International Series in Operations Research & Management Science

Volume 180

Series Editor:
Frederick S. Hillier
Stanford University, CA, USA

Special Editorial Consultant:
Camille C. Price
Stephen F. Austin State University, TX, USA

This book was recommended by Dr. Price

For further volumes:
http://www.springer.com/series/6161

Steven A. Gabriel • Antonio J. Conejo
J. David Fuller • Benjamin F. Hobbs
Carlos Ruiz

Complementarity Modeling in Energy Markets

Springer

Steven A. Gabriel
Department of Civil
and Environmental Engineering
University of Maryland
College Park
MD, USA

Antonio J. Conejo
University of Castilla – La Mancha
Campus Universitario s/n
Ciudad Real
Spain

J. David Fuller
Department of Management Sciences
University of Waterloo
Waterloo
Ontario, Canada

Benjamin F. Hobbs
Department of Geography
and Environmental Engineering
The Johns Hopkins University
Baltimore
MD, USA

Carlos Ruiz
Chair on Systems Science and the Energetic
Challenge
European Foundation for New Energy – EDF
École Centrale Paris and Supélec
Châtenay-Malabry
France

ISSN 0884-8289
ISBN 978-1-4899-8675-7 ISBN 978-1-4419-6123-5 (eBook)
DOI 10.1007/978-1-4419-6123-5
Springer New York Heidelberg Dordrecht London

Printed on acid-free paper

Springer is part of Springer Science+Business Media (www.springer.com)

To our families.

Foreword

Motivation for the Book

This book was motivated by energy markets, particularly electricity and natural gas markets which have been restructured in recent years in a number of countries. The book provides models and tools to understand the functioning of these markets and to identify appropriate decisions for public and private sector energy market participants. Moreover, the book also provides an introduction to market equilibrium models using complementarity formulations and other variations – a particulary active area of research in the last 30 plus years. Thus, the book may be of interest to both energy practitioners as well as researchers in energy and other market equilibria as well as the more broad category of complementarity modeling.

The reader might ask why another book on complementarity models or energy markets? Our rationale is the following. While complementarity models (and generalizations) have been explored in a number of books and articles over the years, we perceived a gap in the treatment of this subject particular to energy that was accessible to a large audience. This observation was key and helped guide the way we decided to present the material. This means that while there are mathematical theorems in the book, we do not focus on proofs of these results but rather give many examples to illustrate the topics. It is our hope that this treatment of the subject will make it widely accessible and provide motivation for research in the future.

What the Book Covers

From the modeling viewpoint, two main issues characterize energy markets (and most markets):

- The need to make decisions (by all market agents) anticipating market outcomes, particularly market-clearing prices.
- From the individual perspective of any market agent, the uncertain behavior of other agents including producers, retailers and consumers.

This book focuses on the first issue and provides models and tools to mathematically represent the hierarchy involved in anticipating market outcomes (bilevel models) and the result of the interaction of all market agents (equilibrium models). Since many energy markets are cleared with the purpose of maximizing social welfare, the description of such clearing mechanism can be conveniently formulated as an optimization problem. On the other hand, any market agent seeking to maximize its profit/utility needs to anticipate market outcomes (which result from the maximum welfare optimization problem). This structure involves a hierarchy: an optimization problem constrained by other optimization problems. The interactions among all the market agents, each one pursuing its maximum profit/utility, result in what we call an equilibrium. Such hierarchical and equilibrium models are of interest in energy markets, and particulary in electricity and natural gas markets. Precise mathematical models can be derived to describe the behavior of market participants and of the market as a whole.

This mathematical characterization of energy markets makes particularly useful the MPEC (mathematical problem with equilibrium constraints) and EPEC (equilibrium problem with equilibrium constraints) as well as the basic MCP (mixed complementarity problem) and VI (variational inequality) approaches described and analyzed in this book.

The chapters of this book are organized as follows:

Chapter 1 provides a friendly introduction to complementarity models, and motivates the interest of such models using real-world examples involving energy markets.

Chapter 2 gives a quick mathematical introduction for most complementarity models, namely, optimization problems constrained by other optimization problems, mathematical programs with equilibrium constraints, equilibrium problems and equilibrium problems with equilibrium constraints.

Chapters 1 and 2 are introductory and reading them allows understanding the concept of complementarity, the structure of complementarity models, and the basic math supporting such models.

Chapter 3 explains introductory microeconomic principles relevant for complementarity problems and market equilibria, and builds the economic base needed in subsequent chapters.

Chapter 4 presents various concepts of equilibria and carefully explores the correspondence between optimization and equilibrium problems.

Chapter 5 describes variational inequality problems as a natural extension/alternative to complementarity problems.

Chapters 6 and 7 provide a detailed description of hierarchical complementarity problems and equilibria involving such problems. This is a very timely topic and these chapters provide an unified and insightful description of such problems.

Chapter 6 analyzes hierarchical problems, particularly, optimization problems with equilibrium constraints, MPECs. Formulation issues are analyzed in detail and solution techniques for some type of problems (linear) are also considered.

Chapter 7 is devoted to equilibrium problems with equilibrium constraints, EPECs, including formulation issues and solution techniques. Relevant practical problems are considered to illustrate the EPEC concept.

Chapters 8 and 9 describe solution algorithms (basic and advanced) for complementarity problems, variational inequality problems, MPECs and EPECs. The MPECs highlighted are MPCCs (mathematical programs with complementarity constraints) which are special instances of MPECs. Since solution techniques to tackle MPECs and EPECs are still being researched, this chapter provides an overview of promising solution techniques.

Chapter 10 provides an overview of natural gas markets as well as an illustrative equilibrium problem with supporting numerical results.

Chapter 11 discusses real-world case studies pertaining to electricity markets including environmental issues.

Chapter 12 analyzes multi-commodity markets: their structures and equilibria. A focus is placed on markets linked through consumption decisions using demand functions with cross-price elasticities.

Appendix A provides a brief overview of convex sets and convex functions.

Appendix B provides GAMS codes for selected examples discussed throughout the book.

Appendix C provides some basic details on modeling electric power flow in a network in ways useful for modeling electricity markets.

Appendix D discusses some engineering considerations useful in modeling gas flow in networks and other non-convex relationships.

Conclusions and Possible Future Research Directions

Perhaps our main conclusion is that MCP/VI, MPEC, and EPEC models for energy markets can be both extraordinarily flexible and useful tools for realistically characterizing these markets. However, in some cases, especially when large-scale or certain non-convex aspects are present, solving these models may also be challenging. Nevertheless, as energy markets become more complicated and distributed (e.g, with the expansion of smart grids), it will be harder to accurately gauge the effects of proposed regulations, policies, or other inputs without tools like the ones proposed and explored in this book. It is our belief that improved models and algorithms in this area will therefore be fruitful and challenging areas of research and we hope that this book serves as a useful guide in that direction.

The authors would like to express their gratitude to Eddie Anderson, Hakob Avetisyan, Cedric de Jonghe, Patricia Josefchak, Saamrat Kasina, Seksun Moryadee, Stephan Nagl, Fernando Oliveira, Yves Smeers, Jeffery Song, Damiano Ting and Sonja Wogrin for their insightful observations and help with the book. At Springer we would like to thank Matthew Amboy, Fred Hillier, Neil Levine, and Camille Price for their support. Lastly, we would like to dedicate the book to our families.

Steven A. Gabriel, University of Maryland, College Park, Maryland, USA

Antonio J. Conejo, University of Castilla - La Mancha, Ciudad Real, Spain

J. David Fuller, University of Waterloo, Waterloo, Canada

Benjamin F. Hobbs, The Johns Hopkins University, Baltimore, Maryland, USA

Carlos Ruiz, École Centrale Paris and Supélec, Paris, France

May 2012

Contents

Chapter 1
Introduction and Motivation

1.1 Introduction

This chapter provides motivation for studying mixed complementarity problems in energy. This class of problems has proven to be especially important in modeling the various liberalized/liberalizing energy markets around the world given its flexibility and ability to directly manipulate both primal (physical) variables as well as dual (price) variables. In this chapter, we introduce complementarity problems and generalizations such as mathematical programs with equilibrium constraints through easy-to-understand and in some cases, well-known energy examples. We will show that one optimization problem, several ones, or optimization problems and nonlinear equations combined are all examples of complementarity problems or extensions, allowing for very general formulations. Upon finishing this chapter, it is anticipated that the reader will have a clearer picture of the modeling advantages of complementarity problems vis-à-vis optimization and other standard models. This chapter is organized as follows: Section 1.2 provides a motivation and description of complementarity models with a number of illustrative examples; Section 1.3 then summarizes the chapter and Section 1.4 provides a computational appendix.

1.2 Complementarity Models: Motivation and Description

In roughly the last twenty years, energy markets in various parts of the world have been liberalizing. The goals of this restructuring in North America and Europe have been to make these markets more competitive and allow greater third party access to energy infrastructure such as pipelines, transmission lines, or the like. In some cases, this restructuring has been perceived as having positive outcomes [52], e.g., PJM, a regional transmission operator that manages wholesale electricity flow in 13 U.S. states and the District of Columbia [46]. On the other hand, this liberalization in California resulted in an energy crisis in the early part of this century and as of January 2010 deregulation has been suspended there [17].

In some cases, due perhaps in part to these trends in energy markets, there have been indications of market power being exerted. One or more players use their influential market position to decrease production or transmission to enhance their own position or profit. Often this sort of action can have cross-border implications and when different nations are involved (as opposed to regions of the same country), can have important geopolitical ramifications. Take for example the curtailment of natural gas supplies from Russia to Ukraine in 2006 [35, 36] and 2009 [28, 45, 53] which had negative effects on downstream clients in Europe and was the impetus, in part, for increasing supply security in Europe. Since roughly 80% of the gas from Russia to Europe passes through Ukraine [8], this has led to a number of alternative pipelines being proposed bypassing Ukraine but transporting Russian gas to Europe (Nord Stream [42], South Stream [51]) or bringing gas from Central Asia and the Middle East to Europe avoiding both Russia and Ukraine (Nabucco [41]). Needless to say, energy regulators in various parts of the world are trying to allow fair access to energy infrastructure [38].

Given the new paradigm for energy markets combining in some cases, an old regulated structure and new, deregulated markets with either imperfect or perfect competition, there is a need for robust models to aid decision-makers. This is where complementarity modeling comes into the picture. Complementarity models generalize linear programs (LP), (convex) quadratic programs (QP), and (convex) nonlinear programs (NLPs). This correspondence is made through the Karush-Kuhn-Tucker (KKT) optimality conditions [6] for these problems. Indeed, the statement of these conditions is a special case of a complementarity problem. The connection with non-convex optimization and complementarity problems breaks down due to the lack of meaningful KKT conditions.

Besides generalizing optimization, complementarity problems also subsume other important classes of modeling including: spatial price equilibria, Wardrop traffic equilibria, Nash-Cournot games, and many more [18]. These problems either cannot be expressed as just one optimization problem (e.g., Nash-Cournot models that have multiple players optimizing their own payoffs), or there is no explicit optimization problem (e.g., spatial price equilibria), just a set of conditions that must hold at an equilibrium. As it turns out, complementarity models are quite appropriate for modeling the mixture of regulated/deregulated, perfect/imperfect competition that characterizes today's energy markets in which both primal variables (e.g., power generation, gas flows) and dual variables (e.g., prices) can be constrained together.

What then is a complementarity problem? Formally, having a function $F : \mathbb{R}^n \to \mathbb{R}^n$, the pure nonlinear complementarity problem denoted NCP(F) is to find an $x \in \mathbb{R}^n$ such that for all i:

1. $F_i(x) \geq 0$
2. $x_i \geq 0$
3. $x_i \cdot F_i(x) = 0$

These conditions can be stated equivalently in vector form as $F(x) \geq 0, x \geq 0, F(x)^T x = 0$ and are often written more compactly as $0 \leq F(x) \perp x \geq 0$ with the "perp" operator $\perp$ denoting the inner product of two vectors equal to zero.[1] The mixed version of the complementarity problem (MCP) is closely related but also allows for both equations with corresponding free variables and inequalities with associated nonnegative variables. The most common form of the MCP is stated as follows: having a function $F : \mathbb{R}^n \to \mathbb{R}^n$, MCP($F$) is to find vectors $x \in \mathbb{R}^{n_1}, y \in \mathbb{R}^{n_2}$ such that for all i:

1. $F_i(x, y) \geq 0, x_i \geq 0, x_i \cdot F_i(x) = 0, i = 1, \ldots, n_1$
2. $F_{j+n_1}(x, y) = 0, y_j$ free $, j = 1, \ldots, n_2$.

Related to the MCP is the variational inequality problem (VI) which is stated as follows: having a vector-valued function $F : \mathbb{R}^n \to \mathbb{R}^n$ and nonempty set $K \subseteq \mathbb{R}^n$, find an $x \in K$ such that:

$$F(x)^T (y - x) \geq 0, \forall y \in K \tag{1.1}$$

This problem is related to the optimality conditions of the nonlinear program with scalar-valued objective function $f : \mathbb{R}^n \to \mathbb{R}$

$$\text{Minimize} \quad f(x) \tag{1.2a}$$

$$\text{s. t.}$$

$$x \in K \tag{1.2b}$$

where "s. t." stands for "subject to". The optimality conditions of this problem are to find a solution $x \in X$ such that

$$\nabla f(x)^T (y - x) \geq 0, \forall y \in K$$

These conditions state that the directional derivative of the objective function $\nabla f(x)^T (y - x)$, should be nonnegative for all feasible directions (i.e., $y - x$ where $y \in K$ assuming that K is convex). Clearly, setting $F(x) = \nabla f(x)$ shows the correspondence with the more general VI. A more complete discussion of the VI and its relationship with optimization problems can be

[1] Similarly, conditions $F(x) \geq 0, x \geq 0, F(x)^T x = 0$ can be written $0 \geq -F(x) \perp x \geq 0$, $0 \leq x \perp -F(x) \leq 0$ and $0 \leq x \perp F(x) \geq 0$ as well.

found in Chapter 5 as well as [18]. Also, it can easily be shown that when K is taken as the nonnegative orthant, that $\mathrm{VI}(\mathbb{R}^n_+)$ corresponds exactly to the pure NCP stated above. Additionally, it can be easily shown that for vectors of lower and upper bounds, respectively, $l, u \in \mathbb{R}^n$, when $K=[l,u]$ that $\mathrm{VI}(K)$ corresponds to a generalization of the mixed complementarity problem shown above. For the MCP stated above, the correspondence with the solution set of the VI is exact when $u_i = \infty$ for all indices and $l_i = 0$ for inequalities and $l_i = -\infty$ for all equations [18]. For more details on the relationship of the VI to optimization and complementarity problems, see Chapter 5.

To clearly illustrate the complementarity problem (mixed or pure forms) and its extensions, we next present some examples most of which relate to energy markets. For concreteness, first consider the following three-variable MCP example.

1.2.1 Illustrative Example. Three-Variable MCP

Let n_1=2, $n_2 = 1$ and $F : \mathbb{R}^3 \to \mathbb{R}^3$ be defined as:

$$F(x_1, x_2, y_1) = \begin{pmatrix} F_1(x_1, x_2, y_1) \\ F_2(x_1, x_2, y_1) \\ F_3(x_1, x_2, y_1) \end{pmatrix} = \begin{pmatrix} x_1 + x_2 \\ x_1 - y_1 \\ x_1 + x_2 + y_1 - 2 \end{pmatrix} \tag{1.3}$$

Then, the corresponding MCP is to find x_1, x_2, y_1 that solve the following conditions:

$$\begin{array}{rl} x_1 + x_2 \geq 0 & x_1 \geq 0 \; (x_1 + x_2) \cdot x_1 = 0 \\ x_1 - y_1 \geq 0 & x_2 \geq 0 \; (x_1 - y_1) \cdot x_2 = 0 \\ x_1 + x_2 + y_1 - 2 = 0 & y_1 \text{ free} \end{array} \tag{1.4}$$

It can be shown that $(x_1, x_2, y_1) = (0, 2, 0)$ is the only solution [20]. To understand MCPs it is instructive to show why this is the case with a simple analysis that we show at the end of this chapter along with other computational issues. More generally, MCPs can have zero solutions, one solution, two solutions, an arbitrary number of finite solutions, or an infinite number of solutions. To see this, note that MCPs generalize both zero-finding problems [2] which can have multiple roots as well as linear equations and convex optimization problems that can have just zero, one, or an infinite number of solutions.

The next result indicates the most basic connection between optimization and complementarity problems.

[2] If only equations are present in the MCP.

1.2.2 Illustrative Example. Nonlinear Program Expressed as an MCP

Consider the following generic optimization problem: [3]

$$\text{Minimize}_x \quad f(x) \tag{1.5a}$$
s. t.
$$g_i(x) \le 0, i = 1, \ldots, m \quad (\alpha_i) \tag{1.5b}$$
$$h_j(x) = 0, j = 1, \ldots, p \quad (\beta_j) \tag{1.5c}$$

where $f, g_i(x), h_j(x) : \mathbb{R}^n \to \mathbb{R}$ are respectively, the objective function and the constraint functions for the problem. The KKT conditions [6] are to find primal variables $x \in \mathbb{R}^n$, and Lagrange multipliers (dual variables) $\alpha \in \mathbb{R}^m_+, \beta \in \mathbb{R}^p$ such that the following conditions hold:

$$0 = \nabla f(x) + \sum_{i=1}^{m} \nabla g_i(x) \alpha_i + \sum_{j=1}^{p} \nabla h_j(x) \beta_j \tag{1.6a}$$
$$0 \le -g_i(x) \perp \alpha_i \ge 0, i = 1, \ldots, m \tag{1.6b}$$
$$0 = h_j(x), j = 1, \ldots, p \quad \beta_j \text{ free} \tag{1.6c}$$

This is an instance of an MCP with the function $F : \mathbb{R}^{n+m+p} \to \mathbb{R}^{n+m+p}$ defined as

$$F(x, \alpha, \beta) = \begin{pmatrix} F_x(x, \alpha, \beta) \\ F_\alpha(x, \alpha, \beta) \\ F_\beta(x, \alpha, \beta) \end{pmatrix} =$$
$$\begin{pmatrix} \nabla f(x) + \sum_{i=1}^{m} \nabla g_i(x) \alpha_i + \sum_{j=1}^{p} \nabla h_j(x) \beta_j \\ -g(x) \\ h(x) \end{pmatrix} \tag{1.7}$$

and complementarity only required between $-g_i(x)$ and the corresponding nonnegative Lagrange multipliers α_i; the other elements of F being equations with associated free variables (x, β). Since for linear and convex quadratic programs these KKT conditions are both necessary and sufficient, we see that the solution sets to these problems correspond exactly to the solutions of (1.6). However, only those nonlinear programming problems that have meaningful KKT conditions will have a correspondence to MCPs, e.g., integer programs do not relate to complementarity problems (at least via KKT conditions).

[3] In this and other optimization problems, Greek letters in parentheses next to constraints are the associated Lagrange multipliers.

It is useful to note the specific form of the corresponding MCP for linear and convex, quadratic programs which both give rise to linear complementarity problems (LCPs). Consider the following convex, quadratic program (which necessarily includes LPs when the matrix $Q = 0$):

$$\text{Minimize}_x \quad \frac{1}{2}x^T Qx + c^T x \tag{1.8a}$$

s. t.

$$Ax \geq b \quad (\alpha) \tag{1.8b}$$

$$Dx = d \quad (\beta) \tag{1.8c}$$

$$x \geq 0 \tag{1.8d}$$

for $Q \in \mathbb{R}^{n\times n}$, a symmetric, positive semi-definite matrix, $c \in \mathbb{R}^n$, $A \in \mathbb{R}^{m\times n}, b \in \mathbb{R}^m, D \in \mathbb{R}^{p\times n}, d \in \mathbb{R}^p$. The KKT conditions for this problem are both necessary and sufficient [6] and are to find vectors $x \in \mathbb{R}^n, \alpha \in \mathbb{R}^m_+, \beta \in \mathbb{R}^p$ such that:

$$0 \leq Qx + c - A^T\alpha - D^T\beta \perp x \geq 0 \tag{1.9a}$$

$$0 \leq Ax - b \perp \alpha \geq 0 \tag{1.9b}$$

$$0 = Dx - d, \ \beta \text{ free} \tag{1.9c}$$

giving rise to the corresponding linear MCP

$$F(x,\alpha,\beta) = \begin{pmatrix} F_x(x,\alpha,\beta) \\ F_\alpha(x,\alpha,\beta) \\ F_\beta(x,\alpha,\beta) \end{pmatrix} = \begin{pmatrix} Q & -A^T & -D^T \\ A & 0 & 0 \\ D & 0 & 0 \end{pmatrix} \begin{pmatrix} x \\ \alpha \\ \beta \end{pmatrix} + \begin{pmatrix} c \\ -b \\ -d \end{pmatrix} \tag{1.10}$$

which is of the form $F(z) = Mz + q$ for $z = \left(x^T, \alpha^T, \beta^T\right)^T$, $q = \left(c^T, -b^T, -d^T\right)^T$. We note that $2z^T Mz = z^T\left(M + M^T\right)z = 2x^T Qx$ so that M is positive semi-definite given the earlier assumptions on Q. Being positive semi-definite, this matrix M has the important property that whenever the corresponding linear MCP is feasible (i.e., $\exists z \geq 0$ such that $q + Mz \geq 0$) then it automatically has a solution (Theorem 3.1.2, [13]). When $Q = 0$ so that (1.8) represents a linear program, the KKT conditions (1.9) represent primal and dual feasibility as well as complementary slackness. This complementarity existence result can be understood to mean that if the primal and dual LPs are feasible, then there is automatically a solution (to both), further illustrating the connection between optimization and complementarity problems.

The next example of an MCP, the Project Independence Evaluation System (PIES) [2], [3] is a famous energy planning model from the 1970's. The more complex, current-day version of PIES is the National Energy Modeling System (NEMS) [4] used by the U.S. Department of Energy to produce their Annual Energy Outlook as well as to respond to a multitude of energy-economic requests from the U.S. government [23], [12]. This example shows that the KKT conditions for a single optimization problem can be combined with additional (econometric) equations linking the primal and dual variables to form an MCP. For more information about the PIES model, see Chapter 5 where it is expressed as a VI and again in Chapter 8 where a PIES algorithm is discussed.

1.2.3 Illustrative Example. PIES Model

The supply side of the energy sector is modeled as the following production cost linear program:

$$\text{Minimize}_x \quad c^T x \tag{1.11a}$$

s. t.

$$Ax \geq q \quad (\pi) \tag{1.11b}$$

$$Bx \geq b \quad (\gamma) \tag{1.11c}$$

$$x \geq 0 \tag{1.11d}$$

where x is the vector of decision variables in energy production. The objective function (1.11a) represents total production costs which are to be minimized. The first set of constraints (1.11b) enforce that the demand q for energy products needs to be met by a combination of energy production types Ax. The second set of constraints (1.11c) are non-demand related. The dual prices are respectively, π and γ in (1.11b) and (1.11c).

The consumption side is modeled by the following econometric equations where p is price:

$$q_i\,(p) = q_i^0 \prod \left(\frac{p_i}{p_i^0} \right)^{e_{ij}} \tag{1.12}$$

here q_i^0, p_i^0 are, respectively, reference demands and prices for energy product i, and e_{ij} is an elasticity between products i and j. To join the supply and demand sides of the market, the equilibrating condition:

$$\pi = p \tag{1.13}$$

is needed. There are several important modeling points to make here. First, it is not possible within the context of an optimization problem to model (1.11),(1.12),(1.13). The reason is that to solve (1.11), one needs to know an optimal value for π to feed into the right-hand side $q = q\,(\pi)$ of (1.11b).

[4] www.eia.doe.gov

But to get such a value for π one needs to first solve (1.11), hence there is a computational difficulty. The solution is to write out the KKT optimality conditions for (1.11) and at the same time, write out the econometric equations (1.12) substituting $\pi = p$ to effectively solve for both the primal and dual variables (x and π, respectively) at the same time. This also brings up a second important point about MCPs, one can combine optimization and econometric equations in a computationally clean way. Lastly, this example also shows that one can handle conditions on both primal and dual variables simultaneously under the MCP format. The corresponding MCP function for PIES is thus the following:

$$F\begin{pmatrix} x \\ \pi \\ \gamma \end{pmatrix} = \begin{pmatrix} c - A^T\pi - B^T\gamma \\ Ax - q(\pi) \\ Bx - b \end{pmatrix}$$

A fourth example of MCPs relevant for energy markets involves a duopoly of energy producers which can easily be extended to an $N-$person Nash-Cournot game [20]. Nash-Cournot games are prevalent in a lot of market equilibrium literature and are also treated in Chapters 3, 4, 5, 7, 10 and 11 for example in different contexts.

1.2.4 Illustrative Example. Nash-Cournot Duopoly in Energy Production, Two Simultaneous Optimization Problems

In this setting, there are two energy producers that compete with each other but have market power to affect prices [50]. The simplest form of this model has $q_i, i = 1, 2$ the amount that player i produces (e.g., electricity or gas). The market power aspects are related to an inverse market demand function $p(q_1 + q_2) = \alpha - \beta(q_1 + q_2)$ where $\alpha, \beta > 0$. This function provides the price for the energy product p given the total market production $(q_1 + q_2)$. For $\gamma_i > 0, i = 1, 2$, a linear production cost function for producer i is given as $c_i(q_i) = \gamma_i q_i$. With market power assumptions in place, each player, taking the other's optimal production as given, can decide to reduce its own production to force higher prices and ultimately higher profits. However, if the market price is too high, demand will also decrease so producer i seeks to balance this tradeoff by solving the following profit-maximization problem:

$$\text{Maximize}_{q_i} \quad p(q_1 + q_2) \cdot q_i - c_i(q_i) \tag{1.14a}$$

s. t.

$$q_i \geq 0 \tag{1.14b}$$

or

$$\text{Maximize}_{q_i} \quad (\alpha - \beta(q_1 + q_2)) \cdot q_i - \gamma_i q_i \tag{1.15a}$$

s. t.

$$q_i \geq 0 \tag{1.15b}$$

The KKT conditions [5] for the two producers' problems but taken together are to find energy production quantities $q_i, i = 1, 2$ such that:

$$0 \leq 2\beta q_1 + \beta q_2 - \alpha + \gamma_1, q_1 \geq 0, \ (2\beta q_1 + \beta q_2 - \alpha + \gamma_1) \cdot q_1 = 0 \text{ (producer 1)}$$
$$0 \leq \beta q_1 + 2\beta q_2 - \alpha + \gamma_2, q_2 \geq 0, (\beta q_1 + 2\beta q_2 - \alpha + \gamma_2) \cdot q_2 = 0 \ \text{ (producer 2)}$$

which need to be solved simultaneously giving rise to the following pure linear complementarity problem with function F:

$$F\begin{pmatrix} q_1 \\ q_2 \end{pmatrix} = \begin{pmatrix} 2\beta q_1 + \beta q_2 - \alpha + \gamma_1 \\ \beta q_1 + 2\beta q_2 - \alpha + \gamma_2 \end{pmatrix} = \begin{pmatrix} 2\beta & \beta \\ \beta & 2\beta \end{pmatrix} \begin{pmatrix} q_1 \\ q_2 \end{pmatrix} + \begin{pmatrix} -\alpha + \gamma_1 \\ -\alpha + \gamma_2 \end{pmatrix}$$

The important point here is that there is more than one optimization problem but to be consistent with a duopoly (Nash-Cournot) perspective, both have to be solved at the same time. In game theory literature, often this simultaneity issue is to establish best reaction or best response functions [43] which are essentially closed-form expressions for q_i as a function of the other production quantity. Then, the two expressions are solved as nonlinear equations to determine the equilibrium quantities. This approach works for small problems and with no constraints besides nonnegativity. Once the number of constraints and players increases, it is much more efficient to use the MCP formulation. Indeed, for realistic energy-planning models, it is not really possible to use this best response approach. Examples of realistic Nash-Cournot and other energy market models that use MCP include [33], [5], [22], [27], [15], [34], [26], [16] to name some selected recent ones. In some of these models, in addition to the standard duopoly/Nash-Cournot approach presented above, additional market-clearing conditions are also needed. This is not a difficulty for MCP given that it can combine optimization problems and equations as long as the system is square; more is said about such an MCP in Illustrative Example 1.2.7 below. Moreover, recent advances in algorithms to solve MCPs in the last twenty years [18], have made solving large-scale versions of these sorts of models computationally feasible and hence relevant. The next example shows a variation on the duopoly when a shared constraint is present leading to a Generalized Nash-Cournot problem.

[5] These conditions are both necessary and sufficient in this case given that the producer problems are convex, quadratic programs since $\beta > 0$.

1.2.5 Illustrative Example. Generalized Nash Equilibria, Energy Production Duopoly

The above discussion concerning Nash equilibria either in pure or mixed form was predicated on the fact that that each player's decisions only affected the other one(s) through the objective function values. The general form of these sort of equilibria allowing for continuous levels of production q_i [6] is the following for player $i = 1, \ldots, p$:

$$\begin{aligned}&\text{Maximize}_{q_i} f_i\left(q_i, q_{-i}\right) = p\left(q_i, q_{-i}\right) q_i - c_i\left(q_i\right)\\&\text{s. t.}\\&\quad q_i \in K_i\end{aligned} \tag{1.16}$$

where K_i is a (convex) feasible region for producer i's decision variable q_i. Previously,
$K_i = \{q_i | 0 \leq q_i\}, p\left(q_i, q_{-i}\right) = \alpha - \beta\left(\sum_i q_i\right)$, and $c_i\left(q_i\right) = \gamma_i q_i$. The optimality conditions for this problem can of course be expressed as either an MCP or the more general VI. In the latter case, the related problem would be to find $q_i, i = 1, \ldots, p$ such that:

$$-\nabla_{q_i} f_i\left(q_i, q_{-i}\right)^T \left(y_i - q_i\right) \geq 0, \forall y_i \in \prod_i K_i$$

A variation on this standard problem that allows for interaction of the players' variables also in the constraint sets is an example of a generalized Nash equilibrium first discussed in Chapter 3. Here we present another example and then the more general statement of this problem.

Consider the energy production duopoly (Example 1.2.4). The associated VI is to find q_1, q_2 such that:

$$\begin{pmatrix}\gamma_1 - \alpha + 2\beta q_1 + \beta q_2\\ \gamma_2 - \alpha + \beta q_1 + 2\beta q_2\end{pmatrix}^T \begin{pmatrix} y_1 - q_1\\ y_2 - q_2\end{pmatrix} \geq 0, \forall \begin{pmatrix} y_1\\ y_2\end{pmatrix} \in K_1 \times K_2 \tag{1.17}$$

where $K_i = R_+, i = 1, 2$. Since there are only nonnegativity constraints, the KKT conditions for the two producers' problems taken together are equivalent to solving this VI (i.e., no dual variables needed) and are:

$$0 \leq 2\beta q_1 + \beta q_2 - \alpha + \gamma_1 \perp q_1 \geq 0 \text{ (producer 1)} \tag{1.18a}$$
$$0 \leq \beta q_1 + 2\beta q_2 - \alpha + \gamma_2 \perp q_2 \geq 0 \text{ (producer 2)} \tag{1.18b}$$

Assuming that both quantities q_1, q_2 are positive leads to solving:

[6] Here q_i is taken as a scalar for player i but could easily be extended to a vector without loss of generality. The former is chosen for notational simplicity.

$$\begin{pmatrix} 2\beta & \beta \\ \beta & 2\beta \end{pmatrix} \begin{pmatrix} q_1 \\ q_2 \end{pmatrix} = \begin{pmatrix} \alpha - \gamma_1 \\ \alpha - \gamma_2 \end{pmatrix} \Leftrightarrow \begin{pmatrix} q_1 \\ q_2 \end{pmatrix} = \begin{pmatrix} \frac{2}{3\beta}(\alpha - \gamma_1) - \frac{1}{3\beta}(\alpha - \gamma_2) \\ \frac{2}{3\beta}(\alpha - \gamma_2) - \frac{1}{3\beta}(\alpha - \gamma_1) \end{pmatrix}$$

For $\alpha = 10, \beta = 5, \gamma_1 = 1, \gamma_2 = 2$,

$$\left[\begin{pmatrix} \frac{2}{3\beta}(\alpha - \gamma_1) - \frac{1}{3\beta}(\alpha - \gamma_2) \\ \frac{2}{3\beta}(\alpha - \gamma_2) - \frac{1}{3\beta}(\alpha - \gamma_1) \end{pmatrix} \right]_{\alpha=10,\beta=5,\gamma_1=1,\gamma_2=2} = \begin{pmatrix} \frac{2}{3} \\ \frac{7}{15} \end{pmatrix} \approx \begin{pmatrix} 0.6677 \\ 0.4667 \end{pmatrix}.$$

In fact, for this value of β, the matrix $\begin{pmatrix} 2\beta & \beta \\ \beta & 2\beta \end{pmatrix}$ is a P-matrix so that by Theorem 3.3.7 in [13], this is a unique solution for any values of $\alpha, \gamma_1, \gamma_2$ (more about this result in Chapter 4).

Now consider a generalization of the above duopoly in which there is a joint constraint related to maximum production. In particular, the new problem for player i is:

$$\text{Maximize}_{q_i} \ (\alpha - \beta(q_1 + q_2)) \cdot q_i - \gamma_i q_i \tag{1.19a}$$

s. t.

$$q_i \geq 0 \tag{1.19b}$$

$$q_1 + q_2 \leq q^{TOT} \tag{1.19c}$$

where q^{TOT} is the maximum total production for the market based on common constraints (e.g., available drilling rigs). As long as this new constraint appears in both players' problems and has the same Lagrange multiplier λ^{TOT}, then there exists an MCP (or VI) solution to the problem [31]. See the Computational Appendix 1.4.3 of this chapter for more details on this problem.

1.2.6 Illustrative Example. Nash-Cournot Duopoly Expressed as a Variational Inequality

Next, the Nash-Cournot duopoly from Illustrative Example 1.2.4 is expressed as a VI as opposed to an MCP. First, since each player's constraint set only involves its own variables, we can express the overall VI set $K = K_1 \times K_2$, i.e., the Cartesian product of sets specific to each player. Then, we take the function $F : \mathbb{R}^2 \to \mathbb{R}^2$ as

$$F\begin{pmatrix} q_1 \\ q_2 \end{pmatrix} = \begin{pmatrix} \frac{df_1(q_1)}{dq_1} \\ \frac{df_2(q_2)}{dq_2} \end{pmatrix}$$

where $f_i, i = 1, 2$ is the negative of the objective function for player i (to convert to a minimization problem) and is given as

$$f_i(q_i) = -\left[(\alpha - \beta(q_1 + q_2)) \cdot q_i - \gamma_i q_i\right].$$

Thus,

$$F\begin{pmatrix} q_1 \\ q_2 \end{pmatrix} = \begin{pmatrix} -\alpha + 2\beta q_1 + \beta q_2 + \gamma_1 \\ -\alpha + \beta q_1 + 2\beta q_2 + \gamma_2 \end{pmatrix}$$

In this case since there are only nonnegativity constraints, $K_1 = K_2 = \mathbb{R}_+$. The increased modeling flexibility of the VI formulation comes when the constraint set is more complicated as might be the case for the duopoly if for example these constraint sets included all sorts of production/operational considerations. There are algorithmic advantages to putting these complexities into these sets as opposed to just expressing the whole problem (complicated constraints and all) as one large MCP and in Chapter 5 we provide more details on VI formulations.

1.2.7 Illustrative Example. Energy Network with Multiple Players

In this example, we consider a two-node energy network (e.g., power or natural gas). Producers A and B operate at node 1 whereas node 2 has producers C and D; all these producers are modeled as price-takers. The amount that is produced by each of these players is given respectively as, $q_1^A, q_1^B, q_2^C, q_2^D$. The amount that is sold by each of these producers at their associated node is denoted respectively as $s_1^A, s_1^B, s_2^C, s_2^D$. Lastly, the two producers at node 1 (A and B) are assumed to have a trading function with f_{12}^A, f_{12}^B representing the amount of energy flow that is shipped by the producer and sold at the other node. Both producers at node 1 have the same sort of optimization problem to solve illustrated by the one for producer A at node 1 given below. Also, node 2 producers C and D have a similar optimization problem but without any flow variables and related terms. Thus, this equilibrium is asymmetric with respect to production possibilities between the two nodes. Producer A's optimization problem is the following:

$$\text{Maximize}_{s_1^A, q_1^A, f_{12}^A} \quad \pi_1 s_1^A + \pi_2 f_{12}^A - c_1^A\left(q_1^A\right) - \left(\tau_{12}^{\text{Reg}} + \tau_{12}\right) f_{12}^A \qquad (1.20\text{a})$$

s. t.

$$q_1^A \le \overline{q}_1^A \quad (\lambda_1^A) \qquad (1.20\text{b})$$

$$s_1^A = q_1^A - f_{12}^A \quad (\delta_1^A) \qquad (1.20\text{c})$$

$$s_1^A, q_1^A, f_{12}^A \ge 0 \qquad (1.20\text{d})$$

Here π_1 is the producer price at node 1 (same idea for π_2) but in contrast to Illustrative Example 1.2.4, each producer cannot control it in their objective function by withholding sales (or production). Rather, π_1 is determined by market-clearing conditions which is an important feature of this example. The function $c_1^A\left(q_1^A\right)$ is the (marginal) production cost function and as in Illustrative Example 1.2.4 we assume for simplicity that it is linear, i.e., $c_1^A\left(q_1^A\right) = \gamma_1^A q_1^A$ with $\gamma_1^A > 0$ (and similarly for the other producers). The terms τ_{12}^{Reg}, τ_{12} represent respectively, the regulated tariff and the congestion tariff for using the network from node 1 to node 2 (e.g., transmission line or pipeline). While τ_{12}^{Reg} is a fixed, exogenous parameter, τ_{12} is a variable which is determined in another part of the formulation as will be shown. Thus, the objective function (1.20a) represents the net profit for the producer (only one time period is considered). The constraints for this problem are an upper bound on production (1.20b), definition of sales at the producer's own node as production less outflow (1.20c), and nonnegativity of all variables. The KKT conditions for each of the producers' problems are both necessary and sufficient [6] in light of the functions chosen. These conditions for each of the four producers are as follows.

Producer A, node 1

$$0 \le -\pi_1 + \delta_1^A \perp s_1^A \ge 0 \tag{1.21a}$$

$$0 \le \gamma_1^A + \lambda_1^A - \delta_1^A \perp q_1^A \ge 0 \tag{1.21b}$$

$$0 \le -\pi_2 + \left(\tau_{12}^{\mathrm{Reg}} + \tau_{12}\right) + \delta_1^A \perp f_{12}^A \ge 0 \tag{1.21c}$$

$$0 \le \overline{q}_1^A - q_1^A \perp \lambda_1^A \ge 0 \tag{1.21d}$$

$$0 - s_1^A - q_1^A + f_{12}^A \text{ , } \delta_1^A \text{ free} \tag{1.21e}$$

Producer B, node 1

$$0 \le -\pi_1 + \delta_1^B \perp s_1^B \ge 0 \tag{1.22a}$$

$$0 \le \gamma_1^B + \lambda_1^B - \delta_1^B \perp q_1^B \ge 0 \tag{1.22b}$$

$$0 \le -\pi_2 + \left(\tau_{12}^{\mathrm{Reg}} + \tau_{12}\right) + \delta_1^B \perp f_{12}^B \ge 0 \tag{1.22c}$$

$$0 \le \overline{q}_1^B - q_1^B \perp \lambda_1^B \ge 0 \tag{1.22d}$$

$$0 = s_1^B - q_1^B + f_{12}^B \text{ , } \delta_1^B \text{ free} \tag{1.22e}$$

Producer C, node 2

$$0 \le -\pi_2 + \delta_2^C \perp s_2^C \ge 0 \tag{1.23a}$$

$$0 \le \gamma_2^C + \lambda_2^C - \delta_2^C \perp q_2^C \ge 0 \tag{1.23b}$$

$$0 \le \overline{q}_2^C - q_2^C \perp \lambda_2^C \ge 0 \tag{1.23c}$$

$$0 = s_2^C - q_2^C \text{ , } \delta_2^C \text{ free} \tag{1.23d}$$

Producer D, node 2

$$0 \leq -\pi_2 + \delta_2^D \perp s_2^D \geq 0 \tag{1.24a}$$
$$0 \leq \gamma_2^D + \lambda_2^D - \delta_2^D \perp q_2^D \geq 0 \tag{1.24b}$$
$$0 \leq \overline{q}_2^D - q_2^D \perp \lambda_2^D \geq 0 \tag{1.24c}$$
$$0 = s_2^D - q_2^D \ , \delta_2^D \text{ free} \tag{1.24d}$$

The market-clearing conditions that stipulate that suppy must equal demand are:

$$0 = \left[s_1^A + s_1^B\right] - D_1\left(\pi_1\right), \ \pi_1 \text{ free} \tag{1.25a}$$
$$0 = \left[s_2^C + s_2^D + f_{12}^A + f_{12}^B\right] - D_2\left(\pi_2\right), \ \pi_2 \text{ free} \tag{1.25b}$$

The terms in square brackets represent the net supply at each node assuming for ease of presentation that no energy is lost between nodes due to network engineering laws. The functions $D_i\left(\pi_i\right), i = 1, 2$ represent the demand at node i as a function of the nodal price π_i.

In addition to the four producer problems and the two market-clearing conditions shown above, this energy network model also considers a transportation system operator (TSO) who manages the congestion and flows. The TSO's problem is to maximize net profit and is as follows (other objectives are also possible):

$$\text{Maximize}_{g_{12}, g_{21}} \quad \left(\tau_{12}^{\text{Reg}} + \tau_{12}\right) g_{12} - c^{TSO}\left(g_{12}\right) \tag{1.26a}$$

s. t.

$$g_{12} \leq \overline{g}_{12} \ \left(\varepsilon_{12}\right) \tag{1.26b}$$
$$g_{12} \geq 0 \tag{1.26c}$$

where g_{12} is the variable for the flow that the TSO manages between nodes 1 and 2 and $c^{TSO}\left(g_{12}\right)$ is a transportation network operations cost function which for simplicity is taken to be linear, i.e., $c^{TSO}\left(g_{12}\right) = \gamma^{TSO} g_{12}$ where $\gamma^{TSO} > 0$. The KKT conditions for this problem are both necessary and sufficient and are the following:

$$0 \leq -\tau_{12}^{\text{Reg}} - \tau_{12} + \gamma^{TSO} + \varepsilon_{12} \perp g_{12} \geq 0 \tag{1.27a}$$
$$0 \leq \overline{g_{12}} - g_{12} \perp \varepsilon_{12} \geq 0 \tag{1.27b}$$

Lastly, the transportation market-clearing conditions also need to be taken into account. They are given as follows:

$$0 = g_{12} - \left[f_{12}^A + f_{12}^B\right] \ \tau_{12} \text{ free} \tag{1.28}$$

The MCP related to this energy network problem is thus the KKT conditions of the producers: (1.21), (1.22), (1.23), (1.24), the nodal market-clearing conditions (1.25), the KKT conditions of the TSO (1.27) and lastly the market-clearing conditions of the transportation market (1.28). One further modeling point for such a system is to note that the prices in the two sets of market-clearing conditions are stated as free. This means that in principle they can be negative which does happen from time to time in power markets. However, in some cases only nonnegative prices are desired. This constraint can be easily incorporated by changing these market-clearing conditions from equalities to inequalities that specify that supply should be greater than or equal to demand. Then, the corresponding prices will be nonnegative necessarily. To insure that there is no excess supply in equilibrium, some additional assumptions on the functions involved (e.g., demand) may be needed [18].

To stop at MCPs and VIs would be premature since many important market features go beyond these formulations. One noteworthy extension includes taking into account stochastic elements combined with equilibrium as was done in, [29], [39], [54], [21]. In this setting, one cannot always just solve the resulting MCP (or related variational inequality) as it is too large to fit into memory in some cases or would take too long to solve. To solve the resulting large-scale stochastic MCPs, borrowing ideas from stochastic optimization, one can separate the problem into variables that need to be determined before the uncertainty is realized and those scenario-dependent variables. Such an approach represents an extension of the L-shaped method [10] or Benders algorithm but applied to stochastic complementarity problems as discussed in [21]. See Chapters 4 and 5 for more details on stochastic equilibrium problems.

Another approach is to take the set of scenarios representing the uncertainty and somehow reduce them to a manageable number. Scenario reduction methods as applied to energy models have appeared in [30], [14], [32], [40], [44]. Such methods approximate the scenario tree by a simpler yet representative tree. Determining such a tree has been shown to be computationally challenging [47] but one can use heuristic procedures. A stochastic complementarity problem approach as applied to energy problems appears in [26].

Another extension to MCPs are bi-level problems also called Mathematical Programs with Equilibrium Constraints (MPEC) [37] given as follows:

$$\text{Minimize} \quad f(x,y) \tag{1.29a}$$

s. t.

$$(x,y) \in \Omega \tag{1.29b}$$

$$y \in S(x) \tag{1.29c}$$

where the vector x represents some upper-level set of decisions (e.g., investments) for a dominant player (company or government) and the vector y some lower-level decisions (e.g., the market). For example, in [48] and [24], a two-level problem is presented in which the profit of a strategic power generator is maximized at the upper-level and the bottom level refers to market-clearing and price formation. Both works convert the associated MPEC to a mixed-integer linear program and use the notion of disjunctive constraints [19] to represent the complementarity aspects of the KKT conditions to the lower-level problem. In [48], to avoid bilinear terms of the form price*quantity, strong duality is applied to the linear programming lower-level problem. In [24], the lower-level is a quadratic program and a different approximation of this bilinear term using binary variables and extra constraints is applied. In [11], a two and three-level model of an intelligent attacker and defender with respect to energy and other areas is modeled. Other MPEC models in energy are considered in Chapters 2, 3, and 6. Other examples in energy can be more complicated based on adequately accounting for the lower-level problem in the upper one. For example, one recent approach [25] is to use a variation on Benders method with domain decomposition.

In the above formulation the constraints (1.29b) represent joint restrictions on both x and y and (1.29c) enforces that the lower-level variables y be in the solution set of a lower-level problem (optimization or MCP) given that the upper-level variables x are fixed. The latter set of constraints make this an especially difficult problem to solve, hence the need for alternative representations such as disjunctive constraints. As a concrete example that shows how the constraint (1.29c) makes the feasible region non-convex, consider the following example from [37].

1.2.8 Illustrative Example. MPEC

Here there are just two variables x, y and we want to solve:

$$\begin{aligned} &\text{Minimize}_{x,y} \quad f(x,y) && (1.30a)\\ &\text{s. t.} \\ &\quad |x| \le 1 && (1.30b)\\ &\quad y \in S(x) && (1.30c)\end{aligned}$$

where the solution set $S(x)$ corresponds to the problem:

$$\begin{aligned} &\text{Minimize}_{y} \quad y && (1.31a)\\ &\text{s. t.} \\ &\quad |y| \le 1 && (1.31b)\\ &\quad xy \le 0 && (1.31c)\end{aligned}$$

As stated in [37] and easily verified,

$$S(x) = \begin{Bmatrix} \{-1\} & x \in [0,1] \\ \{0\} & x \in [-1,0) \end{Bmatrix}$$

so that the feasible region to the overall problem (1.30) is

$$\{(x,-1) : x \in [0,1]\} \cup \{(x,0) : x \in [-1,0)\} \tag{1.32}$$

and is shown in Figure 1.1.

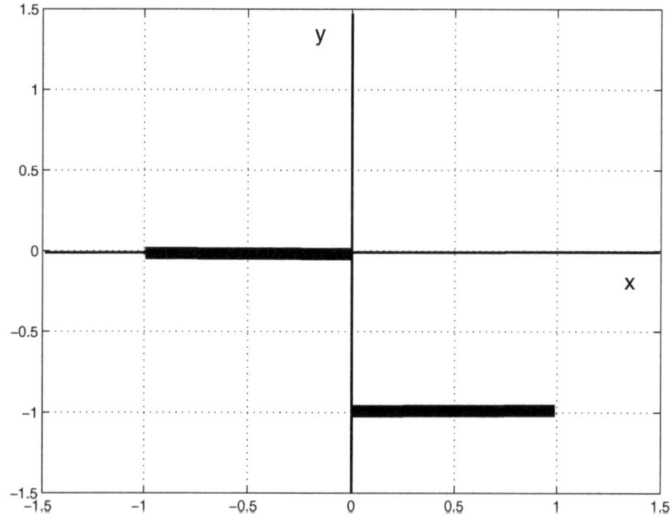

Fig. 1.1 Feasible region for simple MPEC

An Equilibrium Problem with Equilibrium Constraints (EPEC) is an extension to the MPEC when more than one strategic player is allowed at the top level. An early example of an EPEC (although not stated as such) was [4] where a duopoly of two firms was modeled. In the first period they competed in the forward contracts market which then affected the duopolistic competition in the second period's production market. More recently in [49], the authors present a three-period model. In period three, the energy spot market is modeled as an LCP (for generation and transmission). Period 2 then considers for each firm, how much to invest in new generation capacity by maximizing the expected value of the the investment. Thus, each firm is solving an MPEC. Taken together, all firms' problems constitute an EPEC. In period 1, a network planner decides which transmission lines to build/upgrade given the anticipated reactions of the lower two levels. In [7], another EPEC is considered in which the top level represents price-makers for the electric power market that are trying to maximize net profits by choosing appropriate quantity bids. The common lower-level problem is that of market clearing which provides in part, the spot market prices used at the upper level. Many other EPEC models in energy have been presented and this topic is discussed in more detail in Chapters 2, 3, and 7.

1.3 Summary

In this chapter we have introduced and motivated complementarity problems, variational inequality problems, mathematical programs and equilibrium problems with equilibrium constraints. The first two classes are often referred to as one-level equilibrium problems with the third and fourth being two- (or more) level equilibria. These examples were demonstrated to easily show the versatility of equilibrium problems to combine both optimization formulations and nonlinear equations relevant for energy markets. The first example was a simple three-variable model shown both analytically and geometrically to have a unique solution. Illustrative Example 1.2.2 demonstrated that all nonlinear programs via their KKT conditions (if meaningful) are special cases of MCPs (and VIs) and further made the connection between a complementarity existence result and LP duality theory. The third example showed that both optimization problems and nonlinear equations (econometrically-derived perhaps) can be easily combined using the MCP approach. This example further illustrated that optimization problems alone are not sufficient if one wants to manipulate both the primal and dual variables simultaneously. Illustrative Example 1.2.4 showed that multiple optimization problems can be combined (using their KKT conditions) to arrive at one larger complementarity system. This example also indicated that market power considerations (e.g., Nash-Cournot producers) can easily be modeled and extensions of these simple producer problems with more constraints and more players still will be an MCP. The fifth and sixth examples were extensions to the duopoly problem illustrating a generalized Nash problem and a variational inequality, respectively. Illustrative Example 1.2.7, depicting an energy network illustrated that price-taking behavior of players encoded by optimization problems can be combined with market-clearing conditions to model an entire system. Adding more realistic engineering considerations for the players (and market-clearing conditions) would still constitute an MCP as long as the players' problems had valid KKT conditions. Lastly, Illustrative Example 1.2.8 showed a mathematical program with equilibrium constraints and why such problems are difficult to solve in some cases.

This chapter has been an overview of some of the types of energy market problems fitting within the format of complementarity problems or their extensions. In the rest of the book we explore in more detail various aspects of these sorts of problems from both a modeling and algorithmic perspective.

1.4 Appendix: Computational Issues for Selected Problems

In this section we provide computational details on some of the examples described above for further clarification.

1.4.1 Illustrative Example 1.2.1

Here we show with a simple analysis that $(x_1, x_2, y_1) = (0, 2, 0)$ is the unique solution to the problem stated above. First, using the one equation, we can replace the variable y_1 by $2 - x_1 - x_2$ and substitute this value back in to the remaining conditions to get the following after some algebra:

$$\begin{array}{lll} x_1 + x_2 \geq 0 & x_1 \geq 0 & (x_1 + x_2) \cdot x_1 = 0 \\ 2x_1 + x_2 - 2 \geq 0 & x_2 \geq 0 & (2x_1 + x_2 - 2) \cdot x_2 = 0 \end{array} \quad (1.33)$$

Four cases result, namely:

1. $x_1 > 0, x_2 > 0$
2. $x_1 > 0, x_2 = 0$
3. $x_1 = 0, x_2 > 0$
4. $x_1 = 0, x_2 = 0.$

In light of the above conditions (1.33), the first case means that $0 = (x_1 + x_2) = (2x_1 + x_2 - 2)$ or that $x_1 = 2, x_2 = -2$ which is not feasible since both variables have to be positive. The second case implies that $0 = (x_1 + x_2)$ and $x_2 = 0$ so that $x_1 = 0$, which is a contradiction to it being positive. Case 3 reveals that $(x_1, x_2) = (0, 2)$ which is feasible and gives $y_1 = 0$. Lastly, case 4 is not possible since it would violate $2x_1 + x_2 - 2 \geq 0$. This shows that the solution $(x_1, x_2, y_1) = (0, 2, 0)$ is unique but also highlights that an MCP can be solved (in principle) by exploring all the 2^n cases considering if $x_i > 0$ or equal to 0, for $i = 1, \ldots, n$, after substituting out any free variables using their associated equations.

Another way to view MCPs and this example in particular is geometrically. The four conditions $x_1 + x_2 \geq 0, x_1 \geq 0$, $2x_1 + x_2 - 2 \geq 0, x_2 \geq 0$ define the feasible region of this MCP and are depicted in Figure 1.2 as the intersection of half-spaces given by the dashed lines bounded by the bold ones. The complementarity conditions $(x_1 + x_2) \cdot x_1 = 0, (2x_1 + x_2 - 2) \cdot x_2 = 0$ can then be interpreted as the following:

- If the point has a positive first component it must be on the line $x_1 + x_2 = 0$; no such feasible point is possible.
- If the point has a positive second component it must be on the line $2x_1 + x_2 = 2$. This would then leave the line segment from $(1, 0)$ to $(0, 2)$ as

candidate solutions whose points are of the form: $\lambda \cdot (1,0) + (1-\lambda) \cdot (0,2)$ for $\lambda \in [0,1]$. If the weight $\lambda \in (0,1]$, this means that $x_1 > 0$ so by the earlier remark, the point would have to be on the line $x_1 + x_2 = 0$ which is not feasible. This leaves $\lambda = 0$ corresponding to $(0,2)$ as the only solution (considering that $(0,0)$ was indicated to be infeasible).

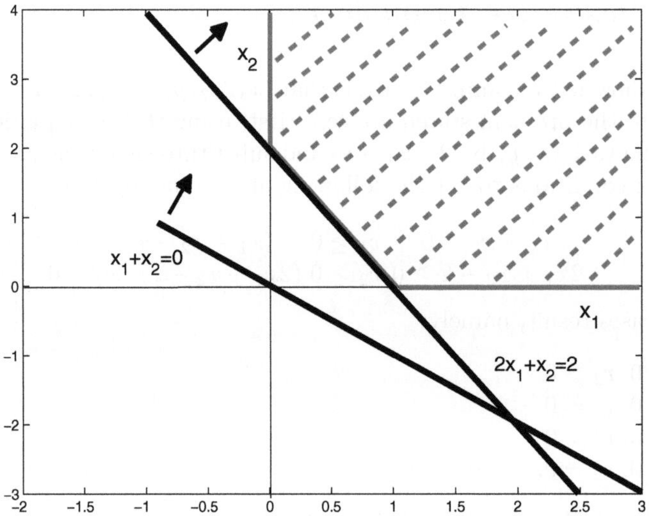

Fig. 1.2 Geometric depiction of Illustrative Example 1.2.1

1.4.2 Illustrative Example 1.2.4

If both production levels q_1 and q_2 are assumed to be positive, then the unique solution to the pure NCP duopoly example is to take q_1 and q_2 to satisfy

$$\begin{pmatrix} 2\beta & \beta \\ \beta & 2\beta \end{pmatrix} \begin{pmatrix} q_1 \\ q_2 \end{pmatrix} + \begin{pmatrix} -\alpha + \gamma_1 \\ -\alpha + \gamma_2 \end{pmatrix} = \begin{pmatrix} 0 \\ 0 \end{pmatrix} \tag{1.34}$$

which leads to

$$\begin{pmatrix} q_1 \\ q_2 \end{pmatrix} = \frac{1}{3\beta} \begin{pmatrix} \alpha - 2\gamma_1 + \gamma_2 \\ \alpha + \gamma_1 - 2\gamma_2 \end{pmatrix} \tag{1.35}$$

In light of the assumptions that $\beta, q_1, q_2 > 0$, this forces

$$\alpha > \max\{2\gamma_1 - \gamma_2, -\gamma_1 + 2\gamma_2\} \tag{1.36}$$

For example, when $\alpha = 10, \beta = 5, \gamma_1 = \gamma_2 = 1$, the solution is $q_1 = q_2 = 0.6$. When (1.36) does not hold, this does not necessarily mean that this NCP does not have a solution. Indeed, when $\alpha = \frac{1}{2}, \beta = 5, \gamma_1 = \gamma_2 = 1$ a solution is $q_1 = q_2 = 0$.

1.4.3 Illustrative Example 1.2.5

Assuming the new joint constraint $q_1 + q_2 \leq q^{TOT}$ appears in both players' problems and has the same Lagrange multiplier λ^{TOT}, then from [31] we know that there exists an MCP (or VI) solution to the problem. In that case, the resulting MCP will be:

$$0 \leq 2\beta q_1 + \beta q_2 - \alpha + \gamma_1 + \lambda^{TOT} \perp q_1 \geq 0 \text{ (producer 1)} \tag{1.37a}$$
$$0 \leq \beta q_1 + 2\beta q_2 - \alpha + \gamma_2 + \lambda^{TOT} \perp q_2 \geq 0 \text{ (producer 2)} \tag{1.37b}$$
$$0 \leq q^{TOT} - q_1 - q_2 \perp \lambda^{TOT} \geq 0 \text{ (joint constraint)} \tag{1.37c}$$

Since the previous problem's solution had $q_1 + q_2 = \frac{17}{15} \approx 1.13$, setting q^{TOT} to 1.06 forces a new solution which turns out to be:

$$\begin{pmatrix} q_1 \\ q_2 \\ \lambda^{TOT} \end{pmatrix} = \begin{pmatrix} 0.63 \\ 0.43 \\ 0.55 \end{pmatrix}$$

If potentially different values for λ^{TOT} were allowed, the resulting KKT system would be:

$$0 \leq 2\beta q_1 + \beta q_2 - \alpha + \gamma_1 + \lambda_1^{TOT} \perp q_1 \geq 0 \text{ (producer 1)} \tag{1.38a}$$
$$0 \leq \beta q_1 + 2\beta q_2 - \alpha + \gamma_2 + \lambda_2^{TOT} \perp q_2 \geq 0 \text{ (producer 2)} \tag{1.38b}$$
$$0 \leq q^{TOT} - q_1 - q_2 \perp \lambda_1^{TOT} \geq 0 \text{ (joint constraint, 1)} \tag{1.38c}$$
$$0 \leq q^{TOT} - q_1 - q_2 \perp \lambda_2^{TOT} \geq 0 \text{ (joint constraint, 2)} \tag{1.38d}$$

The resulting MCP would be of the form:

$$0 \leq \begin{pmatrix} -\alpha + \gamma_1 \\ -\alpha + \gamma_2 \\ q^{TOT} \\ q^{TOT} \end{pmatrix} + \begin{pmatrix} 2\beta & \beta & 1 & 0 \\ \beta & 2\beta & 0 & 1 \\ -1 & -1 & 0 & 0 \\ -1 & -1 & 0 & 0 \end{pmatrix} \begin{pmatrix} q_1 \\ q_2 \\ \lambda_1^{TOT} \\ \lambda_2^{TOT} \end{pmatrix} \perp \begin{pmatrix} q_1 \\ q_2 \\ \lambda_1^{TOT} \\ \lambda_2^{TOT} \end{pmatrix} \geq 0 \tag{1.39a}$$

Assuming $q_1, q_2 > 0$ but with $q_1 + q_2 = q^{TOT}$, would give:

$$\begin{pmatrix} 2\beta & \beta \\ \beta & 2\beta \end{pmatrix} \begin{pmatrix} q_1 \\ q_2 \end{pmatrix} = \begin{pmatrix} \alpha - \gamma_1 \\ \alpha - \gamma_2 \end{pmatrix} - \begin{pmatrix} \lambda_1^{TOT} \\ \lambda_2^{TOT} \end{pmatrix} \tag{1.40a}$$

$$q_1 + q_2 = q^{TOT} \tag{1.40b}$$

There are in general an infinite number of solutions to this system. To see this first note that we can substitute out q_2 via $q_2 = q^{TOT} - q_1$ to get:

$$\begin{pmatrix} 2\beta & \beta \\ \beta & 2\beta \end{pmatrix} \begin{pmatrix} q_1 \\ q^{TOT} - q_1 \end{pmatrix} = \begin{pmatrix} \alpha - \gamma_1 \\ \alpha - \gamma_2 \end{pmatrix} - \begin{pmatrix} \lambda_1^{TOT} \\ \lambda_2^{TOT} \end{pmatrix} \tag{1.41}$$

and solve for q_1 for both rows:

$$q_1 = \frac{\alpha - \gamma_1 - \beta q^{TOT} - \lambda_1^{TOT}}{\beta} \tag{1.42a}$$

$$q_1 = \frac{\alpha - \gamma_2 - 2\beta q^{TOT} - \lambda_2^{TOT}}{-\beta} \tag{1.42b}$$

then equate the right-hand sides to get:

$$\lambda_1^{TOT} + \lambda_2^{TOT} = 2\alpha - (\gamma_1 + \gamma_2) - 3\beta q^{TOT} \tag{1.43}$$

Assuming that the right-hand side is nonnegative (to be consistent with the nonnegative λ values), there are an infinite number of solutions for $\lambda_1^{TOT}, \lambda_2^{TOT}$ since it is one equation in two variables. With the choices for $\alpha = 10, \beta = 5, \gamma_1 = 1, \gamma_2 = 2, q^{TOT} = 1.06$, the right-hand side is 1.1. The solution obtained above where both λ values were equal to 0.55 was just one of a number of infinite choices. If there are an infinite number of values for $\lambda_1^{TOT}, \lambda_2^{TOT}$, then necessarily there are an infinite number of choices for q_1 and q_2 as well. Clearly since there is an equivalent VI to the linear MCP (1.38), the same result can be stated in terms of a variational inequality.

As noted in [31] more generally, the generalized Nash problem is an instance of a quasi-variational inequality problem differing from the classical VI formulation in the ground set (more about QVIs in Chapter 5). More specifically, having a point-to-set mapping $\Omega : R^l \to R^l$ and a function $F : \Omega \to R^l$, the quasi-variational inequality problem (QVI) is to find a vector $x^* \in \Omega(x^*)$ such that:

$$F(x^*)^T (y - x^*) \geq 0, \forall y \in \Omega(x^*)$$

As mentioned in [31], Bensoussan [9] is credited with showing that the generalized Nash equilibrium (an example shown above) is a special case of a QVI. Under certain conditions (see Theorem 4 in [31]), it is shown that there is an equivalent VI but the QVI is the more general formulation. Theorem 6 from [31] indicates that as long as the multipliers to shared constraints are the same (as well as other conditions holding), that this can only happen for VI solutions in the QVI solution set. Since solving VIs (or their associated MCPs) is easier than solving QVIs, this is an important computational result and has been used for example in a recent natural gas market application [1]. Thus, when trying to solve a generalized Nash problem like the one shown above in GAMS, if it is able to provide a solution, it must be the one that relates to the VI (or corresponding MCP) which means that the Lagrange multipliers to the shared constraints are equal.

1.4.4 Illustrative Example 1.2.7

If we examine the conditions for this energy network equilibrium we can see some interesting observations. For example, suppose that the sales for all four producers are positive. Then, by complementarity in (1.21a), (1.22a), (1.23a), and (1.24a), respectively,

$$s_1^A, s_1^B > 0 \Longrightarrow \pi_1 = \delta_1^A = \delta_1^B \tag{1.44a}$$

$$s_2^C, s_2^D > 0 \Longrightarrow \pi_2 = \delta_2^C = \delta_2^D \tag{1.44b}$$

Additionally if the flow variables f_{12}^A and f_{12}^B are assumed to be positive then by complementarity in (1.21c) and (1.22c) and combining the results from (1.44)

$$\pi_2 = \left(\tau_{12}^{\text{Reg}} + \tau_{12}\right) + \pi_1 \tag{1.45}$$

showing that in equilibrium, the delivered price at node 2 (π_2) is the sum of the production price at node 1 (π_1) and the transportation charges to get the energy from node 1 to node 2 ($\tau_{12}^{\text{Reg}} + \tau_{12}$).

For a concrete example of the type of relationships between the variables described above, consider the following input data with $D_i(\pi_i) = a_i - b_i\pi_i$ and:

$$\begin{aligned}
\tau_{12}^{\text{Reg}} &= 0.5\\
\gamma_{1A} &= 10, \gamma_{1B} = 12, \gamma_{2C} = 15, \gamma_{2D} = 18\\
a_1 &= 20, b_1 = 1, a_2 = 40, b_2 = 2\\
\overline{q}_1^A &= 10, \overline{q}_1^B = 10, \overline{q}_2^C = 5, \overline{q}_2^D = 5\\
\overline{g}_{12} &= 5\\
\gamma^{TSO} &= 1
\end{aligned}$$

Here, in order of increasing production costs, the producers are A, B, C, and D. A solution to this model as computed by GAMS is given in Figure

1.3 with the resulting demand of 8 and 10 units, respectively at nodes 1 and 2.

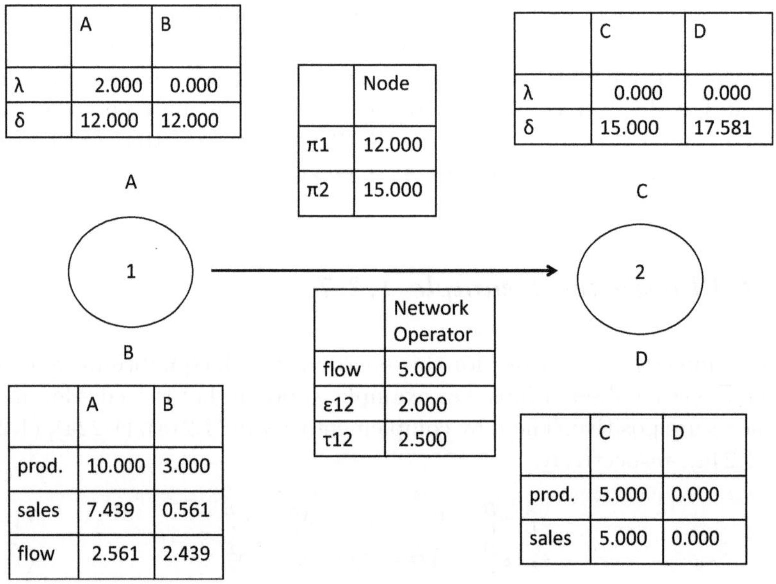

Fig. 1.3 A solution to Illustrative Example 1.2.7

First, we note that using the Producer A and B conditions (1.21) and (1.22), the node 1 price is determined to be $\pi_1 = \$12$. Additionally, from (1.45) we see that

$$\begin{aligned}\pi_2 &= \left(\tau_{12}^{\text{Reg}} + \tau_{12}\right) + \pi_1 \\ &= (0.5 + 2.5) + 12 \\ &= \$15\end{aligned}$$

For every unit produced by the least costly producer A, \$12 is made in revenue while only incurring a production cost of \$10 hence maximum production follows. For every unit produced, it can stay at node 1 (profit is \$2/unit) or can be sent to node 2 where it has a higher selling price of \$15 but transportation costs of $\left(\tau_{12}^{\text{Reg}} + \tau_{12}\right) = (0.5 + 2.5) = \$3$. Thus, all things being equal, this producer should be indifferent to these two markets to sell its energy. Of course all the other conditions need to be maintained. For producer A, this production is broken out into sales to node 1 (7.439) and flow sent to node 2 (2.561). Likewise, producer B also has positive production but only generates 3 out of 10 possible units (0.561 used at node 1 and the rest (2.439) sent on to node 2). The rationale for less production is that this producer is more expensive than A. Every unit produced by B will cost \$12 but sold at node 1 will bring a per unit revenue of only \$12. A similar cancelling out of revenues and costs happens for sending B's flow to node 2. Thus, node 1's demand of 8 units is met by 7.439 from producer A and 0.561 from producer B.

From node 2's point of view, the marginal cost of getting one more unit is due to C's production cost of \$15. It is indifferent between the various ways that this can be achieved (i.e., from node 1 at a cost of \$12+\$3 transportation or from node 2 at \$15). Thus, node 2's demand is met by 5 units transported from node 1 (2.561 from producer A and 2.439 from B) as well as 5 units produced locally by producer C at node 2.

Given the flexibility in some of the above decisions, one might ask is the solution presented unique? In this case it is not. If the variable s_2^C is given an initial guess of 100 instead of what GAMS automatically uses, we see the following differences with the previous solution to this energy network problem: respectively, producer A sales (s_1^A) and flow (f_{12}^A) are now 5.75,4.25; producer B sales (s_1^B) and flow (f_{12}^B) are now 2.25,0.75 (also δ_2^D changes from 17.581 to 18.000). While this is a different solution, it has the essential features of the previous one, namely, same production profile by each producer, same flow of 5 units from node 1 to node 2, same prices and tariffs as well as demands.

References

1. I. Abada, V. Briat, S.A. Gabriel, O. Massol. A generalized Nash-Cournot model for the European natural gas markets with a fuel substitution demand function: The GaMMES model, *Networks and Spatial Economics*, accepted, 2012.
2. B.H. Ahn. *Computation of Market Equilibria for Policy Analysis*, Garland Publishing, Inc., New York, 1979.
3. B.H. Ahn and W.W. Hogan. On convergence of the PIES algorithm for computing equilibria. *Operations Research*, 30:281-300,1982.
4. B. Allaz and J.-L. Vila. Cournot competition, forward markets and efficiency. *Journal of Economic Theory*, 59(1):1-16, 1993.
5. F.R. Aune, R. Golombek, S.A.C. Kittelsen, and K.E. Rosendahl. Liberalizing the energy markets of Western Europe-A computable equilibrium model approach. *Applied Economics*, 36:2137-2149, 2004.
6. M.S. Bazaraa, H.D. Sherali, and C.M. Shetty. *Nonlinear Programming Theory and Algorithms*, Wiley, New York, 1993.
7. L.A. Barroso, R.D. Carneiro, S. Granville, M. V. Pereira, and M.H.C. Fampa. Nash equilibrium in strategic bidding: a binary expansion approach. *IEEE Transactions on Power Systems*, 21(2):629-638, 2006.
8. BBC, http://news.bbc.co.uk/2/hi/7814743.stm, May 2012.
9. A. Bensoussan. Points de Nash dans le cas de fonctionnelles quadratiques et jeux différentiels linéaires à n personnes. *SIAM Journal Control*, 12:460-499, 1974.
10. J.R. Birge and F. Louveaux. *Introduction to Stochastic Programming*, Springer, New York, 1997.
11. G. Brown, M. Carlyle, J. Salmerón, and K. Wood. *Interfaces*, 36(6):530-544, 2006.
12. W. Chung, J.D. Fuller and Y.J. Wu. A new decomposition method for multi-regional economic equilibrium models. *Operations Research*, 54(4):643-655, 2006.
13. R.W. Cottle, J.-S. Pang, and R. E. Stone. *The Linear Complementarity Problem*, Academic Press, Inc., San Diego, 1992.
14. J. Dupačová, N. Gröwe-Kuska and W. Römisch. Scenario reduction in stochastic programming. *Mathematical Programming*, 93(3):493-511, 2003.
15. R. Egging and S.A. Gabriel. Examining market power in the European natural gas market. *Energy Policy*, 34(17):2762-2778, 2006.
16. R. Egging, F. Holz, C. von Hirschhausen, and S.A. Gabriel. Representing GASPEC with the World Gas Model. *The Energy Journal*, 30:97-118, 2009.
17. Energy Information Administration, www.eia.gov, May 2012.
18. F. Facchinei and J.-S. Pang. *Finite-Dimensional Variational Inequalities and Complementarity Problems, Volumes I and II.* Springer, New York, 2003.
19. J. Fortuny-Amat and B. McCarl. A representation and economic interpretation of a two-level programming problem. *Journal of the Operational Research Society*, 32(9):783-792, 1981.

20. S.A. Gabriel. Energy market equilibria expressed as complementarity problems, Manuscript, December, 2008.
21. S.A. Gabriel and J.D. Fuller. A Benders decomposition method for solving stochastic complementarity problems with an application in energy. *Computational Economics*, online publication February, 35:301-329, 2010.
22. S.A. Gabriel, S. Kiet, and J. Zhuang. A mixed complementarity-based equilibrium model of natural gas markets. *Operations Research*, 53(5):799-818, 2005.
23. S.A. Gabriel, A.S. Kydes, and P. Whitman. The National Energy Modeling System: a large-scale energy-economic equilibrium model. *Operations Research*, 49(1):14-25, 2001.
24. S.A. Gabriel and F.U. Leuthold. Solving discretely-constrained MPEC problems with applications in electric power markets. *Energy Economics*, 32:3-14, 2010.
25. S.A. Gabriel, Y. Shim, A.J. Conejo, S. de la Torre, and R. García-Bertrand. A Benders decomposition method for discretely-constrained mathematical programs with equilibrium constraints. *Journal of the Operational Research Society*, 61(9):1404-1419, 2010.
26. S.A. Gabriel, J. Zhuang, and R. Egging. Solving stochastic complementarity problems in energy market modeling using scenario reduction. *European Journal of Operational Research*, 197(3): 1028-1040, 2009.
27. S.A. Gabriel, J. Zhuang, and S. Kiet. A large-scale complementarity model of the North American natural gas market. *Energy Economics*, 27:639-665, 2005.
28. Gasping for gas. *The Economist*, p. 53, January 17, 2009.
29. T.S. Genc, S.S. Reynolds , and S. Sen. Dynamic oligopolistic games under uncertainty: a stochastic programming approach. *Journal of Economic Dynamics and Control*, 31(1):55-80, 2007.
30. N. Gröwe-Kuska, H. Heitsch, and W. Römisch. Scenario reduction and scenario tree construction for power management problems. *In Proceedings IEEE Bologna Power Tech*, 2003.
31. P.T. Harker. Generalized Nash games and quasi-variational inequalities. *European Journal of Operational Research*, 54(1):81-94, 1991.
32. H. Heitsch and W. Römisch. Scenario reduction algorithms in stochastic programming. *Computational Optimization and Applications*, 24:187-206, 2003.
33. B.F. Hobbs. Linear complementarity models of Nash-Cournot competition in bilateral and POOLCO power markets. *IEEE Transactions on Power Systems*, 16(2):194-202, 2001.
34. B.F. Hobbs, G. Drayton, E.B. Fisher, and W. Lise. Improved transmission representations in oligopolistic market models: quadratic losses, phase shifters, and dc lines. *IEEE Transactions on Power Systems,* 23(3):1018-1029,2008.
35. A.E. Kramer. Gazprom builds wealth for itself, but anxiety for others. *The New York Times*, January 13, 2006.
36. M. Landler. Gas halt may produce big ripples in european policy. *The New York Times*, January 4, 2006.
37. Z.Q. Luo, J.-S. Pang, and D. Ralph. *Mathematical Programs with Equilibrium Constraints*, Cambridge University Press, Cambridge, United Kingdom, 1996.
38. S. Meichtry. Eni faces threat to lucrative gas network. *The Wall Street Journal*, p. B8, November 10, 2009.
39. K.T. Midthun. Optimization models for liberalized natural gas markets., Ph.D. Thesis, Norwegian University of Science and Technology, Trondheim, Norway, 2007.
40. J.M. Morales, S. Pineda, A.J. Conejo, and M. Carrión. Scenario reduction for futures market trading in electricity markets. *IEEE Transactions on Power Systems*, 24(2):878-888, 2009.
41. Nabucco, www.nabucco-pipeline.com, May 2012.
42. Nord Stream, www.nord-stream.com, May 2012.
43. M.J. Osborne and A. Rubinstein. *A Course in Game Theory*, The MIT Press, Cambridge, Massachusetts, 1994.

44. S. Pineda and A.J. Conejo. Scenario reduction for risk-averse electricity trading. *IET Generation, Transmission & Distribution*, 4(6):694–705, 2010.
45. Pipe down. *The Economist*, pp. 44-45, January 10, 2009.
46. PJM interconection, www.pjm.com, May 2012.
47. W. Römisch, J. Dupačová, N. Gröwe-Kuska, and H. Heitsch, Approximations of stochastic programs. scenario tree reduction and construction. GAMS Workshop, Heidelberg, DFG Research Center Berlin, 2003.
48. C. Ruiz and A.J. Conejo. Pool strategy of a producer with endogenous formation of locational marginal prices. *IEEE Transactions on Power Systems*, 24(4):1855-1866, 2009.
49. E.E. Sauma and S.S. Oren. Economic criteria for planning transmission investment in restructured electricity markets, *IEEE Transactions on Power Systems*, 22(4):1394-1405, 2007.
50. Shy, O. *Industrial Organization Theory and Applications.* The MIT Press, Cambridge, MA, 1995.
51. South Stream, http://south-stream.info/?L=1, May 2012.
52. Testimony of Glen R. Thomas, On behalf of the PJM power providers group before the Pennsylvania Public Utility Commission public hearing on current and future wholesale electricity markets October 23, 2008, www.puc.state.pa.us/electric/pdf/EnBanc-WEM/Ttmy-PJM_Power_Provider102308.pdf
53. War-war, not jaw-jaw. *The Economist*, pp. 52-53, January 17, 2009.
54. J. Zhuang and S.A. Gabriel. A complementarity model for solving stochastic natural gas market equilibria. *Energy Economics,* 30(1):113-147, 2008.

Chapter 2
Optimality and Complementarity

2.1 Introduction

This chapter provides a friendly introduction to several mathematical structures used in the following chapters. These structures are useful to describe the functioning of markets and the behavior of market agents. Throughout the chapter clarity and simplicity are emphasized. The mathematical structures considered are:

1. Optimization Problems, OP.
2. Equilibrium Problems, EP.
3. Mathematical Programs with Equilibrium Constraints, MPEC.
4. Equilibrium Problems with Equilibrium Constraints, EPEC.

Additionally, the following topics related to the aforementioned mathematical structures are covered:

1. Karush-Kuhn-Tucker conditions, KKT.
2. Constraint qualifications.
3. Sufficiency conditions for optimality.
4. Mixed Linear Complementarity Problems, MLCP.

These mathematical structures are explained, illustrated and clarified through the use of a variety of examples.

The chapter is organized as follows. Section 2.2 describes the basic elements of an optimization problem. Section 2.3 establishes the optimality conditions (KKT conditions) that the solution of an optimization problem needs to meet to be optimal (necessary conditions). Section 2.4 describes the conditions (constraint qualifications) to be met by optimization problems so that optimality conditions can be formulated. Section 2.5 establishes sufficiency conditions to guarantee an optimal solution. Section 2.6 states the standard form of a mixed linear complementarity problem and establishes its relationship with the KKT conditions of one or several interrelated optimization problems. Section 2.7 defines the concept of equilibrium using the

KKT conditions of a set of interrelated optimization problems. Section 2.8 describes a mathematical program with equilibrium constraints using hierarchically related optimization problems. Section 2.9 defines an equilibrium problem with equilibrium constraints using a set of interrelated mathematical programs with equilibrium constraints. Section 2.10 makes several practical observations pertaining to the non-convex nature of some of the mathematical structures considered in this chapter. Finally, Section 2.11 summarizes the chapter providing some relevant conclusions.

Relevant references for optimization problems (and optimality conditions) include [3, 4, 12, 16], for complementarity and equilibrium problems [6, 9–11] and for MPECs and EPECs [2, 13].

2.2 Optimization Problems

An optimization problem or mathematical programming problem is a mathematical entity that allows maximizing or minimizing a certain objective (i.e., objective function) subject to diverse restrictions, typically in the form of equality or inequality constraints. An optimization problem has the general form

$$\text{Minimize}_x \quad f(x) \tag{2.1a}$$

s. t.

$$h(x) = 0 \tag{2.1b}$$

$$g(x) \leq 0 \tag{2.1c}$$

where $x \in \mathbb{R}^n$ is the optimization variable vector, $f(x) : \mathbb{R}^n \rightarrow \mathbb{R}$ is the objective function to be minimized, $h(x) : \mathbb{R}^n \rightarrow \mathbb{R}^{m_E}$ are the functions for the equality constraints and $g(x) : \mathbb{R}^n \rightarrow \mathbb{R}^{m_I}$ for the inequality constraints.

The following observations are in order:

1. Any $x \in \mathbb{R}^n$ is called a *solution*, while a solution meeting (2.1b)-(2.1c) is called a *feasible solution*, and a feasible solution minimizing (2.1a) is called an *optimal solution* or a *minimizer*. The objective function value at the minimizer is the objective function optimal value.
2. The set of solutions meeting both $h(x) = 0$ and $g(x) \leq 0$ constitutes the *feasible region*.
3. Minimizing $f(x)$ is equivalent to maximizing $-f(x)$.
4. A variety of algorithms are available to numerically solve problems of the form (2.1); see, for instance, [12].

Figure 2.1 illustrates the structure of an optimization problem specifying its components.

Optimization Problem (OP)

Objective function (minimize or maximize)

subject to:

Equality constraints
Inequality constraints
Bound on variables

Fig. 2.1 Optimization problem: Structure and components

The three illustrative examples below provide examples of optimization problems: with only equality constraints, unconstrained, and with both equality and inequality constraints.

2.2.1 Illustrative Example. Optimization Problem: Only Equality Constraints

We consider a producer i with the linear production cost

$$c_i = a_i + b_i(s_i + f_i) \tag{2.2}$$

where c_i is the cost function, a_i is the no-production fixed cost, b_i is the linear production cost, and f_i and s_i are the quantities sold by the producer in the futures and the spot markets, respectively. Note that a_i and b_i are constants, f_i and s_i are decisions variables (what the producer must decide), and c_i is a function of the decision variables. The futures and spot markets (two trading floors) are assumed to be cleared sequentially, first the futures market and then the spot market.

Throughout the chapter, we consider two such producers ($i = 1, 2$) and a single price for both the spot and the futures markets, which depends linearly on the total production, i.e.,

$$\pi = \gamma - \beta(s_1 + s_2 + f_1 + f_2) \tag{2.3}$$

where π is the actual price, $\gamma > 0$ is the no-demand price and $\beta > 0$ is the price-demand slope. Note that the assumption above, similar to the one presented in [1], implies no arbitrage (no possibility of advantageous trading) between the spot and the futures market. Note that γ and β are constants, s_1, s_2, f_1 and f_2 are decisions variables, and the price π is a function of the decision variables.

Fixing the production in the futures market and the production of producer 2 (in both the futures and the spot markets), the profit maximization (minus profit minimization) problem of producer 1 in the spot market is

$$\text{Minimize}_{\pi, c_1, s_1} \quad -\pi s_1 + c_1 \tag{2.4a}$$
s. t.
$$\pi = \gamma - \beta(s_1 + s_2 + f_1 + f_2) \tag{2.4b}$$
$$c_1 = a_1 + b_1(f_1 + s_1) \tag{2.4c}$$

The objective function is the negative profit of producer 1: minus revenue plus cost. For clarity, the price and the cost are explicitly stated as first and second constraints, respectively. Note that production s_1 needs to be nonnegative but this constraint is not explicitly imposed for the sake of simplicity. The optimization variables are π, c_1 and s_1.

Incorporating the cost into the objective function (thus eliminating the optimization variable c_1), problem (2.4) can be equivalently written as

$$\text{Minimize}_{\pi, s_1} \quad -\pi s_1 + \left[a_1 + b_1(f_1 + s_1)\right] \tag{2.5a}$$
s. t.
$$-\pi + \gamma - \beta(s_1 + s_2 + f_1 + f_2) = 0 \tag{2.5b}$$

To follow the general format of problem (2.1), the price equality constraint (2.5b) has been rewritten so that its right-hand-side is zero.

Problems (2.4) and (2.5) are instances of equality constrained optimization problems.

For clarity, consider the following data:

1. $\gamma = 100$ \$/MWh,
2. $\beta = \$0.005/(\text{MWh})^2$,
3. $a_1 = \$30,000$,
4. $a_2 = \$50,000$,
5. $b_1 = \$40/\text{MWh}$,
6. $b_2 = \$50/\text{MWh}$,
7. $f_1 = f_2 = 200$ MWh, and
8. $s_2 = 2000$ MWh.

Using the data above, optimization problem (2.4) becomes

$$\text{Minimize}_{\pi, c_1, s_1} \quad -\pi s_1 + c_1 \tag{2.6a}$$
s. t.
$$\pi = 100 - 0.005(s_1 + 2400) \tag{2.6b}$$
$$c_1 = 30,000 + 40(200 + s_1) \tag{2.6c}$$

and optimization problem (2.5) becomes

$$\text{Minimize}_{\pi,s_1} \quad -\pi s_1 + 30,000 + 40(200 + s_1) \tag{2.7a}$$

s. t.

$$-\pi + 100 - 0.005(s_1 + 2400) = 0 \tag{2.7b}$$

2.2.2 Illustrative Example. Optimization Problem: Unconstrained

Incorporating the price function (2.5b) into the objective function (2.5a) of problem (2.5) formulated in Illustrative Example 2.2.1, the equivalent unconstrained problem below is obtained by

$$\text{Minimize}_{s_1}$$
$$-\left[\gamma - \beta(s_1 + s_2 + f_1 + f_2)\right] s_1 + a_1 + b_1(f_1 + s_1) \tag{2.8}$$

The optimal solution of problem (2.8) above is

$$s_1 = \frac{\gamma - b_1 - \beta(s_2 + f_1 + f_2)}{2\beta} \tag{2.9}$$

The reader is invited to check through numerical experiments that (2.9) is the optimal and unique solution of problem (2.8).

Considering the numerical values in Illustrative Example 2.2.1, problem (2.8) becomes

$$\text{Minimize}_{s_1}$$
$$-(100 - 0.005(s_1 + 2000 + 200 + 200))s_1 + 30,000 + 40(200 + s_1) \tag{2.10}$$

Figure 2.2 depicts the objective function of problem (2.10). The negative profit is a quadratic function of s_1 that reaches its minimum for the quantity value provided by equation (2.9), which is $s_1 = 4800$ MWh. This production level corresponds to a profit of \$77,200. From equation (2.5b) we can compute the spot price value, $\pi = 64$ \$/MWh.

2.2.3 Illustrative Example. Optimization Problem: Equality and Inequality Constraints

If a bound is imposed on the production to be sold in the spot market (i.e., a capacity constraint) by producer 1, $s_1 \leq s_1^{\max}$, problem (2.5) becomes

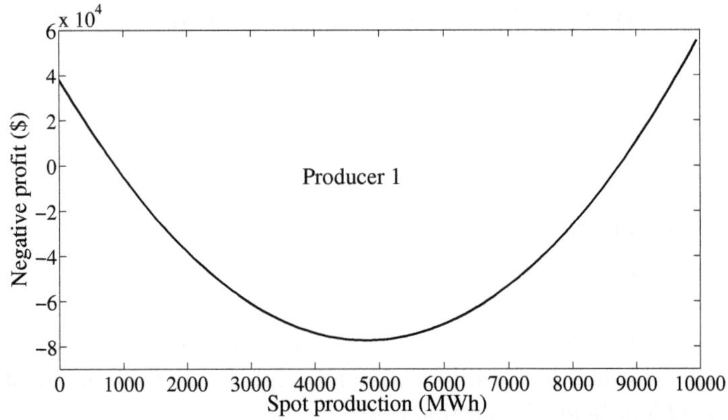

Fig. 2.2 Producer 1: Negative profit vs. spot production

$$\text{Minimize}_{\pi,s_1} \quad -\pi s_1 + \left[a_1 + b_1(f_1 + s_1)\right] \tag{2.11a}$$

s. t.

$$-\pi + \gamma - \beta(s_1 + s_2 + f_1 + f_2) = 0 \tag{2.11b}$$

$$s_1 - s_1^{\max} \leq 0 \tag{2.11c}$$

Objective function (2.11a) is the negative profit of producer 1 (minus revenue plus cost). Constraint (2.11b) expresses the price as a function of total production in both the spot and the futures market (producers 1 and 2), and constraint (2.11c) is a capacity bound on the production of producer 1. Problem (2.11) is an equality and inequality constrained optimization problem.

If the inequality constraint is binding for a particular s_1 value (i.e., it holds as an equality), the optimal solution of problem (2.11) is

$$s_1 = s_1^{\max} \tag{2.12}$$

$$\pi = \gamma - \beta(s_1^{\max} + s_2 + f_1 + f_2) \tag{2.13}$$

If the inequality constraint is not binding, the optimal solution is equal to the one in Illustrative Example 2.2.2 above.

Considering Figure 2.2, it can be concluded that if $s_1^{\max} \geq 4800$ MWh the optimal solution of problem (2.10) does not change with $s_1^{\max}$ and is $s_1 = 4800$ MWh. However, if $s_1^{\max} \leq 4800$ MWh the optimal solution is $s_1 = s_1^{\max}$.

2.2.4 Linear Optimization Problems

A particular class of optimization problems are linear programming (LP) problems. An LP is generally formulated as

$$\text{Minimize}_x \quad c^T x \tag{2.14a}$$

s. t.

$$Ax \geq b \tag{2.14b}$$

$$x \geq 0 \tag{2.14c}$$

where $c \in \mathbb{R}^n$, $b \in \mathbb{R}^m$, $A \in \mathbb{R}^{m \times n}$, and superscript ^{T} indicates the transpose of a vector/matrix. The optimization variable vector is $x \in \mathbb{R}^n$.

Note that constraints $Ax \geq b$ can be alternatively stated (without loss of generality) as $Ax = b$ by adding additional optimization (slack) variables. These are positive variables that make up the difference between "$\geq$" and "$=$".

Directly linked to the primal LP problem (problem (2.14) above) is its dual LP problem that has the form

$$\text{Maximize}_\lambda \quad \lambda^T b \tag{2.15a}$$

s. t.

$$\lambda^T A \leq c^T \tag{2.15b}$$

$$\lambda \geq 0 \tag{2.15c}$$

where the optimization variable vector is $\lambda \in \mathbb{R}^m$.

A relevant result is the Strong Duality Theorem (see [12] for further details) that states that if x is an optimal solution of the primal problem (2.14) and λ is an optimal solution of the dual problem (2.15), then

$$c^T x = \lambda^T b \tag{2.16}$$

Additionally, it can be shown that at an optimal solution,

$$\lambda_j = \frac{\Delta(c^T x)}{\Delta b_j}, \quad \forall j \tag{2.17}$$

for small enough values of Δb_j (change in b_j). This means that λ_j is the sensitivity of the objective function of the primal problem (at the optimal solution) with respect to the right-hand-side parameter b_j of that primal problem.

2.2.5 *Illustrative Example. LP Problem: Primal-Dual Formulation*

We consider the following linear programming problem:

$$\begin{aligned}
&\text{Minimize}_{x_1,x_2} \quad 3x_1 - 2x_2 && \text{(2.18a)}\\
&\text{s. t.}\\
&\quad -x_1 - x_2 \geq -3 && \text{(2.18b)}\\
&\quad 2x_1 - x_2 \geq 1 && \text{(2.18c)}\\
&\quad x_1 \geq 0 && \text{(2.18d)}\\
&\quad x_2 \geq 0 && \text{(2.18e)}
\end{aligned}$$

whose optimal solution is $(x_1, x_2) = (\frac{4}{3}, \frac{5}{3})$, which corresponds to an optimal objective function value of $\frac{2}{3}$.

According to (2.15), the dual LP problem of (2.18) is

$$\begin{aligned}
&\text{Maximize}_{\lambda_1,\lambda_2} \quad -3\lambda_1 + \lambda_2 && \text{(2.19a)}\\
&\text{s. t.}\\
&\quad -\lambda_1 + 2\lambda_2 \leq 3 && \text{(2.19b)}\\
&\quad -\lambda_1 - \lambda_2 \leq -2 && \text{(2.19c)}\\
&\quad \lambda_1 \geq 0 && \text{(2.19d)}\\
&\quad \lambda_2 \geq 0 && \text{(2.19e)}
\end{aligned}$$

The optimal solution of the above problem is $(\lambda_1, \lambda_2) = (\frac{1}{3}, \frac{5}{3})$. The reader is invited to check that, at this optimal solution, the objective function value is $\frac{2}{3}$ and coincides with the objective function value of the primal problem. This is what the Strong Duality Theorem says.

We consider below a modified version of problem (2.18)

$$\begin{aligned}
&\text{Minimize}_{x_1,x_2} \quad 3x_1 - 2x_2 && \text{(2.20a)}\\
&\text{s. t.}\\
&\quad -x_1 - x_2 \geq -4 && \text{(2.20b)}\\
&\quad 2x_1 - x_2 \geq 1 && \text{(2.20c)}\\
&\quad x_1 \geq 0 && \text{(2.20d)}\\
&\quad x_2 \geq 0 && \text{(2.20e)}
\end{aligned}$$

with a different right-hand-side in equation (2.20b), -4 instead of -3. The optimal solution of problem (2.20) is $(x_1, x_2) = (\frac{5}{3}, \frac{7}{3})$, which corresponds to an objective function optimal value of $\frac{1}{3}$.

At the optimal solutions, note that the sensitivity of the objective function (2.18a) with respect to the right-hand-side of equation (2.18b) is $\frac{1}{3}$, i.e, at the optimal solution, a decrease of one unit in the right-hand-side of equation (2.18b) translates into a decrease in the objective function value of $\frac{1}{3}$. This is exactly the value of λ_1, as stated in (2.17).

2.3 Karush-Kuhn-Tucker Conditions

The Karush-Kuhn-Tucker (KKT) conditions are conditions that the optimal solutions of a broad range of optimization problems should satisfy. For some problems the KKT conditions cannot be meaningfully formulated, and thus they cannot characterize optimal solutions for these problems.

Additionally, KKT conditions can be necessary but not sufficient conditions, i.e., solutions meeting them are not necessarily optimal but optimal solutions need to meet them. Also, KKT conditions are first-order conditions, i.e., conditions that are formulated using first derivative vectors and matrices (gradients and Jacobians).

To formulate the KKT conditions it is convenient to define the Lagrangian function. The Lagrangian function of problem (2.1) is

$$\mathcal{L} = f(x) + \lambda^T h(x) + \mu^T g(x) \tag{2.21}$$

where $f(x)$, $h(x)$ and $g(x)$ are continuously differentiable in the *feasible region* $(x \in \{x | h(x) = 0, g(x) \leq 0\})$.

The KKT conditions of problem (2.1) are

$$\nabla_x f(x) + \lambda^T \nabla_x h(x) + \mu^T \nabla_x g(x) = 0 \tag{2.22a}$$

$$h(x) = 0 \tag{2.22b}$$

$$g(x) \leq 0 \tag{2.22c}$$

$$\mu^T g(x) = 0 \tag{2.22d}$$

$$\mu \geq 0 \tag{2.22e}$$

where $\lambda \in \mathbb{R}^{m_E}$ and $\mu \in \mathbb{R}^{m_I}$ are respectively, the equality and inequality Lagrange multiplier vectors, and ∇_x denotes the gradient with respect to x.

Constraint (2.22a) states that the gradient of the Lagrangian function of (2.21) at an optimal solution x should be zero. Constraint (2.22b) states the equality constraints. Constraint (2.22c) enforces the inequality constraints.

Constraint (2.22d) states that inner product of the multiplier vector of the inequality constraints and the inequality constraint vector is zero. Constraint (2.22e) states that the multiplier vector of the inequality constraints is non-negative.

Constraints (2.22c)-(2.22e) are known as complementarity conditions and are equivalently written as

$$0 \leq \mu \perp g(x) \leq 0 \tag{2.23}$$

where $\perp$ indicates complementarity, i.e., $\mu^T g(x) = 0$. Throughout the chapter we often use this compact notation.

In summary, under the appropriate convexity conditions, a solution x of problem (2.1) to be a minimizer needs to meet KKT conditions (2.22).

It can be shown that vector λ is both the sensitivity vector (as in the linear case above, problems (2.14) and (2.15)) and the Lagrange multiplier vector. The interested reader is referred to [12] for further details.

The two illustrative examples below clarify how the KKT conditions work.

2.3.1 Illustrative Example. KKT Conditions: Equality Constraints

The Lagrangian function of problem (2.5) is

$$\mathcal{L}(\pi, s_1, \lambda) = \\ -\pi s_1 + \big[a_1 + b_1(f_1 + s_1)\big] + \lambda\big[-\pi + \gamma - \beta(s_1 + s_2 + f_1 + f_2)\big] \tag{2.24}$$

The KKT conditions for problem (2.5) are

$$-s_1 - \lambda = 0 \tag{2.25a}$$
$$-\pi + b_1 - \lambda\beta = 0 \tag{2.25b}$$
$$-\pi + \gamma - \beta(s_1 + s_2 + f_1 + f_2) = 0 \tag{2.25c}$$

where λ is the multiplier associated with the equality constraint (2.5b). Equations (2.25a), (2.25b) and (2.25c) are obtained by taking derivatives of the Lagrangian function with respect to π, s_1 and λ, respectively.

On the other hand, the KKT condition for problem (2.8) is obtained by differentiating the objective function with respect to s_1 and setting this equal to zero,

$$\beta s_1 - \gamma + \beta(s_1 + s_2 + f_1 + f_2) + b_1 = 0 \tag{2.26}$$

Note that condition (2.26) is equivalent to conditions (2.25), as they correspond to two formulations of the same problem. However, condition (2.26) does not provide Lagrangian multiplier (dual) information.

The KKT conditions (2.25) for the numerical values provided in Illustrative Example 2.2.1 are

$$-s_1 - \lambda = 0 \tag{2.27a}$$
$$-\pi - 0.005\lambda + 40 = 0 \tag{2.27b}$$
$$-\pi - 0.005 s_1 + 88 = 0 \tag{2.27c}$$

which constitute a linear system of equations whose solution is $s_1 = 4800$ MWh, $\pi = 64$ \$/MWh and $\lambda = -4800$ MWh, which are the same values provided in Illustrative Example 2.2.2.

2.3.2 Illustrative Example. KKT Conditions: Equality and Inequality Constraints

The Lagrangian function of problem (2.11) is

$$\begin{aligned}\mathcal{L}(\pi, s_1, \lambda, \mu) = & -\pi s_1 + \left[a_1 + b_1(f_1 + s_1)\right] \\ & + \lambda\left[-\pi + \gamma - \beta(s_1 + s_2 + f_1 + f_2)\right] + \mu(s_1 - s_1^{\max}). \end{aligned} \tag{2.28}$$

The KKT conditions for problem (2.11) are

$$-\pi + b_1 - \lambda\beta + \mu = 0 \tag{2.29a}$$
$$-s_1 - \lambda = 0 \tag{2.29b}$$
$$-\pi + \gamma - \beta(s_1 + s_2 + f_1 + f_2) = 0 \tag{2.29c}$$
$$s_1 \leq s_1^{\max} \tag{2.29d}$$
$$\mu(s_1 - s_1^{\max}) = 0 \tag{2.29e}$$
$$\mu \geq 0 \tag{2.29f}$$

where λ is the multiplier associated with the equality constraint (2.11b), and μ is the multiplier associated with the inequality constraint (2.11c). Constraints (2.29a), (2.29b), (2.29c) and (2.29d) are obtained by differentiating the Lagrangian function with respect to s_1, π, λ and μ, respectively. Constraint (2.29e) enforces that the product of inequality constraint (2.11c) and its multiplier is zero, and constraint (2.29f) enforces the nonnegativity of multiplier μ.

Considering the data provided in Illustrative Example 2.2.1, the system of equations (2.29) becomes

$$-\pi + 40 - 0.005\lambda + \mu = 0 \tag{2.30a}$$
$$-s_1 - \lambda = 0 \tag{2.30b}$$
$$-\pi + 100 - 0.005(s_1 + 2400) = 0 \tag{2.30c}$$

$$s_1 \leq s_1^{\max} \tag{2.30d}$$

$$\mu(s_1 - s_1^{\max}) = 0 \tag{2.30e}$$

$$\mu \geq 0 \tag{2.30f}$$

After some simplifications, system of equations (2.30) becomes

$$(4800 - s_1)(s_1 - s_1^{\max}) = 0 \tag{2.31a}$$

$$s_1 \leq 4800 \tag{2.31b}$$

$$s_1 \leq s_1^{\max} \tag{2.31c}$$

From equations (2.31) note that if $s_1^{\max} > 4800$ MWh then $s_1 = 4800$ MWh, and if $s_1^{\max} \leq 4800$ MWh then $s_1 = s_1^{\max}$. This is the result provided in Illustrative Example 2.2.3.

2.4 Constraint Qualifications

For the KKT conditions to be *meaningful*, the constraints of the considered problem need to meet certain qualifications at each optimal solution. In other words, solutions to KKT conditions correspond to the optimal solutions of those problems whose constraints meet constraint qualifications at the optimal solutions, but not for all optimization problems. Fortunately, many problems of practical interest have constraint qualifications that hold at the corresponding optimal solutions.

If the constraints of the problem under consideration meet a constraint qualification, the KKT conditions are necessary conditions. Being necessary means that at any optimal solution they must hold necessarily. Being sufficient means that a solution to these conditions is a solution to solving the original problem.

Several constraint qualification criteria are available. In particular, we state below the *linear independence constraint qualification* (LICQ), a common and simple one.

If the gradients of (a) the equality constraints $h(x) = 0$ and (b) the binding inequality constraints $g(x) \leq 0$ are linearly independent at a given feasible solution x, this solution is said to be regular, KKT conditions can be *meaningfully* written for x, and thus, the constraints meet the qualification criterion at x. Note that linearity in $h(x)$ and $g(x)$ implies that the constraint qualification criterion is met.

The two illustrative examples below clarify the constraint qualification criterion.

2.4.1 Illustrative Example. Constraint Qualification: Regular Solution

Consider problem (2.11a)-(2.11c). The gradient vectors of constraints (2.11b) and (2.11c) are, respectively,

$$\nabla h_{(\pi,s_1)} = \begin{bmatrix} -1 \\ -\beta \end{bmatrix}, \quad \nabla g_{(\pi,s_1)} = \begin{bmatrix} 0 \\ 1 \end{bmatrix} \tag{2.32}$$

Since these vector are always linearly independent, the constraint qualification is met at any solution.

2.4.2 Illustrative Example. Constraint Qualification: Non-Regular Solution

Consider the following optimization problem

$$\text{Minimize}_{x,y} \quad -y \tag{2.33a}$$

s. t.

$$x^2 + y^2 = 1 \tag{2.33b}$$

$$y + x^2 - 1 \leq 0 \tag{2.33c}$$

A optimal solution for problem (2.33) above is $(x, y) = (0, 1)$. The gradient vectors of constraints (2.33b) and (2.33c) evaluated at the optimal solution are, respectively,

$$\nabla h_{(x,y)} = \begin{bmatrix} 2x \\ 2y \end{bmatrix} = \begin{bmatrix} 0 \\ 2 \end{bmatrix}, \quad \nabla g_{(x,y)} = \begin{bmatrix} 2x \\ 1 \end{bmatrix} = \begin{bmatrix} 0 \\ 1 \end{bmatrix} \tag{2.34}$$

Note that the above gradient vectors are linearly dependent. Therefore the constraints do not meet the LICQ criterion at the optimal solution and the KKT conditions are not necessary for optimal solutions to this problem. In particular, the KKT conditions for problem (2.33) are

$$2\lambda x + 2\mu x = 0 \tag{2.35a}$$

$$-1 + 2\lambda y + \mu = 0 \tag{2.35b}$$

$$x^2 + y^2 = 1 \tag{2.35c}$$

$$\mu(y + x^2 - 1) = 0 \tag{2.35d}$$

$$\mu \geq 0 \tag{2.35e}$$

$$y \leq -x^2 + 1 \tag{2.35f}$$

where λ and μ are, respectively, the Lagrange multipliers for constraints (2.33b) and (2.33c). Note that at the optimal solution, $(x, y) = (0, 1)$, the system of equations (2.35) becomes

$$-1 + 2\lambda + \mu = 0 \tag{2.36a}$$

$$\mu \geq 0 \tag{2.36b}$$

but since λ is unconstrained and μ nonnegative there are an infinite number of solutions. Note also that $(x, y) = (0, -1)$ satisfies the KKT conditions (2.35) despite not being an optimal solution. Thus, the KKT conditions are not *meaningful.*

2.5 Sufficiency Conditions

We state below sufficiency conditions. Solving these conditions is equivalent to solving the problem under consideration.

A solution x of problem (2.1) meeting KKT conditions, where $f(x)$ is a continuously differentiable convex function and $h(x) = 0$ and $g(x) \leq 0$ constitute a convex set, is guaranteed to be an optimal solution.

Additionally, the condition (2.37) below guarantees x to be an optimal solution. This sufficiency condition is

$$\begin{aligned} &\nabla_x^2 f(x) + \lambda^T \nabla_x^2 h(x) + \mu^T \nabla_x^2 g(x) > 0 \text{ on the subspace} \\ &\{y : \nabla_x h(x) y = 0; \nabla_x g_j(x) y = 0, \forall j \in \mathcal{J}\}, \mathcal{J} = j : g_j(x) = 0, \mu_j > 0 \end{aligned} \tag{2.37}$$

Note that

$$\{y : \nabla_x h(x) y = 0; \nabla_x g_j(x) y = 0, \forall j \in \mathcal{J}\}, \mathcal{J} = j : g_j(x) = 0 \mu_j > 0 \tag{2.38}$$

is the tangent hyperplane at x, and that

$$\lambda^T \nabla_x^2 h(x) = \sum_{i=1}^{m_E} \lambda_i^T \nabla_x^2 h_i(x) \tag{2.39}$$

In words, the condition (2.37) implies that the Hessian of the Lagrangian at x should be positive definite on the tangent hyperplane at x.

Observe that sufficiency conditions are second order conditions as they rely on second derivative matrices (Hessians).

The illustrative example below clarifies the sufficiency conditions of an optimization problem.

2.5.1 Illustrative Example. Sufficiency Conditions

Consider problem (2.11) where $f(x) = -\pi s_1 + a_1 + b_1(f_1 + s_1)$, $h(x) = -\pi + \gamma - \beta(s_1 + s_2 + f_1 + f_2)$ and $g(x) = s_1 - s_1^{\max}$ (note that $x = [\pi, s_1]^T$) The gradient vectors and Hessian matrices are provided below.

$$\nabla_x f = \begin{bmatrix} -s_1 \\ -\pi + b_1 \end{bmatrix}, \quad \nabla_x h = \begin{bmatrix} -1 \\ -\beta \end{bmatrix}, \quad \nabla_x g = \begin{bmatrix} 0 \\ 1 \end{bmatrix} \tag{2.40}$$

$$\nabla_x^2 f = \begin{bmatrix} 0 & -1 \\ -1 & 0 \end{bmatrix}, \quad \nabla_x^2 h = \begin{bmatrix} 0 & 0 \\ 0 & 0 \end{bmatrix}, \quad \nabla_x^2 g = \begin{bmatrix} 0 & 0 \\ 0 & 0 \end{bmatrix} \tag{2.41}$$

The Hessian of the Lagrangian becomes

$$\nabla_x^2 f + \lambda^T \nabla_x^2 h + \mu^T \nabla_x^2 g = \begin{bmatrix} 0 & -1 \\ -1 & 0 \end{bmatrix} \tag{2.42}$$

On the other hand, considering that the inequality constraint is not binding, solutions y on the tangent hyperplane meet $y^T \nabla_x h = 0$, thus

$$\begin{bmatrix} y_1 \; y_2 \end{bmatrix} \begin{bmatrix} -1 \\ -\beta \end{bmatrix} = 0 \tag{2.43}$$

and

$$y_1 = -\beta y_2 \tag{2.44}$$

Considering the Hessian of the Lagrangian and solutions belonging to the tangent hyperplane allows showing that this Hessian is definite positive on that subspace, i.e.,

$$\begin{bmatrix} -\beta y_2 \; y_2 \end{bmatrix} \begin{bmatrix} 0 & -1 \\ -1 & 0 \end{bmatrix} \begin{bmatrix} -\beta y_2 \\ y_2 \end{bmatrix} = 2\beta y_2^2 > 0, \; \forall y > 0 \text{ (in subspace)} \tag{2.45}$$

Thus, problem (2.11) meets the second order sufficiency conditions for all solutions.

Using the data provided in Illustrative Example 2.2.1, the gradient vectors are

$$\nabla f_x = \begin{bmatrix} -s_1 \\ -\pi + 40 \end{bmatrix}, \quad \nabla h_x = \begin{bmatrix} -1 \\ -0.005 \end{bmatrix}, \quad \nabla g_x = \begin{bmatrix} 0 \\ 1 \end{bmatrix} \tag{2.46}$$

The Hessian matrices coincide with the ones provided by equations (2.41) and (2.42).

The tangent hyperplane is characterized by

$$y_1 = -0.005y_2 \tag{2.47}$$

Finally, considering the Hessian of the Lagrangian and solutions belonging to the tangent hyperplane, we obtain

$$\begin{bmatrix} -0.005y_2 \ y_2 \end{bmatrix} \begin{bmatrix} 0 & -1 \\ -1 & 0 \end{bmatrix} \begin{bmatrix} -0.005y_2 \\ y_2 \end{bmatrix} = 0.01y_2^2 \tag{2.48}$$

which is always positive for all solutions y, showing that the Hessian of the Lagrangian is positive definite and that second order sufficiency conditions are met for all solutions of problem (2.11).

2.6 Mixed Linear Complementarity Problem, MLCP

A Mixed Linear Complementarity Problem, MLCP, is a system of equalities and inequalities of a certain form that relates to KKT conditions. More general definitions can be established but they are not used in this chapter. The general form of an MLCP is stated below.

Find $z_1 \in \mathbb{R}^n$ and $z_2 \in \mathbb{R}^m$ such that

$$q_1 + M_{11}z_1 + M_{12}z_2 \geq 0 \tag{2.49a}$$

$$q_2 + M_{21}z_1 + M_{22}z_2 = 0 \tag{2.49b}$$

$$z_1 \geq 0 \tag{2.49c}$$

$$z_1^T(q_1 + M_{11}z_1 + M_{12}z_2) = 0 \tag{2.49d}$$

where $q_1 \in \mathbb{R}^n$, $q_2 \in \mathbb{R}^m$, $M_{11} \in \mathbb{R}^{n\times n}$, $M_{22} \in \mathbb{R}^{m\times m}$, $M_{12} \in \mathbb{R}^{n\times m}$, and $M_{21} \in \mathbb{R}^{m\times n}$. Note also that z_2 is unconstrained.

Constraints (2.49a), (2.49c) and (2.49d) allow formulating the complementarity conditions

$$0 \leq z_1 \perp q_1 + M_{11}z_1 + M_{12}z_2 \geq 0 \tag{2.50}$$

while equations (2.49b) are other equality constraints.

Since MLCPs are related to KKT conditions, these conditions are relevant in studying complementarity problems.

The illustrative example below clarifies the structure of an MLCP.

2.6.1 Illustrative Example. MLCP

The KKT conditions (2.29) of problem (2.11) in Illustrative Example 2.2.3 constitute an MLCP where

$$z_1 = \begin{bmatrix} \mu \end{bmatrix}, \ z_2 = \begin{bmatrix} s_1 \\ \pi \\ \lambda \end{bmatrix}, \ q_1 = \begin{bmatrix} s_1^{\max} \end{bmatrix}, \ q_2 = \begin{bmatrix} b_1 \\ 0 \\ \gamma - \beta(s_2 + f_1 + f_2) \end{bmatrix},$$

$$M_{11} = \begin{bmatrix} 0 \end{bmatrix}, \ M_{12} = \begin{bmatrix} -1 \ 0 \ 0 \end{bmatrix}, \ M_{22} = \begin{bmatrix} 0 & -1 & -\beta \\ -1 & 0 & -1 \\ -\beta & -1 & 0 \end{bmatrix}, \ M_{21} = \begin{bmatrix} 1 \\ 0 \\ 0 \end{bmatrix}$$

Note that $n = 1$ and $m = 3$, so that $z_1 \in \mathbb{R}^1$, $z_2 \in \mathbb{R}^3$, $q_1 \in \mathbb{R}^1$, $q_2 \in \mathbb{R}^3$, $M_{11} \in \mathbb{R}^{1\times 1}$, $M_{22} \in \mathbb{R}^{3\times 3}$, $M_{12} \in \mathbb{R}^{1\times 3}$, and $M_{21} \in \mathbb{R}^{3\times 1}$.

2.7 Equilibrium Problems, EP

For the purposes of this chapter, an equilibrium problem is a mathematical entity constituted by the joint consideration of the KKT conditions of several interrelated optimization problems. More general definitions of equilibrium problems are also possible but they are not considered in this chapter but rather in Chapters 3 and 4.

Consider the $i = 1, \dots, n$ problems below

$$\text{Minimize}_{x^i} \quad f^i(x^1, \dots, x^n) \tag{2.51a}$$

s. t.

$$h^i(x^1, \dots, x^n) = 0 \tag{2.51b}$$

$$g^i(x^1, \dots, x^n) \le 0 \tag{2.51c}$$

The vector $x^i \in \mathbb{R}^{n^i}$ includes the set of optimization variables of problem i. Setting $n_T = \sum_{i=1}^{n} n^i$, the objective functions and the constraints of problem i are defined as follows, $f^i : \mathbb{R}^{n_T} \rightarrow \mathbb{R}$, $h^i : \mathbb{R}^{n_T} \rightarrow \mathbb{R}^{m_E^i}$, and $g^i : \mathbb{R}^{n_T} \rightarrow \mathbb{R}^{m_I^i}$.

It is important to note that although the decision variables of problem i are just those in the vector x^i, the objective function and the constraints of problem i, $f^i(x^1, \dots, x^n)$, $h^i(x^1, \dots, x^n) = 0$ and $g^i(x^1, \dots, x^n) \le 0$, depend on the decision variables of all problems, $i = 1, \dots, n$. This fact makes the n problems considered interrelated, which results in a well-posed equilibrium problem. Specifically, this problem is an instance of the Generalized Nash Equilibrium Problem (GNEP), first proposed by Debreu in 1952, [7]. The GNEP is a generalization of the well known Nash Equilibrium Problem (NEP), [14, 15]. In an GNEP each player's set of constraints depends on the rival players' decisions; that is, the actions of other players are not considered fixed for establishing the decisions of each player. The analysis of the properties of the GNEP is out of the scope of this chapter but the interested reader is referred to [8] for further details as well as additional discussion in Chapters 3 and 4.

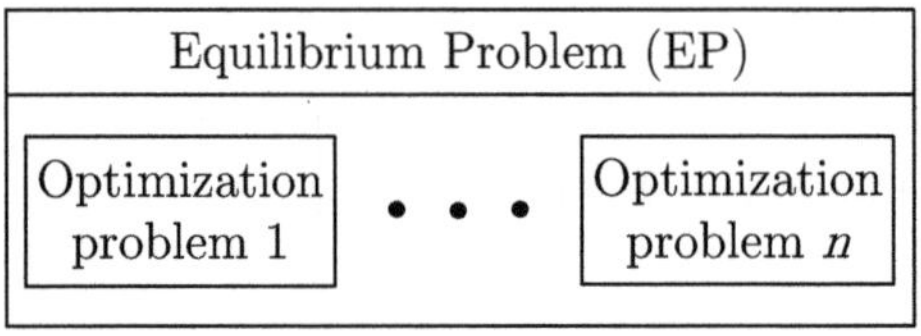

Fig. 2.3 Equilibrium problem: Joint solution of a number of interrelated optimization problems

The corresponding equilibrium problem is defined by jointly solving or considering the KKT conditions of the n problems (2.51) above, i.e.,

$$\nabla_{x^i} f^i(x^1,\dots,x^n) + \lambda^{i^T} \nabla_{x^i} h^i(x^1,\dots,x^n) + \mu^{i^T} \nabla_{x^i} g^i(x^1,\dots,x^n) = 0, \quad i = 1,\dots,n \tag{2.52a}$$

$$h^i(x^1,\dots,x^n) = 0, \quad i = 1,\dots,n \tag{2.52b}$$

$$0 \le \mu^i \perp - g(x^1,\dots,x^n) \ge 0, \quad i = 1,\dots,n \tag{2.52c}$$

where $\lambda^i \in \mathbb{R}^{m_E^i}$ and $\mu^i \in \mathbb{R}^{m_I^i}$ are the Lagrangian multiplier vectors corresponding to the equality and inequality constraints of problem i, $h^i(x^1,\dots,x^n) = 0$ and $g^i(x^1,\dots,x^n) \le 0$, respectively.

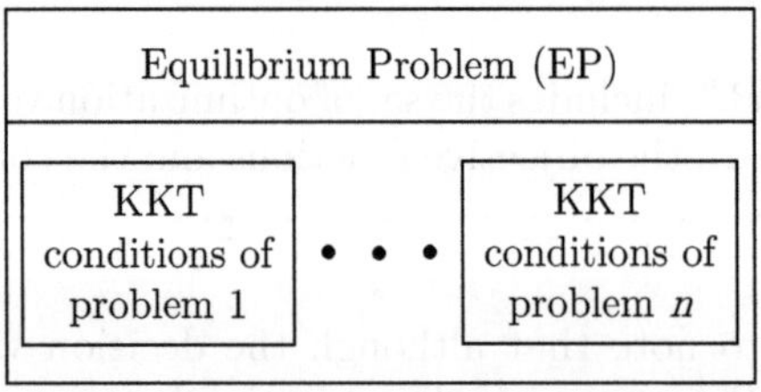

Fig. 2.4 Equilibrium problem (KKT): Joint solution of several systems of KKT conditions

The structure of the equilibrium problem under consideration is illustrated in Figures 2.3 and 2.4. The first of these figures illustrates the structure of the equilibrium problem jointly considering a set of interrelated optimization problems, while the second one considering a set of systems of interrelated KKT conditions.

The three illustrative examples below clarify equilibrium problems involving no constraints, only equality constraints, and both equality and inequality constraints.

2.7.1 Illustrative Example. Equilibrium Conditions: No Constraints

The single optimality condition for the problem of the first producer, problem (2.8) in Illustrative Example 2.2.2, and the single optimality condition of the problem of producer 2 (similar to that of producer 1), stated jointly, constitute an equilibrium problem, i.e.,

$$-\beta s_1 + (\gamma - \beta(s_1 + s_2 + f_1 + f_2)) - b_1 = 0 \tag{2.53a}$$

$$-\beta s_2 + (\gamma - \beta(s_1 + s_2 + f_1 + f_2)) - b_2 = 0 \tag{2.53b}$$

The unknowns of the linear system of equations (2.53) above are s_1 and s_2, and a solution is

$$s_1 = \frac{\gamma + b_2 - 2b_1 - \beta(f_1 + f_2)}{3\beta} \tag{2.54a}$$

$$s_2 = \frac{\gamma + b_1 - 2b_2 - \beta(f_1 + f_2)}{3\beta} \tag{2.54b}$$

which considering (2.3) leads to the price

$$\pi = \frac{\gamma + b_1 + b_2 - \beta(f_1 + f_2)}{3} \tag{2.55}$$

Using the numerical values provided in Illustrative Example 2.2.1, the system of equations (2.53) becomes

$$-0.01s_1 - 0.005s_2 + 58 = 0 \tag{2.56a}$$

$$-0.005s_1 - 0.01s_2 + 48 = 0 \tag{2.56b}$$

whose solution is $s_1 = 4533.3$ MWh, $s_2 = 2533.3$ MWh. We can also compute the equilibrium price from (2.55) resulting in $\pi = 62.66$ \$/MWh. Note that the highest production quantity corresponds to the cheapest producer (producer 1).

2.7.2 Illustrative Example. Equilibrium Conditions: Only Equality Constraints

The KKT conditions for the problem of the first producer (including one equality constraint), problem (2.5) and the KKT conditions of the problem of the second producer (including also an equality constraint), considered jointly, constitute an equilibrium problem, which characterize the equilibrium of the spot market, i.e.,

$$-\pi + \gamma - \beta(s_1 + s_2 + f_1 + f_2) = 0 \tag{2.57a}$$
$$-\pi + b_1 - \lambda_1\beta = 0 \tag{2.57b}$$
$$-s_1 - \lambda_1 = 0 \tag{2.57c}$$
$$-\pi + b_2 - \lambda_2\beta = 0 \tag{2.57d}$$
$$-s_2 - \lambda_2 = 0 \tag{2.57e}$$

Note that equation (2.57a) appears in the optimality conditions of the problems of both producer 1 and producer 2. However, it is not duplicated in the system of equations above. Additionally, note that this system of equations includes no inequalities as the original problem includes no inequalities.

Considering the numerical values in Illustrative Example 2.2.1, the system of equations (2.57) above is equivalent to

$$-\pi - 0.005(s_1 + s_2) + 98 = 0 \tag{2.58a}$$
$$-\pi - 0.005\lambda_1 + 40 = 0 \tag{2.58b}$$
$$-s_1 - \lambda_1 = 0 \tag{2.58c}$$
$$-\pi - 0.005\lambda_2 + 50 = 0 \tag{2.58d}$$
$$-s_2 - \lambda_2 = 0 \tag{2.58e}$$

Note that the solution of the system of equations (2.58) is the one provided in Illustrative Example 2.7.1. Additionally, note that $\lambda_1 = s_1 = 4533.3$ MWh and $\lambda_2 = s_2 = 2533.3$ MWh.

2.7.3 Illustrative Example. Equilibrium Conditions: Equality and Inequality Constraints

The KKT conditions for the problem of the first producer with capacity constraints, problem (2.11) in Illustrative Example 2.2.3, derived in (2.29), and the KKT conditions of the problem of the second producer also with capacity constraints, considered jointly, constitute an equilibrium problem that characterizes the equilibrium of the spot market, i.e.,

$$-\pi + \gamma - \beta(s_1 + s_2 + f_1 + f_2) = 0 \tag{2.59a}$$
$$-\pi + b_1 - \lambda_1\beta + \mu_1 = 0 \tag{2.59b}$$
$$-s_1 - \lambda_1 = 0 \tag{2.59c}$$
$$0 \leq \mu_1 \perp (s_1 - s_1^{\max}) \leq 0 \tag{2.59d}$$
$$-\pi + b_2 - \lambda_2\beta + \mu_2 = 0 \tag{2.59e}$$
$$-s_2 - \lambda_2 = 0 \tag{2.59f}$$
$$0 \leq \mu_2 \perp (s_2 - s_2^{\max}) \leq 0 \tag{2.59g}$$

Note that the KKT conditions of producers 1 and 2 have their own multipliers λ_1 and μ_1, and λ_2 and μ_2, respectively. Multipliers λ_1 and λ_2 are associated with equality constraints (of producers 1 and 2, respectively), while multipliers μ_1 and μ_2 with inequality constraints (of producers 1 and 2, respectively). Equation (2.59a) appears in the optimality conditions of the problems of both producer 1 and producer 2. However, it is not duplicated in the system above.

Using the numerical values provided in Illustrative Example 2.2.1, the system of equalities and inequalities (2.59) becomes

$$-\pi - 0.005(s_1 + s_2) + 98 = 0 \tag{2.60a}$$
$$-\pi - 0.005\lambda_1 + \mu_1 + 40 = 0 \tag{2.60b}$$
$$-s_1 - \lambda_1 = 0 \tag{2.60c}$$
$$0 \leq \mu_1 \perp (s_1 - s_1^{\max}) \leq 0 \tag{2.60d}$$
$$-\pi - 0.005\lambda_2 + \mu_2 + 50 = 0 \tag{2.60e}$$
$$-s_2 - \lambda_2 = 0 \tag{2.60f}$$
$$0 \leq \mu_2 \perp (s_2 - s_2^{\max}) \leq 0 \tag{2.60g}$$

The reader can easily check that if $s_1^{\max} \geq 4533.3$ MWh and $s_2^{\max} \geq 2533.3$ MWh the solution of (2.60) is equal to the one provided in Illustrative Example 2.7.1.

The KKT conditions (2.59), which include equality and inequality constraints, constitute an MLCP where

$$z_1 = \begin{bmatrix} \mu_1 \\ \mu_2 \end{bmatrix}, \; z_2 = \begin{bmatrix} s_1 \\ s_2 \\ \pi \\ \lambda_1 \\ \lambda_2 \end{bmatrix}, \; q_1 = \begin{bmatrix} s_1^{\max} \\ s_2^{\max} \end{bmatrix}, \; q_2 = \begin{bmatrix} b_1 \\ b_2 \\ 0 \\ 0 \\ \gamma - \beta(f_1 + f_2) \end{bmatrix}$$

$$M_{11} = \begin{bmatrix} 0 & 0 \\ 0 & 0 \end{bmatrix}, \; M_{12} = \begin{bmatrix} -1 & 0 & 0 & 0 & 0 \\ 0 & -1 & 0 & 0 & 0 \end{bmatrix}$$

$$M_{21} = \begin{bmatrix} 1 & 0 \\ 0 & 1 \\ 0 & 0 \\ 0 & 0 \\ 0 & 0 \end{bmatrix}, \ M_{22} = \begin{bmatrix} 0 & 0 & -1 & -\beta & 0 \\ 0 & 0 & -1 & 0 & -\beta \\ -1 & 0 & 0 & -1 & 0 \\ 0 & -1 & 0 & 0 & -1 \\ -\beta & -\beta & -1 & 0 & 0 \end{bmatrix}$$

2.7.4 Illustrative Example. Linear Equilibrium Problem

We analyze below a particular instance of the set of problems (2.51). Each of these problems is assumed to be linear and additionally we consider that $n^i = 1, \forall i$ (i.e., each optimization problem involves a single optimization variable). These assumptions render the $i = 1, \ldots, n$ linear problems below

$$\text{Minimize}_{x_i} \quad {c^i}^T x \tag{2.61a}$$
$$\text{s. t.}$$
$$A^i x = b^i \tag{2.61b}$$
$$D^i x \leq e^i \tag{2.61c}$$

where $x = (x_1, \ldots, x_i, \ldots, x^n)^T$, $x \in \mathbb{R}^{n\times 1}$, $c^i \in \mathbb{R}^{n\times 1}$, $A^i \in \mathbb{R}^{m_E\times n}$, $b^i \in \mathbb{R}^{m_E\times 1}$, $D^i \in \mathbb{R}^{m_I\times n}$ and $e^i \in \mathbb{R}^{m_I\times 1}$ Note that the only optimization variable of linear problem i is the *i-th* component of vector x, i.e., x_i.

For the sake of clarity, we consider the particular case in which $n = 2$, $m_E = 1$ and $m_I = 1$. Thus, $x = (x_1, x_2)^T$, $c^i = (c^i_1, c^i_2)^T$, $A^i = (a^i_1, a^i_2)$, $b^i = (b^i)$, $D^i = (d^i_1, d^i_2)$ and $e^i = (e^i)$.

The corresponding equilibrium problem is defined by jointly considering the KKT conditions of the problems (2.61), i.e.,

$$c^1_1 + \lambda^1 a^1_1 + \mu^1 d^1_1 = 0 \tag{2.62a}$$
$$a^1_1 x_1 + a^1_2 x_2 = b^1 \tag{2.62b}$$
$$0 \leq \mu^1 \perp d^1_1 x_1 + d^1_2 x_2 - e^1 \leq 0 \tag{2.62c}$$
$$c^2_2 + \lambda^2 a^2_2 + \mu^2 d^2_2 = 0 \tag{2.62d}$$
$$a^2_1 x_1 + a^2_2 x_2 = b^2 \tag{2.62e}$$
$$0 \leq \mu^2 \perp d^2_1 x_1 + d^2_2 x_2 - e^2 \leq 0 \tag{2.62f}$$

where $\lambda^i \in \mathbb{R}$ and $\mu^i \in \mathbb{R}$ are the Lagrangian multiplier vectors of constraints (2.61b) and (2.61c), respectively. Note that (2.62a) and (2.62d) result from differentiating the Lagrangian of each problem with respect to x_1 and x_2, respectively. Equations (2.62b) and (2.62e) are the equality constraints contained in (2.61b). Finally, (2.62c) and (2.62f) are the complementarity constraints.

Optimality conditions (2.62) constitute a MLCP where

$$z_1 = \begin{bmatrix} \mu^1 \\ \mu^2 \end{bmatrix}, \; z_2 = \begin{bmatrix} x_1 \\ x_2 \\ \lambda^1 \\ \lambda^2 \end{bmatrix}, \; q_1 = \begin{bmatrix} e^1 \\ e^2 \end{bmatrix}, \; q_2 = \begin{bmatrix} -b^1 \\ -b^2 \\ c_1^1 \\ c_2^2 \end{bmatrix}$$

$$M_{11} = \begin{bmatrix} 0\;0 \\ 0\;0 \end{bmatrix}, \qquad M_{12} = \begin{bmatrix} -d_1^1 & -d_2^1 & 0 & 0 \\ -d_1^2 & -d_2^2 & 0 & 0 \end{bmatrix},$$

$$M_{21} = \begin{bmatrix} 0 & 0 \\ 0 & 0 \\ d_1^1 & 0 \\ 0 & d_2^2 \end{bmatrix}, \qquad M_{22} = \begin{bmatrix} a_1^1 & a_2^1 & 0 & 0 \\ a_1^2 & a_2^2 & 0 & 0 \\ 0 & 0 & a_1^1 & 0 \\ 0 & 0 & 0 & a_2^2 \end{bmatrix}$$

Note that $z_1 \in \mathbb{R}^{2\times 1}$, $z_2 \in \mathbb{R}^{4\times 1}$, $q_1 \in \mathbb{R}^{2\times 1}$, $q_2 \in \mathbb{R}^{4\times 1}$, $M_{11} \in \mathbb{R}^{2\times 2}$, $M_{12} \in \mathbb{R}^{2\times 4}$, $M_{21} \in \mathbb{R}^{4\times 2}$ and $M_{22} \in \mathbb{R}^{4\times 4}$.

2.8 Mathematical Programs with Equilibrium Constraints, MPEC

A mathematical program with equilibrium constraints (MPEC) is an optimization problem involving constraints that represent equilibrium conditions; for instance, KKT conditions for one or several interrelated optimization problems modeling the equilibrium in the electricity market. In other words, an MPEC is an optimization problem whose constraints include other interrelated optimization or complementarity problems. A detailed analysis of MPECs is provided in Chapter 6 but with other chapters also mentioning important aspects.

The formulation of a particular case of an MPEC is provided below,

$$\text{Minimize}_{\{x\}\cup\{x^1,x^2,\lambda^1,\lambda^2,\mu^1,\mu^2\}} \quad f(x,x^1,x^2,\lambda^1,\lambda^2,\mu^1,\mu^2) \tag{2.63a}$$

s. t.

$$h(x,x^1,x^2,\lambda^1,\lambda^2,\mu^1,\mu^2) = 0 \tag{2.63b}$$

$$g(x,x^1,x^2,\lambda^1,\lambda^2,\mu^1,\mu^2) \le 0, \tag{2.63c}$$

$$\left\{ \begin{array}{lll} \text{Minimize}_{x^1} & f^1(x,x^1,x^2) & \\ \text{s. t.} & & \\ \quad h^1(x,x^1,x^2) = 0 & & (\lambda^1) \\ \quad g^1(x,x^1,x^2) \le 0 & & (\mu^1) \end{array} \right. \tag{2.63d}$$

$$\left\{\begin{array}{ll} \text{Minimize}_{x^2} & f^2(x, x^1, x^2) \\ \text{s. t.} & \\ \quad h^2(x, x^1, x^2) = 0 & (\lambda^2) \\ \quad g^2(x, x^1, x^2) \leq 0 & (\mu^2) \end{array}\right. \tag{2.63e}$$

Observe the nested structure of problem (2.63) above, i.e., an optimization problem constrained by a set of two optimization problems. Moreover, Figure 6.1 illustrates the structure of an MPEC considering n optimization problems as constraints.

Vector $x \in \mathbb{R}^{n^0}$ includes the set of optimization variables of the upper-level problem, vector $x^i \in \mathbb{R}^{n^i}$ includes the set of optimization variables of constraining problem i, and $\lambda^i \in \mathbb{R}^{m_E^i}$ and $\mu^i \in \mathbb{R}^{m_I^i}$ are the dual variable vectors of problem i associated with the equality and inequality constraints, respectively (indicated in (2.63) in parenthesis). Calling $n_g = \sum_{i=0}^{2} n^i$, and $n_G = n_g + \sum_{i=1}^{2}(m_E^i + m_I^i)$, the objective function and the constraints of the upper level problem are defined as follows $f : \mathbb{R}^{n_G} \rightarrow \mathbb{R}$, $h : \mathbb{R}^{n_G} \rightarrow \mathbb{R}^{m_E}$, and $g : \mathbb{R}^{n_G} \rightarrow \mathbb{R}^{m_I}$; while the the objective function and the constraints of constraining problem i are defined as follows, $f^i : \mathbb{R}^{n_g} \rightarrow \mathbb{R}$, $h^i : \mathbb{R}^{n_g} \rightarrow \mathbb{R}^{m_E^i}$, and $g^i : \mathbb{R}^{n_g} \rightarrow \mathbb{R}^{m_I^i}$.

It is important to note that although the optimization variables of constraining problem i are just those in vector x^i, the objective function and the constraints of constraining problem i, $f^i(x, x^1, x^2)$, $h^i(x, x^1, x^2) = 0$ and $g^i(x, x^1, x^2) \leq 0$, depend on the optimization variables of all problems (the upper-level one and the constraining ones), $x, x^i, i = 1, 2$. Note that under these assumptions the lower-level problems represent a Generalized Nash Equilibria (GNE), as described in Section 2.7. Additionally, the upper-level problem depends on all primal optimization variables, $x, x^i, i = 1, 2$, as well as on the dual variables of the lower-level problems $\lambda^i, \mu^i, i = 1, 2$. This fact makes the upper level problems and the two constraining problems interrelated, which results in a well-posed MPEC. The optimization variables are arranged in two subsets to differentiate those specifically pertaining to the upper-level problem, $\{x\}$, and those influencing the upper-level problem though the constraining problems, $\{x^1, x^2, \lambda^1, \lambda^2, \mu^1, \mu^2\}$.

If the lower-level optimization problems constraining the upper level problem are convex, they can be replaced by their corresponding KKT conditions. Then, problem (2.63) becomes

$$\text{Minimize}_{x, x^1, x^2, \lambda^1, \lambda^2, \mu^1, \mu^2} \quad f(x^1, x^2, \lambda^1, \lambda^2, \mu^1, \mu^2) \tag{2.64a}$$

s. t.

$$h(x, x^1, x^2, \lambda^1, \lambda^2, \mu^1, \mu^2) = 0 \tag{2.64b}$$

$$g(x, x^1, x^2, \lambda^1, \lambda^2, \mu^1, \mu^2) \leq 0, \tag{2.64c}$$

$$\nabla_{x^1} f^1(x, x^1, x^2) + {\lambda^1}^T \nabla_{x^1} h^1(x, x^1, x^2) + {\mu^1}^T \nabla_{x^1} g^1(x, x^1, x^2) = 0, \tag{2.64d}$$

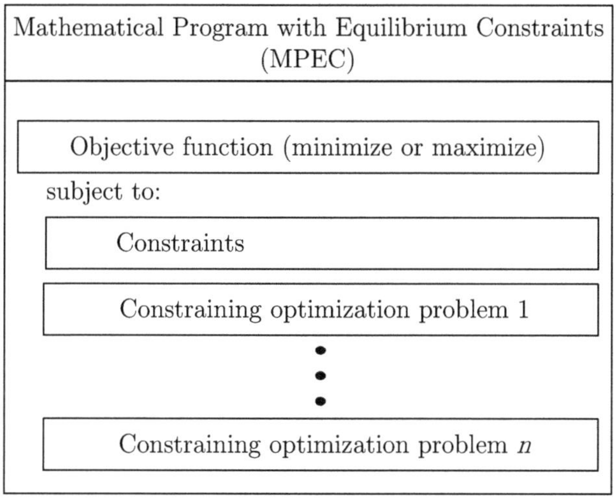

Fig. 2.5 MPEC: Optimization problem constrained by a number of interrelated optimization problems

$$
\begin{aligned}
&\nabla_{x^2} f^2(x, x^1, x^2) + \lambda^{2^T} \nabla_{x^2} h^2(x, x^1, x^2) + \\
&\quad \mu^{2^T} \nabla_{x^2} g^2(x, x^1, x^2) = 0, && \text{(2.64e)} \\
&h^1(x, x^1, x^2) = 0, && \text{(2.64f)} \\
&h^2(x, x^1, x^2) = 0, && \text{(2.64g)} \\
&0 \leq \mu^1 \perp g^1(x, x^1, x^2) \leq 0, && \text{(2.64h)} \\
&0 \leq \mu^2 \perp g^2(x, x^1, x^2) \leq 0 && \text{(2.64i)}
\end{aligned}
$$

Figure 2.6 illustrates the structure of an MPEC considering KKT conditions as constraints, as in (2.64).

Problem (2.64) above is a typical MPEC. It is relevant to note that the feasibility region (2.64b)-(9.22f) is not a convex set. Moreover, the complementarity constraints (9.22f) and (2.64h) do not meet standard constraint qualifications, like LICQ, at any feasible point. It is worth mentioning that there are some variants of constraint qualifications specifically adapted for MPECs, but their analysis is beyond the scope of this chapter.

Finally, note that if an MPEC includes a single, lower-level constraining problem, it is called an MPCC, mathematical program with complementarity (KKT) constraints.

The two examples below provide clarify MPECs, the first one with only equality constraints and the second one including both equality and inequality constraints.

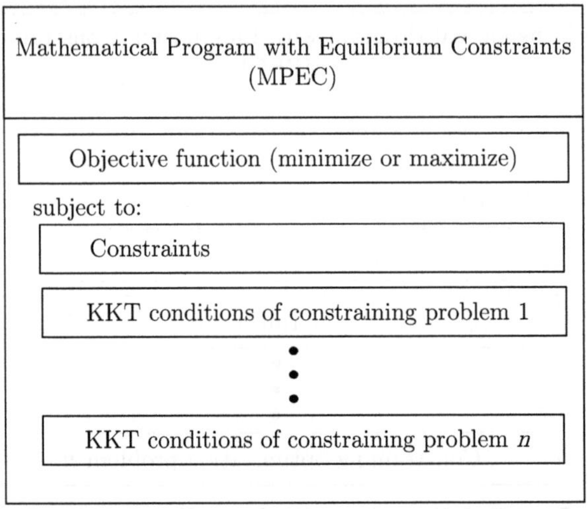

Fig. 2.6 MPEC (KKT): Optimization problem constrained by sets of interrelated KKT conditions

2.8.1 Illustrative Example. MPEC: Only Equality Constraints

We consider that producer 1 of Illustrative Example 2.2.1 maximizes its total profit from selling in the futures market, assuming equilibrium in the spot market, i.e., assuming that the equilibrium conditions in the spot market are satisfied. Thus, its corresponding optimization problem is

$$\text{Minimize}_{\{\pi, f_1\} \cup \{s_1, s_2\}} \quad -\pi(s_1 + f_1) + a_1 + b_1(s_1 + f_1) \tag{2.65a}$$

s. t.

$$\left\{ \begin{array}{l} \text{Minimize}_{s_1} \quad -\pi s_1 + a_1 + b_1(s_1 + f_1) \\ \text{s. t.} \\ \quad \pi = \gamma - \beta(s_1 + s_2 + f_1 + f_2), \end{array} \right. \tag{2.65b}$$

$$\left\{ \begin{array}{l} \text{Minimize}_{s_2} \quad -\pi s_2 + a_2 + b_2(s_2 + f_2) \\ \text{s. t.} \\ \quad \pi = \gamma - \beta(s_1 + s_2 + f_1 + f_2) \end{array} \right. \tag{2.65c}$$

The objective function (2.65a) represents the negative profit of producer 1 (minus revenue plus cost) from selling energy in both the spot and the futures markets, selling respectively s_1 and f_1. It should be noted that s_1 and f_1 may depend on the productions of other producers. This objective function is constrained by two optimization problems, (2.65b) and (2.65c), which when solved jointly represent the equilibrium in the spot market. Note that these two interrelated, lower-level problems represent the profit maximization of producers 1 and 2, respectively.

The optimization variables of the upper-level problem are separated into two subsets, namely $\{\pi, f_1\}$ and $\{s_1, s_2\}$. The first subset includes the optimization variables directly related to the upper-level problem, while the second subset includes optimization variables related to the upper-level problem through the subproblems.

Problem (2.65) can also be expressed as

$$\text{Minimize}_{\pi,f_1,s_1} \quad -\pi(s_1+f_1)+a_1+b_1(s_1+f_1) \tag{2.66a}$$

s. t.

$$s_1=\frac{\gamma+b_2-2b_1-\beta(f_1+f_2)}{3\beta} \tag{2.66b}$$

$$\pi=\frac{\gamma+b_1+b_2-\beta(f_1+f_2)}{3} \tag{2.66c}$$

where problems (2.65b) and (2.65c) have been replaced by their joint solution, represented by equality constraints (2.66b) and (2.66c). Incorporating these equality constraints into the objective function, problems (2.66a)-(2.66c) becomes unconstrained, i.e.,

$$\begin{aligned}&\text{Minimize}_{f_1}\\&\quad -\frac{\gamma+b_1+b_2-\beta(f_1+f_2)}{3}\left(\frac{\gamma+b_2-2b_1-\beta(f_1+f_2)}{3\beta}+f_1\right)\\&\quad +a_1+b_1\left(f_1+\frac{\gamma+b_2-2b_1-\beta(f_1+f_2)}{3\beta}\right)\end{aligned} \tag{2.67}$$

The problem above is solved by differentiating (2.67) with respect to f_1 and setting the resulting expression equal to 0 (optimality condition), i.e.,

$$\begin{aligned}\gamma(-1+2\psi_1)+b_1(2-4\psi_1)+b_2(-1+2\psi_1)&\\+\beta f_1(4+\psi_1)+\beta f_2(1-2\psi_1)&=0\end{aligned} \tag{2.68}$$

where

$$\psi_1=\frac{\partial f_2}{\partial f_1} \tag{2.69}$$

Parameter ψ_1 above represents the reaction of producer 1 in the futures market to the action of producer 2, also in the futures market, and is called the reaction parameter.

The optimal value for f_1 can be obtained solving (2.68) for f_1, namely

$$f_1=\frac{\beta f_2(-1+2\psi_1)+\gamma(1-2\psi_1)+b_1(-2+4\psi_1)+b_2(1-2\psi_1)}{\beta(4+\psi_1)} \tag{2.70}$$

Note that this solution constitutes the optimal solution of MPEC (2.65).
Considering the numerical data provided in Illustrative Example 2.2.1, problem (2.66) becomes

$$\text{Minimize}_{\pi,f_1,s_1} \quad -\pi(s_1+f_1)+30,000+40(s_1+f_1) \tag{2.71a}$$

s. t.

$$s_1=\frac{70-0.005(f_1+f_2)}{0.015} \tag{2.71b}$$

$$\pi=\frac{190-0.005(f_1+f_2)}{3} \tag{2.71c}$$

whose solution (assuming $f_2 = 200$MWh) is provided by (2.70),

$$f_1=\frac{13,800-26,000\psi_1}{4+\psi_1} \tag{2.71d}$$

Results for different values of the reaction parameter ψ_1 are provided in Table 2.1.

Table 2.1 Illustrative Example 2.8.1: Solution for producer 1

ψ_1	Price, π ($/MWh)	Futures production, f_1 (MWh)	Spot production, s_1 (MWh)	Profit ($)
+0.1	58.44	2731.7	3689.4	88,452
0.0	57.25	3450.0	3450.0	89,025
−0.1	55.99	4205.1	3198.3	88,391

2.8.2 Illustrative Example. MPEC: Both Equality and Inequality Constraints

We consider the same two producers as in Illustrative Example 2.8.1 but including production capacity limits. The resulting MPEC for producer 1 is stated below,

$$\text{Minimize}_{\{\pi,f_1\}\cup\{s_1,s_2\}} \quad -\pi(s_1+f_1)+a_1+b_1(s_1+f_1) \tag{2.72a}$$

s. t.

$$\begin{cases} \text{Minimize}_{s_1} & -\pi s_1 + a_1 + b_1(s_1 + f_1) \\ \text{s. t.} & \\ & -\pi + \gamma - \beta(s_1 + s_2 + f_1 + f_2) = 0 \\ & s_1 - s_1^{\max} \leq 0, \end{cases} \tag{2.72b}$$

$$\begin{cases} \text{Minimize}_{s_2} & -\pi s_2 + a_2 + b_2(s_2 + f_2) \\ \text{s. t.} & \\ & -\pi + \gamma - \beta(s_1 + s_2 + f_1 + f_2) = 0 \\ & s_2 - s_2^{\max} \leq 0 \end{cases} \tag{2.72c}$$

As in Illustrative Example 2.8.1, observe that the problem above is constrained by two interrelated optimization problems that jointly solved represent the equilibrium in the spot market.

Using the KKT conditions of the two lower-level problems, MPEC (2.72) above can also be written as

$$\text{Minimize}_{\pi, f_1, s_1, s_2, \lambda_1, \mu_1, \lambda_2, \mu_2} \quad -\pi(s_1 + f_1) + a_1 + b_1(s_1 + f_1) \tag{2.73a}$$

s. t.

$$-\pi + b_1 - \lambda_1 \beta + \mu_1 = 0 \tag{2.73b}$$

$$-s_1 - \lambda_1 = 0 \tag{2.73c}$$

$$-\pi + \gamma - \beta(s_1 + s_2 + f_1 + f_2) = 0 \tag{2.73d}$$

$$0 \leq \mu_1 \perp (s_1 - s_1^{\max}) \leq 0 \tag{2.73e}$$

$$-\pi + b_2 - \lambda_2 \beta + \mu_2 = 0 \tag{2.73f}$$

$$-s_2 - \lambda_2 = 0 \tag{2.73g}$$

$$-\pi + \gamma - \beta(s_1 + s_2 + f_1 + f_2) = 0 \tag{2.73h}$$

$$0 \leq \mu_2 \perp (s_2 - s_2^{\max}) \leq 0 \tag{2.73i}$$

Using the data considered in Illustrative Example 2.2.1 and assuming that $f_2 = 200$ MWh, $\psi_1 = 0$ and $s_1^{\max} = s_2^{\max} = 3000$ MWh, problem (2.73) becomes

$$\text{Minimize}_{\pi, f_1, s_1, s_2, \mu_1, \mu_2} \quad -\pi(s_1 + f_1) + 30{,}000 + 40(s_1 + f_1) \tag{2.74a}$$

s. t.

$$-\pi + 40 + 0.005 s_1 + \mu_1 = 0 \tag{2.74b}$$

$$-\pi + 100 - 0.005(s_1 + s_2 + f_1 + 200) = 0 \tag{2.74c}$$

$$0 \leq \mu_1 \perp (s_1 - 3000) \leq 0 \tag{2.74d}$$

$$-\pi + 50 + 0.005 s_2 + \mu_2 = 0 \tag{2.74e}$$

$$0 \leq \mu_2 \perp (s_2 - 3000) \leq 0 \tag{2.74f}$$

Taking into account equalities (2.73c) and (2.73g), λ_1 and λ_2 have been replaced by $-s_1$ and $-s_2$, respectively, in problem (2.74).

The analysis of the complementarity conditions (2.74d) and (2.74f) allows enumerating the four possible combinations of the final values of s_1 and s_2:

1. $s_1 < 3000$ ($\mu_1 = 0$) and $s_2 = 3000$ ($\mu_2 \geq 0$).
2. $s_1 < 3000$ ($\mu_1 = 0$) and $s_2 < 3000$ ($\mu_2 = 0$).
3. $s_1 = 3000$ ($\mu_1 \geq 0$) and $s_2 = 3000$ ($\mu_2 \geq 0$).
4. $s_1 = 3000$ ($\mu_1 \geq 0$) and $s_2 < 3000$ ($\mu_2 = 0$).

The only feasible solution is achieved for the fourth case with $s_1 = 3000$ MWh ($\mu_1 \geq 0$) and $s_2 < 3000$ MWh ($\mu_2 = 0$). Thus, problem (2.74) can be simplified to

$$\text{Minimize}_{\pi, f_1, s_1, s_2, \mu_1, \mu_2} \quad -\pi(s_1 + f_1) + 30,000 + 40(s_1 + f_1) \tag{2.75a}$$

s. t.

$$\pi = 67 - 0.0025 f_1 \tag{2.75b}$$

$$s_1 = 3000 \tag{2.75c}$$

$$\mu_1 = 12 - 0.005 f_1 \tag{2.75d}$$

$$s_2 = 3400 - 0.5 f_1 \tag{2.75e}$$

$$\mu_2 = 0 \tag{2.75f}$$

The solution of problem (2.75) is $f_1 = 3990$ MWh, $s_1 = 3000$ MWh, $\pi = \$57.025$/MWh, $\mu_1 = \$2.025$/MWh, $s_2 = 1405$ MWh, $\mu_2 = 0$ and the profit for the first producer is \$89,004.

2.9 Equilibrium Problems with Equilibrium Constraints, EPEC

For the purpose of this chapter, an Equilibrium Problem with Equilibrium Constraints, or EPEC, is composed of a set of interrelated MPECs. A more general definition of an EPEC is possible [13], but for the sake of simplicity, it is not used here. More details on EPECs can be found in Chapters 3, 7, and 9. A solution obtained by jointly solving the interrelated MPECs constituting an EPEC is a solution of the EPEC.

A simple EPEC made up by two MPECs is presented below:

$$\text{Minimize}_{X = \{x, x^1, x^2, \lambda_x^1, \lambda_x^2, \mu_x^1, \mu_x^2\}} \quad f_x(X, Y) \tag{2.76a}$$

s. t.

$$h_x(X, Y) = 0 \tag{2.76b}$$

$$g_x(X, Y) \leq 0, \tag{2.76c}$$

$$\nabla_{x^1} f_x^1(x, x^1, x^2, Y) + {\lambda_x^1}^T \nabla_{x^1} h_x^1(x, x^1, x^2, Y) + {\mu_x^1}^T \nabla_{x^1} g_x^1(x, x^1, x^2, Y) = 0, \tag{2.76d}$$

$$\nabla_{x^2} f_x^2(x, x^1, x^2, Y) + {\lambda_x^2}^T \nabla_{x^2} h_x^2(x, x^1, x^2, Y) +$$

$$\mu_x^{2^T} \nabla_{x^2} g_x^2(x, x^1, x^2, Y) = 0, \quad (2.76e)$$
$$h_x^1(x, x^1, x^2, Y) = 0, \quad (2.76f)$$
$$h_x^2(x, x^1, x^2, Y) = 0, \quad (2.76g)$$
$$0 \leq \mu_x^1 \perp g_x^1(x, x^1, x^2, Y) \leq 0, \quad (2.76h)$$
$$0 \leq \mu_x^2 \perp g_x^2(x, x^1, x^2, Y) \leq 0 \quad (2.76i)$$

and

$$\text{Minimize}_{Y=\{y,y^1,y^2,\lambda_y^1,\lambda_y^2,\mu_y^1,\mu_y^2\}} \quad f_y(Y, X) \quad (2.77a)$$

s. t.

$$h_y(Y, X) = 0 \quad (2.77b)$$
$$g_y(Y, X) \leq 0, \quad (2.77c)$$
$$\nabla_{y^1} f_y^1(y, y^1, y^2, X) + \lambda_y^{1^T} \nabla_{y^1} h_y^1(y, y^1, y^2, X) + \mu_y^{1^T} \nabla_{y^1} g_y^1(y, y^1, y^2, X) = 0, \quad (2.77d)$$
$$\nabla_{y^2} f_y^2(y, y^1, y^2, X) + \lambda_y^{2^T} \nabla_{y^2} h_y^2(y, y^1, y^2, X) + \mu_y^{2^T} \nabla_{y^2} g_y^2(y, y^1, y^2, X) = 0, \quad (2.77e)$$
$$h_y^1(y, y^1, y^2, X) = 0, \quad (2.77f)$$
$$h_y^2(y, y^1, y^2, X) = 0, \quad (2.77g)$$
$$0 \leq \mu_y^1 \perp g_y^1(y, y^1, y^2, X) \leq 0, \quad (2.77h)$$
$$0 \leq \mu_y^2 \perp g_y^2(y, y^1, y^2, X) \leq 0 \quad (2.77i)$$

The notation (variable vectors and functions) is analogous to that used to define MPECs in Section 2.8.

Denoting the problem MPEC_x by (2.76) and MPEC_y by (2.77), the EPEC (2.76)-(2.77) can be more compactly written as

$$\text{MPEC}_x \quad (2.78)$$
$$\text{MPEC}_y \quad (2.79)$$

which can be easily generalized to an EPEC including n MPECs,

$$\text{MPEC}_1 \quad (2.80)$$
$$\vdots$$
$$\text{MPEC}_i \quad (2.81)$$
$$\vdots$$
$$\text{MPEC}_n \quad (2.82)$$

Figure 2.7 illustrates the general structure of an EPEC that involves the joint solution of a collection of interrelated MPECs.

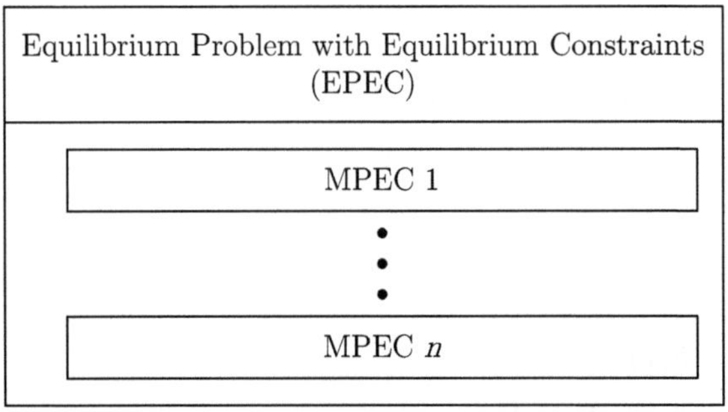

Fig. 2.7 EPEC: Joint solution of a collection of interrelated MPECs

The illustrative example below clarifies the form of an EPEC involving MPECs including only equality constraints.

2.9.1 Illustrative Example. EPEC: Only Equality Constraints

The joint consideration of two producers, both behaving through MPECs similar to that in Illustrative Example 2.8.1 constitutes the EPEC below:

$$\text{Minimize}_{\pi,f_1,s_1,s_2,\lambda_1,\lambda_2} \quad -\pi(s_1+f_1)+a_1+b_1(s_1+f_1) \tag{2.83a}$$

s. t.

$$-\pi+\gamma-\beta(s_1+s_2+f_1+f_2)=0 \tag{2.83b}$$

$$-\pi+b_1-\lambda_1\beta=0 \tag{2.83c}$$

$$-s_1-\lambda_1=0 \tag{2.83d}$$

$$-\pi+b_2-\lambda_2\beta=0 \tag{2.83e}$$

$$-s_2-\lambda_2=0 \tag{2.83f}$$

and

$$\text{Minimize}_{\pi,f_2,s_1,s_2,\lambda_1,\lambda_2} \quad -\pi(s_2+f_2)+a_2+b_2(s_2+f_2) \tag{2.84a}$$

s. t.

$$-\pi+\gamma-\beta(s_1+s_2+f_1+f_2)=0 \tag{2.84b}$$

$$-\pi+b_1-\lambda_1\beta=0 \tag{2.84c}$$

$$- s_1 - \lambda_1 = 0 \tag{2.84d}$$

$$- \pi + b_2 - \lambda_2 \beta = 0 \tag{2.84e}$$

$$- s_2 - \lambda_2 = 0 \tag{2.84f}$$

According to (2.76), the optimization variables of MPEC_x and MPEC_y are $X = \{\pi, f_1, s_1, s_2, \lambda_1, \lambda_2\}$ and $Y = \{\pi, f_2, s_1, s_2, \lambda_1, \lambda_2\}$, respectively. Since the representation of the market is the same for both producers, note that some of the optimization variables are shared. The analytical procedure to solve each one of the two MPECs constituting EPEC (2.83)-(2.84) is similar to the one described in Illustrative Example 2.8.1. The analytical solution of MPEC (2.83) corresponding to producer 1 is given by (2.85a) while the analytical solution of MPEC (2.84) corresponding to producer 2 by (2.85b), i.e.,

$$\begin{aligned}\gamma(-1 + 2\psi_1) + b_1(2 - 4\psi_1) + b_2(-1 + 2\psi_1)&\\ + \beta f_1(4 + \psi_1) + \beta f_2(1 - 2\psi_1) &= 0\end{aligned} \tag{2.85a}$$

$$\begin{aligned}\gamma(-1 + 2\psi_2) + b_2(2 - 4\psi_2) + b_1(-1 + 2\psi_2)&\\ + \beta f_2(4 + \psi_2) + \beta f_1(1 - 2\psi_2) &= 0\end{aligned} \tag{2.85b}$$

where

$$\psi_1 = \frac{\partial f_2}{\partial f_1} \quad \text{and} \quad \psi_2 = \frac{\partial f_1}{\partial f_2} \tag{2.86}$$

Note that f_1 depends on f_2 and, conversely, f_2 depends on f_1. Moreover, parameter ψ_1 represents the reaction of producer 2 in the futures market to the action of producer 1, also in the futures market, while parameter ψ_2 the reaction of producer 1 to the action of producer 2. Therefore, the solution of EPEC (2.83)-(2.84) is obtained by solving the system of equations (2.85). The solution of this system of equations is

$$f_1 = \frac{(2\psi_1 - 1)(\gamma\psi_2 + \gamma - \psi_2 b_2 + 2b_2 - 3b_1)}{\beta(\psi_1\psi_2 - 2\psi_1 - 2\psi_2 - 5)} \tag{2.87a}$$

$$f_2 = \frac{(2\psi_2 - 1)(\gamma\psi_1 + \gamma - \psi_1 b_1 + 2b_1 - 3b_2)}{\beta(\psi_1\psi_2 - 2\psi_1 - 2\psi_2 - 5)} \tag{2.87b}$$

which leads to a price

$$\begin{aligned}\pi = \frac{b_1(\psi_1\psi_2 + \psi_1 - 2\psi_2 - 2) + b_2(\psi_1\psi_2 + \psi_2 - 2\psi_1 - 2)}{-2\psi_2 - 2\psi_1 + \psi_1\psi_2 - 5} -&\\ \frac{\gamma(\psi_1\psi_2 + \psi_1 + \psi_2 + 1)}{-2\psi_2 - 2\psi_1 + \psi_1\psi_2 - 5}&\end{aligned} \tag{2.88}$$

For the particular case of the data provided in Illustrative Example 2.2.1 and assuming the arbitrary values $\psi_1 = 0$ and $\psi_2 = -0.1$, the system of equations (2.85) is equivalent to

$$0.02f_1 + 0.05f_2 - 70 = 0 \tag{2.89a}$$

$$0.006f_1 + 0.0195f_2 - 48 = 0, \tag{2.89b}$$

whose solution is $f_1 = 3125$ MWh and $f_2 = 1500$ MWh. The price is obtained from (2.88) resulting in $\pi = \$55.625$/MWh.

Using (2.66b) for producer 1 and a similar equation for producer 2, we can also compute the production in the spot market, $s_1 = 3125$ MWh and $s_2 = 1125$ MWh. The resulting profits are \$67,656 for producer 1 and −\$35,234 for producer 2. Note that at the equilibrium (instable in this case), the most expensive producer (producer 2) experiences a negative profit, i.e., a loss.

The illustrative example below illustrates the structure of an EPEC involving two MPECs and including both equality and inequality constraints.

2.9.2 Illustrative Example. EPEC: Both Equality and Inequality Constraints

The joint consideration of two producers, both behaving through MPECs similar to that in Illustrative Example 2.8.2 constitutes the EPEC (2.90)-(2.91) below

$$\text{Minimize}_{\pi, f_1, s_1, s_2, \lambda_1, \mu_1, \lambda_2, \mu_2}$$

$$-\pi(s_1 + f_1) + a_1 + b_1(s_1 + f_1) \tag{2.90a}$$

s. t.

$$-\pi + \gamma - \beta(s_1 + s_2 + f_1 + f_2) = 0 \tag{2.90b}$$

$$-s_1 - \lambda_1 = 0 \tag{2.90c}$$

$$-\pi + b_1 - \lambda_1\beta + \mu_1 = 0 \tag{2.90d}$$

$$0 \le \mu_1 \perp (s_1 - s_1^{\max}) \le 0 \tag{2.90e}$$

$$-\pi + b_2 - \lambda_2\beta + \mu_2 = 0 \tag{2.90f}$$

$$-s_2 - \lambda_2 = 0 \tag{2.90g}$$

$$0 \le \mu_2 \perp (s_2 - s_2^{\max}) \le 0 \tag{2.90h}$$

and

$$\text{Minimize}_{\pi, f_2, s_1, s_2, \lambda_1, \mu_1, \lambda_2, \mu_2}$$

$$-\pi(s_2 + f_2) + a_2 + b_2(s_2 + f_2) \tag{2.91a}$$

s. t.

$$-\pi + \gamma - \beta(s_1 + s_2 + f_1 + f_2) = 0 \tag{2.91b}$$
$$-s_1 - \lambda_1 = 0 \tag{2.91c}$$
$$-\pi + b_1 - \lambda_1\beta + \mu_1 = 0 \tag{2.91d}$$
$$0 \leq \mu_1 \perp (s_1 - s_1^{\max}) \leq 0 \tag{2.91e}$$
$$-\pi + b_2 - \lambda_2\beta + \mu_2 = 0 \tag{2.91f}$$
$$-s_2 - \lambda_2 = 0 \tag{2.91g}$$
$$0 \leq \mu_2 \perp (s_2 - s_2^{\max}) \leq 0 \tag{2.91h}$$

Observe that capacity constraints are included. Hence the MPECs involve both equality and inequality constraints.

Note also that MPEC (2.90) corresponds to producer 1, while MPEC (2.91) corresponds to producer 2. They should be jointly solved to attain a solution of the EPEC.

Assuming that $s_1^{\max} = s_2^{\max} = 3000$ MWh and analyzing the complementarity conditions (2.90e) and (2.90h) (or (2.91e) and (2.91h)), we can enumerate the four possible values of variables s_1 and s_2:

1. $s_1 < 3000$ ($\mu_1 = 0$) and $s_2 = 3000$ ($\mu_2 \geq 0$).
2. $s_1 < 3000$ ($\mu_1 = 0$) and $s_2 < 3000$ ($\mu_2 = 0$).
3. $s_1 = 3000$ ($\mu_1 \geq 0$) and $s_2 = 3000$ ($\mu_2 \geq 0$).
4. $s_1 = 3000$ ($\mu_1 \geq 0$) and $s_2 < 3000$ ($\mu_2 = 0$).

Considering the data provided in Illustrative Example 2.2.1 and analyzing each of the above four cases, we can conclude that the solution for the equilibria is reached for $s_1 = 3000$ MWh ($\mu_1 \geq 0$) and $s_2 < 3000$ MWh ($\mu_2 = 0$). For these values, EPEC (2.90)–(2.91) can be simplified as

$$\text{Minimize}_{\pi, f_1, s_1, s_2, \lambda_1, \mu_1, \lambda_2, \mu_2}$$
$$-\pi(s_1 + f_1) + 30{,}000 + 40(s_1 + f_1) \tag{2.92a}$$
s. t.
$$\pi = 67.5 - 0.0025(f_1 + f_2) \tag{2.92b}$$
$$s_1 = 3000 \tag{2.92c}$$
$$s_2 = 3500 - 0.5(f_1 + f_2) \tag{2.92d}$$
$$\mu_1 = 12.5 - 0.0025(f_1 + f_2) \tag{2.92e}$$
$$\mu_2 = 0 \tag{2.92f}$$
$$\lambda_1 = -3000 \tag{2.92g}$$
$$\lambda_2 = -3500 + 0.5(f_1 + f_2) \tag{2.92h}$$

and

$$\text{Minimize}_{\pi, f_2, s_1, s_2, \lambda_1, \mu_1, \lambda_2, \mu_2}$$
$$-\pi(s_2 + f_2) + 50{,}000 + 50(s_2 + f_2) \tag{2.93a}$$

s. t.

$$\pi = 67.5 - 0.0025(f_1 + f_2) \tag{2.93b}$$

$$s_1 = 3000 \tag{2.93c}$$

$$s_2 = 3500 - 0.5(f_1 + f_2) \tag{2.93d}$$

$$\mu_1 = 12.5 - 0.0025(f_1 + f_2) \tag{2.93e}$$

$$\mu_2 = 0 \tag{2.93f}$$

$$\lambda_1 = -3000 \tag{2.93g}$$

$$\lambda_2 = -3500 + 0.5(f_1 + f_2) \tag{2.93h}$$

Incorporating the equality constraints (2.92b)-(2.92h) into the objective functions (2.92a) and (2.93a), problems (2.92) and (2.93) become unconstrained and their corresponding optimality conditions can be reduced to the following linear system of equations,

$$7.5(1 + \psi_1) - 27.5 + 0.0025(2 + \psi_1)f_1 + 0.0025f_2 = 0 \tag{2.94a}$$

$$17.5\psi_2 - 0.0025\psi_2 f_1 + 0.0025f_2 = 0 \tag{2.94b}$$

The solution of (2.94) is

$$f_1 = \frac{7000\psi_2 - 3000\psi_1 + 8000}{\psi_1 + \psi_2 + 2} \tag{2.95a}$$

$$f_2 = -\frac{6000\psi_2 + 10,000\psi_1\psi_2}{\psi_1 + \psi_2 + 2} \tag{2.95b}$$

For $\psi_1 = 0$ and $\psi_2 = -0.1$ we obtain $f_1 = 3842.1$ MWh, $f_2 = 315.8$ MWh, and the resulting price is $\pi = 57.103$ \$/MWh.

The production in the spot market is $s_1 = 3000$ MWh and $s_2 = 1421.1$ MWh, and the profits are \$ 87,036 for producer 1 and $-\$37,659$ for producer 2. Note that at the equilibrium (instable in this case) the most expensive producer (producer 2) experiences a negative profit, i.e., a loss.

2.10 Non-Convexity and Non-Regularity Issues

The constraints of an MPEC include complementarity (i.e., KKT) conditions, which result in two unfortunate facts, [13]:

1. The MPEC feasible region is not generally a convex set.
2. The complementarity constraints partly defining this feasible region are not regular at optimal solutions.

Therefore, generally, no KKT conditions can be formulated meaningfully for MPECs. This fact makes finding the solution of MPECs challenging, and even more so for EPECs, which require jointly solving a collections of interrelated MPECs.

Nonetheless, in some cases a way to solve an MPEC is to formulate and solve its KKT conditions. Analogously, an EPEC can be solved by formulating and jointly solving the sets of KKT conditions corresponding to its associated MPECs.

2.11 Summary

This chapter provides a friendly overview of mathematical structures relevant to the analysis of markets. These structures include: optimization problems, equilibrium problems, mathematical programs with equilibrium constraints and equilibrium problem with equilibrium constraints.

From a market perspective, it was observed that:

1. Optimization problems allow modeling the behavior of market agents participating in competitive markets.
2. Equilibrium problems allow modeling the market interactions of competitive agents and characterizing the market outcomes.
3. Mathematical programs with equilibrium constraints allow characterizing the behavior of strategic agents that influence market outcomes.
4. Equilibrium problems with equilibrium constraints allow characterizing the interaction of several strategic market agents and the resulting market outcomes.

From the computational viewpoint the remarks below are worth mentioning:

1. Optimization problem are relatively *easy* to solve. If they are linear, they are solvable even if including millions of variables and millions of constraints. It they include binary variables, they are easily solvable up to a few thousand binary variables and hundreds of thousands of continuous variables. It they are nonlinear, they can be solved routinely up to dozens of thousands of variables and constraints. If the problem is nonlinear and includes binary variables, it is generally hard to solve. In this last case, decomposition techniques are advisable, [5].
2. The computational complexity of solving equilibrium problems and MLCPs is generally similar to that of solving optimization problems.
3. Solving MPECs is generally complex unless the equilibrium conditions are related to linear or convex problems. Nevertheless several techniques are available to efficiently attack such problems.
4. Since optimality conditions cannot be formulated easily for MPECs that include complementarity (KKT) conditions as constraints, solving EPECs rigorously is generally hopeless, unless a particular structure of

the EPEC considered can be exploited. Heuristic and approximate techniques are available to identify near-optimal solutions.

2.12 Exercises

1. Consider an electricity retailer characterized by a linear utility (profit) function. Formulate an optimization problem describing its optimal strategy in a spot market involving a linear price demand function.
2. Consider two producers similar to the one in Illustrative Example 2.2.1 and one retailer similar to the one is Exercise 1. Formulate the equilibrium problem in the spot market.
3. Formulate the equilibrium conditions in Exercise 2 as a MLCP.
4. Assuming equilibrium in the spot market, formulate and solve an MPEC to establish the strategy in a futures market of the retailer in Exercise 2.
5. Formulate an EPEC considering the two producers and the retailer of Exercise 2. The behavior of each agent should be modeled through an MPEC.
6. If the KKT conditions constraining a generic MPEC correspond to a linear programming problem (Minimize$_x$ $c^T x$ subject to $Ax \geq b, x \geq 0$), show through an example that these conditions can be substituted by the constraints of the primal problem ($Ax \geq b$), the constraints of the dual problem ($\lambda^T A \leq c^T$) and the strong duality theorem ($c^T x = \lambda^T b$).
7. Extend the results in Illustrative Example 2.7.1 (EP) to the case of n producers.
8. Extend the results in Illustrative Example 2.8.1 (MPEC) to the case of n producers.
9. Extend the results in Illustrative Example 2.9.1 (EPEC) to the case of n producers.
10. Consider the problem

$$\begin{aligned} &\text{Minimize}_{x,y} \quad f(x,y) \\ &\text{s. t.} \\ &\qquad 0 \leq y \perp g(x) \leq 0 \end{aligned}$$

Is the constraint regular? Does the constraint constitute a convex set?

References

1. B. Allaz and J.-L. Vila. Cournot competition, futures markets and efficiency. *Journal of Economic Theory*, 59(1):1–16, February 1993.
2. J.F. Bard. *Practical Bilevel Optimization: Algorithms and Applications.* Springer, New York, 1999.
3. D.P. Bertsekas. *Nonlinear Programming.* Athena Scientific, Beltmont, Massachusetts, second edition, 1999.
4. E. Castillo, A.J. Conejo, P. Pedregal, R. García, and N. Alguacil. *Building and Solving Mathematical Programming Models in Engineering and Science.* Wiley, New York, 2001.
5. A.J. Conejo, E. Castillo, R. Mínguez, and R. García Bertrand. *Decomposition Techniques in Mathematical Programming. Engineering and Science Applications.* Springer, Heidelberg, Germany, 2006.
6. R.W. Cottle, J.-S. Pang, and R.E. Stone. *The Linear Complementarity Problem.* Society for Industrial and Applied Mathematics (SIAM), Philadelphia, 2009.
7. G. Debreu. A social equilibrium existence theorem. *Proceedings of the National Academy of Sciences of the United States of America*, 38(10):886–893, 1952.
8. F. Facchinei and C. Kanzow. Generalized Nash equilibrium problems. *4OR*, 5(3):173–210, September 2007.
9. F. Facchinei and J.-S. Pang. *Finite-Dimensional Variational Inequalities and Complementarity Problems, Volume I.* Springer, New York, 2003.
10. F. Facchinei and J.-S. Pang. *Finite-Dimensional Variational Inequalities and Complementarity Problems, Volume II.* Springer, New York, 2003.
11. M.C. Ferris, O.L. Mangasarian, and J.-S. Pang, Editors. *Complementarity: Applications, Algorithms and Extensions.* Springer, New York, 2001.
12. D.G. Luenberger and Y. Ye. *Linear and Nonlinear Programming.* Springer, New York, third edition, 2008.
13. Z.-Q. Luo, J.-S. Pang, and D. Ralph. *Mathematical Programs with Equilibrium Constraints.* Cambridge University Press, Cambridge, UK, 2008.
14. J. Nash. Equilibrium points in n-person games. *Proceedings of the National Academy of Sciences of the United States of America*, 36(1):48–49, 1950.
15. J. Nash. Non-cooperative games. *The Annals of Mathematics. Second Series*, 54(2):286–295, September 1951.
16. J. Nocedal. *Numerical Optimization.* Springer, New York, 2006.

Chapter 3
Some Microeconomic Principles

3.1 Introduction

In this chapter, we explain some useful principles of microeconomics for those readers with little or no background in the subject. Readers who have studied microeconomics may also benefit from this chapter, as we show how to construct several different kinds of models of markets, using optimization and complementarity techniques.

Models are simplifications of reality. Even the complex models discussed in later chapters are, at best, approximations of the aggregate behavior of many people making decisions about buying, producing and selling things. In this chapter, the models are especially simple, in order to focus on the principles that guide the construction of the more complex models. If these simple models were explained through illustrations based in the modern world in which we all live, then the discrepancies between the simple models and the myriad details of our experiences would be distracting. Therefore, we invent an imaginary late 19th century city as the context to explain the principles and models, with illustrations based on the buying and selling of coal (and sometimes wood) for heating and industrial uses.

Microeconomics is a very large subject, and it is impossible to cover all topics in one short chapter. Readers interested in learning more about the subject may wish to read one of the popular textbooks, such as [9] or [13]. The main topics included in this chapter are:

- supply and demand curves;
- single commodity competitive equilibrium as the intersection of supply and demand curves;
- multicommodity equilibrium as the solution of a system of equations;
- single commodity competitive equilibrium as the solution of a social welfare maximization problem;
- complementarity models of equilibrium that explicitly identify individual firms;

- the Nash-Cournot model of imperfect competition, as a complementarity model;
- the generalized Nash extension of the Nash-Cournot model, with constraints that link the firms; and
- the Stackelberg leader-follower model of imperfect competition, as a mathematical program with equilibrium constraints (MPEC).

3.2 Basics of Supply and Demand

3.2.1 Supply Curves

Imagine a late 19th century city, where most buildings used coal for heat, and many industrial processes also used coal. Many firms supplied coal to the city. Each firm was legally free to charge whatever price it wished, but if a firm tried to charge a higher price than the others, then most of its customers would buy from the competition, thus drastically reducing revenue. If a firm charged less than its competitors, then most buyers would try to buy from this firm and not the others, but it could not possibly supply everyone – the firm would be better off to raise the price to match the competitors, and still sell lots of coal. For these reasons, all of the firms normally charged the same price at any given time, with any differences quickly eliminated by managers who were ever watchful of competitors' prices. Therefore, we can speak of *the* price of coal at any time, although the price would vary from month to month, and since no firm tried to set the price any differently from the going price, we say that all the firms were *price-takers.*

Each firm had several coal mines, and the costs of supplying coal differed for different mines: some mines were deeper and had greater costs of labor and of bringing the coal to the surface; some mines were farther from the city, and the cost of transporting the coal to the city was greater. The managers of the various firms were constantly aware of the ever-changing price at which they could sell their coal output in the city, and whenever the selling price per ton[1] was less than a mine's variable cost of producing and transporting a ton of coal (mainly labor and coal transport cost per ton), that mine would stop supplying coal to the city, until the price returned to a level that at least covered the mine's variable cost. Even if sometimes the price did not cover all costs – variable cost and the historical cost of equipment, digging the original shafts and tunnels, etc. – it was worthwhile to operate the mine as long as price exceeded variable cost, because at least some money would be available to help pay for the fixed, historical costs. On the other hand,

[1] We assume that coal from different mines had the same heating value. If some coal had produced more heat per ton than others, then the better quality coal would have sold at a higher price per ton than the lower quality coal.

sometimes the price would rise to a high enough level that it was worthwhile for some distant, high-cost mines to supply coal to the city.

Figure 3.1 illustrates the decisions made by a firm with three mines, when the selling price was $2 per ton. The cheapest of its three mines could produce and transport coal to the city at a variable cost of $0.40 per ton, up to a maximum amount of 4000 tons of coal per month. The height of the lowest step of Figure 3.1 represents the variable cost of coal from the mine, and the width of the step represents the mine's output capacity. The second cheapest mine owned by this firm could supply up to 7000 tons per month of coal to the city at a variable cost of $1.60 per ton, as shown in Figure 3.1 by the height and width (11000 minus 4000) of the second step. The most expensive mine belonging to this firm could deliver up to 4000 tons per day at a variable cost of $2.20 per ton. At a selling price of, say, $2 per ton, the firm of Figure 3.1 would find it worthwhile to operate the cheapest and second cheapest mines, but not the most expensive mine, and it would deliver 4000+7000 = 11000 tons of coal per month to the city. The firm's excess of revenue over variable cost would be $2x11000 - $0.40x4000 - $1.60x7000 = $9200 per month, which is equal to the gray shaded area in Figure 3.1. This $9200 per month was not the firm's profit in the usual accounting sense, because historical costs have not been deducted. The $9200 could be called gross margin, or another accounting term, but here we shall call it the firm's surplus, or the *producer's surplus*, to keep with common economic terminology. [2]

If the selling price were instead $1 per ton, then the firm would produce 4000 tons per month, only from the cheapest mine. At any price between $0.40 per ton and $1.59 per ton, the firm would produce 4000 tons per month; at any price between $1.60 and $2.19 per ton, the firm would produce 11000 tons per month (from two mines); and at any price at or above $2.20 per ton, the firm would operate all three of its mines, and deliver altogether 15000 tons per month. This relationship between the selling price and the firm's delivery quantity is indicated by the solid black lines forming the steps of Figure 3.1. Note that, as the price increases in Figure 3.1, the quantity supplied by the firm either stays the same, or increases; the quantity never decreases when price increases.

The total quantity of coal supplied to the city by all firms in a month is the sum of all the coal supplied by all the mines for which it was profitable to operate, given the price of coal during that month. The relationship between total quantity from all firms and price would lead to an illustration like Figure 3.1, but with a great many steps. In such an all-firms picture, the horizontal axis would stretch from zero to a potential total quantity much larger than in Figure 3.1, making individual step widths appear relatively tiny. If there

[2] Alternatively, the gross margin minus fixed costs, or profit, could be termed "producer's surplus." However, whether or not fixed costs are subtracted when defining producer's surplus would not affect the differences in producer's surpluses when comparing two different runs of a model (e.g., for two government policies). In models for which capital is variable, investment costs should be subtracted when calculating this surplus.

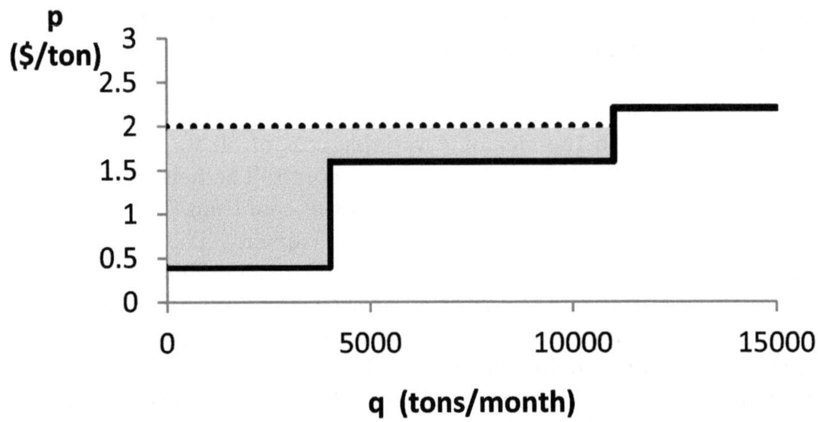

Fig. 3.1 Price-quantity relationship (solid black lines) and producers surplus (gray shaded area) for one firm with three mines

were a great variety of costs per ton among all of the mines of all firms, with relatively small quantities of coal at each marginal cost, then the height of each step (the difference between a mine's cost and the next higher cost for a mine, possibly belonging to another firm) would be relatively small. Such an all-firms picture could therefore be represented by a smooth curve, to a good approximation. Such a smooth curve would have a positive slope (unless a step was especially wide, in which case the slope would be zero over the step). Figure 3.1 illustrates a linear relationship between price and total supply, but keep in mind that other nonlinear relationships are possible. The solid black line is called the *supply curve* for coal in the city. Although we think of the price as the independent variable which causes the quantity supplied to take a value, it is customary in economics to show price on the vertical axis and quantity on the horizontal axis. Extending the illustration of producer's surplus from the one-firm case of Figure 3.1 to the all-firms picture of Figure 3.2, we can see that the total producers' surplus of all firms is represented by the shaded triangular area in Figure 3.2, when the price is $2 per ton.

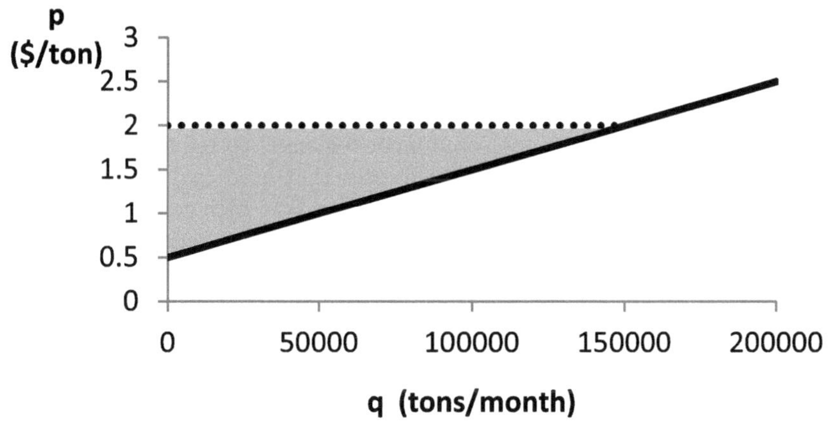

Fig. 3.2 Supply curve for coal (solid black line) and producers' surplus of all firms (gray shaded area)

Whether the supply curve is linear or nonlinear, we see that the total quantity of coal supplied is a function of the price that producers expect to receive; we may write $q = f_s(p)$, and refer to f_s as the *supply function.* We shall often refer to the inverse of the supply function, called the *inverse supply function*, i.e., f_s^{-1}, where $p = f_s^{-1}(q)$. The supply curve is the graph of f_s and the graph of f_s^{-1}. To say that the supply curve has a positive slope is to say that $\frac{df_s^{-1}}{dq} > 0$, and equally that $\frac{df_s}{dp} > 0$. In Figure 3.2, $f_s^{-1}(q) = 0.5 + 0.00001q$, and $f_s(p) = (p - 0.5)/0.00001$.

The same sort of story could be told about *any* good or service that buyers view as *homogenous* – the same regardless of which company supplies it – whether it is wheat, gasoline, standard sized sheets of plywood, legal services, or anything else. The key principles are: (a) there tends to be *one price* at any given time; (b) *a higher selling price tends to induce a larger quantity supplied*, i.e., the supply curve has a nonnegative slope; and (c) the *producers' surplus* can be represented by the area above the supply curve and below the horizontal line at the height of the market price.

3.2.2 Demand Curves

In aggregate, the producers of coal for the imaginary city supplied more coal when the price increased. But there were not only producers in the city's market for coal – there were also purchasers of coal, namely, building owners looking to heat their homes or places of business, and owners of businesses that used coal in their production processes, e.g., for steam. The purchasers also reacted to changing prices, by altering the quantities that they chose to purchase, but their reaction was opposite to that of the coal suppliers. When the coal price went up, some building owners chose to burn less coal in their furnaces by closing off some rooms that were not used much, while others burned less coal but partly made up for it by also burning some cheap firewood in their fireplaces (at some inconvenience) to reduce the total cost of heating. Similarly, some of the businesses that used coal for steam found that their production costs were just too high at the high coal prices, so they shut down and purchased no coal until the coal price returned to a lower level.

Figure 3.3 illustrates the reactions of one household customer to various prices of coal. If the price exceeded \$2.20 per ton, then this customer chose to buy no coal at all; the heated parts of the house were heated entirely by firewood, even though this meant some difficult labor of cutting and splitting the wood. At prices greater than \$1.40 but no more than \$2.20 per ton, this householder cut back on the back-breaking labor for firewood, and purchased 0.1 tons of coal per month. When the price was greater than \$0.60 but no more than \$1.40 per ton, this household stopped using firewood and relied only on coal, buying 0.2 tons of coal per month. If the price fell to \$0.60 or less per ton, then the heating vents to some little-used rooms were opened, to heat them too, and 0.4 tons of coal were purchased per month.

The height of each step in Figure 3.3 can be described as the maximum price that the householder is willing to pay for increments of coal. For any total quantity between 0 and 0.1 tons per month, the householder is willing to pay up to \$2.20 per ton. Thus, \$2.20 per ton is the intrinsic value, for this customer, of each ton of coal when the total amount of coal purchased is between 0 and 0.1 tons per month. If, instead, the customer were to buy between 0.1 and 0.2 tons per month, then the intrinsic value of the first 0.1 tons per month is still \$2.20 per ton, but the intrinsic value of the extra increment of coal (above 0.1 tons per month) is only \$1.40 per ton, the height of the second step. Similarly, the intrinsic value of incremental coal purchases greater than 0.2 tons per month is \$0.60 per ton. If the price that the customer pays to a supplier is, say, \$1.00 per ton, as illustrated in Figure 3.3, then this customer would choose to purchase 0.2 tons per month, with an intrinsic value of \$2.20x0.1+1.40x(0.2-0.1)=\$0.36 per month. However, the customer pays less money to the supplier than this intrinsic value; the total payment is \$1.00x0.2=\$0.20 per month. The difference between the intrinsic value and the actual payment is described as this *consumer's surplus*: \$0.36-0.20=\$0.16 per month, as illustrated by the gray shaded area of Figure 3.3. A consumer's surplus is analogous to a producer's surplus, but with an important difference: in the calculation of a producer's surplus, both the revenue and the variable costs subtracted from it are actual amounts of money that change hands – revenue is paid by customers to the firm, and the firm pays for labor and transportation; but in the calculation of a consumer's surplus, the intrinsic value is a theoretical interpretation, and does not correspond to any money that actually changes hands. Nevertheless, a consumer's surplus measures how much better off the consumer is by purchasing coal, compared to the alternatives such as chopping wood and closing off some heating vents to parts of the house.

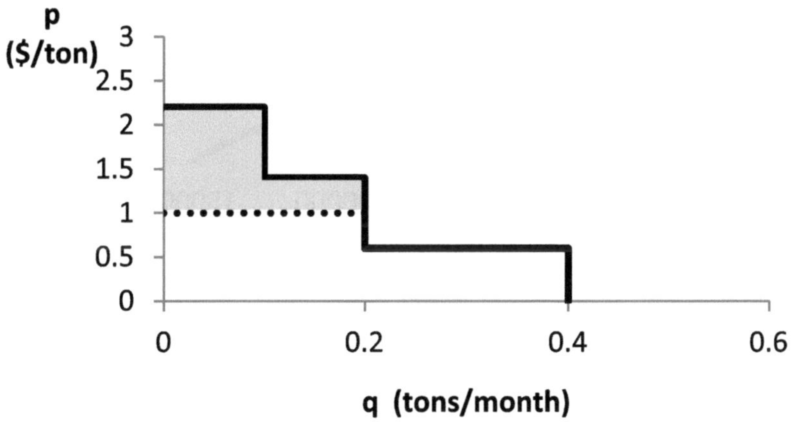

Fig. 3.3 Price-quantity relationship (solid black lines) and consumer's surplus (gray shaded area) for one household

Now consider the total quantity of coal purchased in the city during a month by all customers – the total *quantity demanded* – and how it related to various prices that prevailed in the city. As with the supply curve, this aggregate behavior of all consumers could be represented by a smooth curve as a good approximation to a relationship with many tiny steps. The total quantity demanded would decrease when the price increased. Figure 3.4 illustrates this behavior in a diagram called the *demand curve*; the units on the axes are the same as in the supply curve of Figure 3.2, and once again, price is on the vertical axis and quantity on the horizontal axis. Figure 3.4 shows a straight line demand curve, but nonlinear relationships are possible, too. The triangular gray shaded area in Figure 3.4 represents the *consumers' surplus*, totalled across all consumers, when the price was \$1.00 per ton of coal and consumers purchased 90,000 tons per month.

Corresponding to the demand curve is the *demand function*, f_d, which gives the total demand quantity as a function of price: $q = f_d(p)$. We will often refer to the *inverse demand function*, which gives the price that must correspond to a quantity demanded: $p = f_d^{-1}(q)$. The graph of either function is the demand curve, and the nonpositive slope of the demand curve is reflected in the facts that $\frac{df_d^{-1}}{dq} \leq 0$ and $\frac{df_d}{dp} \leq 0$. In Figure 3.4, $f_d^{-1}(q) = 2.5 - 0.0000167q$, and $f_d(p) = (2.5 - p)/0.0000167$.

The most important feature to note about Figure 3.4 is that *a higher selling price tends to induce a smaller quantity demanded*, i.e., the demand curve has a negative slope (strictly speaking, a non-positive slope). This type of

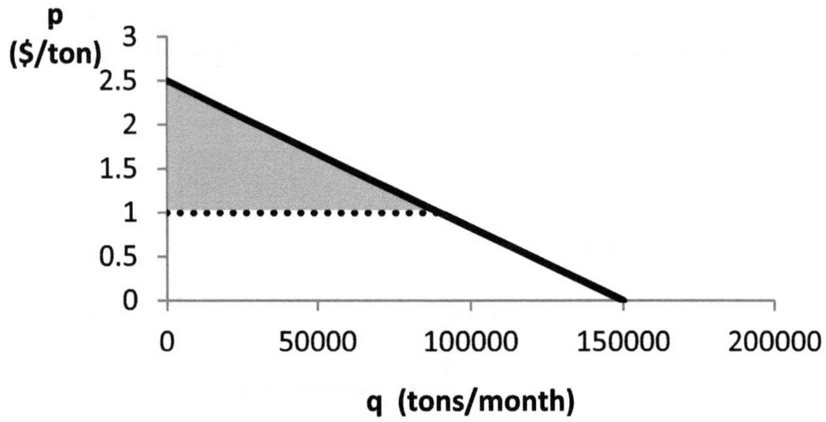

Fig. 3.4 Demand curve for coal (solid black line), and consumers' surplus (gray shaded area) for all consumers

behavior is normally observed for purchasers of *any* good or service, although there are rare exceptions.

3.2.3 Notion of Equilibrium as Intersection of Supply and Demand Curves

With higher prices causing purchasers to buy less, but producers to produce more, there can be only one price at which the quantity produced equals the quantity purchased.[3] Figure 3.5 illustrates this *equilibrium quantity and price*, (q^*, p^*) at the intersection of the supply and demand curves.

Why is this point called "equilibrium"? Suppose that the price charged by producers were higher than p^*: then the quantity produced in the expectation of receiving the higher price would be more than the quantity purchased at that higher price, and producers would therefore have unsold *surplus* coal. With surplus inventory on hand, producers would be eager to reduce the price, in order to encourage more sales of coal, so that they earn some revenue, even though it would be less than they had originally expected. To summarize, then, if producers were to charge a price higher than p^*, then economic forces would tend to reduce the price towards p^*. If, on the other hand, producers were to charge less than p^*, then the amount produced would fall short of the amount demanded at that price – there would be a *shortage* – and some purchasers who could not get enough coal would offer a higher price. Thus, if

[3] Assuming that the demand curve has a strictly negative slope, and that the supply curve has a strictly positive slope.

the price were below p^*, then economic forces would tend to make the price increase towards p^*. Therefore, the term "equilibrium" makes sense for the point (q^*, p^*) because deviations from the price p^* are naturally corrected, and q^* is the quantity that corresponds to p^* (on both curves). It is often plausible to suppose that the prices and quantities that are *observed* in a real world market are equilibrium prices, because non-equilibria probably don't last very long. This is likely to be true for short-run equilibria, defined as an equilibrium given a fixed capital stock (equipment, houses, etc.). However, long run equilibria, in which consumers and producers adjust their capital stock to new conditions, may take years or even decades to achieve. For instance, power plants may last 50 years, and so an obsolete technology might still be in use many years after new investments in such technology become unprofitable.

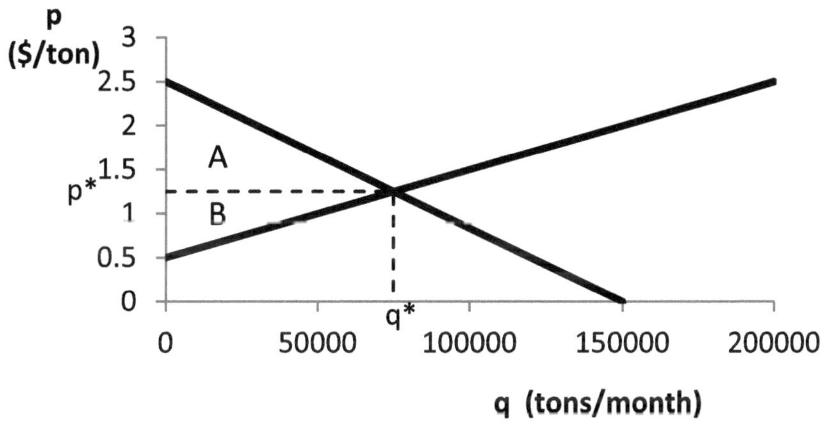

Fig. 3.5 Equilibrium (q^*, p^*) in coal market, consumers' surplus (area A), and producers' surplus (area B)

3.2.3.1 Illustrative Example. Equilibrium in the Coal Market

The demand curve in Figure 3.5 can be described by the inverse demand function

$$p = 2.5 - 0.0000167q \tag{3.1}$$

and the supply curve by the inverse supply function

$$p = 0.5 + 0.00001q \tag{3.2}$$

Therefore, equilibrium is obtained by equating the two expressions in q, solving for q^*, and substituting q^* into either (3.1) or (3.2) to obtain p^*:

$$\begin{aligned}
2.5 - 0.0000167q &= 0.5 + 0.00001q \\
q^* &= (2.5 - 0.5)/(0.0000167 + 0.00001) \\
&= 75000 \text{ tons per month} \\
p^* &= 0.5 + 0.00001q^* \\
&= \$1.25 \text{ per ton}
\end{aligned}$$

The triangular area A in Figure 3.5 represents the consumers' surplus at equilibrium, and the triangular area B represents producers' surplus. Consumers' surplus is calculated by subtracting the amount paid, $p^* \times q^*$ from the area under the demand curve, i.e., from the integral of the inverse demand function:

$$\begin{aligned}
\text{consumers' surplus} &= \int_0^{q^*} (2.5 - 0.0000167q)dq - p^*q^* \qquad (3.3) \\
&= 2.5q^* - \left(\frac{0.0000167}{2}\right)(q^*)^2 - p^*q^* \\
&= \$46,875 \text{ per month}
\end{aligned}$$

Producers' surplus is calculated by subtracting the area under the supply curve (the integral of the inverse supply function) from the amount paid:

$$\begin{aligned}
\text{producers' surplus} &= p^*q^* - \int_0^{q^*} (0.5 + 0.00001q)dq \qquad (3.4) \\
&= p^*q^* - (0.5q^* + \left(\frac{0.00001}{2}\right)(q^*)^2 \\
&= \$28,125 \text{ per month}
\end{aligned}$$

□

Consumers' and producers' surpluses are not usually measured in total amounts, as above. Instead, economists usually measure *changes* in consumers' and producers' surpluses relative to some base case, as a result of interventions in the market.

3.2.3.2 Illustrative Example. Changes in Consumers' and Producers' Surpluses Due to a Cartel

Figure 3.6 illustrates the result of all coal producers forming a cartel, and agreeing to restrict their combined total sales in the city to no more than $q^c = 50,000$ tons per month, less than the equilibrium – the solution of (3.1),(3.2), 75,000 tons per month. The price rises from $p^* = \$1.25$ to $p^c = \$1.67$ per ton. A comparison of Figures 3.6 and 3.5 shows that the consumers' surplus is smaller with the cartel (area C) than in the equilibrium (area A), and it appears that the producers' surplus is larger with the cartel (area D) than in the equilibrium (area B). (This assumes that the colluding producers agree to produce the 50,000 tons as cheaply as possible. This does not necessarily happen in a cartel.) The change in consumers' surplus due to the cartel can be calculated by subtracting area A from area C, and simplifying to a calculation involving only p^c, p^*, q^c and q^*. Note that, by similarity of triangles A and C,

$$\frac{2.5 - p^c}{2.5 - p^*} = \frac{q^c}{q^*}$$

which implies that

$$(2.5 - p^c)q^* = (2.5 - p^*)q^c \tag{3.5}$$

Therefore, the change in consumers' surplus (in the sense of new surplus minus old surplus) is

$$\begin{aligned}
&(\text{Area C}) - (\text{Area A}) \\
&= (2.5 - p^c)q^c/2 - (2.5 - p^*)q^*/2 \\
&= \frac{(2.5 - p^c)q^c - (2.5 - p^*)q^* + (2.5 - p^c)q^* - (2.5 - p^*)q^c}{2}, \text{ by (3.5)} \\
&= \frac{(2.5 - p^c)(q^* + q^c) - (2.5 - p^*)(q^* + q^c)}{2} \\
&= (p^* - p^c) \times (q^* + q^c)/2 \qquad (3.6) \\
&= -\$26,041.67 \text{ per month}
\end{aligned}$$

The increase in producers' surplus due to the cartel is extra revenue that comes from the sale of the amount q^c, offset somewhat by the loss in producers' surplus due to cutting back sales from q^* to q^c: $(p^c - p^*) \times q^c - \{p^* \times (q^* - q^c) - \int_{q^c}^{q^*} (0.5 + 0.00001q)dq\} = \$17,708.33$ per month. We can also see that the cartel causes the total of consumers' plus producers' surpluses to fall by the amount of area E (often called deadweight loss), i.e., $\int_{q^c}^{q^*} (2.5 - 0.0000167q)dq - \int_{q^c}^{q^*} (0.5 + 0.00001q)dq = \$8,333.33$ per month. □

3.2.4 Non-Price Influences: Shifting Supply and Demand Curves

To this point, we have focussed on the influence of price on the quantity supplied or demanded, but it is easy to see that there are many other factors which influence quantities, i.e., the demand and supply functions depend on more than just the price. For example, the weather would have influenced the quantity of coal demanded in the old city of our story: winter weather would have owners of homes and business buildings seeking more coal for heat during these cold months. Weather may also have influenced the quantity that

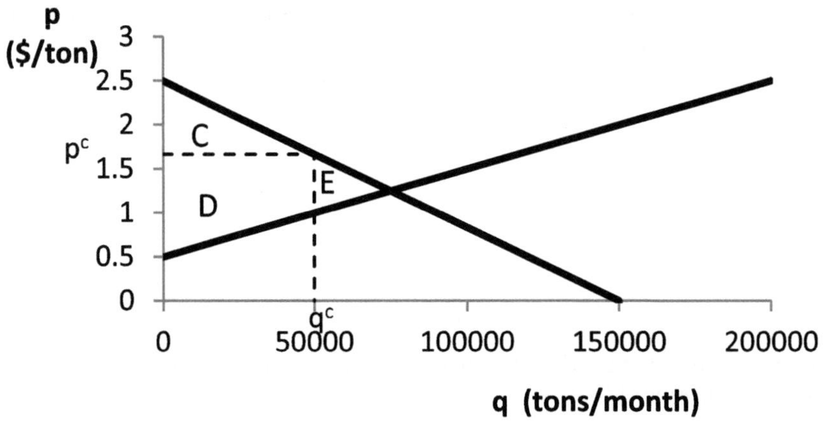

Fig. 3.6 Cartel with output restricted to 50,000, consumers' surplus (area C), producers' surplus (area D), and deadweight loss (area E)

could be supplied: let's suppose that some of the roads from the mines to the city were difficult or impassable during the wet spring and fall months, but that summer roads were good, and winter roads were good for sleighs on the ice and snow. Figure 3.7 illustrates these influences: the winter demand curve is shifted to the right (larger quantity at any given price) compared to the spring and fall demand curve (the demand curve of Figures 3.4 and 3.5); summer demand is shifted to the left; and the supply curve for summer-winter is shifted to the right, compared with the spring and fall supply curve (the supply curve of Figures 3.2 and 3.5). Thus, (q^*, p^*) is the equilibrium during the spring and fall, (q^W, p^W) is the winter equilibrium, and (q^S, p^S) is the summer equilibrium.

There are many other factors that can influence the position of the supply and demand curves. For example, an increase in population, and a corresponding increase in the number of houses, would likely shift the demand curve to the right, as would an increase in average household income.[4] An increase in the price of firewood would make firewood less attractive to households as an alternative to coal for heat, and we would expect the demand curve to shift to the right. So might a clever marketing campaign by coal producers who could, for instance, try to convince consumers to buy more coal and less wood and thereby preserve forests. A new mining technology, that makes it cheaper to produce each ton of coal, would shift the supply

[4] It is also possible that an increase in average income could shift the curve to the left, if some of the extra income were spent on something that reduces the demand for coal, e.g., expensive, more efficient furnaces. Or people might switch to heating with a more expensive but cleaner or otherwise more desirable fuel such as natural gas. Here, we suppose that this does not happen, and instead that people spend some of their extra income to buy more coal, at any price, to stay warmer.

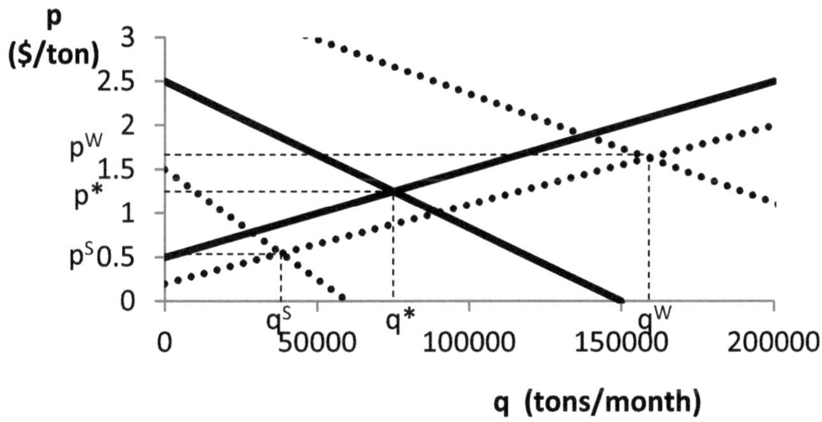

Fig. 3.7 Equilibrium in summer (p^S, q^S), spring & fall (p^*, q^*), winter (p^W, q^W)

curve downwards. To express the point mathematically, the demand function should show dependence on the good's price and also other variables – e.g., we can write $q = f_d(p, \widehat{p}, W, N, I)$ where $p =$ the good's price, $\widehat{p} =$ the price of a substitutable good (like wood), $W =$ a weather variable (e.g., temperature), $N =$ population in the market, and $I =$ average household income. Similarly, we can extend the list of variables that influence supply, and write the supply function as, e.g., $q = f_s(p, W)$, where W measures rainfall. For the models of this book, if supply or demand curves shift, we suppose that a new equilibrium comes about quickly, i.e., that disequilibrium does not last long; we do not try to model the dynamics of change from one equilibrium to another.

3.2.5 Multicommodity Equilibrium

If the price of wood affects the quantity of coal demanded, then the markets for coal and wood interact: fluctuations in the wood market are transmitted into the coal market. We can extend the description of equilibrium in the coal market to a description of simultaneous equilibrium in both markets. This is an example of a multicommodity equilibrium model. See Chapter 12 for more complex models of multicommodity equilibria in energy markets.

3.2.5.1 Illustrative Example. Simultaneous Equilibrium of Coal and Wood Markets

Let p, q^D and q^S represent the price, demand quantity and supply quantity for coal, and let $\widehat{p}, \widehat{q}^D$ and $\widehat{q}^S$ represent the price, demand quantity and supply quantity of wood, with the wood price in dollars per ton, and wood quantity in tons per month. Suppose that the demand for coal depends on both the coal and wood prices, while the supply of coal depends only on the coal price:[5]

$$\begin{aligned} q^D &= 137{,}850 - 60{,}000p + 13{,}500\widehat{p} \\ q^S &= -50{,}000 + 100{,}000p \end{aligned}$$

Suppose also that the demand for wood depends on both prices, while the supply of wood depends only on the wood price:

$$\begin{aligned} \widehat{q}^D &= 66{,}150 + 5000p - 36{,}000\widehat{p} \\ \widehat{q}^S &= -23{,}000 + 70{,}000\widehat{p} \end{aligned}$$

The equilibrium prices can be found by setting the demand and supply expressions equal in each market, and solving the two simultaneous equations:

$$137{,}850 - 60{,}000p + 13{,}500\widehat{p} = -50{,}000 + 100{,}000p \tag{3.7a}$$

$$66{,}150 + 5000p - 36{,}000\widehat{p} = -23{,}000 + 70{,}000\widehat{p} \tag{3.7b}$$

The solution is: $p = 1.25$ and $\widehat{p} = 0.90$. These prices can be substituted into either the demand or supply equations in each market, to determine equilibrium quantities: $q^D = q^S = 75{,}000$ and $\widehat{q}^D = \widehat{q}^S = 40{,}000$.

□

3.2.6 *Estimation of Parameters of Demand and Supply Functions*

3.2.6.1 Top-Down or Statistical Estimation on Observations

Statistical estimation of a demand function begins by gathering observations of variables that are thought to influence demand (the independent vari-

[5] For other models of supply, it may be reasonable to suppose that supply of the good depends not only on its own price, but also on the prices of other goods; for instance, the price of inputs to making the good, such as the prices of axes and foresters' labor in the case of wood. Here, we suppose that the price of wood has no direct effect on the supply of coal, i.e., that if the price of coal stays constant, while the price of wood changes, then coal producers keep coal production constant.

ables), as well as the demand quantities themselves (the dependent variable), for different time periods and/or for different geographical locations. A mathematical form is assumed for the demand function, with initially unknown parameter values, and regression analysis is performed, using the observed values of the independent and dependent variables, to determine the values of the parameters of the demand function.

For example, a demand function for coal in the imaginary city could have been determined by compiling a large number of observations of $(q, p, \widehat{p}, W, N, I)$, over many months of several years, where q = number of tons of coal sold in a month, p = the average coal price during the month, $\widehat{p}$ = the average price of firewood during the month, W = average number of degrees Fahrenheit below 60 degrees during the month, N = population of the city during the month, and I = the average household income during the month. One possible form for the demand function assumes a linear relationship between the demand and the independent variables:

A Linear Demand Function

$$q = f_d(p, \widehat{p}, W, N, I) = a_0 + a_1 p + a_2 \widehat{p} + a_3 W + a_4 N + a_5 I$$

Regression analysis on the observations produces estimates of the parameters a_0, a_1, a_2, a_3, a_4, and a_5. We expect that $a_1 < 0$, $a_2 > 0$, $a_3 > 0$, $a_4 > 0$, and $a_5 > 0$.

It may be thought that a different form of the demand function produces a more accurate representation of the relationship between demand and the independent variables. For example, one nonlinear form is the constant elasticity model:

A Nonlinear Demand Function (Constant Elasticity)

$$q = f_d(p, \widehat{p}, W, N, I) = b_0 p^{b_1} \widehat{p}^{b_2} W^{b_3} N^{b_4} I^{b_5}$$

Regression analysis on the observations produces estimates of the parameters b_0, b_1, b_2, b_3, b_4, and b_5. In this case, we expect that $b_0 > 0$, $b_1 < 0$, $b_2 > 0$, $b_3 > 0$, $b_4 > 0$, and $b_5 > 0$. However, if wood and coal were complements rather than substitutes, then b_2 would also be expected to be negative. For instance, a demand function for wood as a function of the price of wood and the price of wood stoves (a complement of wood use) would have this relationship.

If statistical analysis verifies that there is a good enough fit of an assumed form of the demand function to the observations, then the demand function can be used to estimate the impact of changes in the independent variables on the amount demanded. The impact of an independent variable X (where X is one of $p, \widehat{p}, W, N, I$ in the example) could be expressed with the partial derivative $\partial q / \partial X$. But it is far more common to express the impact with a dimensionless parameter called the *elasticity* of demand with respect to X, that gives the *fractional* change in demand, divided by the *fractional* change in X:

Elasticity of Demand with Respect to an Independent Variable
$\varepsilon_X = (\partial q/q)/(\partial X/X) = (\partial q/\partial X)(X/q)$
In our example, we expect that the *own-price elasticity of demand* $\varepsilon_p < 0$, the *cross-price elasticity of demand* $\varepsilon_{\widehat{p}} > 0$ (for substitutes), and that $\varepsilon_X > 0$ for the other independent variables X. In Chapter 12, we discuss two models that have own-price and cross-price elasticities of demand: an electricity market model with peak and off-peak prices jointly influencing peak and off-peak demand; and a model with demand for coal, light oil and heavy oil each dependent on the three goods' prices.

For the constant elasticity demand function, all elasticities are independent of the values of the independent variables, e.g., $\varepsilon_p = (\partial q/\partial p)(p/q) = (b_1 b_0 p^{b_1 - 1} \widehat{p}^{b_2} W^{b_3} N^{b_4} I^{b_5})(p/b_0 p^{b_1} \widehat{p}^{b_2} W^{b_3} N^{b_4} I^{b_5}) = b_1$. Similarly, $\varepsilon_{\widehat{p}} = b_2$, $\varepsilon_W = b_3$, $\varepsilon_N = b_4$, and $\varepsilon_I = b_5$. However, for the linear demand function, elasticities depend on the values of the independent variables, e.g., $\varepsilon_p = (\partial q/\partial p)(p/q) = (a_1)(p/(a_0 + a_1 p + a_2 \widehat{p} + a_3 W + a_4 N + a_5 I))$.

It is also sometimes possible to use regression techniques to estimate the parameters of a supply function. For example, observations on the monthly supply quantity of coal, the coal price, and W = the amount of rainfall by month could be used in a regression analysis to determine the parameters of a linear or a nonlinear supply function, $q = f_s(p, W)$ to the imaginary city. Elasticities can be computed, such as the *price elasticity of supply* = $(\partial f_s/\partial p)(p/f_s(p, W))$.

The discipline of statistically estimating the parameters that describe the relationships among economic variables is known as *econometrics*. This is a huge subject, and the details are beyond the scope of this book. For further reading, there are many textbooks on econometrics, such as [14].

3.2.6.2 Bottom-Up or Process-Based Estimation

If there are not too many firms that supply a market, and not too many different conditions of supply technology and raw materials availability, then it is sometimes possible to estimate directly the form and parameters of each firm's supply function. For example, in Figure 3.1, the width of a step is the output capacity of one process by which the firm produces output, from a particular combination of technology, raw materials, and other factors such as mode of transportation to market. The height of a step is the per-unit cost of output for this combination of technology, raw materials, etc. The entire market supply function, for all firms, is obtained by combining the steps of all firms into one picture, but ordered from least costly to most costly. The resulting multi-step function is sometimes approximated by a smooth curve, as indicated in the previous discussion about Figure 3.2, but this is not always necessary.

In such direct measurement, it may be observed that the per-unit cost of output for a particular combination of technology, etc., varies with the output rate. For example, as the coal output rate approaches a mine's capacity, it may be observed that some congestion occurs along the rail tracks that carry the carts to the mine mouth, causing a reduction in the number of tons per month *per unit of mine labor*, but the laborers must still be paid for their time; this would cause the per-unit cost of coal to increase as output approaches capacity. The opposite can also happen – e.g., some electric generators operate inefficiently at low output rates, i.e., at low output rates, they burn more fuel per unit of electric output than at higher output rates, which implies that the per-unit cost of electricity decreases as output increases, at least over the low range of output. When the per-unit cost varies with output, the steps of the supply function will not have horizontal tops, and the construction of the supply function picture becomes complicated if the range of variation of per-unit cost of one combination of technology, etc., overlaps the range of another.

For many models of this book, it is not necessary to construct an explicit supply function for the whole market; it is sufficient to know the supply functions of individual firms.

In order to be able to measure a firm's supply function directly, it is necessary to have access to information on the firm's production function and input costs. A "production function" is an economist's term for the relationship between amount of inputs to a production process and the amount of outputs. A production function for power, for instance, expresses the relationship between the inputs fuel and generation equipment, and the output electricity production, reflecting capacity limits and efficiencies of energy conversion. However, the firm may be very reluctant to release such information, because it could be exploited by the firm's competitors. Nevertheless, it is sometimes possible to make reasonable estimates of a firm's supply function, from publicly available information. For example, in many jurisdictions, an electric generation company must present information on its generation equipment (e.g., technology, fuel, capacity) before public hearings. This information, together with publicly available information on prices of fuels (e.g., coal, natural gas, or uranium) can allow an outsider to estimate the firm's supply function.

3.2.6.3 Auctions

There is another way in which firms' supply functions are sometimes revealed – when the market is run as an *auction*. For example, in many electricity markets, a non-profit organization (often called the Independent System Operator (ISO)) receives *offers* from all generating companies to generate electric power during a near-future time period of a short duration (e.g., an hour). Each offer usually consists of the parameters of a step function – (offer price, capacity) pairs, by which the firm declares that it is willing to produce up to the capacity amount as long as the market price is above or equal to the offer price. The ISO considers all firms' offers, together with the forecast of electricity demand in the near-future period, and performs a calculation that, in effect, produces a step function supply curve for the whole market. There is a step at which the total supply of all steps adds up to the forecast demand.[6] The price that all generators receive for their output is the offer price of the most expensive generator that is called upon to provide electricity. The auction mechanism reveals supply function information to the ISO, which allows the ISO to operate the market for electricity, with data that are as accurate as the firms themselves know.

Such an auction can be solved using an optimization model. Constraints other than market clearing (supply = demand) can be added to such models, such as transmission limits between subareas of a market or requirements for spare generation capacity to be on line. Examples are shown in several chapters, with the most detailed in Chapter 11.

However, it is important to note that firms do not necessarily reveal their true costs and capacities to the ISO. A firm may have very good reasons to submit (offer price, capacity) pairs that are not accurate statements of per-unit costs and capacities. See Illustrative Example 6.2.2 in Chapter 6 for an example of a model of such behavior.

3.3 Social Welfare Maximization

3.3.1 Definition of Social Welfare in Single Commodity Models: Consumers' Plus Producers' Surpluses

In the discussion of Figure 3.5, the consumers' surplus and the producers' surplus were defined at the equilibrium (q^*, p^*), in equations (3.3) and (3.4). Note that if the two expressions are added together, the revenue terms cancel, and the result is *consumers' plus producers' surplus* = $\int_0^{q^*} (2.5 - 0.0000167q)dq - \int_0^{q^*} (0.5 + 0.00001q)dq$, which can be calculated without direct reference to p^*. This quantity represents the net gain of all participants in the market – consumers and producers – due to the buying and selling which takes place in the market. The consumers' plus producers' surplus is also called *social welfare* or the *gain from trade*, and is a standard measure of market efficiency.

It is possible to evaluate consumers' plus producers' surplus, or social welfare, at *any* value of the quantity – there is no need to be restricted to the equilibrium quantity, q^*. In general, then, we may define the social welfare function, $SW(q)$ as the difference between the integral of the inverse demand

[6] Assuming that all transmission lines have power flows within their rated capacities to carry power.

function, $f_d^{-1}(q)$, and the integral of the inverse supply function, $f_s^{-1}(q)$, as follows (the variable of integration is shown by the symbol q', to distinguish it from the upper limit of integration, q):

$$\text{social welfare} \quad SW(q) = \int_0^q f_d^{-1}(q')dq' - \int_0^q f_s^{-1}(q')dq' \tag{3.8}$$

The function $SW(q)$ represents the value to consumers of having the amount q, without deducting the amount of money paid for q (the first term in equation (3.8)), minus the variable costs incurred by producers to supply the amount q.

3.3.2 Equilibrium as Maximization of Social Welfare in Single Commodity Models

The equilibrium (q^*, p^*) can be calculated by finding the intersection of the demand and supply curves, i.e., by solving the simultaneous equations $p = f_d^{-1}(q)$ and $p = f_s^{-1}(q)$, but there is another way to do the calculation, which opens the door to a great variety of modeling techniques. The alternative way, due to Samuelson [11], is to find the value of q which maximizes the social welfare function, $SW(q)$. This can be justified by noting that, for unconstrained maximization, the function is maximized at the point where the derivative of the function is zero:[7]

$$\begin{aligned} &\text{Maximize} \quad SW(q) \\ &\text{which implies that} \quad \frac{dSW}{dq} = 0 \ , \\ &\text{which implies that}^8 \quad f_d^{-1}(q) - f_s^{-1}(q) = 0 \ . \\ &\text{Therefore,} \quad f_d^{-1}(q) = f_s^{-1}(q) \ . \end{aligned} \tag{3.9}$$

The last condition requires that, at the maximum point, the price on the demand curve equals the price on the supply curve, i.e., the maximum point is the equilibrium quantity q^*.

---

[7] Assuming that the inverse demand function has a negative slope, and that the inverse supply function has a positive slope, it follows that $SW(q)$ is strictly concave. Thus, if there is a maximum point, then it is unique, and defined by the condition that the derivative is zero.

3.3.2.1 Illustrative Example. Equilibrium in Coal Market as Social Welfare Maximization

With the linear demand (3.1) and supply (3.2) curves for the coal market, the social welfare function is

$$\int_0^q (2.5 - 0.0000167q')dq' - \int_0^q (0.5 + 0.00001q')dq'$$
$$= 2.5q - \frac{0.0000167}{2}q^2 - 0.5q - \frac{0.00001}{2}q^2$$

and the maximum is achieved where the derivative is zero:

$$2.5 - 0.0000167q - 0.5 - 0.00001q = 0$$
$$\text{i.e.,}\quad 2.5 - 0.0000167q = 0.5 + 0.00001q$$

which is the condition for intersection of the demand and supply curves, with the solution $q = 75000$ tons per month.

□

3.3.3 Pareto Efficiency Versus Social Welfare Optimization

It appears that a market acts as if it maximizes a single objective function, i.e., social welfare. This is very strange at first sight, because the market is actually composed of many different firms and consumers, each trying to optimize their own profit (for firms) or well being (for consumers, usually called utility). There is a concept of optimality, due to the economist Vilfredo Pareto, that can be applied to economic situations (i.e., production and consumption decisions, and the prices at which things trade) with many different agents: Pareto optimality, also known as Pareto efficiency.

Definition 3.1. An economic situation is said to be Pareto optimal (or Pareto efficient) if it is impossible to make an improvement for any agent without making things worse for another agent.

In Section 3.4, we present an example which is Pareto optimal – the coal market when all firms and consumers behave as price-takers – and several examples of market outcomes which are not Pareto optimal – some firms do not act as price-takers.

3.3.4 Social Welfare in Multicommodity Models

Can the social welfare maximization approach be used to define the simultaneous equilibrium of several commodities? For example, could it be done for the example of the coal and wood markets? The answer is: not in general, unless some unusual conditions are satisfied for the demand functions. The difficulty lies with the attempt to generalize the integration of a one-dimensional inverse demand function to the integration of a vector of inverse demand functions of several variables, as we now explain for the case of two commodities. The result is easily extended to more than two commodities and is discussed also in Chapter 4, under the name Principle of Symmetry (Integrability).

3.3.4.1 Possible Difficulty to Integrate Inverse Demand Functions in Multicommodity Models

Consider a two commodity equilibrium model, with p and q the price and quantity of one good, and $\widehat{p}$ and $\widehat{q}$ the price and quantity in the other market. Suppose, as in the coal and wood example, that the demand for a good depends on both prices, but that the supply of a good depends only on that good's own price. The system of equations for simultaneous equilibrium can be written as follows, using the inverse demand and supply functions:

$$
\begin{aligned}
p &= f_d^{-1}(q, \widehat{q}) \\
p &- f_s^{-1}(q) \\
\widehat{p} &= \widehat{f^{-1}}_d(q, \widehat{q}) \\
\widehat{p} &= \widehat{f^{-1}}_s(\widehat{q})
\end{aligned}
$$

To define a social welfare function in a manner that is analogous to the one commodity case, we would need to extend the notion of integrating one inverse demand function to integration of two inverse demand functions. That is, we would need to find a consumer value function, $CV(q, \widehat{q})$ from which we would subtract producers' variable costs to define the two-variable social welfare function, $SW(q, \widehat{q}) = CV(q, \widehat{q}) - \int_0^q f_s^{-1}(q')dq' - - \int_0^{\widehat{q}} \widehat{f^{-1}}_s(\widehat{q}')d\widehat{q}'$. In order to have maximization of social welfare define the equilibrium, the function $CV(q, \widehat{q})$ must have the property that its two partial derivatives equal the two demand functions, i.e.,

$$
\begin{aligned}
\frac{\partial CV}{\partial q} &= f_d^{-1}(q, \widehat{q}) \\
\frac{\partial CV}{\partial \widehat{q}} &= \widehat{f^{-1}}_d(q, \widehat{q})
\end{aligned}
$$

We can take second cross partial derivatives by differentiating each equation with respect to the *other* quantity variable:

$$\frac{\partial^2 CV}{\partial \widehat{q} \partial q} = \frac{\partial f_d^{-1}}{\partial \widehat{q}}$$
$$\frac{\partial^2 CV}{\partial q \partial \widehat{q}} = \frac{\partial \widehat{f^{-1}}_d}{\partial q}$$

However, it is well known that reversing the order of differentiation in the second cross partial derivative does not affect the result, i.e.,

$$\frac{\partial^2 CV}{\partial \widehat{q} \partial q} = \frac{\partial^2 CV}{\partial q \partial \widehat{q}}$$

and therefore,

$$\frac{\partial f_d^{-1}}{\partial \widehat{q}} = \frac{\partial \widehat{f^{-1}}_d}{\partial q}$$

To summarize, if the notion of integration of a single inverse demand function can be extended to two inverse demand functions, each of which depends on both quantities q and $\widehat{q}$, then the partial derivatives of the inverse demand functions, with respect to the *other* quantity must be equal, for *all* values of q and $\widehat{q}$. This symmetry condition is seldom true for econometrically estimated demand functions, unless that condition is imposed as a constraint during the estimation. When the symmetry condition is not satisfied, we say that demand is *asymmetric*.

3.3.4.2 Illustrative Example. Impossibility of Integrating Inverse Demand Functions for Coal and Wood

The linear demand functions for the coal and wood markets are:

$$q = 137,850 - 60,000p + 13,500\widehat{p}$$
$$\widehat{q} = 66,150 + 5000p - 36,000\widehat{p}$$

Solving for prices in terms of the quantities, i.e., for the inverse demand functions gives:

$$p = 2.7984 - 0.0000172q - 0.00000645\widehat{q}$$
$$\widehat{p} = 2.2262 - 0.00000239q - 0.0000287\widehat{q}$$

and it is clear that the symmetry condition is violated, i.e., $-0.00000645 \neq -0.00000239$.

□

Social welfare maximization does not, in general, work for calculation of equilibrium in multicommodity models. However, there are other approaches that do work, namely, complementarity and variational inequality modeling methods, as discussed in various chapters of this book.

3.3.4.3 Measuring Changes in Social Welfare in Multicommodity Models

Although it is not possible, in general, to define a consumer value function in a multicommodity setting, and therefore consumers' surplus cannot be defined, it is nevertheless possible to estimate the *change* in consumer value when an external influence shifts equilibrium, using a result due to [3]. This is so even if the demand functions are not integrable. The change in consumers' surplus can therefore be estimated as the change in consumer value, minus the change in the amount of money that consumers pay.

3.3.4.4 Illustrative Example. Changes in Consumers' Surplus, for Wood and Coal, Due to a Tax on Coal

In the coal and wood market, the original equilibrium was $p = 1.25$, $\widehat{p} = 0.90$, $q = 75{,}000$ and $\widehat{q} = 40{,}000$, which is the solution of (3.7a)-(3.7b). Suppose that the government imposes a tax of \$0.25 on each ton of coal, to be collected from the coal producers. This has the effect of adding \$0.25 to the per-ton cost of coal, for any quantity of coal output, thus changing the inverse supply function for coal from $p = 0.5 + 0.00001q$ to $p = 0.75 + 0.00001q$. The new simultaneous equilibrium of the coal and wood markets shifts to $p' = 1.41$, $\widehat{p}' = 0.91$, $q' = 65{,}687$ and $\widehat{q}' = 40{,}518$. Based on analysis (beyond the scope of this book) by [3], it is possible to approximate the *change* in consumer value for each commodity separately, and the change in consumers' surplus is obtained by subtracting the change in consumers' payment for the good. The approximation for the change in consumers' surplus for coal is the single commodity formula (3.6), with the *new* surplus corresponding to the with-tax case here (and price p^c in (3.6) corresponding to p' here):

$$\begin{aligned} \text{change in consumers' surplus for coal} &= (p - p')(q + q')/2 \\ &= (1.25 - 1.41)(75{,}000 + 65{,}687)/2 \\ &= -\$11{,}255 \text{ per month} \end{aligned}$$

The change in consumers' surplus for coal is negative because consumers pay a higher price after the tax is imposed. The same calculation can be done for the change in consumers' surplus for wood purchases:

$$\text{change in consumers surplus for wood} = (\widehat{p} - \widehat{p}')(\widehat{q} + \widehat{q}')/2$$

$$=(0.90-0.91)(40,000+40,518)/2$$
$$=-\$403 \text{ per month}$$

The change in consumers' surplus for wood is negative because the wood price is higher after the tax is imposed.

□

3.4 Modeling Individual Players in Single Commodity Markets

This section introduces methods to construct equilibrium models by combining models of each player in a market. The term "player" typically means a producing firm, a consumer, or an aggregation of consumers (e.g., the representation of total consumer demand in a demand function). Often, the term *agent* is used synonymously for player. We examine some simple one-commodity markets in this section. Chapter 12 examines multicommodity models, constructed from considerations of individual players, using complementarity methods.

3.4.1 Profit-Maximization Problem for Price-Taking Firms, and Form of Equilibrium Problem

Let $VC_i(q_i)$ represent the variable cost of firm i, as a function of the firm's output q_i. Assume for the moment that each VC_i function is continuously differentiable and strictly convex, which implies that its derivative $\frac{dVC_i(q_i)}{dq_i}$ is increasing in q_i. If a price-taking firm i expects to receive a price p for its output q_i, then it will choose the output rate q_i in order to maximize its profit, $pq_i - VC_i(q_i)$. The profit maximum occurs where $p - \frac{dVC_i(q_i)}{dq_i} = 0$, i.e. $\frac{dVC_i(q_i)}{dq_i} = p$. (The point q_i which satisfies this condition is unique, due to strict convexity of VC_i.) Thus, $\frac{dVC_i(q_i)}{dq_i}$, called the *marginal cost function*, is the definition of the firm's inverse supply function, because it traces out the relationship between price and the quantity that the firm would choose to produce. Note that this also equals the derivative of the total cost function $TC_i(q_i)$, which equals $VC_i(q_i)$ plus any fixed cost.

For the models in this book, it is usually somewhat complicated to estimate the supply functions of all firms, combine them into a single market supply function, and then integrate the market supply function as in the second term of equation (3.8). It is often more straightforward to estimate $VC_i(q_i)$, and add the variable costs of all firms, to get the second term of equation (3.8). Thus, maximization of social welfare (3.9) can be represented as (the dual variable for the constraint is to the right of the constraint in parentheses)

$$\text{Maximize} \quad \int_0^q f_d^{-1}(q')dq' - \sum_i VC_i(q_i) \tag{3.10a}$$

s. t.

$$q - \sum_i q_i = 0 \quad (\lambda) \tag{3.10b}$$

where the constraint requires that the sum of all firms' outputs must equal the demand. Let us check that the maximum point of (3.10) satisfies the notion of equilibrium. The variables of the problem (3.10) are total demand q, and individual firm output q_i (one such variable for each firm i), and the problem has one constraint with associated dual variable λ. Assuming that the Karush-Kuhn-Tucker (KKT) conditions (see Chapter 2) are valid for the problem, and recalling that maximization of an objective produces the same optimal point as minimization of the negative of the objective, we can conclude that, at the optimal point,

$$-f_d^{-1}(q) + \lambda = 0$$
$$\frac{dVC_i(q_i)}{dq_i} - \lambda = 0 \ \forall i$$

and therefore,

$$\lambda = f_d^{-1}(q)$$
$$\lambda = \frac{dVC_i(q_i)}{dq_i} \quad \forall i$$

It can be seen that the dual variable, λ plays the role of equilibrium price, because it is equal to the price on the demand curve, at the maximum point, and for each firm i, it is also equal to the marginal cost $\frac{dVC_i(q_i)}{dq_i}$, i.e., the value of the inverse supply function for firm i. The optimal point which maximizes (3.10) is the equilibrium.

It is quite restrictive to require that $VC_i(q_i)$ be strictly convex and differentiable. For instance, these requirements are not met by one common way to represent a firm's costs: the representation of the step functions in Figure 3.1. The step function height gives the derivative of variable cost with respect to output, but this derivative is not defined at the end of a step (it could take any value between the height of one step and the height of the next step). Now we show how to define a profit maximization model, and an equilibrium model, for which the firms' variable cost functions $VC_i(q_i)$ are based on the step function idea of Figure 3.1.

Each step represents a production process defined by a combination of a production technology, raw materials, and geographical location-dependent factors such as costs of transportation to the market. Let us index the steps/processes by the symbol b (for block), and let x_{ib} represent the output of firm i from process b, which can be produced at a per-unit cost of c_{ib} (the height of step b), up to the capacity K_{ib} of the process (i.e., we must require that $x_{ib} \leq K_{ib}$, or equivalently, $x_{ib} - K_{ib} \leq 0$). To be realistic, we must require that $x_{ib} \geq 0$ (which appears below in the form $-x_{ib} \leq 0$ to facilitate computation of the KKT conditions, following the description of KKT conditions in Chapter 2). The firm's total output, from all processes, is $q_i = \sum_b x_{ib}$. If the firm expects to receive a price p for its output, then in order to maximize its profit (i.e., minimize the negative of profit), the firm will choose x_{ib} for all processes b in order to

$$\text{Minimize} \quad \sum_b (c_{ib} - p)x_{ib} \tag{3.11a}$$

s. t.

$$x_{ib} - K_{ib} \leq 0 \quad (\gamma_{ib}) \ \forall b \tag{3.11b}$$

$$-x_{ib} \leq 0 \quad (\delta_{ib}) \ \forall b \tag{3.11c}$$

If γ_{ib} is the dual variable associated with (3.11b), and δ_{ib} is the dual variable associated with (3.11c), then the KKT conditions are:

$$0 \geq x_{ib} - K_{ib} \perp \gamma_{ib} \geq 0 \ \ \forall b \tag{3.12a}$$

$$0 \geq -x_{ib} \perp \delta_{ib} \geq 0 \ \ \forall b \tag{3.12b}$$

$$c_{ib} - p + \gamma_{ib} - \delta_{ib} = 0 \ \ \forall b \tag{3.12c}$$

If $p > c_{ib}$, then by (3.12c) and nonnegativity of γ_{ib} and δ_{ib}, it follows that $\gamma_{ib} > 0$; thus by complementarity in (3.12a) and (3.12b), $x_{ib} = K_{ib}$, and so $\delta_{ib} = 0$; therefore, $c_{ib} + \gamma_{ib} = p$. Summarizing this in words, if the price p exceeds the unit cost c_{ib} of process b (and the excess is $\gamma_{ib} > 0$, often called the *rent* per unit of output), then it is optimal to set $x_{ib} = K_{ib}$, the capacity of process b. On the other hand, if $p < c_{ib}$, then $\delta_{ib} > 0$ and by complementarity in (3.12b), $x_{ib} = 0$. In words, if the price p is less than the unit cost c_{ib}, then it is optimal to produce nothing. Finally, if it is optimal to set x_{ib} strictly between its limits, $0 < x_{ib} < K_{ib}$, then by complementarity, $\gamma_{ib} = \delta_{ib} = 0$, and it must be true that $c_{ib} - p = 0$ – the only condition for which it can be optimal to have $0 < x_{ib} < K_{ib}$ is when $p = c_{ib}$.

To summarize these mathematical derivations, the firm will choose to produce from any process whose unit cost is less than or equal to the price, which is just what is illustrated in Figure 3.1.

The maximum social welfare problem (3.10) can be modified to accommodate the step function form of firms' costs, as follows (converting to minimization form, to ease the derivation of the KKT conditions):

$$\text{Minimize} \quad \sum_i \sum_b c_{ib} x_{ib} - \int_0^q f_d^{-1}(q')dq' \tag{3.13a}$$

s. t.

$$q - \sum_i \sum_b x_{ib} = 0 \tag{3.13b}$$

$$x_{ib} - K_{ib} \leq 0 \quad (\gamma_{ib}) \quad \forall i, b \tag{3.13c}$$

$$-x_{ib} \leq 0 \quad (\delta_{ib}) \quad \forall i, b \tag{3.13d}$$

The KKT conditions necessarily hold at an optimum:

$$\lambda = f_d^{-1}(q) \tag{3.14a}$$

$$q - \sum_i \sum_b x_{ib} = 0 \tag{3.14b}$$

$$0 \geq x_{ib} - K_{ib} \perp \gamma_{ib} \geq 0 \quad \forall i, b \tag{3.14c}$$

$$0 \geq -x_{ib} \perp \delta_{ib} \geq 0 \quad \forall i, b \tag{3.14d}$$

$$c_{ib} - \lambda + \gamma_{ib} - \delta_{ib} = 0 \quad \forall i, b \tag{3.14e}$$

and it can be seen that λ is the equilibrium price, as it is the price on the inverse demand curve that corresponds to demand q, and it relates to each firm's step function as described above. Because f_d^{-1} has a negative first derivative, the objective function (3.13a) is convex, which, together with the linearity of the constraints (3.13b)-(3.13d), ensures that the KKT conditions are sufficient: any solution to (3.14) is an optimal solution to (3.13). Therefore, the optimization form of the model (3.13) is equivalent to the KKT conditions (3.14), which form a complementarity model.

The complementarity version of the model can be derived by a line of reasoning that does not refer directly to the social welfare optimization form. From this perspective, an equilibrium model of a market is the collection of all behavioral conditions of participants in the market, together with conditions that express the interactions of their decisions. The behavioral condition for consumers is the price-quantity relationship (3.14a). Each producer i follows the profit-maximizing behavior given by (3.14c), (3.14d), and (3.14e), i.e., the same as (3.12a), (3.12b) and (3.12c), with the substitution of λ for p. The condition that expresses the interaction among producers and consumers is (3.14b), the *market clearing condition.*

3.4.2 Perfect Versus Imperfect Competition

If all producers and consumers are price-takers, then the market is said to experience *perfect competition.* If instead any producers try to anticipate the results of their actions on the market price (or if any consumers do), then the market experiences imperfect competition. In this section, we discuss a market with three producing firms, under seven different behavioral assumptions: in the first case, all firms are price-takers; in the second case, the three

Table 3.1 Data for price-taker, monopoly and Cournot models

Producer Data	Firm and Block					
	$i = 1$		$i = 2$		$i = 3$	
	$b = 1$	$b = 2$	$b = 1$	$b = 2$	$b = 1$	$b = 2$
c_{ib} (\$/ton)	0.55	0.81	0.62	1.25	0.78	1.35
K_{ib} (tons/month)	21000	16000	17000	22000	18000	14000
Demand Data	$\alpha = 2.5$		$\beta = 0.0000166666667$			

firms have merged into one firm, a *monopoly*. In the other cases, we illustrate several "oligopoly" models. In the third case, the three firms are separate, but each tries to anticipate the effect of its production (sales) decisions on the market price, through knowledge of the demand curve; in the fourth case, only one firm anticipates the market price, while the other two firms are price-takers; and in the fifth case, two firms anticipate the market price, with the third a price-taker. In case six, the three firms are assumed to choose their selling prices and then live with the consequences for their sales, instead of choosing their sales quantities as in the other cases. Finally, in case seven, we look at a model with separate firms that coordinate their decisions in a "cartel". For all seven models, consumers are assumed to be price-takers, and the inverse demand function is assumed to be linear with the data given in Table 3.1:

$$f_d^{-1}(q) = \alpha - \beta q \tag{3.15}$$

Each of the three firms has two production processes, with each process represented by a unit cost c_{ib} and capacity K_{ib}. We use the same data for all five models, with values chosen to continue the story about coal supplies and demands; see Table 3.1.

3.4.2.1 Illustrative Example. Three Price-Taking Firms: Social Welfare Maximization Model

When all three firms are price-takers, equilibrium can be calculated by social welfare maximization, (3.16).

$$\text{Minimize} \quad \sum_i \sum_b c_{ib} x_{ib} - (\alpha q - \frac{1}{2} \beta q^2) \tag{3.16a}$$

s. t.

$$q - \sum_i \sum_b x_{ib} = 0 \tag{3.16b}$$

$$x_{ib} - K_{ib} \leq 0 \;\; \forall i, b \tag{3.16c}$$

$$- x_{ib} \leq 0 \;\; \forall i, b \tag{3.16d}$$

Table 3.2 Firm-related solution values for five models

Firm Related Values		Firm i and Mine b					
		$i=1$		$i=2$		$i=3$	
		$b=1$	$b=2$	$b=1$	$b=2$	$b=1$	$b=2$
Price Taker	x_{ib}	21000	16000	17000	3000	18000	0
	Revenue	46250		25000		22500	
	Variable Cost	24510		14290		14040	
	Surplus	21740		10710		8460	
Monopoly	x_{ib}	21000	0	17000	0	13600	0
	Revenue	34440		27880		22304	
	Variable Cost	11550		10540		10608	
	Surplus	22890		17340		11696	
All Cournot	x_{ib}	21000	12220	17000	0	18000	0
	Revenue	45263		23177		24540	
	Variable Cost	21432		10540		14040	
	Surplus	23831		12637		10500	
One Cournot ($i=1$)	x_{ib}	21000	5400	17000	13600	18000	0
	Revenue	33000		38250		22500	
	Variable Cost	15924		27540		14040	
	Surplus	17076		10710		8460	
Two Cournot ($i=1,2$)	x_{ib}	21000	16000	17000	0	18000	0
	Revenue	48100		22100		23400	
	Variable Cost	24510		10540		14040	
	Surplus	23590		11560		9360	

Table 3.3 Other solution values for five models

Other Values	q	p	Consumer Payment	Consumer Surplus	Producer Surplus	Social Welfare
Price-Takers	75000	1.25	93750	46875	40910	87785
Monopoly	51600	1.64	84624	22188	51926	74114
All Cournot	68200	1.36	92979	38760	46967	85728
1 Cournot	75000	1.25	93750	46875	36246	83121
1, 2 Cournot	72000	1.30	93600	43200	44510	87710

The optimal solution[9] of (3.16) is shown in the price-taker rows of Tables 3.2 and 3.3.

□

[9] It is possible to use the KKT conditions of this model to show that there is no other optimal solution, for other values of x_{ib}, q. The argument depends on the fact that, for this set of data, there are no ties in the unit cost values x_{ib}, for different i, b combinations.

3.4.2.2 Illustrative Example. Three Price-Taking Firms: Complementarity Model

The same solution is found by solving the corresponding complementarity problem (3.14), with the linear inverse demand function (3.15).

□

Notice that the complementarity model is more difficult to code than the optimization model: the complementarity model contains more variables and constraints than the optimization model, because the complementarity model must explicitly represent the dual variables and the expressions involving derivatives with respect to the primal variables x and q. The coding of the extra variables and constraints takes more time, but it also introduces more possibilities for coding errors and subsequent debugging. The difference in difficulty is not great for this very small example, but in larger, more complex models, the difference can be substantial. For this reason, if a model can be represented as an optimization model, then this is often the preferred choice. However, for many of the models in this book, the complementarity approach (or equivalent variational inequality approach, explained in Chapter 5) is the only approach that can be taken – e.g., a multicommodity model cannot normally be represented as a social welfare maximization model. Chapter 12 provides some detailed examples.

The price-taker solution is Pareto efficient. This can be seen by considering changes in total production, or in the amounts produced by each firm: any change causes at least one producer, or a consumer, to be worse off, i.e., to have a smaller surplus. If total production were less than the 75,000 tons per month shown in Table 3.3, and if price stayed at \$1.25 per ton, then consumers' surplus would decrease, making at least some individual consumers worse off; smaller production could also reduce the surplus of one or more producers, unless the reduction in total production is 3000 tons or less, in which case the reduction could be borne entirely by producer 2, in its $b = 2$ mine, where the unit cost is the same as the price of \$1.25. If we try to repair the damage to consumers by reducing the price below \$1.25 while reducing total production below 75,000, in such a way as to preserve consumers' surplus, then all three firms would receive less revenue per ton sold, and would therefore have less producers' surplus.

What about an increase in total production? Assuming that the extra, above 75,000 tons, is purchased by consumers, and the price stays at \$1.25, it would be possible to increase production from the $b = 2$ mine of firm 2 without reducing any firm's surplus (because price equals unit cost of the extra production), but the extra coal would have to be purchased by some consumers who value the extra coal at less than the \$1.25 price that they are forced to pay (i.e., further down the inverse demand curve), causing a loss in the surplus of some consumers. If the price of the extra production is reduced below \$1.25, to eliminate the loss in consumers' surplus, then at least one producer must produce the extra coal at a unit cost that exceeds the price received, causing a loss in producer's surplus for at least one producer.

3.4.2.3 Monopoly Model

Suppose that all three firms merge into one single firm. All of the original three firms' production facilities still exist as before. If this one firm acted as a price-taker, then the resulting market equilbrium would be the same as with three separate firms. However, if the one firm can estimate the inverse demand function, then it would be wise to anticipate the effect of its total production (which equals demand) on the market price. This is the essence of the monopoly model, which maximizes total profit (below, minimizes the negative of profit), with the revenue term recognizing the price-quantity relationship of the inverse demand function:

$$\text{Minimize} \quad \sum_i VC_i(q_i) - f_d^{-1}(q)q \tag{3.17}$$

s. t.

$$q - \sum_i q_i = 0 \quad (\lambda) \tag{3.18}$$

Applying the KKT conditions results in

$$-f_d^{-1}(q) - \frac{df_d^{-1}(q)}{dq}q + \lambda = 0$$
$$\frac{dVC_i(q_i)}{dq_i} - \lambda = 0 \quad \forall i$$
$$q - \sum_i q_i = 0$$

and therefore,

$$\lambda = f_d^{-1}(q) + \frac{df_d^{-1}(q)}{dq}q \tag{3.19a}$$

$$\lambda = \frac{dVC_i(q_i)}{dq_i} \quad \forall i \tag{3.19b}$$

The dual variable λ is no longer the price; instead it is the *marginal revenue* as defined in (3.19a). The marginal revenue is less than the price because $\frac{df_d^{-1}(q)}{dq} < 0$. A key condition for optimality is that marginal revenue (3.19a) equals marginal cost (3.19b).

3.4.2.4 Illustrative Example. Three Firms Merged as One Firm: Monopoly Model

To implement the step function model with the monopoly behavioral assumption, we need only change the objective function to one that maximizes profit for the single firm, with price given by the inverse demand function:

$$\text{Minimize} \quad \sum_i \sum_b c_{ib} x_{ib} - (\alpha - \beta q) q \tag{3.20a}$$

s. t.

$$(3.16b), (3.16c) \text{ and } (3.16d) \tag{3.20b}$$

The monopoly results[10] are shown in Tables 3.2 and 3.3. Comparison with the price-taker solution reveals that in the monopoly solution, demand (which is equal to total production) is smaller, price is greater, producers' surplus is greater, and both consumers' surplus and social welfare are smaller. Compared with perfect competition, the monopolist restricts production in order to drive the price up and reap a greater profit.

□

Comparing the competitive and monopoly solutions, we see that they are apparently Pareto optimal with respect to each other, as producers would be worse off if the market is competitive while consumerrs would be worse off under Monopoly. However, economists also use the criterion of Potential Pareto Optimality, also called the Kaldor-Hicks criterion. That criterion asks whether one solution could potentially make all agents better off (or at least one better off and no one worse off) if side-payments were made among the players. In this case, a potential Pareto improvement is possible by moving from the Monopoly to the competitive solution if consumers made a side payment of, say, \$12,000 to producers. Then producer surplus would be \$52910 (=\$40910+\$12000) in the competitive case, while consumer surplus would be \$34875 (= \$46875-\$12000); these surpluses are strictly greater than the respective surpluses in the Monopoly solution. Economists use the potential Pareto improvement criterion to evaluate market efficiency. If one alternative would represent a potential Pareto improvement over another, then the former is said to be more efficient. This is equivalent to saying that the first alternative has positive net benefits relative to the second, where net benefits equal the social surplus. Indeed, maximizing net benefits by this measure is equivalent to choosing a potentially Pareto optimal alternative. We use these concepts in the case studies of Chapters 11 and 12.

It may be impractical actually to collect such side payments from consumers, but there are sometimes more plausible mechanisms to achieve a Pareto improvement. For example, production by all firms could be increased

[10] As with the price-taker model (3.16), it can be shown that the solution to the monopoly model is unique.

to the levels of the price-taker solution (75,000 tons per month), with the monopoly-level of production (51,600 tons per month) sold at the monopoly price of $1.64 per ton, and the *extra* production (75,000-51,600 tons per month) sold at the price-taker equilibrium price of $1.25 per ton. Such a *two-price* arrangement (also called price discrimination) would increase consumers' surplus, and also the surplus of the monopoly (as readers can easily check). This discussion leaves out an important practical issue: can such a two-price arrangement actually be implemented? Some consumers would be tempted to buy more than they need for their own use, at the $1.25 price, and try to sell it at a profit to other consumers who are asked to pay the monopoly price of $1.64. Such *arbitrage* activity could quickly destroy the two-price system.

3.4.2.5 Nash-Cournot Model

Even if the three firms remain separate, it may be possible for each firm to use its knowledge of the inverse demand function to anticipate its own effect on the market price. However, in order to be able to predict the market price, a firm would have to make a guess about the production (equal to sales quantity) of each of the other firms, because the inverse demand function depends on *total* demand, not just on one firm's production. A model based on these assumptions – that each firm knows the inverse demand function for the entire market, and each firm assumes fixed values for the production of the other firms – was originally formulated by Cournot in 1838 [2]. Much later, the idea that each firm takes the other firms' decisions as beyond its control was investigated by Nash in 1951 [8] in a much more general setting, as a key foundation of modern game theory. Thus, the market model with these assumptions is often referred to as the *Nash-Cournot* model.

In general terms, firm i chooses its production quantity q_i in order to maximize its profit, under the assumption that other firms j will choose production quantities $\overline{q}_j$. For firm i, the quantities $\overline{q}_j$ are merely guesses, or forecasts, and are beyond the control of firm i. It is possible to create a Nash-Cournot model with total demand q as a variable, and with a constraint as in (3.10b), to ensure that total demand equals the total production over all firms, but such a *linking constraint* (a constraint having terms under the control of different agents in the game model) raises technical issues that are discussed later in Section 3.4.3. We avoid these issues here by substituting $q_i + \sum_{j \neq i} \overline{q}_j$ for firm i's estimate of total demand. Therefore, firm i's problem is to choose q_i [11]in order to

[11] In this introduction to the key ideas of the Cournot model, we assume for simplicity that we can ignore nonnegativity requirement on q_i, i.e., that the solution to the model satisfies nonnegativity without explicitly enforcing it. The simplicity comes in avoiding dual variables and complementarity conditions.

$$\text{Maximize}_{q_i} \quad f_d^{-1}(q_i + \sum\nolimits_{j \neq i} \overline{q}_j)q_i - VC_i(q_i) \tag{3.21}$$

Applying the KKT condition, for firm i, we have

$$-f_d^{-1}(q_i + \sum\nolimits_{j \neq i} \overline{q}_j) - \frac{df_d^{-1}(q_i + \sum_{j \neq i} \overline{q}_j)}{dq} q_i + \frac{dVC_i(q_i)}{dq_i} = 0$$

and therefore,

$$f_d^{-1}(q_i + \sum\nolimits_{j \neq i} \overline{q}_j) + \frac{df_d^{-1}(q_i + \sum_{j \neq i} \overline{q}_j)}{dq} q_i = \frac{dVC_i(q_i)}{dq_i}$$

This is the familiar *marginal revenue equals marginal cost* condition (3.19) of the monopoly model, with the exception that only the firm i production is multiplied by the slope of the inverse demand function.

If any firm i makes guesses $\overline{q}_j$ about other firms, but some or all of the other firms choose production values different from $\overline{q}_j$, then firm i will notice and have to revise its guesses $\overline{q}_j$, and recompute its own best q_i. If, instead, all firms make guesses about the other firms' production, and these guesses are confirmed, then there is no need for any firm to recompute its best production quantity; we are justified to say that the correct guess condition defines an equilibrium, because no firm has any incentive to change its decision. The model of such a market could be represented by the collection of all firms' KKT conditions, substituting guesses $\overline{q}_j$ by actual market outcomes, leading to a system of equations, with as many equations as there are unknowns, q_i:

$$f_d^{-1}(\sum\nolimits_j q_j) + \frac{df_d^{-1}(\sum_j q_j)}{dq} q_i = \frac{dVC_i(q_i)}{dq_i} \quad \forall i \; . \tag{3.23}$$

Equation-solving techniques could be used to solve the Nash-Cournot model of the above form, but if the models of the firms contain inequality conditions (e.g., nonnegativity and capacity constraints), then the KKT conditions must contain inequalities and complementarity conditions: the collection of all firms' KKT conditions is no longer a set of equations, and is instead a complementarity model.

3.4.2.6 Illustrative Example. Nash-Cournot Model of Three Firms: Complementarity Model

Consider the three-firm illustration, with step functions that describe their costs. For now, we keep the general form of the inverse demand function, $f_d^{-1}(q)$, i.e., it is not necessarily linear in form. Each firm i makes a guess about the decisions $\overline{x}_{jb}$ of other firms $j \neq i$, and chooses its own x_{ib} in order to maximize its profit. In the following model for firm i, total demand q, is understood by firm i to be equal to $\sum_b x_{ib} + \sum_{j \neq i} \sum_b \overline{x}_{jb}$.

$$\text{Minimize}_{x_{ib}} \quad \sum_b c_{ib} x_{ib} - f_d^{-1}(\sum_b x_{ib} + \sum_{j \neq i} \sum_b \overline{x}_{jb}) \sum_b x_{ib} \tag{3.24a}$$

s. t.

$$x_{ib} - K_{ib} \leq 0 \quad (\gamma_{ib}) \quad \forall b \tag{3.24b}$$

$$- x_{ib} \leq 0 \quad (\delta_{ib}) \quad \forall b \tag{3.24c}$$

The KKT conditions for firm i are:

$$c_{ib} - f_d^{-1}(\sum_{b'} x_{ib} + \sum_{j \neq i} \sum_{b'} \overline{x}_{jb'}) \tag{3.25a}$$

$$- \frac{df_d^{-1}(\sum_{b'} x_{ib'} + \sum_{j \neq i} \sum_{b'} \overline{x}_{jb'})}{dq} \sum_{b'} x_{ib'} + \gamma_{ib} - \delta_{ib} = 0 \;\; \forall b$$

$$0 \geq x_{ib} - K_{ib} \perp \gamma_{ib} \geq 0 \;\; \forall b \tag{3.25b}$$

$$0 \geq -x_{ib} \perp \delta_{ib} \geq 0 \;\; \forall b \tag{3.25c}$$

where the alias index b' is used to index summations in (3.25a) to avoid confusion with "$\forall b$" which indicates that this condition holds for every element of b. At equilibrium, the guesses $\overline{x}_{jb}$ are confirmed as the actual decisions made by the other players. The whole complementarity model is therefore all of the KKT conditions of all firms, with the guesses $\overline{x}_{jb}$ represented as variables of the overall model. In order to make comparisons with the perfect competition model, we re-introduce the total demand variable $q = \sum_i \sum_b x_{ib}$ into firm i's KKT conditions, resulting in the following complementarity representation of the Nash-Cournot model:

$$q - \sum_i \sum_b x_{ib} = 0 \tag{3.26a}$$

$$0 \geq x_{ib} - K_{ib} \perp \gamma_{ib} \geq 0 \;\; \forall i, b \tag{3.26b}$$

$$0 \geq -x_{ib} \perp \delta_{ib} \geq 0 \;\; \forall i, b \tag{3.26c}$$

$$c_{ib} - f_d^{-1}(q) - \frac{df_d^{-1}(q)}{dq} \sum_{b'} x_{ib'} + \gamma_{ib} - \delta_{ib} = 0 \;\; \forall i, b \tag{3.26d}$$

There is only one key difference between the Nash-Cournot model (3.26) and the perfect competition model (3.14). Comparing the last conditions listed in each model, (3.14e) and (3.26d), we can see that the terms $-f_d^{-1}(q) - \frac{df_d^{-1}(q)}{dq} \sum_b x_{ib}$ are present in the Nash-Cournot model, but only the term $-f_d^{-1}(q)$ (i.e., λ) appears in the perfect competition model. That is, price in the perfect competition model is replaced by marginal revenue in the oligopoly model.

With the linear inverse demand function (3.15), (3.26d) becomes

$$c_{ib} - (\alpha - \beta q) + \beta \sum_{b'} x_{ib'} + \gamma_{ib} - \delta_{ib} = 0 \;\; \forall i, b \tag{3.27}$$

Tables 3.2 and 3.3 present the results of the Nash-Cournot model with linear inverse demand (3.15), in the All Cournot rows. By all the measures in Table 3.3 , the Nash-Cournot equilibrium lies *between* the price-taker and monopoly solutions. This is expected, intuitively, because the price-taker model assumes no attempt by producers to exert control over the price, the monopoly model allows the most control over the price, and the Nash-Cournot model allows each firm to have some control over price while being at the mercy of other producers' decisions. We can also observe, in Table 3.2 , that each firm enjoys a larger surplus in the Nash-Cournot solution than in the price-taker solution.

Comparing the monopoly and all Cournot rows of Table 3.3, we see that, as the large monopoly is split into several competing firms, the social welfare increases. This does not always happen, for reasons not represented in this coal market illustration. Sometimes, a large firm can have lower per-unit costs than a small firm, due to "economies of scale," so splitting a large firm into smaller firms can sometimes increase costs enough to cause a decrease in social welfare.

The Nash-Cournot solution is not Pareto efficient – e.g., a two-price solution (if enforceable by producers – see the discussion above, after the illustrative example in Section 3.4.2.4) with greater production could improve some producers' surpluses and also consumers' surplus, as in the monopoly case. Also, the competitive solution represents a potential Pareto improvement over the Nash-Cournot solution, since the total surplus is greater in the former case.

□

The Nash-Cournot model represents the interaction of optimizing behaviors of several agents, each with its own separate objective function. Therefore, we would not normally expect that a model of their equilibrium behavior could be constructed as a *single* optimization model, with only one objective function. However, if the inverse demand function is linear in demand, then it is possible to represent the Nash-Cournot model as the result of a single optimization model (i.e., only one model, not one for each firm),using an idea often attributed to [6], but that also appeared earlier, e.g. in [10]. In the following model, *if it is assumed that the inverse demand function is linear* – i.e., that the second derivative $\frac{d^2 f_d^{-1}(q)}{dq^2} = 0$ – then the KKT conditions lead to the same set of equations (3.23). Continuing with the general case which leads to the Cournot model (3.23), and again assuming that we can ignore nonnegativity conditions on the variables q_i, we define the unconstrained optimization problem as

$$\text{Maximize}_{q_i} \quad \int_0^q f_d^{-1}(q')dq' + \frac{1}{2}\sum_i \frac{df_d^{-1}(q)}{dq} q_i^2 - \sum_i VC_i(q_i) \tag{3.28}$$

where the symbol q is used as shorthand for $\sum_i q_i$. Because the derivative of the inverse demand function is negative, it is *as if* the market is maximizing social welfare, but there are extra quadratic cost terms, $\frac{1}{2}\sum_i \frac{df_d^{-1}(q)}{dq} q_i^2$. The KKT conditions amount to setting the partial derivatives of the objective function equal to zero:

$$f_d^{-1}(q) + \frac{df_d^{-1}(q)}{dq} q_i + \frac{1}{2}\sum_i \frac{d^2 f_d^{-1}(q)}{dq^2} q_i^2 - \frac{dVC_i(q_i)}{dq_i} = 0 \ \ \forall i \qquad (3.29)$$

The equations (3.29) are equivalent to the Cournot model's marginal revenue equals marginal cost conditions (3.23) if the second derivative of the inverse demand function, $\frac{d^2 f_d^{-1}(q)}{dq^2}$, is zero, i.e., if the inverse demand function is linear. Therefore, the single optimization model (3.28) can represent the Cournot model when the inverse demand function is linear.[12]

3.4.2.7 Illustrative Example. Nash-Cournot Model of Three Firms: Optimization Model if Demand is Linear

For the coal market, the inverse demand function is linear, and so the Nash-Cournot model can be represented by a single optimization model, using the idea in (3.28), to insert extra cost-like terms into the objective function of the social welfare maximization model. Here, total production of firm i, which is q_i in (3.28), is represented by $\sum_b x_{ib}$, and the derivative of the inverse demand function is $\frac{df_d^{-1}(q)}{dq} = -\beta$. The Cournot model, in optimization form is:

$$\begin{aligned} \text{Minimize}_{x_{ib}} \quad & \sum_i \sum_b c_{ib} x_{ib} + \frac{1}{2}\sum_i \beta \left(\sum_b x_{ib} \right)^2 && (3.30a) \\ & - (\alpha(\sum_i \sum_b x_{ib}) - \frac{1}{2}\beta(\sum_i \sum_b x_{ib})^2) && \\ \text{s. t.} \quad & && \\ & x_{ib} - K_{ib} \le 0 \ \ \forall i, b && (3.30b) \\ & - x_{ib} \le 0 \ \ \forall i, b && (3.30c) \end{aligned}$$

The reader can easily check that the KKT conditions of (3.30) are the same as the complementarity model (3.26), when the inverse demand function

[12] Another way to explain this point relies on the discussion of symmetry and integrability in Chapter 4. The Cournot model (3.23), as derived from the several optimization models of the firms, is integrable when f_d^{-1} is linear, because linearity implies symmetry of the Jacobian matrix of the expressions on the left of (3.23).

$f_d^{-1}(q) = \alpha - \beta q$. The optimization version of the model (3.30) therefore produces the same solution as the complementarity version of the model.

□

It is conceivable that some firms may behave in the Nash-Cournot fashion, while others act as price-takers.

3.4.2.8 Illustrative Example. Mixed Behaviors: Firm 1 as Cournot, Firms 2 and 3 as Price-Takers

If firm 1 in the coal market behaves as a Nash-Cournot player, while the other two firms are price-takers, then we could modify the objective function (3.30a) to have the extra quadratic cost-like term only for firm 1:

$$\text{Minimize} \quad \sum_i \sum_b c_{ib} x_{ib} + \frac{1}{2}\beta \left(\sum_b x_{1b} \right)^2 - (\alpha q - \frac{1}{2}\beta q^2)$$

Alternatively, the complementarity version of the model could be modified to have the extra β term in (3.27) only for firm 1:

$$c_{1b} - (\alpha - \beta q) + \beta \sum_{b'} x_{1b'} + \gamma_{1b} - \delta_{1b} = 0 \;\; \forall b$$

$$c_{ib} - (\alpha - \beta q) + \gamma_{ib} - \delta_{ib} = 0 \;\; i = 2, 3, \forall b$$

The results of this model are displayed in Tables 3.2 and 3.3 , in the rows labelled One Cournot. From the consumers' point of view, the outcome is the same as in the price-taker case: q, p, consumer payments, and consumer surplus are all identical to the price-taker case. The only difference between the one-Cournot and price-taker cases is that firm 1 (which tries to influence the price) enjoys *smaller* profit than in the price-taker case, with some of firm 1's production transferred to the marginal supplier, firm 2. Firm 2's unit cost of the extra production is the same as the price, so firm 2 sees no change in profit.

Clearly, this result is not Pareto efficient: if the extra production were taken back from firm 2 and allocated once again to firm 1, then firm 1 would enjoy a larger profit, while the profits of firms 2 and 3 are unchanged, as is the consumer's surplus.

□

3.4.2.9 Illustrative Example. Mixed Behaviors: Firms 1 and 2 as Cournot, Firm 3 as Price-Taker

Another model that mixes Cournot and price-taking behavior is shown in Tables 3.2 and 3.3, in the rows labelled Two Cournot. In this model, firms 1

and 2 behave in the Cournot fashion, while firm 3 is a price-taker. There are extra quadratic cost-like terms for firms 1 and 2 in the objective function. It can be seen that firms 1 and 2 enjoy surpluses that are larger than in the One Cournot case, but smaller than the All Cournot case.
□

3.4.2.10 Bertrand Game

In a Cournot model, the *strategic variables* are the firms' production quantities. In a review and criticism of the Cournot model in the late nineteenth century, Joseph Bertrand [1] formulated a model in which firms' strategic variables are the prices that each firm charges. The Bertrand model is a type of Nash game, because each firm takes the other firms' strategic price variables as fixed and beyond its control. Bertrand pointed out that if the commodity is homogeneous – i.e., customers believe that the product of one firm is physically identical to the product of any other firm – then all customers will try to buy from the firm offering the lowest price. Therefore, each firm has an incentive to undercut other firms' prices, even if only by a small amount, as long as the price does not fall below the firm's marginal cost of production. If the price does fall below the marginal cost of production, then the firm would cut its output until its marginal cost is less than the price. The equilibrium resulting from such a process is the same as the price-taker equilibrium, even if the Bertrand oligopoly has as few as two firms.[13] There has been much discussion of the merits of the Cournot versus Bertrand models. In some circumstances, the Cournot assumption is realistic – e.g., if it is difficult or slow for firms to change their production quantity decisions – and in other circumstances, the Bertrand assumption seems more realistic.

3.4.2.11 Illustrative Example. Bertrand Model of Coal Market

We start by assuming that each firm i chooses a price p_i for coal that it sells, and it also decides how much to produce from each of its mines, x_{ib}. Consumers choose how much coal to buy from each firm, q_i, given the price p_i set by firm i, in order to maximize consumers' surplus. Since the total quantity purchased by consumers is $\sum_i q_i$, the expression for consumers' surplus becomes $\alpha(\sum_i q_i) - \frac{1}{2}\beta(\sum_i q_i)^2 - \sum_i p_i q_i$. Therefore, the model of consumer behavior is

[13] The Nash-Bertrand game does not always yield the same solution as price-taking. In other circumstances, e.g., producers located in different but connected geographic markets, the Nash-Bertrand equilibrium can be different from the price taking equilibrium. See [7] for an example.

$$\text{Minimize}_{q_i} \quad \sum_i p_i q_i - \alpha(\sum_i q_i) + \frac{1}{2}\beta(\sum_i q_i)^2 \tag{3.31a}$$

s. t.

$$q_i \geq 0 \ \ \forall i \tag{3.31b}$$

From the KKT conditions of the consumer model, we may conclude that

$$0 \leq p_i - \alpha + \beta(\sum_j q_j) \perp q_i \geq 0 \ \ \forall i$$

where the index j is an alias for i. Therefore, for any $q_i > 0, p_i = \alpha - \beta(\sum_j q_j)$, which implies that, at equilibrium, any firms that sell coal must charge the same price – the price as given by the inverse demand function.

Each firm i chooses outputs x_{ib} and the price to charge p_i in order to maximize its profit, while taking as given the decision by consumers to purchase q_i tons of coal:

$$\text{Minimize}_{x_{ib}, p_i} \quad \sum_b c_{ib} x_{ib} - p_i q_i \tag{3.32a}$$

s. t.

$$x_{ib} \leq K_{ib} \ \ \forall b \tag{3.32b}$$

$$\sum_b x_{ib} = q_i \tag{3.32c}$$

$$x_{ib} \geq 0, \forall b, \ p_i \geq 0 \tag{3.32d}$$

At equilibrium, if the parameter $q_i > 0$ in firm i's model, then the firm's price p_i must be the same as every other firm that has positive sales, as shown above – the firm's belief that it controls its selling price becomes an illusion, at equilibrium. If, on the other hand, $q_i = 0$ in firm i's model, then the firm's choice of price p_i is irrelevant because there will be no revenue, and $q_i = 0$ implies that $x_{ib} = 0, \forall b$.

The overall result of this analysis is that, in spite of the firms' initial beliefs that they control the prices that they can charge, at equilibrium, all firms are price-takers. Thus, the equilibrium is the same as the price-taker equilibrium in Tables 3.2 and 3.3.

□

3.4.2.12 Cartels

In a cartel arrangement, the individual firms are separate (unlike a monopoly) and have their separate interests. The cartel typically tries to strike an agreement among its member firms to restrict their production, in order to drive the price higher than it would otherwise be. An example is given for the coal market in Illustrative Example 3.2.3.2, without specifics about how much each individual firm restricts production to meet the overall target of 50,000 tons per month. Another example might be an agreement to restrict production to achieve the monopoly solution of Tables 3.2 and 3.3. However, firms 1 and 3 might object to this particular agreement because it increases the surplus of firm 2 much more than the other two firms, compared with the price-taker solution. One resolution of such a disagreement could be for firm 2 to transfer some of its surplus increase to firms 1 and 3 – a Nash cooperative game with exchangeable utility (the side payments). Another type of agreement could be based on the idea of a Nash cooperative game without exchangeable utility – to share the restriction of production more fairly, making such an agreement more likely to be struck than the monopoly solution of Tables 3.2 and 3.3.

Another typical problem with any cartel agreement is the incentive for individual firms to cheat. For example, if the agreement were the monopoly solution of Tables 3.2 and 3.3, then every firm would be tempted to produce some extra amounts, secretly, from any of their mines not already at capacity under the agreement, and try to sell the extra, secret production at the high price of \$1.64 per ton, which is above the unit cost of production of every mine. For this reason, cartel agreements usually include mechanisms to discourage cheating, e.g., inspections, etc.

More elaborate cartel models have been formulated that explicitly include so-called "incentive compatibility" constraints. These constrain the solutions so that subsets of agents have no incentive to split off or "cheat". The models often make sometimes complicated assumptions about how such firms would be punished by the non-cheating firms should deviation from the cartel agreement be detected (see, e.g., [5]).

Cartels are usually illegal in most countries, under laws that prohibit anti-competitive behavior. However, governments sometimes allow, or even organize, cartels for exports from their countries. A well-known example is the Organization of Petroleum Exporting Countries (OPEC). Chapter 10 also discusses the Gas Exporting Countries Forum (GECF).

3.4.3 Nash Versus Generalized Nash Equilibria

The Nash-Cournot model is an example of a more general concept, i.e., the Nash equilibrium concept of game theory, due to Nash [8]. Suppose that the different players in the game are indexed by i, and that the vector of *strategic variables* under the control of player i is denoted by x_i. There may be constraints on x_i, but to simplify the following discussion, we assume that there are no constraints. Suppose further that each player is interested in maximizing its utility U_i, and that this utility depends not only on the values of the player's own variables x_i, but also on the values of other players' variables, x_j for $j \neq i$. For example, in the Nash-Cournot model, the utility of a firm is its profit, but its profit depends not only on its own production decisions, but also on the production decisions of the other firms, because the other firms' sales influence the market price and therefore the revenue of the firm. To show the dependence of utility on all variables, we can write $U_i(x_i; x_{-i})$ where the symbol x_{-i} is shorthand for "x_j, for all $j \neq i$." The Nash equilibrium is defined as values for the x_i for all players i such that no player could improve its utility by changing the variables under its control. Another way of stating this is that at a Nash equilibrium, the equilibrium values for x_i maximize the utility of player i, given the equilibrium values of all other players, x_{-i}. Thus, the Nash equilibrium can be characterized as a solution to the following set of optimization problems, one for each player i:

$$\text{choose } x_i \text{ in order to } \max U_i(x_i; x_{-i}) \quad . \tag{3.33}$$

Under the simplifying assumption of no constraints, and also assuming that U_i is concave in x_i, the KKT conditions for all of these optimization problems take the simple form of a set of equations:

$$\nabla_i U_i(x_i; x_{-i}) = 0 \quad \forall i \tag{3.34}$$

where the subscripted gradient symbol ∇_i means partial derivatives with respect to x_i. Therefore, a Nash equilibrium can be calculated as a solution of a set of equations, with as many equations as there are unknowns x_i.

Matters become more complicated if there are constraints that involve relationships among the variables of different players, leading to a *generalized Nash model.* One example is presented in Chapter 1, Section 1.2.5, and in Chapter 5, the generalized Nash model is presented in the context of the variational inequality form, as a quasi-variational inequality. In the present chapter's example which follows, the basic Nash-Cournot model of the coal market is modified under the assumption that there is only one coal yard as a depot in the city, and that the size of the coal yard and its accessibility effectively limit the number of tons of coal per month that can be sold in total, by all firms. To keep the discussion as simple as possible, let us suppose that there is just one constraint, and that it is linear in form: $\sum_i a_i x_i \leq l$, where a_i is a row vector of the same dimension as the column vector x_i, and l is the limit that cannot be exceeded (e.g., the maximum amount of coal that can be moved through the coal yard per month). Thus, the Nash equilibrium model (3.33) or (3.34) is modified to become the generalized Nash model: for each player i,

$$\text{choose } x_i \text{ in order to } \max U_i(x_i; x_{-i})$$

$$\text{subject to } \; a_i x_i \leq l - \sum_{j \neq i} a_j x_j \quad (\phi_i)$$

which has KKT conditions (with the dual variable ϕ_i for player i's problem) that lead to a complementarity-like version of a model:

$$\nabla_i U_i(x_i; x_{-i}) - a_i^T \phi_i = 0 \ \forall i \tag{3.35a}$$

$$0 \leq l - \sum_{j \neq i} a_j x_j - a_i x_i \perp \phi_i \geq 0 \ \forall i \tag{3.35b}$$

where the superscript T indicates the transpose (the row vector becomes a column vector). The dual variable ϕ_i is player i's extra optimal utility per extra unit of l, i.e., the *marginal utility* of l. There is something strange about the complementarity-like model (3.35): the number of equations in (3.35a) equals the number of x_i variables, but are there are as many equations in (3.35b) as there are ϕ_i variables? The equations denoted by the $\perp$ symbol in (3.35b) may be written explicitly as

$$(l - \sum_j a_j x_j)\phi_i = 0 \ \forall i$$

Now consider solutions for which the linking constraint is binding, i.e., $l - \sum_j a_j x_j = 0$ – the many ϕ_i variables may take different values, but there is only one equation, which makes the system (3.35) *nonsquare*, with more variables than equations and thus many possible solutions. If the modeler were to force a special case that all ϕ_i are equal – i.e., if there were only one variable ϕ, not indexed by i – then the system (3.35) would be square, and it would be a proper complementarity model. The excessive number of ϕ_i variables in the nonsquare system (3.35) introduces extra degrees of freedom, and in general implies that there are many solutions, even if the single-ϕ square system has only one solution. See the paper by Harker [4] for further discussion of this point.

In the real world, if trading is allowed of whatever the limit l represents (e.g., if the right to use some of the capacity of the coal yard can be bought and sold among firms), then players will have an incentive to trade their rights to l if there is any difference in the players' marginal utilities ϕ_i. Therefore, if trading is allowed and no players exercise market power in such trading, then an equilibrium will have equality of all marginal utilities, and so the square, single-ϕ version of the generalized Nash model is applicable (see Illustrative Example 3.4.3.2 for further discussion of this point):

$$\nabla_i U_i(x_i; x_{-i}) - a_i^T \phi = 0 \ \forall i \tag{3.36a}$$

$$0 \leq l - \sum_{j \neq i} a_j x_j - a_i x_i \perp \phi \geq 0 \tag{3.36b}$$

Suppose that the 19th century city government wanted to control the dust and noise due to coal handling done previously at various places in the city. The government decided to require the coal firms to unload their deliveries from the mine, and load onto local delivery vehicles, all in one special field, just outside the city. The chosen spot for this coal yard had various limitations, due to the limited area of the field, and the traffic limits on the access roads. This led to an effective upper limit on the total amount of coal that could be delivered to consumers in the city of no more than $l =$ 60,000 tons per month. This capacity limit had to be shared by all three firms. Suppose that the market structure had been like the all-Cournot model whose solution is in Tables 3.2 and 3.3; then the coal yard limit of $l = 60,000$ would not allow the all Cournot solution with total production equal to 68,200 tons per month. How are the coal yard shares to be determined? This question is addressed in the next three illustrative examples shown below.

3.4.3.1 Illustrative Example. Generalized Nash Model for Coal Market: Limit on Coal Yard, with Government Allocation of Coal Yard Shares

As one solution, the city government could decide to allocate shares l_i to the three companies in proportion to their shares of total production prior to the imposition of the coal yard requirement. From Tables 3.2 and 3.3, for the all Cournot solution, the shares for the three firms would be:

$$\text{Firm 1 share } l_1 = \frac{21,000 + 12,200}{68,200} \times 60,000 \approx 29,208.2 \text{ tons per month}$$

$$\text{Firm 2 share } l_2 = \frac{17,000}{68,200} \times 60,000 \approx 14,956.0 \text{ tons per month}$$

$$\text{Firm 3 share } l_3 = \frac{18,000}{68,200} \times 60,000 \approx 15,835.8 \text{ tons per month.}$$

To represent this, the model of firm i, (3.24) can be modified to include the constraint on the coal yard share for firm i:

$$\sum_b x_{ib} \leq l_i \tag{3.37}$$

The dual variable for (3.37), ϕ_i, is in units of dollars per ton of coal. Assuming linear inverse demand, taking the KKT conditions for firm i, assembling all firms' KKT conditions, and re-introducing the variable q for total demand (as in (3.26)) yields a square complementarity model that no longer has any linking constraints:

$$q - \sum_i \sum_b x_{ib} = 0 \tag{3.38a}$$

$$0 \geq x_{ib} - K_{ib} \perp \gamma_{ib} \geq 0 \ \ \forall i, b \tag{3.38b}$$

$$0 \geq -x_{ib} \perp \delta_{ib} \geq 0 \ \ \forall i, b \tag{3.38c}$$

$$c_{ib} - (\alpha - \beta q) + \beta \sum_{b'} x_{ib'} + \gamma_{ib} + \phi_i - \delta_{ib} = 0 \ \ \forall i, b \qquad (3.38d)$$

$$0 \leq l_i - \sum_b x_{ib} \perp \phi_i \geq 0 \ \ \forall i \qquad (3.38e)$$

The solution of (3.38) is given in Table 3.4, in the rows Share Allocation. Note that the marginal utilities ϕ_i are different for different firms. For example, firm 1 places a lower marginal value on its share of the coal yard than either of the other two firms. This suggests that firm 1 might be interested in selling some of its coal yard share to the other firms, and the other two firms might be interested in buying, if the price is right.

□

3.4.3.2 Illustrative Example. Generalized Nash Model for Coal Market: Limit on Coal Yard, with Trading of Shares and Equal Marginal Utilities of Yard Shares

If the city government decides to allow the firms to trade their allocated shares with one another, then the end result of such trading would be a common marginal utility ϕ for all three firms. To see this, we introduce another variable for firm i, s_i – amount of i's yard share that is sold (if $s_i > 0$) or purchased from other firms (if $s_i < 0$). Then the constraint (3.37) must account for a reduction in space available to the firm, if $s_i > 0$, or an increase in space, if $s_i < 0$: $\sum_b x_{ib} + s_i \leq l_i$. The objective function of firm i must be modified to include a term for revenue from sales of some of its allocated yard space (if $s_i > 0$) at a price ϕ (common to all firms, and assumed to be beyond the control of any one firm), or a cost of purchase of extra yard space (if $s_i < 0$). Thus, the modified model of firm i becomes:

$$\text{Minimize}_{x_{ib}, s_i} \quad \sum_b c_{ib} x_{ib} - p \sum_b x_{ib} - \phi s_i \qquad (3.39a)$$

s. t.

$$x_{ib} \leq K_{ib} \ \ (\gamma_{ib}) \ \ \forall b \qquad (3.39b)$$

$$\sum_b x_{ib} + s_i \leq l_i \ \ (\phi_i) \qquad (3.39c)$$

$$-x_{ib} \leq 0 \ \ (\delta_{ib}) \ \ \forall b \qquad (3.39d)$$

where p is shorthand for the inverse demand function $p = \alpha - \beta \sum_j \sum_b x_{jb}$. The overall equilibrium model must have a market clearing constraint for yard space trading, $\sum_i s_i = 0$, which together with the KKT conditions of (3.39) for all firms leads to a square complementarity model:

$$q - \sum_i \sum_b x_{ib} = 0 \qquad (3.40a)$$

$$0 \geq x_{ib} - K_{ib} \perp \gamma_{ib} \geq 0 \;\; \forall i, b \tag{3.40b}$$

$$0 \geq -x_{ib} \perp \delta_{ib} \geq 0 \;\; \forall i, b \tag{3.40c}$$

$$c_{ib} - (\alpha - \beta q) + \beta \sum_{b'} x_{ib'} + \gamma_{ib} + \phi - \delta_{ib} = 0 \;\; \forall i, b \tag{3.40d}$$

$$0 \leq l_i - \sum_b x_{ib} \perp \phi_i \geq 0 \;\; \forall i \tag{3.40e}$$

$$\sum_i s_i = 0 \tag{3.40f}$$

$$-\phi + \phi_i = 0 \;\; \forall i \tag{3.40g}$$

Thus, as claimed, all firms have the same marginal utility, $\phi_i = \phi$, the equilibrium price in the market for trading yard space. Any solution to (3.40) must also satisfy a simpler complementarity model that has no s_i or ϕ_i variables. The simpler model is derived by replacing ϕ_i by ϕ in (3.40e), dropping (3.40g), and noting that, by (3.40e) and (3.40f),

$$\sum_i \sum_b x_{ib} + \sum_i s_i = \sum_i \sum_b x_{ib} \leq \sum_i l_i \text{ , i.e.,}$$

$$\sum_i \sum_b x_{ib} \leq l$$

because the government-allocated shares are assumed to sum to the total size of the yard, l.

Therefore, any solution to (3.40) also solves the following square complementarity model:

$$q - \sum_i \sum_b x_{ib} = 0 \tag{3.41a}$$

$$0 \geq x_{ib} - K_{ib} \perp \gamma_{ib} \geq 0 \;\; \forall i, b \tag{3.41b}$$

$$0 \geq -x_{ib} \perp \delta_{ib} \geq 0 \;\; \forall i, b \tag{3.41c}$$

$$c_{ib} - (\alpha - \beta q) + \beta \sum_{b'} x_{ib'} + \gamma_{ib} + \phi - \delta_{ib} = 0 \;\; \forall i, b \tag{3.41d}$$

$$0 \leq l - \sum_i \sum_b x_{ib} \perp \phi \geq 0 \tag{3.41e}$$

Note that this model makes no reference to the particular government allocation of yard shares. A solution to (3.41) is shown in Table 3.4, in the rows Share Trading. Note that the firms' surpluses have been calculated without any accounting for revenue to firm 1 or costs to the other firms, due to trading of shares; we leave it as an exercise to modify the profit calculations to account for these revenues or costs. The production of firm 1 is less than in the Share Allocation case, because it has traded away some of its share, down to 25,000 tons per month, while the production quantities of the other two

Table 3.4 Three generalized Nash solutions: Limit on coal yard

		Firm		
		$i = 1$	$i = 2$	$i = 3$
Share Allocation	Production	29208	14956	15835
	Surplus	25614	13161	11402
	ϕ_i	0.203	0.631	0.456
Share Trading	Production	25000	17000	18000
	Surplus	22710	14960	12960
	ϕ	0.273		
Weighted Price Auction	Production	27533	17000	15467
	Surplus	24458	14960	11136
	ϕ	0.231		

firms are greater than in the Share Allocation case. The marginal utility is common to all three firms, at a value greater than ϕ_1 and less than both ϕ_2 and ϕ_3 of the Share Allocation case.

There is another interpretation of the model (3.41): it could also represent the end result of the city government *auctioning* the coal yard space l in such a way that all firms pay the same price ϕ which is determined by the competitive auction. In this interpretation, the government takes all the revenue from sales of yard space to firms – there is no trading among firms – but the resulting values of x_{ib}, q and ϕ are the same as in the share-trading interpretation of (3.41). We leave it as an exercise to compare the firms' profits (including accounting for revenues and costs of yard space) in the two interpretations.

□

Many other solutions can be found for the coal yard sharing situation. For example, the city government could allocate different tradeable shares l_i than in the above example, leading to different profits for the firms, but this would not affect the equilibrium values of x_{ib}. As another example, the government could auction the coal yard shares, but give a preference to some firms over others. Let us suppose that firm 1 has its head office in the city, but firms 2 and 3 have their head offices in another city, and that the government has decided that the price to be charged for coal yard shares is twice as much for firms 2 and 3 as for firm 1. This leads to our final example.

3.4.3.3 Illustrative Example. Generalized Nash Model for Coal Market: Limit on Coal Yard, with Auctioning of Shares and Unequal Marginal Utilities

We introduce new parameters, w_i which are the weights to multiply the dual variable ϕ. To implement the aim of twice the price for firms 2 and 3 as for firm 1, we set $w_1 = 1$, $w_2 = 2$, and $w_3 = 2$. The model is similar to (3.41), except that (5.35d) is modified with the weights:

$$(5.35a), (5.35b), (5.35c), (5.35e)$$

$$c_{ib} - (\alpha - \beta q) + \beta \sum_{b'} x_{ib'} + \gamma_{ib} + w_i \phi - \delta_{ib} = 0 \ \ \forall i, b \qquad (3.42)$$

A solution is shown in Table 3.4, in the rows Weighted Price Auction.
□

We return to a discussion of generalized Nash equilibria in Chapter 5, in the context of quasi-variational inequality (QVI) models.

3.5 Multi-Level Games

3.5.1 Stackelberg Leader-Follower Games (MPECs)

Cournot behavior has a firm using its knowledge of the inverse demand curve, but assuming that it has no influence over other firms' output decisions. In contrast, Stackelberg [12] supposed that a *leader* firm could use its knowledge of the inverse demand curve *and* it could also anticipate the reactions of other *follower* firms to the leader's decisions. That is, in a Stackelberg model, the leader firm knows both the inverse demand curve and the followers' behavior. In a mathematical programming formulation of the Stackelberg idea, we specify the objective function as the leader's profit function, with the revenue term as the product of the inverse demand function and the leader's quantity produced. The constraints include the usual constraints on the leader's ability to produce its own output, and in addition, the KKT conditions of the follower firms. The mathematical form of such a model is usually called a Mathematical Program with Equilibrium Constraints (MPEC), but an alternative name is Mathematical Program with Complementarity Constraints (MPCC); see Chapter 6 for a more thorough discussion of various optimization problems that are constrained by other optimization problems, including MPCCs. The follower firms could behave in the Cournot fashion (using knowledge of the inverse demand curve, and assuming that the other players' production quantity decisions are given, i.e., beyond the firm's control), or as price-takers; different assumptions lead to different models. For illustrations here, we continue the example of three firms, with step functions that describe their costs, and we suppose that the follower (or followers) act as price-takers.

3.5.1.1 Illustrative Example. Stackelberg MPEC with Firm 2 as Leader

In this Stackelberg model, we suppose that firm 2 is the leader, and that firms 1 and 3 are price-taking followers. Here is the model.

$$\text{Minimize} \quad \sum_b c_{2b}x_{2b} - f_d^{-1}(q)\sum_b x_{2b} \tag{3.43a}$$

s. t.

$$q - \sum_i \sum_b x_{ib} = 0 \tag{3.43b}$$

$$x_{2b} - K_{2b} \leq 0 \ \ \forall b \tag{3.43c}$$

$$-x_{2b} \leq 0 \ \ \forall b \tag{3.43d}$$

$$0 \geq x_{ib} - K_{ib} \perp \gamma_{ib} \geq 0 \ \ i = 1, 3, \forall b \tag{3.43e}$$

$$0 \geq -x_{ib} \perp \delta_{ib} \geq 0 \ \ i = 1, 3, \forall b \tag{3.43f}$$

$$c_{ib} - f_d^{-1}(q) + \gamma_{ib} - \delta_{ib} = 0 \ \ i = 1, 3, \forall b \tag{3.43g}$$

Note that the constraints include the complementarity conditions that represent the optimal behavior of firms 1 and 3 in the face of a price $f_d^{-1}(q)$. □

3.5.1.2 Illustrative Example. Stackelberg MPEC with Firms 1 and 2 Merged as One Leader

In another Stackelberg model, firms 1 and 2 have merged into one large firm which acts as the leader, and firm 3 acts as a price-taking follower. The formulation is as follows.

$$\text{Minimize} \quad \sum_{i=1,2} \sum_b c_{ib}x_{ib} - f_d^{-1}(q) \sum_{i=1,2} \sum_b x_{ib} \tag{3.44a}$$

s. t.

$$q - \sum_i \sum_b x_{ib} = 0 \tag{3.44b}$$

$$x_{ib} - K_{ib} \leq 0 \ \ i = 1, 2, \forall b \tag{3.44c}$$

$$-x_{ib} \leq 0 \ \ i = 1, 2, \forall b \tag{3.44d}$$

$$0 \geq x_{3b} - K_{3b} \perp \gamma_{3b} \geq 0 \ \ \forall b \tag{3.44e}$$

$$0 \geq -x_{3b} \perp \delta_{3b} \geq 0 \ \ \forall b \tag{3.44f}$$

$$c_{3b} - f_d^{-1}(q) + \gamma_{3b} - \delta_{3b} = 0 \ \ \forall b \tag{3.44g}$$

□

Table 3.5 Firm-related solution values for Stackelberg models

Firm Related Values		Firm i and Mine b					
		$i = 1$		$i = 2$		$i = 3$	
		$b = 1$	$b = 2$	$b = 1$	$b = 2$	$b = 1$	$b = 2$
Stackel-berg Leader $i = 2$	x_{ib}	21000	16000	17000	0	18000	0
	Revenue	48100		22100		23400	
	Variable Cost	24510		10540		14040	
	Surplus	23590		11560		9360	
Stackel-berg Leader $i = 1\&2$	x_{ib}	21000	13000	17000	0	18000	0
	Revenue	45900		22950		24300	
	Variable Cost	22080		10540		14040	
	Surplus	23820		12410		10260	

Table 3.6 Other solution values for Stackelberg models

Other Values	q	p	Cons. Payment	Cons. Surplus	Prod. Surplus	Social Welfare
Leader $i = 2$	72000	1.30	93600	43200	44510	87710
Leader i=1&2	69000	1.35	93150	39675	46490	86165

The results of the two Stackelberg models are presented in Tables 3.5 and 3.6. Comparing with the price-taker solution in Tables 3.2 and 3.3, we can see that when firm 2 is the leader, greater profits result for all three firms than in the price-taker case, but consumers' surplus and social welfare are reduced by the market manipulation strategy. When firms 1 and 2 are merged and act as a single leader, these tendencies are even greater, but the solution turns out to be the same as the solution to the model of Section 3.4.2.9 with firms 1 and 2 acting independently in the Cournot fashion, with firm 3 as a price-taker.

3.5.2 Multi-Leader Games (EPECs)

It is possible that more than one firm may attempt to act as a leader. In a duopoly of only two firms, however, it is not possible to determine a consistent solution: if each firm assumes that it is the leader, and that it can manipulate the other, then both firms' expectations will not be confirmed when they see how the other firm actually behaves. It is possible to have a more limited kind of multiple leadership, though.

For example, in our three-firm coal market example, suppose that firms 1 and 2 both anticipate the reaction of consumers to different market prices, and both of them also anticipate the reaction of firm 3 to different market prices. Suppose further that firm 1 does not presume to be able to anticipate the reaction of firm 2 to firm 1's production decisions, nor does firm 2 anticipate firm 1's reaction to firm 2's production decisions. Instead, we suppose that firms 1 and 2 behave in a Nash fashion with respect to each other: they just take the other player's production decisions as given, i.e., they make a guess of the other firm's production, and equilibrium is defined to be when the guesses are confirmed. Thus, the whole model is composed of two MPECs: the MPEC for firm 1 contains the guesses of firm 2's production as parameters, and similarly, the MPEC for firm 2 contains the guesses of firm 1's production, with firm 3's behavior modeled in the lower levels of both MPECs.

Such a model is called an Equilibrium Program with Equilibrium Constraints (EPEC), which is discussed in much more detail in Chapter 7.

3.6 Summary

This chapter presents some fundamental economic principles and their application in several models of economic behavior. The main points are summarized below.

When all agents (consumers and producing firms) are price-takers (i.e., they assume that their decisions do not affect the equilibrium price in the market), equilibrium in a single commodity market can be modeled as an optimal solution of a mathematical program that maximizes a social welfare function. Social welfare is measured as consumers' surplus (area under the inverse demand curve, minus money paid by consumers) plus producers' surplus (revenue paid to producers, minus producers' variable costs). In a multicommodity model of equilibrium, it may not be possible to define a social welfare function, because there may not be a well defined integral of the several inverse demand functions. Without a social welfare function, it is not possible to represent equilibrium as an optimal solution of a mathematical program like that for the single commodity case. But in that case, formulation of the equilibrium problem as a complementarity problem can come to the rescue.

However, the price-taker assumption is not always valid. Some agents, e.g., large firms, may recognize that their decisions do influence the market price and the decisions of other agents.

For example, in a Nash-Cournot model, it is assumed that firms choose their production quantities to maximize profit, and that they have knowledge of the inverse demand curve – they recognize the effect that an increase in their own sales quantity has on depressing the market price, due to the reactions of consumers. However, in the Nash-Cournot model, a firm does not assume that its decisions affect the decisions made by other firms. A Nash-Cournot model can be represented as a complementarity model, essentially by writing out all firms' KKT conditions for optimality.

If there are constraints that link the decisions of different firms in a Nash-Cournot model, then the model is an example of a Generalized Nash Equilibrium (GNE) model. A GNE can have many different solutions, depending on how the linking constraints are handled. Different variations of a GNE can correspond to different ways in which an agent in the real world actually enforces the linking constraints.

The Stackelberg model goes beyond the Nash-Cournot model by representing a greater degree of influence by a firm over the actions of other agents in the market. In a Stackelberg model, there is a single leader firm which not only uses its knowledge of the inverse demand function, but also anticipates the reactions of the other follower firms to its own decisions about the quantity to produce. A Stackelberg model is an optimization model with the objective to maximize the profit of the leader, but the constraints include the reactions of the followers to the leader's decisions by representing the KKT conditions of the followers' profit maximization problems. The KKT conditions of the followers include the leader's decisions as parameters, from the point of view of the followers, but the leader's decisions are variables in the overall leader's problem. The mathematical form of the Stackelberg model is known as a Mathematical Program with Equilibrium Constraints (MPEC), or a Mathematical Program with Complementarity Constraints (MPCC).

3.7 Exercises

1. In the price-taker model of the three firms, change one parameter of the inverse demand function to represent a 20% increase in the quantity demanded at any given price. [Hint: The demand curve should rotate counterclockwise, with the price intercept not moving.] Is the new equilibrium price larger or smaller? Explain.
2. The demands for gas and electricity, G and E respectively, are related to their prices, p_G and p_E respectively, according to the following demand functions:
   $G = 15 - 3p_G + 1p_E$
   $E = 20 + 0.5p_G - 2p_E$
   The government imposes taxes on gas and electricity which changes the prices from $p_G = 4$ and $p_E = 6$ to $p_G = 5$ and $p_E = 7$.
   (a) Calculate the quantities of G and E demanded, both before and after the taxes are imposed.
   (b) On a diagram with p_G on the vertical axis and G on the horizontal axis, graph:
   (i) the demand curve for G, before the taxes are imposed (holding p_E constant at its before-tax value);
   (ii) the demand curve for G, after the taxes are imposed; and
   (iii) the straight line through the before and after points on the graph.

(c) Using Harberger's approximation (as in Section 3.3.4.4), estimate the change in the total use value to consumers of their purchases of gas and electricity, i.e., the difference between the after and before situations, based on treating the straight line through the before and after points as if it is a single commodity demand curve. State whether this is a loss or a gain in total use value.
(d) Estimate the change in consumers' surplus by deducting the change in exchange value (amount of money paid by consumers) from your answer to part (c). State whether this is a loss or a gain in consumers' surplus.

3. Change the GAMS social welfare maximization model of the three firms (see Appendix B) to represent the following change in technology at firm $i = 3$. Firm 3 has just installed improved water pumps at its mine $b = 2$, resulting in a lower variable cost of production, and a larger capacity, namely \$1.12 per ton, and 20,000 tons per month, respectively.
(a) Does the equilibrium price increase or decrease? Explain.
(b) Calculate the change in consumers' surplus, and in producers' surplus, as a result of this technological improvement.
4. Modify the GAMS social welfare maximization model (see Appendix B) to create GAMS code for the monopoly model of Illustrative Example 3.4.2.4. Check that the solution is the same as given in the text.
5. Modify the GAMS social welfare maximization model (see Appendix B) to create GAMS code for the Nash-Cournot model in optimization form, of Illustrative Example 3.4.2.7. Check that the solution is the same as the solution of the complementarity form of the model.
6. Modify your GAMS code from the previous question to represent the model of Illustrative Example 3.4.2.8, i.e., with firm 1 behaving in the Cournot fashion, and firms 2 and 3 as price-takers. Check that the solution is the same as in the text.
7. Modify the GAMS complementarity model (with all three firms behaving in the Nash-Cournot fashion – see Appendix B for this GAMS model) to represent the model of Illustrative Example 3.4.2.8, i.e., with firm 1 behaving in the Cournot fashion, and firms 2 and 3 as price takers. Check that the solution is the same as that of the optimization model of the previous question.
8. Modify the GAMS complementarity model (with all three firms behaving in the Nash-Cournot fashion – see Appendix B for this GAMS model) to represent the model of Illustrative Example 3.4.3.1, i.e., with government allocation of coal yard shares. Check that your solution is the same as in the text.
9. Use the yard shares l_i given in Illustrative Example 3.4.3.1, and the results in Table 3.4 to determine the values of the share trade variables s_i in Illustrative Example 3.4.3.2. Also calculate the profits of each firm, accounting for the revenue or cost of trading yard shares (the surpluses in Table 3.4 do not account for the yard share trading revenues or costs). Also, use your GAMS model of the previous question to redo the calcula-

tions of s_i and profits after changing the l_i shares by reducing l_1 by 2000 tons per month, and increasing each of l_2 and l_3 by 1000 tons per month, and compare with the results of the original shares l_i.

10. For the auction interpretation of the solution to Illustrative Example 3.4.3.2 (i.e., the government sells the yard space), calculate the profits of the firms, including the value of yard space as a cost. Compare with the profits under the share trading interpretation of the previous question.
11. Compute the profits of the three firms for the weighted price auction of Illustrative Example 3.4.3.3, and compare with the profits under the equal price auction of the previous question.
12. Modify the GAMS model that has firm 2 as the Stackelberg leader (see Appendix B for this GAMS model) to represent firms 1 and 2 merged as one Stackelberg leader, as in Illustrative Example 3.5.1.2. Check that your solution is the same as in the text.

References

1. J. Bertrand. Book review of Walras's "Théorie mathématique de la richesse sociale" and of Cournot's "Recherches sur les principles mathématiques de la théorie des richesses." *Journal de Savants*, 67:499-508, 1883.
2. A.A. Cournot. *Researches into the Mathematical Principles of the Theory of Wealth.* (translated by Nathaniel T. Bacon, 1927) Macmillan, New York, 1838.
3. A. Harberger. Three basic postulates for applied welfare economics. *Journal of Economic Literature*, 9(3):785-797, 1971.
4. P.T. Harker. Generalized Nash games and quasi-variational inequalities. *European Journal of Operational Research*, 54(1):81-94, 1991.
5. J.E. Harrington, B.F. Hobbs, J.-S. Pang, A. Liu and G. Roch. Collusive game solutions via optimization. *Mathematical Programming, Series B*, 104(2-3):407-435, 2005.
6. H. Hashimoto. A spatial Nash equilibrium model. *In Spatial Price Equilibria: Advances in Theory, Computation, and Application*, P. T. Harker, Springer, 1985.
7. B.F. Hobbs. Network models of spatial oligopoly with an application to deregulation of electricity generation. *Operations Research*, 34(3):395-409, 1986.
8. J.F. Nash. Non-cooperative games. *The Annals of Mathematics*, 54(2):286-295, 1951.
9. R.S. Pindyck and D.L. Rubinfeld. *Microeconomics.* Prentice-Hall, Upper Saddle River, New Jersey, 2008.
10. S.W. Salant. Imperfect competition in the international energy market: a computerized Nash-Cournot model. *Operations Research*, 30(2):252-280, 1982.
11. P.A. Samuelson. Spatial price equilibrium and linear programming. *The American Economic Review*, 42:283-303, 1952.
12. H. von Stackelberg. *Marktform und Gleichgewicht.* (*Market Structure and Equilibrium*). J. Springer, Vienna, 1934.
13. H.R. Varian. *Intermediate Microeconomics: A Modern Approach.* Norton, New York, 2006.
14. J.M. Wooldridge. *Econometrics.* South Western Cengage Learning, Mason, Ohio, 2009.

Chapter 4
Equilibria and Complementarity Problems

4.1 Introduction

In this chapter, we explore the notions of equilibria and optimization and show how in some cases they are related. The notion of an equilibrium is a fundamental concept that has been used in a variety of disciplines such as economics, engineering, and science to name just a few. At its core, an equilibrium is a state of the system being modeled for which the system has no "incentive" to change. These incentives can be monetary in the case of economics or based on natural forces and scientific laws such as total input equals total output. Some well-known engineering examples include: conservation of energy, conservation of mass, conservation of momentum [8], steady-state probabilities in Markov chains such as birth-and-death processes [53] to name a few. These and other engineering examples are typified by a balancing of forces or conditions so that the state once reached will not easily (if at all) be left.

This is not to say that all systems under study will have an equilibrium or steady-state. Take for example the predator-prey models for population growth and decay also known as Lotka-Volterra equations [8] in which the number of predator and prey vary according to two linear ordinary differential equations. The numbers of each of these two populations can cycle through several values based on dynamics of the populations being studied with no equilibrium state in some cases.

From a modeling perspective, identifying the equilibrium state(s) allows one to predict where the system might be at a future time or under reasonable conditions. This is not to say that the system will definitely be in such a state in the future, in part due to the potential multiple equilibria that might exist, but also because it may take a certain amount of time to transition to one of these equilibria. Nevertheless, knowledge of such equilibria can be instrumental in determining various characteristics of the system under consideration and serve as benchmarks.

Equilibria can be viewed in many different ways and we concentrate on those that have direct relevance to energy markets and complementarity or other related formulations. This direction necessarily leads to the notions of equilibria in dominant actions (EDA), Nash equilibria (NE), pure and mixed forms, equivalence between bimatrix games and linear complementarity problems (LCPs), the Principle of Symmetry and other equilibrium concepts (see also Chapter 3 for a related discussion). EDA refers to actions that each "player" can take that are best regardless of what the other players choose. It is not always the case that such equilibria can be found and a relaxation of this concept is the NE. Here each player chooses an action that has the highest payoff for themselves assuming that the other players are choosing actions that are best for themselves at the same time. A further distinction in equilibrium modeling is made as to the number of players and the number of actions that each player can exert. When just two players are present, equilibria can be zero-sum (what one player wins is what the other loses), variants of this in which the sum of the players' payoffs are a constant sum, as well as the so-called "bimatrix games". This last concept involves a finite set of actions for each player and two separate payoff matrices related to the actions chosen by each player. While zero-sum games are special cases of primal and dual linear programs, bimatrix games are associated with a certain linear complementarity problem. We explore these different concepts of equilibria and how they relate to optimization problems (this is where the Principle of Symmetry fits in).

In the rest of this chapter, we concentrate on equilibrium concepts for economics or engineering that are relevant for the current energy markets context. The concept of an equilibrium is very much related to that of game theory and in what follows, notions from the latter discipline will be used. In Section 4.2 we focus on economics and engineering equilibria presenting various notions of equilibria (e.g., EDA, NE). In Section 4.3 we explore the connection between optimization problems and equilibria showing that duality theory can be viewed as an equilibria between two objectives. This connection between optimization and equilibrium is further explored in Section 4.4 via generalizations of the optimality conditions to linear and nonlinear programs (e.g., Spatial Price Equilibrium, complementarity problems). In this section, we also present the Principle of Symmetry, a well-known concept that tells exactly when a certain equilibrium problem corresponds to the KKT conditions of an optimization problem. Section 4.5 gives basic existence and uniqueness results for equilibrium problems expressed as complementarity or the related variational inequality problem. Section 4.6 presents equilibria with certain key non-convexities as well as combining stochasticity and equilibria for energy markets. And lastly, Section 4.7 summarizes the chapter and Section 4.8 provides a Computational Appendix.

4.2 Economics and Engineering Equilibria

4.2.1 Equilibria in Dominant Actions

The first concept considered is that of an equilibrium in dominant actions (EDA) which is a decision by each player which is best independent of what the other player has chosen. Clearly, this is a reasonable concept when trying to predict how each player will act. Consider the following illustrative example.

4.2.1.1 Illustrative Example. Energy Production Duopoly

Suppose that as discussed in Chapter 1, there are two energy producers, $i = 1, 2$ each of whom is trying to maximize their own profit given as the objective function in the optimization problem for player i below.[1]

$$\text{Maximize}_{q_i} \quad p \cdot q_i - c_i(q_i) \tag{4.1a}$$

s. t.

$$q_i \geq 0 \tag{4.1b}$$

One can imagine that the context is production of electric power, natural gas, oil, or other energy product. Producer i is trying to select the level of production q_i that will give it the highest possible profit derived from the revenue (price p times quantity) less costs ($c_i(q_i)$). Here, the profit function is $f_i(q_i, q_{-i}) = p \cdot q_i - c_i(q_i)$ where $-i$ denotes the other player (producer). So $f_1(q_1, q_2) = -\beta q_1^2 + (\alpha - \gamma_1 - \beta q_2)q_1$ and $f_2(q_1, q_2) = -\beta q_2^2 + (\alpha - \gamma_2 - \beta q_1)q_2$ assuming that as presented in Chapter 1, $p = p(q_1 + q_2) = \alpha - \beta(q_1 + q_2)$ is the inverse demand function depending on what both players produce and $c_i(q_i) = \gamma_i q_i$ the linear cost functions where $\alpha, \beta, \gamma_i, i = 1, 2$ are positive parameters.

Let $\alpha = 100, \beta = 1.5, \gamma_1 = 10, \gamma_2 = 5$. Also, for simplicity, suppose that just three levels of production are available to each producer as given in Table 4.1 with the appropriate profit levels indicated as (profit for producer 1, profit for producer 2).

Notice that for producer 1, a level of $q_1 = 20$ always results in higher profits than the other choices regardless of what producer 2 does (e.g., \$1200 vs. \$750, \$0 if player 2 produces nothing). Similarly, $q_2 = 20$ also dominates other choices for the second player regardless of what the first player chooses. Thus, the optimal energy production level of one player is independent of whatever the other producer did and represents a *dominant action* for that

[1] While only two players are listed here, the concepts extend easily to $N \geq 2$ players.

Table 4.1 Example of an equilibrium in dominant actions, $\alpha = 100, \beta = 1.5, \gamma_1 = 10, \gamma_2 = 5$

$q_1 \backslash q_2$	0	10	20
0	(\$0, \$0)	(\$0, \$800)	(\$0, \$1300)
10	(\$750, \$0)	(\$600, \$650)	(\$450, \$1000)
20	(\$1200, \$0)	(\$900, \$500)	(\$600, \$700)

Table 4.2 Example showing no equilibrium in dominant actions, $\alpha = 100, \beta = 2, \gamma_1 = 10, \gamma_2 = 5$

$q_1 \backslash q_2$	0	10	20
0	(\$0, \$0)	(\$0, \$750)	(\$0, \$1100)
10	(\$700, \$0)	(\$500, \$550)	(\$300, \$700)
20	(\$1000, \$0)	(\$600, \$350)	(\$200, \$300)

player. In terms of an equilibrium for both players, the production level pair of $(q_1^{EDA}, q_2^{EDA}) = (20, 20)$ is called an *equilibrium in dominant actions* [50]. More generally, such an equilibrium satisfies:

$$f_i(q_i^{EDA}, q_{-i}) \geq f_i(q_i, q_{-i}), \forall q_i \in F_i, q_{-i} \in F_{-i}, i = 1, 2 \tag{4.2}$$

where F_i is the feasible region for player i (nonnegative production levels in the example above) and $f_i(q_i, q_{-i})$ is the objective to be maximized $(p \cdot q_i - c_i(q_i)$ above) noting that the other player's production level q_{-i} is a constant.

4.2.1.2 Illustrative Example. Energy Production Duopoly, β Changed from 1.5 to 2)

Now if β changes from 1.5 to 2 related to a steeper slope for the demand curve, the payoffs in Table 4.2 result. Here neither player has a dominant action. To see this note that $q_1 = 20$ produces the higher player 1 profits when $q_2 \in \{0, 10\}$ but not so for a production level of 20 for player 2. Likewise, a similar pattern holds for player 2 indicating no dominant action. The end result is that there is no equilibrium in dominant actions.[2] What to do in this case then to predict the actions of the two producers? The next concept, that of a Nash equilibrium is a relaxation of EDA to handle this sort of situation.

[2] To not have an equilibrium in dominant actions, it would suffice that one player didn't have a dominant action at the production pair in question.

4.2.2 Nash Equilibria

The definition of an EDA (in the current context) specifies that the best production levels (q_1^{EDA}, q_2^{EDA}) should satisfy (4.2). A relaxation of this, called a Nash equilibrium (NE) is to only assert the same sort of inequality but for equilibrium values of the other player [50]. That is, production levels (q_1^{NE}, q_2^{NE}) such that:

$$f_i(q_i^{NE}; q_{-i}^{NE}) \geq f_i(q_i; q_{-i}^{NE}), \forall q_i \in F_i, q_{-i} \in F_{-i}, i = 1, 2 \tag{4.3}$$

This definition implies that there there should be no unilateral incentive to deviate from such an equilibrium (q_1^{NE}, q_2^{NE}).

4.2.2.1 Illustrative Example. Energy Production Duopoly, Nash Equilibrium

Using the payoff matrices from Table 4.2 shows that there are exactly two Nash equilibria for this energy production problem: $(q_1^{NE}, q_2^{NE}) \in \{(10, 20), (20, 10)\}$. Moreover, assuming $\beta > 0$, as described in the Computational Appendix to this chapter, it can be shown for $\alpha = 100, \gamma_1 = 10, \gamma_2 = 5$ that:[3]

$$\begin{cases} (0,0) \text{ is a NE if } & 9.5 < \beta \\ (0,10) \text{ is a NE if } & 4.5 < \beta < 9.5 \\ (0,20) \text{ is a NE if } & 3 < \beta < 3.1\overline{6} \\ (10,0) \text{ is a NE if } & 4.75 < \beta < 9 \\ (10,10) \text{ is a NE if } & 2.375 < \beta < 4.5 \\ (10,20) \text{ is a NE if } & 1.8 < \beta < 2.375 \\ (20,0) \text{ is a NE } & \text{not possible} \\ (20,10) \text{ is a NE if } & 1.9 < \beta < 2.25 \\ (20,20) \text{ is a NE if } & \beta < 1.8 \end{cases} \tag{4.4}$$

so for this problem the NE depends on the value of β. The equilibrium $(0, 0)$ when $\beta > 9.5$ indicates that no production should occur by either energy producer consistent with a very steep demand curve.

From the definitions above, clearly a Nash equilibrium is the less restrictive concept and in fact every EDA is a NE but the converse is not necessarily true (e.g., Proposition 2.1, [50]). However, a given problem may not even have any Nash equilibria. Depending on the data, if a NE exists, it may be unique. For example, for $\alpha = 100, \gamma_1 = 10, \gamma_2 = 5, \beta = 1$, the above analysis shows that only (20, 20) is a NE with payoffs of (1000, 1100), respectively, for producers 1 and 2.

[3] The related analysis assumes that payoffs for Nash equilibria need to be strictly greater than alternative choices. To include a tie, all $<$ signs can be replaced by $\leq$.

Table 4.3 Example of an equilibrium in dominant actions, $\alpha = 100, \beta = 1, \gamma_1 = 10, \gamma_2 = 5$, additional costs

$q_1 \backslash q_2$	0	10	20
0	(\$0, \$0)	(\$0, \$850)	(\$0, \$1500)
10	(\$800, \$0)	(\$700, \$750)	(\$600, \$1300 − \$800 = \$500)
20	(\$1400, \$0)	(\$1200, \$650)	(\$1000 − \$500 = \$500, \$1100 − \$100 = \$1000)

4.2.2.2 Illustrative Example. Energy Production Duopoly, $\beta = 1$, Additional Costs

Suppose that only for certain high levels of production, there are additional costs. As a practical example, one might think of a rig availability for oil and gas drilling when one or both players are demanding as much drilling equipment as possible to handle high production. The extra cost might be the expense of importing additional rigs from neighboring regions, building new ones from scratch, overtime for workers, etc. Moreover, due to geographical and workforce differences, this extra cost might not be the same for each player nor at each level of individual production. Assume that when both players produced 20, player 1 faced additional costs of \$500 but player 2 only \$100. When the production pair was (10, 20), only producer 2 faced this extra charge and this time it was \$800 since they were "going it alone" in terms of maximal production and having to bear all the costs of new machines from the manufacturer (or the like). Thus, the new payoff matrix is as follows:

Thus, (20, 20) is no longer a NE since player 1 can do better by selecting $q_1 = 10$ when $q_2 = 20$. In fact for this modified problem there is no Nash equilibrium.

4.2.3 Types of Game Theory Problems Considered

In the current energy production context, the game being played is not zero-sum since what one player wins is not what the other loses [51]. Nor is it a constant-sum game [51] which can be seen by examining the payoff matrices and checking that the sum of both players' payoffs is not the same for each possible production pair. Moreover, in energy markets with competing players, it is less usual to have zero-sum games (with just two players) since the payoff is not from the other player but from the market-based profit. Also, the feasible region for each player is often much more complicated than just a small number of possible production levels. There may be engineering constraints (e.g., Kirchoff's Laws) that need to be considered, economic restrictions in terms of available budgets, or a host of other operational or investment-related aspects. Moreover, energy markets usually have many more than just two players. In summary, the example above is only illustrative.

Those qualifications having been said, while two-person, zero-sum games (and more generally matrix games) are an important part of game theory and discussion about equilibria in general, we will skip it here to concentrate more on games that relate directly to complementarity problems and energy markets. The interested reader can find excellent descriptions of two-person, zero-sum games in a number of textbooks such as [51] and [53].

Also, the games considered above and what we will consider below are static, that is they are one-shot calculations made by the players to get the highest payoffs possible and are also called games in *normal form* [51]. In reality, energy producers would make a series of production decisions sequentially learning from previous time periods and what the other player(s) and market did. A tree of possible actions over the time horizion would then constitute a repeated game also called a game in *extensive form* [51]. Since these latter games can be reduced to ones in normal form by spelling out the information available and the set of instructions to follow [51], we concentrate just on the normal form games in the rest of this chapter and relate them to optimization and complementarity problems keeping in mind the application to energy markets.

4.2.4 Mixed Versus Pure Equilibria

To expand the notion of an equilibrium for those situations when none can be computed like the Nash equilibrium example above, it is instructive to widen the definition of what an equilibrium means. Instead of the two energy producers choosing production quantities (or in more general terms "actions"), one can suppose that with probability p_i producer 1 selects a level i and similarly for producer 2 and probability q_j. Such problem are called *mixed games.* If the players were competing in a two-person, zero-sum game in which one player's payoff was the other player's payout, then by the famous *Minimax Theorem* by John von Neumann from 1928, a solution would always exist and furthermore a linear programming problem could be used to obtain it for both players [51], [53].

In the more general two-person games such as the energy production example above, there is not an equivalent linear program to be solved but rather a corresponding linear complementarity problem [41], [42]. Consider the energy production example above but in the context of a more general bimatrix game [12]. Energy producer 1 has m production levels with the vector $x = [x_{1,\ldots,}x_m]^T$ representing the probability of choosing these levels, i.e., $x_i = P\left(\text{Producer 1 selects level } i\right).$ For the second energy producer, suppose that there are n production levels with $y = [y_{1,\ldots,}y_n]^T$ the vector of associated probabilities and $y_j = P\left(\text{Producer 2 selects level } j\right).$ As shown above, in the

Nash perspective, each player is trying to maximize their own profit taking the other player's optimal decisions as given. In the mixed setting where the players are selecting not the production levels but just the probabilities of choosing those levels, we see that the respective optimization problems are as follows. Producer 1 is trying to select the probability values x to maximize his expected profit given that producer 2 has already optimally selected y, i.e.,[4]

$$\text{Maximize}_x \quad \sum_{i=1}^{m}\sum_{j=1}^{n} x_i y_j m_{ij} = x^T M y$$

s. t.

$$1 - e_m^T x = 0 \quad (\lambda) \tag{4.5a}$$

$$x_i \geq 0, \quad i = 1, \ldots, m \tag{4.5b}$$

where the constraints stipulate that x must be a valid probability with e_m a vector of m ones and $M = [m_{ij}]$ is the profit to producer 1. Analogously, having an optimal value for x, producer 2 is trying to select the values for y that optimize the following program:

$$\text{Maximize}_y \quad \sum_{i=1}^{m}\sum_{j=1}^{n} x_i y_j n_{ij} = x^T N y$$

s. t.

$$1 - e_n^T y = 0 \quad (\gamma) \tag{4.6a}$$

$$y_j \geq 0, \quad j = 1, \ldots, n \tag{4.6b}$$

where $N = [n_{ij}]$ is producer 2's profit and with e_n a vector of n ones. We note that each of these problems are linear programs whose KKT conditions are both necessary and sufficient [2] and are the following: find primal variable vectors x, y and dual scalar variables λ, γ so that

$$0 \leq -My - e_m\lambda \perp x \geq 0 \tag{4.7a}$$

$$0 = 1 - e_m^T x, \lambda \text{ free} \tag{4.7b}$$

$$0 \leq -N^T x - e_n\gamma \perp y \geq 0 \tag{4.7c}$$

$$0 = 1 - e_n^T y, \gamma \text{ free} \tag{4.7d}$$

One solution for the multipliers is to take $\lambda, \gamma = 1$ and explicity take out the equality constraints but add them in later in a normalization step. Also and for consistency with [12] we set $-M = A$ and $-N = B$. To make the correspondence with the linear complementarity form of the problem, both the payoff matrices A, B will be assumed to have strictly positive entries

[4] Greek letters in parentheses besides constraints are the associated dual variables (Lagrange multipliers).

although this is not restrictive since a large positive value can be added to each element to ensure this without affecting the solution set.

We can write the corresponding LCP as follows: find x, y such that

$$0 \leq -e_m + Ay \perp x \geq 0 \tag{4.8a}$$
$$0 \leq -e_n + B^T x \perp y \geq 0 \tag{4.8b}$$

Clearly, a solution (x, y) of the LCP (4.8) may not have the elements of x and y sum to one so a normalization is needed. Specifically, if (x^*, y^*) solve (4.5) and (4.6), i.e., they are a Nash equilibrium, then (x', y') solves the LCP (4.8) where [12]:

$$x' = \frac{1}{(x^*)^T By^*} x^* \tag{4.9a}$$
$$y' = \frac{1}{(x^*)^T Ay^*} y^* \tag{4.9b}$$

Also, if (x', y') solves the LCP (4.8), a corresponding Nash equilibrium (x^*, y^*) can be recovered via the following [12]:

$$x^* = \frac{1}{(e_m)^T x'} x' \tag{4.10a}$$
$$y^* = \frac{1}{(e_n)^T y'} y' \tag{4.10b}$$

4.2.4.1 Illustrative Example. Energy Production Bimatrix Game, Version 1

As a practical example, consider the following two-person, bimatrix game in which energy producer 1 can choose $m = 3$ levels of production: 10, 20, 30 and for producer 2 $n = 2$ levels: 15 , 30 with the following payoff matrices:

$$A = \begin{bmatrix} 50 & 2 \\ 10 & 3 \\ 20 & 4 \end{bmatrix}, B = \begin{bmatrix} 10 & 5 \\ 7 & 11 \\ 130 & 17 \end{bmatrix}$$

Note that given the conversion from M, N to A, B, the entries of A, B should be interpreted as costs which are to be minimized. A solution to this specific LCP is:

$$x' = [0, 0.14285714, 0]^T$$
$$y\prime = [0.1, 0]^T$$

with corresponding Nash equilibrium

$$x^* = \frac{1}{(e_m)^T x'} x' = \begin{bmatrix} 0 \\ 1 \\ 0 \end{bmatrix} \tag{4.11a}$$

$$y^* = \frac{1}{(e_n)^T y'} y' = \begin{bmatrix} 1 \\ 0 \end{bmatrix} \tag{4.11b}$$

indicating that producer 1 should produce 20 and producer 2, 15 with 100% probability. Thus, this solution is both a mixed and a pure Nash equilibrium.

4.2.4.2 Illustrative Example. Energy Production Bimatrix Game, Version 2

Now we return to the payoff matrices from Table 4.3 for which it was shown that no pure Nash solution existed. In this case, $m = 3, n = 3$ and the payoff matrices were:

$$M = \begin{bmatrix} 0 & 0 & 0 \\ 800 & 700 & 600 \\ 1400 & 1200 & 500 \end{bmatrix}, N = \begin{bmatrix} 0 & 850 & 1500 \\ 0 & 750 & 500 \\ 0 & 650 & 1000 \end{bmatrix}$$

To convert these to the matrices A, B described above, we multiply each entry by -1 and add a sufficiently large constant to each entry (e.g., 1600) to obtain:

$$A = \begin{bmatrix} 1600 & 1600 & 1600 \\ 800 & 900 & 1000 \\ 200 & 400 & 1100 \end{bmatrix}, B = \begin{bmatrix} 1600 & 750 & 100 \\ 1600 & 850 & 1100 \\ 1600 & 950 & 600 \end{bmatrix}$$

A solution to the corresponding LCP (4.8) is to take

$$x' = [0, 0.00065421, 0.00046729]^T$$
$$y\prime = [0, 0.00016949, 0.00084746]^T$$

which after normalizing gives:

$$x^* = \frac{1}{(e_m)^T x'} x' = \begin{bmatrix} 0 \\ 0.583334819 \\ 0.416665181 \end{bmatrix} \tag{4.12a}$$

$$y^* = \frac{1}{(e_n)^T y'} y' = \begin{bmatrix} 0 \\ 0.166665028 \\ 0.833334972 \end{bmatrix} \tag{4.12b}$$

This solution indicates that producer 1 should produce 10 units with probability about 58% and 20 units with the remaining 42%. For producer 2, it is better to produce 10 units with probability about 17% and 20 units with the remaining 83%.

How can we interpret this solution in terms of real-world energy production? Different from a pure Nash solution in which the meaning of the solution is just the production level (or other strategy being modeled), for mixed problems it is the probability of selecting the associated production level. It is as if the producer has a "probability dial" that he can set which controls a simulation. The output of the simulation is the decision on the production level (0, 10, 20 in this example).

4.3 Duality in Optimization Versus Equilibria

The distinction between optimization and equilibrium problems is an important one especially when modeling energy markets. In these markets, optimization will be restricted to a single problem with one or more players whose goals are all aligned. That is to say, the one objective such as cost minimization which if optimized, will satisfy all the parties involved. By contrast, market equilibrium problems may or may not have players' goals consistent with one overall objective. Thus, to assume an optimization formulation to model an energy market may be inappropriate. In this section, we explore the differences between optimization and equilibrium problems and begin with well-known linear programs. Duality theory will be seen to be the common concept between optimization and equilibrium problems.

4.3.1 Linear Programs as Equilibrium Problems

Linear programs have been well studied for a number of decades and their application in a variety of areas is well-documented (e.g., [9], [43], [53]). In some sense, the primal and dual aspects of linear programs are perhaps the first example of an equilibrium problem but generally not stated as such. Consider the following primal linear program:

$$\text{Minimize}_x \quad c^T x \tag{4.13a}$$

$$\text{s. t.}$$

$$Ax \geq b \tag{4.13b}$$

$$x \geq 0 \tag{4.13c}$$

with corresponding dual given as:

$$\text{Maximize}_y \quad b^T y \tag{4.14a}$$
s. t.
$$A^T y \leq c \tag{4.14b}$$
$$y \geq 0 \tag{4.14c}$$

The Weak Duality Theorem for LPs [43] states that for the primal and dual feasible pair of vectors x, y that:

$$c^T x \geq b^T y$$

so that one objective is a bound on the other. The Strong Duality Theorem [43] shows that if there is a feasible x to the primal and y to the dual that:

$$c^T x = b^T y$$

In some sense, one can imagine that two players with different objectives are competing against each other where player 1 controls the x variables and player 2 the y variables. Then, these duality results show that at optimality, there is a balance between these competing players' programs.

4.3.1.1 Illustrative Example. Energy Production Optimization Problem, One Player

Consider a variation on the energy production problem from before (Example 4.2.1.1. In this setting there is just one producer who decides on q, the level of energy production which must be in $[0, q^{\max}]$ where $q^{\max}$ is the maximum production possible. Further, assume that the price of the energy is an exogenous factor for this producer and is given by p with γ the unit cost and it is assumed that $p > \gamma$ so that the producer has an incentive to produce. Thus, the energy production problem for this producer can be stated as the following linear program:

$$\text{Maximize}_q \quad pq - \gamma q \tag{4.15a}$$
s. t.
$$q \leq q^{\max} \quad (\lambda) \tag{4.15b}$$
$$q \geq 0 \tag{4.15c}$$

Clearly the producer is trying to maximize profit. But suppose that this producer is having capacity problems and would like to increase $q^{\max}$. How much should he pay for one unit more of production capacity?[5] Assuming

[5] This unit cost is not the same for all levels of $q^{\max}$ but depends on the basis (see [9], [43] for more details on bases in linear programming).

that all the extra capacity $\Delta q^{\max}$ is used, the increase in the profit would be from $(p-\gamma)\,q^{\max}$ to $(p-\gamma)\left(q^{\max}+\Delta q^{\max}\right)$. A rational producer would only pay at most the difference in the profit levels, i.e.,$(p-\gamma)\left(\Delta q^{\max}\right)$. Clearly paying less than this amount would be advantageous but may not be possible. The dual problem is consequently:

$$\text{Minimize}_{\lambda} \quad \lambda q^{\max} \tag{4.16a}$$

$$\text{s. t.}$$

$$\lambda \geq (p-\gamma) \tag{4.16b}$$

$$\lambda \geq 0 \tag{4.16c}$$

but what is the interpretation of this problem? First, this dual problem is one of determing a best capacity price λ with the capacity total costs (at the current capacity level) being the objective function value. Similar to the "Dakota production problem" in [53], one can imagine another business which is considering buying the energy production capacity but wants to determine the price at which the takeover costs $\lambda q^{\max}$ would be minimized. The constraints are that the price should not be less than the per unit profit and it should be nonnegative. If for example, $\lambda < (p-\gamma)$, then the producer would have no incentive to sell since he could make $(p-\gamma)$ per unit which would be superior to λ. Thus, this dual problem is one of finding the lowest possible price and payout to take over the energy producer's business.

Considering both the primal and dual problems together, a solution pair (q,λ) will only be optimal for the producer and takeover company if both cannot do any better, i.e., if they have no incentive to change. The Weak Duality result interpreted for this problem amounts to

$$(p-\gamma)\,q \leq \lambda q^{\max} \tag{4.17}$$

so the profit will never be higher than $\lambda q^{\max}$ and the takeover cost will never be lower than the profit $(p-\gamma)\,q$ from (4.15). If there is a (q,λ) that is optimal for both entities, then by the Strong Duality Theorem, equality must hold in (4.17). Thus, the equilibrium aspects are made apparent only when one considers both the primal and dual problems together. This is made clear through the KKT optimality conditions which are to find a production level q and dual price λ such that:

$$0 \leq -(p-\gamma)+\lambda \perp q \geq 0 \tag{4.18a}$$

$$0 \leq q^{\max}-q \perp \lambda \geq 0 \tag{4.18b}$$

Assuming that an optimal production $q^* > 0$, there are two cases of interest to consider. First, if an optimal $q^* = q^{\max} > 0$,these conditions imply that in equilibrium an optimal $\lambda^* = p-\gamma$ or that $(p-\gamma)\,q^{\max} = \lambda^* q^{\max} > 0$. That is, the optimal producer profit is positive and the same as the total buyout cost so the producer is indifferent between the options of continuing to produce or selling his operations. Second, if $q^* \in (0, q^{\max})$, then by complementarity,[6] an optimal price λ^* of the resource should be equal to zero (no incentive for the producer to sell his capacity since it's not being fully utilized) but also by complementarity $(p-\gamma) = \lambda^* = 0$ which is not possible since $p > \gamma$. Lastly, suppose that $q^* = 0$. By complementarity this would imply that $\lambda^* = 0$. But then from the first set of conditions this would mean that $p - \gamma \leq 0$ which is not possible since $(p-\gamma) > 0$. Thus, only the first case is possible, that of producing at capacity.

4.3.2 Nonlinear Programs as Equilibrium Problems

The above discussion showed that a linear program, by considering both the primal and dual problems simultaneously is an example of an equilibrium in which the objectives of two "players" are balanced. A natural question might be, how far does this concept extend beyond linear programs? That is, is there a similar connotation for nonlinear ones? The answer is both yes and no with the key property being convexity of the problem as well as certain constraint qualifications. First assume that the primal nonlinear program under consideration is given by:

$$\text{Minimize}_x \quad f(x) \tag{4.19a}$$

s. t.

$$g(x) \leq 0 \quad (\lambda) \tag{4.19b}$$

$$x \in X \tag{4.19c}$$

where X can be R^n or some subset to be handled separately and $g: R^n \to R^m, \lambda \in R^m_+$. The Lagrangian function is defined as $L(x,\lambda) = f(x) + \lambda^T g(x)$ for $x \in X, \lambda \geq 0$. Then, we can define the primal function for any $x \in X$ as follows [44]:

$$L^*(x) = \underset{\lambda \geq 0}{\text{Maximize}} \; L(x,\lambda) \tag{4.20}$$

For x satisfying $g(x) \leq 0$, $\lambda^T g(x) \leq 0$ since $\lambda \geq 0$. Thus, in (4.20), the best choice will be to take $\lambda = 0$ so that $L^*(x) = f(x)$. If for a given $x, g_i(x) > 0$, for some i, then λ_i can take on arbitrarily large values while holding the other values of λ constant to make the primal function equal to positive infinity. Thus,

$$L^*(x) = \left\{ \begin{array}{c} f(x) \text{ if } g(x) \leq 0 \\ \infty \quad \text{otherwise} \end{array} \right\}$$

Consequently, the primal problem is then:

---

[6] For linear programs, complementarity is usually referred to as "complementarity slackness" the former term also applying to settings more general than LPs.

$$\underset{x \in X}{\text{Minimize}} \;\; L^* (x) \tag{4.21}$$

which corresponds to the original problem (4.19) if we ignore the region where the primal function is positive infinity [44]. For any nonnegative λ, a corresponding nonlinear dual function is defined as:

$$L_* (\lambda) = \underset{x \in X}{\text{Minimize}} \;\; L (x, \lambda) \tag{4.22}$$

and the associated dual problem as [44]:

$$\underset{\lambda \geq 0}{\text{Maximize}} \;\; \underset{x \in X}{\text{Minimize}} \;\; L (x, \lambda)$$

Similar to the case of linear programming, there is a weak duality result relating the above primal and dual problems.

Theorem 4.1. *(Theorem 14.5, [44]) Let x be a feasible solution to the primal problem (4.19), and (x^*, λ^*) be a feasible solution to its dual problem. Then,*

$$f(x) + \lambda^T g(x) \leq f(x)$$

When f is a convex function and each g_i is also convex with all functions being continuously differentiable with a regularity condition holding [2], then the optimal primal and dual objectives are equal (Theorem 14.6, [44]). This is the analog with linear programming of two competing problems and a balance sought between them.

4.4 More About the Connection Between Optimization and Equilibrium Problems

From the above discussion, optimization problems viewed from the perspective of considering both the primal and dual problems can be thought of an equilibrium between these two problems. This is not the only way to connect optimization and equilibrium. In what follows we will describe several more important connections between these two broad classes of problems.[7] The first one involves the classical Spatial Price Equilibrium (SPE) problem [47], [30], [31].

[7] An important point when considering optimization and equilibrium problems is the "squareness" of the system. For a complementarity problem, there needs to be the same number of variables as constraints (equations or inequalities). This is not a problem when considering the Karush-Kuhn-Tucker optimality conditions for an optimization problem but this issue may arise in the model formuation aspects of more general complementarity problems.

4.4.1 Spatial Price Equilibrium Problem

Imagine that there are a number of separate regions some producing energy products and some consuming them (a region could do both as well). An important question is what will be the flows and prices that will balance the supply and demand of these energy products (e.g., electricity, coal, natural gas)? Following the formulation from [30], every region $r \in R$ would first need to balance its internal supply S_r with its own demand D_r while also taking into account the flow F_{ij} between regions i and j. Or more specifically, for each region r,

$$S_r - D_r + \sum_i F_{ir} - \sum_j F_{rj} = 0 \tag{4.23}$$

where $\sum_i F_{ir}$ is the total flow into region r and $\sum_j F_{rj}$ is the total flow out of the region making the assumption that all regional pairs are possible.[8] Additionally, the total demand in the system should equal the total supply or

$$\sum_r S_r - \sum_r D_r = 0 \tag{4.24}$$

which will be achieved in (4.23) if this condition is summed over all regions [30].

Of course, in the case of energy products that are transported, there are usually losses that are accrued. For example, in natural gas, some of the fuel can be used to feed the compressors in the pipeline or could be lost due to cracks in the pipes. In electricity, the losses can be realized as part of the transmission and distribution system. In that case, usually the losses are modeled at the destination region (node) so that the inflows would be adjusted to $\sum_i F_{ir} (1 - loss_{ir})$ where $loss_{ir}$ is the loss rate (percentage) along the link connecting region i and r. In reality, the losses could also be a function of the level of flow. For our purposes without loss of generality, we will ignore these losses.

The spatial price equilibrium conditions can be very simply stated as: if there is flow between a pair of regions, then the supply price $\psi_i (S_i)$ plus the transportation cost $c_{ij} (F_{ij})$ must equal the price in the demand region $\theta_j (D_j)$. These items are written as a function of their arguments, S_i, F_{ij}, D_j to show their possible dependence. When there is no flow between the pair of regions, then the cost of getting the energy to the destination region j should be strictly greater than the price that the energy transporter could sell it for. Mathematically speaking, this means the following set of conditions:

$$0 \leq \psi_i (S_i) + c_{ij} (F_{ij}) - \theta_j (D_j) \perp F_{ij} \geq 0 \tag{4.25}$$

---

[8] This is not restrictive as an infinite cost could be applied to non-existent regional pairs and is notationally simpler than spelling out only the valid regional flow pairs.

Table 4.4 Transportation unit costs

c_{ij} =

from\to	1	2	3	4	5
1	10	20	30	5	6
2	3	15	12	5	6
3	17	40	5	3	7
4	50	23	28	12	4

Table 4.5 Maximum supply amounts

S_i =

Supply Region	
1	20
2	30
3	35
4	15

Table 4.6 Demand amounts

D_j =

Demand Region	
1	20
2	25
3	10
4	10
5	35

which are clearly stating a complementarity between the flows and the delivered prices [30]. In the case when $\psi_i(S_i) = \psi_i, c_{ij}(F_{ij}) = c_{ij}, \theta_j(D_j) = \theta_j$, then the above conditions relate to the classical transportation problem [31].

4.4.1.1 Illustrative Example. Spatial Price Equilibrium for Energy Products

As an example of the transportation problem in an energy setting, consider four petroleum refinery plants (supply nodes) supplying products to five cities (demand nodes) with all 20 supply, demand pairs possible. No engineering phenomena, line capacities, nor other details are considered in this example to keep it simple but certainly could be in a more fleshed out model. Consider that a company is managing this nine-node system and would like to minimize the total transportation costs subject to supply and demand balancing. The costs are given as per unit amounts between a supply region i and a demand region j. Table 4.4 gives these costs and the supply demand levels are indicated in the Tables 4.5 and 4.6 noting that total supply equals total demand (i.e., this is a balanced system).

The related transportation problem [43] can then be stated as the following:

Table 4.7 Optimal flows for transportation problem

$x_{ij} =$

from\to	1	2	3	4	5	supply
1	0	15	0	0	5	**20**
2	20	10	0	0	0	**30**
3	0	0	10	10	15	**35**
4	0	0	0	0	15	**15**
demand	**20**	**25**	**10**	**10**	**35**	**100**

$$\text{Minimize}_{x_{ij}} \quad \sum_{i=1}^{4}\sum_{j=1}^{5} c_{ij}x_{ij}$$

s. t.

$$\sum_{j=1}^{5} x_{ij} \leq S_i, \quad i = 1, 2, 3, 4 \quad (\psi_i) \tag{4.26a}$$

$$\sum_{i=1}^{4} x_{ij} \geq D_j, \quad j = 1, 2, 3, 4, 5 \quad (\theta_j) \tag{4.26b}$$

$$x_{ij} \geq 0, \quad \forall i, j \tag{4.26c}$$

The dual problem is:

$$\text{Maximize}_{\psi_i,\theta_j} \quad \sum_{i=1}^{4} -\psi_i S_i + \sum_{j=1}^{5} \theta_j D_j$$

s. t.

$$-\psi_i + \theta_j \leq c_{ij}, \quad \forall i, j \tag{4.27a}$$

$$\psi_i, \theta_j \geq 0, \quad \forall i, j \tag{4.27b}$$

One interpretation of the dual problem is another player that is trying to maximize the total net payments (demand payments less supply costs) subject to the delivered costs $c_{ij} + \psi_i$ being greater than or equal to the selling price θ_j. One solution of this problem is shown in Table 4.7 and indicates that all supply and demand are met exactly as must be the case given that this is a balanced transportation problem.

Besides the flows, it is important to consider the dual prices, one for each equation corresponding to a supply or demand node as shown in Table 4.8.

We note that the SPE equilibrium complementarity conditions are met, that is:

$$0 \leq \psi_i + c_{ij} - \theta_j \perp x_{ij} \geq 0 \tag{4.28}$$

For example, consider the region pairs (1,1) and (4,5). Accordingly, the complementarity conditions are:

Table 4.8 Optimal prices for transportation problem

Supply Node Prices	ψ
1	1
2	6
3	0
4	3
Demand Node Prices	θ
1	9
2	21
3	5
4	3
5	7

$$(1,1) : 0 < 2 = \psi_1 + c_{11} - \theta_1 = 1 + 10 - 9, x_{11} = 0$$
$$(4,5) : 0 = \psi_4 + c_{45} - \theta_5 = 3 + 4 - 7, x_{45} = 15 > 0$$

This is another example of the correspondence between optimization and equilibrium problems. In equilibrium, the complementarity condition (4.28) enforces that if there is a positive flow of energy products between the supply and demand regions, no profits can be made. Clearly (4.28) is a special case of (4.25).

One could of course start with just the optimality conditions of the transportation problem and make the prices ψ_i, θ_j not just constant but a function of supply and demand [31].

A natural question to ask is whether there is an optimization problem (not necessarily linear) that would give rise to the more general SPE complementarity conditions as well as market balance described above? The answer is yes as described in [30] in which $r \in R$ indicates the regions (nodes) and $(i,j) \in A$ corresponds to the inter-regional arcs.

$$\text{Minimize} \quad \sum_r \int_0^{S_r} \psi_r(s)\, ds - \sum_r \int_0^{D_r} \theta_r(s)\, ds + \sum_{(i,j)} \int_0^{F_{ij}} c_{ij}(s)\, ds \tag{4.29a}$$

s. t.

$$S_r - D_r + \sum_i x_{ir} - \sum_j x_{rj} = 0, \quad \forall r \tag{4.29b}$$

$$\sum_r S_r - \sum_r D_r = 0 \tag{4.29c}$$

$$S_r, D_r \geq 0, \quad \forall r \tag{4.29d}$$

$$x_{ij} \geq 0, \quad \forall i,j \tag{4.29e}$$

whose KKT conditions will be the SPE equilibrium conditions cited above. The objective function requires some explanation. The left-most term describes the total production cost summed over all regions r. Similarly, the second term from the left is the (negative of) the area under the demand curve summed across the regions, i.e., the "consumer satisfaction" [52]. The last term is the total distribution costs summed over the arcs in the network. As such, the objective function represents (the negative of) the net social pay-off which is to be optimized [47]. Thus, it seems like an optimization problem can be converted into an equilibrium one and vice-versa. This is not the case however. What is needed is that the equilibrium conditions (e.g., for complementarity) need to be the KKT conditions of an appropriate optimization problem which may not be the case for a general complementarity problem.

Conversely, if one starts with an optimization problem, one can always end up with an MCP. Recall that as mentioned in Chapter 1, every nonlinear program of the form:

$$\text{Minimize}_x \quad f(x) \tag{4.30a}$$
$$\text{s. t.}$$
$$g_i(x) \leq 0, \quad i = 1, \ldots, m \quad (\lambda) \tag{4.30b}$$
$$h_j(x) = 0, \quad j = 1, \ldots, p \quad (\gamma) \tag{4.30c}$$

gives rise to a mixed complementarity problem via its KKT conditions which are to find x, λ, γ such that:

$$0 = \nabla f(x) + \sum_i \nabla g_i(x)\lambda_i + \sum_j \nabla h_j(x)\gamma_j \tag{4.31a}$$
$$0 \leq -g_i(x) \perp \lambda_i \geq 0, i = 1, \ldots, m \tag{4.31b}$$
$$0 = h_j(x), \gamma_j \text{ free } j = 1, \ldots, p \tag{4.31c}$$

The corresponding MCP function is $F : R^n \times R^m_+ \times R^p \rightarrow R^n \times R^m_+ \times R^p$ is:

$$F(x, \lambda, \gamma) = \begin{cases} \nabla f(x) + \sum_i \nabla g_i(x)\lambda_i + \sum_j \nabla h_j(x)\gamma_j \\ -g_i(x), i = 1, \ldots, m \\ h_j(x), j = 1, \ldots, p \end{cases}$$

4.4.2 Optimization Problems from Equilibrium Conditions?

A next question then is what equilibrium problems translate back to an optimization problem like the SPE example shown above? Or are there some equilibrium problems for which there is no natural underlying optimization problem? As it turns out, this question can be answered in a number of ways with the first one relying on the notion of the variational inequality (VI) problem. The VI will be explored in more detail in Chapter 5 but for now we just state its definition and relevance to complementarity problems.

Definition 4.1. Having a function $F : R^n \to R^n$, and a set $K \subseteq R^n$, the variational inequality problem VI(K, F) is to find a vector $x^* \in K$ such that

$$F(x^*)^T (y - x^*) \geq 0, \forall y \in K$$

Recall that, having a function $F : R^n \to R^n$, the pure nonlinear complementarity problem NCP(F) is to find an $x \in R^n$ such that for all i:

1. $F_i(x) \geq 0$
2. $x_i \geq 0$
3. $x_i \cdot F_i(x) = 0$

or more compactly as $0 \leq F(x) \perp x \geq 0$. The mixed version of the complementarity problem (MCP) has several forms with perhaps the most general one being the following. Have lower and upper bound vectors, respectively, $l, u \in R^n \cup \{+\infty, -\infty\}$, the mixed complementarity problem is to find a vector x such that the following conditions hold:

1. if $x_i = l_i \Rightarrow F_i(x) \geq 0$
2. if $x_i = u_i \Rightarrow F_i(x) \leq 0$
3. if $x_i \in (l_i, u_i) \Rightarrow F_i(x) = 0$

If we divide the indices $\{1, \ldots, n\}$ into two sets with the first one S_1 relating to greater than or equal to inequalities and the second one S_2 equations, then these three conditions can be specialized to:

1. $F_i(x_{S_1}, x_{S_2}) \geq 0, x_i \geq 0, x_i \cdot F_i(x) = 0, \forall i \in S_1$
2. $F_j(x_{S_1}, x_{S_2}) = 0, x_j$ free $, \forall j \in S_2$.

by choosing $l_i = 0, u_i = +\infty, \forall i \in S_1, l_j = -\infty, u_i = +\infty, \forall j \in S_2$.

By selecting appropriate values for $y \in K$ in the definition of the VI, it is not hard to show that the pure NCP(F) is a special case of the the VI in which the set $K = R^n_+$. Additionally, one can show that the mixed complementarity problem is a special case of the VI in which the set $K = [l, u]$. These two results and a more general connection between the VI and mixed complementarity problem are discussed in Chapter 5.

The equivalence between optimization and the VI is given by the famous Principle of Symmetry (or Integrability) stated below in which one is considering the following optimization problem in which K is assumed to be a convex subset of R^n:

$$\text{Minimize}_x \quad \theta(x) \tag{4.32}$$

s. t.

$$x \in K \tag{4.33}$$

Theorem 4.2. *[Theorem 1.3.1 from [15]]Let $F : U \to R^n$ be continuously differentiable on the open convex set $U \subseteq R^n$. The following three statements are equivalent: a. there exists a real-valued function θ such that $F(x) = \nabla\theta(x)$ for all $x \in U$; b. the Jacobian matrix $JF(x)$ is symmetric for all $x \in U$; c. F is integrable on U.*

Moreover, if any one of these equivalent conditions is in force that the function θ can be generated according to the following formula:

$$\theta(x) = \int_0^1 F(x^0 + t(x - x^0))^T (x - x^0) dt \tag{4.34}$$

where $x^0 \in U$ is an arbitrary vector. The essence of this result says that there will be an equivalent VI as long as the VI function F has a symmetric Jacobian. This makes sense since if one started with an optimization problem of the form (4.32), the gradient of the objective function $\nabla\theta(x)$ would be vector-valued like F but would necessarily have a symmetric Jacobian since it would be the Hessian matrix $\nabla^2\theta(x)$. Thus, this result is instrumental in saying when an optimization problem is associated with an equilibrium problem and when it can't. Moreover, it gives a formula to let one produce the corresponding optimization functions (objective or constraints) as described below.

4.4.2.1 Illustrative Example. Extended Energy Production Optimization Problem

As a way to relate this result to energy markets, consider a variation on the energy production problem (4.15). Specifically, now assume that there are two energy products to be produced whose levels are given by q_1 and q_2 along with associated prices p_1 and p_2. Also, assume that there are now quadratic costs for the two products given by $\frac{1}{2}\gamma_1 q_1^2$ and $\frac{1}{2}\gamma_2 q_2^2$ for $\gamma_1, \gamma_2 > 0$ as well as product-specific upper bounds $q_1^{\max}, q_2^{\max}$. Then, the new optimization problem expressed as a minimization of negative profits is given as:

$$\text{Minimize}_{q_1,q_2} \quad \theta(q_1, q_2) = \frac{1}{2}\gamma_1 q_1^2 + \frac{1}{2}\gamma_2 q_2^2 - p_1 q_1 - p_2 q_2 \tag{4.35a}$$

s. t.

$$q_1 \le q_1^{\max} \quad (\lambda_1) \tag{4.35b}$$

$$q_2 \le q_2^{\max} \quad (\lambda_2) \tag{4.35c}$$

$$q_1, q_2 \ge 0 \tag{4.35d}$$

An optimal solution (q_1^*, q_2^*) will necessarily satisfy:

$$\nabla\theta\left(q_1^*, q_2^*\right)^T \begin{pmatrix} y_1 - q_1^* \\ y_2 - q_2^* \end{pmatrix} \geq 0, \forall \begin{pmatrix} y_1 \\ y_2 \end{pmatrix} \in K =$$

$$= \left\{(q_1, q_2) \,| 0 \leq q_1 \leq q_1^{\max}, 0 \leq q_2 \leq q_2^{\max}\right\}$$

which is exactly the VI(K, F) with

$$F(q_1, q_2) = \nabla\theta\left(q_1, q_2\right) = \begin{pmatrix} \gamma_1 q_1 - p_1 \\ \gamma_2 q_2 - p_2 \end{pmatrix}$$

and Jacobian

$$JF = \begin{pmatrix} \gamma_1 & 0 \\ 0 & \gamma_2 \end{pmatrix}$$

which is clearly symmetric since it is diagonal. Now suppose that we started with a VI whose function F had a Jacobian that was symmetric but not diagonal, for example with $\gamma_3 > 0$:

$$JF = \begin{pmatrix} \gamma_1 & \gamma_3 \\ \gamma_3 & \gamma_2 \end{pmatrix}$$

What would be the equivalent energy production optimization problem? In this case, we could take $F(q_1, q_2) = \begin{pmatrix} \gamma_1 q_1 + \gamma_3 q_2 - p_1 \\ \gamma_3 q_1 + \gamma_2 q_2 - p_2 \end{pmatrix}$ (although other constant terms could also be added). By the Principle of Symmetry, setting $x^0 = 0$, the choice for the objective function of the corresponding optimization problem would be:

$$\begin{aligned}
\theta(q_1, q_2) &= \int_0^1 F\left(t\begin{pmatrix} q_1 \\ q_2 \end{pmatrix}\right)^T \begin{pmatrix} q_1 \\ q_2 \end{pmatrix} dt \\
&= \int_0^1 \begin{pmatrix} t\gamma_1 q_1 + t\gamma_3 q_2 - p_1 \\ t\gamma_3 q_1 + t\gamma_2 q_2 - p_2 \end{pmatrix}^T \begin{pmatrix} q_1 \\ q_2 \end{pmatrix} dt \\
&= \int_0^1 \left\{t\left[\gamma_1 q_1^2 + \gamma_3 q_2 q_1 + \gamma_3 q_1 q_2 + \gamma_2 q_2^2\right] - \left[p_1 q_1 + p_2 q_2\right]\right\} dt \\
&= \left[\gamma_1 q_1^2 + \gamma_3 q_2 q_1 + \gamma_3 q_1 q_2 + \gamma_2 q_2^2\right] \int_0^1 t dt - \int_0^1 \left[p_1 q_1 + p_2 q_2\right] dt \\
&= \left[\gamma_1 q_1^2 + \gamma_3 q_2 q_1 + \gamma_3 q_1 q_2 + \gamma_2 q_2^2\right] \frac{1}{2}\left[t^2\right]_0^1 - \left[p_1 q_1 + p_2 q_2\right] \\
&= \frac{1}{2}\left[\gamma_1 q_1^2 + \gamma_3 q_2 q_1 + \gamma_3 q_1 q_2 + \gamma_2 q_2^2\right] - \left[p_1 q_1 + p_2 q_2\right] \\
&= \frac{1}{2}\gamma_1 q_1^2 + \frac{1}{2}\gamma_2 q_2^2 - p_1 q_1 - p_2 q_2 + \gamma_3 q_1 q_2
\end{aligned}$$

which is the original objection function plus an additional cost term of the form $\gamma_3 q_1 q_2$ representing interaction costs relating perhaps to congestion in the energy production, rig availabiity, or the like. This example shows that if one started with a equilibrium problem expressed as a VI with symmetric

Jacobian, then an equivalent energy production optimization problem could be found.

4.4.2.2 Illustrative Example. Extended Energy Production Optimization Derived from MCP

A similar correspondence can be made with the complementarity form of the equilibrium. We start with some complementarity problem and try to relate it to (4.31) identifying a $\nabla f(x)$ along the way in order to end up with a nonlinear program of the form (4.30). Consider the necessary and sufficient KKT conditions for (4.35):

$$0 \le \gamma_1 q_1 - p_1 + \lambda_1 \perp q_1 \ge 0 \tag{4.36a}$$
$$0 \le \gamma_2 q_2 - p_2 + \lambda_2 \perp q_2 \ge 0 \tag{4.36b}$$
$$0 \le q_1^{\max} - q_1 \perp \lambda_1 \ge 0 \tag{4.36c}$$
$$0 \le q_2^{\max} - q_2 \perp \lambda_2 \ge 0 \tag{4.36d}$$

Now consider the following variation on these conditions with quadratic terms in the first two sets of conditions as well as $\delta_1 q_2$ and $\delta_2 q_1$ terms added to the first and second conditions, respectively; here δ_1 and δ_2 are coefficients. Also there is an additional multiplier λ_3 and extra associated constraint of the form: $q_1 + q_2 \le q^{up}$.

$$0 \le \gamma_1 q_1^2 + \delta_1 q_2 - p_1 + \lambda_1 + \lambda_3 \perp q_1 \ge 0 \tag{4.37a}$$
$$0 \le \gamma_2 q_2^2 + \delta_2 q_1 - p_2 + \lambda_2 + \lambda_3 \perp q_2 \ge 0 \tag{4.37b}$$
$$0 \le q_1^{\max} - q_1 \perp \lambda_1 \ge 0 \tag{4.37c}$$
$$0 \le q_2^{\max} - q_2 \perp \lambda_2 \ge 0 \tag{4.37d}$$
$$0 \le q^{up} - q_1 - q_2 \perp \lambda_3 \ge 0 \tag{4.37e}$$

Clearly the last three sets of conditions are of the form $0 \le -g_i(q) \perp \lambda_i \ge 0$ so these will be the constraints and associated multipliers of an optimization problem to be constructed. For the first two sets of conditions, given that the (nonnegative) primal variables are q_1, q_2, the task is to identify the function f for which the terms in these conditions make up $\nabla f(q)$. Specifically, we have:

$$F(q_1, q_2) = \nabla f(q_1, q_2) = \begin{pmatrix} \gamma_1 q_1^2 + \delta_1 q_2 - p_1 \\ \gamma_2 q_2^2 + \delta_2 q_1 - p_2 \end{pmatrix}$$

for some function f to be determined noting that the terms $\lambda_1 + \lambda_3, \lambda_2 + \lambda_3$ would represent $\sum_i \nabla g_i(q) \lambda_i$. Thus, it comes down to the Principle of Symmetry so we need to check if $JF(q_1, q_2)$ is a symmetric matrix. But

$$JF(q_1, q_2) = \begin{pmatrix} 2\gamma_1 q_1 & \delta_1 \\ \delta_2 & 2\gamma_2 q_2 \end{pmatrix}$$

which is symmetric if and only if $\delta_1 = \delta_2 = \delta$. Assuming that this is the case, the Principle of Symmetry shows that setting $x^0 = 0$, the choice for the objective function of the corresponding optimization problem would be:

$$\begin{aligned}
f(q_1, q_2) &= \int_0^1 F\left(t\begin{pmatrix} q_1 \\ q_2 \end{pmatrix}\right)^T \begin{pmatrix} q_1 \\ q_2 \end{pmatrix} dt \\
&= \int_0^1 \begin{pmatrix} \gamma_1 t^2 q_1^2 + \delta t q_2 - p_1 \\ \gamma_2 t^2 q_2^2 + \delta t q_1 - p_2 \end{pmatrix}^T \begin{pmatrix} q_1 \\ q_2 \end{pmatrix} dt \\
&= \int_0^1 \left(\gamma_1 t^2 q_1^3 + \delta t q_1 q_2 - p_1 q_1 + \gamma_2 t^2 q_2^3 + \delta t q_1 q_2 - p_2 q_2\right) dt \\
&= \gamma_1 q_1^3 \left[\frac{t^3}{3}\right]_0^1 + 2\delta q_1 q_2 \left[\frac{t^2}{2}\right]_0^1 - p_1 q_1 + \gamma_2 q_2^3 \left[\frac{t^3}{3}\right]_0^1 - p_2 q_2 \\
&= \frac{\gamma_1 q_1^3}{3} + \delta q_1 q_2 - p_1 q_1 + \frac{\gamma_2 q_2^3}{3} - p_2 q_2
\end{aligned}$$

Thus, the corresponding nonlinear program would be:

$$\text{Minimize}_{q_1, q_2} \quad \frac{\gamma_1 q_1^3}{3} + \delta q_1 q_2 - p_1 q_1 + \frac{\gamma_2 q_2^3}{3} - p_2 q_2 \tag{4.38a}$$

s. t.

$$q_1 \le q_1^{\max} \quad (\lambda_1) \tag{4.38b}$$

$$q_2 \le q_2^{\max} \quad (\lambda_2) \tag{4.38c}$$

$$q_1 + q_2 \le q^{up} \quad (\lambda_3) \tag{4.38d}$$

relating to more complicated costs and an additional joint upper bound on the sum of the two quantities q_1 and q_2.

4.4.3 Equilibria with No Corresponding KKT-Based Optimization Problem

From the analysis above, to make the correspondence between the optimization problem (4.38) and the equilibrium conditions (4.37), it was assumed that $\delta_1 = \delta_2 = \delta$. But what if this is not the case? Can we still form some optimization problem? Since (4.37) is an example of a complementarity problem, there is always an optimization problem related to it but not necessarily one that is derived from associated KKT conditions. To be more precise, suppose that we started with the following MCP:

$$0 \leq F_{S_1}(x_{S_1}, x_{S_2}) \perp x_{S_1} \geq 0 \tag{4.39a}$$
$$0 = F_{S_2}(x_{S_1}, x_{S_2}),\ x_{S_2},\ \text{free} \tag{4.39b}$$

where as before, $S_1 \cup S_2$ is a partition of the indices $\{1,\ldots,n\}$ and $x = \left(x_{S1}^T, x_{S_2}^T\right)^T$. Then, we can always form the following related optimization problem:

$$\text{Minimize}_x \quad F_{S_1}(x_{S_1}, x_{S_2})^T x_{S_1} \tag{4.40a}$$
s. t.
$$F_{S_1}(x_{S_1}, x_{S_2}) \geq 0 \tag{4.40b}$$
$$F_{S_2}(x_{S_1}, x_{S_2}) = 0 \tag{4.40c}$$
$$x_{S_1} \geq 0 \tag{4.40d}$$
$$x_{S_2},\ \text{free} \tag{4.40e}$$

A solution of this problem for which the objective function value was zero would correspond exactly to an MCP solution. If the objective function were not equal to zero, this correspondence would break down. While this optimization problem is related to the MCP, it is not derived from it in the sense of it relating to the KKT conditions of that problem.

As a way to see the advantage of the more general MCP vs. optimization formulations, let's return to the energy market spatial price equilibrium problem of Example 4.4.1.1. Written out in the most general form, such a problem is to find flows x_{ij} and prices ψ_i, θ_j for all supply nodes i and demand nodes j that satisfy the following conditions:

$$0 \leq \psi_i \left(\sum_j x_{ij}\right) + c_{ij}(x) - \theta_j \left(\sum_i x_{ij}\right) \perp x_{ij} \geq 0, \forall i, j \tag{4.41}$$

where the total supply is given by $S_i = \sum_j x_{ij}$ and the total demand by $D_j = \sum_i x_{ij}$.

Here the MCP function $F : R^{20} \rightarrow R^{20}$ is given by

$$F(x) = \psi_i \left(\sum_j x_{ij}\right) + c_{ij}(x) - \theta_j \left(\sum_i x_{ij}\right), i = 1,2,3,4; j = 1,2,3,4,5 \tag{4.42}$$

If there were an equivalent optimization problem with objective function f, $F(x)$ would have to represent $\nabla f(x)$ (since there are no constraints other than nonnegativity in this problem). By the Principle of Symmetry this would mean that $JF(x)$ would have to be symmetric but:

$$JF(x)_{(i,j),(k,l)} = \frac{\partial F_{ij}(x)}{\partial x_{kl}} = \frac{\partial \psi_i \left(\sum_j x_{ij}\right)}{\partial x_{kl}} + \frac{\partial c_{ij}(x)}{\partial x_{kl}} - \frac{\partial \theta_j \left(\sum_i x_{ij}\right)}{\partial x_{kl}} \tag{4.43a}$$

$$JF(x)_{(k,l),(i,j)} = \frac{\partial F_{kl}(x)}{\partial x_{ij}} = \frac{\partial \psi_k \left(\sum_j x_{ij}\right)}{\partial x_{ij}} + \frac{\partial c_{kl}(x)}{\partial x_{ij}} - \frac{\partial \theta_l \left(\sum_i x_{ij}\right)}{\partial x_{ij}} \tag{4.43b}$$

But in general, the right-hand sides of (4.43a) and (4.43b) will not be equal so there will not be an optimization problem whose KKT conditions give rise to this more general version of the SPE. From that perspective, the MCP is more flexible since it subsumes optimization problems (via KKT conditions) but also includes formulations not arising from an optimization setting.

4.4.3.1 Illustrative Example. Spatial Price Equilibrium, Version 2

A next question might be what would the differences in the answers given this more general formulation? To answer that assume that that the inverse price and inverse demand functions have the following form:

$$\psi_i(x) = \kappa_i^S + \rho_i^S \cdot \left(\sum_j x_{ij}\right) \tag{4.44a}$$

$$\theta_j(x) = \kappa_j^D - \rho_j^D \cdot \left(\sum_i x_{ij}\right) \tag{4.44b}$$

where $\rho_i^S, \rho_i^D > 0$. Also, assume for concreteness that the transportation cost function is given as:

$$c_{ij}(x) = c_{ij} \cdot (x_{ij}) \tag{4.45}$$

We would like to pick the various parameter values to match the solution from Example 4.4.1.1 shown above for better comparison purposes (details are shown in the Computational Appendix). This means that for the supply prices we have eight unknowns in four equations so we arbitrarily set $\kappa_1^S = \kappa_2^S = \kappa_4^S = 0, \rho_3^S = \rho_2^S$ to obtain $\kappa_3^S = -7$, and $\rho_1^S = \frac{1}{20}, \rho_2^S = \rho_3^S = \rho_4^S = \frac{1}{5}$. Similarly for the demand prices, by setting $\rho_1^D = \rho_2^D = \rho_3^D = \rho_4^D = \rho_5^D = 1$ we get: $\kappa_1^D = 29, \kappa_2^D = 46, \kappa_3^D = 15, \kappa_4^D = 13, \kappa_5^D = 42$. Lastly, we use the same unit transportation costs as in the transportation problem stated above. By design, the set of prices and flows are the same as in Example 4.4.1.1. The advantage of the complementarity form of the problem (4.41) is that it has direct feedback between the flows and the prices that can be manipuated by the modeler. For example, if we wanted to make the inverse supply function more steep at node 1, we could just adjust the value for ρ_1^S and make it larger.

Table 4.9 Optimal flows for SPE

$\left(x_{ij}^{old}, x_{ij}^{new}\right) =$

from\to	1	2	3	4	5	**supply**
1	(0, 0)	(15, 9.167)	(0, 0)	(0, 0)	(5, 0)	**(20, 9.167)**
2	(20, 19.167)	(10, 15)	(0)	(0)	(0)	**(30, 34.167)**
3	(0, 0)	(0, 0)	(10, 9.615)	(10, 9.615)	(15, 17.692)	**(35, 36.923)**
4	(0)	(0)	(0)	(0)	(15, 16.923)	**(15, 16.923)**
demand	**(20, 19.167)**	**(25, 24.167)**	**(10, 9.615)**	**(10, 9.615)**	**(35, 34.615)**	**(100, 97.180)**

Table 4.10 Optimal prices for transportation problem

Supply Node Prices	ψ^{old}, ψ^{new}
1	1, 1.833
2	6, 6.833
3	0, 0.385
4	3, 3.385
Demand Node Prices	$\theta^{old}, \theta^{new}$
1	9, 9.833
2	21, 21.833
3	5, 5.385
4	3, 3.385
5	7, 7.385

Suppose that we want it to have the same slope as at the other nodes, i.e., $\frac{1}{5}$ increasing it from $\frac{1}{20}$. The resulting new equilibrium flows and prices are shown in Tables 4.9 and 4.10, respectively.

As shown in the Computational Appendix to this chapter, the solutions for version of the SPE are unique. The first thing to notice is that by adjusting the slope of the inverse supply function only at node 1, the effects are felt network-wide. Indeed, the total supply-demand balance of 100 units of energy products from before drops to 97.180 due to the higher supply price at node 1 since ρ_1^S increased also increasing all supply prices in the network. These higher prices have a dampening effect on what is sent out from node 1 (previously 20 now only 9.167) with nodes 2 , 3 and 4 increasing their supply as a result. From an energy consumer perspective, the destination prices θ_j all go up due to the higher costs of getting energy products just from node 1. This example shows the important feedback aspects of the complementarity form of the equilibrium problem in which various parts of the model (e.g., prices) can be directly manipulated as compared to optimization.

While the total supply and demand have decrease from 100 to 97.180, they are still equal to each other. This is always the case since $\text{supply}_i = \sum_j x_{ij}$, and $\text{demand}_j = \sum_i x_{ij}$

$$\Rightarrow \sum_i \text{supply}_i = \sum_i \sum_j x_{ij} = \sum_j \sum_i x_{ij} = \sum_j \text{demand}_j$$

Also, note that the version of the SPE considered here is a specialization of (4.29) with $S_r = D_r = 0$ in (4.29b) for all regions r. This concludes the illustrative example.

4.5 Selected Existence/Uniqueness Results for Equilibrium Problems

In this section we describe some basic existence and uniquness results for equilibrium problems and relate them to optimization to highlight the differences. Consider the following generic optimization problem:

$$\text{Minimize}_x \quad f(x) \tag{4.46a}$$

s. t.

$$g_i(x) \le 0, \quad i = 1, \dots, m \quad (\lambda_i) \tag{4.46b}$$

$$h_j(x) = 0, \quad j = 1, \dots, p \quad (\gamma_j) \tag{4.46c}$$

where $f, g_i, h_j : S^F \to R$ where the feasible region $S^F \subseteq R^n$ is defined as $S = \{x | g_i(x) \le 0, i = 1, \dots, m, h_j(x) = 0, j = 1, \dots, p\}$. Clearly such a general formulation includes all the energy production and other optimization problems discussed so far in this chapter. Perhaps the most basic existence result for (4.46) is the Weirstrass Theorem which relies on the notion of a compact set [2].

Definition 4.2. A set $S \subseteq R^n$ is closed if it contains all its limit points. That is, for all convergent sequences $\{x^k\} \subseteq S$, and $\{x^k\} \to \bar{x}$, $\bar{x} \in S$.

Definition 4.3. A set $S \subseteq R^n$ is bounded if there exists a $\beta > 0$ such that $\|x\| \le \beta$ for all $x \in S$.

Definition 4.4. A set $S \subseteq R^n$ is compact if it is both closed and bounded.

All the optimization problem illustrative examples discussed so far in this chapter had closed feasible regions since they were polyhedral. That is, they were of the form $\{x | Ax \le b\}$ for some matrix A and vector b. The boundedness condition however is problem-specific. For example, for the network equilibrium problem (Example 1.7 from Chapter 1), bounds on the primal variables in producer A's optimization problem are:

$$q_1^A \le \bar{q}_1^A$$
$$s_1^A = q_1^A - f_{12}^A \le \bar{q}_1^A$$
$$f_{12}^A \le \bar{g}_{12}$$

where $\beta \ge \{\bar{q}_1^A, \bar{g}_{12}\}$. This boundedness of the feasible region is a key point which reappears in the existence results for MCPs and VIs to be discussed below.

Then, the Weirstrass Theorem [2] can be stated as the following.

Theorem 4.3. *[2] Suppose that S^F is nonempty and compact and the function $f{:}S^F \to R$ is continuous on S^F. Then (4.46) has a solution.*

Another way to consider the existence of a solution to (4.46) is through its KKT conditions: find $(x, \lambda, \gamma) \in R^n \times R^m_+ \times R^p$ such that:

$$0 = \nabla f(x) + \sum_i \nabla g_i(x)\lambda_i + \sum_j \nabla h_j(x)\gamma_j \tag{4.47a}$$

$$0 \leq -g_i(x) \perp \lambda_i \geq 0, i = 1, \ldots, m \tag{4.47b}$$

$$0 = h_j(x), \ \gamma_j \text{ free}, j = 1, \ldots, p \tag{4.47c}$$

Clearly this system includes both the primal and dual variables variables, x and (λ, γ), respectively, taken together. The next two results indicate when the KKT conditions are sufficient or necessary at an optimal solution.

Theorem 4.4. *[2] Let $\bar{x}$ be a feasible solution to (4.46). Suppose that $(\bar{x}, \bar{\lambda}, \bar{\gamma}) \in R^n \times R^m_+ \times R^p$ is a KKT triple, i.e., it solves (4.47). If f, g_i are convex and h_j is an affine function, then $\bar{x}$ is a global optimum to (4.46).*[9]

As for uniqueness of the solution x, the next theorem presents a simplified result from the more general version in [2].

Theorem 4.5. *[2] Suppose that the function f is strictly convex. Consider (4.46) where S^F is assumed convex . Then, if $\bar{x}$ is a local optimal solution, then it is the unique global optimal solution.*

It is illustrative to relate these existence and uniqueness results to the energy examples shown above. Consider Example 4.3.1.1 in which a single energy producer is trying to maximize its profit. Here $S^F = \{q | 0 \leq q \leq q^{\max}\}$ where $q^{\max} > 0$. Since $0 \in S^F$, clearly this set is nonempty. Moreover S^F is closed since it is polyhedral, i.e., the intersection of a finite number of linear inequalities. It is also clearly bounded with $\beta \geq q^{\max}$, hence S^F is compact. The objective function $f(q) = pq - \gamma q = (p - \gamma)\, q$ is continuous in q since it is linear. Hence, by the Weirstrass Theorem, there must be a solution which previously was shown to be equal to $q^{\max}$. The objective function is convex since it is linear but not strictly convex. So the above theorem cannot be directly used. It is important to note that this theorem only provides sufficient conditions for uniqueness, not necessary ones as evidenced in this example.

The KKT conditions to this problem are to find q, λ such that:

$$0 \leq \gamma - p + \lambda \perp q \geq 0$$

[9] Note that Theorem 4.3.8 in [2] states this result in a slightly more general context. Also, the KKT conditions are necessary for (4.46) as long as certain constraint qualifications are present [2].

$$0 \leq q^{\max} - q \perp \lambda \geq 0$$

Since f and g are both convex and there are no equality constraints to consider, a solution to these KKT conditions solves the original problem (4.15) [2]]. There are four cases to consider:

Case 1: $q = 0, \lambda = 0$
Case 2: $q > 0, \lambda = 0$
Case 3: $q = 0, \lambda > 0$
Case 4: $q > 0, \lambda > 0$

Under Case 1, we must have $p \leq \gamma$ but this violates the assumption made above that $p > \gamma$ so this case is not possible. Under Case 2, by complementarity $p = \gamma$ which is not possible for the same reason. Under Case 3, by complementarity $0 < q^{\max} = q = 0$ which is a contradiction. Only Case 4 is possible for which the unique solution is: $q = q^{\max} > 0$, $\lambda = p - \gamma > 0$. Note that both the primal and dual variables are considered with this KKT approach.

To motivate a similar analysis but for complementarity and variational inequality problems, lets consider this simple energy production problem recast with the following formulation. A VI form of this problem would be to take

$$F^{VI,1}(q) - \nabla f(q) - \gamma - p \text{ and } X = \{q | q \in [0, q^{\max}]\} \tag{4.48}$$

so that the corresponding variational inequality problem would be:

$$(\gamma - p)(q - q^*) \geq 0, \forall q \in X \tag{4.49}$$

where q^* is a solution to this problem. Clearly, $q^{\max} \in X$ which implies that $(\gamma - p)(q^{\max} - q^*) \geq 0$ but since $\gamma - p < 0$ this means that $q^{\max} = q^*$. Note that in this case $F^{VI,1}(q) = \nabla f(q)$ is continuous and X is nonempty, compact, and convex. This example then is motivation for the general result shown in Theorem 4.6 below.

An alternative form for the VI is to take the KKT conditions and form the associated variational inequality from them. In this case, $X = R^2_+$ and

$$F^{VI,2}\begin{pmatrix} q \\ \lambda \end{pmatrix} = \begin{pmatrix} 0 & 1 \\ -1 & 0 \end{pmatrix}\begin{pmatrix} q \\ \lambda \end{pmatrix} + \begin{pmatrix} \gamma - p \\ q^{\max} \end{pmatrix} \tag{4.50}$$

$F^{VI,2}$ is a continuous mapping from $R^2 \to R^2$, since its linear but $X = R^2_+$ is only closed and convex but not bounded. Hence, Theorem 4.6 cannot be used. This is not to say that this refutes the earlier existence result but only that the premise of this theorem doesn't apply for this second version of the VI form of the problem.

The existence and uniqueness conditions for MCPs and VIs are a generalization of those stated for the optimization problem (4.46). One of the more basic existence results for the VI is the following [32]:

Theorem 4.6. *[14], [33] Let X be a nonempty, compact and convex subset of R^n and let F be a continuous mapping from X to R^n. Then, there exists a solution to the problem VI(X, F).*

Theorem 4.6 does not assert anything about uniqueness of the solution, only when the VI has a solution. As noted in [32], the boundedness part of the compactness assumption on the set X is essential and for the MCP is not generally guaranteed, an example of which was just shown. To show uniqueness we can introduce several important definitions for the vector-valued functon F that appears in the VI and the MCP; these all can be found in [46] as noted in [32] which provides many other useful functional definitions for equilibrium problems. For each of the following definitions, the function $F : R^n \to R^n$ is given. From an intuitive point of view, the notions of vector-valued monotonicity described below are variations of their scalar-valued analogs. In the latter case, a scalar-valued function is monotone if it is always non-decreasing or always non-increasing. A strictly monotone scalar-valued function is a more restrictive version but allows for things like inverses to be established and corresponds to always strictly increasing or always strictly decreasing. When dealing with vector-valued functions that have more than one component, clearly the same definition cannot be used. For example, if the function is always increasing in one component but not so in the other ones then the scalar-valued definition will not make sense. Thus, a generalization of these notions is needed and that is what is provided below.

Definition 4.5. F is monotone over a set X if $\left[F(x) - F(y)\right]^T (x - y) \geq 0, \forall x, y \in X$.

Monotonicity of a vector-valued function is analogous to convexity of a scalar-valued function $f : R^n \to R$.

Consider $F^{VI,2}$ relative to the monotonicity definition. Letting $x = \left(q^1\, \lambda^1\right)^T$,$y = \left(q^2\, \lambda^2\right)^T$:

$$\left[F(x) - F(y)\right]^T (x - y) \geq 0, \forall x, y \in X = R^2_+ \Leftrightarrow$$

$$\left[\begin{pmatrix} 0 & 1 \\ -1 & 0 \end{pmatrix}\begin{pmatrix} q^1 \\ \lambda^1 \end{pmatrix} + \begin{pmatrix} \gamma - p \\ q^{\max} \end{pmatrix} - \begin{pmatrix} 0 & 1 \\ -1 & 0 \end{pmatrix}\begin{pmatrix} q^2 \\ \lambda^2 \end{pmatrix} - \begin{pmatrix} \gamma - p \\ q^{\max} \end{pmatrix}\right]^T \begin{pmatrix} q^1 - q^2 \\ \lambda^1 - \lambda^2 \end{pmatrix}$$

$$= \begin{pmatrix} q^1 - q^2 \\ \lambda^1 - \lambda^2 \end{pmatrix}^T \begin{pmatrix} 0 & 1 \\ -1 & 0 \end{pmatrix}^T \begin{pmatrix} q^1 - q^2 \\ \lambda^1 - \lambda^2 \end{pmatrix} \geq 0, \forall q^1, q^2, \lambda^1, \lambda^2 \in R_+$$

Now consider an affine $F = q + Mx$ with $X = R^n$ which relates to the linear and quadratic programming examples from this chapter (though expressed as MCPs); here M is a matrix and q is a vector. It is important to see what these definitions mean. First, monotonicity of such an F means:

$$\left[q + Mx - (q + My)\right]^T (x - y) \geq 0, \forall x, y \in R^n$$

$$\Leftrightarrow [M(x-y)]^T (x-y) \geq 0, \forall x, y \in R^n$$
$$\Leftrightarrow (x-y)^T M^T (x-y) \geq 0, \forall x, y \in R^n$$

which means that M^T is a positive semi-definite (psd) matrix hence so is M. This is because M^T is psd if and only if $(M + M^T)$ is since $0 \leq z^T (M + M^T) z = 2z^T M^T z = 2z^T M z$ for all z [12]. But note that from the example above $M = \begin{pmatrix} 0 & 1 \\ -1 & 0 \end{pmatrix}$ is psd since $M + M^T = \begin{pmatrix} 0 & 0 \\ 0 & 0 \end{pmatrix}$ which is psd and by the above argument. Thus, $F^{VI,2}$ is a monotone function over $X = R^2_+$. A more restrictive definition that leads to stronger results is the next one.

Definition 4.6. F is strictly monotone over a set X if $[F(x) - F(y)]^T (x-y) > 0, \forall x, y \in X, x \neq y$.

In the affine case, if F is strictly monotone, then:

$$[q + Mx - (q + My)]^T (x-y) > 0, \forall x, y \in R^n \quad, x \neq y$$
$$\Leftrightarrow (x-y)^T M^T (x-y) > 0, \forall x, y \in R^n \quad, x \neq y$$

which means that M^T hence M (by similar reasoning) is positive definite. An even more restrictive yet stronger definition is the following.

Definition 4.7. F is strongly monotone over a set X if there exists an $\alpha > 0$ such that $[F(x) - F(y)]^T (x-y) \geq \alpha \|x-y\|^2, \forall x, y \in X$.

Also, strong monotonicity would lead to the same conclusion in this affine case, namely that M is positive definite. Another definition concerns the notion of coercivity.

Definition 4.8. F is coercive with respect to X if there exists a vector $x^0 \in X$ such that $\lim_{x \in X, \|x\| \to \infty} \frac{F(x)^T (x - x^0)}{\|x\|} = \infty$ for any vector norm $\|\cdot\|$ in R^n.

An important result concerning the uniqueness of solutions to the VI is the following (Corollary 3.2, [32]) :

Theorem 4.7. *Let X be a nonempty, closed, convex subset of R^n and let F be a continuous mapping from X into R^n. If F is strongly monotone with respect to X, then there exists a unique solution to the problem VI(X, F).*

If the function F is only strictly monotone, then the associated VI is only guaranteed to have at most one solution; it may not have any solutions as noted in [32].

It is important to consider the function F derived from the KKT conditions to optimization problems of the form (4.46) but with $f(x) = \frac{1}{2} x^T Q x + c^T x$ (Q symmetric and psd) and $g(x) = Ax - b, h(x) = Dx - d$, i.e., a quadratic program, such functions will never be strictly nor strongly monotone. To see this note that the KKT conditions (4.47) can be expressed as the affine VI with $X = R^n \times R^m_+ \times R^p$ where

$$F(x,\lambda,\gamma) = \begin{pmatrix} Qx + c + A^T\lambda + D^T\gamma \\ b - Ax \\ d - Dx \end{pmatrix} = \begin{pmatrix} Q & A^T & D^T \\ -A & 0 & 0 \\ -D & 0 & 0 \end{pmatrix} \begin{pmatrix} x \\ \lambda \\ \gamma \end{pmatrix} + \begin{pmatrix} c \\ b \\ d \end{pmatrix}$$

Let $x = \left(\left(x^1\right)^T \left(\lambda^1\right)^T \left(\gamma^1\right)^T \right)^T, y = \left(\left(x^2\right)^T \left(\lambda^2\right)^T \left(\gamma^2\right)^T \right)^T$ be two vectors in $R^n \times R^m \times R^p$. Then,

$$\begin{aligned} &[F(x) - F(y)]^T (x - y) \\ &= \begin{pmatrix} x^1 - x^2 \\ \lambda^1 - \lambda^2 \\ \gamma^1 - \gamma^2 \end{pmatrix}^T \begin{pmatrix} Q & A^T & D^T \\ -A & 0 & 0 \\ -D & 0 & 0 \end{pmatrix}^T \begin{pmatrix} x^1 - x^2 \\ \lambda^1 - \lambda^2 \\ \gamma^1 - \gamma^2 \end{pmatrix} \\ &= \left(x^1 - x^2\right)^T Q^T \left(x^1 - x^2\right) + \left(x^1 - x^2\right)^T \left(-A^T\right) \left(\lambda^1 - \lambda^2\right) \\ &\quad + \left(x^1 - x^2\right)^T \left(-D^T\right) \left(\gamma^1 - \gamma^2\right) + \left(\lambda^1 - \lambda^2\right)^T A \left(x^1 - x^2\right) \\ &\quad + \left(\gamma^1 - \gamma^2\right)^T D \left(x^1 - x^2\right) \\ &\not> 0 \text{ for } x \neq y \end{aligned}$$

To see this let $x^1 = x^2$ but $\lambda^1 \neq \lambda^2$ or $\gamma^1 \neq \gamma^2$ so that $[F(x) - F(y)]^T (x - y) = 0$. Thus, requiring strict or strong monotonicity on optimization-derived KKT VIs (or corresponding MCPs) is too restrictive in general.

There are also results that are specific to complementarity problems two of which are given below. The first one is due to Cottle ([10], [11]).

Theorem 4.8. *[Theorem 3.6, [32]] Let $F : R^n_+ \to R^n$ be continuously differentiable and suppose that there exists a $\delta \in (0,1)$ such that all principal minors of the Jacobian matrix $JF(x)$ are bounded between δ and δ^{-1} for all $x \in R^n_+$. Then the NCP(F) has a unique soluton.*

If we consider the (pure) linear complementarity problem derived from Example 4.3.1.1, i.e., (4.50), the Jacobian of the associated function F is just the matrix:

$$A = \begin{pmatrix} 0 & 1 \\ -1 & 0 \end{pmatrix}$$

Its principal minors by definition are the determinants of its principal submatrices (i.e., $A_{\alpha\alpha}$ where α is a subset of the rows (columns) [12]). In this case there are three principal submatrices:

$$A_{\{1\}\{1\}} = (0), A_{\{2\}\{2\}} = (0), A_{\{1,2\}\{1,2\}} = A$$

with corresponding principal minors: $0, 0, 1$, respectively. Thus, the above result is not directly applicable given that at least one of the principal minors equals zero. However, if we return to the energy production duopoly from

Chapter 1 (Illustrative Example 1.2.4), we see that the Jacobian of the related function F is given by:

$$A = \begin{pmatrix} 2\beta & \beta \\ \beta & 2\beta \end{pmatrix} \tag{4.51}$$

with three principal submatrices:

$$A_{\{1\}\{1\}} = (2\beta)\,, A_{\{2\}\{2\}} = (2\beta)\,, A_{\{1,2\}\{1,2\}} = A$$

with principal minors: $2\beta, 2\beta, 3\beta^2$ which are all positive (given that $\beta > 0$). For example, if $\beta \geq 1$, then we can take $\delta = \frac{1}{3\beta^2} \in (0,1)$ with

$$\delta^{-1} = 3\beta^2 \geq 2\beta > \frac{1}{2\beta} > \frac{1}{3\beta^2} = \delta$$

so that the theorem above applies. In fact, Example 1.4 is a linear complementarity problem so that stronger existence and uniqueness results can be obtained. Since $\beta > 0$, the above analysis shows that all principal minors are positive so that it's a P-Matrix [12]. Theorem 3.3.7 from this reference indicates that there is a unique solution in fact not just for the constant vector

$$q = \begin{pmatrix} -\alpha + \gamma_1 \\ -\alpha + \gamma_2 \end{pmatrix}$$

but for all $q \in R^2$. For more details on existence and uniqueness results for equilibrium problems, the interested reader can consult [32], [12]. and [15].

4.6 Extensions to Equilibrium Problems: Non-Convex Side Constraints and Incorporating Stochasticity

4.6.1 Overview

So far we have considered equilibrium problems expressed as MCPs, VIs, or QVIs. These problems inherently possess non-convexities via complementarity or other similar conditions. Additionally, for the purposes of more realistic models, there may be other non-convex restrictions to impose beyond the standard ones. In this section, we review some of these extensions that have a bearing on energy markets as well as considering stochastic aspects to equilibrium problems. We first consider a variation on the spatial price equilibrium formulation presented earlier.

4.6.1.1 Illustrative Example. Integer-Constrained Spatial Price Equilibrium

As motivation, consider an extension of the SPE version 2 (Example 4.4.3.1) but with integer restrictions. Specifically, we wish to find flows x_{ij} such that:

$$0 \leq \psi_i \left(\sum_j x_{ij} \right) + c_{ij}(x) - \theta_j \left(\sum_i x_{ij} \right) \perp x_{ij} \geq 0, \forall i, j \tag{4.52}$$

where additionally we want

$$x_{ij} \in Z_+, \; \forall (k,l) \in D \subseteq \{(i,j) \,| i = 1,2,3,4; j = 1,2,3,4,5\}.$$

Essentially, we want an SPE of energy products where a subset of the flows indexed by D (possibly all of them) are integer-valued. Such a restriction might relate to practical considerations such as only a certain, discrete number of truckloads of coal.

From Example 4.4.3.1 we know that the solution was unique but did include some non-integer flows (e.g., $x_{21} = 19.167$). Thus, clearly imposing integer restrictions on top of an equilibrium solution might be infeasible. In that case, coming up with a compromise solution which relaxed integrality, complementarity or both would be one approach and was the subject of [19] to be discussed in more detail below. First we present the general version of a discretely-constrained mixed linear complementarity problem (DC-MLCP) as a unifying formulation [19].

4.6.2 Discretely-Constrained Mixed Linear Complementarity Problem

Given the vector $q = \begin{pmatrix} q_1 \\ q_2 \end{pmatrix}$ and matrix $A = \begin{pmatrix} A_{11} & A_{12} \\ A_{21} & A_{22} \end{pmatrix}$, the DC-MLCP is to find $z = (z_1, z_2) \in R^{n_1} \times R^{n_2}$ such that:

$$0 \leq q_1 + \begin{pmatrix} A_{11} & A_{12} \end{pmatrix} \begin{pmatrix} z_1 \\ z_2 \end{pmatrix} \perp z_1 \geq 0 \tag{4.53a}$$

$$0 = q_2 + \begin{pmatrix} A_{21} & A_{22} \end{pmatrix} \begin{pmatrix} z_1 \\ z_2 \end{pmatrix}, z_2 \text{ free} \tag{4.53b}$$

$$(z_1)_c \in R_+, \; c \in C_1, (z_1)_d \in Z_+, d \in D_1 \tag{4.53c}$$

$$(z_2)_c \in R, \; c \in C_2, (z_2)_d \in Z, d \in D_2 \tag{4.53d}$$

where $C_k \cup D_k$ is a partition of the indices for $z_k, k = 1,2$ into continuous-valued and discrete-valued variables, respectively. Note that the complementarity relationship and nonnegativity for z_1 (4.53a) can be reformulated via disjunctive constraints [16] in the following way:

$$0 \leq q_1 + \begin{pmatrix} A_{11} & A_{12} \end{pmatrix} \begin{pmatrix} z_1 \\ z_2 \end{pmatrix} \leq M_1 (u) \tag{4.54a}$$

$$0 \leq z_1 \leq M_1 (1 - u) \tag{4.54b}$$

$$u_j \in \{0, 1\}, \forall j \tag{4.54c}$$

where M_1 is a suitably large, positive constant and u is a vector of binary variables. As proposed in [19], the following mixed-integer linear program can be used to solve the DC-MLCP with $\sigma = (\sigma_1, \ldots, \sigma_{n_1})^T$ and $\varepsilon_{1ri}, \varepsilon_{2rj}$ representing deviations from complementarity and integrality, respectively. The integrality approximation comes from the constraints (4.55f) and (4.55g) which target the integers $i \in \{0, 1, \ldots, N\}$ and $j \in \{-N_1, \ldots, -1, 0, 1, \ldots, N_2\}$, respectively for $(z_1)_r, r \in D_1$ and $(z_2)_r, r \in D_2$. In this mixed-integer, linear program there are two positive constants, M_1 and M_2 that need to be computed. The former one relates to the relaxation of complementarity and is generally application-specific but can be obtained by trial and error or problem-specific information (see [22] and Example 4.6.2.1 below). The other constant relates to the relaxation of integrality and the first theorem below shows valid values for M_2.

Minimize

$$\omega_1 \left[\sum_{r \in D_1} \sum_{i=0}^{N} (\varepsilon_{1ri})^+ + (\varepsilon_{1ri})^- + \sum_{r \in D_2} \sum_{i=0}^{N} (\varepsilon_{2ri})^+ + (\varepsilon_{2ri})^- \right] + \omega_2 \left[1^T \sigma \right] \tag{4.55a}$$

s. t.

$$0 \leq q_1 + (A_{11} \; A_{12}) \begin{pmatrix} z_1 \\ z_2 \end{pmatrix} \leq M_1 (u) + M_1 \sigma \tag{4.55b}$$

$$0 \leq z_1 \leq M_1 (1 - u) + M_1 \sigma \tag{4.55c}$$

$$0 = q_2 + (A_{21} \; A_{22}) \begin{pmatrix} z_1 \\ z_2 \end{pmatrix} \tag{4.55d}$$

$$u_j \in \{0, 1\}, \forall j \tag{4.55e}$$

$$- M_2 (1 - w_{1ri}) \leq (z_1)_r - i - \varepsilon_{1ri} \leq M_2 (1 - w_{1ri}), \quad i = 0, 1, \ldots, N, \quad r \in D_1 \tag{4.55f}$$

$$- M_2 (1 - w_{2ri}) \leq (z_2)_r - i - \varepsilon_{2ri} \leq M_2 (1 - w_{2ri}), \quad i = -N_1, \ldots, -1, 0, 1, \ldots, N_2, \quad r \in D_2 \tag{4.55g}$$

$$\varepsilon_{1ri} = (\varepsilon_{1ri})^+ - (\varepsilon_{1ri})^-, \quad i = 0, 1, \ldots, N, \quad r \in D_1 \tag{4.55h}$$

$$\varepsilon_{2ri} = (\varepsilon_{2ri})^{+} - (\varepsilon_{2ri})^{-}, \quad i = 0, 1, \ldots, N, \quad r \in D_1 \tag{4.55i}$$

$$\sum_{i=0}^{N} w_{1ri} = 1 \tag{4.55j}$$

$$\sum_{i=0}^{N} w_{2ri} = 1 \tag{4.55k}$$

$$w_{1ri} \in \{0,1\}, \quad i = 0, 1, \ldots, N, \quad r \in D_1 \tag{4.55l}$$

$$w_{2ri} \in \{0,1\}, \quad i = 0, 1, \ldots, N, \quad r \in D_2 \tag{4.55m}$$

$$\sigma \geq 0 \tag{4.55n}$$

$$(\varepsilon_{1ri})^{+}, (\varepsilon_{1ri})^{-} \geq 0, \quad i = 0, 1, \ldots, N, \quad r \in D_1 \tag{4.55o}$$

$$(\varepsilon_{2ri})^{+}, (\varepsilon_{2ri})^{-} \geq 0, \quad i = 0, 1, \ldots, N, \quad r \in D_2 \tag{4.55p}$$

Theorem 4.9. *[19] Let $M_2 \geq \max\{N, N_1 + N_2\}$. Then, this value will be valid for the constraints (4.55f) and (4.55g).*

To prove when (4.55) will have a solution, the following assumption is made in which the discrete restrictions are dropped to form the relaxation of the original DC-MLCP (4.53).

Assumption 1 *Define the set*

$$S = \left\{ (z_1, z_2) \,|\, 0 \leq q_1 + \begin{pmatrix} A_{11} & A_{12} \end{pmatrix} \begin{pmatrix} z_1 \\ z_2 \end{pmatrix}, 0 = q_2 + \begin{pmatrix} A_{21} & A_{22} \end{pmatrix} \begin{pmatrix} z_1 \\ z_2 \end{pmatrix}, z_1 \geq 0 \right\}$$

Then, assume that:

i. S is nonempty and
ii. There exists a constant M^* such that $M^* \geq \max\{\|z_1\|_\infty, \|z_2\|_\infty\} = \left\| \begin{pmatrix} z_1 \\ z_2 \end{pmatrix} \right\|_\infty$ for all $(z_1, z_2) \in S$.

Theorem 4.10. *[19] If Assumption 1 holds, and $M_2 \geq \max\{N, N_1 + N_2\}$ then problem (4.55) always has a solution.*

4.6.2.1 Illustrative Example. Integer-Constrained Network Equilibrium

We return to the energy network equilibrium problem (Example 1.2.7) initially discussed in Chapter 1. In this example, one can explicitly show what the value of M_1 should be by examining the problem specifics and based on a coercivity-type condition [15] on the demand functions and also related to the needed compactness stated in the earlier existence theorems.

Theorem 4.11. *[19]. Assume that the demand functions $D_n(\pi_n)$ satisfy the following condition:*

Table 4.11 Description of problem variations

Variation	σ–Complementarity?	ϵ–Integrality?	Description
1	No	No	MLCP
2	No	No	Integer Variables
3	Yes	No	Integer Variables
4	Yes	No	Continuous Variables
5	No	Yes	Continuous Variables
6	Yes	Yes	Continuous Variables

$$[|D_n(\pi_n)| \text{ is bounded} \Rightarrow |\pi_n| \text{ is bounded}], n = 1, 2$$

Then, there exists a constant M^* for the network equilibrium problem that satisfies Assumption 1.

In the case of linear demand, a specific value of M_1 can be obtained as shown in [19].

To analyze the relaxation of either complementarity, integrality or both, several variations were used as indicated in Table 4.11 with specific parameter values described in [19].

Variation 1 corresponds to solving the MCP without any integer restrictions (e.g., just continuous variables). Variation 2 involved converting the MCP to a formulation with disjunctive constraints but restricting the variables of production and sales to be integer. Variation 3 relaxes complementarity but still enforces integrality for the relevant variables. Variation 4 is the continuous analog of Variation 3. Variation 5 allows only for relaxing integrality and lastly, Variation 6 allows for relaxing both integrality and complementarity.

The sales, $s_1^A, s_1^B, s_2^C, s_2^D$ and production quantities $q_1^A, q_1^B, q_2^C, q_2^D$ are the variables that are integer-constrained in Variations 2 and 3 with the objective of getting integer values for these variables in a given solution (there are multiple integer solutions). The values of M_1, and M_2 were set equal to 100 but since there are multiple solutions, different values of M_1 may lead to different solutions.

Table 4.12 from [19] shows the results and indicate that Variations 3 and 6 give an integer solution. An interesting observation is that Variation 6 targets integers via ε-complementarity while Variation 3 directly enforces integer values. As might be expected, Variation 1 produced a non-integer solution and Variation 2 was infeasible. These results indicate that σ-complementarity is crucial for a feasible integer solution (cf. Variations 3 and 6). However, just σ-complementarity by itself is not enough to get integer solutions (Variation 4) nor is only ε-complementarity (Variation 5).

Table 4.12 Network example: Results

Variations	1	2	3	4	5	6
s_1^A	7.440	Infeasible	8.000	8.000	8.000	8.000
s_1^B	0.560	Infeasible	0	0	0	0
s_2^C	4.500	Infeasible	4.000	4.500	4.500	4.000
s_2^D	0	Infeasible	0	0	0	0
q_1^A	10.000	Infeasible	10.000	10.000	10.000	10.000
q_1^B	3.000	Infeasible	3.000	3.000	3.000	3.000
q_2^C	4.500	Infeasible	4.000	4.500	4.500	4.000
q_2^D	0	Infeasible	0	0	0	0
f_{12}^A	2.560	Infeasible	2.000	2.000	2.000	2.000
f_{12}^B	2.440	Infeasible	3.000	3.000	3.000	3.000
λ_1^A	2.000	Infeasible	2.000	2.000	2.000	2.000
λ_1^B	0	Infeasible	0	0	0	0
λ_2^C	0.250	Infeasible	0.500	0.250	0.250	0.500
λ_2^D	0	Infeasible	0	0	0	0
g_{12}	5.000	Infeasible	5.000	5.000	5.000	5.000
ϵ_{12}	2.250	Infeasible	2.500	2.250	2.250	2.500
δ_1^A	12.000	Infeasible	12.000	12.000	12.000	12.000
δ_2^B	12.000	Infeasible	12.000	12.000	12.000	12.000
δ_2^C	15.250	Infeasible	15.500	15.250	15.250	15.500
δ_2^D	17.581	Infeasible	18.000	18.000	18.000	18.000
π_1	12.000	Infeasible	12.000	12.000	12.000	12.000
π_2	15.250	Infeasible	15.500	15.250	15.250	15.500
τ_{12}	2.750	Infeasible	3.000	2.750	2.750	3.000
Sum σ	n/a	n/a	0.005	0	n/a	0.005
Sum ϵ	n/a	n/a	n/a	n/a	1.000	0

4.6.3 Stochastic Equilibria

In an equilibrium problem (MCP or VI), in which multiple players are competing against each other for market or system resources, one can expand on the examples cited above (e.g., Example 1.2.7) to include stochasticity. This uncertainty may be in the form of random demand, capacity, costs or the like. An interesting problem is to see how incorporating stochasticity might adjust equilibrium solutions. Such an approach is the analog of mixed (bimatrix) games presented earlier in this chapter in which two energy producers sought to select the best probabilities for choosing production levels (see Example 4.2.4.2). The problem of solving a stochastic MCP or VI has been mostly unstudied until recently. Some earlier important works in this area include: [36] who considered a stochastic dynamic Nash-Cournot model for the European gas market; [13] who presented a stochastic Stackelberg game for the European gas market; [29] who considered the European gas market but from a simulation, sample-path perspective; [3] who analyzed the stochastic MCP/VI with a gradient-based, quasi-Monte Carlo simulation technique.

More recently, Haurie and Moresino [34] and Haurie and Zaccour [35] considered S-adapted equilibria under uncertainty; Cabero et al. [6], [7] developed an approach for solving stochastic complementarity problems in electricity basedon on a Benders decomposition approach [4]; Shanbhag et al. [48], [49] presented a matrix splitting-based decomposition method for a stochastic complementarity problem in electricity; Genc et al. [28] analyzed oligopolistic markets under uncertainty; Zhuang and Gabriel [54] who came up with and analyzed a natural gas market equilibrium expressed as a stochasic complementarity problem and Gabriel et al. [25] who proposed scenario-reduction methods for solving stochastic complementarity problems in energy; Gabriel and Fuller [20] who developed a Benders variant for solving stochastic VIs and MCPs and applied it to a power market equilibrium problem.

4.6.3.1 Generator *f*'s Problem

As a representative energy example, we present below a stochastic MCP by Gabriel and Fuller [20] based on the earlier work by Hobbs [37]. As originally presented in [37], we consider an electric power market model depicting the decisions of generating firms (f), a grid operator, and market-clearing conditions. The firms compete with each other in a non-cooperative Nash-Cournot form in which a linear, inverse demand function is given and is of the form $a - b\,$(total sales) where $a, b > 0$. The competition occurs in that each firm can adjust its production to lower total sales and hence raise the market price to ultimately maximize its own expected profit (5.30). This optimization is done separately for each generator while keeping both slow (5.30b) and rapid ramping (5.30c) limits within their bounds as well as an equality between sales and generation (5.30d). Here, no power losses are considered but this feature could easily be added. Each player thus solves a convex, quadratic program shown below in which the intercept of the demand curve at node i, a_{im} is allowed to be random. The index m refers to the scenario which occurs with probabiity $prob(i,m)$ at node i.

$$
\begin{aligned}
&\text{Maximize}_{s_f, x_f, r_f} \\
&\quad \sum_m \sum_i prob(i,m) \left(a_{im} - b_i \left(\sum_g s_{gim} \right) - w_{im} \right) s_{fim} \\
&\quad - \sum_m \sum_{i,h} prob(i,m) \left(C_{fih} - w_{im} \right) x_{fih} \\
&\quad - \sum_m \sum_i prob(i,m) \left(RC_{fih} - w_{im} \right) r_{fihm} \qquad (4.56a)
\end{aligned}
$$

s. t.

$$x_{fih} - X_{fih} \le 0, \quad \forall i, h \quad (\rho_{fih}) \qquad (4.56b)$$

$$r_{fihm} - R_{fih} \le 0 \quad \forall i, m \quad (\sigma_{fihm}) \qquad (4.56c)$$

$$\sum_{i} s_{fim} - \sum_{i,h} x_{fih} - \sum_{i,h} r_{fihm} = 0, \quad \forall m \quad (\theta_{fm}) \tag{4.56d}$$

$$x_{fih} \geq 0, \quad \forall i, h \tag{4.56e}$$

$$r_{fihm}, s_{fim} \geq 0, \quad \forall i, h, m \tag{4.56f}$$

The primal variables are:

- s_{fim} is sales for firm f at node i for scenario m
- x_{fih} is slow-ramping generation for firm f at node i for generator type h
- r_{fihm} is rapid-ramping generation for firm f at node i for generator type h for scenario m

The dual variables are:

- ρ_{fih} for the generation capacity constraint (5.30b)
- σ_{fihm} for the rapid generation capacity constraint (5.30c)
- θ_{fm} for the sales and generation energy balance constraint (5.30d)

and

- w_{im} is the exogenous wheeling fee (an endogenous variable in the overall equilibrium)
- a_{im},b_i are coefficients for the inverse demand function
- C_{fih}, RC_{fih} are the marginal costs of power generation for respectively, slow-ramping and rapid-ramping generation
- X_{fih},R_{fih} are the generation capacities for respectively, slow-ramping and rapid-ramping generation

The KKT conditions for firm f are both necessary and sufficient in this case [2] and are the following:

$$0 \leq prob(i,m)\left[-a_{im} + b_i\left(s_{fim+} \sum_{g} s_{gim}\right) + w_{im}\right] + \theta_{fm} \perp s_{fim} \geq 0,$$
$$\forall f, i, m \tag{4.57a}$$

$$0 \leq C_{fih} - \sum_{m} prob(i,m) w_{im} - \sum_{m} \theta_{fm} + \rho_{fih} \perp x_{fih} \geq 0 \quad \forall f, i, h \tag{4.57b}$$

$$0 \leq prob(i,m) RC_{fih} + \sigma_{fihm} - prob(i,m) w_{im} - \theta_{fm} \perp r_{fihm} \geq 0,$$
$$\forall f, i, h, m \tag{4.57c}$$

$$0 \leq X_{fih} - x_{fih} \perp \rho_{fih} \geq 0 \quad \forall f, i, h \tag{4.57d}$$

$$0 \leq R_{fih} - r_{fihm} \perp \sigma_{fihm} \geq 0 \quad \forall f, i, h, m \tag{4.57e}$$

$$0 = \sum_{i} s_{fim} - \sum_{i,h} x_{fih} - \sum_{i} r_{fihm}, \quad \theta_{fm} \text{ free} \quad \forall f, m \tag{4.57f}$$

4.6.3.2 Grid Owner's Problem

In addition to the generators competing at each of the nodes, there is a grid operator whose job it is to keep the network running smoothly. There are many possible objectives for such a market player, for example maximizing social welfare, minimizing total costs, but for ease of presentation with some realism, the grid operator is modeled as maximizing expected wheeling (transportation) fees subject to obeying line limits. The parameter $PTDF_{il}$ is the amount of power flow on line l for each unit of power injection y_{im} at node i.[10] The resulting linear program is thus:

$$\text{Maximize}_y \quad \sum_m \sum_i prob(i,m) w_{im} y_{im} \tag{4.58a}$$

s. t.

$$-T_{l-} - \sum_i PTDF_{il} y_{im} \leq 0, \quad \forall l, m \quad (\lambda_{lm-}) \tag{4.58b}$$

$$-T_{l+} + \sum_i PTDF_{il} y_{im} \leq 0, \quad \forall l, m \quad (\lambda_{lm+}) \tag{4.58c}$$

Note that the primal variable are: y_{im} which is the inflow (if positive) or the outflow (if negative) at node i for scenario m; the dual variables are $\lambda_{lm-}, \lambda_{lm+}$ corresponding to the lower and upper line limits, respectively. The wheeling fees, w_{im} are exogenous to the grid owner but are variables in the larger MCP as will be indicated below. Since (5.31) is a linear program, the KKT conditions are both necessary and sufficient [2] and are the following:

$$0 = -prob(i,m) w_{im} - \sum_l PTDF_{il} \lambda_{lm-} + \sum_l PTDF_{il} \lambda_{lm+},$$

$$y_{im} \text{ free} \quad \forall i, m \tag{4.59a}$$

$$0 \leq T_{l-} + \sum_i PTDF_{il} y_{im} \perp \lambda_{lm-} \geq 0 \quad \forall l, m \tag{4.59b}$$

$$0 \leq T_{l+} - \sum_i PTDF_{il} y_{im} \perp \lambda_{lm+} \geq 0 \quad \forall l, m \tag{4.59c}$$

4.6.3.3 Market Clearing

Lastly, the following market-clearing conditions are also included:

[10] The acronym PTDF is commonly used for Power Transfer Distribution Factor.

$$0 = -\sum_{f} s_{fim} + \sum_{f,h} x_{fih} + \sum_{f} r_{fihm} + y_{im} \tag{4.60}$$

with free dual variable $\hat{w}_{im} \equiv prob(i,m)w_{im}$.

Note that the wheeling fees w_{im} take the values that would be observed if scenario m were realized; they are not expected values, i.e., they are not scaled by probabilities $prob(i,m)$ but need to be consistent with the units in the other parts of the system being modeled. This is the rationale for the scaling of the variable $\hat{w}_{im} \equiv prob(i,m)w_{im}$.

Putting together the KKT conditions (4.57) and (4.59) as well as the market-clearing (5.32) produces a potentially very large linear complementarity problem if the number of scenarios $|M|$ is big.

4.7 Summary

In this chapter we have discussed the connection between optimization and equilibrium problems. The latter are seen to generalize optimization and include: complementarity, variational inequality, and quasi-variational inequality formulations. Optimization as examples of equilibrium were discussed relative to Weak and Strong Duality theorems for both linear and convex programs. The Principle of Symmetry was seen to be another concept that tied together these two classes of problems. Lastly, variations on the standard equilibrium problems in which certain non-convexities or stochastic elements was introduced to show the wide range of possible extensions.

4.8 Appendix: Computational Issues for Selected Problems

4.8.1 Computation of Nash Equilibrium Based on the Range for the Parameters

We are given $f_1(q_1,q_2) = -\beta q_1^2 + (\alpha - \gamma_1 - \beta q_2)q_1$ and $f_2(q_1,q_2) = -\beta q_2^2 + (\alpha - \gamma_2 - \beta q_1)q_2$. To investigate the range on the parameters β given $\alpha = 100, \gamma_1 = 10, \gamma_2 = 5$, and the resulting Nash equilibria, consider the payoff matrix given below:

$q_1 \backslash q_2$	0	10	20
0	(\$0, \$0)	(\$0, A)	(\$0, B)
10	(C, \$0)	(D, E)	(F, G)
20	(H, \$0)	(I, J)	(K, L)

(4.61)

First we compute the values for $A, \dots, L$.

For Player 1:

- $C = \left[-\beta q_1^2 + (\alpha - \gamma_1 - \beta q_2)q_1\right]_{\alpha=100,\gamma_1=10,q_1=10,q_2=0} = 900 - 100\beta$
- $D = \left[-\beta q_1^2 + (\alpha - \gamma_1 - \beta q_2)q_1\right]_{\alpha=100,\gamma_1=10,q_1=10,q_2=10} = 900 - 200\beta$
- $F = \left[-\beta q_1^2 + (\alpha - \gamma_1 - \beta q_2)q_1\right]_{\alpha=100,\gamma_1=10,q_1=10,q_2=20} = 900 - 300\beta$
- $H = \left[-\beta q_1^2 + (\alpha - \gamma_1 - \beta q_2)q_1\right]_{\alpha=100,\gamma_1=10,q_1=20,q_2=0} = 1800 - 400\beta$
- $I = \left[-\beta q_1^2 + (\alpha - \gamma_1 - \beta q_2)q_1\right]_{\alpha=100,\gamma_1=10,q_1=20,q_2=10} = 1800 - 600\beta$
- $K = \left[-\beta q_1^2 + (\alpha - \gamma_1 - \beta q_2)q_1\right]_{\alpha=100,\gamma_1=10,q_1=20,q_2=20} = 1800 - 800\beta$

For Player 2:

- $A = \left[-\beta q_2^2 + (\alpha - \gamma_2 - \beta q_1)q_2\right]_{\alpha=100,\gamma_2=5,q_1=0,q_2=10} = 950 - 100\beta$
- $B = \left[-\beta q_2^2 + (\alpha - \gamma_2 - \beta q_1)q_2\right]_{\alpha=100,\gamma_2=5,q_1=0,q_2=20} = 1900 - 400\beta$
- $E = \left[-\beta q_2^2 + (\alpha - \gamma_2 - \beta q_1)q_2\right]_{\alpha=100,\gamma_2=5,q_1=10,q_2=10} = 950 - 200\beta$
- $G = \left[-\beta q_2^2 + (\alpha - \gamma_2 - \beta q_1)q_2\right]_{\alpha=100,\gamma_2=5,q_1=10,q_2=20} = 1900 - 600\beta$
- $J = \left[-\beta q_2^2 + (\alpha - \gamma_2 - \beta q_1)q_2\right]_{\alpha=100,\gamma_2=5,q_1=20,q_2=10} = 950 - 300\beta$
- $L = \left[-\beta q_2^2 + (\alpha - \gamma_2 - \beta q_1)q_2\right]_{\alpha=100,\gamma_2=5,q_1=20,q_2=20} = 1900 - 800\beta$

The point $(0, 0)$ is a NE if $0 > C, 0 > H, 0 > A, 0 > B \Leftrightarrow \beta > \max\{9, 4.5, 9.5, 4.75\} = 9.5$

The point $(0, 10)$ is a NE if $0 > D, 0 > I, A > 0, A > B \Leftrightarrow 9.5 > \beta > \max\left\{4.5, 3, 3.1\overline{6}\right\} = 4.5$

The point $(0, 20)$ is a NE if $0 > F, 0 > K, B > A, B > 0 \Leftrightarrow 3.1\overline{6} = \min\left\{3.1\overline{6}, 4.75\right\} > \beta > \max\{3, 2.25\} = 3$

The point $(10, 0)$ is a NE if $C > 0, C > H, 0 > E, 0 > G \Leftrightarrow 9 > \beta > \max\left\{3, 4.75, 3.1\overline{6}\right\} = 4.75$

The point $(10, 10)$ is a NE if $D > 0, D > I, E > 0, E > G \Leftrightarrow 4.5 = \min\{4.5, 4.75\} > \beta > \max\{2.25, 2.375\} = 2.375$

The point $(10, 20)$ is a NE if $\Leftrightarrow F > 0, F > K, G > 0, G > E \Leftrightarrow 2.375 = \min\left\{3, 3.1\overline{6}, 2.375\right\} > \beta > 1.8$

The point $(20, 0)$ is a NE if $\Leftrightarrow H > C, H > 0, 0 > J, 0 > L \Leftrightarrow 3 = \min\{3, 4.5\} > \beta > \max\left\{3.1\overline{6}, 2.375\right\} = 3.1\overline{6}$, not possible

The point $(20, 10)$ is a NE if $\Leftrightarrow I > 0, I > D, J > 0, J > L \Leftrightarrow 2.25 = \min\left\{3, 2.25, 3.1\overline{6}\right\} > \beta > 1.9$

The point $(20, 20)$ is a NE if $\Leftrightarrow K > F, K > 0, L > J, L > 0 \Leftrightarrow 1.8 = \min\{1.8, 2.25, 1.9, 2.375\} > \beta$

$$\begin{cases} (0,0) \text{ is a NE} & 9.5 < \beta \\ (0,10) \text{ is a NE} & 4.5 < \beta < 9.5 \\ (0,20) \text{ is a NE} & 3 < \beta < 3.1\overline{6} \\ (10,0) \text{ is a NE} & 4.75 < \beta < 9 \\ (10,10) \text{ is a NE} & 2.375 < \beta < 4.5 \\ (10,20) \text{ is a NE} & 1.8 < \beta < 2.375 \\ (20,0) \text{ is a NE} & \text{not possible} \\ (20,10) \text{ is a NE} & 1.9 < \beta < 2.25 \\ (20,20) \text{ is a NE} & \beta < 1.8 \end{cases} \tag{4.62}$$

4.8.2 Computations for Price Functions in Spatial Price Equilibrium- Version 2

To match the supply prices obtained as output from the transportation problem, there are eight unknowns ($\kappa_i^S, \rho_i^S, i = 1, 2, 3, 4$) and four equations (one for each node). To be able to solve this system, we arbitrarily set $\kappa_1^S = \kappa_2^S = \kappa_4^S = 0$ and $\rho_3^S = \rho_2^S$ to obtain the following square system:

$$1 = \psi_1^* = \rho_1^S \cdot \left(\sum_j x_{1j}^* \right) = \rho_1^S (20) \Leftrightarrow \rho_1^S = \frac{1}{20} \tag{4.63a}$$

$$6 = \psi_2^* = \rho_2^S \cdot \left(\sum_j x_{2j}^* \right) = \rho_2^S (30) \Leftrightarrow \rho_2^S = \frac{1}{5} \tag{4.63b}$$

$$0 = \psi_3^* = \kappa_3^S + \rho_3^S \cdot \left(\sum_j x_{3j}^* \right) = \kappa_3^S + \frac{1}{5} \cdot (35) \Leftrightarrow \kappa_3^S = -7 \tag{4.63c}$$

$$3 = \psi_4^* = \rho_4^S \cdot \left(\sum_j x_{4j}^* \right) = \rho_4^S (15) \Leftrightarrow \rho_4^S = \frac{1}{5} \tag{4.63d}$$

We can do a similar analysis for the inverse demand function parameters setting the slopes equal to 1 to get a square system:

$$9 = \theta_1^* = \kappa_1^D - \rho_1^D \cdot \left(\sum_i x_{i1}^* \right) = \kappa_1^D - 1\,(20) \Leftrightarrow \kappa_1^D = 29 \tag{4.64a}$$

$$21 = \theta_2^* = \kappa_2^D - \rho_2^D \cdot \left(\sum_i x_{i2}^* \right) = \kappa_2^D - (25) \Leftrightarrow \kappa_2^D = 46 \tag{4.64b}$$

$$5 = \theta_3^* = \kappa_3^D - \rho_3^D \cdot \left(\sum_i x_{i3}^* \right) = \kappa_3^D - (10) \Leftrightarrow \kappa_3^D = 15 \qquad (4.64c)$$

$$3 = \theta_4^* = \kappa_4^D - \rho_4^D \cdot \left(\sum_i x_{i4}^* \right) = \kappa_4^D - (10) \Leftrightarrow \kappa_4^D = 13 \qquad (4.64d)$$

$$7 = \theta_5^* = \kappa_5^D - \rho_5^D \cdot \left(\sum_i x_{i5}^* \right) = \kappa_5^D - (35) \Leftrightarrow \kappa_5^D = 42 \qquad (4.64e)$$

4.8.3 Uniqueness of Spatial Price Equilibrium Version 2 Solution

The analysis above comparing the equilibrium flows x_{ij} and prices ψ_i, θ_j assumed implicitly that there was just a unique answer. Clearly, if there were multiple solutions then the same conclusions might not be true. In this section we show that there will always be a unique solution to the linear complementarity SPE given by:

$$0 \leq \psi_i \left(\sum_j x_{ij} \right) + c_{ij} (x) - \theta_j \left(\sum_i x_{ij} \right) \perp x_{ij} \geq 0, \forall i, j$$

with the prices and costs as described above. First, we rewrite the function $F : R^{20} \to R^{20}$ in terms of the given choices for ψ_i, θ_j, c_{ij} :

$$F_{ij} (x) = \kappa_i^S + \rho_i^S \left(\sum_j x_{ij} \right) + c_{ij} x_{ij} - \kappa_j^D + \rho_j^D \left(\sum_i x_{ij} \right)$$

We will show that this function F is strictly monotone which implies that there is at most one solution to the corresponding MCP. This is because Karmardian showed that this property implies at most one solution to the pure NCP and the SPE under consideration is a special case of such a problem [40]. Combined with the actual solution presented earlier, this analysis will show uniqueness of that solution.

Recall that strict monotonicity is defined as

$$0 < [F(x) - F(y)]^T (x - y), \forall x \neq y$$

But in this case

$$[F(x) - F(y)]^T (x - y)$$

$$
\begin{aligned}
&= \sum_i \sum_j \left[\rho_i^S \left(\sum_j (x_{ij} - y_{ij}) \right) (x_{ij} - y_{ij}) + c_{ij} (x_{ij} - y_{ij}) (x_{ij} - y_{ij}) \right] \\
&\quad + \sum_i \sum_j \left[\rho_j^D \left(\sum_i (x_{ij} - y_{ij}) \right) (x_{ij} - y_{ij}) \right] \\
&= \sum_i \sum_j \left[\rho_i^S \left(\sum_j (x_{ij} - y_{ij}) \right) (x_{ij} - y_{ij}) \right] \\
&\quad + \sum_i \sum_j \left[c_{ij} (x_{ij} - y_{ij}) (x_{ij} - y_{ij}) \right] \\
&\quad + \sum_i \sum_j \left[\rho_j^D \left(\sum_i (x_{ij} - y_{ij}) \right) (x_{ij} - y_{ij}) \right] \\
&= \sum_i \rho_i^S \left(\sum_j (x_{ij} - y_{ij}) \right) \sum_j \left[(x_{ij} - y_{ij}) \right] \\
&\quad + \sum_i \sum_j \left[c_{ij} (x_{ij} - y_{ij})^2 \right] \\
&\quad + \sum_j \rho_j^D \left(\sum_i (x_{ij} - y_{ij}) \right) \sum_i \left[(x_{ij} - y_{ij}) \right] \\
&= \sum_i \rho_i^S \left(\sum_j (x_{ij} - y_{ij}) \right)^2 + \sum_i \sum_j \left(c_{ij} (x_{ij} - y_{ij})^2 \right) \\
&\quad + \sum_j \rho_j^D \left(\sum_i (x_{ij} - y_{ij}) \right)^2
\end{aligned}
$$

The left-most term in the final line is strictly positive since $x \neq y$ and $\rho_i^S > 0$ for all i; the next term is also strictly positive since $c_{ij} > 0$ for all i, j and the third term is strictly positive since $\rho_j^D > 0$ for all j. Thus, the related MCP has a unique solution as shown above.

4.9 Exercises

1. Solve the following transportation problem with

$c_{ij} =$

from\to	1	2	3	4	5
1	50	19	10	25	37
2	30	5	110	76	36
3	80	4	50	44	10
4	100	7	27	10	5

$S_i =$

Supply Region	
1	50
2	10
3	10
4	20

, $D_j =$

Demand Region	
1	18
2	18
3	18
4	18
5	18

and verify for each of the 20 region pairs that (4.25) holds.

2. Consider the following MCP: find x_1, x_2 such that

$$0 \leq x_1^3 + \delta_1 x_1 x_2 + 7 + \lambda \perp x_1 \geq 0$$
$$0 \leq x_2^2 + \delta_2 x_1 x_2 + 15 + \lambda \perp x_2 \geq 0$$
$$0 \leq 25 - x_1 - x_2 \perp \lambda \geq 0$$

For what values of δ_1, δ_2 will there be an equivalent optimization problem whose KKT conditions give rise to this MCP? If there are any such values, what will be the associated optimization problem?

3. Solve the following two-person, bimatrix game with $A = \begin{bmatrix} 30 & 12 \\ 30 & 31 \\ 25 & 14 \end{bmatrix}, B = \begin{bmatrix} 8 & 3 \\ 9 & 12 \\ 100 & 15 \end{bmatrix}$ or show that no solution exists.

4. Consider the following LCP :

$$0 \leq 1 - 2\lambda \perp x_1 \geq 0$$
$$0 \leq 1 - 3\lambda \perp x_2 \geq 0$$
$$0 \leq 2x_1 + 3x_2 - 4 \perp \lambda \geq 0$$

Using the Principle of Symmetry, find an objective function $f(x_1, x_2)$ and a constraint function $g(x_1, x_2)$ with multiplier λ so that this LCP represents the KKT conditions of the optimization problem

$$\begin{aligned} &\text{Minimize}_{x_1,x_2} \quad f(x_1, x_2) \\ &\text{s. t.} \\ &\quad g(x_1, x_2) \leq 0 \\ &\quad x_1, x_2 \geq 0 \end{aligned}$$

or show that no such optimization problem exists.

5. Consider the following LCP:

$$
\begin{aligned}
&0 \leq 0 + 2x_1 - 7\lambda \perp x_1 \geq 0 \\
&0 \leq 1 - 6\lambda \perp x_2 \geq 0 \\
&0 \leq 7x_1 + 6x_2 - 5 \perp \lambda \geq 0
\end{aligned}
$$

Using the Principle of Symmetry, find an objective function $f(x_1, x_2)$ and a constraint function $g(x_1, x_2)$ with multiplier λ so that this LCP represents the KKT conditions of the optimization problem

$$
\begin{aligned}
&\text{Minimize}_{x_1,x_2} \quad f(x_1, x_2) \\
&\text{s. t.} \\
&\quad g(x_1, x_2) \leq 0 \\
&\quad x_1, x_2 \geq 0
\end{aligned}
$$

or show that no such optimization problem exists.

6. Consider the following LCP:

$$
\begin{aligned}
&0 \leq 0 + 2x_1 + 2x_2 - 7\lambda \perp x_1 \geq 0 \\
&0 \leq 1 - 6\lambda \perp x_2 \geq 0 \\
&0 \leq 7x_1 + 6x_2 - 5 \perp \lambda \geq 0
\end{aligned}
$$

Using the Principle of Symmetry, find an objective function $f(x_1, x_2)$ and a constraint function $g(x_1, x_2)$ with multiplier λ so that this LCP represents the KKT conditions of the optimization problem

$$
\begin{aligned}
&\text{Minimize}_{x_1,x_2} \quad f(x_1, x_2) \\
&\text{s. t.} \\
&\quad g(x_1, x_2) \leq 0 \\
&\quad x_1, x_2 \geq 0
\end{aligned}
$$

or show that no such optimization problem exists.

7. Consider the computations and GAMS code for the SPE problem of Section 4.8.2. Use another approach for matching the supply prices from the transportation problem. Specifically, you will need to convert the eight unknowns to just four to match the four equations in a way different from what is shown in that section. Revise the GAMS code accordingly and report on your results.
8. Perform a similar analysis to what is done in Section 4.8.1 to determine the range of values for β and the resulting Nash equilibria. This time use $\alpha = 10$, $\gamma_1 = 1$, $\gamma_2 = 1$ and provide mathematical justification for your answer.

References

1. B.H. Ahn and W.W. Hogan. On Convergence of the PIES algorithm for computing equilibria. *Operations Research*, 30(2):281-300, 1982.
2. M.S. Bazaraa, H.D. Sherali, and C.M. Shetty. *Nonlinear Programming Theory and Algorithms*, Wiley, New York, 2nd edition, 1979.
3. M.H. Belknap, C.H. Chen , P.T. Harker. *A Gradient-Based Method for Analyzing Stochastic Variational Inequalities with One Uncertain Parameter.* OPIM Working Paper 00–03–13. Department of Operations and Information Management, Wharton School. March, 2000.
4. J.F. Benders. Partitioning procedures for solving mixed-variables programming problems. *Numerische Mathematik*, 4:238-252, 1962.
5. A. Bensoussan. Points de Nash dans le cas de fonctionnelles quadratiques et jeux différentiels linéaires à n personnes. *SIAM Journal Control*, 12:460-499, 1974.
6. J. Cabero, A. Baillo, S. Cerisola, M. Ventosa. *Electricity market equilibrium model with risk constraints via Benders decomposition.* INFORMS Annual Conference. November, San Francisco, 2005.
7. J. Cabero, A. Baillo, M. Ventosa, S. Cerisola. *Application of Benders decomposition to an equilibrium problem.* 15th PSCC, Liege. August, 2005.
8. S.C. Chapra and R. P. Canale. *Numerical methods for engineers with software and programming applications*, McGraw Hill, Boston, Massachusetts, 4th edition, 2002.
9. V. Chvátal. *Linear programming*, W.H. Freeman and Company, New York, 1983.
10. R.W. Cottle. Nonlinear programs with positively bounded Jacobians. *SIAM Journal on Applied Mathematics*, 14:147-158, 1966.
11. R.W. Cottle and G.B. Dantzig. Complementary pivot theory of mathematical programming. *Linear Algebra and Its Applications*, 1:103-125, 1968.
12. R.W. Cottle, J.-S. Pang, R.E. Stone. *The linear complementarity problem.* SIAM, Philadelphia, 1992.
13. D. De Wolf and Y. Smeers. A stochastic version of a Stackelberg-Nash-Cournot equilibrium model, *Management Science*, 43(2):190-197, 1997.
14. B.C. Eaves. On the basic theorem of complementarity. *Mathematical Programming*, 1:68-75, 1971.
15. F. Facchinei and J.-S. Pang. *Finite-dimensional variational inequalities and complementarity problems, Volume 1*, Springer, New York, 2003.
16. J. Fortuny-Amat, B. McCarl. A representation and economic interpretation of a two-level programming problem. *The Journal of the Operational Research Society* 32(9):783-792, 1981.
17. J.D. Fuller. Market equilibrium models with continuous and binary variables, working paper (personal communication), 2008.

18. J.D. Fuller, Capacity pricing in unit commitment, working paper (personal communication), 2010.
19. S.A. Gabriel, A.J. Conejo, C. Ruiz, and S. Siddiqui. Solving discretely-constrained, mixed linear complementarity problems with applications in energy, in review at *Computers and Operations Research*, February 2011.
20. S.A. Gabriel and J.D. Fuller, A Benders decomposition method for solving stochastic complementarity problems with an application in energy. *Computational Economics*, 35(4):301-329, 2010.
21. S.A. Gabriel, S. Siddiqui, A.J. Conejo, C. Ruiz, Discretely-constrained, Nash-Cournot games in energy, in review at *Networks and Spatial Economics*, February 2011.
22. S.A. Gabriel, F.U. Leuthold. Solving discretely-constrained MPEC problems with applications in electric power markets. *Energy Economics* 32:3-14, 2010.
23. S.A. Gabriel, S. Kiet, J. Zhuang. A mixed complementarity-based equilibrium model of natural gas markets. *Operations Research*, 53(5):799-818, 2005.
24. S.A. Gabriel, Y. Shim, A.J. Conejo, S. de la Torre, R. García-Bertrand. A Benders decomposition method for discretely-constrained mathematical programs with equilibrium constraints. *Journal of the Operational Research Society*, 61:1404-1419, 2010.
25. S.A. Gabriel , J. Zhuang, R. Egging. Solving stochastic complementarity problems in energy market modeling using scenario reduction. *Euopean Journal of Operational Research*, 197(3):1028-1040, 2009.
26. S.A. Gabriel, J. Zhuang, S. Kiet. A large-scale complementarity model of the North American natural gas market. *Energy Economics*, 27:639-665, 2005.
27. F.D. Galiana, A. L. Motto, and F. Bouffard. Reconciling social welfare, agent profits, and consumer payments in electricity pools. *IEEE Transactions on Power Systems*, 18:452-459, 2003.
28. T.S. Genc, S.S. Reynolds, and S. Sen. Dynamic oligopolistic games under uncertainty: a stochastic programming approach. *Journal of Economic Dynamics and Control*, 31(1):55-80, 2007.
29. G. Gürkan, A.Y. Özge, S.M. Robinson. Sample-path solution of stochastic variational inequalities. *Mathematical Programming*, 84:313-333, 1999.
30. P.T. Harker. Alternative models of spatial competition. *Operations Research*, 34(3):410-425, 1986.
31. P. T. Harker. *Lectures on computation of equilibria with equation-based methods*, CORE Lecture Series, Université catholique de Louvain, Louvain-la-Neuve, Belgium, 1993.
32. P.T. Harker and J.-S. Pang. Finite-dimensional variational inequality and nonlinear complementarity problems: A survey of theory, algorithms and applications. *Mathematical Programming*, 48:161-220, 1990.
33. P. Hartman and G. Stampacchia. On some nonlinear elliptic differential functional equations, *Acta Mathematica*, 115:153-188, 1966.
34. A. Haurie and F. Moresino. S-adapted oligopoly equilibria and approximations in stochastic variational inequalities. *Annals of Operation Research*, 114:183-201, 2002.
35. A. Haurie and G. Zaccour. S-Adapted equilibria in games played over event trees: An overview, in A.S. Nowak et al. eds. Advances in Dynamic Games, *Annals of the International Society of Dynamic Games* 7:417-444, 2005.
36. A. Haurie , G. Zaccour, J. Legrand , Y. Smeers. A stochastic dynamic Nash-Cournot model for the European gas market. Technical Report G-86-24, GERAD, École des Hautes Etudes Commerciales, Montréal, Québec, Canada, 1987.
37. B.F. Hobbs. Linear complementarity models of Nash-Cournot competition in bilateral and POOLCO power markets. *IEEE Transaction on Power Systems*, 16(2):194-202, 2001.
38. W.W. Hogan and B.J. Ring. On Minimum-uplift pricing for electricity markets. Working Paper, John F. Kennedy School of Government, Harvard University, 2003.
39. S. Karamardian. The Complementarity Problem, *Mathematical programming*, 2:107-129, 1972.

40. S. Karamardian. Complementarity problems over cones with monotone and pseudomonotone maps, *Journal of Optimization Theory and Applications*, 18(4):445-454, 1976.
41. C.E. Lemke and J.T. Howson, Jr. Equilibrium points of bimatrix games, *SIAM Journal on Applied Mathematics* 12:413-423, 1964.
42. C.E. Lemke. Bimatrix equilbrium points and mathematical programming, *Management Science* 11:681-689, 1965.
43. K. G. Murty. *Linear programming*, Wiley, New York, 1983.
44. S.G. Nash and A. Sofer. *Linear and nonlinear programming*, McGraw-Hill, New York, 1996.
45. R.P O'Neill, P.M. Sotkiewicz, B.F. Hobbs, M.H. Rothkopf, and W.R. Stewart, Jr. Efficient market-clearing prices in markets with nonconvexities. *European Journal Operational Research*, 164(1):269-285, 2005.
46. J.M. Ortega and W.C. Rheinboldt. *Iterative solution of nonlinear equations in several variables*, Academic Press, New York, 1970.
47. P.A. Samuelson. Spatial price equilibrium and linear programming, *American Economic Review*, 42(3):283-303, 1952.
48. U. Shanbhag, P. Glynn P, G. Infanger. A complementarity framework for forward contracting under uncertainty. INFORMS Annual Conference. November, San Francisco, 2005.
49. U. Shanbhag, G. Infanger, P. Glynn . A complementarity framework for forward contracting under uncertainty, *Operations Research*, 59(4):810-834, 2011.
50. O. Shy. *Industrial organization theory and applications*, MIT Press, Cambridge, Massachusetts, 1996.
51. P.D. Straffin. *Game theory and strategy*, The Mathematical Association of America, Washington, 1993.
52. T. Takayama and W.C. Labys. Spatial equilibrium analysis, Chapter 5 of *Handbook of Regional and Urban Economics*, Vol. I, edited by P. Nijkamp, Elsevier Science Publishers BV, 173-199. 1986.
53. W.L. Winston. *Introduction to probability models*, Thomson Brooks/Cole, Belmont, California, 2004.
54. J. Zhuang and S.A. Gabriel, A complementarity model for solving stochastic natural gas market equilibria. *Energy Econ*, 30(1):113-147, 2008

Chapter 5
Variational Inequality Problems

5.1 Introduction

The purpose of this chapter is to explain variational inequality (VI) formulations of equilibrium problems, and the close connection of a VI problem to an equivalent complementarity problem. There are sometimes advantages to a VI formulation compared to a complementarity formulation: the complementarity formulation has primal decision variables, *and* dual variables that arise, e.g., when specifying the KKT conditions of individual agents; but a VI formulation has the same primal variables, with few, or no dual variables, which can considerably ease the coding of the model in GAMS. This coding advantage is particularly evident when implementing large complex models, or decomposition algorithms as discussed in Chapter 9. However, the derivation of a complementarity model formulation is usually easier than the derivation of an equivalent VI model: e.g., many complementarity models in this book are derived by writing down the KKT conditions of the agents, together with market-clearing conditions, but the procedure to write down a VI model with few or no dual variables is not as easily stated. In this chapter, we alleviate this difficulty by showing how to arrive quickly at the formulation of a VI model for a large class of Nash equilibrium and generalized Nash equilibrium settings.

We begin with a discussion of how to formulate a VI problem for the cases of a single optimizing agent, and of a Nash equilibrium of several optimizing agents without any constraints that link the agents together in their feasible sets. Next, we consider a VI formulation of a special case of the generalized Nash equilibrium problem, with constraints that link the agents. Following this, we discuss the relations between a VI formulation and a complementarity formulation of equilibrium. Finally, we examine the quasi-variational inequality problem as a way to represent a generalized Nash equilibrium.

Many VI models of energy markets can be found in the literature, often together with an equivalent complementarity model. For example, the large modeling system used by the U.S. Department of Energy for its forecasts is described in [11] as a VI. There have been numerous papers with VI formulations of electricity markets, e.g., [1], [2], [14], [15], [21]. A natural gas market model is described as a complementarity model and a VI in [10].

Throughout this chapter, we frequently refer to the following general statement of the form of a VI problem, denoted by VI(K, G), where $G(x) \in \mathbb{R}^n$ is a vector-valued mapping, and $K \subseteq \mathbb{R}^n$ is the feasible set.

$$VI(K,G): \text{ find } x^* \in K \text{ such that } G(x^*)^T(x - x^*) \geq 0 \text{ for all } x \in K \quad (5.1)$$

Note that VIs are discussed in Chapter 4 where basic mathematical results such as existence and uniqueness of solutions are presented, for VIs and complementarity problems. In contrast, this chapter is dedicated to VIs and therefore contains many more details on that topic.

5.2 Formulation of Variational Inequality Problems

We start by formulating a single objective, constrained optimization problem as a variational inequality (VI) problem, and then we extend the idea to formulate a Nash equilibrium of several agents as a VI problem. Compared to Chapter 4, in this chapter we provide more details on the general VI formulation and also relate it to optimization in a different way (via the Taylor Series approximation).

5.2.1 Optimization Problem as a VI Problem

Consider an optimization problem to choose $x \in \mathbb{R}^n$ in order minimize a single-valued objective function $f(x)$, assumed to be continuously differentiable, but x is restricted to be in a closed, convex set K, where typically, K is defined by a set of constraints. We may write the problem as

$$\text{Minimize}_{x \in K} \quad f(x) \quad (5.2)$$

Let x^* stand for an optimal solution to (5.2). For any other feasible point $y \in K$ that is close to x^*, the linear approximation of f based on the first order Taylor series is fairly accurate:

$$f(y) \approx f(x^*) + \nabla f(x^*)^T(y - x^*)$$

Because x^* is optimal, it follows that $f(y) \geq f(x^*)$, and so, to a good approximation,

$$\nabla f(x^*)^T(y - x^*) \geq 0 \quad (5.3)$$

for any $y \in K$ that is close to x^*.

Next, consider *any* $x \in K$, but not necessarily close to the optimal solution x^*. By convexity of K, any point on the line between x^* and x is also in K, i.e.,

$$x^* + \alpha(x - x^*) \in K \text{ for any } \alpha \in [0, 1]$$

Now choose $\alpha > 0$ small enough so that $y = x^* + \alpha(x - x^*)$ is close enough to x^* that (5.3) is true, to a good approximation. Then, to a good approximation,

$$\nabla f(x^*)^T \alpha(x - x^*) \geq 0$$

which, after dividing by α, implies that, to a good approximation,

$$\nabla f(x^*)^T (x - x^*) \geq 0 \text{ , for all } x \in K \tag{5.4}$$

It is possible to prove that any optimal solution x^* of problem (5.2) must satisfy the condition (5.4) precisely, without any qualification such as "to a good approximation". See, e.g., [19], Proposition 1.2.

Theorem 5.1. *([19], Proposition 1.2) Let x^* be a solution to (5.2). Then x^* solves (5.4).*

The converse is also true, if f is convex.

Theorem 5.2. *([19], Proposition 1.3) If $f(x)$ is convex, and x^* is a solution to (5.4), then x^* solves (5.2).*

The problem of finding a vector $x^* \in K$ such that condition (5.4) is true is a VI problem of the form (5.1), with $G(x) = \nabla f(x)$. However, not all VI problems have the property that the vector-valued mapping G is the gradient of a single-valued function, as illustrated next by the derivation of a Nash equilibrium as a VI problem and highlighted earlier via the Principle of Symmetry discussed in Chapter 4.

5.2.2 VI Formulation of Nash Equilibrium: No Linking Constraints

Now consider several agents, indexed by $i = 1, 2, ..., I$, each with its own objective function and constraints. Agent i controls a vector of decision variables, x_i, and the vector of all agents' variables is

$$x = \begin{pmatrix} x_1 \\ \vdots \\ x_I \end{pmatrix}$$

The objective function of agent i depends not only on its own decision variables x_i, but also on the other agents' decisions, which we group together into a vector x_{-i}. To begin, let us suppose that there are no *linking constraints* – a linking constraint involves variables under the control of more than one agent. The problem of agent i is

$$\text{Minimize}_{x_i \in K_i} \quad f_i(x_i, x_{-i}) \tag{5.5}$$

where $K_i \subset \mathbb{R}^{n_i}$ is a closed convex set, and f_i is convex and continuously differentiable in x_i. A Nash equilibrium is the result of the simultaneous optimization of all agents, with all agents correctly guessing the decisions of other agents; a more precise definition follows.

Definition 5.1. A Nash equilibrium x^* has the property that, for every i, x_i^* solves (5.5) given $x_{-i} = x_{-i}^*$.

Applying the result (5.4) and the two theorems to each problem (5.5) leads to an equivalent statement of Nash equilibrium: for every agent i,

$$\nabla_i f_i(x_i^*, x_{-i}^*)^T (x_i - x_i^*) \geq 0 \text{ for all } x_i \in K_i \tag{5.6}$$

where the symbol ∇_i represents the gradient with respect to only the x_i variables. Define the mapping $G(x)$ by concatenating $\nabla_i f_i(x)$ for all agents i, i.e.,

$$G(x) = \begin{pmatrix} \nabla_1 f_1(x) \\ \vdots \\ \nabla_I f_I(x) \end{pmatrix} \tag{5.7}$$

Restricting $x_i \in K_i$ for all agents i can be described as a restriction that x should be in the Cartesian product of all the feasible sets K_i: define $K = K_1 \times K_2 \times \cdots \times K_I \equiv \prod_i K_i$, and require that $x \in K$, with $x^* \in K$, in particular. We can add the I inequalities described by (5.6), to arrive at the requirement

$$G(x^*)^T (x - x^*) \geq 0 \text{ for all } x \in K \tag{5.8}$$

which, along with $x^* \in K$, is the VI representation of the Nash equilibrium problem.

The above discussion shows that any solution to the Nash equilibrium problem (5.5) for all i is also a solution to the VI problem (5.8) with G defined as in (5.7) and $K = \prod_i K_i$. Assuming, as we are, that $f_i(x_i, x_{-i})$ is convex in x_i, the converse can also be shown, with the help of Theorem 5.2; we leave this is an exercise for the reader. These results are summarized in the following theorem.

Theorem 5.3. *Suppose that, for* $i = 1, 2, ..., I$, $K_i \subset \mathbb{R}^{n_i}$ *is a closed convex set, and* $f_i : K_i \longrightarrow \mathbb{R}$ *is convex and continuously differentiable. Then* x^* *is a Nash equilibrium if and only if* $x^* \in K = \prod_i K_i$ *solves the VI (5.8) with G defined as in (5.7).*

5.2.2.1 Illustrative Example. Nash-Cournot Model of Coal Market from Chapter 3

In the coal market model of Chapter 3, the variables under the control of firm i are the outputs from its mines, x_{ib} indexed by b. There is also another variable that is not directly under the control of any of the firms – the demand q – but there is a constraint that equates q with the sum of all mine outputs, $q - \sum_i \sum_b x_{ib} = 0$, so here we eliminate the variable q using this constraint, in order to fit the form of the Nash model with no linking constraints, described in general, above. With this minor variation, the Nash-Cournot coal market model can be described by writing down the form of each firm's optimization problem. For firm i, the problem is

$$\text{Minimize}_{x_{ib}} \quad \sum_b c_{ib} x_{ib} - \left(\alpha - \beta \sum_{j,b} x_{jb} \right) \sum_b x_{ib}$$

$$\text{s. t.}$$

$$x_{ib} - K_{ib} \leq 0 \quad \forall b$$

$$x_{ib} \geq 0 \quad \forall b$$

Note that, in the expression for price in the inverse demand function, the summation over j includes variables x_{ib} under the control of firm i, and also other variables x_{jb} with $j \neq i$ that are treated by firm i as parameters because they are not under the control of firm i.

To formulate the VI problem, we need to define the feasible set K, and we need the concatenation of all firms' gradients into the mapping G, as in (5.7):

$$K = \{ x = (x_{ib})_{\forall i,b} \mid x_{ib} - K_{ib} \leq 0, \; x_{ib} \geq 0, \;\; \forall i, b \} \tag{5.9a}$$

$$G_{ib}(x) = c_{ib} - \left(\alpha - \beta \sum_{j,b'} x_{jb'} \right) + \beta \sum_{b'} x_{ib'} \tag{5.9b}$$

where G_{ib} indicates the i, b component of G, i.e., the partial derivative of the objective function of firm i, with respect to x_{ib}.

In a GAMS representation of this model, the modeler must (1) represent the constraints in the same way as in an optimization model, (2) give the form of all G_{ib}, and (3) indicate the pairing of G_{ib} expressions with the appropriate variables x_{ib}. Compared with the coding of a complementarity model, the task of representing a VI model is somewhat easier, because there is no need to deal with dual variables – neither declarations of dual variables, nor expressions involving dual variables.

□

5.2.3 VI Formulation of Generalized Nash Equilibrium With Linking Constraints: A Special Case

Now let us consider a Nash equilibrium model in which there is at least one constraint that involves decision variables of more than one agent; we call such constraints *linking constraints.* (Other terms for such constraints are *coupling constraints* or *complicating constraints.*) In the example which follows below, from a Chapter 3 illustration, the linking constraint among the Cournot players in the coal market is a limit on the size of the coal storage yard, which effectively limits the total production and sale of coal in the city.

As mentioned in Chapters 3 and 4, and in more detail in Section 5.4, a *generalized Nash equilibrium* model is like a Nash equilibrium model except that the generalized Nash model has linking constraints, whereas the traditional Nash model does not. As explained in Chapter 4 and below in Section 5.4, the generalized Nash equilibrium model may have many different solutions, if the agents are allowed to have different marginal valuations of the resources of the linking constraints (e.g., coal yard space), but in this subsection, we are interested in the special case where the agents have the same marginal value vector of the resources, at the equilibrium. They arrive at the same marginal valuation of resources typically because they buy or sell each resource among themselves, at a common equilibrium price (the marginal valuation).

To proceed with the general setting, the Nash model (5.5), of I simultaneous optimizations, allows for links among the agents' objective functions, but no linking constraints. Let us consider inclusion of one or more linking constraints that are linear in form:

$$\sum_{i=1}^{I} A_i x_i \leq L \tag{5.10}$$

where A_i is a matrix with m_0 rows and $L \in \mathbb{R}^{m_0}$ is a vector of known parameters. Note that, if a modeler wishes to have an equality for a linking constraint, then the equality constraint can be represented equivalently as two inequality constraints (one in the "$\leq$" form, and the other in the "$\geq$" form, which can itself be transformed into "$\leq$" form by multiplying through by -1). Therefore, (5.10) is general enough to represent any constraints that are linear in the variables x_i, after this slight transformation.

The linear form of (5.10) is actually less restrictive than it may appear at first sight. A more general form would have agent i using amounts of the limited resources in the vector L as specified in a vector of functions $r_i(x_i)$, with the constraint (5.10) modified to the following more general form (termed separable in the subvectors x_i)

$$\sum_{i=1}^{I} r_i(x_i) \leq L \tag{5.11}$$

This can be converted to the linear form (5.10) by introducing new decision variables w_i where each vector w_i represents the amounts of the limited resources purchased by agent i for its use. The definition of the feasible set for agent i is modified to include the new variables w_i and also constraints

$$r_i(x_i) \leq w_i \tag{5.12}$$

The constraints (5.11) will tend to be binding at an optimal solution of firm i's problem if we assume that agent i must purchase w_i at positive prices – it would be nonoptimal for agent i to purchase any more than is needed. If the functions in the vector $r_i(x_i)$ are convex, then the constraints (5.12) define a convex region, which is the reason to define (5.12) as inequalities instead of equalities. Therefore, the linking constraints can be written in the linear form

$$\sum_{i=1}^{I} w_i \leq L$$

Now proceeding with the linking constraints in the form (5.10), in order to extend the adding up procedure that led from (5.6) to (5.8), we temporarily reformulate (5.10) by introducing a vector of variables, $y \in \mathbb{R}^{m_0}$, and writing an equivalent system of two sets of constraints:

$$\sum_{i=1}^{I} A_i x_i - y = 0 \tag{5.13a}$$

$$y \leq L \tag{5.13b}$$

The vector y is interpreted as the variables under the control of an agent, numbered $I+1$, who enforces the limit (5.13b). We also temporarily introduce a vector of prices $\mu \in \mathbb{R}^{m_0}$(the common valuations of the linking resources), with $\mu \geq 0$. Both y and μ will disappear in the final formulation of the VI. We assume that the prices μ are beyond the control of any of the agents.

The feasible set for the equilibrium problem of $(I+1)$ agents is

$$\widehat{K} = \left\{ (x,y) | x_i \in K_i,\, y \leq L,\, \sum_{i=1}^{I} A_i x_i - y = 0 \right\} \tag{5.14}$$

A generalized Nash equilibrium, (x^*, y^*) must be feasible with respect to all constraints, i.e., $(x^*, y^*) \in \widehat{K}$. The objective function of agent i now has to take account of the price on the part $A_i x_i$ that agent i takes in the linking constraint (5.13a), i.e., to minimize $f_i(x_i, x^*_{-i}) + \mu^T A_i x_i$. Agent $I+1$ is assumed to choose y in order to maximize revenue from the sale of y at prices μ, i.e., to minimize $-\mu^T y$. An equilibrium must simultaneously optimize all agents' objectives, with respect to any feasible points, i.e., using (5.4), (x^*, y^*) must satisfy

$$\{\nabla_i f_i(x_i^*, x_{-i}^*)^T + \mu^T A_i\}(x_i - x_i^*) \geq 0, \; i = 1, ..., I \tag{5.15a}$$

$$-\mu^T(y - y^*) \geq 0 \tag{5.15b}$$

for any feasible $(x, y) \in \widehat{K}$.

The joint VI problem of all $I+1$ agents is obtained by adding (5.15a) over all $i = 1, ..., I$, plus (5.15b): find $(x^*, y^*) \in \widehat{K}$ such that

$$\sum_{i=1}^{I} \{\nabla_i f_i(x_i^*, x_{-i}^*)^T + \mu^T A_i\}(x_i - x_i^*) \tag{5.16}$$

$$- \mu^T(y - y^*) \geq 0 \text{ for all } (x, y) \in \widehat{K}$$

However, y can be eliminated by using the equality in (5.14), and both μ and y can be eliminated from (5.16) due to cancellations among the terms involving μ:

$$\begin{aligned}
&\sum_{i=1}^{I} \mu^T A_i(x_i - x_i^*) - \mu^T(y - y^*) \\
&= \sum_{i=1}^{I} \mu^T A_i(x_i - x_i^*) - \mu^T \left(\sum_{i=1}^{I} A_i x_i - \sum_{i=1}^{I} A_i x_i^* \right) \\
&= 0
\end{aligned}$$

Note that the cancellations involving μ depend on the fact that all agents agree on the same value of μ. This results in a VI problem only in the original x variables: with the feasible set defined as

$$\overline{K} = \{x \,|\, x_i \in K_i, \sum_{i=1}^{I} A_i x_i \leq L\} \tag{5.17}$$

and G defined as in (5.7), find $x^* \in \overline{K}$ such that

$$G(x^*)^T(x - x^*) \geq 0 \text{ for all } x \in \overline{K} \tag{5.18}$$

Note that the VI statement (5.18) refers to the linking constraints in the definition of the feasible set $\overline{K}$, but the vector valued mapping G is the same as in the Nash equilibrium model *without* linking constraints: the concatenation of the gradients of the individual agents' objectives, without any reference to the prices associated with the linking constraints.

To summarize, we have shown that a Nash-like equilibrium among the $I+1$ agents, with the additional linking constraints (5.10), also solves the VI defined by (5.17), (5.18), with G given by (5.7). A key assumption is that all agents have the same marginal valuation μ of linking resources. These ideas are explained in Section 5.4 with more precision, and for the more general case of different marginal valuations.

5.2.3.1 Illustrative Example. Nash-Cournot Model of Coal Market with Coal Yard Limit from Chapter 3

Compared with the coal market model that has no coal yard limit, the new element of the model with the coal yard limit is the linking constraint that relates the decision variables x_{ib} of the output of firm i from its mine b,

$$\sum_i \sum_b x_{ib} \leq l$$

which places an upper limit $l \in \mathbb{R}^1$ on the total output of all firms that can sit in the city's coal yard. Application of the general results (5.17), (5.18) to extend the model (5.9) to include the coal yard limit leads to the following definitions of a feasible set K and elements G_{ib} of the mapping G that defines a VI:

$$K = \left\{ (x_{ib}) \,|\, x_{ib} - K_{ib} \leq 0,\ x_{ib} \geq 0,\ \ \forall i, b,\ \sum_{i,b} x_{ib} \leq l \right\} \tag{5.19a}$$

$$G_{ib}(x) = c_{ib} - \left(\alpha - \beta \sum_{j,b'} x_{jb'} \right) + \beta \sum_{b'} x_{ib'} \tag{5.19b}$$

Note that the only difference between the models with and without the coal yard is in the definitions of the feasible sets, (5.9a) and (5.19a).

If the goal is simply to formulate the VI model, then it is unnecessary to introduce explicitly the agent controlling variable y as in (5.13) and the developments that follow it. However, we do so here, in order to illustrate the general development.

We suppose that the city government owns the coal yard, and that it sells the space l in the yard to the coal firms, at a price μ that is beyond the city's control – the price will be determined in the market, e.g., perhaps by an auction, and the city government is a price-taker in that market. The city government wants to maximize its revenue from sale of up to l units of coal yard space:

$$\begin{aligned} &\text{Minimize}_y \quad -\mu y \\ &\text{s. t.} \\ &\quad y \leq l \end{aligned}$$

In VI form (5.4) for a single optimization model, this is:

$$\text{find } y^* \leq l \text{ such that } \ -\mu(y - y^*) \geq 0 \text{ for all } y \leq l \tag{5.20}$$

Each firm i faces its usual costs that are included in $\sum_b c_{ib} x_{ib}$, but it must also pay for coal yard space, for an additional cost of $\mu \sum_b x_{ib}$:

$$\text{Minimize}_{x_{ib}} \sum_b c_{ib}x_{ib} + \mu \sum_b x_{ib} - \left(\alpha - \beta \sum_{j,b} x_{jb}\right) \sum_b x_{ib}$$

s. t.

$$x_{ib} - K_{ib} \le 0 \quad \forall b$$
$$x_{ib} \ge 0 \quad \forall b$$

In VI form for a single optimization model (for firm i only), this is:

with the feasible set $K_i = \{x_{ib}\forall b | \, x_{ib} - K_{ib} \le 0, \, x_{ib} \ge 0, \ \forall b\}$ find $x_{ib}^* \in K_i$ such that

$$\sum\nolimits_b \left(c_{ib} + \mu - \left(\alpha - \beta \sum_{j,b} x_{jb}^*\right) + \beta \sum\nolimits_b x_{ib}^* \right) (x_{ib} - x_{ib}^*)$$
$$\ge 0 \text{ for all } x_{ib} \in K_i \qquad (5.21)$$

The constraint that links the firms and the city government is that the total amount that the city sells must equal the total that the firms buy:

$$y = \sum_i \sum_b x_{ib} \qquad (5.22)$$

which can be used to eliminate the variable y in the definition of the feasible region K for the whole equilibrium model, i.e., (5.19a). Now sum the inequality in (5.21) over all firms i, add $-\mu(y - y^*) \ge 0$ from (5.20) to the result, and use (5.22) to eliminate all terms involving μ through cancellations that simply mean that the revenue collected by the government equals the total money paid by the firms for coal yard space. The result is

$$\sum_i \sum_b \left(c_{ib} - \left(\alpha - \beta \sum_{j,b} x_{jb}^*\right) + \beta \sum_b x_{ib}^* \right) (x_{ib} - x_{ib}^*) \ge 0 \ \forall x = (x_{ib}) \in K$$

i.e., the mapping G for the VI is exactly as given in (5.19b).

□

For further illustration of the formulation of VIs with linking constraints and common marginal valuations, we now present three more examples from earlier chapters.

5.2.3.2 Illustrative Example. Competitive Equilibrium of Two Related Markets: Coal and Wood from Chapter 3

In this model, p and q represent the price and quantity for coal, and $\widehat{p}$ and $\widehat{q}$ represent the price and quantity of wood. The demand for coal depends on both p and $\widehat{p}$, and similarly, the demand for wood also depends on both prices. The supply of of each commodity depends only on its own price. The whole model is a set of four simultaneous equations in the four unknowns – we present the two demand equations followed by the two supply equations:

$$q = 137,850 - 60,000p + 13,500\widehat{p} \tag{5.23a}$$
$$\widehat{q} = 66,150 + 5000p - 36,000\widehat{p} \tag{5.23b}$$
$$q = -50,000 + 100,000p \tag{5.23c}$$
$$\widehat{q} = -23,000 + 70,000\widehat{p} \tag{5.23d}$$

From Chapter 3, a solution is $p^* = 1.25$, $\widehat{p}^* = 0.90$, $q^* = 75,000$ and $\widehat{q}^* = 40,000$.

We formulate a VI version of this model, by thinking of the situation *as if* it involves the simultaneous optimization of four agents that are linked in their objective functions and through some linking constraints. The four agents are not individual firms or consumers, but instead are mathematical aggregations that are chosen in order to fit the general form in (5.17) and (5.18). The agents are: the coal supply industry, the wood supply industry, the coal consumers and the wood consumers. Each agent controls a variable, which is the amount of coal or wood that is supplied or consumed. To fit the form in (5.17) and (5.18), we need to distinguish supply variables from consumption variables: we denote supplies by subscript-s and demands by subscript-d, i.e., q_s, q_d, $\widehat{q}_s$ and $\widehat{q}_d$.

It is useful to invert the two demand functions (5.23a) and (5.23b) to give prices in terms of quantities. This is done in Chapter 3, and is repeated here for convenience:

$$p = 2.7984 - 0.0000172q_d - 0.00000645\widehat{q}_d \tag{5.24a}$$
$$\widehat{p} = 2.2262 - 0.00000239q_d - 0.0000287\widehat{q}_d \tag{5.24b}$$

Treating (5.24a) as a marginal value function for coal consumption, parameterized by the amount of wood consumption $\widehat{q}_d$, we can integrate over q_d to get a total value function for coal consumption, $TV_c(q_d; \widehat{q}_d)$, which is also parameterized by $\widehat{q}_d$:

$$TV_c(q_d; \widehat{q}_d) = (2.7984 - 0.00000645\widehat{q}_d)q_d - \frac{0.0000172}{2}q_d^2$$

Similarly, we can integrate (5.24b) to get a total value function for wood consumption, $TV_w(\widehat{q}_d; q_d)$, which is parameterized by coal consumption q_d:

$$TV_w(\widehat{q}_d; q_d) = (2.2262 - 0.00000239q_d)\widehat{q}_d - \frac{0.0000287}{2}\widehat{q}_d^{\,2}$$

We also need to invert the two supply relationships (5.23c) and (5.23d):

$$p = 0.5 + 0.00001 q_s \tag{5.25a}$$

$$\widehat{p} = \frac{23}{70} + \frac{1}{70000}\widehat{q_s} \tag{5.25b}$$

Treating the inverse supply functions as marginal cost functions, we can integrate each one to get total cost functions for coal and wood, $TC_c(q_s)$ and $TC_w(\widehat{q_s})$:

$$TC_c(q_s) = 0.5q_s + \frac{0.00001}{2}q_s^2$$

$$TC_w(\widehat{q_s}) = \frac{23}{70}\widehat{q_s} + \frac{1/70000}{2}\widehat{q_s}^2$$

The linking constraints are the two market clearing conditions:

$$q_s - q_d = 0 \tag{5.26a}$$

$$\widehat{q_s} - \widehat{q_d} = 0 \tag{5.26b}$$

Recall that these equality constraints can be represented as pairs of inequality constraints – e.g., (5.26a) can be replaced by $q_s - q_d \leq 0$ and $-q_s + q_d \leq 0$ – so the setup fits the form of (5.10).

In the general development above, in Section 5.2.3, agent i chooses x_i to minimize $f_i(x_i, x_{-i}^*) + \mu^T A_i x_i$, where $\mu^T A_i$ represents the negative of the vector of prices paid to agent i for the vector x_i. In the present example, x_i for the coal supply agent corresponds to the scalar variable q_s, and the price paid to the coal supply agent is the dual variable associated with the linking constraint (5.26a). The coal supply agent incurs costs $TC_c(q_s)$, which corresonds to $f_i(x_i, x_{-i}^*)$ in the general development of Section 5.2.3 above. Similarly, the function f_i for the wood supply agent is $TC_w(\widehat{q_s})$.

The coal consumer agent buys q_d from the market and obtains its value, $TV_c(q_d; \widehat{q_d})$; it chooses q_d to maximize net value, which is total value minus what it has to pay in the market. However, the amount that it pays corresponds to $\mu^T A_i x_i$ in the general development above because it pays into the market, at a price determined by the dual variable of the linking constraint (5.26a). Therefore, interpreting the general development above, the function f_i for the coal consumer agent is just the negative of total value, $-TV_c(q_d; \widehat{q_d})$. Similarly, the function f_i for the wood consumer agent is just the negative of total value, $-TV_w(\widehat{q_d}; q_d)$.

Therefore, the mapping G of the VI formulation of equilibrium involves the partial derivatives of the TC and TV functions, i.e., the inverse supply and demand functions:

$$G(q_s, \widehat{q_s}, q_d, \widehat{q_d}) = \begin{pmatrix} \partial TC_c / \partial q_s \\ \partial TC_w / \partial \widehat{q_s} \\ -\partial TV_c / \partial q_d \\ -\partial TV_w / \partial \widehat{q_d} \end{pmatrix}$$

$$= \begin{pmatrix} 0.5 + 0.00001 q_s \\ \frac{23}{70} + \frac{1}{70000} \widehat{q_s} \\ -2.7984 + 0.0000172 q_d + 0.00000645 \widehat{q_d} \\ -2.2262 + 0.00000239 q_d + 0.0000287 \widehat{q_d} \end{pmatrix}$$

The only constraints are the market clearing constraints, so the feasible region is

$$K = \{(q_s, \widehat{q_s}, q_d, \widehat{q_d}) | q_s - q_d = 0, \widehat{q_s} - \widehat{q_d} = 0\}$$

Therefore, the VI problem is to

$$\text{find } (q_s^*, \widehat{q_s^*}, q_d^*, \widehat{q_d^*}) \in K \text{ such that}$$

$$G(q_s^*, \widehat{q_s^*}, q_d^*, \widehat{q_d^*})^T \begin{pmatrix} q_s - q_s^* \\ \widehat{q_s} - \widehat{q_s^*} \\ q_d - q_d^* \\ \widehat{q_d} - \widehat{q_d^*} \end{pmatrix} \geq 0 \text{ for all } (q_s, \widehat{q_s}, q_d, \widehat{q_d}) \in K$$

If equilibrium prices $p, \widehat{p}$ are desired, then they can be calculated from (5.24a), (5.24b), using $q_d^*, \widehat{q_d^*}$ from the VI solution. (The same values of $p, \widehat{p}$ can be calculated from (5.25a), (5.25b).)

□

5.2.3.3 Illustrative Example. PIES Multicommodity Competitive Equilibrium Model from Chapter 1

Chapter 1 introduced the Project Independence Evaluation System (PIES) model. Supply side decisions are in the vector x, and demand quantities are in $q \in \mathbb{R}^n$. Under competitive conditions, at equilibrium, x has the property that the total cost (across all firms) of meeting demand q is minimized; in the PIES model, this minimum cost model is assumed to be a linear program in the variables x. Demands q depend on prices $p \in \mathbb{R}^n$ we write $q = q(p)$. However, for our purpose here, it is more useful to use the inverse demand function: $p = p(q)$. The PIES model can be written as follows (π and γ are vectors of dual variables):

$$\text{Minimize}_x \quad c^T x \tag{5.27a}$$

s. t.

$$Ax \geq q \qquad (\pi) \tag{5.27b}$$

$$Bx \geq b \qquad (\gamma) \tag{5.27c}$$

$$x \geq 0 \tag{5.27d}$$

together with

$$\pi = p(q) \tag{5.28}$$

In order to derive the VI form of the PIES model, we need to identify several agents, their objective functions and constraints, and any linking constraints. Here, we again look for mathematical forms that appear *as if* they represent behaviors of several agents. We use the following agents:

- the supply sector, which receives revenue $\pi^T Ax$ and incurs costs $c^T x$ in choosing x;
- n consumer groups, one for each commodity, where the *marginal* value (derivative of total value) of commodity i is given by the ith component, $p_i(q)$ of the inverse demand function, and the cost of purchasing q_i is $\pi_i q_i$.

Therefore, the linking constraints are (5.27b), which connect the supply sector to each of the consumer groups, and the corresponding prices are in the vector π. Each consumer group i chooses q_i in order to maximize its net value, which is total value of q_i minus cost of purchase $\pi_i q_i$; the optimal choice is characterized by setting the derivative equal to zero, $\pi_i - p_i(q) = 0$, giving (8.46), and this is the justification for identifying consumer group i as if it is an agent. In the VI formulation, the element of the mapping G corresponding to consumer group i's objective terms that do *not* involve π is therefore $-p_i(q)$. The elements of G corresponding to the supply sector are in the cost vector c.

Using the general development (5.17), (5.18), the VI formulation can now be presented:

$$\text{Let } K = \{x, q |\, Ax \geq q\,,\, Bx \geq b\,,\, x \geq 0\}. \tag{5.29a}$$

$$\text{Find } (x^*, q^*) \in K \text{ such that} \tag{5.29b}$$

$$c^T(x - x^*) - p(q^*)^T(q - q^*) \geq 0 \quad \forall (x, q) \in K \tag{5.29c}$$

□

5.2.3.4 Illustrative Example. Stochastic Equilibrium Model from Chapter 4

In this example originally introduced in Chapter 4, Section 4.6.3, where a stochastic complementarity model is developed, the agents are: several electricity generation firms, indexed by f; and the owner of the transmission grid. The linking constraints require that, at each node (bus) i of the transmission grid, for each stochastic scenario m, the net injection of power (a grid owner decision variable) must equal the sum of generation minus sales of all firms at that node. The price associated with the linking constraint at node i in scenario m is denoted by w_{im}, and is known as the wheeling fee for transmission services. Each generation firm maximizes its expected profit, and the grid owner maximizes its expected revenue from wheeling fees, where expected values are calculated with the scenario probabilities $prob(i, m)$; this implies, in particular, that the expected value of the wheeling fee at node i in scenario m is $prob(i,m)w_{im}$.

From Chapter 4, we repeat the firms' and grid owner's optimization problems below, along with the linking constraints; some additional symbol definitions follow the problem statements. Then we derive the VI formulation of the model, using the general developments (5.17), (5.18).

Each firm f has the following optimization problem

$$\underset{s_f,x_f,r_f}{\text{Maximize}} \left[\begin{array}{c} \sum_{m,i} prob(i,m)\Big(a_{im}-b_i\left(\sum_g s_{gim}\right)-w_{im}\Big)s_{fim} \\ -\sum_{m,i,h} prob(i,m)\left(C_{fih}-w_{im}\right)x_{fih} \\ -\sum_{m,i} prob(i,m)\left(RC_{fih}-w_{im}\right)r_{fihm} \end{array}\right] \quad (5.30a)$$

s. t.

$$x_{fih} - X_{fih} \le 0 \;\forall i,h \quad (5.30b)$$

$$r_{fihm} - R_{fih} \le 0 \;\forall i,m \quad (5.30c)$$

$$\sum_i s_{fim} - \sum_{i,h} x_{fih} - \sum_{i,h} r_{fihm} = 0 \;\forall m \quad (5.30d)$$

$$x_{fih} \ge 0 \;\forall i,h \quad (5.30e)$$

$$r_{fihm}, s_{fim} \ge 0 \;\forall i,h,m \quad (5.30f)$$

the grid owner's problem is

$$\text{Maximize}_y \quad \sum_m \sum_i prob(i,m)w_{im}y_{im} \quad (5.31a)$$

s. t.

$$-T_{l-} - \sum_i PTDF_{il}y_{im} \le 0 \;\forall l,m \quad (5.31b)$$

$$-T_{l+} + \sum_i PTDF_{il}y_{im} \le 0 \;\forall l,m \quad (5.31c)$$

and the linking constraints are the market clearing conditions

$$0 = -\sum_f s_{fim} + \sum_{f,h} x_{fih} + \sum_f r_{fihm} + y_{im} \;\forall i,m \quad (5.32)$$

with free dual variable $\hat{w}_{im} \equiv prob(i,m)w_{im}$

where s_{fim} is sales, x_{fih} is slow-ramping generation, r_{fihm} is rapid-ramping generation, and y_{im} is the net injection of power at node i. Definitions of all symbols can be found in Chapter 4, Section 4.6.3.1.

To formulate the VI model, we must define the feasible set K, and the components of the mapping G. Recall that G is the concatenation of certain gradients of all agents. Each such gradient for an agent is composed of the partial derivatives of the agent's objective function with respect to that agent's decision variables, but terms involving the dual variables associated with linking constraints are ignored.

The feasible set of the VI is defined for primal variables only, by all of the constraints in all agents' problems, together with the linking constraints:

$$K = \{s_{fim}, x_{fih}, r_{fihm}, y_{im} \mid x_{fih} - X_{fih} \leq 0, r_{fihm} - R_{fih} \leq 0,$$
$$\sum_i s_{fim} - \sum_{i,h} x_{fih} - \sum_{i,h} r_{fihm} = 0 \; \forall f, m,$$
$$x_{fih}, r_{fihm}, s_{fim} \geq 0, \forall f, i, h, m,$$
$$-T_{l-} - \sum_i PTDF_{il} y_{im} \leq 0, -T_{l+} + \sum_i PTDF_{il} y_{im} \leq 0, \forall l, m,$$
$$0 = -\sum_f s_{fim} + \sum_{f,h} x_{fih} + \sum_f r_{fihm} + y_{im}, \forall i, m\}$$

The mapping G must list its elements in the order which corresponds properly with the elements of the variable vector, i.e., if an element of G is the partial derivative of a function with respect to a particular variable, then this element of G must appear in the same position in the vector G as the variable appears in the variable vector. The variable vector is

$$\begin{pmatrix} s_{fim}, \forall f, i, m \\ x_{fih}, \forall f, i, h \\ r_{fihm}, \forall f, i, h, m \\ y_{im}, \forall i, m \end{pmatrix}$$

and the mapping is (note that the coefficient of x_{fih} is $\sum_m prob(i,m) C_{fih} = C_{fih} \sum_m prob(i,m) = C_{fih}$)

$$G(s_{fim}) = \begin{pmatrix} prob(i,m)\left[-a_{im} + b_i\left(s_{fim} + \sum_g s_{gim}\right)\right], \forall f, i, m \\ C_{fih}, \forall f, i, h \\ prob(i,m) RC_{fih}, \forall f, i, h, m \\ 0, \forall i, m \end{pmatrix}$$

Note that, in general, G may depend on all variables. However, in this model, G depends only on the sales variables s_{fim}, and only for the elements of G that correspond to the sales variables. All other elements of G are constant, and in particular, the elements that correspond to the net injections y_{im} are zero because *all* terms in the grid owner's objective (5.31a) involve the prices associated with the linking constraints, i.e., there is no cost function of the grid owner, there is only revenue to the grid owner.

□

5.3 Relations between Variational Inequality and Complementarity Problems

The class of VI problems is very closely related to the class of complementarity problems. This close relationship is exploited in many of the theoretical developments for both classes of problems, e.g., existence, uniqueness, and the design of algorithms for the solution of the problems, as discussed for example in [8]. However, in this section, we focus on the relationship between the problem classes for the purposes of model formulation and analysis of the properties of a model.

In Section 5.3.1, we first show that any complementarity problem can be represented as a variational inequality problem whose feasible set is very simple in form – e.g., the nonlinear complementarity problem (NCP) has a corresponding VI with the nonnegative orthant as the feasible set. This is actually not very useful for model formulation or analysis of the properties of a model, but it is presented here for completeness.

The second result, in Section 5.3.2, is more useful for model formulation and analysis of properties of a model: we show that (under some assumptions) at the solution of a VI problem, dual variables exist for each constraint that defines the feasible region, and that the VI solution and these dual variables satisfy certain relations that extend the Karush-Kuhn-Tucker (KKT) conditions from the class of single objective optimization models to the class of VI models. Furthermore, and most important for formulation and analysis of models, the KKT conditions for the VI form a complementarity problem that is equivalent to the VI problem (under some assumptions, of course).

From the second result about the KKT conditions for a given VI model, it is apparent that it is useful to use notation that distinguishes between the primal variables of a VI model, and the dual variables that are defined in the KKT conditions. For this reason, we use the symbol x for the vector of primal variables of a VI model, different symbols for the dual variables, and the symbol z for the vector of *all* variables (primal and dual). Because a VI model has a corresponding complementarity problem (second result below) which has both primal and dual variables, we use the symbol z for the vector of all variables in a complementarity problem.

In Section 5.3.3, we show how some functions that define constraints for the feasible set K may be moved into the mapping G, while the (former) constraints' dual variables become variables of an alternative, equivalent VI model. For example, if the original VI represents a Nash equilibrium (5.8) in the primal variables x (or a generalized Nash equilibrium (5.18) with equal valuations of linking resources), then this process of moving constraints into the mapping G and introducing dual variables produces a mixed formulation of the model, with both primal and dual variables, in a VI form. Taken to the extreme, with all constraints moved into the mapping (except, perhaps, nonnegativity conditions), we end up with a VI like the one discussed in Section 5.3.1.

5.3.1 Any Complementarity Problem Has an Equivalent Variational Inequality Problem

A very general form of a complementarity problem is the mixed complementarity problem, for which there may be both upper and lower bounds on every variable, and some equality constraints. However, for our purposes here, the simpler nonlinear complementarity problem (NCP) is preferable, because the derivation of the corresponding VI is easier to present.

We first define the NCP that we use. The variables are in the vector $z \in \mathbb{R}^n$, with individual elements z_i for $i = 1, 2, ..., n$. Given a mapping $F : \mathbb{R}^n \longrightarrow \mathbb{R}^n$, with individual elements F_i for $i = 1, 2, ..., n$, the NCP is defined as follows. **NCP**: find a vector $z \in \mathbb{R}^n$ such that $0 \leq z_i,\ \ F_i(z) \geq 0$ and $z_i F_i(z) = 0$ for all $i = 1, ..., n$,
or, in vector form,

$$0 \leq F(z) \perp z \geq 0 \tag{5.33}$$

The result has no clear economic intuition behind it, so we simply present it as a theorem. The proof is omitted here, but is similar to the proof of Proposition 1.4 in [19].

Theorem 5.4. *Define a VI as follows: for* $K = \{z | z \geq 0\}$

$$\textit{find } z^* \in K \textit{ such that } F(z^*)^T(z - z^*) \geq 0\ \forall z \in K \tag{5.34}$$

The set of all solutions z^* *of the VI (5.34) equals the set of all solutions of the NCP (8.29).*

5.3.1.1 Illustrative Example. NCP and Two VI Forms for Coal Yard Model

We begin this example with a reminder of the complementarity form of the coal yard model, from Chapter 3, for the case of equal marginal utilities of yard shares:

$$q - \sum_i \sum_b x_{ib} = 0 \tag{5.35a}$$

$$0 \geq x_{ib} - K_{ib} \perp \gamma_{ib} \geq 0\ \ \forall i, b \tag{5.35b}$$

$$0 \geq -x_{ib} \perp \delta_{ib} \geq 0\ \ \forall i, b \tag{5.35c}$$

$$c_{ib} - (\alpha - \beta q) + \beta \sum_{b'} x_{ib'} + \gamma_{ib} + \phi - \delta_{ib} = 0\ \ \forall i, b \tag{5.35d}$$

$$0 \geq -l + \sum_i \sum_b x_{ib} \perp \phi \geq 0 \tag{5.35e}$$

For the present example, we convert (5.35) to the NCP form, by eliminating δ_{ib} through the use of (5.35d), by removing the equality constraint (5.35a) in the course of eliminating q, to be consistent with the VI form developed in Section 5.2, and by converting all inequalities to "≥ 0" form. Thus, the NCP form that we use here is:

$$0 \leq c_{ib} - (\alpha - \beta q) + \beta \sum_{b'} x_{ib'} + \gamma_{ib} + \phi \perp x_{ib} \geq 0 \ \forall i, b \tag{5.36a}$$

$$0 \leq K_{ib} - x_{ib} \perp \gamma_{ib} \geq 0 \ \ \forall i, b \tag{5.36b}$$

$$0 \leq l - \sum_i \sum_b x_{ib} \perp \phi \geq 0 \tag{5.36c}$$

To connect this with the general form of the NCP (8.29),

$$z = \begin{pmatrix} x_{ib}, \forall i, b \\ \gamma_{ib}, \forall i, b \\ \phi \end{pmatrix}$$

and

$$F(z) = \begin{pmatrix} c_{ib} - (\alpha - \beta q) + \beta \sum_{b'} x_{ib'} + \gamma_{ib} + \phi, \forall i, b \\ K_{ib} - x_{ib}, \forall i, b \\ l - \sum_i \sum_b x_{ib} \end{pmatrix} \tag{5.37}$$

The corresponding VI model, from the theorem above, is defined over the feasible set

$$K = \left\{ z = \begin{pmatrix} x_{ib}, \forall i, b \\ \gamma_{ib}, \forall i, b \\ \phi \end{pmatrix} | x_{ib} \geq 0, \gamma_{ib} \geq 0, \forall i, b, \phi \geq 0 \right\} \tag{5.38}$$

and the mapping is given by (5.37).

Note that the feasible set and the mapping for this form of the VI model are different from (5.19), which were developed in Section 5.2.3.1: the feasible region (5.38) is simply the nonnegative orthant of the combined vector of the primal variables x_{ib} *and* the dual variables γ_{ib}, ϕ; the mapping (5.37) contains expressions which appear in the definition of the feasible region (5.19a); and the elements of the mapping (5.37) which correspond most closely to the mapping (5.19b) contain some extra terms involving the dual variables γ_{ib}, ϕ. In the next subsection, we continue this example and show that the KKT conditions for the VI (5.19) of Section 5.2.3.1 are precisely the NCP (5.36). Thus, there are two VI formulations of the coal yard model – one with only primal variables, and the other with both primal and dual variables.

□

5.3.2 Any Variational Inequality Problem Has an Equivalent Complementarity Problem

Let $G : \mathbb{R}^n \to \mathbb{R}^n$, let $g : \mathbb{R}^n \to \mathbb{R}^m$ be such that g_i is convex for all $i = 1, \ldots, m$, and let $h : \mathbb{R}^n \to \mathbb{R}^p$ be such that h_j is affine (i.e., a constant term plus a linear term) for all $j = 1, \ldots, p$. Assume further that g is continuously differentiable. Define the set

$$K = \{x \in \mathbb{R}^n | g(x) \leq 0, h(x) = 0\} \tag{5.39}$$

The assumptions about g_i and h_j ensure that K is a convex set. We define the variational inequality problem

$$VI(K, G)\text{: find } x^* \in K \text{ such that } G(x^*)^T(x - x^*) \geq 0 \; \forall x \in K \tag{5.40}$$

We now show that at a solution x^* of $VI(K, G)$, we can define Karush-Kuhn-Tucker (KKT) conditions which are the same as KKT conditions for a constrained optimization problem, except that the gradient of the objective function is replaced by the mapping $G(x)$. To make it clear where we are headed, we first present the result. The KKT conditions for VI(K, G) are in the form of the mixed complementarity problem denoted MCP-VI below. The variable vectors $v^* \in \mathbb{R}^m$ and $\beta^* \in \mathbb{R}^p$.

MCP-VI:

$$G(x^*) + \nabla g(x^*)^T v^* + \nabla h(x^*)^T \beta^* = 0\,,\ x \text{ is free} \tag{5.41a}$$
$$0 \geq g(x^*) \perp v^* \geq 0 \tag{5.41b}$$
$$h(x^*) = 0\,,\ \beta^* \text{ is free} \tag{5.41c}$$

Now we use a standard argument [8] to show that at any solution x^* of $VI(K, G)$, there exist vectors $v^* \in \mathbb{R}^m$ and $\beta^* \in \mathbb{R}^l$ such that the conditions (5.41) are true. Let $c = G(x^*)$, and define the following constrained optimization problem (note that c is treated as a constant in this problem):

$$\text{Minimize}_x \quad c^T x \tag{5.42a}$$

s. t.

$$g(x) \leq 0 \tag{5.42b}$$
$$h(x) = 0 \tag{5.42c}$$

A solution x^* of $VI(K, G)$ is feasible in the optimization problem (5.42), and for any other feasible point x, $c^T(x - x^*) = G(x^*)^T(x - x^*) \geq 0$, by (5.40), i.e., $c^T x \geq c^T x^*$, which says that the solution x^* of $VI(K, G)$ is an optimal solution of the optimization problem (5.42). At a solution of the optimization problem (5.42), such as x^*, the KKT conditions must hold (see Chapter 2), i.e., there exist vectors $v^* \in \mathbb{R}^m$ and $\beta^* \in \mathbb{R}^p$ such that

$$c + \nabla g(x^*)^T v^* + \nabla h(x^*)^T \beta^* = 0\,,\ x \text{ is free}$$
$$0 \geq g(x^*) \perp v^* \geq 0$$
$$h(x^*) = 0\,,\ \beta^* \text{ is free}$$

Replacing c by $G(x^*)$ gives the desired result (5.41).

The same argument works in the reverse direction. Any vectors x^*, v^* and β^* which solve the MCP-VI (5.41) can be viewed as vectors which satisfy the KKT conditions of the optimization problem (5.42) with $c = G(x^*)$. This implies that $c^T(x - x^*) = G(x^*)^T(x - x^*) \geq 0$ for any other feasible point x, which means that x^* solves $VI(K, G)$ (5.40).

The relations between the VI and MCP-VI are summarized in the following theorem.

Theorem 5.5. *Assume that $G : \mathbb{R}^n \to \mathbb{R}^n$, that $g : \mathbb{R}^n \to \mathbb{R}^m$ is continuously differentiable, and that $h : \mathbb{R}^n \to \mathbb{R}^p$ is affine. (a) If x^* solves the VI defined by (5.39), (5.40), then there exist $v^* \in \mathbb{R}^m, \beta^* \in \mathbb{R}^p$ such that MCP-VI (5.41) is satisfied. (b) If (x^*, v^*, β^*) satisfies (5.41), then x^* solves the VI defined by (5.39), (5.40).*

Because of this equivalence of the VI and MCP-VI, one simple procedure to follow for solution of VI(K, G) is to solve MCP-VI – the robust Path solver is available in GAMS. Furthermore, even the conversion of VI(K, G) into MCP-VI is automated by the Extended Mathematical Programming (EMP) interface in GAMS [9]. The coding in GAMS of the VI formulation of a problem is easier and less error prone than the coding of the same problem as the equivalent MCP, because the MCP coding requires the explicit declaration of the dual variables and, more importantly, the entry of expressions for all terms in (5.41a) involving the dual variables. However, derivation of the MCP form of a model is relatively straightforward – typically, one writes down the KKT conditions of each agent and includes market clearing constraints to formulate the MCP model – while derivation of the VI form of the same model is not as easy. This chapter is intended to present methods to ease the formulation of VI models.

However, it is important to note that the conditions in (5.41) involving dual variables can give important theoretical insights into a VI model. Therefore, when creating a VI model, we usually write out the KKT conditions, even if the computer coding is done without explicit mention of dual variables.

The VI and MCP formulations are compared in the next example of a model from Chapter 3.

5.3.2.1 Illustrative Example. Comparison of MCP and VI Forms of Coal Market Model with Coal Yard Limits

The VI form of the coal yard model was given in Section 5.2.3.1. We repeat it here, for ease of comparison with the MCP-VI form, but we slightly rearrange the constraints to fit the general form in (5.39). The feasible set K and mapping G are:

$$K=\left\{(x_{ib})\,|\,x_{ib}-K_{ib}\leq 0,-x_{ib}\leq 0,\forall i,b,-l+\sum_{i,b}x_{ib}\leq 0\right\} \tag{5.44a}$$

$$G_{ib}(x)=c_{ib}-\left(\alpha-\beta\sum_{j,b}x_{jb}\right)+\beta\sum_{b}x_{ib} \tag{5.44b}$$

There are no equality constraints $h(x)=0$, and so the MCP-VI (5.41) for this model is:

$$c_{ib}-\left(\alpha-\beta\sum_{j,b}x_{jb}\right)+\beta\sum_{b}x_{ib}+v_{ib}-w_{ib}+\mu=0,\ x_{ib}\text{ is free }\forall i,b$$
$$0\geq x_{ib}-K_{ib}\perp v_{ib}\geq 0\ \forall i,b$$
$$0\geq -x_{ib}\perp w_{ib}\geq 0\ \forall i,b$$
$$0\geq -l+\sum_{i,b}x_{ib}\perp\mu\geq 0$$

which can be simplified by elimination of the w_{ib} dual variables, to become

$$0\leq c_{ib}-\left(\alpha-\beta\sum_{j,b}x_{jb}\right)+\beta\sum_{b}x_{ib}+v_{ib}+\mu\perp x_{ib}\geq 0\ \forall i,b \tag{5.45a}$$

$$0\geq x_{ib}-K_{ib}\perp v_{ib}\geq 0\ \forall i,b \tag{5.45b}$$

$$0\geq -l+\sum_{i,b}x_{ib}\perp\mu\geq 0 \tag{5.45c}$$

The first line (5.45a) of the MCP-VI statement contains the VI mapping expression (5.44b) plus terms involving dual variables and gradients of constraint expressions. The remaining lines of (5.45) contain the constraints that define the VI feasible region (5.44a), sign restrictions on dual variables, and statements of complementarity with dual variables.

□

5.3.3 Alternative Equivalent Forms of Variational Inequality Problems

For a given VI model, it is possible to drop some constraints from the definition of the feasible set K, include the dual variables of the dropped constraints among the variables of the VI model, and modify the mapping G, to obtain a different-looking, but equivalent VI model. We can show this by proving that the KKT conditions of the new VI are the same as the KKT conditions of the original VI. We follow the notation in the VI and MCP-VI definitions, (5.39), (5.40) and (5.41), but suitably modified.

To begin, we need to distinguish between the constraints that will remain in the definition of the feasible set, $g_1(x) \leq 0, h_1(x) = 0$, and those which will be dropped from the feasible set, $g_2(x) \leq 0, h_2(x) = 0$. Thus, in (5.39) and (5.41),

$$g(x) = \begin{pmatrix} g_1(x) \\ g_2(x) \end{pmatrix} \quad h(x) = \begin{pmatrix} h_1(x) \\ h_2(x) \end{pmatrix}$$
$$v = \begin{pmatrix} v_1 \\ v_2 \end{pmatrix} \qquad \beta = \begin{pmatrix} \beta_1 \\ \beta_2 \end{pmatrix}$$

Therefore, the VI (5.40) is described, with the distinction between the two sets of constraints, as follows,

$$\text{for } K=\{x \in \mathbb{R}^n | g_1(x) \leq 0, h_1(x)=0, g_2(x) \leq 0, h_2(x)=0\}, \quad (5.46a)$$
$$\text{find } x^* \in K \text{ such that } G(x^*)^T(x - x^*) \geq 0 \text{ for all } x \in K, \quad (5.46b)$$

and the equivalent MCP VI (5.41), with the distinction between constraint sets, is

$$G(x^*) + \nabla g_1(x^*)^T v_1^* + \nabla h_1(x^*)^T \beta_1^*$$
$$+ \nabla g_2(x^*)^T v_2^* + \nabla h_2(x^*)^T \beta_2^* = 0\,, \ x \text{ is free} \quad (5.47a)$$
$$0 \geq g_1(x^*) \perp v_1^* \geq 0 \quad (5.47b)$$
$$0 \geq g_2(x^*) \perp v_2^* \geq 0 \quad (5.47c)$$
$$h_1(x^*) = 0\,, \ \beta_1^* \text{ is free} \quad (5.47d)$$
$$h_2(x^*) = 0\,, \ \beta_2^* \text{ is free} \quad (5.47e)$$

The alternative VI is defined over the variables x, v_2 and β_2. The mapping $G(x)$, which has the same dimension as x, is replaced by the following mapping, which has dimension equal to the sum of the dimensions of x, v_2 and β_2:

$$\text{mapping for alternative VI} \ = \begin{pmatrix} G(x) + \nabla g_2(x)^T v_2 + \nabla h_2(x)^T \beta_2 \\ -g_2(x) \\ -h_2(x) \end{pmatrix} \quad (5.48)$$

The definition of the alternative VI is:

$$\text{for } K'=\{x, v_2, \beta_2 | g_1(x) \leq 0, h_1(x)=0, -v_2 \leq 0, \beta_2 \,\text{is free}\} \quad (5.49a)$$
$$\text{find } (x^*, v_2^*, \beta_2^*) \in K'$$

$$\text{s.t. } [G(x^*) + \nabla g_2(x^*)^T v_2^* + \nabla h_2(x^*)^T \beta_2^*]^T (x - x^*)$$
$$- g_2(x^*)^T (v_2 - v_2^*) - h_2(x^*)^T (\beta_2 - \beta_2^*) \geq 0 \ \ \forall\, x \in K' \quad (5.49b)$$

Now we write out the KKT conditions for (5.49), and then show that these KKT conditions are equivalent to (5.47). Since (5.49) and (5.46) are equivalent to the same MCP (5.47), they are equivalent to each other.

To derive the KKT conditions for (5.49), we follow the route from (5.39), (5.40) to (5.41), but applied to the alternative VI (5.49). Expression (5.41a) involves the mapping of the VI, gradients of the constraint functions, and dual variables of the constraint functions. In (5.49), the mapping is (5.48), the constraint functions are $g_1(x), h_1(x)$,and -v_2, and for dual variables of the constraint functions, we use the symbols v_1, β_1 and w, respectively. Therefore, expression (5.41a), applied to (5.49) is:

$$\begin{aligned} &G(x^*) + \nabla g_1(x^*)^T v_1^* + \nabla h_1(x^*)^T \beta_1^* \\ &\qquad + \nabla g_2(x^*)^T v_2^* + \nabla h_2(x^*)^T \beta_2^* = 0\,,\ x \text{ is free} \\ &-g_2(x^*) - w^* = 0,\ v_2^* \text{ is "free"} \\ &-h_2(x^*) = 0,\ \beta_2^* \text{ is free} \end{aligned} \quad (5.50)$$

and the expressions (5.41b), (5.41c), applied to (5.49a) become

$$\begin{aligned} &0 \geq g_1(x^*) \perp v_1^* \geq 0 \\ &h_1(x^*) = 0\,,\ \beta_1^* \text{ is free} \\ &0 \geq v_2^* \perp w^* \geq 0 \end{aligned} \quad (5.51)$$

Using (5.50) to eliminate w^* in (5.51) produces the same set of conditions as in (5.47).

If desired, a modeler can take this process to the point that the only constraints remaining in the definition of the feasible set are nonnegativity conditions on variables, with all other constraint expressions appearing as entries in the mapping function. An example of this can be found in [17].

5.3.3.1 Alternative Form of VI for Nash Equilibrium with Linking Constraints

In Section 5.2.3, we saw that the Nash equilibrium among agents $i = 1, \ldots, I$, when there are linking constraints (and all agents agree on the marginal value of the linking constraint) can be represented as a VI in this way:

$$\text{for } \overline{K} = \{x\,|x_i \in K_i,\ -L + \sum_{i=1}^{I} A_i x_i \leq 0\}$$
$$\text{find } x^* \in \overline{K} \text{ such that} \quad (5.52)$$

$$G(x^*)^T(x - x^*) \geq 0 \text{ for all } x \in \overline{K} \tag{5.53}$$

$$\text{where } G(x) = \begin{pmatrix} \nabla_1 f_1(x) \\ \vdots \\ \nabla_I f_I(x) \end{pmatrix}$$

Now let us derive the alternative form of the VI which arises from dropping the linking constraints from the feasible set, introducing the dual variables μ of the linking constraints as variables in the alternative VI, and modifying the mapping G. We suppose that "$x_i \in K_i$" in the definition of $\overline{K}$ stands for a list of constraints that correspond to $g_1(x) \leq 0, h_1(x) = 0$ in (5.46a), that the linking constraints $-L + \sum_{i=1}^{I} A_i x_i \leq 0$ correspond to $g_2(x) \leq 0$, and that there are no equality constraints $h_2(x) = 0$. Following (5.48) and (5.49),

$$\text{the alternative mapping is } \begin{pmatrix} \nabla_1 f_1(x) + A_1^T \mu \\ \vdots \\ \nabla_I f_I(x) + A_I^T \mu \\ L - \sum_{i=1}^{I} A_i x_i \end{pmatrix}$$

and the alternative VI is described as follows:

$$\text{for } \overline{K}' = \{x, \mu \,|\, x_i \in K_i,\ \mu \geq 0\}$$

$$\text{find } (x^*, \mu^*) \in \overline{K}' \text{ such that}$$

$$\sum_{i=1}^{I} [\nabla_i f_i(x^*) + A_i^T \mu^*]^T (x - x^*) \tag{5.54}$$

$$+ (L - \sum_{i=1}^{I} A_i x_i^*)^T (\mu - \mu^*) \geq 0 \ \forall (x, \mu) \in \overline{K}' \tag{5.55}$$

The original form (5.52) is a problem in the primal decision variables of the agents, x, but the alternative form (5.55) also includes some dual variables μ, which are the prices of the resources that are limited in the linking constraints. Examples of this approach to linking constraints can be found in [19] (Chapter 8) and [15].

5.3.3.2 Illustrative Example. Alternative VI for Nash-Cournot Model of Coal Market with Yard Limit

The original VI form (5.19) of this model is

$$\text{for} \quad K = \left\{ x = (x_{ib}) \mid x_{ib} - K_{ib} \le 0,\ x_{ib} \ge 0,\ \forall i, b \ \ -l + \sum_{i,b} x_{ib} \le 0 \right\},$$

$$\text{find } x^* \in K \text{ such that } G(x^*)^T (x - x^*) \ge 0 \text{ for all } x \in K$$

$$\text{where } G_{ib}(x) = c_{ib} - \left(\alpha - \beta \sum_{j,b} x_{jb} \right) + \beta \sum_{b} x_{ib}$$

Following (5.55), the alternative VI is

$$\text{for } K' = \{ x = (x_{ib}), \mu \mid x_{ib} - K_{ib} \le 0,\ x_{ib} \ge 0,\ \ \forall i, b,\ \mu \ge 0 \}$$

$$\text{find } (x^*, \mu^*) \in K' \text{ s.t.} \sum_{i,b} [c_{ib} - \left(\alpha - \beta \sum_{j,b} x^*_{jb} \right) + \beta \sum_{b} x^*_{ib} + \mu^*]^T (x - x^*)$$

$$+ (l - \sum_{i,b} x^*_{ib})^T (\mu - \mu^*) \ge 0 \text{ for all } (x, \mu) \in K'$$

Some modelers may prefer the original form, because it is a problem in only the primal variables. Others may prefer the alternative form because the expression for the negative of the marginal profit of the x_{ib} decision variable, $c_{ib} - \left(\alpha - \beta \sum_{j,b} x^*_{jb} \right) + \beta \sum_b x^*_{ib} + \mu^*$, includes the equilibrium price that must be paid for coal yard space.
□

5.4 Generalized Nash Equilibrium as Quasi-variational Inequality Problem

The notion of a generalized Nash equilibrium is introduced in Chapters 1 and 3. Here, we discuss generalized Nash equilibrium in the context of VI models, which leads to the quasi-variational inequality (QVI) problem.

Let us return to the Nash equilibrium problem for agents $i = 1, \ldots, I$, with linking constraints $\sum_{i=1}^{I} A_i x_i \le L$, but with a different twist. Instead of having each agent i purchase amounts $A_i x_i$ (or sell them, depending on the signs of coefficients in A_i) at prices μ that are common to all agents, we explore the consequences of applying the Nash idea to the linking constraints: agent i takes the view that the decisions of the other players are beyond the control of i, and therefore the linking constraints are

$$A_i x_i \le L - \sum_{j=1,\, j \ne i}^{I} A_j x_j$$

where the right side is treated as a constant, from the point of view of agent i. This approach is consistent with the treatment of the objective function of agent i: $f_i(x_i, x_{-i})$ depends on the decisions of the other agents, but agent i views the other agents' variables as parameters, beyond the control of i. However, this consistency in approach brings a complication when we explore the KKT conditions of the optimization problems of all agents.

Each agent i is assumed to have a feasible set that is parameterized by the decisions of the other agents, x_{-i},

$$K_i(x_{-i}) = \{x_i \in \mathbb{R}^{n_i} | g_i(x_i) \leq 0, A_i x_i \leq L - \sum_{j=1,\, j\neq i}^{I} A_j x_j\} \tag{5.57}$$

where $g_i : \mathbb{R}^{n_i} \longrightarrow \mathbb{R}^{m_i}$, each function in g_i is convex, A_i has m_0 rows, and agent i treats x_{-i} variables as parameters. (There could also be affine equality constraints in the definition of K_i, but an affine equality constraint can be represented by two affine inequality constraints, which fits the form (5.57) and leaves the following expressions a little simpler.) Each agent i wishes to optimize its objective subject to the constraints that it faces,

$$\text{Minimize}_{x_i \in K_i(x_{-i})} \quad f_i(x_i, x_{-i}) \tag{5.58}$$

which leads to the definition of generalized Nash equilibrium.

Definition 5.2. A generalized Nash equilibrium (GNE) x^* solves (5.58) simultaneously for all $i = 1, \ldots, I$, i.e. for every agent i,

$$f_i(x_i, x^*_{-i}) \geq f_i(x^*_i, x^*_{-i}), \forall x_i \in K_i(x^*_{-i})$$

With dual variables v_i for $g_i(x_i) \leq 0$ constraints, and μ_i for the linking constraints (note the i subscript on μ_i, which means that each agent i can have different dual variables for the linking constraints), the KKT conditions are equivalent to (5.58):

$$\nabla_i f_i(x_i, x_{-i}) + \nabla_i g_i(x_i)^T v_i + A_i^T \mu_i = 0 \tag{5.59a}$$

$$0 \geq g_i(x_i) \perp v_i \geq 0 \tag{5.59b}$$

$$0 \geq -L + \sum_{j=1,\, j\neq i}^{I} A_j x_j + A_i x_i \perp \mu_i \geq 0 \tag{5.59c}$$

The conditions (5.59) form a well-defined, square NCP for a single agent i – the variables are x_i, v_i, μ_i, and there are as many conditions as there are variables. However, if we collect (5.59) for all the agents into one overall NCP, to represent a GNE problem, then we encounter a difficulty: the linking constraints in (5.59c) are identical in form for every agent, but there is a separate vector of dual variables μ_i for each agent, which implies that the overall NCP is nonsquare, with more variables than conditions. There could be more than one solution to the overall NCP, and therefore more than one equilibrium. This is troublesome for a modeler, because it is not clear which solution is preferred as a prediction of the model. It seems that we need to specify some extra conditions, to have a well-defined model.

Before we explore different types of extra conditions, let us first express the overall NCP of (5.59) for all i in a form as close as possible to a variational inequality problem. Given the optimal choices x^*_{-i} of the other agents, the VI form of the optimization problem (5.58) for agent i is

$$\begin{aligned}&\text{find } x_i^* \in K_i(x^*_{-i}) \text{ such that}\\&\nabla_i f_i(x_i^*, x^*_{-i})^T(x_i - x_i^*) \geq 0, \forall x_i \in K_i(x^*_{-i})\end{aligned} \tag{5.60}$$

When we put all of the VIs together for all agents, we add up the inequality in (5.60) over all agents i, but the tricky part is to define the overall feasible set as a Cartesian product of the sets $K_i(x^*_{-i})$. We must recognize that the very definition of the overall feasible set depends on the solution x^*, because its constituent parts do:

$$K(x^*) = K_1(x^*_{-1}) \times K_2(x^*_{-2}) \times \ldots \times K_I(x^*_{-I})$$

It is this point to set mapping $K(x^*)$ which causes the overall model of all agents to deviate from the VI form. The overall model of all agents is called a quasi-variational inequality, or QVI for short, and is defined as

$$\begin{aligned}&\textbf{QVI: } \text{find } x^* \in K(x^*) \text{ such that}\\&\qquad G(x^*)^T(x - x^*) \geq 0, \forall x \in K(x^*)\end{aligned} \tag{5.61}$$

$$\text{where } G(x) = \begin{pmatrix} \nabla_1 f_1(x) \\ \vdots \\ \nabla_I f_I(x) \end{pmatrix} \tag{5.62}$$

The multiplicity of solutions to (5.61) arises because different values of x^* define different feasible sets $K(x^*)$. From the point of view of agent i, a different x^* means a different x^*_{-i}, which implies a different right hand side of the linking constraint in (5.57), and this means that the optimal solution for i is different.

The multiplicity of solutions to a QVI is not just an abstract theoretical issue; for a modeler, the different solutions correspond to different policy or regulatory mechanisms by which the agents in the game can share the common resources of the linking constraints. For example, Boucher and Smeers [1] discuss a QVI of a generalized Nash equilibrium in which the agents are electric generating companies, electricity consumers, an electricity system operator, power marketers, and owners of transmission capacity. They present several different ways in which the markets for electric energy and transmission services can conceivably be organized – e.g., one way has the system operator merely checking feasibility of power flows on the transmission lines but playing no economic role, while another has the system operator also acting as the sole seller of transmission services. The authors point out that, without a well-defined agent in charge of allocating tradable transmission services, the market is incomplete, and it "is bound to have unpredictable results." [1]

Other papers that use QVI formulations of energy markets include [4] and [14]. An early formulation of generalized Nash equilibrium was introduced by Rosen [20], before finite-dimensional VI or QVI models were known in the literature. Rosen referred to the linking constraints as "coupling constraints". There have been several applications of Rosen's approach to energy market models and to equilibrium in international climate policy, including [3], [5], [13] and [18].

A review of generalized Nash equilibrium models may be found in [7], including examples of energy and non-energy applications, and mathematical properties.

5.4.1 Some Important Properties of Quasi-variational Inequality Problems

The QVI problem (5.61) can be defined in a general way, without direct reference to the GNE as the motivation, but for all applications of interest here, we are only concerned with the special case of the GNE, with each agent i interested to solve (5.58). There are three important theorems that allow us to understand QVI solutions as solutions of related, ordinary VI problems. Therefore, these theorems allow us to explore the multiple solutions to QVI problems, and to compute QVI solutions, using the machinery developed for solving ordinary VI problems. We present these three theorems, for the case of our GNE problem (in particular, we assume an affine form for the linking constraints). For notational convenience, the vector $x = (x_i)_{i=1}^I$, the $n = \sum_i n_i$ dimensional concatenation of the vectors x_i, and similarly, $g(x) = (g_i(x_i))_{i=1}^I$.

5.4.1.1 The VI Solution is a QVI Solution: Linking Duals are Equal

The first result is due to Harker [12]. The VI in this theorem is the VI (5.17), (5.18) derived in Section 5.2.3.

Theorem 5.6. *[12] Let $K = \{x \in \mathbb{R}^n | g(x) \leq 0, \sum_{i=1}^I A_i x_i \leq L\}$, and let $G(x)$ be defined as in (5.62). Any solution x^* to the variational inequality problem VI(K, G) is also a solution to the QVI problem (5.61). Furthermore,* *for this solution to the QVI problem, the dual variable vectors μ_i^* in (5.59) are the same for all i, i.e., $\mu_i^* = \mu^*$ for all i.*

5.4.1.2 Illustrative Example. Simple Electric Capacity Market Model with High Cost Green Energy and Equal Prices for All

Suppose that the government wants electric generation capacity in its jurisdiction of D megawatts in a future year. The firms that own and can build capacity are indexed by f, and there are three types of generation, indexed by h: existing capacity, which pollutes, and which needs a small expense now to continue to be available in the plan year; new conventional capacity, which also pollutes; and new green capacity, which does not pollute. The variable for the amount of capacity of firm f, type h, is x_{fh}. The total cost to firm f of providing capacity x_{fh} is given by

$$c_{fh}(x_{fh}) = a_{fh}x_{fh} + \frac{b_{fh}}{2}x_{fh}^2$$

The function $\sum_h c_{fh}(x_{fh})$ corresponds to $f_i(x)$ in the general theory above, i.e., f corresponds to i. Alternatively, we could view each type of capacity, h, as a separate agent, which means that each combination (f, h) corresponds to i, and therefore $c_{fh}(x_{fh})$ corresponds to $f_i(x)$. This latter view is adopted in the remainder of this example, and also in subsequent examples.

Each capacity variable x_{fh} is nonnegative and bounded above by X_{fh}. The linking constraint requires that the total of all capacity variables is at least as great as the government target of D megawatts. Therefore, the feasible set of the VI is

$$K = \{x_{fh} | 0 \leq x_{fh} \leq X_{fh} \, \forall f, h; \sum_{f,h} x_{fh} \geq D\}$$

and the mapping G is given by the derivatives of the functions c_{fh}:

$$G(x) = (a_{fh} + b_{fh}x_{fh}) \forall f, h$$

At a solution x^* of VI(K, G), the common price paid to all firms, for every type of capacity is given by the dual variable, μ^*, associated with the linking constraint. Therefore, the total payment to the firms is $\mu^* D$, the total cost of all capacity is $\sum_{f,h}(a_{fh}x_{fh}^* + \frac{b_{fh}}{2}x_{fh}^{*2})$ and the profit earned by firm f for its capacity of type h is

$$\mu^* x_{fh}^* - (a_{fh}x_{fh}^* + \frac{b_{fh}}{2}x_{fh}^{*2})$$

Data for a two-firm example are given in Table 5.1.

Table 5.1 Data for electric capacity model

	a_{fh}, b_{fh}, X_{fh}		
Firm	Existing	Conventional	Green
1	180, 0, 140	1000, 15, 90	1800, 90, 50
2	110, 0, 100	900, 18, 75	2000, 70, 80
	Capacity Requirement $D = 400$		

Table 5.2 Capacity model results: VI with equal prices

	Existing	Conventional	Green
Capacity	240	152.6	7.4
Cost	36,200	240,812.5	15,069.3
Profit	492,574.2	95,436.1	1198.3
Total Payment = 881,290.3		Total Cost = 292,081.7	

Output from the solution to VI(K, G) is given in Table 5.2. Capacity, Cost and Profit are each summed over the two firms, but broken down by type of generation capacity.

□

5.4.1.3 Modified VI: First Price-Directed Search for QVI Solutions

The next result, from [16], allows us to explore different solutions to the QVI problem, for which the dual variables μ_i^* may be different for different agents i. The modeler provides some non-negative parameters in vectors ω_i, of the same dimension as the linking dual vector μ, then solves a related VI problem that treats ω_i as a vector of extra prices faced by agent i for the amounts $A_i x_i$.

Theorem 5.7. *[16] Let* $K = \{x \in \mathbb{R}^n | g(x) \le 0, \sum_{i=1}^I A_i x_i \le L\}$, *let* $\omega = (\omega_i)_{i=1}^I$ *where* $\omega_i \in \mathbb{R}^{m_0}$ *and* $\omega_i \ge 0$, *let*

$$G^\omega(x) = (\nabla_i f_i(x) + A_i^T \omega_i)_{i=1}^I$$

and let (x^*, v^*, μ^*) *be a solution to the variational inequality* $VI(K, G^\omega)$. *If* $(L - \sum_{i=1}^I A_i x_i^*)^T \omega_i = 0$ *for* $i = 1, \ldots, I$, *then* x^* *is a solution to the* ***QVI*** *problem (5.61) and the corresponding GNE problem, with* $\mu_i^* = \mu^* + \omega_i$.

Proof. See [16], Theorem 3.3.

The theorem says that if every ω_i is complementary to the linking constraint at the solution x^* of the related VI problem, then x^* is a solution to the QVI problem, with $\mu_i^* = \mu^* + \omega_i$, where μ^* is the dual vector for the linking constraint from the related VI solution. In other words, provided the modeler correctly guesses zero values for components of ω_i that correspond

Table 5.3 Capacity model results: VI with subsidy

	Existing	Conventional	Green
Capacity	240	142.1	17.9
Cost	36,200	218,100.5	40,409.6
Profit	471,929.0	82,760.0	6,431.1
Total Payment = 855,830.4		Total Cost = 294,710.2	

to nonbinding linking constraints at the solution to the related VI problem, this procedure finds a QVI solution with unequal linking duals that equal the common dual μ^* plus ω_i. In [16], it is also shown that any QVI solution is also a solution to a variational inequality problem VI(K, G^ω), with $\omega = (\mu_i^*)_{i=1}^I$ from the QVI solution.

5.4.1.4 Illustrative Example. Electric Capacity Market Model with Subsidized Green Energy

Continuing with the same example, we suppose that the government is not pleased with the small amount of nonpolluting green electric capacity as in the solution to the VI, based on every type of capacity receiving the same price. The government therefore decides to offer a subsidy for green capacity, but no subsidy for existing or conventional capacity. Therefore, the subsidy symbol is indexed by h in this example: $\omega_h = 0$ for $h \in \{existing, conventional\}$, and $\omega_h = 500$ for $h = green$. The feasible set K is unchanged from the previous example. The mapping changes to

$$G^\omega(x) = (a_{fh} + b_{fh}x_{fh} - \omega_h)\forall f, h$$

Output from the solution to VI(K, G^ω) is given in the Table 5.3. The requirement of the theorem – that $(L - \sum_{i=1}^I A_i x_i^*)^T \omega_i = 0$ – here means that the product of the subsidy times the slack in the linking constraint must be zero. This is true of the solution because the slack is zero.

Comparison with Table 5.2 shows that the subsidy has increased the amount of green capacity, at the expense of new conventional capacity. The total cost of capacity has increased, and the total payment for capacity, $(\mu^* + \omega_h)D$, has decreased.

□

5.4.1.5 Modified VI: Second Price-Directed Search for QVI Solutions

A different price-directed way to find QVI solutions is most easily explained using the alternative form (5.55) of VI representation of Nash equilibrium with linking constraints. The idea is to make the different agents' linking dual variable vectors proportional to each other, according to proportionality factors that are guessed in advance. In the following theorem, the proportionality factors (one for each linking constraint) are in the vector $\phi_i = (\phi_{ik})_{k=1}^{m_0}$, and the symbol "$\circ$" in $\phi_i \circ \mu$ stands for the Hadamard product, i.e., the vector whose k^{th} component is $\phi_{ik}\mu_k$, the product of the corresponding components in the two vectors. The idea is originally due to Rosen [20], but we have a different notation and approach here: Rosen's proportionality factors, $r_{ik} > 0$ are the inverse of ours, $r_{ik} = \phi_{ik}^{-1}$.

Theorem 5.8. *Let* $K' = \{x \in \mathbb{R}^n, \mu \in \mathbb{R}^{m_0} | g(x) \leq 0, \mu \geq 0\}$, *let* $\phi = (\phi_i)_{i=1}^I$ *where* $\phi_i \in \mathbb{R}^{m_0}$ *and* $\phi_i > 0$, *let*

$$G^\phi(x) = \begin{pmatrix} (\nabla_i f_i(x) + A_i^T(\phi_i \circ \mu))_{i=1}^I \\ L - \sum_{i=1}^{I} A_i x_i \end{pmatrix}$$

and let (x^*, v^*, μ^*) *be a solution to the variational inequality VI*(K', G^ϕ). *If* $\mu_i^* = \phi_i \circ \mu^*$, *then* $(x^*, v^*, (\mu_i^*)_{i=1}^I)$ *is a solution to the QVI problem (5.61).*

Remark. *Readers can easily check that, with* $\mu_i^* = \phi_i \circ \mu^*$, *the KKT conditions for VI*(K', G^ϕ) *are the same as the set of KKT conditions (5.59) for all agents in the generalized Nash equilibrium.*

5.4.1.6 Illustrative Example. Electric Capacity Market Model with Green Price a Multiple of Conventional Price

Suppose that the government wishes to explore a different mechanism to encourage green capacity – a guarantee that green capacity will be paid 25% more than existing or new conventional capacity. Therefore, for this example, $\phi_h = 1$ for $h \in \{existing, conventional\}$, and $\phi_h = 1.25$ for $h = green$. The feasible set is

$$K' = \{x_{fh}, \mu | 0 \leq x_{fh} \leq X_{fh} \, \forall f, h; \, \mu \geq 0\}$$

and the mapping is

$$G^\phi(x) = \begin{pmatrix} (a_{fh} + b_{fh}x_{fh} - \phi_h \mu) \forall f, h \\ -D + \sum_{f,h} x_{fh} \end{pmatrix}$$

Output from the solution to VI(K', G^ϕ) is given in Table 5.4.

This price multiple policy has an effect similar to the subsidy of the previous example, but with slightly more green capacity.

□

Table 5.4 Capacity model results: VI with proportional prices

	Existing	Conventional	Green
Capacity	240	141.5	18.5
Cost	36,200	216,851.3	41,962.5
Profit	470,769.1	82,074.6	6,854.3
Total Payment = 854,711.8		Total Cost = 295,013.8	

5.4.1.7 Modified VI: Resource-Directed Search for QVI Solutions

Another result by [16] involves a different type of guess to explore various solutions to the QVI problem: the right side vector of the linking constraint, L, is allocated to each agent i as $L = \sum_i \beta_i$. Each β_i becomes a limitation directly on agent i as $A_i x_i \leq \beta_i$, instead of the linking constraints, and the corresponding VI problem is solved with no linking constraints. To put it another way, a guess is made of $\beta_i = L - \sum_{j \neq i} A_j x_j$ for each i, thereby replacing the linking constraints by ordinary, nonlinking constraints for each agent.

Theorem 5.9. *[16] Let $\beta_i \in \mathbb{R}^{m_0}, i = 1, \ldots, I$, be vector parameters such that $\sum_i \beta_i = L$, let $K^\beta = \{x \in \mathbb{R}^n | g(x) \leq 0, A_i x_i \leq \beta_i \, \forall i\}$ where $\beta = (\beta_i)_{i=1}^I$, let $G(x) = (\nabla_i f_i(x))_{i=1}^I$, and let x^* be a solution to VI(K^β, G). If*

$$\text{for each } k = 1, \ldots, m_0, \quad \left\{ \begin{array}{lll} \text{either} & A_{ik} x_i^* = \beta_{ik}, & i = 1, \ldots, I \\ \text{or} & A_{ik} x_i^* < \beta_{ik}, & i = 1, \ldots, I \;, \end{array} \right\}$$

*where A_{ik} is the kth row of A_i and β_{ik} is the kth entry of β_i, then x^* is a solution to the **QVI** problem (5.61) and the corresponding GNE problem.*

Proof. *See [16], Theorem 3.6.*

The theorem gives a way to recognize that any particular guess of the β_i vectors leads to a valid solution of the QVI problem: for each linking constraint, either all agents experience it as a binding constraint, or all agents experience it as a nonbinding constraint. Another theorem in [16] says that any QVI solution can be obtained as the solution to a variational inequality problem VI(K^β, G) for a suitably chosen β.

5.4.1.8 Illustrative Example. Electric Capacity Market Model with Quotas for Green and Conventional

In this form of intervention, the government decides to split the total capacity target $D = 400$ into three parts that must be met separately by existing, new conventional and green capacity: $\beta_h = 230, 140, 30$, for existing, new conventional and green, respectively. The feasible set is

Table 5.5 Capacity model results: VI with quotas

	Existing	Conventional	Green
Capacity	230	140	30
Cost	34,400	213,666.7	74,968.8
Profit	7,000	80,333.3	17,843.8
Total Payment = 428,212.5		Total Cost = 323,035.4	

$$K^{\beta} = \{x_{fh} | 0 \leq x_{fh} \leq X_{fh} \forall f, h; \sum_f x_{fh} \geq \beta_h \forall h\}$$

and the mapping is just as in the first example,

$$G(x) = (a_{fh} + b_{fh} x_{fh}) \forall f, h$$

Output from the solution to VI(K^{β}, G) is given in Table 5.5.

There is more green capacity than in the other two types of intervention (subsidy and price multiple), as required, but at the highest total cost of all solutions. The total payment is dramatically lower than any of the other solutions, because there are different prices for each of the three capacity types, and in particular, the price for existing capacity takes the value of the low marginal cost of firm 1.

□

5.5 Summary

In this chapter, we have shown how to formulate Nash equilibrium problems as variational inequality (VI) models for situations with or without constraints that link the agents in the problem. When there are linking constraints, i.e., we have a generalized Nash equilibrium (GNE), the VI model is a quasi-variational inequality (QVI) model, which in general has many possible solutions because such a model is incompletely defined. The VI form of a model can be easier to code in GAMS than the corresponding complementarity form of the same equilibrium situation; but because the VI form is more difficult to formulate than the complementarity form of an equilibrium situation, this chapter stresses methods of formulation.

We have also discussed the relation between the complementarity formulation of a Nash equilibrium or GNE, and corresponding VI or QVI formulations. One key relation is that KKT conditions can be defined for a given VI, and these KKT conditions are the complementarity version of the model. There can be several different-looking, but equivalent forms of a VI model, depending on whether or not some dual variables are included among the variables of the VI model.

The chapter ends with a discussion of several different VI solutions to a QVI problem of planning for energy production capacity, together with economic interpretations of the different VI solutions.

5.6 Exercises

1. Prove the "only if" part of Theorem 5.4. (Hint: in (5.8), consider an x which differs from x^* only in the subvector x_i, and use Theorem 5.2.)
2. Write out the complementarity problem form of the PIES multicommodity competitive equilibrium model of Section 5.2.3.3. Identify the expressions which correspond to the mapping and the constraint functions of the VI formulation of the same model.
3. Write out the VI form of the PIES multicommodity competitive equilibrium model (Section 5.2.3.3) which has all primal and dual variables of the complementarity problem form, and a feasible set defined as the nonnegative values of all variables.
4. Write out the alternative VI form (with linking prices as variables) of the PIES multicommodity competitive equilibrium model (Section 5.2.3.3).
5. Write out the complementarity problem form of the stochastic equilibrium model (Section 5.2.3.4). Identify the expressions which correspond to the mapping and the constraint functions of the VI formulation of the same model.
6. Write out the VI form of the stochastic equilibrium model (Section 5.2.3.4) which has all primal and dual variables of the complementarity problem form, and a feasible set defined as the nonnegative values of all variables.
7. Write out the alternative VI form (with linking prices as variables) of the stochastic equilibrium model (Section 5.2.3.4).
8. Modify the GAMS model for the Illustrative Example in Section 5.4.1.6 (see Appendix B for the GAMS code) to create VI models in GAMS that reproduce the results for

   a. the Illustrative Example in Section 5.4.1.2, the electric capacity market model with equal prices;
   b. the Illustrative Example in Section 5.4.1.4, the electric capacity market model with subsidized green energy; and
   c. the Illustrative Example in Section 5.4.1.8, the electric capacity market model with quotas for green and conventional.

9. Using coding methods for a VI, from the GAMS model for Illustrative Example 5.4.1.6 (see Appendix B for the GAMS code), create VI models in GAMS for

a. the Illustrative Example in Section 5.2.2.1, the Nash-Cournot model for the coal market from Chapter 3 (Hint: a good starting point is the GAMS complementarity model for this, in Appendix B);
b. the Illustrative Example in Section 5.2.3.1, the Nash-Cournot model for the coal market, with the coal yard limit, from Chapter3; and
c. the Illustrative Example in Section 5.3.3.2, the alternative VI for the Nash-Cournot model for the coal market, with the coal yard limit.

References

1. J. Boucher and Y. Smeers. Alternative models of restructured electricity systems, part 1: no market power. *Operations Research*, 49(6):821-838, 2001.
2. E. Celebi and J.D. Fuller. Time of use pricing in electricity markets under different market structures. *IEEE Transactions on Power Systems*, to appear, 2012.
3. J. Contreras, M. Klusch and J.B. Krawczyk. Numerical solutions to Nash-Cournot equilibria in coupled constraint electricity markets. *IEEE Transactions on Power Systems,* 19(1):195-206, 2004.
4. O. Daxhelet and Y. Smeers. Variational inequality models of restructured electricity systems. In *Applications and Algorithms of Complementarity*, ed. M.C. Ferris, O.L. Mangasarian, J.-S. Pang. Kluwer, New York, 2001.
5. L. Drouet, A. Haurie, F. Moresino, J.-P. Vial, M. Vielle and L. Viguier. An oracle based method to compute a coupled equilibrium in a model of international climate policy. *Computational Management Science*, 5:119-140, 2008.
6. F. Facchinei, A. Fischer and V. Piccialli. On generalized Nash games and variational inequalities. *Operations Research Letters*, 35:159-164, 2007.
7. F. Facchinei and C. Kanzow. Generalized Nash equilibrium problems. *Annals of Operations Research,* 175:177-211, 2010.
8. F. Facchinei and J.-S. Pang. *Finite-Dimensional Variational Inequalities and Complementarity Problems, Volumes I and II*, Springer, New York, 2003.
9. M.C. Ferris, S.P. Dirkse, J.-H. Jagla and A. Meeraus. An extended mathematical programming framework. *Computers and Chemical Engineering*, 33:1973-1982, 2009.
10. S.A. Gabriel, S. Kiet, J. Zhuang. A mixed complementarity-based equilibrium model of natural gas markets. *Operations Research*, 53(5), 799-818,2005.
11. S.A. Gabriel, A.S. Kydes and P. Whitman. The National Energy Modeling System: a large-scale energy-economic equilibrium model. *Operations Research*, 49(1):14–25, 2001.
12. P.T. Harker. Generalized Nash games and quasi-variational inequalities. *European Journal of Operational Research*, 54(1):81-94, 1991.
13. A. Haurie. Environmental coordination in dynamic oligopolistic markets. *Group Decision and Negotiation*, 4:39-57, 1995.
14. B.F. Hobbs and J.-S. Pang. Nash-Cournot equilibria in electric power markets with piecewise linear demand functions and joint constraints. *Operations Research*, 55(1):113-127, 2007.
15. Z. Liu and A. Nagurney. An integrated electric power supply chain and fuel market network framework: theoretical modeling with empirical analysis for New England. *Naval Research Logistics*, 56:600-624.
16. K. Nabetani, P. Tseng and M. Fukushima. Parametrized variational inequality approaches to generalized Nash equilibrium problems with shared constraints. *Computational Optimization and Applications*, 48:423-452, 2011.

17. A. Nagurney and K.K. Dhanda. Marketable pollution permits in oligopolistic markets with transaction costs. *Operations Research*, 48(3):424-435, 2000.
18. F. Nasiri and G. Zaccour. Renewable portfolio standard policy: a game-theoretic analysis. *INFOR*, 48(4):251-260, 2010.
19. A. Nagurney. *Network Economics: A Variational Inequality Approach*, Kluwer, Boston, 1993.
20. J.B. Rosen. Existence and uniqueness of equilibrium points for concave n-person games. *Econometrica*, 33(3):520-534, 1965.
21. J.-Y. Wei and Y. Smeers. Spatial oligopolistic electricity models with Cournot generators and regulated transmission prices. *Operations Research*, 47(1):102-112, 1999.

Chapter 6
Optimization Problems Constrained by Complementarity and Other Optimization Problems

6.1 Introduction

This chapter provides a friendly introduction to the analysis of optimization problems constrained by complementarity and other optimization problems. These problems are also known as bilevel problems [3], and the field of study to which they belong, hierarchical optimization. Throughout this chapter, we refer to them using the acronym OPcOPs, Optimization Problems constrained by other Optimization Problems, which explicitly indicates a hierarchy.

We first consider in Section 6.2 general optimization problems constrained by general optimization problems. Then, in Section 6.3 we analyze the particularly relevant case in which lower-level problems are linear. Next in Section 6.4 we consider the case of both upper-and lower-level problems being linear, and provide mixed-integer linear programming equivalents for certain types of these problems. In Section 6.5 we state and analyze the first-order optimality conditions for linear problems constrained by linear problems. Finally, Section 6.6 summarizes the chapter and provides conclusions.

6.1.1 Practical Interest

Optimization problems constrained by other optimization problems are common in practice, and particularly if dealing with markets.

In general, any market agent has to make decisions anticipating (or subject to) the equilibrium of the market. Anticipating the market equilibrium is necessary, for instance, to estimate the market-clearing prices, which in turn are used to establish the strategy of the agent (e.g., maximizing profit). The decision-making problem of such an agent can be generally formulated as an optimization problem (profit maximization) that is constrained by the market equilibrium. And the market equilibrium can also be conveniently formulated as a complementarity or optimization problem. Therefore, the resulting problem is an optimization problem (agent decision-making) subject to another optimization problem (market equilibrium). Note that an explicit hierarchy is assumed in the above reasoning: an upper-level problem subject to a lower-level one or to a collection of lower-level ones.

Some relevant examples of OPcOPs are listed below:

1. A strategic producer may construct its offer curves for the electricity pool by modeling within its maximum profit decision-making problem (optimization problem) the actual market clearing in the pool (also an optimization problem). Note that the market clearing is influenced by the actions of the strategic producer.
2. A producer may expand its production facilities to maximize its long-term profit (optimization problem) subject to the clearing of the market for many market conditions (a set of optimization problems).
3. An attacker may seek to maximize the damage it can inflict to a power system (optimization problem) subject to the actions of the system operator pursuing to minimize the adverse impact of such intentional attack (an optimization problem).
4. A transmission operator may expand its transmission network to minimize the unserved energy (optimization problem) subject to the clearing of the market for many load and offering conditions (a set of optimization problems).

6.1.2 Structure and Basic Classification

We consider that a structural hierarchy exists in the models we propose. That is, the upper-level problem is subject to a lower-level problem or to multiple lower-level problems, but not the other way around. This assumption is consistent with the design and functioning of markets, but this may not be the case in other areas of application such as transportation systems, where the identification of the upper-level problem might not be straightforward, [3]. The hierarchy considered is related to Stackelberg games in microeconomic analysis [12].

A variety of formats characterizing optimization problems constrained by complementarity or other optimization problems can be established, among them:

1. If the constraining problems are expressed as a set of complementarity conditions ($0 \leq x \perp y \geq 0$), i.e., an MCP, the resulting problem is called an MPCC, mathematical program with complementarity conditions. Note that "mathematical program" is other way to refer to an optimization problem.

2. If the constraining problems are transformed into a set of equilibrium constraints (e.g., an MCP, a mixed complementarity problem), the resulting problem is called an MPEC, mathematical program with equilibrium constraints. In some applications, equilibrium constraints in addition to including the optimality conditions of optimization problems include other constraints. In such situation the bilevel problem is still called an MPEC.
3. If the constraining problems are expressed by primal constraints, dual constraints and equalities enforcing the strong duality theorem, the resulting problem is an MPPDC, mathematical program with primal and dual constraints.

6.2 Optimization Problems Constrained by Other Optimization Problems, OPcOP

We characterize in this section the general form of the considered hierarchical optimization problems.

6.2.1 General Formulation

The formulation of the rather general OPcOP considered in this chapter is provided below,

$$\begin{aligned}
&\text{Minimize}_{\{x\}\cup\{x^1,\dots,x^n\}\cup\{\lambda^1,\dots,\lambda^n,\mu^1,\dots,\mu^n\}} \\
&\quad f(x,x^1,\dots,x^n,\lambda^1,\dots,\lambda^n,\mu^1,\dots,\mu^n) && \text{(6.1a)}\\
&\text{s. t.}\\
&\quad h(x,x^1,\dots,x^n,\lambda^1,\dots,\lambda^n,\mu^1,\dots,\mu^n) = 0 && \text{(6.1b)}\\
&\quad g(x,x^1,\dots,x^n,\lambda^1,\dots,\lambda^n,\mu^1,\dots,\mu^n) \le 0, && \text{(6.1c)}\\
&\quad \left\{\begin{array}{l}\text{Minimize}_{x^1} \quad f^1(x,x^1,\dots,x^n)\\ \text{s. t.}\\ \quad h^1(x,x^1,\dots,x^n) = 0 \quad (\lambda^1)\\ \quad g^1(x,x^1,\dots,x^n) \le 0 \quad (\mu^1),\end{array}\right. && \text{(6.1d)}\\
&\qquad \vdots\\
&\quad \left\{\begin{array}{l}\text{Minimize}_{x^i} \quad f^i(x,x^1,\dots,x^n)\\ \text{s. t.}\\ \quad h^i(x,x^1,\dots,x^n) = 0 \quad (\lambda^i)\\ \quad g^i(x,x^1,\dots,x^n) \le 0 \quad (\mu^i),\end{array}\right. && \text{(6.1e)}
\end{aligned}$$

$$\vdots$$

$$\begin{cases} \text{Minimize}_{x^n} \quad f^n(x, x^1, \ldots, x^n) \\ \text{s. t.} \\ \quad h^n(x, x^1, \ldots, x^n) = 0 \quad (\lambda^n) \\ \quad g^n(x, x^1, \ldots, x^n) \leq 0 \quad (\mu^n). \end{cases} \tag{6.1f}$$

Problem (6.1a)-(6.1c) is the upper-level problem, while problems (6.1d)-(6.1f) are the constraining lower-level problems.

Vector $x \in \mathbb{R}^{n^0}$ includes the set of optimization variables specifically belonging to the upper-level problem, while vector $x^i \in \mathbb{R}^{n^i}$ includes the set of optimization variables of constraining lower-level problem i. Vectors $\lambda^i \in \mathbb{R}^{m_{\text{E}}^i}$ and $\mu^i \in \mathbb{R}^{m_{\text{I}}^i}$ are respectively the equality and inequality dual variable vectors pertaining to constraining lower-level problem i.

Note the hierarchical structure of problem (6.1) above, i.e., an optimization problem constrained by its own constraints and a set of other optimization problems. Note also that the dual variables of the constraining problems (indicated in parentheses after the corresponding constraints) affect the upper-level problem, but not the other way around. That is, the dual variables of the upper-level problem do not directly affect the lower-level problems.

Note that hierarchical model (6.1) is rather general and can be adapted to quite a few practical settings.

An important observation is that the objective functions of the upper-level and the lower-level problems should be conflicting, otherwise the OPcOP reduces to a single-level optimization problem, which is obtained by simply replacing each constraining optimization problem with its constraints and adding the upper- and lower-level objective functions together.

Figure 6.1 visualizes the structure of an OPcOP considering n optimization problems as constraints.

The number of variables specific to the upper-level problem is n^0, while the number of variables of constraining problem i is n^i. m_E and m_I are, respectively, the numbers of equality and inequality constraints of the upper-level problem, while m_E^i and m_I^i are, respectively, the numbers of equality and inequality constrains of constraining problem i. Calling $n_{\text{G}} = \sum_{i=0}^{n} n^i$, $m_{\text{GE}} = \sum_{i=1}^{n} m_{\text{E}}^i$ and $m_{\text{GI}} = \sum_{i=1}^{n} m_{\text{I}}^i$, the objective function and the constraints of the upper-level problem are mathematically characterized as follows

1. $f(x, x^1, \ldots, x^n, \lambda^1, \ldots, \lambda^n, \mu^1, \ldots, \mu^n) : \mathbb{R}^{n_{\text{G}}} \times \mathbb{R}^{m_{\text{GE}}} \times \mathbb{R}^{m_{\text{GI}}} \to \mathbb{R}$,
2. $h(x, x^1, \ldots, x^n, \lambda^1, \ldots, \lambda^n, \mu^1, \ldots, \mu^n) : \mathbb{R}^{n_{\text{G}}} \times \mathbb{R}^{m_{\text{GE}}} \times \mathbb{R}^{m_{\text{GI}}} \to \mathbb{R}^{m_{\text{E}}}$, and
3. $g(x, x^1, \ldots, x^n, \lambda^1, \ldots, \lambda^n, \mu^1, \ldots, \mu^n) : \mathbb{R}^{n_{\text{G}}} \times \mathbb{R}^{m_{\text{GE}}} \times \mathbb{R}^{m_{\text{GI}}} \to \mathbb{R}^{m_{\text{I}}}$;

while the the objective function and the constraints of constraining problem i are defined as follows,

1. $f^i(x, x^1, \ldots, x^n) : \mathbb{R}^{n_{\text{G}}} \to \mathbb{R}$,

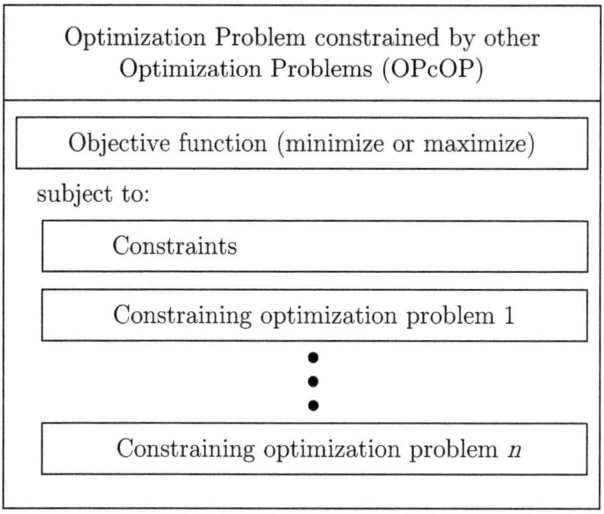

Fig. 6.1 OPcOP: Optimization problem constrained by a number of interrelated optimization problems

2. $h^i(x, x^1, \ldots, x^n) : \mathbb{R}^{n_G} \to \mathbb{R}^{m^i_E}$, and
3. $g^i(x, x^1, \ldots, x^n) : \mathbb{R}^{n_G} \to \mathbb{R}^{m^i_I}$.

It is important to note that although the optimization variables of constraining problem i are just those in vector x^i, the objective function and the constraints of this constraining problem i, $f^i(x, x^1, \ldots, x^n), h^i(x, x^1, \ldots, x^n) = 0$ and $g^i(x, x^1, \ldots, x^n) \leq 0$, depend on the primal optimization variables of all problems (the upper-level one and the constraining ones), x and $x^i, i = 1, \ldots, n$.

On the other hand, the upper-level problem depends on its optimization variables x, and of all the (primal and dual) variables of the lower-level problems x^i, λ^i and $\mu^i, i = 1, \ldots, n$. This fact makes the upper-level problem and the n constraining problems interrelated, which results in a well-posed and rather general OPcOP. The above dependencies are similar to those exhibited by Nash and Generalized Nash equilibrium problems [7, 14, 15].

Note that the primal optimization variables are arranged in two subsets to distinguish those specifically pertaining to the upper-level problem, $\{x\}$, and those influencing the upper-level problem though the constraining problems, $\{x^1, \ldots, x^n\}$.

In the following, three illustrative examples provide instances of OPcOPs, including a strategic offering model, a vulnerability assessment problem and a transmission investment model.

Specifically, the illustrative example below describes a strategic offering model.

6.2.2 Illustrative Example. Strategic Offering, OPcOP

We analyze in this example the problem faced by a strategic producer that seeks to derive hourly offers to maximize its profit subject to equilibrium in the pool [16], which is cleared through a network-constrained auction.

For the sake of simplicity we consider a three-node network including a demand and a generating unit per bus. Thus, the same index i is used to refer to buses, generating units and loads. Three lines interconnect these tree nodes as shown in Figure. 6.2.

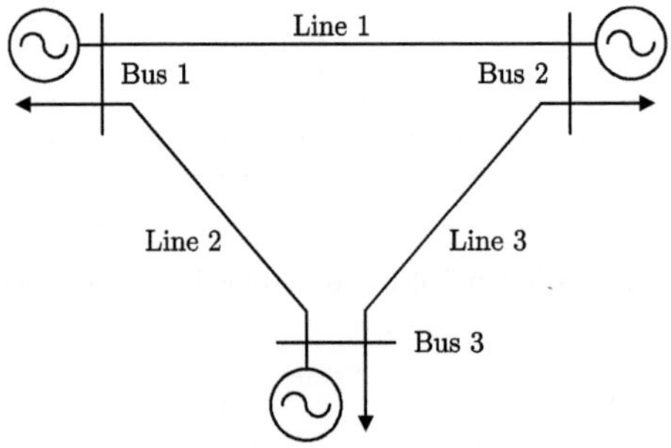

Fig. 6.2 Illustrative Example 6.2.2: Three-bus network

Data for generation units and demands are provided in Table 6.1, while data for the lines are in Table 6.2.

Table 6.1 Illustrative Example 6.2.2: Generating unit and demand data in MW

Bus #	Capacity (MW)	Production cost ($/MWh)	Demand (MW)	Demand utility ($/MWh)
1	20	16	5	18
2	10	19	20	20
3	25	15	15	21

The OPcOP profit maximization problem of generator 1 is formulated below.

$$\text{Minimize}_{\{\hat{c}_1^{\text{g}}\}\cup\{g_1,g_2,g_3,d_1,d_2,d_3,\theta_1,\theta_2,\theta_3\}\cup\Xi_{\text{SO}}}$$

$$c_1^{\text{g}} g_1 - \alpha_1 g_1 \tag{6.2a}$$

s. t.

$$\text{Minimize}_{g_1,g_2,g_3,d_1,d_2,d_3,\theta_1,\theta_2,\theta_3}$$

Table 6.2 Illustrative Example 6.2.2: Line data

Line #	From	To	Susceptance (S)	Capacity (MW)
1	1	2	100	5
2	1	3	125	10
3	2	3	150	10

$$\hat{c}_1^{\mathrm{g}} g_1 + \sum_{i=2}^{3} c_i^{\mathrm{g}} g_i - \sum_{i=1}^{3} c_i^{\mathrm{d}} d_i \tag{6.2b}$$

s. t.

$$g_1 - d_1 - B_1(\theta_1 - \theta_2) - B_2(\theta_1 - \theta_3) = 0 \quad (\alpha_1) \tag{6.2c}$$
$$g_2 - d_2 - B_1(\theta_2 - \theta_1) - B_3(\theta_2 - \theta_3) = 0 \quad (\alpha_2) \tag{6.2d}$$
$$g_3 - d_3 - B_2(\theta_3 - \theta_1) - B_3(\theta_3 - \theta_2) = 0 \quad (\alpha_3) \tag{6.2e}$$
$$0 \leq g_i \leq g_i^{\max} \quad (\phi_i^{\mathrm{gmin}}, \phi_i^{\mathrm{gmax}}), i = 1, 2, 3 \tag{6.2f}$$
$$0 \leq d_i \leq d_i^{\max} \quad (\phi_i^{\mathrm{dmin}}, \phi_i^{\mathrm{dmax}}), i = 1, 2, 3 \tag{6.2g}$$
$$- f_1^{\max} \leq B_1(\theta_1 - \theta_2) \leq f_1^{\max} \quad (\rho_1^{\min}, \rho_1^{\max}) \tag{6.2h}$$
$$- f_2^{\max} \leq B_2(\theta_1 - \theta_3) \leq f_2^{\max} \quad (\rho_2^{\min}, \rho_2^{\max}) \tag{6.2i}$$
$$- f_3^{\max} \leq B_3(\theta_2 - \theta_3) \leq f_3^{\max} \quad (\rho_3^{\min}, \rho_3^{\max}) \tag{6.2j}$$
$$\theta_3 = 0 \quad (\gamma). \tag{6.2k}$$

The optimization variables are described below. Variables g_i, d_i and θ_i are the generation, demand and angle at bus i, ($i = 1, 2, 3$), $\hat{c}_1^{\mathrm{g}}$ is the strategic price offer of unit 1, and $\Xi_{\mathrm{SO}} = \{\alpha_i, \phi_i^{\mathrm{gmin}}, \phi_i^{\mathrm{gmax}}, \rho_i^{\min}, \rho_i^{\max}, i = 1, 2, 3, \gamma\}$ is the set of dual variables. Note that each dual variable is indicated in parenthesis to the right of its associated equation.

Constants include the susceptance of each line B_k ($k = 1, 2, 3$), its transmission capacity $f_k^{\max}$ ($k = 1, 2, 3$), the minimum power output and the capacity of each generating unit, $g_i^{\min}$ and $g_i^{\max}$ ($i = 1, 2, 3$), respectively, the production cost of each generating unit c_i^{g} ($i = 1, 2, 3$) and the utility of each demand c_i^{d} ($i = 1, 2, 3$).

Upper-level problem (6.2a) seeks to maximize the profit of producer 1. The objective function includes cost minus revenue. The revenue is computed by multiplying the energy sold in the pool and the clearing price obtained from the lower-level problem (6.2b)-(6.2k). Note that this clearing price is the dual variable of the balance equation at bus 1, the bus in which generator 1 is located (recall that this is the problem faced by generator 1). The upper-level problem has no constraints other than those from the lower-level problem.

Lower-level problem (6.2b)-(6.2k) pursues clearing the market minimizing the minus declared social welfare, and provides productions, consumptions and prices. Objective function (6.2b) is the negative of declared social welfare (see Chapter 3 for a detailed definition of social welfare). Constraints (6.2c), (6.2d) and (6.2e) are power balance at buses 1, 2 and 3, respectively. Constraints (6.2f) and (6.2g) are generation and demand bounds, respectively. Constraints (6.2h), (6.2i) and (6.2j) are transmission capacity limits for lines 1, 2 and 3, respectively, and constraint (6.2k) sets the reference angle.

The solution of problem (6.2), obtained as explained in Illustrative Example 6.4.3.1, provides an optimal price offer for unit 1 of $c_1^{g*} = \$20$ per MWh and a profit of \$33.636. Productions, consumptions and locational marginal prices are provided in Table 6.3.

Table 6.3 Illustrative Example 6.2.2: Results for unit 1 offering as a strategic producer

Bus #	Production (MWh)	Consumption (MWh)	Price ($/MW)
1	8.409	0	20
2	10	18.409	20
3	15	15	20
Total	33.409	33.409	–

Table 6.4 Illustrative Example 6.2.2: Results for unit 1 offering as a competitive producer

Bus #	Production (MWh)	Consumption (MWh)	Price ($/MW)
1	13.409	5	16
2	10	18.409	20
3	15	15	18.182
Total	38.409	38.409	–

If unit 1 decides not to act strategically and offers its marginal cost, the solution of problem (6.2) provides a profit of \$0 (lower than that in the case of strategic behavior since its marginal cost is equal to the locational marginal price at bus 1). Productions, consumptions and locational marginal prices are provided in Table 6.4.

The example below considers a totally different problem and illustrates how a vulnerability assessment problem can be formulated as an OPcOP [17].

6.2.3 Illustrative Example. Vulnerability Assessment, OPcOP

Pursuing a disruptive strategy, the objective of an attacker (upper-level agent) is to maximize the unserved energy throughout an electric energy system, while the objective of the system operator (lower-level agent) is to minimize that unserved energy [2,13,17].

The solution of this vulnerability problem allows quantifying the level of damage than an attacker can inflict to a system, and to identify the most critical components of such system, with the ultimate goal of protecting these critical components and enhancing the security of the system.

We consider the three-bus network in Figure 6.2 (Illustrative Example 6.2.2). Generating unit and demand data are provided in Table 6.5, while line data are provided in Table 6.6.

Table 6.5 Illustrative Example 6.2.3: Generating unit and demand data

Bus #	Minimum power output (MW)	Capacity (MW)	Demand (MW)
1	10	20	5
2	5	10	20
3	12.5	25	15

Table 6.6 Illustrative Example 6.2.3: Line data

Line #	From	To	Susceptance (S)	Capacity (MW)
1	1	2	100	5
2	1	3	125	10
3	2	3	150	10

We assume that the attacker has resources to destroy just two components: being these components generating units or lines, not buses.

The OPcOP formulation of the above vulnerability assessment problem is as follows:

$$\text{Minimize}_{\{x_1,x_2,x_3\}\cup\{r_1,r_2,r_3,g_1,g_2,g_3,f_1,f_2,f_3,\theta_1,\theta_2,\theta_3,\}\cup\Xi_{\text{VA}}}$$

$$-\sum_{i=1}^{3} r_i \tag{6.3a}$$

s. t.

$$\sum_{k=1}^{3} x_k + \sum_{i=1}^{3} y_i = 2 \tag{6.3b}$$

$$x_k \in \{0,1\}, k = 1,2,3; \; y_i \in \{0,1\}, i = 1,2,3 \tag{6.3c}$$

$$\text{Minimize}_{g_1,g_2,g_3,f_1,f_2,f_3,r_1,r_2,r_3,\theta_1,\theta_2,\theta_3}$$

$$\sum_{i=1}^{3} r_i \tag{6.3d}$$

s. t.

$$g_1 + r_1 - d_1 - f_1 - f_2 = 0 \quad (\alpha_1) \tag{6.3e}$$

$$g_2 + r_2 - d_2 + f_1 - f_3 = 0 \quad (\alpha_2) \tag{6.3f}$$

$$g_3 + r_3 - d_3 + f_2 + f_3 = 0 \quad (\alpha_3) \tag{6.3g}$$

$$f_1 - (1 - y_1)B_1(\theta_1 - \theta_2) = 0 \quad (\beta_1) \tag{6.3h}$$

$$f_2 - (1 - y_2)B_2(\theta_1 - \theta_3) = 0 \quad (\beta_2) \tag{6.3i}$$

$$f_3 - (1 - y_3)B_3(\theta_2 - \theta_3) = 0 \quad (\beta_3) \tag{6.3j}$$

$$\theta_3 = 0 \quad (\gamma) \tag{6.3k}$$

$$(1 - x_i)g_i^{\min} \leq g_i \leq (1 - x_i)g_i^{\max} \quad (\phi_i^{\min}, \phi_i^{\max}), i = 1,2,3 \tag{6.3l}$$

$$0 \leq r_i \leq d_i \quad (\nu_i^{\min}, \nu_i^{\max}), i = 1,2,3 \tag{6.3m}$$

$$-f_k^{\max} \leq f_k \leq f_k^{\max} \quad (\rho_k^{\min}, \rho_k^{\max}), k = 1,2,3 \tag{6.3n}$$

Optimization variables are described as follows. Binary variable x_k ($k = 1,2,3$) is 1 if line k is destroyed by the upper-level agent (attacker) and 0 otherwise, binary variable y_i ($i = 1,2,3$) is 1 if generating unit i is destroyed by the upper-level agent and 0 otherwise, variables r_i, g_i and θ_i ($i = 1,2,3$) are the unserved energy, the production and the angle, respectively, corresponding with bus i, and f_k ($k = 1,2,3$) is the power flow trough line k. $\Xi_{\text{VA}} = \{\gamma, \alpha_i, \beta_i, \phi_i^{\min}, \phi_i^{\max}, \nu_i^{\min}, \nu_i^{\max}, \rho_i^{\min}, \rho_i^{\max}, i = 1,2,3\}$ is the set of dual variables of the lower-level problem.

Constants include a fixed demand in each bus d_i ($i = 1,2,3$), the susceptance of each line B_k ($k = 1,2,3$), the minimum power output and the capacity of each generating unit, $g_i^{\min}$ and $g_i^{\max}$ ($i = 1,2,3$), respectively, and the transmission capacity of each line $f_k^{\max}$ ($k = 1,2,3$).

The upper-level problem (6.3a)-(6.3c) is constrained by the lower-level one, (6.3d)-(6.3n). The objective function (6.3a) to be maximized is the unserved energy throughout the system, and the constraints include condition (6.3b) enforcing that only two components can be destroyed (lines, generators or a combination of both) and binary variable declarations (6.3c).

The objective of the lower-level problem (6.3d)-(6.3n) is minimizing the unserved energy. Constraints (6.3e)-(6.3g) are bus energy balances, constraints (6.3h)-(6.3j) are flow definitions, and constraint (6.3k) defines the reference bus. Inequalities (6.3l) enforce minimum power outputs and capacities of the production units, (6.3m) state per bus that the unserved energy should be positive and below the demand, and (6.3n) enforce line capacity limits in both directions.

Observe that problem (6.3) is an OPcOP, whose solution is easily reached by inspection as indicated in Table 6.7.

Table 6.7 Illustrative Example 6.2.3: Solution by inspection

Strategy #	x_1	x_2	x_3	y_1	y_2	y_3	Unserved energy (MW)
1	1	1	0	0	0	0	27.5
2	1	0	1	0	0	0	35
3	0	1	1	0	0	0	20
4	0	0	0	1	1	0	12.5
5	0	0	0	1	0	1	12.5
6	0	0	0	0	1	1	12.5
7	1	0	0	1	0	0	22.5
8	1	0	0	0	1	0	22.5
9	1	0	0	0	0	1	22.5
10	0	1	0	1	0	0	17.5
11	0	1	0	0	1	0	17.5
12	0	1	0	0	0	1	17.5
13	0	0	1	1	0	0	25
14	0	0	1	0	1	0	25
15	0	0	1	0	0	1	25

Inspecting Table 6.7, it can be concluded that an optimal solution is strategy 2 involving 35 MW of unserved energy.

Note the highly combinatorial nature of OPcOP (6.3), which makes impossible a solution by inspection in larger systems. Efficient solution techniques are illustrated in some of the following sections of this chapter and in Chapter 8.

The next illustrative example describes a transmission investment model that properly represents the functioning of the market [9].

6.2.4 Illustrative Example. Transmission Investment, OPcOP

We consider the problem of a transmission planner that pursues the expansion of its transmission grid with the objective of minimizing unserved energy, and subject to a limited investment budget and the clearing of the market. The market is cleared by minimizing the social cost.

The system considered is depicted in Figure. 6.2 (Illustrative Example 6.2.2), with its generation and demand data provided in Table 6.8. Existing line data are given in Table 6.9, while data pertaining to new lines are provided in Table 6.10.

Table 6.8 Illustrative Example 6.2.4: Generating unit and demand data

Bus #	Minimum power output (MW)	Capacity (MW)	Production cost ($/MWh)	Demand (MW)	Unserved energy cost ($/MWh)
1	10	25	16	8	20
2	15	30	19	10	18
3	5	10	15	45	25

Table 6.9 Illustrative Example 6.2.4: Existing line data

Line #	From	To	Susceptance (S)	Capacity (MW)
1	1	2	100	5
2	1	3	125	8
3	2	3	150	5

Table 6.10 Illustrative Example 6.2.4: New line data

New line #	From	To	Susceptance (S)	Capacity (MW)	Building cost ($)
1	1	2	100	10	500
2	1	3	125	15	1000
3	2	3	150	15	1000

The conversion factor e and the investment budget $c^{\max}$ are 10 and \$2500, respectively.

The OPcOP formulation of the transmission investment problem is as follows:

$$\text{Minimize}_{\{x_1,x_2,x_3\}\cup\{g_1,g_2,g_3,r_1,r_2,r_3,f_1,f_2,f_3,t_1,t_2,t_3,\theta_1,\theta_2,\theta_3\}\cup\Xi_{\text{TI}}}$$

$$\sum_{k=1}^{3} c_k x_k + e\sum_{i=1}^{3} c_i^{\text{U}} r_i \tag{6.4a}$$

s. t.

$$\sum_{k=1}^{3} c_k x_k - c^{\max} \leq 0 \tag{6.4b}$$

$$x_k \in \{0,1\}, \quad k = 1,2,3 \tag{6.4c}$$

$$\text{Minimize}_{g_1,g_2,g_3,r_1,r_2,r_3,f_1,f_2,f_3,t_1,t_2,t_3,\theta_1,\theta_2,\theta_3}$$

$$\sum_{i=1}^{3} c_i^{\mathrm{g}} g_i + \sum_{i=1}^{3} c_i^{\mathrm{U}} r_i \tag{6.4d}$$

s. t.

$$g_1 + r_1 - d_1 - f_1 - f_2 - t_1 - t_2 = 0 \quad (\alpha_1) \tag{6.4e}$$

$$g_2 + r_2 - d_2 + f_1 - f_3 + t_1 - t_3 = 0 \quad (\alpha_2) \tag{6.4f}$$

$$g_3 + r_3 - d_3 + f_2 + f_3 + t_2 + t_3 = 0 \quad (\alpha_3) \tag{6.4g}$$

$$f_1 - x_1 B_1^{\mathrm{N}}(\theta_1 - \theta_2) = 0 \quad (\beta_1) \tag{6.4h}$$

$$f_2 - x_2 B_2^{\mathrm{N}}(\theta_1 - \theta_3) = 0 \quad (\beta_2) \tag{6.4i}$$

$$f_3 - x_3 B_3^{\mathrm{N}}(\theta_2 - \theta_3) = 0 \quad (\beta_3) \tag{6.4j}$$

$$t_1 - B_1^{\mathrm{E}}(\theta_1 - \theta_2) = 0 \quad (\delta_1) \tag{6.4k}$$

$$t_2 - B_2^{\mathrm{E}}(\theta_1 - \theta_3) = 0 \quad (\delta_2) \tag{6.4l}$$

$$t_3 - B_3^{\mathrm{E}}(\theta_2 - \theta_3) = 0 \quad (\delta_3) \tag{6.4m}$$

$$\theta_3 = 0 \quad (\gamma) \tag{6.4n}$$

$$g_i^{\min} \leq g_i \leq g_i^{\max} \quad (\phi_i^{\min}, \phi_i^{\max}), i = 1,2,3 \tag{6.4o}$$

$$0 \leq r_i \leq d_i \quad (\nu_i^{\min}, \nu_i^{\max}), i = 1,2,3 \tag{6.4p}$$

$$-t_i^{\max} \leq t_i \leq t_i^{\max} \quad (\psi_i^{\min}, \psi_i^{\max}), i = 1,2,3 \tag{6.4q}$$

$$-f_i^{\max} \leq f_i \leq f_i^{\max} \quad (\rho_i^{\min}, \rho_i^{\max}), i = 1,2,3. \tag{6.4r}$$

The optimization variables are explained in what follows. Binary variable x_k ($k = 1,2,3$) is 1 if new line k is built by the upper-level agent (transmission planner) and 0 otherwise, variables r_i, g_i and θ_i ($i = 1,2,3$) are the unserved energy, the production and the angle, respectively, associated with bus i, t_k ($k = 1,2,3$) is the power flow through existing line k, while f_k ($k = 1,2,3$) is the power flow through new line k. Set $\Xi_{\mathrm{TI}} = \{\gamma, \alpha_i, \beta_i, \delta_i, \phi_i^{\min}, \phi_i^{\max}, \nu_i^{\min}, \nu_i^{\max}, \psi_i^{\min}, \psi_i^{\max}, \rho_i^{\min}, \rho_i^{\max}, i = 1,2,3\}$ include the dual variables of the lower-level problem.

Constants include the fixed demand in each bus d_i ($i = 1,2,3$), the susceptance of each existing line B_k^{E} ($k = 1,2,3$) and of each new line B_k^{N} ($k = 1,2,3$), the minimum power output and the capacity of each generating unit, $g_i^{\min}$ and $g_i^{\max}$ ($i = 1,2,3$), respectively, and the transmission capacity of each new and existing line $f_k^{\max}$ and $t_k^{\max}$ ($k = 1,2,3$), respectively. Cost data include line k building cost c_k, unit i generating cost c_i^{G}, bus i unserved energy cost c_i^{U}, and the investment budget $c^{\max}$. Constant e makes comparable unserved energy and investment costs, i.e., they are referred to a yearly basis.

The upper-level problem (6.4a)-(6.4c) is constrained by the lower-level one (6.4d)-(6.4r). The objective of the upper-level problem is to minimize the investment costs plus the unserved energy costs throughout the system, and the constraints include condition (6.19b) enforcing a limited budget for transmission investment and binary variable declarations (6.19c).

Lower-level problem (6.4d)-(6.4r) represents the clearing of the market. Its objective is minimizing the generation cost. Constraints (6.4e)-(6.4g) are bus energy balances, constraints (6.4h)-(6.4j) are flow definitions for new lines, constraints (6.4k)-(6.4m) are flow definitions for existing lines, and constraint (6.4n) provides the definition of the reference bus. Inequalities (6.4o) enforce minimum power outputs and capacities of production units, constraints (6.4p) state per bus that the unserved energy should be positive and below the demand, constraints (6.4q) enforce line capacity limits of existing lines in both directions, and finally constraints (6.4r) enforce line capacity limits of new lines in both directions.

Table 6.11 Illustrative Example 6.2.4: Solution by inspection

Investment strategy #	x_1	x_2	x_3	Investment cost ($)	Unserved energy cost ($)	Generation cost ($)	Upper-level obj. function ($)
1	0	0	0	0	645.833	629.667	6458.333
2	1	0	0	500	645.833	629.667	6958.333
3	0	1	0	1000	541.667	696.333	6416.667
4	0	0	1	1000	459.800	740.067	5598.000
5	1	1	0	1500	541.667	696.333	6916.667
6	1	0	1	1500	442.708	759.667	5927.083
7	0	1	1	2000	301.133	835.000	5011.333
8	1	1	1	2500	312.083	835.000	5620.833

Observe that problem (6.4) is an OPcOP. Its solution is easily reached by inspection as indicated in Table 6.11. From Table 6.11, it can be concluded that an optimal solution is investment strategy 7.

Note that a solution by inspection is not viable in realistic systems. Solution techniques are illustrated in some of the following sections of this chapter and in Chapter 8.

6.2.5 Basic Assumption: Constraining Problems are Convex

It is important to note that constraining problems need to be convex for the complete hierarchical problem to be tractable. Moreover, if they are not, the analytical tools to address OPcOP became almost non-existent. For this reason, in the following we only consider convex lower-level problems. In practice, a particular but very relevant class of convex constraining problems is the class of linear problems.

6.2.6 Mathematical Program with Complementarity Constraints, MPCC

If the lower-level optimization problems (i.e., problems (6.1d)-(6.1f)) constraining the upper-level problem (i.e., problem (6.1a)-(6.1c)) are convex, they can be replaced by their corresponding Karush-Kuhn-Tucker (KKT) conditions [1]. Note that for convex problems meeting appropriate constraints qualifications, KKT conditions are both necessary and sufficient optimality conditions. Then, problem (6.1) becomes

$$
\begin{aligned}
&\text{Minimize}_{\{x\}\cup\{x^1,\ldots,x^n\}\cup\{\lambda^1,\ldots,\lambda^n,\mu^1,\ldots,\mu^n\}} \\
&\quad f(x,x^1,\ldots,x^n,\lambda^1,\ldots,\lambda^n,\mu^1,\ldots,\mu^n) && \text{(6.5a)}\\
&\text{s. t.} \\
&\quad h(x,x^1,\ldots,x^n,\lambda^1,\ldots,\lambda^n,\mu^1,\ldots,\mu^n) = 0 && \text{(6.5b)}\\
&\quad g(x,x^1,\ldots,x^n,\lambda^1,\ldots,\lambda^n,\mu^1,\ldots,\mu^n) \le 0, && \text{(6.5c)}\\
&\quad \nabla_{x^i} f^i(x,x^1,\ldots,x^n) + {\lambda^i}^T \nabla_{x^i} h^i(x,x^1,\ldots,x^n) + \\
&\qquad {\mu^i}^T \nabla_{x^i} g^i(x,x^1,\ldots,x^n) = 0, \quad i=1,\ldots,n && \text{(6.5d)}\\
&\quad h^i(x,x^1,\ldots,x^n) = 0, \quad i=1,\ldots,n && \text{(6.5e)}\\
&\quad 0 \le \mu^i \perp -g^i(x,x^1,\ldots,x^n) \ge 0, \quad i=1,\ldots,n. && \text{(6.5f)}
\end{aligned}
$$

Problem (9.22) above is called an MPCC, a mathematical program with complementarity constraints. This is so because constraints (9.22f) are complementarity constraints. It is relevant to note that the feasible region (9.22d)-(9.22f) is not generally a convex set. Moreover, the specific complementarity constraints (9.22f) do not generally meet constraint qualifications. Thus, solving problem (9.22) might not be easy.

The illustrative example below characterizes the vulnerability assessment problem described in Illustrative Example 6.2.3 as an MPCC.

6.2.7 Illustrative Example. Vulnerability Assessment, MPCC

OPcOP (6.3) can be recast as an MPCC as explained in what follows. Since x_i and y_i ($i = 1, 2, 3$) are given (constants) for the lower-level problem, such a problem is linear (thus convex), and therefore it can be replaced by its KKT conditions. This way the original OPcOP (6.3) becomes the MPCC below:

$$\text{Minimize}_{\{x_1,x_2,x_3\}\cup\{r_1,r_2,r_3,g_1,g_2,g_3,f_1,f_2,f_3,\theta_1,\theta_2,\theta_3\}\cup\Xi_{\text{VA}}}$$

$$-\sum_{i=1}^{3} r_i \tag{6.6a}$$

s. t.

$$\sum_{i=1}^{3} x_k + \sum_{i=1}^{3} y_i = 2 \tag{6.6b}$$

$$x_k \in \{0,1\}, \quad k = 1,2,3 \tag{6.6c}$$

$$y_i \in \{0,1\}, \quad i = 1,2,3 \tag{6.6d}$$

$$g_1 + r_1 - d_1 - f_1 - f_2 = 0 \tag{6.6e}$$

$$g_2 + r_2 - d_2 + f_1 - f_3 = 0 \tag{6.6f}$$

$$g_3 + r_3 - d_3 + f_2 + f_3 = 0 \tag{6.6g}$$

$$f_1 - (1 - y_1)B_1(\theta_1 - \theta_2) = 0 \tag{6.6h}$$

$$f_2 - (1 - y_2)B_2(\theta_1 - \theta_3) = 0 \tag{6.6i}$$

$$f_3 - (1 - y_3)B_3(\theta_2 - \theta_3) = 0 \tag{6.6j}$$

$$\theta_3 = 0 \tag{6.6k}$$

$$\alpha_i - \phi_i^{\min} + \phi_i^{\max} = 0, \quad i = 1,2,3 \tag{6.6l}$$

$$-\alpha_1 + \alpha_2 + \beta_1 - \rho_1^{\min} + \rho_1^{\max} = 0 \tag{6.6m}$$

$$-\alpha_1 + \alpha_3 + \beta_2 - \rho_2^{\min} + \rho_2^{\max} = 0 \tag{6.6n}$$

$$-\alpha_2 + \alpha_3 + \beta_3 - \rho_3^{\min} + \rho_3^{\max} = 0 \tag{6.6o}$$

$$1 + \alpha_i - \nu_i^{\min} + \nu_i^{\max} = 0, \quad i = 1,2,3 \tag{6.6p}$$

$$-\beta_1(1 - y_1)B_1 - \beta_2(1 - y_2)B_2 = 0 \tag{6.6q}$$

$$+\beta_1(1 - y_1)B_1 - \beta_3(1 - y_3)B_3 = 0 \tag{6.6r}$$

$$\beta_2(1 - y_2)B_2 + \beta_3(1 - y_3)B_3 + \gamma = 0 \tag{6.6s}$$

$$0 \le \phi_i^{\min} \perp g_i - (1 - x_i)g_i^{\min} \ge 0, \quad i = 1,2,3 \tag{6.6t}$$

$$0 \le \phi_i^{\max} \perp (1 - x_i)g_i^{\max} - g_i \ge 0, \quad i = 1,2,3 \tag{6.6u}$$

$$0 \le \nu_i^{\min} \perp r_i \ge 0, \quad i = 1,2,3 \tag{6.6v}$$

$$0 \le \nu_i^{\max} \perp d_i - r_i \ge 0, \quad i = 1,2,3 \tag{6.6w}$$

$$0 \le \rho_i^{\min} \perp f_i^{\max} + f_i \ge 0, \quad i = 1,2,3 \tag{6.6x}$$

$$0 \le \rho_i^{\max} \perp f_i^{\max} - f_i \ge 0, \quad i = 1, 2, 3. \tag{6.6y}$$

Note that the optimization problem (6.6) is single-level, mixed-integer and nonlinear. Such problems can be solved using mathematical programming procedures. However, the combination of nonlinearities and binary variables may require ad hoc solution techniques.

An optimal solution of problem (6.6) is that provided in Illustrative Example 6.2.3.

6.2.8 Mathematical Program with Equilibrium Constraints, MPEC

If the lower-level optimality conditions (9.22d)-(9.22f) constitute an equilibrium problem, then (9.22) is known as an MPEC, a mathematical program with equilibrium constraints. In some applications, equilibrium constraints include not only optimality conditions but other conditions as well. In such situations the bilevel problem is still called an MPEC.

The illustrative example below provides the MPEC formulation of the strategic offering problem explained in Illustrative Example 6.2.2.

6.2.9 Illustrative Example. Strategic Offering, MPEC

Consider OPcOP (6.2). Since the lower-level problem (6.2c)-(6.2k) is linear and thus convex, OPcOP (6.2) can be recast as an MPEC as indicated below.

$$\text{Minimize}_{\{\hat{c}_1^{\text{g}}\} \cup \{g_1, g_2, g_3, d_1, d_2, d_3, \theta_1, \theta_2, \theta_3\} \cup \Xi_{\text{SO}}}$$

$$c_1^{\text{g}} g_1 - \alpha_1 g_1 \tag{6.7a}$$

s. t.

$$g_1 - d_1 = B_1(\theta_1 - \theta_2) + B_2(\theta_1 - \theta_3) \tag{6.7b}$$

$$g_2 - d_2 = B_1(\theta_2 - \theta_1) + B_3(\theta_2 - \theta_3) \tag{6.7c}$$

$$g_3 - d_3 = B_2(\theta_3 - \theta_1) + B_3(\theta_3 - \theta_2) \tag{6.7d}$$

$$\theta_3 = 0 \tag{6.7e}$$

$$\hat{c}_1^{\text{g}} - \alpha_1 - \phi_1^{\text{gmin}} + \phi_1^{\text{gmax}} = 0 \tag{6.7f}$$

$$c_i^{\text{g}} - \alpha_i - \phi_i^{\text{gmin}} + \phi_i^{\text{gmax}} = 0, \quad i = 2, 3 \tag{6.7g}$$

$$-c_i^{\text{d}} + \alpha_i - \phi_i^{\text{dmin}} + \phi_i^{\text{dmax}} = 0, \quad i = 1, 2, 3 \tag{6.7h}$$

$$B_1(\alpha_1 - \alpha_2 - \rho_1^{\min} + \rho_1^{\max}) + B_2(\alpha_1 - \alpha_3 - \rho_2^{\min} + \rho_2^{\max}) = 0 \tag{6.7i}$$

$$B_1(-\alpha_1 + \alpha_2 + \rho_1^{\min} - \rho_1^{\max}) + \\ B_3(\alpha_2 - \alpha_3 - \rho_3^{\min} + \rho_3^{\max}) = 0 \tag{6.7j}$$

$$B_2(-\alpha_1 + \alpha_3 + \rho_2^{\min} - \rho_2^{\max}) + \\ B_3(-\alpha_2 + \alpha_3 + \rho_3^{\min} - \rho_3^{\max}) + \gamma = 0 \tag{6.7k}$$

$$0 \le g_i \perp \phi_i^{\text{gmin}} \ge 0, i = 1, 2, 3 \tag{6.7l}$$

$$0 \le g_i^{\max} - g_i \perp \phi_i^{\text{gmax}} \ge 0, \quad i = 1, 2, 3 \tag{6.7m}$$

$$0 \le d_i \perp \phi_i^{\text{dmin}} \ge 0, \quad i = 1, 2, 3 \tag{6.7n}$$

$$0 \le d_i^{\max} - d_i \perp \phi_i^{\text{dmax}} \ge 0, \quad i = 1, 2, 3 \tag{6.7o}$$

$$0 \le B_1(\theta_1 - \theta_2) + f_1^{\max} \perp \rho_1^{\min} \ge 0 \tag{6.7p}$$

$$0 \le f_1^{\max} - B_1(\theta_1 - \theta_2) \perp \rho_1^{\max} \ge 0 \tag{6.7q}$$

$$0 \le B_2(\theta_1 - \theta_3) + f_2^{\max} \perp \rho_2^{\min} \ge 0 \tag{6.7r}$$

$$0 \le f_2^{\max} - B_2(\theta_1 - \theta_3) \perp \rho_2^{\max} \ge 0 \tag{6.7s}$$

$$0 \le B_3(\theta_2 - \theta_3) + f_3^{\max} \perp \rho_3^{\min} \ge 0 \tag{6.7t}$$

$$0 \le f_3^{\max} - B_3(\theta_2 - \theta_3) \perp \rho_3^{\max} \ge 0 \tag{6.7u}$$

Observe that problem (9.23) is a continuous, single-level, nonlinear problem. Nonlinearities are due to the complementarity conditions (9.23l)-(9.23u) and the objective function (9.23a).

The illustrative example below provides an MPEC formulation for the transmission investment problem explained in Illustrative Example 6.2.4.

6.2.10 Illustrative Example. Transmission Investment, MPEC

OPcOP (6.4) can be recast as an MPEC as shown below. Since x_i $(i = 1, 2, 3)$ are given values (constants) for the lower-level problem, such problem is linear (and thus convex), and therefore it can be replaced by its KKT conditions. This way the original OPcOP (6.4) becomes the MPEC below:

$$\text{Minimize}_{\{x_1,x_2,x_3\} \cup \{g_1,g_2,g_3,r_1,r_2,r_3,f_1,f_2,f_3,t_1,t_2,t_3,\theta_1,\theta_2,\theta_3\} \cup \Xi_{\text{TI}}}$$

$$\sum_{k=1}^{3} c_k x_k + e \sum_{i=1}^{3} c_i^{\text{U}} r_i \tag{6.8a}$$

s. t.

$$\sum_{k=1}^{3} c_k x_k - c^{\max} \le 0 \tag{6.8b}$$

$$x_k \in \{0, 1\}, \quad k = 1, 2, 3 \tag{6.8c}$$

$$(6.4e)\text{-}(6.4n) \tag{6.8d}$$
$$c_i^{\mathrm{g}} + \alpha_i - \phi_i^{\mathrm{min}} + \phi_i^{\mathrm{max}} = 0, \quad i = 1,2,3 \tag{6.8e}$$
$$c_i^{\mathrm{U}} + \alpha_i - \nu_i^{\mathrm{min}} + \nu_i^{\mathrm{max}} = 0, \quad i = 1,2,3 \tag{6.8f}$$
$$-\alpha_1 + \alpha_2 + \beta_1 - \rho_1^{\mathrm{min}} + \rho_1^{\mathrm{max}} = 0 \tag{6.8g}$$
$$-\alpha_1 + \alpha_3 + \beta_2 - \rho_2^{\mathrm{min}} + \rho_2^{\mathrm{max}} = 0 \tag{6.8h}$$
$$-\alpha_2 + \alpha_3 + \beta_3 - \rho_3^{\mathrm{min}} + \rho_3^{\mathrm{max}} = 0 \tag{6.8i}$$
$$-\alpha_1 + \alpha_2 + \delta_1 - \psi_1^{\mathrm{min}} + \psi_1^{\mathrm{max}} = 0 \tag{6.8j}$$
$$-\alpha_1 + \alpha_3 + \delta_2 - \psi_2^{\mathrm{min}} + \psi_2^{\mathrm{max}} = 0 \tag{6.8k}$$
$$-\alpha_2 + \alpha_3 + \delta_3 - \psi_3^{\mathrm{min}} + \psi_3^{\mathrm{max}} = 0 \tag{6.8l}$$
$$-\beta_1 x_1 B_1^{\mathrm{N}} - \beta_2 x_2 B_2^{\mathrm{N}} - \delta_1 B_1^{\mathrm{E}} - \delta_2 B_2^{\mathrm{E}} = 0 \tag{6.8m}$$
$$\beta_1 x_1 B_1^{\mathrm{N}} - \beta_3 x_3 B_3^{\mathrm{N}} + \delta_1 B_1^{\mathrm{E}} - \delta_3 B_3^{\mathrm{E}} = 0 \tag{6.8n}$$
$$\beta_2 x_2 B_2^{\mathrm{N}} + \beta_3 x_3 B_3^{\mathrm{N}} + \delta_2 B_2^{\mathrm{E}} + \delta_3 B_3^{\mathrm{E}} + \gamma = 0 \tag{6.8o}$$
$$0 \le \phi_i^{\mathrm{min}} \perp - g_i^{\mathrm{min}} + g_i \ge 0, \quad i = 1,2,3 \tag{6.8p}$$
$$0 \le \phi_i^{\mathrm{max}} \perp g_i^{\mathrm{max}} - g_i \ge 0, \quad i = 1,2,3 \tag{6.8q}$$
$$0 \le \nu_i^{\mathrm{min}} \perp r_i \ge 0 \quad i = 1,2,3 \tag{6.8r}$$
$$0 \le \nu_i^{\mathrm{max}} \perp d_i - r_i \ge 0, \quad i = 1,2,3 \tag{6.8s}$$
$$0 \le \psi_i^{\mathrm{min}} \perp t_i^{\mathrm{max}} + t_i \ge 0, \quad i = 1,2,3 \tag{6.8t}$$
$$0 \le \psi_i^{\mathrm{max}} \perp t_i^{\mathrm{max}} - t_i \ge 0, \quad i = 1,2,3 \tag{6.8u}$$
$$0 \le \rho_i^{\mathrm{min}} \perp f_i^{\mathrm{max}} + f_i \ge 0, \quad i = 1,2,3 \tag{6.8v}$$
$$0 \le \rho_i^{\mathrm{max}} \perp f_i^{\mathrm{max}} - f_i \ge 0, \quad i = 1,2,3 \tag{6.8w}$$

Note that MPEC (6.8) is single-level, mixed-integer and nonlinear. Solution techniques are available to attack such problem. Nonlinearities are due to complementarity conditions (6.8p)-(6.8w) and equalities (6.8m)-(6.8o).

An optimal solution of MPEC (6.8) is that provided in Illustrative Example 6.2.4.

6.2.11 Stochastic OPcOPs

If the constraining problems constitute a collection of optimization problems where each one is associated with a particular realization of a stochastic process (characterized by a set of scenarios indexed by $\omega = 1, \ldots, s$), the bilevel problem is called a stochastic OPcOP, in short sOPcOP. Its formulation is as follows

$$\text{Minimize}_{\{x\} \cup \{x^1, \ldots, x^s\} \cup \{\lambda^1, \ldots, \lambda^s, \mu^1, \ldots, \mu^s\}}$$

$$f(x, x^1, \dots, x^s, \lambda^1, \dots, \lambda^s, \mu^1, \dots, \mu^s) \tag{6.9a}$$

s. t.

$$h(x, x^1, \dots, x^s, \lambda^1, \dots, \lambda^s, \mu^1, \dots, \mu^s) = 0 \tag{6.9b}$$

$$g(x, x^1, \dots, x^s, \lambda^1, \dots, \lambda^s, \mu^1, \dots, \mu^s) \leq 0, \tag{6.9c}$$

$$\left\{\begin{array}{ll} \text{Minimize}_{x^\omega} & f^\omega(x, x^1, \dots, x^\omega, \dots, x^s) \\ \text{s. t.} & \\ \quad h^\omega(x, x^1, \dots, x^\omega, \dots, x^s) = 0 & (\lambda^\omega) \\ \quad g^\omega(x, x^1, \dots, x^\omega, \dots, x^s) \leq 0 & (\mu^\omega), \end{array}\right. \tag{6.9d}$$

$\omega = 1 \dots, s$

For the sake of simplicity, note that we include only one *family* of constraining lower-level problems in the above formulation, i.e., a single constraining lower-level problem is associated with each scenario. However, a more general formulation can be obtained by considering scenarios in each of the lower-level problems of OPcOP (6.1).

The illustrative example below provides a stochastic instance of the strategic offering problem explained in Illustrative Example 6.2.2.

6.2.12 Illustrative Example. Strategic Offering, sOPcOP

Considering two scenarios for the lower-level problem, the OPcOP (6.2) becomes the sOPcOP below.

$$\text{Minimize}_{\{\hat{c}_1^{\text{g}}\}\cup\{g_1^\omega, g_2^\omega, g_3^\omega, d_1^\omega, d_2^\omega, d_3^\omega, \theta_1^\omega, \theta_2^\omega, \theta_3^\omega\}\cup\Xi_{\text{SO}}^\omega, \omega=1,2}$$

$$\sum_{\omega=1}^{2} \Pi^\omega \left(c_1^{\text{g}} g_1^\omega - \alpha_1^\omega g_1^\omega\right) \tag{6.10a}$$

s. t.

$$\text{Minimize}_{g_1^\omega, g_2^\omega, g_3^\omega, d_1^\omega, d_2^\omega, d_3^\omega, \theta_1^\omega, \theta_2^\omega, \theta_3^\omega} \tag{6.10b}$$

$$\hat{c}_1^{\text{g}} g_1^\omega + \sum_{i=2}^{3} c_i^{\text{g}} g_i^\omega - \sum_{i=1}^{3} c_i^{\text{d}} d_i^\omega \tag{6.10c}$$

$$\text{s. t.} \tag{6.10d}$$

$$g_1^\omega - d_1^\omega = B_1(\theta_1^\omega - \theta_2^\omega) + B_2(\theta_1^\omega - \theta_3^\omega) \quad (\alpha_1^\omega) \tag{6.10e}$$

$$g_2^\omega - d_2^\omega = B_1(\theta_2^\omega - \theta_1^\omega) + B_3(\theta_2^\omega - \theta_3^\omega) \quad (\alpha_2^\omega) \tag{6.10f}$$

$$g_3^\omega - d_3^\omega = B_2(\theta_3^\omega - \theta_1^\omega) + B_3(\theta_3^\omega - \theta_2^\omega) \quad (\alpha_3^\omega) \tag{6.10g}$$

$$0 \leq g_i^\omega \leq g_i^{\max} \quad (\phi_i^{\text{gmin}^\omega}, \phi_i^{\text{gmax}^\omega}), i = 1, 2, 3 \tag{6.10h}$$

$$0 \leq d_i^\omega \leq d_i^{\max} \quad (\phi_i^{\text{dmin}^\omega}, \phi_i^{\text{dmax}^\omega}), i = 1, 2, 3 \tag{6.10i}$$

$$- f_1^{\max} \leq B_1(\theta_1^\omega - \theta_2^\omega) \leq f_1^{\max} \quad (\rho_1^{\min^\omega}, \rho_1^{\max^\omega}) \tag{6.10j}$$

$$- f_2^{\max} \leq B_2(\theta_1^\omega - \theta_3^\omega) \leq f_2^{\max} \quad (\rho_2^{\min^\omega}, \rho_2^{\max^\omega}) \tag{6.10k}$$

$$- f_3^{\max} \leq B_3(\theta_2^\omega - \theta_3^\omega) \leq f_3^{\max} \quad (\rho_3^{\min^\omega}, \rho_3^{\max^\omega}) \tag{6.10l}$$

$$\theta_3^\omega = 0 \quad (\gamma^\omega) \tag{6.10m}$$

$$\omega = 1, 2.$$

6.3 Optimization Problems Constrained by Linear Problems, OPcLP

The general formulation of an optimization problem constrained by a collection of linear programming problems is as follows,

$$\text{Minimize}_{\{x\} \cup \{x^1, \ldots, x^n\} \cup \{\mu^1, \ldots, \mu^n\}}$$

$$f(x, x^1, \ldots, x^n, \mu^1, \ldots, \mu^n) \tag{6.11a}$$

s. t.

$$h(x, x^1, \ldots, x^n, \mu^1, \ldots, \mu^n) = 0 \tag{6.11b}$$

$$g(x, x^1, \ldots, x^n, \mu^1, \ldots, \mu^n) \leq 0, \tag{6.11c}$$

$$\left\{ \begin{array}{l} \text{Minimize}_{x^1} \quad c^1(z^1)^T x^1 \\ \text{s. t.} \\ \quad A^1(z^1) x^1 \geq b^1(z^1) \quad (\mu^1) \\ \quad x^1 \geq 0, \end{array} \right. \tag{6.11d}$$

$$\vdots$$

$$\left\{ \begin{array}{l} \text{Minimize}_{x^i} \quad c^i(z^i)^T x^i \\ \text{s. t.} \\ \quad A^i(z^i) x^i \geq b^i(z^i) \quad (\mu^i) \\ \quad x^i \geq 0, \end{array} \right. \tag{6.11e}$$

$$\vdots$$

$$\left\{ \begin{array}{l} \text{Minimize}_{x^n} \quad c^n(z^n)^T x^n \\ \text{s. t.} \\ \quad A^n(z^n) x^n \geq b^n(z^n) \quad (\mu^n) \\ \quad x^n \geq 0, \end{array} \right. \tag{6.11f}$$

$$\text{where } z^i = [x^T {x^1}^T \ldots {x^{i-1}}^T {x^{i+1}}^T \ldots {x^n}^T]^T. \tag{6.11g}$$

Functions $f(x, x^1, \ldots, x^n, \mu^1, \ldots, \mu^n)$, $h(x, x^1, \ldots, x^n, \mu^1, \ldots, \mu^n)$ and $g(x, x^1, \ldots, x^n, \mu^1, \ldots, \mu^n)$ and variable vectors x, x^i and μ^i are characterized in Problem (6.1). Vector $c^i(z^i) \in \mathbb{R}^{n^i}$, matrix $A^i(z^i) \in \mathbb{R}^{m_\text{I}^i \times n^i}$ and vector

$b^i(z^i) \in \mathbb{R}^{m^i_\mathrm{I}}$ are the cost vector, the constraint matrix and the right-hand-side vector, respectively, of problem i, which include variables pertaining to other problems playing the role of constants for problem i.

Note that each constraining linear programming problem is formulated using its canonical form, so that its dual problem is straightforwardly written.

It should be noted that Illustrative Examples 6.2.2, 6.2.3 and 6.2.4 are examples of optimization problems constrained by linear problems.

6.3.1 Mathematical Program with Primal and Dual Constraints, MPPDC

If constraining problems (6.11d)-(6.11f) are replaced in problem (6.11) with their corresponding primal constraints, dual constraints, and strong duality theorem equalities, problem (6.11) renders:

$$\text{Minimize}_{\{x,x^1,\ldots,x^n,\mu^1,\ldots,\mu^n\}}$$
$$f(x,x^1,\ldots,x^n,\mu^1,\ldots,\mu^n) \tag{6.12a}$$
s. t.
$$h(x,x^1,\ldots,x^n,\mu^1,\ldots,\mu^n) = 0 \quad (\alpha) \tag{6.12b}$$
$$g(x,x^1,\ldots,x^n,\mu^1,\ldots,\mu^n) \leq 0 \quad (\beta), \tag{6.12c}$$
$$\left\{\begin{array}{ll} c^1(z^1)^T x^1 = b^1(z^1)^T \mu^1 & (\delta^1) \\ A^1(z^1)x^1 \geq b^1(z^1) & (\phi^1) \\ x^1 \geq 0 & (\gamma^1) \\ A^1(z^1)^T \mu^1 \leq c^1(z^1) & (\eta^1) \\ \mu^1 \geq 0 & (\rho^1), \end{array}\right. \tag{6.12d}$$
$$\vdots$$
$$\left\{\begin{array}{ll} c^i(z^i)^T x^i = b^i(z^i)^T \mu^i & (\delta^i) \\ A^i(z^i)x^i i \geq b^i(z^i) & (\phi^i) \\ x^i \geq 0 & (\gamma^i) \\ A^i(z^i)^T \mu^i \leq c^i(z^i) & (\eta^i) \\ \mu^i \geq 0 & (\rho^i), \end{array}\right. \tag{6.12e}$$
$$\vdots$$
$$\left\{\begin{array}{ll} c^n(z^n)^T x^n = b^n(z^n)^T \mu^n & (\delta^n) \\ A^n(z^n)x^n \geq b^n(z^n) & (\phi^n) \\ x^n \geq 0 & (\gamma^n) \\ A^n(z^n)^T \mu^n \leq c^n(z^n) & (\eta^n) \\ \mu^n \geq 0 & (\rho^n), \end{array}\right. \tag{6.12f}$$

$$\text{where } z^i = [x^T x^{1^T} \ldots x^{i-1^T} x^{i+1^T} \ldots x^{n^T}]^T. \tag{6.12g}$$

Note that the problem above in a MPPDC, that is, a mathematical program with primal and dual constraints.

Dual variables pertaining to both the upper-level problem and the constraining lower-level problems are indicated in parentheses. The dimension of the dual variable vectors of problem i are indicated in the following: $\delta^i \in \mathbb{R}$, $\phi^i \in \mathbb{R}^{m_I^i}$, $\gamma^i \in \mathbb{R}^{n^i}$, $\eta^i \in \mathbb{R}^{n^i}$ and $\rho^i \in \mathbb{R}^{m_I^i}$.

Since, in general, KKT conditions are numerically difficult to handle, it is relevant to note that an MPPDC might be easier to solve than its associated MPCC.

For simplicity Problem (6.11) only includes constraints of the type "$\geq$". Thus, if the problem under study contains constraints of the type "$\leq$", these constraints need to be converted into constraints of the type "$\geq$" before applying formulations (6.12). This is equivalent to considering constraints of the type "$\leq$" as such and changing the sign of the corresponding multipliers. This is in accordance with the KKT formulations used in Subsections 6.2.6 and 6.2.8. (see Illustrative Examples 6.2.7 - 6.2.10). It is important to note that this consideration is relevant to interpret correctly the illustrative examples below. Additionally, note that equality constraints are easily converted to constraints of the type "$\geq$" if required, [11].

The illustrative example below clarifies the structure of an MPPDC using the strategic offering problem formulated in Illustrative Example 6.2.2.

6.3.2 *Illustrative Example. Strategic Offering, MPPDC*

Since the lower-level problem of OPcOP (6.2) is linear, and thus convex, the whole OPcOP (6.2) can be recast as an MPPDC, as shown below:

$$\text{Minimize}_{\{\hat{c}_1^g\} \cup \{g_1, g_2, g_3, d_1, d_2, d_3, \theta_1, \theta_2, \theta_3\} \cup \Xi_{SO}}$$

$$c_1^g g_1 - \alpha_1 g_1 \tag{6.13a}$$

s. t.

Primal constraints:

$$g_1 - d_1 = B_1(\theta_1 - \theta_2) + B_2(\theta_1 - \theta_3) \quad (\sigma_1) \tag{6.13b}$$

$$g_2 - d_2 = B_1(\theta_2 - \theta_1) + B_3(\theta_2 - \theta_3) \quad (\sigma_2) \tag{6.13c}$$

$$g_3 - d_3 = B_2(\theta_3 - \theta_1) + B_3(\theta_3 - \theta_2) \quad (\sigma_3) \tag{6.13d}$$

$$0 \leq g_i \leq g_i^{\max}, \quad (\beta_i^{\text{gmin}}, \beta_i^{\text{gmax}}), i = 1, 2, 3 \tag{6.13e}$$

$$0 \leq d_i \leq d_i^{\max}, \quad (\beta_i^{\text{dmin}}, \beta_i^{\text{dmax}}), i = 1, 2, 3 \tag{6.13f}$$

$$-f_1^{\max} \leq B_1(\theta_1 - \theta_2) \leq f_1^{\max} \quad (\beta_1^{\text{fmin}}, \beta_1^{\text{fmax}}) \tag{6.13g}$$

$$-f_2^{\max} \leq B_2(\theta_1 - \theta_3) \leq f_2^{\max} \quad (\beta_2^{\text{fmin}}, \beta_2^{\text{fmax}}) \tag{6.13h}$$

$$- f_3^{\max} \leq B_3(\theta_2 - \theta_3) \leq f_3^{\max} \quad (\beta_3^{\text{fmin}}, \beta_3^{\text{fmax}}) \tag{6.13i}$$

$$\theta_3 = 0 \quad (\sigma^\theta) \tag{6.13j}$$

Dual constraints:

$$\hat{c}_1^{\text{g}} - \alpha_1 - \phi_1^{\text{gmin}} + \phi_1^{\text{gmax}} = 0 \quad (\rho_1^{\text{g}}) \tag{6.13k}$$

$$c_i^{\text{g}} - \alpha_i - \phi_i^{\text{gmin}} + \phi_i^{\text{gmax}} = 0, \quad (\rho_i^{\text{g}}), i = 2, 3 \tag{6.13l}$$

$$- c_i^{\text{d}} + \alpha_i - \phi_i^{\text{dmin}} + \phi_i^{\text{dmax}} = 0, \quad (\rho_1^{\text{d}}), i = 1, 2, 3 \tag{6.13m}$$

$$B_1(\alpha_1 - \alpha_2 - \rho_1^{\min} + \rho_1^{\max}) + B_2(\alpha_1 - \alpha_3 - \rho_2^{\min} + \rho_2^{\max}) = 0 \quad (\rho_1^\theta) \tag{6.13n}$$

$$B_1(-\alpha_1 + \alpha_2 + \rho_1^{\min} - \rho_1^{\max}) + B_3(\alpha_2 - \alpha_3 - \rho_3^{\min} + \rho_3^{\max}) = 0 \quad (\rho_2^\theta) \tag{6.13o}$$

$$B_2(-\alpha_1 + \alpha_3 + \rho_2^{\min} - \rho_2^{\max}) + B_3(-\alpha_2 + \alpha_3 + \rho_3^{\min} - \rho_3^{\max}) + \gamma = 0 \quad (\rho_3^\theta) \tag{6.13p}$$

$$\phi_i^{\text{gmin}}, \phi_i^{\text{gmax}} \geq 0, \quad (\eta_i^{\text{gmin}}, \eta_i^{\text{gmax}}), i = 1, 2, 3 \tag{6.13q}$$

$$\phi_i^{\text{dmin}}, \phi_i^{\text{dmax}} \geq 0, \quad (\eta_i^{\text{dmin}}, \eta_i^{\text{dmax}}), i = 1, 2, 3 \tag{6.13r}$$

$$\rho_i^{\min}, \rho_i^{\max} \geq 0, \quad (\eta_i^{\min}, \eta_i^{\max}), i = 1, 2, 3 \tag{6.13s}$$

Strong duality theorem:

$$\hat{c}_1^{\text{g}} g_1 + \sum_{i=2}^{3} c_i^{\text{g}} g_i - \sum_{i=1}^{3} c_i^{\text{d}} d_i = - \sum_{i=1}^{3} g_i^{\max} \phi_i^{\text{gmax}} - \sum_{i=1}^{3} d_i^{\max} \phi_i^{\text{dmax}} - \sum_{i=1}^{3} f_i^{\max} \rho_i^{\max} - \sum_{i=1}^{3} f_i^{\max} \rho_i^{\min}. \quad (\delta) \tag{6.13t}$$

Note that the MPPDC above is single-level and nonlinear due to equality (6.13t) enforcing the strong duality theorem. Solution algorithms for such problems are available [1]. Its solution is that provided in Illustrative Example (6.2.2).

The next illustrative example is intented to clarify the MPPDC formulation for the vulnerability assessment problem stated in Illustrative Example 6.2.3.

6.3.3 Illustrative Example. Vulnerability Assessment, MPPDC

OPcOP (6.3) can be recast as an MPPDC as explained below. Since x_i and y_i $(i = 1, 2, 3)$ are given values (constants) for the lower-level problem, such problem is linear, and therefore it can be replaced by its constraints (primal problem), the constraints of its dual problem and the equality corresponding to the strong duality theorem. This way the original OPcOP (6.3) becomes the MPPDC below:

$$\text{Minimize}_{\{x_1,x_2,x_3\}\cup\{r_1,r_2,r_3,g_1,g_2,g_3,f_1,f_2,f_3,\theta_1,\theta_2,\theta_3\}\cup\Xi_{\text{VA}}}$$

$$-\sum_{i=1}^{3} r_i \tag{6.14a}$$

s. t.

$$\sum_{i=1}^{3} x_k + \sum_{i=1}^{3} y_i = 2 \tag{6.14b}$$

$$x_k \in \{0,1\},\ k=1,2,3; \quad y_i \in \{0,1\},\ i=1,2,3 \tag{6.14c}$$

Primal constraints:

$$g_1 + r_1 - d_1 - f_1 - f_2 = 0 \tag{6.14d}$$

$$g_2 + r_2 - d_2 + f_1 - f_3 = 0 \tag{6.14e}$$

$$g_3 + r_3 - d_3 + f_2 + f_3 = 0 \tag{6.14f}$$

$$f_1 - (1-y_1)B_1(\theta_1 - \theta_2) = 0 \tag{6.14g}$$

$$f_2 - (1-y_2)B_2(\theta_1 - \theta_3) = 0 \tag{6.14h}$$

$$f_3 - (1-y_3)B_3(\theta_2 - \theta_3) = 0 \tag{6.14i}$$

$$\theta_3 = 0 \tag{6.14j}$$

$$(1-x_i)g_i^{\min} \le g_i \le (1-x_i)g_i^{\max}, \quad i=1,2,3 \tag{6.14k}$$

$$0 \le r_i \le d_i, \quad i=1,2,3 \tag{6.14l}$$

$$-f_i^{\max} \le f_i \le f_i^{\max}, \quad i=1,2,3 \tag{6.14m}$$

Dual constraints:

$$\alpha_i - \phi_i^{\min} + \phi_i^{\max} = 0, \quad i=1,2,3 \tag{6.14n}$$

$$-\alpha_1 + \alpha_2 + \beta_1 - \rho_1^{\min} + \rho_1^{\max} = 0 \tag{6.14o}$$

$$-\alpha_1 + \alpha_3 + \beta_2 - \rho_2^{\min} + \rho_2^{\max} = 0 \tag{6.14p}$$

$$-\alpha_2 + \alpha_3 + \beta_3 - \rho_3^{\min} + \rho_3^{\max} = 0 \tag{6.14q}$$

$$1 + \alpha_i - \nu_i^{\min} + \nu_i^{\max} = 0, \quad i=1,2,3 \tag{6.14r}$$

$$-\beta_1(1-y_1)B_1 - \beta_2(1-y_2)B_2 = 0 \tag{6.14s}$$

$$+\beta_1(1-y_1)B_1 - \beta_3(1-y_3)B_3 = 0 \tag{6.14t}$$

$$\beta_2(1-y_2)B_2 + \beta_3(1-y_3)B_3 + \gamma = 0 \tag{6.14u}$$

$$\phi_i^{\min}, \phi_i^{\max}, \nu_i^{\min}, \nu_i^{\max}, \rho_i^{\min}, \rho_i^{\max} \ge 0, \quad i=1,2,3 \tag{6.14v}$$

Strong duality theorem:

$$\sum_{i=1}^{3} r_i = -\sum_{i=1}^{3} \alpha_i d_i + \sum_{i=1}^{3} \phi_i^{\min}(1-x_i)g_i^{\min} - \sum_{i=1}^{3} \phi_i^{\max}(1-x_i)g_i^{\max}$$

$$- \sum_{i=1}^{3} \nu_i^{\max} d_i - \sum_{i=1}^{3} f_i^{\max} \rho_i^{\min} - \sum_{i=1}^{3} f_i^{\max} \rho_i^{\max} \tag{6.14w}$$

Note that the MPPDC (6.14) is single-level, mixed-integer and nonlinear. Nonlinearities are due to the objective function (6.14a) and the equality pertaining to the strong duality theorem, (6.14w). Solution algorithm for such problems are available [1,4]. An optimal solution is that provided in Illustrative Example 6.2.3.

To further clarify the structure of MPPDCs, the illustrative example below provides the MPPDC formulation of the transmission investment problem explained in Illustrative Example 6.2.4.

6.3.4 Illustrative Example. Transmission Investment, MPPDC

OPcOP (6.4) can be recast as an MPPDC as shown below. Since x_i ($i = 1, 2, 3$) are given values (constants) for the lower-level problem, such problem is linear, an therefore it can be replaced by its constraints (primal problem), its dual constraints and the equality enforcing the strong duality theorem. This way the original OPcOP (6.4) becomes the MPPDC below:

$$\text{Minimize}_{\{x_1,x_2,x_3\} \cup \{g_1,g_2,g_3,r_1,r_2,r_3,f_1,f_2,f_3,t_1,t_2,t_3,\theta_1,\theta_2,\theta_3\} \cup \Xi_{\text{TI}}}$$

$$\sum_{k=1}^{3} c_k x_k + e \sum_{i=1}^{3} c_i^{\text{U}} r_i \tag{6.15a}$$

s. t.

$$\sum_{k=1}^{3} c_k x_k - c^{\max} \leq 0 \tag{6.15b}$$

$$x_k \in \{0, 1\}, \quad k = 1, 2, 3 \tag{6.15c}$$

Primal contraints:

$$\text{(6.4e)-(6.4r)} \tag{6.15d}$$

Dual constraints:

$$c_i^{\text{g}} + \alpha_i - \phi_i^{\min} + \phi_i^{\max} = 0, \quad i = 1, 2, 3 \tag{6.15e}$$

$$c_i^{\text{U}} + \alpha_i - \nu_i^{\min} + \nu_i^{\max} = 0, \quad i = 1, 2, 3 \tag{6.15f}$$

$$- \alpha_1 + \alpha_2 + \beta_1 - \rho_1^{\min} + \rho_1^{\max} = 0 \tag{6.15g}$$

$$- \alpha_1 + \alpha_3 + \beta_2 - \rho_2^{\min} + \rho_2^{\max} = 0 \tag{6.15h}$$

$$- \alpha_2 + \alpha_3 + \beta_3 - \rho_3^{\min} + \rho_3^{\max} = 0 \tag{6.15i}$$

$$- \alpha_1 + \alpha_2 + \delta_1 - \psi_1^{\min} + \psi_1^{\max} = 0 \tag{6.15j}$$

$$- \alpha_1 + \alpha_3 + \delta_2 - \psi_2^{\min} + \psi_2^{\max} = 0 \quad (6.15k)$$

$$- \alpha_2 + \alpha_3 + \delta_3 - \psi_3^{\min} + \psi_3^{\max} = 0 \quad (6.15l)$$

$$- \beta_1 x_1 B_1^{\mathrm{N}} - \beta_2 x_2 B_2^{\mathrm{N}} - \delta_1 B_1^{\mathrm{E}} - \delta_2 B_2^{\mathrm{E}} = 0 \quad (6.15m)$$

$$\beta_1 x_1 B_1^{\mathrm{N}} - \beta_3 x_3 B_3^{\mathrm{N}} + \delta_1 B_1^{\mathrm{E}} - \delta_3 B_3^{\mathrm{E}} = 0 \quad (6.15n)$$

$$\beta_2 x_2 B_2^{\mathrm{N}} + \beta_3 x_3 B_3^{\mathrm{N}} + \delta_2 B_2^{\mathrm{E}} + \delta_3 B_3^{\mathrm{E}} + \gamma = 0 \quad (6.15o)$$

$$\phi_i^{\min}, \phi_i^{\max}, \nu_i^{\min}, \nu_i^{\max}, \psi_i^{\min}, \psi_i^{\max}, \rho_i^{\min}, \rho_i^{\max} \geq 0, i = 1,2,3 \quad (6.15p)$$

Strong duality theorem:

$$\sum_{i=1}^{3} c_i^{\mathrm{g}} g_i + \sum_{i=1}^{3} c_i^{\mathrm{U}} r_i = - \sum_{i=1}^{3} \alpha_i d_i + \sum_{i=1}^{3} \phi_i^{\min} g_i^{\min} - \sum_{i=1}^{3} \phi_i^{\max} g_i^{\max}$$
$$- \sum_{i=1}^{3} \nu_i^{\max} d_i - \sum_{i=1}^{3} t_i^{\max} \psi_i^{\min} - \sum_{i=1}^{3} t_i^{\max} \psi_i^{\max}$$
$$- \sum_{i=1}^{3} f_i^{\max} \rho_i^{\min} - \sum_{i=1}^{3} f_i^{\max} \rho_i^{\max}. \quad (6.15q)$$

Note that MPPDC (6.15) is single-level, mixed-integer and nonlinear. Algorithms for solving such problems are available. An optimal solution is that provided in Illustrative Example 6.2.4.

6.3.5 Mathematical Program with Complementarity Constraints, MPCC

Reformulating the constraining linear problems i ($i = 1 \ldots n$) of problem (6.11) as

$$\text{Minimize}_{x^i} \quad c^i(z^i)^T x^i \quad (6.16a)$$

s. t.

$$- A^i(z^i)x^i + b^i(z^i) \leq 0 \quad (\mu^i) \quad (6.16b)$$

$$- x^i \leq 0 \quad (\xi^i), \quad (6.16c)$$

and using KKT conditions to characterize the optimality region of each of these constraining problems, problem (6.11) becomes

$$\text{Minimize}_{\{x, x^1, \ldots, x^n, \mu^1, \ldots, \mu^n, \xi^1, \ldots, \xi^n\}}$$
$$f(x, x^1, \ldots, x^n, \mu^1, \ldots, \mu^n, \xi^1, \ldots, \xi^n) \quad (6.17a)$$

s. t.

$$h(x, x^1, \ldots, x^n, \mu^1, \ldots, \mu^n, \xi^1, \ldots, \xi^n) = 0 \quad (6.17b)$$

$$g(x, x^1, \ldots, x^n, \mu^1, \ldots, \mu^n, \xi^1, \ldots, \xi^n) \leq 0, \tag{6.17c}$$

$$\begin{cases} c^1(z^1) - A^1(z^1)^T \mu^1 - \xi^1 = 0 \\ 0 \leq \mu^1 \perp A^1(z^1)x^1 - b^1(z^1) \geq 0 \\ 0 \leq \xi^1 \perp x^1 \geq 0. \end{cases} \tag{6.17d}$$

$$\vdots$$

$$\begin{cases} c^i(z^i) - A^i(z^i)^T \mu^i - \xi^i = 0 \\ 0 \leq \mu^i \perp A^i(z^i)x^i - b^i(z^i) \geq 0 \\ 0 \leq \xi^i \perp x^i \geq 0. \end{cases} \tag{6.17e}$$

$$\vdots$$

$$\begin{cases} c^n(z^n) - A^n(z^n)^T \mu^n - \xi^n = 0 \\ 0 \leq \mu^n \perp A^n(z^n)x^n - b^n(z^n) \geq 0 \\ 0 \leq \xi^n \perp x^n \geq 0. \end{cases} \tag{6.17f}$$

$$\text{where } z^i = [x^T {x^1}^T \ldots {x^{i-1}}^T {x^{i+1}}^T \ldots {x^n}^T]^T. \tag{6.17g}$$

Note that the problem above is an MPCC, that is, a mathematical program with complementarity constraints.

An instance of the MPCC (6.17) is provided in Illustrative Example 6.3.3 pertaining to the vulnerability assessment problem.

6.3.6 Stochastic OPcLPs

We describe below a stochastic OPcLP. Note that we consider a collection of lower-level problems indexed by scenarios. This is not the most general formulation but it is appropriate to characterize the structure of a stochastic OPcLP. A more general formulation can be obtained by converting each lower-level problem of problem (6.11) into a family of problems indexed by scenarios.

The stochastic OPcLP has the form shown below:

$$\text{Minimize}_{\{x\} \cup \{x^1, \ldots, x^s\} \cup \{\mu^1, \ldots, \mu^s\}}$$

$$f(x, x^1, \ldots, x^s, \mu^1, \ldots, \mu^s) \tag{6.18a}$$

s. t.

$$h(x, x^1, \ldots, x^s, \mu^1, \ldots, \mu^s) = 0 \tag{6.18b}$$

$$g(x, x^1, \ldots, x^s, \mu^1, \ldots, \mu^s) \leq 0, \tag{6.18c}$$

$$\begin{cases} \text{Minimize}_{x^\omega} \quad c^\omega(x, x^1, \ldots, x^{\omega-1}, x^\omega, \ldots, x^s)^T x^\omega \\ \text{s. t.} \\ \quad A^\omega(x, x^1, \ldots, x^{\omega-1}, x^\omega, \ldots, x^s) x^\omega \geq \\ \quad\quad b^\omega(x, x^1, \ldots, x^{\omega-1}, x^\omega, \ldots, x^s) \quad (\mu^\omega) \end{cases} \tag{6.18d}$$

$$\omega = 1 \ldots, s. \tag{6.18e}$$

To clarify the structure of a sOPcLP, the illustrative example below provides a stochastic version of the transmission investment problem described in Illustrative Example 6.2.4.

6.3.7 Illustrative Example. Transmission Investment, sOPcLP

A stochastic version of the OPcLP described in Illustrative Example 6.2.4 is provided below. For the sake of simplicity, just two scenarios are considered.

$$\text{Minimize}_{\{x_1,x_2,x_3\}\cup\{g_1^\omega,g_2^\omega,g_3^\omega,r_1^\omega,r_2^\omega,r_3^\omega,f_1^\omega,f_2^\omega,f_3^\omega,t_1^\omega,t_2^\omega,t_3^\omega,\theta_1^\omega,\theta_2^\omega,\theta_3^\omega\}\cup\Xi_{\text{TI}}^\omega}$$

$$\sum_{\omega=1}^{2} \Pi^\omega \left(\sum_{k=1}^{3} c_k x_k + e \sum_{i=1}^{3} c_i^{\text{U}} r_i^\omega \right) \tag{6.19a}$$

s. t.

$$\sum_{k=1}^{3} c_k x_k - c^{\max} \leq 0 \tag{6.19b}$$

$$x_k \in \{0,1\}, k = 1,2,3 \tag{6.19c}$$

$$\text{Minimize}_{g_1^\omega,g_2^\omega,g_3^\omega,r_1^\omega,r_2^\omega,r_3^\omega,f_1^\omega,f_2^\omega,f_3^\omega,t_1^\omega,t_2^\omega,t_3^\omega,\theta_1^\omega,\theta_2^\omega,\theta_3^\omega} \tag{6.19d}$$

$$\sum_{i=1}^{3} c_i^{\text{g}} g_i^\omega + \sum_{i=1}^{3} c_i^{\text{U}} r_i^\omega \tag{6.19e}$$

s. t. (6.19f)

$$g_1^\omega + r_1^\omega - d_1^\omega - f_1^\omega - f_2^\omega - t_1^\omega - t_2^\omega = 0 \quad (\alpha_1^\omega) \tag{6.19g}$$

$$g_2^\omega + r_2^\omega - d_2^\omega + f_1^\omega - f_3^\omega + t_1^\omega - t_3^\omega = 0 \quad (\alpha_2^\omega) \tag{6.19h}$$

$$g_3^\omega + r_3^\omega - d_3^\omega + f_2^\omega + f_3^\omega + t_2^\omega + t_3^\omega = 0 \quad (\alpha_3^\omega) \tag{6.19i}$$

$$f_1^\omega - x_1^\omega B_1^{\text{N}}(\theta_1^\omega - \theta_2^\omega) = 0 \quad (\beta_1^\omega) \tag{6.19j}$$

$$f_2^\omega - x_2^\omega B_2^{\text{N}}(\theta_1^\omega - \theta_3^\omega) = 0 \quad (\beta_2^\omega) \tag{6.19k}$$

$$f_3^\omega - x_3^\omega B_3^{\text{N}}(\theta_2^\omega - \theta_3^\omega) = 0 \quad (\beta_3^\omega) \tag{6.19l}$$

$$t_1^\omega - B_1^{\text{E}}(\theta_1^\omega - \theta_2^\omega) = 0 \quad (\delta_1^\omega) \tag{6.19m}$$

$$t_2^\omega - B_2^{\text{E}}(\theta_1^\omega - \theta_3^\omega) = 0 \quad (\delta_2^\omega) \tag{6.19n}$$

$$t_3^\omega - B_3^{\text{E}}(\theta_2^\omega - \theta_3^\omega) = 0 \quad (\delta_3^\omega) \tag{6.19o}$$

$$\theta_3^\omega = 0 \quad (\gamma^\omega) \tag{6.19p}$$

$$g_i^{\min} - g_i^\omega \leq 0 \quad (\phi_i^{\min^\omega}), i = 1,2,3 \tag{6.19q}$$

$$g_i^\omega - g_i^{\max} \leq 0 \quad (\phi_i^{\max^\omega}), i = 1,2,3 \tag{6.19r}$$

$$- r_i^\omega \leq 0 \quad (\nu_i^{\min^\omega}), i = 1,2,3 \tag{6.19s}$$

$$r_i^\omega - d_i^\omega \leq 0 \quad (\nu_i^{\max^\omega}), i = 1,2,3 \tag{6.19t}$$

$$- t_i^{\max} - t_i^\omega \leq 0 \quad (\psi_i^{\min^\omega}), i = 1,2,3 \tag{6.19u}$$

$$t_i^\omega - t_i^{\max} \leq 0 \quad (\psi_i^{\max^\omega}), i = 1,2,3 \tag{6.19v}$$

$$- f_i^{\max} - f_i^\omega \leq 0 \quad (\rho_i^{\min^\omega}), i = 1,2,3 \tag{6.19w}$$

$$f_i^\omega - f_i^{\max} \leq 0 \quad (\rho_i^{\max^\omega}), i = 1,2,3, \tag{6.19x}$$

$$\omega = 1,2.$$

6.4 Transforming an MPCC/MPEC/MPPDC into a MILP

Diverse linearization techniques are available to transform an OPcLP expressed as a MPCC, MPEC or MPPDC into a mixed-integer linear programming problem, MILP. They are briefly reviewed below.

Such linearizations are important to achieve solvable instances of the considered MPcLPs. In other words, such linearizations are relevant in practice because MILP problems can be straightforwardly solved provided that the number of integer variables is not *too* large.

6.4.1 Fortuny-Amat McCarl Linearization

Complementarity constraints of the type

$$0 \leq x \perp y \geq 0, \tag{6.20}$$

can be replaced by the following set of linear constraints

$$0 \leq x \leq Mu \tag{6.21a}$$

$$0 \leq y \leq M(1-u) \tag{6.21b}$$

being M large enough and u binary.

This linearization is due to Fortuny-Amat and McCarl [8]. Care should be exercised to select constant M to avoid numerical ill-conditioning.

6.4.2 SOS1 and Penalty Function Linearization

An alternative technique to linearize conditions (6.20) relies on using Special Ordered Sets of Type 1 variables (SOS1 variables) [18]. SOS1 variables are defined so that at most one of them can take a strictly positive value while all others are equal to 0.

Condition (6.20) can be linearized by first reformulating it in terms of variables u and v as follows

$$x \geq 0 \tag{6.22a}$$
$$y \geq 0 \tag{6.22b}$$
$$u - |v| = 0 \tag{6.22c}$$
$$u = \frac{x+y}{2} \tag{6.22d}$$
$$v = \frac{x-y}{2} \tag{6.22e}$$

The variable v can in turn be decomposed into its positive and negative parts, i.e., $v = v^+ - v^-$, where v^+ and v^- are SOS1 variables. Hence, an exact formulation for (6.20) is

$$x \geq 0 \tag{6.23a}$$
$$y \geq 0 \tag{6.23b}$$
$$u - \left(v^+ + v^-\right) = 0 \tag{6.23c}$$
$$u - \frac{x+y}{2} \tag{6.23d}$$
$$v^+ - v^- = \frac{x-y}{2} \tag{6.23e}$$
$$v^+, v^- \text{ are SOS1 variables} \tag{6.23f}$$

Note that the absolute value function in (6.22c) can also be approximated using a penalty approach when solving MPCCs and MPECs. In such approach, a suitable penalty constant multiplied by positive variables v^+ and v^- is added into the objective function of the MPCC or MPEC, which then forces either of v^+ or v^- to be zero. Further details are provided in [18].

6.4.3 Other Linearizations

There are certain problems that at a first-look appear to be nonlinear but that can be linearized through exact algebraic transformations. There is no general rule to achieve such linearization, but given the advantage of working with linear problems, it is always worthy to try. We show below an example of an apparently nonlinear problem that can be linearized using simple algebraic transformations.

6.4.3.1 Illustrative Example. Strategic Offering: Exact Linear Transformation

Problem (6.10) is non-linear due the complementarity constraints and the term $\alpha_1 g_1$ in the objective function. The complementarity constraints can be linearized using the Fortuny-Amat and McCarl linearization described in Subsection 6.4.1 above. To find a linear expression for the product $\alpha_1 g_1$, we use the strong duality theorem and some of the KKT conditions as indicated in the following.

From (9.23f)

$$\hat{c}_1^{\mathrm{g}} = \alpha_1 + \phi_1^{\mathrm{gmin}} - \phi_1^{\mathrm{gmax}}, \tag{6.24}$$

and

$$\hat{c}_1^{\mathrm{g}} g_1 = \alpha_1 g_1 + \phi_1^{\mathrm{gmin}} g_1 - \phi_1^{\mathrm{gmax}} g_1. \tag{6.25}$$

From (9.23l) and (9.23m)

$$\phi_1^{\mathrm{gmin}} g_1 = 0, \tag{6.26}$$

$$\phi_1^{\mathrm{gmax}} g_1 = \phi_1^{\mathrm{gmax}} g_1^{\mathrm{max}}. \tag{6.27}$$

Substituting (6.26) and (6.27) in (6.25) renders

$$\hat{c}_1^{\mathrm{g}} g_1 = \alpha_1 g_1 - \phi_1^{\mathrm{gmax}} g_1^{\mathrm{max}}. \tag{6.28}$$

On the other hand, from (6.13t) (strong duality theorem),

$$\begin{aligned} \hat{c}_1^{\mathrm{g}} g_1 = &- \sum_{i=2}^{3} c_i^{\mathrm{g}} g_i + \sum_{i=1}^{3} c_i^{\mathrm{d}} d_i - \sum_{i=1}^{3} g_i^{\mathrm{max}} \phi_i^{\mathrm{gmax}} \\ &- \sum_{i=1}^{3} d_i^{\mathrm{max}} \phi_i^{\mathrm{dmax}} - \sum_{i=1}^{3} f_i^{\mathrm{max}} \rho_i^{\mathrm{max}} - \sum_{i=1}^{3} f_i^{\mathrm{max}} \rho_i^{\mathrm{min}} \end{aligned} \tag{6.29}$$

Substituting (6.28) in (6.29) results in

$$\alpha_1 g_1 = - \sum_{i=2}^{3} c_i^{\mathrm{g}} g_i + \sum_{i=1}^{3} c_i^{\mathrm{d}} d_i - \sum_{i=2}^{3} g_i^{\mathrm{max}} \phi_i^{\mathrm{gmax}}$$

$$-\sum_{i=1}^{3} d_i^{\max}\phi_i^{\text{dmax}} - \sum_{i=1}^{3} f_i^{\max}\rho_i^{\max} - \sum_{i=1}^{3} f_i^{\max}\rho_i^{\min}, \tag{6.30}$$

which provides a linear expression for $\alpha_1 g_1$.

Using (6.30) and linearizing the complementarity conditions through (6.21), problem (6.10) can be recast as a mixed-integer linear programming problem, easily solved using an appropriate MILP solver, such as CPLEX [5].

6.5 Writing and Solving the KKTs of an MPPDC

MPPDC (6.12) is nonlinear but continuous and its constraints are generally regular. Thus, its solutions can be obtained solving its KKT conditions. Moreover, this formulation opens a way for the solution of EPECs. This approach is considered in this section.

6.5.1 KKTs of an MPPDC

The KKTs of problem (6.12), which is an MPPDC, are as follows:

$$\nabla_x f + \alpha^T \nabla_x h + \beta^T \nabla_x g = 0 \tag{6.31a}$$

$$\nabla_{x^i} f + \alpha^T \nabla_{x^i} h + \beta^T \nabla_{x^i} g + \delta^i c^i - {\phi^i}^T A^i - \gamma^i = 0, \quad i = 1, ..., n \tag{6.31b}$$

$$\nabla_{\mu^i} f + \alpha^T \nabla_{\mu^i} h + \beta^T \nabla_{\mu^i} g - \delta^i b^i + {\eta^i}^T {A^i}^T - \rho^i = 0, \quad i = 1, ..., n \tag{6.31c}$$

$$0 \leq \beta \perp - g \geq 0 \tag{6.31d}$$

$$0 \leq A^i x^i - b^i \perp \phi^i \geq 0, \quad i = 1, ..., n \tag{6.31e}$$

$$0 \leq x^i \perp \gamma^i \geq 0, \quad i = 1, ..., n \tag{6.31f}$$

$$0 \leq c^i - {A^i}^T \mu^i \perp \eta^i \geq 0, \quad i = 1, ..., n \tag{6.31g}$$

$$0 \leq \mu^i \perp \rho^i \geq 0, \quad i = 1, ..., n. \tag{6.31h}$$

The functions and constraints of the original upper-level problem are defined in (6.1) as $f = f(x, x^1, \ldots, x^n, \mu^1, \ldots, \mu^n)$, $h = h(x, x^1, \ldots, x^n, \mu^1, \ldots, \mu^n)$ and $g = g(x, x^1, \ldots, x^n, \mu^1, \ldots, \mu^n)$.

For simplicity, to obtain (6.31) problem (6.12) is simplified by considering $c^i(z^i) = c^i$, $A^i(z^i) = A^i$ and $b^i(z^i) = b^i$, i.e., by considering that the constraining lower-level problems are not interrelated.

The above formulation, which constitutes and MCP, is generally nonconvex.

It is relevant to note that the above methodology can be useful for solving EPECs because it allows formulating the optimality conditions of each of the MPECs constituting an EPEC.

6.5.2 Illustrative Example. Strategic Offering, KKTs

The KKTs of the strategic offering MPPDC (6.13) are provided below. Differentiations with respect to the optimization variables are indicated to enhance clarity.

For the sake of clarity, we split up the formulation into equality and complementarity constraints. Moreover, we indicate how the equality constraints are obtained by differentiating the Lagrangian of problem (6.13).
Equality constraints:

$$\frac{\partial L}{\partial g_1} = c_1^{\rm g} - \alpha_1 + \sigma_1 - \beta_1^{\rm gmin} + \beta_1^{\rm gmax} + \delta \hat{c}_1^{\rm g} = 0 \tag{6.32a}$$

$$\frac{\partial L}{\partial g_i} = \sigma_i - \beta_i^{\rm gmin} + \beta_i^{\rm gmax} + \delta c_i^{\rm g} = 0, \quad i = 2,3 \tag{6.32b}$$

$$\frac{\partial L}{\partial d_i} = -\sigma_i - \beta_i^{\rm dmin} + \beta_i^{\rm dmax} - \delta c_i^{\rm d} = 0, \quad i = 1,2,3 \tag{6.32c}$$

$$\frac{\partial L}{\partial \theta_1} = -\sigma_1(B_1 + B_2) + \sigma_2 B_1 + \sigma_3 B_2 +$$
$$B_1(-\beta_1^{\rm fmin} + \beta_1^{\rm fmax}) - B_2(\beta_2^{\rm fmin} + \beta_2^{\rm fmax}) = 0 \tag{6.32d}$$

$$\frac{\partial L}{\partial \theta_2} = -\sigma_2(B_1 + B_3) + \sigma_1 B_1 + \sigma_3 B_3 +$$
$$B_1(\beta_1^{\rm fmin} - \beta_1^{\rm fmax}) + B_3(-\beta_3^{\rm fmin} + \beta_3^{\rm fmax}) = 0 \tag{6.32e}$$

$$\frac{\partial L}{\partial \theta_3} = -\sigma_3(B_2 + B_3) + \sigma_1 B_1 + \sigma_2 B_3 +$$
$$B_2(\beta_2^{\rm fmin} - \beta_2^{\rm fmax}) + B_3(\beta_3^{\rm fmin} - \beta_3^{\rm fmax}) = 0 \tag{6.32f}$$

$$\frac{\partial L}{\partial \alpha_1} = -g_1 + \rho_1^{\rm g} + \rho_1^{\rm d} + \rho_1^{\theta}(B_1 + B_2) - \rho_2^{\theta} B_1 - \rho_3^{\theta} B_2 = 0 \tag{6.32g}$$

$$\frac{\partial L}{\partial \alpha_2} = -\rho_2^{\rm g} + \rho_2^{\rm d} - \rho_1^{\theta} B_1 + \rho_2^{\theta}(B_1 + B_3) - \rho_3^{\theta} B_3 = 0 \tag{6.32h}$$

$$\frac{\partial L}{\partial \alpha_3} = -\rho_3^{\rm g} + \rho_3^{\rm d} - \rho_1^{\theta} B_2 - \rho_2^{\theta} B_3 + \rho_3^{\theta}(B_2 + B_3) = 0 \tag{6.32i}$$

$$\frac{\partial L}{\partial \phi_i^{\rm gmin}} = -\rho_i^{\rm g} - \eta_i^{\rm gmin} = 0, \quad i = 1,2,3 \tag{6.32j}$$

$$\frac{\partial L}{\partial \phi_i^{\rm gmax}} = \rho_i^{\rm g} - \eta_i^{\rm gmax} + \delta g_i^{\rm max} = 0, \quad i = 1,2,3 \tag{6.32k}$$

$$\frac{\partial L}{\partial \phi_i^{\text{dmin}}} = -\rho_i^{\text{d}} - \eta_i^{\text{dmin}} = 0, \quad i = 1,2,3 \tag{6.32l}$$

$$\frac{\partial L}{\partial \phi_i^{\text{dmax}}} = \rho_i^{\text{d}} - \eta_i^{\text{dmax}} + \delta d_i^{\text{max}} = 0, \quad i = 1,2,3 \tag{6.32m}$$

$$\frac{\partial L}{\partial \hat{c}_1^{\text{g}}} = \rho_1^{\text{g}} + \delta g_1 = 0 \tag{6.32n}$$

$$\frac{\partial L}{\partial \rho_1^{\text{min}}} = B_1(-\rho_1^{\theta} + \rho_2^{\theta}) - \eta_1^{\text{min}} + \delta f_1^{\text{max}} = 0 \tag{6.32o}$$

$$\frac{\partial L}{\partial \rho_2^{\text{min}}} = B_2(-\rho_1^{\theta} + \rho_3^{\theta}) - \eta_2^{\text{min}} + \delta f_2^{\text{max}} = 0 \tag{6.32p}$$

$$\frac{\partial L}{\partial \rho_3^{\text{min}}} = B_3(-\rho_2^{\theta} + \rho_3^{\theta}) - \eta_3^{\text{min}} + \delta f_3^{\text{max}} = 0 \tag{6.32q}$$

$$\frac{\partial L}{\partial \rho_1^{\text{max}}} = B_1(\rho_1^{\theta} - \rho_2^{\theta}) - \eta_1^{\text{max}} + \delta f_1^{\text{max}} = 0 \tag{6.32r}$$

$$\frac{\partial L}{\partial \rho_2^{\text{max}}} = B_2(\rho_1^{\theta} - \rho_3^{\theta}) - \eta_2^{\text{max}} + \delta f_2^{\text{max}} = 0 \tag{6.32s}$$

$$\frac{\partial L}{\partial \rho_3^{\text{max}}} = B_3(\rho_2^{\theta} - \rho_3^{\theta}) - \eta_3^{\text{max}} + \delta f_3^{\text{max}} = 0 \tag{6.32t}$$

$$\frac{\partial L}{\partial \gamma} = \rho_3^{\theta} = 0 \tag{6.32u}$$

(6.13b)-(6.13d), (6.13j)-(6.13p) and (6.13t). (6.32v)

Complementarity constraints:

$$0 \le g_i \perp \beta_i^{\text{gmin}} \ge 0, \quad i = 1,2,3 \tag{6.33a}$$

$$0 \le g_i^{\text{max}} - g_i \perp \beta_i^{\text{gmax}} \ge 0, \quad i = 1,2,3 \tag{6.33b}$$

$$0 \le d_i \perp \beta_i^{\text{dmin}} \ge 0, \quad i = 1,2,3 \tag{6.33c}$$

$$0 \le d_i^{\text{max}} - d_i \perp \beta_i^{\text{dmax}} \ge 0, \quad i = 1,2,3 \tag{6.33d}$$

$$0 \le B_1(\theta_1 - \theta_2) + f_1^{\text{max}} \perp \beta_1^{\text{fmin}} \ge 0 \tag{6.33e}$$

$$0 \le B_2(\theta_1 - \theta_3) + f_2^{\text{max}} \perp \beta_2^{\text{fmin}} \ge 0 \tag{6.33f}$$

$$0 \le B_3(\theta_2 - \theta_3) + f_3^{\text{max}} \perp \beta_3^{\text{fmin}} \ge 0 \tag{6.33g}$$

$$0 \le f_1^{\text{max}} - B_1(\theta_1 - \theta_2) \perp \beta_1^{\text{fmax}} \ge 0 \tag{6.33h}$$

$$0 \le f_2^{\text{max}} - B_2(\theta_1 - \theta_3) \perp \beta_2^{\text{fmax}} \ge 0 \tag{6.33i}$$

$$0 \le f_3^{\text{max}} - B_3(\theta_2 - \theta_3) \perp \beta_3^{\text{fmax}} \ge 0 \tag{6.33j}$$

$$0 \le \phi_i^{\text{gmin}} \perp \eta_i^{\text{gmin}} \ge 0, \quad i = 1,2,3 \tag{6.33k}$$

$$0 \le \phi_i^{\text{gmax}} \perp \eta_i^{\text{gmax}} \ge 0, \quad i = 1,2,3 \tag{6.33l}$$

$$0 \le \phi_i^{\text{dmin}} \perp \eta_i^{\text{dmin}} \ge 0, \quad i = 1,2,3 \tag{6.33m}$$

$$0 \le \phi_i^{\text{dmax}} \perp \eta_i^{\text{dmax}} \ge 0, \quad i = 1,2,3 \tag{6.33n}$$

$$0 \leq \rho_i^{\min} \perp \eta_i^{\min} \geq 0, \quad i = 1, 2, 3 \tag{6.33o}$$

$$0 \leq \rho_i^{\max} \perp \eta_i^{\max} \geq 0, \quad i = 1, 2, 3. \tag{6.33p}$$

Note that the KKT conditions (6.32)-(6.33) constitute a MCP.

6.5.3 Reformulating an MCP as an Optimization Problem

Recent improvements in nonlinear optimization solvers motivate the reformulation of MCP (6.31) as a nonlinear optimization problem. State-of-the-art nonlinear solvers include CONOPT [6] and SNOPT [10].

KKT conditions (6.31) can be recast as the following optimization problem

$$\text{Minimize}_{\{x\} \cup \{x^1,\ldots,x^n\} \cup \{\mu^1,\ldots,\mu^n\} \cup \Xi_{\text{NLP}}}$$

$$- \beta^T g + \sum_i^n {\phi^i}^T \left(A^i x^i - b^i\right) + \sum_i^n {\gamma^i}^T x^i$$

$$+ \sum_i^n {\eta^i}^T \left(c^i - {A^i}^T \mu^i\right) + \sum_i^n {\rho^i}^T \mu^i \tag{6.34a}$$

s. t.

$$\text{(6.31a)-(6.31c)} \tag{6.34b}$$

$$g \leq 0 \tag{6.34c}$$

$$A^i x^i - b^i \geq 0, \quad i = 1, ..., n \tag{6.34d}$$

$$x^i \geq 0, \quad i = 1, ..., n \tag{6.34e}$$

$$c^i - {A^i}^T \mu^i \geq 0, \quad i = 1, ..., n \tag{6.34f}$$

$$\lambda^i, \delta^i, \phi^i, \gamma^i, \eta^i, \rho^i \geq 0 \quad i = 1, ..., n. \tag{6.34g}$$

Observe that the complementarity conditions (6.33) have been transferred to the objective function, and that the set $\Xi_{\text{NLP}} = \{\delta^i, \phi^i, \gamma^i, \eta^i, \rho^i, i = 1, \ldots, n\}$ includes the dual variables of the lower-level problems in (6.12). Note that the objective function of problem (6.34) is non-linear and highly non-convex, whereas the constraints remain linear (this fact makes problem (6.34) generally well-behaved). The objective function is always positive (every term is positive) and therefore has a minimum value equal to zero. It is important to note that at this value the complementarity conditions (6.31d)-(6.31h) are satisfied. Note also that an optimal solution of problem (6.34) with zero-valued objective function is also a solution of problem (6.31).

6.5.4 Illustrative Example. Strategic Offering: MCP Optimization Problem

Considering the reformulation explained in Subsection 6.5.3, the strategic offering MCP (6.32)-(6.33) can be recast as

$$
\begin{aligned}
&\text{Minimize}_{\{\hat{c}_1^{\mathrm{g}}\}\cup\{g_1,g_2,g_3,d_1,d_2,d_3,\theta_1,\theta_2,\theta_3\}\cup\Xi_{\mathrm{SO}}\cup\Xi_{\mathrm{SOd}}}\\
&\sum_{i=1}^{3} g_i\beta_i^{\mathrm{gmin}} + \sum_{i=1}^{3}(g_i^{\max} - g_i)\beta_i^{\mathrm{gmax}} + \sum_{i=1}^{3} d_i\beta_i^{\mathrm{dmin}}\\
&+\sum_{i=1}^{3}(d_i^{\max} - d_i)\beta_i^{\mathrm{dmax}} + \left[B_1(\theta_1-\theta_2)+f_1^{\max}\right]\beta_1^{\mathrm{fmin}}\\
&+\left[B_2(\theta_1-\theta_3)+f_2^{\max}\right]\beta_2^{\mathrm{fmin}} + \left[B_3(\theta_2-\theta_3)+f_3^{\max}\right]\beta_3^{\mathrm{fmin}}\\
&+\left[f_1^{\max}-B_1(\theta_1-\theta_2)\right]\beta_1^{\mathrm{fmax}} + \left[f_2^{\max}-B_2(\theta_1-\theta_3)\right]\beta_2^{\mathrm{fmax}}\\
&+\left[f_3^{\max}-B_3(\theta_2-\theta_3)\right]\beta_3^{\mathrm{fmax}} + \sum_{i=1}^{3}\phi_i^{\mathrm{gmin}}\beta_i^{\mathrm{gmin}} + \sum_{i=1}^{3}\phi_i^{\mathrm{gmax}}\beta_i^{\mathrm{gmax}}\\
&+\sum_{i=1}^{3}\phi_i^{\mathrm{dmin}}\beta_i^{\mathrm{dmin}} + \sum_{i=1}^{3}\phi_i^{\mathrm{dmax}}\beta_i^{\mathrm{dmax}}\\
&+\sum_{i=1}^{3}\rho_i^{\min}\beta_i^{\min} + \sum_{i=1}^{3}\rho_i^{\max}\beta_i^{\max} \qquad (6.35a)
\end{aligned}
$$

s. t.

$$
\begin{aligned}
&\text{(6.32a)-(6.32v)} && (6.35b)\\
&g_i \geq 0, \quad i=1,2,3 && (6.35c)\\
&g_i^{\max} - g_i \geq 0, \quad i=1,2,3 && (6.35d)\\
&d_i \geq 0, \quad i=1,2,3 && (6.35e)\\
&d_i^{\max} - d_i \geq 0, \quad i=1,2,3 && (6.35f)\\
&B_1(\theta_1-\theta_2) + f_1^{\max} \geq 0 && (6.35g)\\
&B_2(\theta_1-\theta_3) + f_2^{\max} \geq 0 && (6.35h)\\
&B_3(\theta_2-\theta_3) + f_3^{\max} \geq 0 && (6.35i)\\
&f_1^{\max} - B_1(\theta_1-\theta_2) \geq 0 && (6.35j)\\
&f_2^{\max} - B_2(\theta_1-\theta_3) \geq 0 && (6.35k)\\
&f_3^{\max} - B_3(\theta_2-\theta_3) \geq 0 && (6.35l)\\
&\phi_i^{\mathrm{gmin}}, \phi_i^{\mathrm{gmax}}, \phi_i^{\mathrm{dmin}}, \phi_i^{\mathrm{dmax}}, \rho_i^{\min}, \rho_i^{\max} \geq 0, \quad i=1,2,3. && (6.35m)\\
&\beta_i^{\mathrm{gmin}}, \beta_i^{\mathrm{gmax}}, \beta_i^{\mathrm{dmin}}, \beta_i^{\mathrm{dmax}} \geq 0, \quad i=1,2,3 && (6.35n)\\
&\beta_1^{\mathrm{fmin}}, \beta_2^{\mathrm{fmin}}, \beta_3^{\mathrm{fmin}}, \beta_1^{\mathrm{fmax}}, \beta_2^{\mathrm{fmax}}, \beta_3^{\mathrm{fmax}} \geq 0 && (6.35o)\\
&\beta_i^{\mathrm{gmin}}, \beta_i^{\mathrm{gmax}}, \beta_i^{\mathrm{dmin}}, \beta_i^{\mathrm{dmax}}, \beta_i^{\min}, \beta_i^{\max} \geq 0, \quad i=1,2,3, && (6.35p)
\end{aligned}
$$

where $\Xi_{\mathrm{SOd}} = \{\sigma_i,\ \beta_i^{\mathrm{gmin}},\ \beta_i^{\mathrm{gmax}},\ \beta_i^{\mathrm{dmin}},\ \beta_i^{\mathrm{dax}}, \beta_i^{\mathrm{fmin}},\ \beta_i^{\mathrm{fmax}},\ \rho_i^{\mathrm{g}},\ \rho_i^{\mathrm{d}},\ \rho_i^{\theta},\ \rho_i^{\mathrm{g}}, \eta_i^{\mathrm{gmin}},\ \eta_i^{\mathrm{gmax}}, \eta_i^{\mathrm{dmin}}, \eta_i^{\mathrm{dmax}}, \eta_i^{\mathrm{min}}, \eta_i^{\mathrm{max}}, i = 1, 2, 3, \delta\}$ is the set of dual variables.

Solving nonlinear problem (6.35) for the data provided in Illustrative Example 6.2.2 renders the solution reported in that illustrative example.

6.6 Summary

This chapter considers optimization problems constrained by complementarity and other optimization problems, which are of particular interest in energy markets. The chapter motivates the practical interest of these problems and classifies them according to their structure. Solution technique consisting in deriving equivalent mixed-integer linear problems, nonlinear problmes or mixed complementarity problems are described. A wealth of illustrative examples clarify throughout the chapter the interest and practical applications of OPcOPs. They also illustrate some solution techniques.

The conclusion below are in order:

1. Optimization problems constrained by complementarity and other optimization problems (OPcOP) are of great interest to describe the functioning of energy markets and the interactions among market participants.
2. If the constraining problems of an OPcOP are convex, its optimality region can be efficiently characterized and the OPcOP conveniently transformed into a single-level optimization problem that is tractable and can generally be solved.
3. If the constraining problems are linear, an array to solution technique are available to efficiently solve such OPcLP problems.
4. Modeling stochastic data in the constraining problems is possible but leads to large-scale models that may not be tractable.

6.7 Exercises

1. Reformulate the strategic offering model in Subsection 6.2.2 to include several units per node, and piecewise linear production costs. Consider also that the strategic producer owns more than one unit.
2. Consider the lower-level problem of the the strategic offering model in Subsection 6.2.2. Write the dual problem corresponding to this lower-level problem and its KKT conditions. Verify that the solutions of the primal problem, the dual problem and the optimality conditions are identical.
3. Reformulate the vulnerability assessment model in Subsection 6.2.3 to consider an attacker able to destroy network nodes. Is this problem sub-

stantially more complicated than the problem of an attacker unable to destroy nodes?

4. Reformulate the transmission investment model in Subsection 6.2.4 for a transmission planner that minimizes investment cost while maintaining the unserved energy below a given level. Consider also inelastic demands. How can inelastic demands be modeled?
5. Analyze the characteristics of the bilevel model in Subsection 6.2.4 if the target of the transmission planner is maximizing social welfare (minimizing social cost).
6. Is the numerical behavior of the Fortuny-Amat linearization better than that of the SOS1 linearization? Analyze this issue considering the strategic offering model in Subsection 6.2.2, and the transmission investment model in Subsection 6.2.4.
7. Consider the vulnerability assessment model in Subsection 6.2.3 and verify numerically that the primal-dual formulation of the optimality conditions of the lower-level problem provides the same solutions as its KKT conditions.
8. Using the transmission investment model describe in Subsection 6.2.4, numerically analyze the KKT conditions of its corresponding MPPDC, written in Subsection 6.2.9. Does this system of equality and inequalities have many solutions?
9. Sketch the structure of a bilevel model to be used by a strategic investor in production facilities. Hint: the upper level target is the maximum profit of the investor while at the lower level the market is cleared under different load conditions with the objective of maximizing social welfare.
10. Sketch the structure of a bilevel model to be used by a transmission operator to decide its transmission maintenance plan. Hint: the upper level target is maximizing transmission security while the producers maximize their respective profits at the lower level.

References

1. D.P. Bertsekas. *Nonlinear Programming*. Athena Scientific, Belmont, Massachusetts, second edition, 1999.
2. G. Brown, M. Carlyle, J. Salmerón, and K. Wood. Defending critical infrastructure. *Interfaces*, 36(6):530–544, November 2006.
3. B. Colson, P. Marcotte, and G. Savard. Bilevel programming: A survey. *4OR*, 3(2):87–108, June 2005.
4. A.J. Conejo, E. Castillo, R. Mínguez, and R. García Bertrand. *Decomposition Techniques in Mathematical Programming. Engineering and Science Applications.* Springer, Heidelberg, Germany, 2006.
5. CPLEX. *GAMS. The Solver Manuals. GAMS/CPLEX*, 2010. Available at www.gams.com.
6. A.S. Drud. *GAMS. The Solver Manuals. GAMS/CONOPT.* ARKI Consulting and Development, Bagsvaerdvej 246A, DK-2880 Bagsvaerd, Denmark, 2010. Available at www.gams.com.
7. F. Facchinei and C. Kanzow. Generalized Nash equilibrium problems. *4OR*, 5(3):173–210, September 2007.
8. J. Fortuny-Amat and B. McCarl. A representation and economic interpretation of a two-level programming problem. *The Journal of the Operational Research Society*, 32(9):783–792, September 1981.
9. L. Garcés, A.J. Conejo, R. García-Bertrand, and R. Romero. A bilevel approach to transmission expansion planning within a market environment. *IEEE Transactions on Power Systems*, 24(3):1513–1522, August 2009.
10. P.E. Gill, W. Murray, M.A. Saunders, and A. Drud. *GAMS. The Solver Manuals. GAMS/SNOPT*, 2010. Available at www.gams.com.
11. D.G. Luenberger and Y. Ye. *Linear and Nonlinear Programming*. Springer, New York, third edition, 2008.
12. A. Mas-Colell, M.D. Whinston, and J.R. Green. *Microeconomic Theory.* Oxford University Press, New York, 1995.
13. A.L. Motto, J.M. Arroyo, and F.D. Galiana. A mixed-integer LP procedure for the analysis of electric grid security under disruptive threat. *IEEE Transactions onPower Systems*, 20(3):1357–1365, August 2005.
14. J. Nash. Equilibrium points in n-person games. *Proceedings of the National Academy of Sciences of the United States of America*, 36(1):48–49, 1950.
15. J. Nash. Non-cooperative games. *The Annals of Mathematics. Second Series*, 54(2):286–295, September 1951.
16. C. Ruiz and A.J. Conejo. Pool strategy of a producer with endogenous formation of locational marginal prices. *IEEE Transactions on Power Systems*, 24(4):1855–1866, November 2009.

17. J. Salmerón, K. Wood, and R. Baldick. Analysis of electric grid security under terrorist threat. *IEEE Transactions on Power Systems*, 19(2):905–912, May 2004.
18. S. Siddiqui and S.A. Gabriel. An SOS1-based approach for solving MPECs with a natural gas market application, accepted at *Networks and Spatial Economics*, May 2012.

Chapter 7
Equilibrium Problems with Equilibrium Constraints

7.1 Introduction

The previous chapter presented mathematical programs for solving leader-follower (Stackelberg) games when a single leader correctly anticipates the equilibrium reaction of followers, who in turn naively believe that the leader's decisions are exogenous and fixed. This chapter introduces a type of mathematical program that is useful for modeling such games when there is more than one leader, and one wants to find an equilibrium among them: Equilibrium Problems with Equilibrium Constraints (EPECs). First, we present a general EPEC formulation (Section 7.2) and the basic diagonalization approach to solving EPECs, including a simple example. We then summarize some of their many applications to energy markets (Section 7.3), including the three examples of energy market EPECs that we feature in this chapter.

The detailed examples of Sections 7.4-7.6 include both formulations and simple numerical examples that the reader can replicate. These EPECs represent two types of hierarchical (leader-follower) relationships in the energy supply chain: between oligopolistic power generators and a transmission system operator (Sections 7.4 and 7.5) and among oligopolistic producers who play successive capacity expansion—spot market games (Section 7.6). The models in Sections 7.4 and 7.5 represent short-run market operations situations in which generators submit MW quantity schedules and price offers, respectively, to the grid operator, who then operates the system optimally subject to those schedules or offers. In these multiple leader-follower games, the generators anticipate correctly how the grid operator's calculation of market prices will react to those schedules and offers. However, the generator leaders play a Nash game among themselves.

In the long-run investment game of Section 7.6, we disregard the operation of the transmission grid and instead consider a situation in which generators can invest in capacity and anticipate correctly how those investments affect the short-run operations of the market. However, the generators are Nash players with respect to each other's capacity. The interaction of the multiple investors (leaders) with the short-run operations (followers) yields an EPEC model structure.

Diagonalization is used to solve all the examples in the chapter, in the case where pure strategy solutions exist; another, more advanced algorithm that can be used to solve EPECs is presented in Section 9.4.2. Where pure strategy solutions do not exist in our examples, we use the bimatrix approach of Section 4.2.4 to obtain mixed strategy equilibria.

7.2 The EPEC Problem

7.2.1 Problem Statement and Diagonalization Algorithm

Chapter 6 introduced several closely related mathematical programming problems that involve optimization of an objective (upper-level problem) subject to the solution of another problem (lower-level problem). If the lower-level is an optimization problem, then the overall problem is an OPcOP in that chapter's terminology. Alternatively, the lower-level might instead be a complementarity problem that represents an equilibrium among the followers and which might not necessarily be expressible as the solution of an optimization problem, in which case the overall two-level problem is referred to as an MPEC in Chapter 6. If the lower-level problem is an optimization problem, it is usually represented by its first-order conditions (in the form of a complementarity problem). Chapter 6 calls the overall problem an MPCC. Less commonly, the solution might be represented as three sets of constraints: one for primal variables of the subproblem, one for its dual variables, and finally a zero duality gap condition that sets the primal and dual objectives equal (termed a MPPDC in Chapter 6). If the first-order conditions or primal-dual conditions of an OPcOP lower-level optimization problem are viewed as equilibrium conditions for that single optimization, then the OPcOP is itself an MPEC. Therefore, for the purposes of this chapter, we will refer to all of those problems as MPECs, even though, strictly speaking, some readers might prefer to use the term MPEC for only a subset of the two-level problems of Chapter 6.

This chapter is concerned with the formulation and calculation of an equilibrium among several MPECs that share the same lower-level problem, whether an optimization or an equilibrium problem. In game theoretic terms, the problem is to obtain a Nash equilibrium among multiple leaders of a Stackelberg leader-follower problem, where there is a shared single follower or set of followers. A Stackelberg game represents a situation of asymmetric information in which the leader(s) correctly anticipate the reaction of followers to the leader(s)' decisions, but the followers take the leader(s)' decisions as exogenous, naively acting as if they have no influence over what the leader(s) do. The followers' reactions are captured, as in the MPECs of Chapter 6, by embedding the followers' equilibrium conditions and variables in the constraint set of each leader's MPEC. In particular, the lower level of each MPEC in the EPEC represents the followers' reactions to the leaders' decisions either as an optimization problem (so each follower's reaction is captured by the first-order conditions or primal-dual conditions for its optimization problem) or an equilibrium problem (representing the equilibrium response of the followers to the leaders' decisions).

Adapting slightly the notation of Chapter 6, we can express the EPEC problem as follows. Let vector $\boldsymbol{y}^j$ be the vector of decision variables for the MPEC problem of leader $j = 1, \ldots, J$, while vector $\boldsymbol{x}^i$ and $\boldsymbol{\lambda}^i$ are the vectors of primal and dual decision variables, respectively, for follower $i = 1, \ldots, I$. Let $\text{EP}(\boldsymbol{x}^1, \boldsymbol{\lambda}^1, \ldots, \boldsymbol{x}^I, \boldsymbol{\lambda}^I | \boldsymbol{y}^1, \ldots, \boldsymbol{y}^J)$ be the equilibrium problem for the followers, which is conditioned on the leaders' decisions. For instance, the followers' problem may consist of I optimization problems linked by market-clearing constraints (such as supply = demand) (for instance, follower problems (6.1d)-(6.1f) plus a market-clearing constraint (6.1b) in Chapter 6). Or, EP( ) may be an equilibrium problem in which some of the equilibrium conditions do not derive from optimization problems (such as regulatory rules that determine the price of energy transport services or that ration energy supplies in times of shortage). Define $f_j(\boldsymbol{y}^j; \boldsymbol{x}^1, \boldsymbol{\lambda}^1, \ldots, \boldsymbol{x}^I, \boldsymbol{\lambda}^I | \boldsymbol{y}^1, \ldots, \boldsymbol{y}^{j-1}, \boldsymbol{y}^{j+1}, \ldots, \boldsymbol{y}^J)$ to be the objective function that the jth leader wishes to maximize. Notice that it is conditioned on the other leaders' decisions, indicating that j is a Nash player with respect to the leaders' strategies. A semicolon separates j's decision variable from the followers' decision variables, which are indirectly determined by the leaders' choices. Leader j must choose its $\boldsymbol{y}^j$ from a feasible region $\boldsymbol{Y}^j$. Leader j's MPEC can therefore be stated as:

$$\text{Maximize}_{\boldsymbol{y}^j; \boldsymbol{x}^1, \boldsymbol{\lambda}^1, \ldots, \boldsymbol{x}^I, \boldsymbol{\lambda}^I}$$

$$f_j(\boldsymbol{y}^j; \boldsymbol{x}^1, \boldsymbol{\lambda}^1, \ldots, \boldsymbol{x}^I, \boldsymbol{\lambda}^I | \boldsymbol{y}^1, \ldots, \boldsymbol{y}^{j-1}, \boldsymbol{y}^{j+1}, \ldots, \boldsymbol{y}^J) \tag{7.1a}$$

s. t.

$$\boldsymbol{y}^j \in \boldsymbol{Y}^j \tag{7.1b}$$

$$\text{EP}(\boldsymbol{x}^1, \boldsymbol{\lambda}^1, \ldots, \boldsymbol{x}^I, \boldsymbol{\lambda}^I | \boldsymbol{y}^1, \ldots, \boldsymbol{y}^J) \tag{7.1c}$$

The EPEC can now be expressed as the problem of finding a Nash equilibrium among the leaders $j$whose strategy variables are $\boldsymbol{y}^j$:

$$\text{Find: } \{\boldsymbol{y}^1, \ldots, \boldsymbol{y}^J; \boldsymbol{x}^1, \boldsymbol{\lambda}^1, \ldots, \boldsymbol{x}^I, \boldsymbol{\lambda}^I\} \text{ that solve:}$$

$$\text{MPEC (7.1) for } j = 1, \ldots, J \tag{7.2}$$

There are other solution concepts for EPECs, for instance the local Nash equilibrium proposed by Hu and Ralph [24]. There, a local EPEC solution is defined as one that represents a local, but not necessarily global optimum for each of the constituent MPECs in (7.2).

It is difficult to determine whether a solution exists for an EPEC (7.2), primarily because the feasible region of each leader's (Nash player's) MPEC is, in general, non-convex. In this chapter, we show examples where one EPEC has no pure strategy solution, while slightly changing its parameter values results in such a solution. There is a significant literature on the analysis of EPEC and solution characteristics. As just some examples, Leyffer and Munson [28] characterize solution sets for certain EPECs, deMiguel and Xu [14] address existence and uniqueness, while Sherali [39] proves existence of a unique equilibrium for certain simple cases. Ehrenmann [15] demonstrates, however, that the latter results depend on the assumption of symmetry in strategies among identical players, but that in general, an equilibrium among MPECs for identical leaders might involve different decisions for different leaders.

The most common strategy in the literature for attempting to find a solution to an EPEC is diagonalization. This is a variant of the Gauss-Seidel method for numerical solution of simultaneous equations. The steps are as follows:

1. Diagonalization starts by defining an initial set of values for the leaders' strategies $\boldsymbol{y}^{j0}$ for all j, a convergence criterion ε, and a maximum number of iterations N. The second superscript refers to the iteration of the procedure. Set $\boldsymbol{y}^{j1} = \boldsymbol{y}^{j0}$ for all j.
2. $n \leftarrow 1$.
3. Do for $j = 1, \ldots, J$:

   a. Leader j's MPEC is solved for its $\boldsymbol{y}^j$, given the saved values of the other leaders' decisions $\boldsymbol{y}^{j'1}$, $j' \neq j$.
   b. Reset $\boldsymbol{y}^{j1}$ equal to the $\boldsymbol{y}^j$ just obtained.

4. If $|\boldsymbol{y}^{j1} - \boldsymbol{y}^{j0}| < \varepsilon$ for all j,then quit (where $|\ |$ is some chosen norm), and report the last solution $\{\boldsymbol{y}^{11}, \ldots, \boldsymbol{y}^{J1}; \boldsymbol{x}^1, \boldsymbol{\lambda}^1, \ldots, \boldsymbol{x}^I, \boldsymbol{\lambda}^I\}$ from $j = J$ in Step 3. Else,
5. If $n = N$, quit and report that the algorithm has failed to converge. Else,
6. $n \leftarrow n + 1$, set $\boldsymbol{y}^{j0} \leftarrow \boldsymbol{y}^{j1}$ for all j, and return to Step 3.

This is the solution approach taken in the numerical applications of this chapter. A variant that might increase the probability or speed of convergence involves relaxation, with Step 3c instead calculating $\boldsymbol{y}^{j1} \leftarrow \theta \boldsymbol{y}^j + (1-\theta)\boldsymbol{y}^{j0}$. The parameter θ is a relaxation parameter with $0 < \theta < 1$ implying under relaxation and $\theta > 1$ implying over relaxation. Other approaches to solving EPECs have been proposed, and an example is summarized in Section 9.4.2 (see also [28]).

There are several possible outcomes of applying this algorithm to an EPEC. One is that it converges (quickly or slowly!) to an equilibrium. The illustrative example in Section 7.2.2 below is an example where the algorithm converges nicely to the unique pure strategy equilibrium among the leaders' MPECs.

However, if there are more than one equilibrium solution, then which one is found (if any) will generally depend on the initial set of values (see Exercises 1 and 2 at the end of the chapter). Also, the point it converges to may not be an equilibrium, if the leader's MPEC is solved by an algorithm that doesn't guarantee a global optimum. For instance, if a local method (such as a reduced gradient approach) is used to solve the MPEC in Step 3a, then it may converge to a local equilibrium in the sense of Hu and Ralph [24]. There, each player's MPEC is at a local optimum (so second-order conditions are satisfied for each). But such a solution is not a true solution to the EPEC if it is not a global optimum for at least one player. A local algorithm may fail to discover the alternative, better optimum for that leader in Step 3a, and so the procedure may converge at Step 4, even though that solution is not actually an equilibrium among the MPECs (i.e., not a solution to the EPEC) (see Exercise 3).

Another outcome is that the algorithm does not converge, which would be the case if there is no pure strategy equilibrium among the leaders. We show some examples of such cases in Sections 7.4 and 7.5. However, failure to converge does not necessarily mean that there is no such equilibrium.

7.2.2 Illustrative Example. Diagonalization Applied to EPEC with Two Producers Facing Piecewise Linear Demand

As a simple example of an EPEC formulation and the use of diagonalization to solve it, consider the following situation. A duopoly consists of two leaders (energy producers) j = A and B who have marginal production costs of \$10/unit and no capacity limits. Using the above notation, the nonnegative quantities of energy they sell are scalars y_A and y_B, respectively, which are bought in a single market by two groups of followers (consumers) $i = 1$ and 2. The producers are the leaders, and the consumers are the followers in this game.

The consumers' demand for energy x_i is defined by the followers' equilibrium problem EP(x_1, x_2 ,$\lambda | y_A$, y_B):

$$y_A + y_B - (x_1 + x_2) = 0 \quad (\lambda \text{ unrestricted}) \tag{7.3a}$$
$$0 \leq x_1 \perp 100 - x_1 - \lambda \leq 0 \tag{7.3b}$$
$$0 \leq x_2 \perp 35 - x_2 - \lambda \leq 0 \tag{7.3c}$$

where λ is the market price. The first condition is market-clearing (supply minus demand is zero, with the energy price λ as the dual variable). The two complementary conditions represent a linear demand curve if price is below the "choke" level ($100/unit for consumer 1 and $35/unit for consumer 2), with quantity demanded being zero if price is above the choke level. Such a demand curve can result, for instance, from an optimization such as the following for consumer 1:

$$\begin{aligned} &\text{Maximize}_{x_1} \quad \int_0^{x_1} (100 - x)dx - \lambda x_1 \\ &\text{s. t.} \\ &\quad x_1 \geq 0 \end{aligned}$$

In words, consumer 1 maximizes its consumer surplus (the value of consumption, equal to the integral of the linear demand curve, minus its payment for energy) subject to nonnegativity. Its solution is x_1 = 100-λ, unless $\lambda \geq 100$, in which case $x_1 = 0$. This is, of course, the solution to the KKT condition for that problem, which is the first of the complementarity conditions in EP(x_1, x_2 ,$\lambda|y_A$, y_B), equation (7.3). The overall demand curve for electricity (x = x_1+x_2 as a function of λ) is therefore piecewise linear, consisting of three pieces with knots at $x = 0$ ($\lambda = 100$) and 65 ($\lambda = 35$) (Figure 7.1).[1]

If each leader (producer) maximizes its profit subject to those equilibrium conditions (demand relationships and market clearing) while assuming that the other producer doesn't change its production (Nash-Cournot), producer A's problem then becomes the following MPEC:

$$\text{Maximize}_{y_A, x_1, x_2, \lambda} \quad \lambda y_A - 10 y_A \tag{7.4a}$$

s. t.

$$y_A \geq 0 \tag{7.4b}$$

$$\text{EP}(x_1, x_2, \lambda | y_A, y_B) \tag{7.4c}$$

That is, A maximizes its net revenue subject to nonnegativity and the response of the followers (the consumer-price system) to A's and B's decisions.

So there are two MPECs, one each for leaders (producers) A and B. The equilibrium problem between them is for each to (a) choose their strategic variable (their outputs), while (b) believing that the other's strategic variable won't change (Nash assumption) and (c) correctly anticipating the responses of the follower consumers. This is an EPEC that can be phrased as:

---

[1] Such kinked demand curves were considered in the original version of the GASTALE gas market model for Europe [7], in which Cournot gas producers faced effective demand curves of gas traders, who are in turn playing a Cournot game among themselves in competing to sell power to retail customers. These demand curves were derived from the traders' sales decisions in separate industrial, utility, and residential/commercial retail markets with distinct linear demand curves.

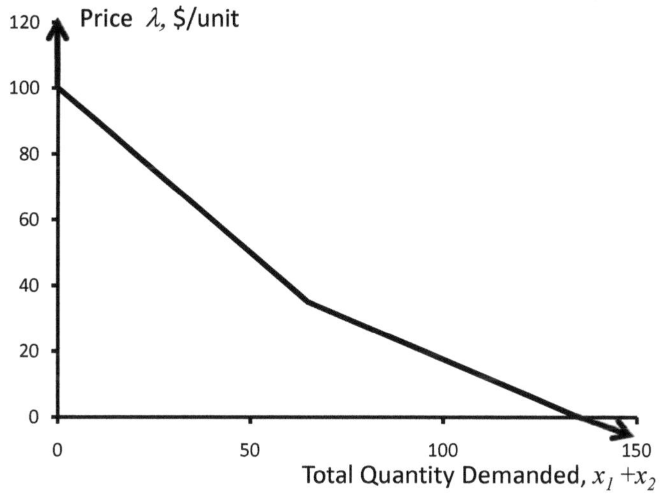

Fig. 7.1 Inverse demand function implied by follower equilibrium problem in EPEC example

$$\text{Find: } \{y_A, y_B; x_1, x_2, \lambda\} \text{ that solve:}$$
$$\text{MPEC (7.4) for} j = \text{A,B}$$

To solve that EPEC, we now apply the six-step diagonalization procedure of Section 7.2.1 as follows:

Initialization:

1. The initial y_{A0} and y_{B0} are assumed to be 20 units apiece, as are y_{A1} and y_{B1}. We impose a convergence criterion $\varepsilon = 0.001$, and a maximum number of iterations $N = 10$.
2. $n \leftarrow 1$

First Iteration n =1:

3. For producer A:
   a. A's MPEC (7.4) is solved for its y_A, given its rival's assumed production $y_{B1} = 20$. This yields $y_A = 35$ units, as well as $\lambda = 45$, $x_1 = 55$, $x_2 = 0$, and a profit of \$1225 for A. Note that this is on the upper, steeper segment of the inverse demand curve above consumer 2's choke price of \$35/unit (Figure 7.1).
   b. Set $y_{A1} = y_A = 35$.
3. For producer B:

a. B's MPEC (7.4) is solved for y_B, given $y_{A1} = 35$. This yields $y_B =$ 40 units, as well as $\lambda = 30$, $x_1 = 70$, $x_2 = 5$, and a profit of \$800 for B. This outcome is on the lower, shallower segment of the inverse demand curve below consumer 2's \$35/unit choke price (Figure 7.1).
b. Set $y_{B1} = y_B = 40$.

4. Since $|y_{A1} - y_{A0}| = |35-20|$ and $|y_{B1} - y_{B0}| = |40-20|$, both exceed ε, so we continue to Step 5.
5. $n = 1$ which is not equal to $N = 10$, so we continue to Step 6.
6. $n \leftarrow n+1 = 2$; $y_{A0} \leftarrow y_{A1} = 35$; and $y_{B0} \leftarrow y_{B1} = 40$.

Second Iteration $n = 2$:

3. For producer A:

a. A's MPEC is solved given $y_{B1} = 40$, yielding $y_A = 37.5$.
b. $y_{A1} \leftarrow 37.5$.

3. For producer B:

a. B's MPEC is solved given $y_{A1} = 37.5$, giving $y_B = 38.75$.
b. $y_{B1} \leftarrow 38.75$.

4. Since $|y_{A1} - y_{A0}| = |37.5-35|$ and $|y_{B1} - y_{B0}| = |38.75-40|$, both exceed ϵ.
5. $n = 2 < N = 10$.
6. $n \leftarrow n+1 = 3$; $y_{A0} \leftarrow y_{A1} = 37.5$; and $y_{B0} \leftarrow y_{B1} = 38.75$.

Third Iteration $n = 3$:
(Yields $y_A = 38.125$, $y_B = 38.438$)
Fourth Iteration $n = 4$:
(Yields $y_A = 38.281$, $y_B = 38.359$)
Fifth Iteration $n = 5$:
(Yields $y_A = 38.3203$, $y_B = 38.3398$)
Sixth Iteration $n = 6$:

3. For producer A:

a. A's MPEC is solved given $y_{B1} = 38.3398$, yielding $y_A = 38.3301$.
b. $y_{A1} \leftarrow 38.3301$.

3. For producer B:

a. B's MPEC is solved given $y_{A1} = 38.3301$, giving $y_B = 38.3350$.
b. $y_{B1} \leftarrow 38.3350$.

4. Since $|y_{A1} - y_{A0}| = |38.3301-38.3203|$ and $|y_{B1} - y_{B0}| = |38.3350\text{-}38.3398|$ are both below $\varepsilon = 0.001$, the diagonalization procedure stops, and the results are reported. The last solution (from solving B's MPEC) is $\lambda = 29.1675$, $x_1 = 70.8325$, $x_2 = 5.8325$, with a profit of \$734.785 for B. Based on that price and $y_{A1} = 38.3301$, A's profit is then \$734.691.

The procedure has converged to what, in this case, is the only equilibrium solution for this EPEC. However, under other demand parameters, there are more than one equilibria, and which it converges to depends on the starting point (see Exercise 7.1). For other problems, this procedure may not converge at all, either because there is not a pure strategy equilibrium (as for some examples in the Sections 7.4 and 7.5) or because there is such an equilibrium but the algorithm is unable to find it.

In the next section, we summarize some uses of EPECs in the energy modeling literature before presenting some example EPEC formulations and numerical applications.

7.3 Energy Applications of EPECs

The sort of two-level Stackelberg game structure that can be formulated as an EPEC can arise in many ways. The different stages can correspond to different players in the game, or a succession of time periods. We illustrate each of these in this chapter. For instance, as noted in Chapter 6, a successive stage structure is a natural way to describe Stackelberg games in which one (or more) market players are leaders who make decisions in the first stage, anticipating how a follower player (or set of players) will react to those decisions. Thus there is an information asymmetry: leaders correctly foresee the reaction of followers, but followers take the leader decisions as given (à la Nash), even if in fact they could influence the leader's decisions. If there is a single leader, this naturally leads to an MPEC formulation; but the presence of multiple leaders who play a simultaneous game against each other in the first stage yields an EPEC problem. Such games naturally arise in supply chains where players in one stage of the chain act as leaders with respect to later stages—for instance, producers making quantity or pricing decisions anticipating how transporters or marketers of their products will react to those decisions.

In this chapter, we consider three such situations in detail. In Section 7.4, we model a short-run power pricing game in which several generating companies are playing a Nash-Cournot game against each other in the first stage. Thus, each acts as if other generators will not alter their output in response to a change in its production and sales. However, a two-stage structure comes about because each generator does recognize that the TSO's transmission charges (reflected in locational marginal prices, or LMPs for short) will depend on everyone's output decisions, so that if the generator worsens (or relieves) congestion, price differences between locations will increase (or decrease). The EPEC is formulated by deriving an MPEC for each generator who optimizes its outputs subject to other generator outputs and the locational prices set by TSOs, the latter determined by the first-order conditions for the grid operator's optimization problem. These conditions enter as constraints in the optimization problems of the generators, making each generator's problem an MPEC.

In the second such situation (Section 7.5), we reconsider the generator-TSO game, but instead assume that the generators' strategic variables are \$/MWh offers into an auction market run by the TSO. This can give very different market equilibria (when they exist) than the previous Cournot producer EPEC. However both EPECs share a leader-follower formulation of the game between generators and the TSO.

A different class of EPECs arises if instead all players make decisions simultaneously in each stage, so each stage is a Nash game, but the stages represent different times or types of decisions that account for effects in subsequent decision stages. A familiar example is the Allaz-Vila [1] game, in which a group of producers each make two types of decisions: the amount of output to sell in a financial forward contract, and then later the amount of output to sell in the spot market. This becomes an EPEC if the game requires that the solution be time consistent (a Markov perfect equilibrium)[2] in that the first stage (contracting) decisions are made simultaneously by all players, with each player anticipating correctly how its decisions would affect the prices in the subsequent spot market game. The spot market game too is a simultaneous game involving only decisions about how much in total to produce, where each player's spot market quantity is just the difference between its production and its forward contracted quantity. This type of game is sometimes called a "closed loop" game, as opposed to "open loop" games in which all decisions are made at once [30, 43, 45], rather than successively. (Yao et al. [47] present classic versions of this closed loop game; Murphy and Smeers [30] also present one involving a succession of capacity, contracting, and spot market decisions.)

In Section 7.6, we present an example of another closed loop game involving all players in each of the stages. In our case, players invest in and subsequently operate electric generation plants. This model is related to the classic model of Kreps and Scheinkman [27], in which players choose capacity in stage one and then compete à la Nash-Bertrand in the second, single-output stage. They show that even though players compete using price (Bertrand) strategies in stage two, the result is the same as a simple one-stage Cournot model. This is because the capacity problem in the first stage essentially imposes a quantity (Cournot) strategy upon the second stage, as players would never build capacity that they wouldn't operate, so that each plant built is fully used in stage 2. The story changes for electricity markets, and other markets in which a good is nonstorable, in that there are multiple operating periods, some of which are likely to result in outputs below capacity [46]. As a result,

---

2 Fudenberg and Tirole [18] provide a more precise definition of a Markov perfect equilibrium as "a profile of Markov strategies that yields a Nash equilibrium in every proper subgame." In this case, the first-stage game must be a Nash equilibrium in the first stage's contracting decisions, and furthermore the second-stage game itself must be a Nash game in the second-stage operating decisions.

some odd things happen, including the result that a model with Bertrand (price) competition in stage two actually can result in less capacity in stage 1, higher prices in stage 2 in some periods, and overall lower consumer surplus and market efficiency compared to Cournot (quantity) competition in stage two. Counter-intuitively, this implies that market regulations designed to force marginal cost bidding in spot markets can perversely harm consumers. This two stage model is posed and solved as an EPEC in Section 7.6, as each investor in the first stage solves an MPEC that chooses its capacity anticipating how its investments will affect the equilibrium in the second stage.

These examples are by no means exhaustive of the possible games in energy markets that can be formulated as EPECs. We cite many others in the footnotes to this chapter. Here are three more examples. For instance, there may be a competitive fringe of smaller energy suppliers whose reaction to price is predictable by the larger suppliers. The large suppliers could then anticipate how the supply from the fringe is deducted from demand in their model, rendering the effective demand more elastic [32]. Thus, there is a Stackelberg game between leaders (large suppliers) and the followers (the fringe). However, since the fringe's exact response (a step function) would be represented by the fringe's KKT conditions, modeling that response would also convert each of the larger generator's problem into an MPEC. Then the equilibrium problem among several large suppliers would become an EPEC.[3]

As another example, leaders and followers in an EPEC can correspond to different levels and sets of market actors in a supply chain. For instance, a natural gas supply chain might be structured as a successive oligopoly, as in Boots et al. [7] (but see the critique by Smeers [40]). In a successive oligopoly, oligopolistic players at the beginning of a supply chain anticipate how oligopolistic players further down the supply chain will react to changes in prices. In the case of Boots et al.'s natural gas market model, oligopolistic gas producers sell gas to oligopolistic local gas suppliers, who in turn sell to three consumer segments–industrial, commercial, and residential. If local suppliers take international gas prices as fixed, but gas producers correctly anticipate the effective demand for gas by suppliers, then suppliers and producers can be viewed as followers and leaders, respectively, in a Stackelberg game. That effective demand is described by the equilibrium conditions for retail gas competition among the suppliers; inserting those conditions in the optimization problem for a producer yields an MPEC for that producer. Finding an equilibrium among the solutions of the producers' MPECs constitutes an EPEC. An additional set of EPECs arising from competition in natural gas markets is described in Chapter 9.

As a third example, Daxhelet and Smeers [12] describe an EPEC in which the leaders are electric utility regulators from different countries (Stackelberg

[3] Section 3.5, above, also discusses a game with one or more Stackelberg leaders with respect to a competitive fringe of other producers. A variant of this game was analyzed by deMiguel and Xu [14] in a more abstract market context.

leaders) who can redistribute fixed costs among generators and consumers, and who attempt do this in a way that maximizes their country's net benefits from the electricity market. In each regulator's model, equilibrium constraints represent the response of the electricity market (so the market is a Stackelberg follower). Meanwhile, each of the regulators assume that the other regulator's decisions won't change (Nash game among regulators).

In the next two sections, we present simple instances of two of the most popular EPEC formulations: that of electricity generators (Stackelberg leaders) competing against each other in a Nash game, while anticipating the prices that result from the reaction of the grid operator and consumers (Stackelberg followers). The two differ in their strategic variables: quantity (MW) in Section 7.4 and price offers ($/MWh) in Section 7.5.

7.4 EPEC Power Market Model 1: Strategic Quantity Decisions by Generators

7.4.1 Model Formulation

7.4.1.1 Model Structural Assumptions

In Chapter 11, we introduce an MCP-based model of producers supplying energy in a network that is operated by a transmission system operator (TSO) that sets prices for transportation services in order to clear the market for transport capacity. That is, prices are adjusted until the demand for transport on each arc in the network (in the form of flow on that arc) is less than or equal to the arc's capacity. The specific example in that case is a power transmission network governed by Kirchhoff's laws, but other networks governed by other physical laws (natural gas, oil, captured CO_2) can be similarly modeled (e.g., the natural gas models in Chapter 10). Under the assumptions of that electricity model, the price of transport is just the difference in their nodal prices, so that the shipping of energy from i to j is just the negative of the cost from j to i. Transportation cost structures other than this "congestion pricing" structure can be represented as well (e.g., [44]).

The MCP formulation that we will present in Section 11.2 follows from an assumption that all producers are naive (price-taking) with respect to the TSO's transport prices. That is, they believe that their decisions about how much to produce and where or what offers to submit to an auction do not affect the market-clearing prices for transportation services. Simultaneously, the TSO also assumes that its decisions about how much transmission service it provides cannot affect prices. From a game theoretical perspective, this is a Nash-Bertrand game between players at different stages in the supply chain. Meanwhile, the specific model formulation in Section 11.2 simultaneously assumes that the producers are Nash-Cournot players with respect to each other's output and sales decisions, although other games among the producers (including Nash-Bertrand) could instead be represented. Thus, that section's model is a mix–horizontal competition among producers is one type of game (Cournot), while vertical interactions are represented by another (Bertrand).

However, other vertical game structures can lead to an EPEC model rather than an MCP. In particular, a Stackelberg games structure between competing producers (leaders) and the TSO (the follower). A number of electricity market models, for instance, represent a set of sophisticated electricity generators who correctly anticipate how the TSO will adjust transmission prices and flows in response to changes in producers' outputs or offers (e.g., [22,47]). For instance, consider a two-node market with a single generator at each node and a capacitated line between them. If the line is uncongested, the prices $\{p_1, p_2\}$ at both nodes will be equal. A clever firm at node 2 will realize that if it expands its output, then both prices will decrease by the same amount; if demand is at all elastic at node 1, then the TSO will need to flow more power from 2 to 1. If the clever firm continues to expand production, at some point it will congest the line in the direction 2 to 1. Further expansion beyond that point cannot increase exports to 1; as a result, no further change in p_1 is possible, and all the additional output will be absorbed by demand at 2. Then p_2 will fall below p_1. The resulting accelerated decreases in p_2 will precisely equal the increases in the price of transmission from 2 to 1 ($= p_1 - p_2$). On the other hand, if the firm at node 2 shrinks its output enough, eventually the line will become congested in the other (importing) direction. At that point, p_2 will rise above p_1, and only p_2 will continue to respond to producer 2's withdrawal of supply. In contrast, the MCP model we will present in Section 11.2 instead has firm 2 anticipating (incorrectly) that changes in its output would leave the price of transmission unaffected i.e., that the prices at different locations change in lockstep ($p_1 - p_2$ does not change).

The more sophisticated model of this section has the following structure: each of several generating firms maximize its own profit subject to (a) the sales of the other generators (Cournot conjecture) and (b) the response of the TSO and consumers (Stackelberg followers), represented by the market-clearing conditions together with the KKT conditions of the followers' problems. Unlike the MPC of Section 11.2, the generator "sees" the hard transmission constraints, knows how the TSO will react to them, and can take advantage of congesting or decongesting them. Unfortunately (or interestingly, depending on whether you welcome computational challenges!), the generator's problem becomes an MPEC, and the collection of those MPECs constitutes an EPEC. Each MPEC's constraint set defines a non-convex feasible region, creating the possibility of local but not necessarily global optima for the supplier's problem. Consequently, it is not possible to define KKT conditions for the generator's problem that will necessarily yield the firm's profit-maximizing solution. As we will see, unlike the MCP model, nothing in general can be said about the existence or uniqueness of a solution to this EPEC, i.e., an equilibrium among the generators' MPEC profit-maximization problems. Depending on the parameterization, there can be a unique pure strategy solution, several such solutions, or none. In addition to the inability to establish analytical conclusions about the nature of an EPEC solution, EPECs are also much more difficult to solve numerically than MCPs such as those in Section 11.2.

Below we outline the formulation of an EPEC that allows generators to anticipate transmission price changes for the type of market considered in the MCP model above.[4] This EPEC can readily be extended so that large generators also anticipate the response of a competitive "fringe" of small suppliers to changes in price. We apply the EPEC as well as the MCP of Section 11.2 to the same simple example, and compare the solutions.

We develop the model as follows. First we describe the price-taking followers' problems; then we embed their KKTs and market-clearing conditions in the generator leader's problem, creating an MPEC. The notation is the same as that in Section 11.2.

7.4.1.2 Consumer

The consumer's response to price is represented by a demand function $q_i = f_{d,i}(p_i)$. Our example below will use an affine demand function.

A richer representation would be an optimization problem in which the consumer maximizes profit (if an industrial or commercial firm) or utility (if a residential customer) subject to: prices of inputs needed to make its product (if a firm) or to provide energy services (if a residence); production functions that translate inputs into products or energy services, respectively; and other constraints. For instance, a steel mill might choose the amounts of ore, scrap metal, coking coal, electricity, labor, and equipment that maximize its profit, subject to prices of those inputs, the production functions that represent the "recipe" for steel, and any capacity or other constraints. The demand function for electricity is then implicit in the optimization, rather than explicit as below. (See [5, 36] for examples.) For a residential customer, it might choose the mix of investment in energy-saving equipment or home insulation (say) and power purchases that maximize the value of heating, lighting, and other services consumed minus the expenditures on those inputs.[5]

[4] There are several models that adopt this EPEC structure, with slight variations. Numerical models that have been implemented with transmission networks of moderate size (dozens of nodes) is that of Cardell et al. [10] and Yao et al. [47]. The economic implications of solutions to simple 2-4 node models have been analyzed extensively [8, 33, 41].

[5] For an example of a MPEC where a generator chooses a social welfare maximizing mix of generation and demand-side management (DSM) conservation subsidies subject to such a consumer model, see Hobbs and Nelson [23]. That model is used to explore the effect of different company objectives upon the resulting choice of "megawatts" and "negawatts."

7.4.1.3 Transmission Provider

The formulation used in the MCP of Section 11.2 is also applied here. A price-taking TSO maximizes the profit from flowing power transactions over its transmission system. We assume a "single-buyer" system, in which generators sell all their output to the TSO at the local price; a generalization that also allows bilateral transactions between generators and consumers that the TSO simply transmits is straightforward [20]. In essence, the TSO is maximizing profit from spatial arbitrage, buying $a_i < 0$ at one node i and reselling some positive a_j at another node j.

Transmission is represented using a lossless linearized dc load flow, captured by "power transmission distribution factors" $PTDF_{ik}$, also called "swing factors." Swing factors describe the MW flow through constraint k resulting from a 1 MW withdrawal at node i, assuming that the matching 1 MW injection takes place at an arbitrary hub node. The swing factors constitute an alternative but equivalent method for expressing network constraints to the representation used in Chapter 6. See Appendix C for a development of the linearized dc load flow model. Section 11.2 presents other numerical examples of swing factors.

The TSO's optimization problem is:

$$\text{Maximize}_{a_i} \quad \sum_i p_i a_i \tag{7.5a}$$

s. t.

$$\sum_i a_i = 0 \quad (\gamma) \tag{7.5b}$$

$$\sum_i PTDF_{ik} a_i < T_k \quad (\lambda_k), \quad \forall k \tag{7.5c}$$

with a_i being the net MW of power withdrawn from the network by consumers and producers at i (and thus sold by the TSO), and T_k being the upper bound on the flow through transmission constraint k.

7.4.1.4 Follower Equilibrium

Let the set of power plant outputs by producers f at nodes at which they have plants $i \in I_f$ be $\{x_{fi}, \forall f, i \in I_f\}$. Given those outputs, the follower equilibrium problem (expressed as MCP) combines the demand functions, KKT conditions for the transmission provider, and market-clearing constraints.

$$\text{Find } \{q_i, a_i, \gamma, \lambda_k, p_i\} \text{ satisfying:} \tag{7.6a}$$

$$q_i = f_{d,i}(p_i) \quad (q_i \text{ unrestricted}), \quad \forall i \tag{7.6b}$$

$$\sum_i a_i = 0 \quad (\gamma \text{ unrestricted}) \tag{7.6c}$$

$$p_i - \gamma - \sum_k PTDF_{ik}\lambda_k = 0 \quad (a_i \,\text{unrestricted}), \quad \forall i \tag{7.6d}$$

$$0 \leq \lambda_k \perp \sum_i PTDF_{ik} a_i - T_k \leq 0, \quad \forall k \tag{7.6e}$$

$$q_i - a_i - \sum_{f \in F_i} x_{fi} = 0 \quad (p_i \text{ unrestricted}), \; \forall i \tag{7.6f}$$

where F_i is the set of firms that have a power plant at node i. We assume that a firm has, at most, one power plant at a given i, which is not a restrictive assumption. The last condition states that consumption q_i equals the net supply at the node from the TSO and generators sited at that location. The problem (7.6) is an MCP which, if no mistakes have been made in setting it up, should be square (i.e., as many variables as conditions). It is a linear or nonlinear MCP, depending on the form of $f_{d,i}(\,)$.

7.4.1.5 Generator (Leader) MPEC

Each generator f chooses the output of each of its power plants $\{x_{fi}, \forall i \in I_f\}$ to maximize its profit subject to a capacity constraint on its output and the follower equilibrium. A more sophisticated representation based on the unit commitment problem could be used, which would involve complexities such as ramp rates, start up costs, and minimum run conditions. Many of these add additional non-convexities (either as concave terms in the objective function or non-convex constraints) to the MPEC; we omit them for simplicity. Even this simple model can result in multiple or no equilibria; the complexities of unit commitment could make such outcomes more likely.

Generator f's MPEC is:

$$\text{Maximize}_{x_{fi}, a_i, p_i, \gamma, \lambda_k} \quad \sum_{i \in I_f} p_i x_{fi} - \sum_{i \in I_f} C_{fi}(x_{fi}) \tag{7.7a}$$

s. t.

$$x_{fi} \leq K_{fi} \quad (\mu_{fi}), \quad \forall i \in I_f \tag{7.7b}$$

$$x_{fi} \geq 0, \quad \forall i \in I_f \tag{7.7c}$$

$$\text{MCP (7.6)} \tag{7.7d}$$

There are several things worth noting about this problem. As mentioned, the feasible region is non-convex, because of the complementarity nature of the KKT conditions for the TSO's problem within (7.6). Also, unlike the MCP in Section 11.2, all the flow variables and price are now decision variables in the generating firm's model, not just its generation. Finally, the objective function is bilinear, involving a product of the price and output decision variables. As a result, the Hessian is neither negative semi-definite nor positive semi-definite, so the objective is non-concave.

7.4.1.6 EPEC

The EPEC may now be written as:

$$\begin{aligned} &\text{Find } \{x_{fi}, \forall f; q_i, a_i, \gamma, \lambda_k, p_i\} \text{ such that:} \\ &\quad \{x_{fi}, q_i, a_i, \gamma, \lambda_k, p_i\} \text{ solves MPEC (7.7)}, \quad \forall f \end{aligned} \tag{7.8}$$

We can define a market equilibrium solution as occurring when each and every generating firm's problem (7.7) yields the same value of all the variables. In that case, firm 1's optimal sales response to firm 2's sales decisions are the same sales that firm 2 assumed when it optimized its response (a Nash-Cournot equilibrium), and the resulting prices and flows that each anticipates the same. Zero, one, or many such pure strategy equilibria may exist.

For very simple cases (e.g., the three-node model in Oren [33]), MPEC (7.7) can be reduced in size to the point that equilibrium solutions can be analytically identified. For situations in which there are two or three firms and only one strategic variable for each (e.g., total output), reaction functions can be calculated and plotted that show each firm's optimal response to the decisions of the other firms (as in [11]). If the reaction functions intersect, then there is a solution. If they don't, then no pure strategy equilibrium exists, although Nash's famous theorem guarantees existence of a mixed strategy equilibrium under certain circumstances [31].[6]

Below, we use the diagonalization algorithm of Section 7.2 in an attempt to find an equilibrium for a two-node example; one assumption about transmission capacity results in an equilibrium, while another yields none. We also contrast that solution to others, including a competitive market model as well as a Cournot model we will present in Section 11.2, in which producers naively think that their actions do not affect transportation prices.

7.4.2 Illustrative Example

7.4.2.1 Assumptions

Figure 7.2 shows the example that we will use to illustrate the EPEC model of generator quantity competition subject to the reactions of the TSO and

6 A mixed strategy for a firm is its choice of a set of probabilities that it will use to choose randomly from among possible values of its strategic variables, and is discussed in Chapter 4. In the present chapter, two examples of the calculation of a mixed strategy equilibrium are presented.

consumers (adapted from [21]). The example has two markets ($i = 1, 2$), each having one power plant along with consumers with a linear demand curve. The hub node is arbitrarily assigned to i=1. There are two generating firms (f = A,B), with A owning the plant at i=1, and B owning the plant at i=2. So $F_1 = \{A\}$, $F_2 = \{B\}$, $I_A = \{1\}$, and $I_B = \{2\}$. As a result of the technological innovations motivated by liberalization of the power sector, each plant miraculously has zero marginal cost and no capacity limit. A transmission line with a 50 MW capacity connects the two markets.[7]

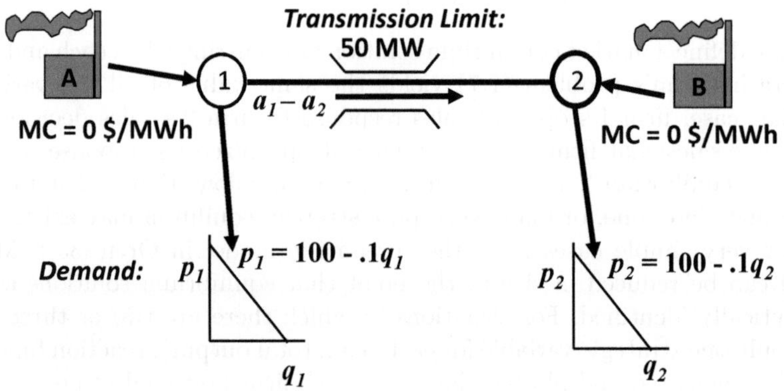

Fig. 7.2 A two-node example of generator competition

7.4.2.2 Follower Problem

First, we derive the follower equilibrium problem (7.6), given $\{x_{A1}, x_{B2}\}$ by first considering the TSO's problem (7.5), in which we assume that the hub node for calculating swing factors is node 2. That assumption leads to zero swing factors for a_2 (since any withdrawal at 2 will have a matching injection at 2, and vice versa).

$$
\begin{aligned}
&\text{Maximize}_{a_1,a_2} \quad p_1 a_1 + p_2 a_2 \\
&\text{s. t.} \\
&\quad a_1 + a_2 = 0 \quad (\gamma) \\
&\quad 1a_1 + 0a_2 \le 50 \quad (\lambda_1) \\
&\quad -1a_1 + 0a_2 \le 50 \quad (\lambda_2)
\end{aligned}
$$

[7] This example is a variation on the simple two-node problem originally introduced in Chapter 1 (Illustrative Example 1.2.7).

The first inequality represents the constraint that flow from 2 to 1 must be less than the transmission capacity of 50 MW (since a positive a_1 means that power is being provided to node 1 from the grid), while the second represents the constraint on flow in the other direction. The optimal solution will be to flow 50 MW from the lower-priced node to the higher priced one; if the prices are equal, then any flow within the [−50 MW, + 50 MW] bounds is optimal.

The resulting follower equilibrium problem is:

Find $\{q_1, q_2, a_1, a_2, p_1, p_2, \gamma, \lambda_1, \lambda_2\}$ satisfying:

$$q_1 = 1000 - 10p_1 \tag{7.9a}$$

$$q_2 = 1000 - 10p_2 \tag{7.9b}$$

$$a_1 + a_2 = 0 \tag{7.9c}$$

$$p_1 - \gamma - \lambda_1 + \lambda_2 = 0 \tag{7.9d}$$

$$p_2 - \gamma = 0 \tag{7.9e}$$

$$0 \geq 1a_1 + 0a_2 - 50 \perp \lambda_1 \geq 0 \tag{7.9f}$$

$$0 \geq -1a_1 + 0a_2 - 50 \perp \lambda_2 \geq 0 \tag{7.9g}$$

$$q_1 - a_1 - x_{A1} = 0 \tag{7.9h}$$

$$q_2 - a_2 - x_{B2} = 0 \tag{7.9i}$$

That MCP is square (9 variables, 9 conditions).

7.4.2.3 Leader Problems

Firm A's MPEC, based on (7.7), is then:

$$\text{Maximize}_{x_{A1}, q_1, q_2, a_1, a_2, p_1, p_2, \gamma, \lambda_1, \lambda_2} \quad p_1 x_{A1} \tag{7.10a}$$

s. t.

$$x_{A1} \geq 0 \tag{7.10b}$$

$$\text{MCP (7.9)} \tag{7.10c}$$

Note that x_{B2} is *not* a decision variable in (7.10), consistent with the Cournot game assumption for players at the same level in the electricity supply chain; that is, each producer takes the output of rival generators as fixed. Unfortunately, the bilinear (and thus non-concave) nature of the objective of (7.10) is not eliminated by substituting the inverse demand function $p_1 = 1000 - 0.1q_1$ for price (which leaves a bilinear term $q_1 x_{A1}$), or by further using market clearing $q_1 = a_1 + x_{A1}$ to eliminate q_1 from the objective (unhappily leaving a bilinear $a_1 x_{A1}$ term).

Meanwhile, B's MPEC is symmetric to A's:

$$\text{Maximize}_{x_{B2}, q_1, q_2, a_1, a_2, p_1, p_2, \gamma, \lambda_1, \lambda_2} \quad p_2 x_{B2} \tag{7.11a}$$

s. t.

$$x_{B2} \geq 0 \tag{7.11b}$$
$$\text{MCP (7.9)} \tag{7.11c}$$

B's problem is conditioned on A's generation x_{A1}, as the latter is not a decision variable in (7.11).

7.4.2.4 EPEC Statement and Analysis

A pure strategy equilibrium, if it exists, solves the following EPEC problem:

$$\begin{aligned}&\text{Find } \{x_{A1}, x_{B2}, q_1, q_2, a_1, a_2, p_1, p_2, \gamma, \lambda_1, \lambda_2\} \text{ such that:}\\&\quad \text{A: } \{x_{A1}, q_1, q_2, a_1, a_2, p_1, p_2, \gamma, \lambda_1, \lambda_2\} \text{solves (7.10) and}\\&\quad \text{B: } \{x_{B2}, q_1, q_2, a_1, a_2, p_1, p_2, \gamma, \lambda_1, \lambda_2\} \text{solves (7.11).}\end{aligned}$$

It turns out that *no* such set of values for the variables exists under Figure 7.2's demand and cost assumptions. This can be most readily seen by examining the two firms' reaction functions and noticing that they do not intercept. To do this, we first note that the above formulations imply that A's generation x_{A1} and profit are unique given a value of B's output x_{B2}, and vice versa. Therefore, we can simply plot A's optimal reaction to B's sales as $x_{A1}{=}x_{A1}(x_{B2})$ (dashed line in Figure 7.3), while B's optimal reaction to A's sales is $x_{B2}{=}x_{B2}(x_{A1})$ (solid line). Note that one function is the mirror image of the other; this symmetry results because each firm has the same supply cost (zero) and the same demand function at its node.

To interpret these functions, consider for instance a situation in which B has chosen $x_{B2}{=}$ 475 MW. The dashed line in the figure shows that A's optimal reaction is to generate 575 MW (which is the optimal x_{A1} from MPEC (7.10), given that A believes that B won't budge from $x_{B2} = 475$).[8] At that point, $p_1 = p_2 = 47.5$ \$/MWh, and $a_2{-}a_1 = 50$ MW. (Note that only the difference between the arbitrage amounts a_i is unique, and not the individual values of a_1 and a_2.) Thus, A aggressively moves into B's home market. It is precisely at this level of A's output that the line becomes congested. Although the flow is at its upper bound, there is no economic congestion since the price difference is zero.[9] If A sells less than 575 MW, the line becomes uncongested

[8] This MPEC is solved by a line search over the variable x_{A1}, and calculating the resulting prices and profits by solving the followers' equilibrium problem (7.6). That equilibrium problem could be solved as an MCP, but we instead applied a quadratic programming solver to a QP whose KKT conditions correspond to MCP (7.6).

[9] Oren [33] notes that Cournot firms in his model, which is for a different network topology, also choose outputs so that lines are just barely congested, and no revenues are received by the TSO. However, this is not a general result for oligopoly models. For example, on average, the TSO may earn a positive congestion revenue in a mixed strategy Cournot equilibrium, as we see in the results later in this section, or even in pure strategy equilibria if the producers' strategic variables are supply functions rather than quantities (as in the results in Section 7.5).

(flow is less than capacity) and price falls at both locations 1 and 2. But if A sells more, the congestion means that node 1 must absorb all the additional sales, which rapidly decreases the price there.

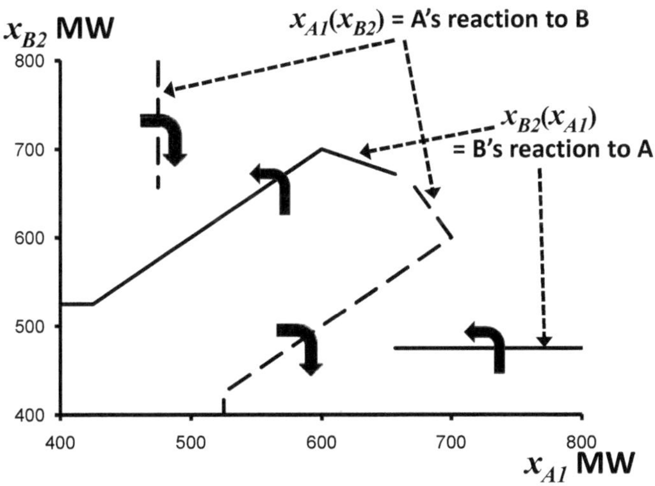

Fig. 7.3 Reaction functions for EPEC model in which Cournot generators are Stackelberg leaders with respect to TSO and consumer followers

Now consider what happens if B has instead picked x_{B2}= 667 MW. A's reaction function in Figure 7.3 indicates that it should sell only 475 MW, which results in 50 MW of flow from node 2 to node 1. Thus, it is optimal for A to concede market 2 to B. Figure 7.4 shows the profit for A as a function of its total sales decision, given that A believes that B will maintain its output at 667 MW. This profit function is neither convex nor concave, as a result of the non-convex feasible region of (7.10). There are three concave sections, each corresponding to a different resolution of the complementarity conditions associated with the two transmission bounds in the constraint set of (7.10) (which are the first-order conditions for the TSO's transmission duals λ_k, k=1,2). The section on the left occurs when the TSO flows the maximum possible amount from 2 to 1, while the right section is the opposite case. The middle concave section corresponds to the situation when transmission flow is basic (between the two bounds). The borders of the region correspond to when a complementarity condition for the TSO's transmission "flips", in that which of two inequalities in a condition is binding changes. The asterisks show two local optima, one at x_{A1} = 475 MW (A's profit = 22,563 \$/hr) and the other at x_{A1} = 667 MW (A's profit = 22,222 \$/hr). Inspection of all possible solutions shows that the lower sales value represents the global optimum. The existence of a local optimum that is not the global optimum implies that traditional gradient-based optimization techniques could find the wrong local optimum when attempting to solve A's profit-maximization MPEC (7.10).

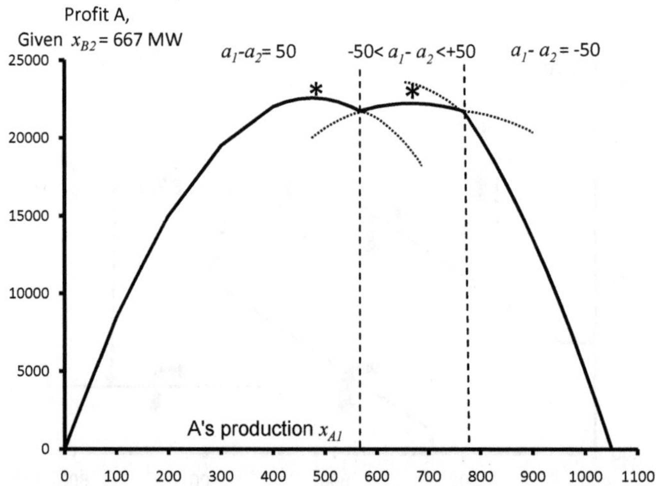

Fig. 7.4 Firm A's profit, given $x_{B2} = 667$ MW, showing two local optima and the global optimum is $x_{1A} = 475$ MW

When x_{B2} instead equals 656.5 MW, the middle curve in Figure 7.4 shifts upwards enough so that the profits from the two local optima are precisely equal. This is the point corresponding to the gap in A's reaction function in Figure 7.3. At that level of x_{B2}, firm A would be indifferent between $x_{A1} =$ 475 MW or 671.75 MW. However, any level of sales that falls strictly between those levels would yield a lower profit for A, which is the reason for the gap in Figure 7.3.

The gaps in these reaction functions are located such that there is no intersection. This implies that there is no Cournot pure strategy equilibrium among the two producers. That is, there is no point at which neither firm wishes to adjust its sales unilaterally. If one attempts to find an equilibrium by the diagonalization approach of Section 7.2, the process would never converge, as we demonstrate next.[10]

[10] For another simple two-node Cournot case, Borenstein et al. [8] also found that pure strategy did not exist.

7.4.2.5 Attempted Solution by Diagonalization

Table 7.1 shows the results of applying the diagonalization procedure of Section 7.2 to the EPEC. It shows a sequence of six solutions (labeled "MPEC_B1" through "MPEC_A3") of A's and B's MPECs ((7.10) and (7.11), respectively).

Starting with an assumption of relatively low sales by A ($x_{A1} = 475$ MW), MPEC_B1 shows that the optimal solution to B's problem (7.11) is $x_{B2} = 575$ MW. MPEC_A1 then shows A's response to that x_{B2}; solving (7.10) given x_{B2} = 575 MW yields x_{A1} = 675 MW. In both of these solutions, the supplier is reacting to relatively low generation by its rival (and thus high prices) by expanding its supply, cutting prices, and exporting 50 MW to its rival's market. However in MPEC_B3, the reverse happens; B reacts to A's sales of 675 MW by cutting production and conceding part of its home market at node 2, allowing A to export 50 MW. Table 7.1 shows the first six iterations; MPEC_B4 turns out to be identical to MPEC_B1, and the cycle starts again. There is no convergence of this diagonalization—not because of a flaw in the algorithm, but because there is no pure strategy equilibrium.

Table 7.1 Comparison of two-node solutions for generator strategic operation, 50 MW transmission limit

Solution	x_{A1} MW	x_{B2} MW	p_1 \$/MWh	p_2 \$/MWh	Cons. Surplus \$/hr	Profit \$/hr	Welfare \$/hr
Cournot MPEC_B1	475	575*	47.5	47.5	27563	49875	77438
Cournot MPEC_A1	675*	575	37.5	37.5	39063	46875	85938
Cournot MPEC_B2	675	475*	37.5	47.5	33313	47875	81688[a]
Cournot MPEC_A2	575*	475	47.5	47.5	27563	49875	77438
Cournot MPEC_B3	575	675*	37.5	37.5	39063	46875	85938
Cournot MPEC_A3	475*	675	47.5	37.5	33313	47875	81688[a]
Average Cournot MPEC	575	575	42.5	42.5	33313	48208	81688[b]
Mixed Strategy Cournot EPEC (Expected Value)	643.1	643.1	35.7	35.7	41533	45498	87090[c]
Cournot MCP (Section 11.2)	666.7	666.7	33.3	33.3	44444	44444	88889
Competitive	1000	1000	0	0	100000	0	100000
EPEC: A Cournot B Competitive,	475	1050	47.5	0	63781	22563	88719[d]

* Indicates optimal x_{fi} from MPEC in reaction to the other firm's production

a. Transmission (congestion) surplus nonzero = 500 \$/hr (price difference times flow); welfare = consumer surplus + profit + congestion surplus.

b. Average congestion surplus is 133 \$/hr.

c. Expected value of congestion surplus is 60 \$/hr.

d. Congestion surplus is 2375 \$/hr.

Of course, the absence of a pure strategy equilibrium may indeed be the reality in power markets; there is no reason a priori that we should expect such an equilibrium. As an example, in Day and Bunn's [13] agent-based simulations of the UK market, prices often cycled without converging. But a challenge facing the analyst using this model is how to characterize the market outcomes. One way is to describe the range of prices that occur during iterations of the diagonalization [29], and calculate averages over that distribution. For instance, in the seventh row of Table 7.1, we show the average sales, prices, consumer surplus, profit, and welfare (the sum of profit, consumer surplus, and congestion revenue, the latter equaling the difference between expenditures by consumers and revenue to producers) over the six solution cycle MPEC_B1 through MPEC_A3. This procedure has the practical problem that the computational cost of doing enough iterations to characterize the distribution of prices may be prohibitive.

Meanwhile, a conceptual difficulty with diagonalization is that the range and distribution of prices it yields will in general differ from the probability distribution that would result if a Nash (mixed strategy) equilibrium for the game were calculated. Therefore, if practical, it is preferable to calculate the mixed strategy equilibrium, which we do next.

7.4.2.6 Mixed Strategy Solution

The absence of a pure strategy solution for the above EPEC leads us to turn to calculation of a mixed strategy equilibrium, which, by Nash's famous theorem, must exist if the strategy space is finite. We use the classic MCP approach to calculating a mixed strategy for a bimatrix game. As explained in Section 4.2.4 above, this consists of populating payoff matrices with A's and B's profits under each combination of strategies considered and then formulating a linear MCP that represents the equilibrium problem for the two players.[11] However, calculating mixed equilibria can be challenging even for simple problems, and requires highly simplified strategy spaces for more realistic problems (e.g., [2]). Furthermore, the bimatrix approach is not practical for games with more than two players; that approach leads to a difficult to solve nonlinear MCP formulation. Other approaches, such as the fictitious-play learning algorithm (Ch. 2 in [17]), are employed instead.

We consider the seven possible strategies for leader A, $x_{1A} = \{425\text{ MW}, 475\text{ MW}, \ldots, 725\text{ MW}\}$, and the same strategies for B. This leads to 49 possible combinations of strategies. The payoff matrix for A is shown in Table 7.2 Since the players are symmetric, the payoff matrix for B is just the transpose of A's matrix.

---

[11] Stoft [41] presents an example of the application of the bimatrix approach in the case of two generators competing à la Cournot, as here, on a three bus network.

Table 7.2 Payoff matrix for leader A in Cournot EPEC ($/hr)

$x_{1A} \backslash x_{2B}$	425	475	525	575	625	675	725
425	24438	23375	22313	22313	22313	22313	22313
475	26125	24938	23750	22563	22563	22563	22563
525	27563	26250	24938	23625	22313	22313	22313
575	27313	27313	25875	24438	23000	21563	21563
625	26563	26563	26563	25000	23438	21875	20313
675	25313	25313	25313	25313	23625	21938	20250
725	23563	23563	23563	23563	23563	21750	19938

Solving the MCP (4.7) for the mixed strategy equilibrium yields the following solution: A chooses 475 MW with probability 0.07846, 625 MW with probability 0.32391, and 675 MW with probability 0.59763. B adopts the same strategy. So out of the 49 possible combinations, only 9 (=3^2) have positive probabilities. From these probabilities, we can calculated expected outcomes in terms of the primal variables and prices, profits, consumer surplus, etc. (Table 7.1), which we discuss below.

However, even if distributions of prices and other market outcomes can be described in this manner for a mixed strategy equilibrium solution of the EPEC, there still remains the practical challenge of communicating the results concisely and clearly. If no pure strategy exists (or, alternatively, several exist), then it becomes difficult to compare solutions for alternative market designs or structures. The ability of policy makers and other users of market simulations to understand the results and the methodology lying behind them becomes harder if they cannot be assured that model solutions exist and are unique.

7.4.2.7 Comparison of Outcomes of Alternative Game Formulations

To illustrate how the model formulation and solution approach can affect the calculated equilibrium, we contrast the prices and surpluses resulting from five different models that use the same cost and demand data (Figure 7.2), as shown. We consider the following games and solution approaches:

1. The average of the six Cournot EPEC solutions MPEC_B1-MPEC_A3 encountered in the application of the deterministic diagonalization approach.
2. The expected value of the mixed strategy solution to the Cournot EPEC.
3. The solution from a game in which the generation firms still play a Cournot game against each other, but play a Bertrand (price-taking) game against the grid operator (yielding the MCP of Section 11.2, below) instead of the Stackelberg EPEC game of this section. In this model,

generators naively assume that price differences among nodes will not be affected by their output decisions, which leads to an easier-to-compute but arguably less realistic model.
4. The competitive (price-taking) equilibrium.
5. An asymmetric EPEC in which only A is a Cournot player, while B is a price-taker.

The results of this "comparative statics" exercise are shown in Table 7.1.

The lowest welfare, which is an indicator of overall market efficiency, results from the average of the six deterministic EPEC solutions. Surprisingly, those solutions paint an appreciably more dismal picture of the market outcomes for this game than the actual equilibrium for this EPEC problem, which is the mixed strategy equilibrium. Equilibrium prices are 15% lower on average than the average of the iterations of the diagonalization. This shows that if diagonalization fails to find a pure strategy equilibrium for an EPEC, averaging the solutions encountered (as in [29]) may be a poor indication of the actual equilibrium. However, we are unaware of any analysis that has established a predictable direction or magnitude of this possible bias.

In contrast, the alternative, simpler Cournot model of electricity competition (Section 11.2) in which generators do not anticipate how the grid operator will alter congestion costs in response to production decisions yields a still more competitive outcome: lower prices, less profit, more consumer surplus, and more welfare. Thus, the nature of the "vertical" game between producers and transporters of energy affects market outcomes. In particular, a savvy set of producers who correctly anticipate the lower level's (TSO's) response appear to be able to exploit that to increase their profit at the expense of consumer surplus and TSO surplus, with the consumers bearing the brunt of the loss. We note, however, that this has not been established as a general result, and furthermore the difference is less if those results are compared to the mixed strategy equilibrium.

The lowest prices – in fact, zero under the assumptions of Figure 7.2 – result if producers behave competitively (price-taking) with respect to each other. Zero profits are earned, and the entire social surplus accrues to consumers.

7.4.2.8 Sensitivity Case: Single Oligopolist

An additional case we consider in Table 7.1 is where only A exercises market power as a Cournot player in our EPEC, while B is a price-taker. This is a different sort of EPEC, where A's MPEC (7.10) is the same but B's problem instead reduces to maximizing profit subject to a fixed price and no capacity limit:

$$\text{Maximize}_{x_{B2}} \quad p_2 x_{B2} \tag{7.12a}$$

s. t.

$$x_{B2} \geq 0 \tag{7.12b}$$

whose KKT condition is simply $0 \leq x_{B2} \perp p_2 \leq 0$; this implies that if prices are nonnegative, the only equilibrium price at node 2 is zero, and then any nonnegative production by B is an equilibrium. The EPEC consists of finding values of all the variables that simultaneously solve A's problem (7.10) and B's price-taking problem (7.12). Diagonalization (or just plain inspection) converges to the pure strategy solution shown in the last line of Table 7.1, where A concedes part of its market in order to raise price at its node to 47.5 \$/MWh. The resulting welfare is better than the average outcome of having both firms play a Cournot game with respect to each other and a Stackelberg game with respect to the TSO and consumers (the average of MPEC_B1-MPEC_A3, or the mixed strategy equilibrium in Table 7.1), as one would expect. However, its welfare is surprisingly worse than when both A and B are Cournot players, but are price-taking with respect to transmission (the MCP solution from the Cournot model of Section 11.2). This indicates that vertical strategic relationships (i.e., among entities at different levels of the supply chain, here, between producers and transmission) can be as or more damaging to markets than horizontal ones (among entities at the same level, here producers).

7.4.2.9 Sensitivity Case: Transmission Expansion

Finally, we consider what happens if transmission capacity is expanded by a mere 10 MW to 60 MW. This changes the entire dynamic of the game. The profit function in Figure 7.4 (given $x_{2B} = 667$ MW) still has the same general shape, but now the profit for A from responding with 667 MW (the center segment, corresponding to no binding transmission constraint) is higher than responding by conceding and shrinking x_{1A} to 475 MW (the local optimum in the left segment). In a diagonalization, B's best response x_{2B} to A's 667 MW would be to stay put at $x_{2B} = 667$ MW, and we have a pure strategy equilibrium. Starting from elsewhere (for instance, from $x_{2B} = 470$ MW, which is B's best response if it knows that A will produce so much that the line will be congested from 1 to 2) converges to the same solution (Table 7.3).

This pure equilibrium is identical to the MCP equilibrium from Section 11.2 (the third line from the bottom in Table 7.1). Thus, adding just 20% (or 10 MW) to the transmission line has significantly lowered prices and increased market efficiency (as measured by welfare). This large change happened because the dynamics switched from a game without a pure strategy equilibrium (in which prices might cycle or randomly vary, with some very high values occurring) to a game with such an equilibrium. This shows that under oligopoly conditions a small change in market structure to increase the ability of rivals to enter markets (here, a 20% increase in transmission) can dramatically change market dynamics and have large market efficiency benefits.

Table 7.3 Diagonalization solution to two-node Cournot generator EPEC, 60 MW transmission limit

Solution	x_{A1} MW	x_{B2} MW	p_1 \$/MWh	p_2 \$/MWh
Cournot MPEC_B1	590*	470	47	47
Cournot MPEC_A1	590	705*	35.25	35.25
Cournot MPEC_B2	647.5*	705	32.375	32.375
Cournot MPEC_A2	647.5	676.25*	33.8125	33.8125
Cournot MPEC_B3	661.875*	676.25	33.0938	33.0938
Cournot MPEC_*fn*, large n	666.67*	666.67*	33.33	33.33

* Indicates optimal x_{fi} from MPEC in reaction to the other firm's production

7.4.2.10 Summary of Cournot EPEC Example

This example has illustrated how an EPEC can represent the interaction of strategic energy producers with each other (à la Cournot) and with the downstream transportation system (as Stackelberg leaders and follower). In general, such an EPEC is more difficult to solve than an MCP model that instead represents the vertical game as one between producers and transporters who are both price-taking with respect to transportation costs (Section 11.2). Drastically different behaviors can also result. Even an extremely simple version of the Cournot EPEC model can result in complex behavior. In the above case, a slight change in transmission capacity can result in either no pure strategy solution or (at least) one such solution, with very different price and market efficiency outcomes.

7.5 EPEC Power Market Model 2: Strategic Offering by Generators

7.5.1 Model Formulation[12]

7.5.1.1 Structural Assumptions

Many energy markets, especially in power, are structured as auctions in which producers submit offers to provide energy, and the auctioneer determines what offers to accept in order to maximize net surplus in the market subject to transportation constraints. A producer's offer consist of a function describing how much it is willing to sell at each location as a function of the price at that location. The function can be as simple as "I am willing to sell up to 100 MW in hour 23 if the price is at least \$35/MWh." In the simplest auctions, the auctioneer minimizes the "as bid" cost of providing energy (the sum of the integrals under the offer curves) subject to meeting demand at various locations as well as the transport constraints.

Such "smart auctions" are used not only in energy, but for other commodities, such as water [35]. The most elaborate versions are in power markets, which often clear several commodities at once (energy and various types of operating reserves) over a multiperiod horizon, consider complex offers (for instance with prohibited operating zones, fixed start-up costs, and alternative configurations for combined cycle plants), and solicit bids to buy from consumers as well as offers to sell from producers [3]. These auction-based electricity markets are sometimes referred to as "pool-based" or "single-buyer" power market designs.

In this section, we consider an EPEC for such an auction-based electricity market that is closely related to the EPEC of the previous section, but differs in two important ways. First, it instead represents a game in which the producer's strategic variable is its offer function rather than quantity produced. Second, the TSO is not a price taking profit maximizer, but instead maximizes apparent market surplus, equal to the integrals of the demand curves (representing demand bids) minus the integrals of the offer functions. We say "apparent" because actual market surplus will differ from the TSO's objective function because producers can submit bid functions that differ from their marginal cost functions, perhaps in an attempt to exercise market power and increase prices.[13]

[12] Illustrative Examples 6.2.2, 6.2.9, and 6.2.12 in Chapter 6 present MPEC formulations for a single producer who strategically offers bids into an auction-based market. Thus the model of this section can be viewed as a generalization of that example to multiple producers, each of whom strategically offers bids into an auction-based market operated by a TSO.

[13] Several EPEC power market models structured in this manner have been proposed, including Hobbs et al. [22], which this section's model is based upon. That reference in-

However, it turns out that if producers are restricted to offering fixed quantities (that is, a vertical offer curve at the quantity they wish to provide) rather than a sloped offer curve, the model of this section reduces to the model of the previous section. This will be shown below. But we begin by presenting the auctioneer's (TSO's) model and its resulting equilibrium (KKT) conditions that are then embedded in the MPEC of each producer, who is a Stackelberg leader with respect to the following TSO (and demand). The next subsection presents a two-node application, showing, as above, that pure strategy equilibria may or may not exist, depending on the precise parameters.

7.5.1.2 Auctioneer (Transmission Provider)

Consumers are assumed to bid their true inverse demand functions $f_{d,i}^{-1}(q_i)$to the TSO, which treats their integrals as the aggregate benefit of consumption. The TSO's objective is then to maximize the sum of those integrals minus the sum of the accepted supply offers. Its decision variables are the quantities supplied at each node q_i and the amount of supply bought from each producer x_{fi}. We simplify the supply offers, representing them as consisting of a maximum quantity that a producer is willing to supply K_{fi} and the minimum price that the producer is willing to accept b_{fi}. Furthermore, we assume that producers offer their capacity truthfully (perhaps because of audit rules), and so their only degree of freedom is the minimum price.

A final simplification we make in this generator model in that we assume that deviation penalties are such that producers will indeed provide the amounts they are asked to. Of course, actual power markets often have two or more settlements. Most common is a combination of a day-ahead market followed by a real-time balancing market in which deviations from day-ahead schedules are settled at real-time prices that generally differ from (and are more volatile than) day-ahead prices.

The auctioneer's model is:

$$\text{Maximize}_{x_{fi},q_i,a_i} \quad \sum_i \int_0^{q_i} f_{d,i}^{-1}(q)dq - \sum_f \sum_{i \in I_f} b_{fi} x_{fi} \tag{7.13a}$$

s. t.

$$q_i - a_i - \sum_{f \in F_i} x_{fi} = 0 \quad (p_i), \quad \forall i \tag{7.13b}$$

cludes a medium size application (a 30 bus network with two firms, each owning three generating units). For closely related models, see Berry et al. [6], who solve their models by diagonalization as well, and Hu and Ralph [24], who establish certain existence results and find local Nash equilibria of the EPEC by solving a complementarity formulation of the simultaneous first-order conditions of all the producers' MPECs.

$$\sum_i a_i = 0 \quad (\gamma) \tag{7.13c}$$

$$\sum_i PTDF_{ik} a_i \le T_k \quad (\lambda_k), \quad \forall k \tag{7.13d}$$

$$x_{fi} \le K_{fi} \quad (\mu_{fi}), \quad \forall f, i \in I_f \tag{7.13e}$$

$$x_{fi} \ge 0 \quad \forall f, i \in I_f \tag{7.13f}$$

7.5.1.3 Producer MPEC

Each producer is a leader with respect to the auctioneer and consumers. To include the auctioneer as a follower in a leader producer's MPEC, we first derive the KKTs for (7.13). We then assume that each producer f chooses its offer in order to maximize its actual profit subject to the auctioneer's reaction to the offers, as captured by the KKTs. Producer f's MPEC is then:

$$\text{Maximize}_{b_{fi}; x_{f'i}, q_i, a_i, p_i, \gamma, \lambda_k, \mu_{f'i}} \quad \sum_{i \in I_f} p_i x_{fi} - \sum_{i \in I_f} C_{fi}(x_{fi}) \tag{7.14a}$$

s. t.

The TSO's KKTs:

$$0 \le x_{f'i} \perp p_i - b_{f'i} - \mu_{f'i} \le 0, \quad \forall f', i \in I_{f'} \tag{7.14b}$$

$$p_i = f_{d,i}^{-1}(q_i) \quad (q_i \text{ unrestricted}), \quad \forall i \tag{7.14c}$$

$$p_i - \gamma - \sum_k PTDF_{ik} \lambda_k = 0 \quad (a_i \text{ unrestricted}), \quad \forall i \tag{7.14d}$$

$$q_i - a_i - \sum_{f' \in F_i} x_{f'i} = 0 \quad (p_i \text{ unrestricted}), \quad \forall i \tag{7.14e}$$

$$\sum_i a_i = 0 \quad (\gamma \text{ unrestricted}) \tag{7.14f}$$

$$0 \le \lambda_k \perp \sum_i PTDF_{ik} a_i - T_k \le 0, \quad \forall k \tag{7.14g}$$

$$0 \le \mu_{f'i} \perp x_{f'i} - K_{f'i} \le 0, \quad \forall f', i \in I_{f'} \tag{7.14h}$$

Notice that f and f' are different subscripts, since a particular firm f is optimizing subject to the TSO's first-order conditions that determine the outputs of plants for all firms f'. Notice also the semicolon in the list of decision variables separating the offer variable from the others; this is to indicate that, in a sense, the producer directly chooses its offer and thereby indirectly determines all the other variables through the offer's effect on the solution to the TSO's auction problem (defined by the TSO's KKTs in the producer's constraint set).

7.5.1.4 EPEC

The EPEC that finds the equilibrium offers for the competing producers, as well as the equilibrium prices, production, consumption, and flows, can be phrased as follows:

$$\begin{aligned}&\text{Find } \{b_{fi}, \forall f; x_{f'i}, q_i, a_i, p_i, \gamma, \lambda_k, \mu_{f'i}\} \text{ such that:}\\&\{b_{fi}; x_{f'i}, q_i, a_i, p_i, \gamma, \lambda_k, \mu_{f'i}\}\text{solves (7.14)}, \quad \forall f\end{aligned} \tag{7.15}$$

We note that (7.15) is overly restrictive – it is not necessary that the producers' MPECs in (7.15) have the same values of $\{q_i, a_i, \gamma, \lambda_k, \mu_{f'i}\}$ because all that the producers care about is the price p_i. It is possible, for instance, that a diagonalization procedure used to solve (7.15) will come up with different values of the TSO's duals $\{\gamma, \lambda_k, \mu_{f'i}\}$ for each of the producer's MPECs because there are multiple dual solutions due to degeneracy in the TSO's problem. However, this is unimportant as long as the prices received by the producers are consistent among the MPECs [28].

Before proceeding to an example, we show that this reduces to the Cournot model of Section 7.4 in which the TSO acts as a price-taker rather than a social welfare maximizer if instead producers are allowed to choose a fixed quantity x_{fi} to offer (or "schedule") into this auctioneer market. If this is the case, x_{fi} is no longer a decision variable for the TSO's problem, but is instead directly decided upon by the producer. This means that the TSO's problem simplifies: the last two constraints of (7.13) can be omitted as well as the x_{fi} offer cost term in the objective function. Hence, the KKT conditions associated with x_{fi} and its upper bound are no longer needed. Then if we substitute $q_i = f_{d,i}(p_i)$ for its inverse $p_i = f_{d,i}^{-1}(q_i)$ in (7.13) (which is permissible if the demand function is monotonically decreasing and continuous), the resulting set of five KKT conditions is the same as (7.6), the equilibrium conditions for the Cournot producer/price-taking TSO model of Section 7.4. The bounds on x_{fi} that were dropped from (7.13) must be reimposed in the producer's problem (which becomes the MPEC (7.7)), since now the producer can freely choose its value (rather than having to accept the value that results from the auctioneer's processing of its price offer b_{fi}).[14]

□

[14] Yao et al. [47] use the welfare formulation of the auctioneer's (TSO) problem in their Cournot/Stackelberg producer EPEC, rather than our price-taking TSO version. However, as just shown, equivalent MPECs for producers result from either formulation, and so the EPECs are also equivalent.

7.5.2 Illustrative Example

7.5.2.1 Assumptions

We adapt the example of Figure 7.2 slightly so that the results are more interesting. In particular, the two plants now have positive (and differing) marginal costs: the marginal operating cost for x_{A1} is \$20/MWh, while that for x_{B2} is \$35/MWh. Also new are capacity constraints on the generators, 550 MW in each case. (Unfortunately, restructuring hasn't yielded so much innovation that plants cost nothing to run and have infinite capacity!) We assume that the market rules are such that each generator has to offer in all 550 MW of its capacity into the market, but the price at which it offers that capacity (b_{A1} for firm A, b_{B2} for firm 2) is not restricted. The demand functions and transmission capacity are the same.

7.5.2.2 EPEC Formulation

The TSO's market auction problem (7.13) can therefore be phrased as maximization of the sum of the integral of the two inverse demand curves minus the (as bid) costs of accepted generation offers, subject to the physical transmission and generation constraints:

$$\text{Maximize}_{x_{A1},x_{B2},q_1,q_2,a_1,a_2} \quad (100q_1 - 0.1q_1^2/2) + (100q_2 - 0.1q_2^2/2) - b_{A1}x_{A1} - b_{B2}x_{B2} \tag{7.16a}$$

s. t.

$$q_1 - a_1 - x_{A1} = 0 \quad (p_1) \tag{7.16b}$$

$$q_2 - a_2 - x_{B2} = 0 \quad (p_2) \tag{7.16c}$$

$$a_1 + a_2 \tag{7.16d}$$

$$1a_1 + 0a_2 \leq 50 \quad (\lambda_1) \tag{7.16e}$$

$$-1a_1 + 0a_2 \leq 50 \quad (\lambda_2) \tag{7.16f}$$

$$x_{A1} \leq 550 \quad (\mu_{A1}) \tag{7.16g}$$

$$x_{B2} \leq 550 \quad (\mu_{B2}) \tag{7.16h}$$

$$x_{A1}, x_{B2} \geq 0 \tag{7.16i}$$

Then A's MPEC (7.14) is to maximize its revenue minus its actual cost, subject to the solution of (7.16) (defined by (7.16)'s KKTs):

$$\text{Maximize}_{b_{A1};x_{A1},x_{B2},q_1,q_2,a_1,a_2,p_1,p_2,\gamma,\lambda_1,\lambda_2,\mu_{A1},\mu_{B2}} \quad p_1x_{A1} - 20x_{A1} \tag{7.17a}$$

s. t.

$$0 \leq x_{A1} \perp p_1 - b_{A1} - \mu_{A1} \leq 0 \tag{7.17b}$$

$$0 \leq x_{B2} \perp p_2 - b_{B2} - \mu_{B2} \leq 0 \tag{7.17c}$$

$$q_1 = 1000 - 10p_1 \tag{7.17d}$$
$$q_2 = 1000 - 10p_2 \tag{7.17e}$$
$$p_1 - \gamma - \lambda_1 + \lambda_2 = 0 \tag{7.17f}$$
$$p_2 - \gamma = 0 \tag{7.17g}$$
$$a_1 + a_2 = 0 \tag{7.17h}$$
$$q_1 - a_1 - x_{A1} = 0 \tag{7.17i}$$
$$q_2 - a_2 - x_{B2} = 0 \tag{7.17j}$$
$$0 \geq 1a_1 + 0a_2 - 50 \perp \lambda_1 \geq 0 \tag{7.17k}$$
$$0 \geq -1a_1 + 0a_2 - 50 \perp \lambda_2 \geq 0 \tag{7.17l}$$
$$0 \leq \mu_{A1} \perp x_{A1} - 550 \leq 0 \tag{7.17m}$$
$$0 \leq \mu_{B2} \perp x_{B2} - 550 \leq 0 \tag{7.17n}$$

Notice that A has only its offer variable b_{A1} as a decision variable, and not B's – this is consistent with a Nash game between the producers in offers. Meanwhile, B's MPEC has the same constraint set, but a different objective function:

$$\text{Maximize}_{b_{B2};x_{A1},x_{B2},q_1,q_2,a_1,a_2,p_1,p_2,\gamma,\lambda_1,\lambda_2,\mu_{A1},\mu_{B2}} \quad p_2 x_{B2} - 35x_{B2}$$
s. t.
$$\text{Same constraint set as (7.17)} \tag{7.18}$$

The EPEC whose solution (if it exists) is a Nash equilibrium in offers between the producers, and a Stackelberg equilibrium between the set of producers and the TSO is then:

Find $\{b_{A1}, b_{B2}, x_{A1}, x_{B2}, q_1, q_2, a_1, a_2, p_1, p_2, \gamma, \lambda_1, \lambda_2, \mu_{A1}, \mu_{B2}\}$ such that:

A: $\{b_{A1}; x_{A1}, x_{B2}, q_1, q_2, a_1, a_2, p_1, p_2, \gamma, \lambda_1, \lambda_2, \mu_{A1}, \mu_{B2}\}$ solves (7.17), and

B: $\{b_{B2}; x_{A1}, x_{B2}, q_1, q_2, a_1, a_2, p_1, p_2, \gamma, \lambda_1, \lambda_2, \mu_{A1}, \mu_{B2}\}$ solves (7.18).

7.5.2.3 Application of Diagonalization

We attempt to solve the EPEC by the diagonalization procedure of Section 7.2, alternating between solving MPECs (7.17) and (7.18), using the other producer's offer from the previous iteration in each case. The MPECs are solved by a line search over the producer's own offer; for each candidate offer, we solved the TSO's auction problem (7.16) using a quadratic program, and noted the resulting profit for the firm. The candidate offer with the highest profit was then chosen as the best response. We then insert that response in the constraint set for the other company's MPEC in the next iteration of the diagonalization.

However, the diagonalization procedure runs into problems when the optimal response of one producer to an offer by another producer is to just undercut the other offer by epsilon, and steal part of its market. Then in the diagonalization procedure, the second producer gets its turn, and drops its price by two times epsilon, taking back the market. The smaller epsilon is, the slower the diagonalization procedure will be. Therefore, in our example, if the producer drops its offer in a round of the diagonalization procedure, it must drop it by a finite amount (we assume at least 5 \$/MWh). However, if it raises it, it can raise it by any amount. (Needless to say, it is nontrivial to implement a restriction that the MPEC's choice of offer is not allowed to be in the range$(b^o_{fi} - 5, b^o_{fi})$, where b^o_{fi} is the previous round's offer. A binary variable can be introduced to do this, but at the cost of significantly complicating the model.[15])

Table 7.4 shows the results of diagonalization, starting with b_{B2} = \$65/MWh. This happens to be B's optimal "concede" offer—that is, its best response if A had previously made a very low offer so as to export 50 MW into B's node 2. B's best option is then to offer its output at a price much higher than its marginal cost and settle for supplying only part of its node's demand and none of node 1's. Then A's optimal response to B's \$65/MWh offer is to undercut by offering \$62.5/MWh (solution MPEC_A1); this maintains A's exports to node 2. B then responds by undercutting in turn to \$60, taking back not only node 2's load, but also stealing 50 MW of load from node 1 (solution MPEC_B1). Thus flow reverses in direction, as shown in the last column. It is then A's turn again, and it undercuts again to an offer of \$57.5/MWh. This back and forth continues until B decides that it is better to concede by raising its offer rather than under cut again – this happens in response to A's offer of \$52.5, when B decides to raise its offer back up to \$65/MWh (solution MPEC_B3). This is because B earns 9000 \$/hr if it concedes, whereas if it instead undercuts A by offering 50 \$/MWh, its profit would have been only 8250 \$/hr. Once B raises its offer in this manner, the price cycle starts again, with A moving its offer up to \$62.5 (first row), then B undercutting (second row), and so forth.

[15] In our implementation of the EPEC for a large system [22] we did not include such variables. Fortunately, in this case, diagonalization did converge to a pure strategy equilibrium.

Table 7.4 Sequence of solutions to generator offering MPECs when K_{fi} = 550 MW, transmission limit = 50 MW, and comparative statics

Solution	b_{A1} $=p_1$	b_{B2} $=p_2$	x_{A1}	x_{B2}	Flow 1→2	Profit 1	Profit 2	Consu. Surplus	Social Welfare[a]
MPEC_A1	62.5*	65	425	300	50	18063	9000	13156	40344
MPEC_B1	62.5	60*	325	450	-50	13813	11250	15031	40219
MPEC_A2	57.5*	60	475	350	50	17813	8750	17031	43719
MPEC_B2	57.5	55*	375	500	-50	14063	10000	19156	43344
MPEC_A3	52.5*	55	525	400	50	17063	8000	21406	46594
MPEC_B3	52.5	65*	525	300	50	17063	9000	17406	44094
Average Offering MPEC	57.5	60	442	383	17	16313	9333	17198	43052
Mixed Strategy Offering EPEC (Expected Value)	55.5	62.0	478	347	33	16902	9000	17262	43534
Competitive	45	45	550	550	0	13750	5500	30250	49500

* Optimal response of producer to other producer's offer

a. Welfare = Profits + Consumer Surplus + TSO Surplus. Transmission surplus = 125 in all MPEC solutions except MPEC_B3, where it equals 625. In mixed strategy solution, expected TSO Surplus = 370.

Economists term this type of price cycling "Edgeworth competition," in which a period of successive undercutting is followed by one firm conceding and raising its price to serve a residual market, which then motivates its rivals to raise their prices, touching off a new cycle of price cutting. Edgeworth competition is more commonly a characteristic of markets in which producers compete on the basis of prices or bids (Bertrand type competition), rather than in quantities (Cournot competition). Usually, there needs to be some spatial or product differentiation that means that outputs of different firms cannot be perfect substitutes (in our case, a transmission bottleneck prevents A from completely replacing B's output, and vice versa). As a result, a conceding firm does not lose all of its market to its lower price rivals.[16]

Figure 7.5 illustrates this Edgeworth-like cycling with a pair of reaction curves, similar to Figure 7.3 above. The 45° sloped portion of the curves represent the "undercut" strategy, in which the reacting generator cuts shaves its price to just less than the other's. (We've assumed that such undercutting must be at least 2.5 \$/MWh smaller than the price being undercut; other assumptions would result in 45° lines that are closer or further apart.) However, as the reaction curves show, if the first generator's price is too low, the reacting generator instead "concedes" with a much higher price offer. This occurs where there is a vertical discontinuity in B's reaction function, and a horizontal discontinuity in A's. As in Figure 7.3, the two reaction curves do not intersect, which implies that there is no pure strategy equilibrium, and instead a mixed strategy equilibrium must be sought.

[16] For another example in a spatial competition context, see Schuler and Hobbs [38].

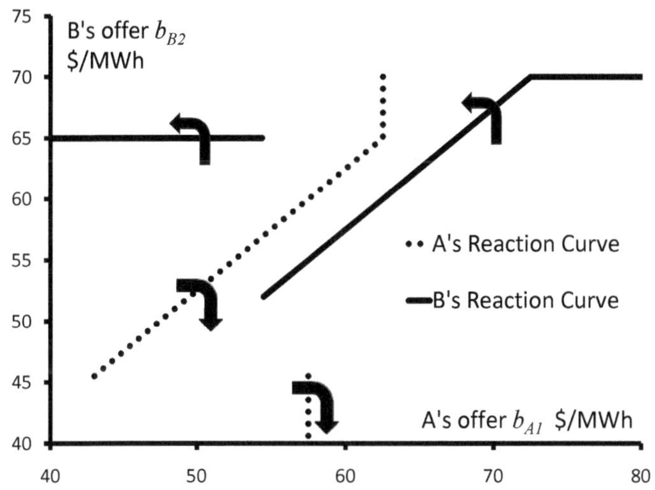

Fig. 7.5 Reaction curves: B's offer in reaction to A's offer, and vice versa

7.5.2.4 Mixed Strategy Equilibrium Computation

Indeed there is no pure strategy equilibrium for this EPEC. However, there is a mixed strategy equilibrium (by Nash's theorem [31]), which can be approximated by discretizing the strategy space and implementing the bimatrix MCP method of Section 4.2.4, as we did for the Cournot EPEC of Section 7.4.

In calculating the mixed strategy equilibrium, we consider eight possible price offers for each leader, $b_{fi} = \{50, 52.5, \ldots, 67.5\}$, yielding the two payoff matrices in Tables 7.5 and 7.6. Along the diagonals of these tables, where the two firms' bids match each other, there are multiple possible solutions that are equally optimal from the operator's point of view. Of all the possible solutions, we assume that the market operator chooses to accept an equal number of MW from each offerer.[17]

The resulting mixed strategy equilibrium is shown in Table 7.7. A notable feature is that the most likely strategy for B is to make a high offer (69% chance of b_{B2}= 62.5 $/MWh), while A's most likely strategy is to bid considerably lower (5% chance of b_{A1}= 55 $/MWh). On average, B's offer is about 5 $/MWh more than A's, which is to be expected since B's marginal cost ($35/MWh) exceeds A's ($20/MWh). Nevertheless, there is a 13% (=

[17] The other allocations between A and B that are equally optimal from the operator's point of view do yield different profits for the two producers that what is shown here. If alternative solutions are used for the diagonals of these tables, then the mixed strategy equilibrium would be slightly different than what we calculated here.

(0.2+0.0143)(0.1439+0.0991) + (0.2+0.0143+0.5714)(0.0991)) chance that B will actually make an offer that is actually less than A's.

Table 7.5 Electricity offer EPEC: A's payoff matrix

Firm A	Firm B's offer b_{B2}, \$/MWh							
b_{A1}, \$/MWh	50	52.5	55	57.5	60	62.5	65	67.5
50	15000	16500	16500	16500	16500	16500	16500	16500
52.5	13813	15438	17063	17063	17063	17063	17063	17063
55	14000	14000	15750	17500	17500	17500	17500	17500
57.5	14063	14063	14063	15938	17813	17813	17813	17813
60	14000	14000	14000	14000	16000	18000	18000	18000
62.5	13813	13813	13813	13813	13813	15938	18063	18063
65	13500	13500	13500	13500	13500	13500	15750	18000
67.5	13063	13063	13063	13063	13063	13063	13063	15438

Table 7.6 Electricity offer EPEC: B's payoff matrix

Firm A	Firm B's offer b_{B2}, \$/MWh							
b_{A1}, \$/MWh	50	52.5	55	57.5	60	62.5	65	67.5
50	7500	7438	8000	8438	8750	8938	9000	8938
52.5	8250	8313	8000	8438	8750	8938	9000	8938
55	8250	9188	9000	8438	8750	8938	9000	8938
57.5	8250	9188	10000	9563	8750	8938	9000	8938
60	8250	9188	10000	10688	10000	8938	9000	8938
62.5	8250	9188	10000	10688	11250	10313	9000	8938
65	8250	9188	10000	10688	11250	11688	10500	8938
67.5	8250	9188	10000	10688	11250	11688	12000	10563

Table 7.7 Electricity offer EPEC: Mixed strategy equilibrium

Offer b_{fi}	Mixed Strategy Probabilities	
\$/MWh	A	B
50	0	0
52.5	0.2143	0.0991
55	0.5714	0.1439
57.5	0.0143	0
60	0.2000	0.0634
62.5	0	0
65	0	0.6937
67.5	0	0

7.5.2.5 Comparison of Average MPEC Results, EPEC Mixed Equilibrium, and Competitive Equilibrium

Table 7.4 presents the comparative statics for our offer models. Unlike the Cournot EPEC model of Section 7.4, the average outcomes of the MPEC solutions from the diagonalization closely resemble the expected values from the EPEC mixed strategy equilibrium. Indeed the average prices across the nodes (unweighted by quantitites) as well as the total supply are precisely equal (58.8 \$/MWh and 825 MW, respectively). Interestingly, however, the values of total profits, consumer surplus, and social welfare differ between the two models. The mixed strategy equilibrium favors consumers and has higher market efficiency by a small but nonnegligible amount. The reason the equilibrium is more efficient may be because there is more price differentiation between the nodes, with higher prices at the higher cost location (B). Also, the mixed strategy equilibrium expects a significantly greater amount of production from the cheaper plant, which is A's. Just as in Section 7.4.2, use of average results over a series of solutions from diagonalization can paint a misleading picture of what the true mixed strategy equilibrium would look like.

However, when compared to the competitive solution (last row of Table 7.4), both offer EPECs tell the same basic story: exercise of market power by the duopoly significantly raises prices and hurts consumers. Interestingly, the competitive prices at the two nodes are equal (\$45/MWh) even though B's marginal cost is higher. The reason is that both producers exhaust their capacity, so that price needs to be set high enough to restrain demand to match the available production.

7.5.2.6 Pure Strategy Equilibrium

Whether or not there is a pure strategy equilibrium for the EPEC describing a game between generators whose strategic variable is the \$/MWh and MW offers they make into a market is highly parameter dependent. Others have reported obtaining such equilibria (e.g., [22]), and we get one here if the parameters are slightly adjusted.

In particular, if B's capacity K_{B2} is 327 MW or less, then it is no longer profitable to undercut A, because B cannot expand its output enough to make up for the profit it would lose by abandoning the "concede" strategy. No matter what offer A starts with below \$62.5/MWh, B's response is to concede and offer \$65/MWh, to which A responds with \$62.5/MWh, and the pure strategy equilibrium is reached. If A starts with an offer higher than \$65/MWh, then B instead undercuts, and A responds with a offer of \$62.5/MWh, at which point B finally concedes (b_{B2} = \$65/MWh), and the same pure strategy equilibrium is attained.

This equilibrium is of course unfavorable to consumers, since its prices correspond to the highest prices experienced in the original EPEC cycle (solution MPEC_A1, Table 7.4). One reason is that the generators are not fully using their capacity, even though their marginal costs are well below the marginal value of power to the consumers (i.e., prices). In particular, A produces 425 out of its possible 550 MW, while B generates 300 out of its capacity of 327 MW. In contrast, if these duopolists were instead price-takers, both producers would fully use their capacity, and prices would fall to $50/MWh at node 1 and $62.3/MWh at node 2. Consumer surplus then would rise by half (from $13,156 to $19,606), and market efficiency would also increase (from $40,344 to $45,649).

7.6 Closed Loop Multistage Nash Equilibrium: Capacity Expansion

7.6.1 Introduction

In most organizations, including power generation companies, different organizational subunits are responsible for capital budgeting and for operations. In capital budgeting, returns from investments, in terms of gross margins from spot and forward markets, are projected over time horizons that may be as long as several decades, and are then balanced against the immediate capital expense. Operators have a much shorter time horizon; in electricity markets, dispatch, unit commitment, and spot market offering decisions are revisited daily or more frequently, while maintenance planning might have time horizons measured in dozens of months at most. To appropriately evaluate investments, the capital budgeting function must consider how those investments will affect outcomes in the commodity market (the operations stage). Will aggressive construction of new plants result in over capacity, causing spot market prices and the resulting profits to crash? Or will demand growth or retirements of old plants take up the slack, and shore up gross margins? Thus, a company analyst who evaluates potential investments must solve an MPEC-type of problem, in which she or he anticipates the reaction of the players in the commodity markets to the present of additional capacity. This is a "closed loop" information structure.

Among multiple companies, this investment game can be viewed as a kind of prisoners' dilemma [42]. An individual company might be able to unilaterally improve its profit by expanding capacity, with higher volumes making up for lower prices. But if all companies do that, then everyone's profits could be lower than if all companies instead refrained from building. (Of course, the prisoners' dilemma metaphor is not completely apt, as there is another set of players that is better off if the companies all build. These are the consumers, who enjoy lower prices and more consumption; as a result, overall market efficiency as measured by total market surplus may improve.) The EPEC model of this section represents this two-stage investment-operation game and provides some surprising results. In particular, less intensive competition in the commodity market (quantity competition, à la Cournot) can result in more investment and more consumer benefits than if competition in the commodity market is intense (price competition, la Bertrand).

This EPEC model is highly simplified in several respects. Each company f specializes in one type of power plant which can be built in any (non-negative) size K_f MW in the first stage. Its overall objective is to maximize its profit in a typical year, accounting for both the annualized capital cost of the investment and the gross margin from operation in that year. The firm's plant operation in a typical year is simulated by considering the equilibrium in the commodity market (or spot market) in each hour; hours are grouped into demand periods (or blocks) b with similar demand levels, each have H_b hours/yr. We assume that the commodity equilibrium problems in different blocks are separable, with decisions in one not affected by decisions in others. Each firm chooses its production level x_{fb} MW in each block respecting its capacity limit, and the marginal cost of operation C_f \$/MWh is a constant. In deciding on x_{fb}, the firm either takes the market price p_b \$/MWh as exogenous (Bertrand competition) or takes the other firms' output decisions as fixed (Cournot competition), in which case $p_b = f_{d,b}^{-1}(\sum_f x_{fb})$, the inverse demand function in period b. More general formulations could consider, for instance, multiyear time horizons, with later capital investment decision stages; unit commitment and other constraints and costs that cause equilibria among different periods to interact; and conjectural variations between the highly competitive Bertrand and oligopolistic Cournot extremes [30, 34, 46].

In the rest of this section, we start with the second-stage commodity market (operations) problem. We then outline each company's MPEC representing its investment problem in the first stage, and follow with the resulting EPEC among multiple companies. We conclude with a proof by example that indeed Cournot competition can be more efficient than Bertrand competition.

7.6.2 Stage 2 Equilibrium: The Commodity Market

7.6.2.1 Perfect Competition

We follow the procedure of other chapters in building a complementarity model of the short run operation of the market: formulate player models, take first-order conditions, and formulate the overall equilibrium problem. We start with the perfect competition case. Starting with the generation firm f and using the notation just introduced, f maximizes its short run profit (gross margin) subject to its capacity and a naïve assumption that price is immutable:

$$\text{Maximize}_{x_{fb}} \quad \sum_b H_b \left(p_b x_{fb} - C_f x_{fb}\right) \tag{7.19a}$$

s. t.

$$x_{fb} - K_f \leq 0 \quad (\mu_{fb}), \quad \forall b \tag{7.19b}$$

$$x_{fb} \geq 0, \quad \forall b \tag{7.19c}$$

The assumption of price-taking behavior is reasonable not only for unconcentrated markets (i.e., markets with many small firms), but also in situations where regulation compels marginal-cost offers. The latter is the case in Ireland for all generators, and also in certain ISO markets in the US where generators with local market power due to transmission constraints can have their offers set to the ISO's estimate of short run marginal cost.[18]

We skip the derivation of the first-order conditions of the generator's model and proceed directly to the equilibrium model by gathering the first-order conditions for the above generator problem (7.19) for all f and combining them with the inverse demand function:

Find $\{x_{fb}, \mu_{fb}, p_b\}$ satisfying:

$$0 \leq x_{fb} \perp H_b(p_b - C_f) - \mu_{fb} \leq 0, \quad \forall f, b \tag{7.20a}$$

$$0 \leq \mu_{fb} \perp x_{fb} - K_f \leq 0, \quad \forall f \tag{7.20b}$$

$$p_b = f_{d,b}^{-1}\left(\sum_f x_{fb}\right) \quad (p_b \text{ unrestricted}), \quad \forall b \tag{7.20c}$$

The first condition is analogous to the familiar marginal revenue (MR) equals marginal cost (MC) condition of elementary microeconomics (see Chapter 3), with two wrinkles: if output is zero, marginal revenue can be less than marginal cost; and marginal cost includes the shadow price of capacity. The second condition is complementary slackness for the capacity constraint, and the third condition represents the demand response of consumers. Price is unrestricted; indeed, real power markets frequently show negative prices due to network constraints, ramp limitations, and the desire of generators to keep their plants on-line through the night so as to avoid costly startups the next day. This equilibrium problem can be solved directly using, for example, the PATH method [16], discussed in Chapter 8. It happens that problem (7.20) is also the set of first-order conditions for the following quadratic program, which can also be solved to yield the short run equilibrium:

$$\text{Maximize}_{x_{fb}} \quad \sum_b H_b \int_0^{\sum_f x_{fb}} f_{d,b}^{-1}(q)dq - \sum_b H_b \sum_f C_f x_{fb} \tag{7.21a}$$

s. t.

[18] Plus 10%, in the case of California.

$$x_{fb} - K_f \leq 0 \quad (\mu_{fb}), \quad \forall f, b \tag{7.21b}$$

$$x_{fb} \geq 0, \quad \forall f, b \tag{7.21c}$$

This is a specific instance of the general Principle of Symmetry, which describes when an optimization problem can be defined whose KKT conditions also define the market equilibrium (Chapters 3 and 4).

7.6.2.2 Cournot Competition

For the Cournot case, generator f's problem instead recognizes that a change in output will change price via the inverse demand function:

$$\text{Maximize}_{x_{fb}} \quad \sum_b H_b(f_{d,b}^{-1}(\sum_{f'} x_{f'b})\, x_{fb} - C_f x_{fb}) \tag{7.22a}$$

s. t.

$$x_{fb} - K_f \leq 0 \quad (\mu_{fb}), \quad \forall b \tag{7.22b}$$

$$x_{fb} \geq 0, \quad \forall b \tag{7.22c}$$

where f and f' are distinct indices, the former a specific number referring to the particular firm that is optimizing, and the latter a general index used to represent any firm. Such a Cournot game is more realistic where markets are concentrated and regulation does not force offers to equal marginal cost. The resulting equilibrium problem among the generators' problems (7.22) becomes:

$$\text{Find } \{x_{fb}, \mu_{fb}, p_b\} \text{ satisfying:} \tag{7.23a}$$

$$0 \leq x_{fb} \perp H_b((p_b + x_{fb}\, dp_b/dx_{fb}) - C_f) - \mu_{fb} \leq 0, \quad \forall f, b \tag{7.23b}$$

$$0 \leq \mu_{fb} \perp x_{fb} - K_f \leq 0, \quad \forall f, b \tag{7.23c}$$

$$p_b = f_{d,b}^{-1}(\sum_f x_{fb}) \quad (p_b \text{ unrestricted}), \quad \forall b \tag{7.23d}$$

where $(p_b + x_{fb}\, dp_b/dx_{fb})$ is marginal revenue, the derivative of revenue $f_{d,b}^{-1}(\sum_{f'} x_{f'b})\, x_{fb}$ with respect to f's output, given everyone else's output. Like the Bertrand/perfect competition case, the first condition is the generalization of our old friend, the marginal revenue equals marginal cost condition, but this time accounting for how expanded output will depress prices (the derivative term).

Also like the Bertrand case, this complementarity problem can be solved directly, or, if demand curves are linear, we can instead formulate and solve a quadratic program whose KKT conditions are equivalent to the above CP. Let the linear inverse demand function be: $p_b = f_{d,b}^{-1}(\sum_f x_{fb}) = P_b^o - (P_b^o/Q_b^o)\sum_f x_{fb}$, where P_b^o and Q_b^o are the price and quantity intercepts, respectively, of the function. This quadratic program is the following optimization problem [19] (see also [4, 37], as well as Section 3.4.2.6 of this book):

$$\text{Maximize}_{x_{fb}} \quad \sum_b H_b\{P_b^o \sum_f x_{fb} - (P_b^o/Q_b^o)(\sum_f x_{fb})^2/2\}$$
$$- \sum_b H_b\{\sum_f (P_b^o/Q_b^o)x_{fb}^2/2\} - \sum_b H_b \sum_f C_f x_{fb}$$

s. t.

$$x_{fb} - K_f \leq 0 \quad (\mu_{fb}), \quad \forall f, b$$
$$x_{fb} \geq 0, \quad \forall f, b$$

The first bracketed term in the objective is the integral of the inverse demand functions (analogous to the integral term in the social welfare maximization objective (7.21)), which represents the benefits to consumers of consumption. The second bracketed term is a trick that results in the derivative of the objective with respect to x_{fb} becoming MR-MC rather than $p_b - MC$, where the marginal revenue MR includes the downward effect that an increment in x_{fb} has upon price and, thus, the revenues received for inframarginal sales. The KKT conditions for this optimization are equivalent to the CP (7.23) formulation of the equilibrium problem in the linear demand case. The advantage of using the optimization formulation is the ability to use widely available nonlinear optimization software if a CP solver is unavailable.

7.6.3 Stage 1 EPEC Problem

In the first stage, each firm f chooses its capacity to maximize annualized profit, subject to the second-stage equilibrium, which yields an MPEC [15].[19] Then the EPEC is obtained by attempting to solve the MPECs for all f simultaneously, so that their second-stage equilibria are identical. We start with the MPEC formulations, which depend on whether the second stage is a Bertrand or Cournot model. In the Bertrand case, firm f's investment problem is the following MPEC:

$$\text{Maximize}_{\{K_f; x_{f'b}, \mu_{f'b}, \forall f', b\}} \quad -D_f K_f + \sum_b H_b(p_b x_{fb} - C_f x_{fb}) \tag{7.24a}$$

s. t.

$$K_f \geq 0 \tag{7.24b}$$

[19] An alternative formulation has the strategic generator making investment and production decisions simultaneously, subject to the reaction of the market [26]. For a formulation of a MPEC in which the generator investor instead benevolently maximizes net market efficiency, while also making transmission and fuel investments as well, see [25].

$$0 \le x_{f'b} \perp H_b(p_b - C_{f'}) - \mu_{f'b} \le 0, \quad \forall f', b \tag{7.24c}$$

$$0 \le \mu_{f'b} \perp x_{f'b} - K_{f'} \le 0, \quad \forall f' \tag{7.24d}$$

$$p_b = f_{d,b}^{-1}(\sum_{f'} x_{f'b}) \quad (p_b \text{ unrestricted}), \forall b \tag{7.24e}$$

D_f is the annualized capital cost of f's generator in \$/MW/yr. This MPEC has a non-convex feasible region because of the complementarity conditions for the production and capacity shadow price variables. In this simple case, it can be solved readily by, for instance, a line search over K_f, solving the complementarity problem (7.24c)-(7.24e) for each value.

For the case of Cournot competition, we obtain the firm's MPEC by substituting the Cournot second-stage equilibrium conditions for the Bertrand ones in (7.24):

$$\text{Maximize}_{\{K_f; x_{f'b}, \mu_{f'b}, \forall f', b\}} \quad -D_f K_f + \sum_b H_b(p_b x_{fb} - C_f x_{fb}) \tag{7.25a}$$

s. t.

$$K_f \ge 0 \tag{7.25b}$$

$$0 \le x_{f'b} \perp (p_b + x_{f'b} dp_b/dx_{f'b}) - C_{f'} - \mu_{f'b} \le 0, \quad \forall f', b \tag{7.25c}$$

$$0 \le \mu_{f'b} \perp x_{f'b} - K_{f'} \le 0, \quad \forall f'b \tag{7.25d}$$

$$p_b = f_{d,b}^{-1}(\sum_{f'} x_{f'b}) \quad (p_b \text{ unrestricted}), \quad \forall b \tag{7.25e}$$

Of course, the feasible region for this MPEC is also non-convex, but again, a line search strategy can be used to solve it. Note that for simplicity of presentation, we've put p_b in the revenue part of the objective of (7.25) rather than the inverse demand curve $f_{d,b}^{-1}(\sum_{f'} x_{f'b})$ used in Cournot second-stage optimization (7.22). This slight simplification is allowed because of the last constraint of the MPEC.

The equilibrium problem for the first stage can now be defined. The EPEC for the competitive/Bertrand case solves MPEC (7.24) simultaneously for all f, constraining the second-stage solution to be the same for all f. That is, our Bertrand EPEC is:

$$\text{Find } \{K_f, \forall f; x_{f'b}, \mu_{f'b}, \forall f', b; p_b, \forall b\} \text{ that solve (7.24)}, \ \forall f$$

For the Cournot case, the EPEC is instead:

$$\text{Find } \{K_f, \forall f; x_{f'b}, \mu_{f'b}, \forall f', b; p_b, \forall b\} \text{ that solve (7.25)}, \ \forall f$$

Finding a solution to these EPECs (if one exists) can be pursued by any of the solution strategies outlined earlier. In the example in the next section, we used diagonalization.

7.6.4 Illustrative Example: Consumers prefer Cournot to Bertrand Competition

Consider two firms $f = 1{,}2$. The first specializes in baseload plants (with capital cost $D_1 = 120{,}000$ \$/MW/yr, and operating cost $C_1 = 50$ \$/MWh) and the other builds only peaking generators ($D_2 = 80{,}000$, $C_2 = 90$ \$/MWh/yr). There are twenty equal length periods per year ($H_b = 438$ hours/yr, b=1,..,20). The inverse demand curve in each is:

$$p_b = f_{d,b}^{-1}(\sum_f x_{fb}) = P_b^o - (P_b^o/Q_b^o)(\sum_f x_{fb}) \tag{7.26}$$

with the price and quantity intercepts being, respectively,P_b^o= 300 \$/MWh and Q_b^o= 2050−50b, b = 1,..,20. Thus, the quantity intercept is 2000 MW for the peak period (b=1) and 1050 MW for the lowest load period (b=20).

First, we consider Bertrand competition. Given $\{K_1,K_2\}$, the second-stage solution to (7.20) is the intercept of the inverse demand curve (7.26) with a supply curve based on the short run marginal cost of production (Figure 7.6). The supply curve shows that price is \$50/MWh for load less than K_1 (= 800 MW in the figure), and \$90/MWh for load strictly within (K_1, K_1+K_2) (= (800 MW, 1300 MW) in the figure). Price can be between \$50 and \$90 if quantity demanded equals exactly K_1, and for demand equal to K_1+K_2, the supply curve is vertical and price will be at least \$90/MWh. In the last situation, this is the case of "shortage" or "scarcity" pricing in which price exceeds the marginal cost of the last unit supplied because of capacity limits. It is during such times that the peaking generator earns a positive gross margin, or scarcity rent, contributing to paying off its capital cost. This is the situation shown in the figure, where the price is just over \$100/MWh, in excess of the marginal operating cost of the costliest generator.

These scarcity rents are key to incenting investment in markets. In a perfectly competitive market in which generators make not only operating but also investment decisions under the assumption that they have no effect on price, a generator will expand capacity until the marginal MW of capacity earns just enough scarcity rent to cover its capital cost [9]. This is equivalent to the so-called "open loop" Bertrand equilibrium, which could be modeled by a complementarity problem constructed by assuming that each firm f chooses both its capacity and production decisions at the same time. This can be shown to maximize market efficiency, measured by the total surplus (consumer surplus plus producer profit) [9].

For Bertrand competition, the second-stage model (7.20), with data inserted, is as follows:

Find $\{x_{fb}, \mu_{fb}, f = 1, 2,\ b = 1, ..., 20; p_b, b = 1, ..., 20\}$ satisfying:

$0 \le x_{1,1} \perp p_1 - 50 - \mu_{1,1} \le 0$

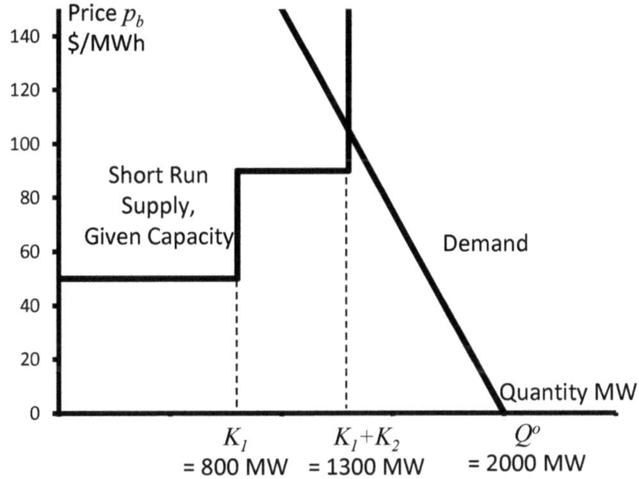

Fig. 7.6 Supply-demand equilibrium in second-stage Bertrand solution of capacity-operation EPEC (assuming plant capacities of 800 and 500 MW, and demand curve for peak period $b = 1$)

$$
\begin{aligned}
&\ldots\\
&0 \leq x_{1,20} \perp p_{20} - 50 - \mu_{1,20} \leq 0\\
&0 \leq x_{2,1} \perp p_1 - 90 - \mu_{2,1} \leq 0\\
&\ldots\\
&0 \leq x_{2,20} \perp p_{20} - 90 - \mu_{2,20} \leq 0\\
&0 \leq \mu_{1,1} \perp x_{1,1} - K_1 \leq 0\\
&\ldots\\
&0 \leq \mu_{1,20} \perp x_{1,20} - K_1 \leq 0\\
&0 \leq \mu_{2,1} \perp x_{2,1} - K_2 \leq 0\\
&\ldots\\
&0 \leq \mu_{2,20} \perp x_{2,20} - K_2 \leq 0\\
&p_1 = 300 - (300/2000)(x_{1,1} + x_{2,1})\\
&\ldots\\
&p_{20} = 300 - (300/1050)(x_{1,20} + x_{2,20})
\end{aligned}
\tag{7.27}
$$

The EPEC for solving the first-stage problem in the Bertrand case is then:

Find $\{K_f, f = 1, 2; x_{fb}, \mu_{fb}, f = 1, 2 \text{ and } b = 1, ..., 20; p_b, b = 1, ..., 20\}$ satisfying:

$$\text{Maximize}_{\{K_1; x_{fb}, \mu_{fb}, f=1,2, b=1,\ldots,20\}} \quad -120000K_1 + 438(p_1 - 50)x_{1,1} + \cdots + 438(p_{20} - 50)x_{1,20}$$

s. t.

$K_1 \geq 0$ and (7.27)

$$\text{Maximize}_{\{K_2; x_{fb}, \mu_{fb}, f=1,2, b=1,\ldots,20\}} \quad -80000K_2 + 438(p_1 - 90)x_{2,1} + \cdots + 438(p_{20} - 90)x_{2,20}$$

s. t.

$K_2 \geq 0$ and (7.27)

We solve this Bertrand EPEC by diagonalization. Table 7.8 shows the iteration results. First, we arbitrarily set firm 2's initial capacity to 350 MW of peaking capacity, and allow firm 1 to adjust its capacity in the first iteration. We solve firm 1's MPEC in that case by a line search over K_1, solving the complementarity problem for the short run Bertrand equilibrium (7.27) for each value of K_1 tested. (We solved that complementarity problem by solving the quadratic program for which that problem is the KKT conditions.) This search yields 436 MW of baseload plant. In the second iteration, firm 2 responds by adjusting its capacity to 274 MW. The equilibrium is shown in the last row; other starting points yield the same equilibrium.

Table 7.8 Diagonalization iterations among leaders' MPECs for Bertrand problem

Iteration	K_1 MW	K_2 MW	Profit f=1	Profit f=2
Starting Capacity		350		
1st solution (1 moves)	436	350	293001002	126567226
2nd (2 moves)	436	274	348670327	134068909
3rd (1 moves)	441	274	348723762	131617771
4th (2 moves)	441	271	351090809	131631281
...	...	...	...	...
Equilibrium	443.5	270.0	351890887	130420950

The quantities and prices resulting in each period in the second-stage equilibrium solution are shown in Table 7.9. What is particularly noticeable is that capacity is binding in every period, and prices are always at scarcity levels, above the marginal cost of operating any of the facilities. This is a result of the particular cost and demand parameters; other parameters can result in slack capacity in some periods. However, the general outcome of this EPEC is that strategic investors in stage 1, realizing that prices will be low in stage 2 unless capacity is tight, will have only a weak incentive to invest. The resulting equilibrium capacity from the first-stage EPEC, K_1+K_2, is 713.5

MW. This is appreciably lower, as we shall see, than the capacity that results if instead Cournot competition occurs in stage two—which allows commodity prices to be higher and increases the incentive to invest. In terms of the prisoners' dilemma outlined earlier in this section, the higher prices in stage two—and their decreased sensitivity to the amount of capacity relative to the Bertrand case—increases the incentive to "cheat" by expanding capacity, as high margins are earned on the additional capacity without depressing prices quite as much as in the Bertand case.

Table 7.9 Commodity (second-stage) prices from Bertrand capacity-operations EPEC solution

Period b:	1	2	3	4	5	6	7
x_{1b} MW	443.5	443.5	443.5	443.5	443.5	443.5	443.5
x_{2b} MW	270	270	270	270	270	270	270
p_b \$/MWh	193.0	190.2	187.3	184.3	181.1	177.7	174.1

Period b:	8	9	10	11	12	13	14
x_{1b} MW	443.5	443.5	443.5	443.5	443.5	443.5	443.5
x_{2b} MW	270	270	270	270	270	270	270
p_b \$/MWh	170.3	166.2	161.9	157.3	152.4	147.1	141.4

Period b:	15	16	17	18	19	20
x_{1b} MW	443.5	443.5	443.5	443.5	443.5	443.5
x_{2b} MW	270	270	270	270	270	270
p_b \$/MWh	135.3	128.8	121.6	113.9	105.4	96.1

Let's now formulate and solve the Cournot EPEC to show that indeed more capacity results, at least under our particular parameterization. First, we insert the data into the second-stage Cournot model (7.23), giving us the following commodity market equilibrium problem, parameterized in the capacity variables K_1 and K_2:

$$
\begin{aligned}
&\text{Find } \{x_{fb}, \mu_{fb}, f = 1,2,\ b = 1, ..., 20; p_b, b = 1, ..., 20\} \text{ satisfying:}\\
&\quad 0 \le x_{1,1} \perp (p_1 + x_{1,1}(-300/2000)) - 50 - \mu_{1,1} \le 0\\
&\quad \ldots\\
&\quad 0 \le x_{1,20} \perp (p_{20} + x_{1,20}(-300/1050)) - 50 - \mu_{1,20} \le 0\\
&\quad 0 \le x_{2,1} \perp (p_1 + x_{2,1}(-300/2000)) - 90 - \mu_{2,1} \le 0\\
&\quad \ldots\\
&\quad 0 \le x_{2,20} \perp (p_{20} + x_{2,20}(-300/1050) - 90 - \mu_{2,20} \le 0\\
&\quad 0 \le \mu_{1,1} \perp x_{1,1} - K_1 \le 0 \qquad (7.28)\\
&\quad \ldots\\
&\quad 0 \le \mu_{1,20} \perp x_{1,20} - K_1 \le 0
\end{aligned}
$$

$$0 \le \mu_{2,1} \perp x_{2,1} - K_2 \le 0$$

$$\ldots$$

$$0 \le \mu_{2,20} \perp x_{2,20} - K_2 \le 0$$

$$p_1 = 300 - (300/2000)(x_{1,1} + x_{2,1})$$

$$\ldots$$

$$p_{20} = 300 - (300/1050)(x_{1,20} + x_{2,20})$$

Note the appearance of the negative demand derivative term $x_{fb}dp_b/dx_{fb}$ in the first set of conditions (the generalization of the marginal revenue equals marginal cost condition). Its negative sign indicates that the strategic Cournot player understands that expanding production depresses price.

The EPEC corresponding to the first-stage Cournot problem is as follows:

Find $\{K_f, f = 1,2;\ x_{fb}, \mu_{fb}, f = 1,2 \text{ and } b = 1, ..., 20; p_b, b = 1, ..., 20\}$ satisfying:

$$\text{Maximize}_{\{K_1; x_{fb}, \mu_{fb}, f=1,2, b=1,...,20\}} \quad -120000K_1 + 438(p_1 - 50)x_{1,1} + \cdots + 438(p_{20} - 50)x_{1,20}$$

s. t.

$K_1 \ge 0$ and (7.28)

$$\text{Maximize}_{\{K_2; x_{fb}, \mu_{fb}, f=1,2, b=1,...,20\}} \quad -80000K_2 + 438(p_1 - 90)x_{2,1} + \cdots + 438(p_{20} - 90)x_{2,20}$$

s. t.

$K_2 \ge 0$ and (7.28)

The only difference between this EPEC and the Bertrand EPEC is the substitution of the second-stage Cournot equilibrium (7.28) for the second-stage Bertrand equilibrium (7.27) in the constraint set in each of the two firms' MPECs.

As in the case of the Bertrand EPEC, we solve the Cournot EPEC by diagonalization. Table 7.10 shows the iterations. This time, we arbitrarily start with firm 2's capacity equaling 200 MW, and firm 1 moves in the first iteration. A line search solves firm 1's MPEC in that case, solving the stage one Cournot complementarity problem (7.28) for over the range of K_1's tested. (We solved each complementarity problem using the equivalent quadratic program whose KKT conditions are the equilibrium conditions for stage 2.) Firm 1's best response is $K_1 = 555$ MW. Then firm 2 responds by adjusting its capacity K_2 from 200 to 311 MW. Continuing in this manner yields the equilibrium in the last row.

Table 7.10 Diagonalization iterations for Cournot problem

Iteration	K_1	K_2	Profit f=1	Profit f=2
Initial Capacity		200		
1st Solution (1 moves)	555	200	420750528	109152659
2nd (2 moves)	555	311	358602951	123383313
3rd (1 moves)	515	311	360253192	129969260
4th (2 moves)	515	309.6	360670136	129970466
5th (1 moves)	515.5	309.6	360671250	129868532
6th (2 moves)	515.5	309.3	360760684	129868579
Equilibrium	515.4	309.3	360760469	129888950

The second-stage equilibrium solution is shown in Table 7.11. Compared to the Bertrand quantities, we see that in the Cournot case, there is more capacity of both types (a total of 824.7 MW, 16% more than the Bertrand case) and capacity constrains far less often (only during periods 1-8). With more capacity and, thus, output during those constrained periods, Cournot prices must therefore be lower than Bertrand prices during those periods. This at first blush is very surprising, because a well-known result of simple oligopoly models is that Cournot oligopolists restrict output and raise prices compared to Bertrand oligopolists. This does not happen here because of the two stage (EPEC) nature of this game–because the second-stage (short-run) Cournot prices are higher than Bertrand prices when capacity is slack, there is more of an incentive for Cournot producers to expand capacity in the first stage. The greater capacity intensifies competition in stage 2, yielding lower prices during most periods (1-13). However, Cournot prices are higher than Bertrand prices in the lowest demand periods (14-20).

Table 7.11 Commodity (second-stage) prices from Cournot capacity-operations EPEC solution

Period b:	1	2	3	4	5	6	7
x_{1b} MW	515.4	515.4	515.4	515.4	515.4	515.4	515.4
x_{2b} MW	309.3	309.3	309.3	309.3	309.3	309.3	309.3
p_b $/MWh	176.3	173.1	169.8	166.3	162.6	158.6	154.5

Period b:	8	9	10	11	12	13	14
x_{1b} MW	515.4	515.4	499.4	483.3	467.2	451.1	435.0
x_{2b} MW	309.3	302.3	292.8	283.3	273.9	264.4	255.0
p_b $/MWh	150.1	146.7	146.7	146.7	146.7	146.7	146.7

Period b:	15	16	17	18	19	20
x_{1b} MW	418.9	402.8	386.7	370.6	354.4	338.3
x_{2b} MW	245.6	236.1	226.7	217.2	207.8	198.3
p_b $/MWh	146.7	146.7	146.7	146.7	146.7	146.7

On balance, the average prices of the two equilibria (revenue/MWh supplied) are almost exactly equal (Cournot 154.4 \$/MWh, Bertrand price 154.3). (This is not a necessary outcome; alternative parameter assumptions could cause them to diverge.) Consumption is higher in the Cournot case, as prices are lower during times of highest demand (average load of 736.2 MW, vs. 713.5 MW under Bertrand competition). Consumers are better off (as measured as consumer surplus) by \$15M/yr under Cournot competition compared to Bertrand competition (\$470M/yr - \$455M/yr).

We illustrate the consumer surplus calculations as follows. As explained in Chapter 3, consumer surplus is the integral of the demand curve—a measure of the total economic value of consumption—minus what the consumer pays for the commodity. For instance, in period $b = 1$ under Cournot competition, the quantity consumed is 824.7 MW, for which \$176.3/MWh is paid. The area under the demand curve is the trapezoid whose area is (438 hours/yr)(824.7 MW)(300+176.3\$/MWh)/2, or \$86M/yr. (Note that \$300/MWh is the price intercept or "choke price" for the demand curve.). Subtracting expenditures of (824.7 MW)(\$176.3/MWh) = \$64M/yr results in a consumer surplus of \$22M for that period. Summing that quantity over all periods yields the total Cournot consumer surplus of \$470M/yr.

Overall market efficiency—as gauged by consumer surplus plus profit ("social surplus" or "social welfare", as economists often term it)—is also higher under the two-stage Cournot equilibrium than under the two-stage Bertrand equilibrium. In the former case, the social surplus is \$960M/yr, while in the latter it is \$938M/yr, a difference of \$22M. Intriguingly, both producers and consumers are better off. Profit is \$8M/yr higher (although the peak producer is actually slightly worse off, as a comparison of equilibrium solutions in the last rows of Tables 7.8 and 7.10 show), and as just mentioned, consumers too experience a higher surplus. This is the result of complex interactions of capacity limits and relative price levels and consumption in different periods.

The above results are not general, and depend on the particular numerical assumptions. Under other parameter values or model structures (e.g., nonlinear demand), Bertrand two-stage competition results in higher market efficiency than Cournot. However, this example has served to show how a multiple stage structure in markets can lead to surprising and counter-intuitive results, in this case Cournot being better for consumers than Bertrand. The example has also illustrated the formulation of an EPEC that represents a closed loop game in which the same market parties are involved in all decision stages. In this case, the parties each make investment decisions in the first stage that take into account how all parties' production decisions in the second stage will be affected.

7.7 Summary

EPEC models represent the interaction of two or more Stackelberg leaders, each of whom solves an MPEC (Chapter 6) whose lower level represents the reactions of followers to the leaders' decisions. A solution to an EPEC is a Nash equilibrium among the leaders such that, given the strategies of other leaders, each leader is content with its choice of strategy. The leaders' MPECs themselves are generally non-convex optimization problems that are difficult to solve. So it should not surprise the reader that EPECs would even be more challenging to solve and, indeed, may have no solution in pure strategies or instead might have multiple solutions.

In this chapter, we have summarized the easiest EPEC solution algorithm—diagonalization. Then we have introduced a series of simple examples of applications of EPECs in energy markets. These models represent a succession of decisions in time or along supply chains, and their results illustrate the kinds of insights that can be obtained when comparing outcomes of models that are based on different market structures or games. We think that the EPEC's hierarchical relationship between leaders and followers, in many circumstances, more realistically represents the economic power and information relationships among players in a market compared to simpler complementarity formulations that assume that all players take decisions simultaneously and with the same information.

But finding pure strategy solutions to energy EPECs can be challenging and, as we show in our examples, may even be impossible when only mixed strategy equilibria exist. However, the basic diagonalization algorithm yields economic insights about the interaction of the leaders, while often being an effective way of identifying a pure strategy equilibrium, if any. For simple two-player problems, mixed strategy equilibria can be calculated using Section 4.2.4's bimatrix procedure with the results of the constituent MPECs of the EPECs as inputs.

There is active research both in using EPECs to represent interactions of players in energy markets and supply chains, and in analyzing EPEC solution properties and devising algorithms to solve them (see Section 9.4). These research topics will likely fascinate researchers for decades to come because of EPECs' practical usefulness and their fascinating mathematics for which general analytical and algorithmic approaches are unlikely to exist.

7.8 Exercises

1. Consider the EPEC problem in Section 7.2.2, but a situation in which the producers A and B have \$10/unit marginal cost as before, but the equilibrium problem for the two consumers are as follows:

$$y_A + y_B - (x_1 + x_2) = 0 \quad (\lambda \text{ unrestricted})$$

$$0 \leq x_1 \perp 100 - 3x_1 - \lambda \leq 0$$

$$0 \leq x_2 \perp 21.25 - (3/8)x_2 - \lambda \leq 0$$

Apply the diagonalization procedure of Section 7.2 in an attempt to solve the resulting EPEC, with a starting point of $y_{A0} = y_{B0} = 5$ $(= y_{A1} = y_{B1})$. Note whether the algorithm converges and, if so, to what equilibrium values of $\{y_A,\ y_B, x_1,\ x_2,\ \lambda\}$. Repeat the application, but instead use a starting point of $y_{A0} = y_{B=} = 16$ $(= y_{A1} = y_{B1})$. If they both converge to an equilibrium, do they converge to the same equilibrium? If not, which equilibrium do you think would be more likely to actually emerge from the market, and why?

2. Graphically analyze Exercise 1 by plotting reaction curves for the two producers in the form of Figure 7.3. That is, start by putting y_A on the X axis and y_B on the Y axis. Then plot A's optimal reaction y_A to a given value of y_B (i.e., a function $y_A(y_B)$), and on the same plot show B's optimal reaction to A's decision (i.e., a function $y_B(y_A)$). Where the reaction functions intersect (if they do), there will be a pure strategy equilibrium. Please locate any and all equilibria, and if there are more than one, please discuss why this happens. Do the intersections (if any) correspond with the equilibria found in Exercise 1?
3. Repeat Exercise 1, but this time assume the following equilibrium problem for demand:

$$y_A + y_B - (x_1 + x_2) = 0 \quad (\lambda \text{ unrestricted})$$

$$0 \leq x_1 \perp 100 - 3x_1 - \lambda \leq 0$$

$$0 \leq x_2 \perp 34.375 - (3/8)x_2 - \lambda \leq 0$$

Using a starting point of $y_{A0} = y_{B0} = 5$ $(= y_{A1} = y_{B1})$, use the global optimum to each MPEC in Step 3c as the solution in that step. Then repeat the procedure, but this time using a local algorithm (such as a gradient algorithm) that would terminate at a local (but not necessarily global) optimum where the second order conditions are satisfied. Since solving the MPEC is a one dimensional search, you can graph the MPEC's objective function as a function of the leader's single strategic variable y_A or y_B (depending on whether A or B's MPEC is being solved) from which you can identify the local optimum that would result from applying a gradient method to solve the MPEC.

Verify that the second application of diagonalization using the local algorithm converges to the following local equilibrium$\{y_A = 10,\ y_B = 10\}$, and confirm this is a local equilibrium (in the sense of Hu and Ralph [24]) by verifying the second order conditions for each leader at that point. Verify also that this local equilibrium is not a true EPEC solution because at least one leader would choose another value for its strategic variable y_j if it instead used an algorithm that guaranteed a global optimum for its MPEC. What is that value, and how does its profit for that leader compare to that under $\{y_A = 10, y_B = 10\}$?

4. Consider the numerical examples in Sections 7.4.2 (Cournot EPEC) and 7.5.2 (Offer EPEC). Assume transmission capacity of 50 MW. Imagine that at node 2 there is 40 MW of distributed generation capacity. Of that 40 MW, 25 MW has a marginal cost of \$60/MWh and 15 MW has a marginal cost of \$80/MWh. The owners of this distributed capacity are price-takers; that is, they are not strategic and exercise no market power. Imagine that leaders A and B both recognize the existence of this distributed capacity, and correctly anticipate its reaction to price. As a result, leaders A and B include the reactions of this competitive fringe into the equilibrium subproblem of each of their MPECs. Show the revised formulations of both Cournot MPECs (corresponding to Section 7.4.2) and both Offer MPECs (corresponding to Section 7.5.2) that account for the reactions of the fringe.
5. For the Cournot MPECs defined in problem 4, execute the diagonalization algorithm, using the same starting point as the examples. Repeat for the Offer MPECs. Do enough iterations to determine whether diagonalization is likely to converge to a pure strategy equilibrium or not.
6. Formulate a capacity expansion EPEC using the data of Section 7.6, except consider two periods whose loads correspond to those assumed in $b - 1$ and 10 in the example, and which are 4380 hours in duration apiece. Use the diagonalization algorithm to determine equilibrium solutions for both the Cournot and Bertrand second-stage spot market cases.

References

1. B. Allaz and J.-L. Vila. Cournot competition, forward markets and efficiency. *Journal of Economic Theory*, 59:1-16, 1993.
2. C. Babayigit, P. Rocha, and T.K. Das. A two-tier matrix game approach for obtaining joint bidding strategies in FTR and energy markets. *IEEE Transactions on Power Systems*, 25(3):1211-1219, 2010.
3. R. Baldick, U. Helman, B.F. Hobbs, and R.P. O'Neill. Design of efficient generation markets. *Proceedings of the IEEE*, 93(11):1998-2012, 2005.
4. J. Barqun, E. Centeno, and J. Reneses. Medium-term generation programming in competitive environments: a new optimisation approach for market equilibrium computing, generation, transmission and distribution. *IEE Proceedings*, 151(1):119-126, 2004.
5. C. Berger, D. Fuller, A. Haurie, R. Loulou, D. Luthra, and J.-P. Waaub. Modelling energy use in the mineral processing industries of Ontario with MARKAL-Ontario. *Energy* 15(9):741-758, 1990.
6. C.A. Berry, B.F. Hobbs, W.A. Meroney, R.P. O'Neill, and W.R. Stewart, Jr. Analyzing strategic bidding behavior in transmission networks. *Utilities Policy*, 8(3):139-158, 1999.
7. M. Boots, F.A.M. Rijkers, and B.F. Hobbs. A two-level oligopoly analysis of the European gas market. *The Energy Journal*, 25(3):73-102, 2004.
8. S. Borenstein, J. Bushnell, and S. Stoft. The competitive effects of transmission capacity in a deregulated electricity industry. *Rand Journal of Economics*, 31(2):294-325, 2000.
9. M. Caramanis. Investment decisions and long-term planning under electricity spot pricing, *IEEE Transactions on Power Apparatus and Systems*, PAS-101(12):4640-4648, 1982
10. J. Cardell, C. C. Hitt, and W. W. Hogan. Market power and strategic interaction in electricity networks. *Resource and Energy Economics,* 19(1–2):109-137, 1997.
11. L. Cunningham, R. Baldick, and M. Baughman. An empirical study of applied game theory: transmission constrained Cournot behavior. *IEEE Transactions on Power Systems*, 17(1):166-172, 2002.
12. O. Daxhalet and Y. Smeers. The EU regulation on cross-border trade of electricity: A two-stage equilibrium model. *European Journal of Operational Research*, 181:1396-141, 2007.
13. C. Day and D. Bunn. Divestiture of generation assets in the electricity pool of England and Wales: A computational approach to analyzing market power. *Journal of Regulatory Economics*, 19(2):123-141, 2001.
14. V. DeMiguel, and H. Xu. A stochastic multiple-leader Stackelberg model: analysis, computation, and application. *Operations Research*, 57(5):1220-1235, 2009.

15. A. Ehrenmann. Manifolds of multi-leader Cournot equilibria. *Operations Research Letters*, 32:121-125, 2004.
16. M. Ferris and T. Munson. Complementarity problems in GAMS and the PATH solver. *Journal Economic Dynamics and Control*, 24:165-188, 2000.
17. D. Fudenberg and D.K. Levine. *Theory of Learning in Games*. MIT Press, Cambridge, MA, 1998.
18. D. Fudenberg and J. Tirole, *Game Theory*. MIT Press, 1991.
19. H. Hashimoto. A spatial Nash equilibrium model. In P. Harker, Editor. *Spatial Price Equilibrium: Advances in Theory, Computation, and Application*. Springer, Berlin, 1982.
20. B.F. Hobbs. Linear complementarity models of Nash-Cournot competition in bilateral and POOLCO power markets. *IEEE Transactions on Power Systems*, 16(2):194-202, 2001.
21. B.F. Hobbs and U. Helman. Complementarity-based equilibrium modeling for electric power markets. In. D. Bunn, Editor. *Modeling Prices in Competitive Electricity Markets*, Wiley, London, 2004.
22. B.F. Hobbs, C.B. Metzler, and J.-S. Pang. Strategic gaming analysis for electric power networks: an MPEC approach. *IEEE Transactions on Power Systems*, 15(2):638-645, May 2000.
23. B.F. Hobbs and S.K. Nelson. A nonlinear bilevel model for analysis of electric utility demand-side planning issues. *Annals of Operational Research*, 34:255-274, 1992.
24. X.Hu and D. Ralph. Using EPECs to model bilevel games in restructured electricity markets with locational prices. *Operations Research*, 55(5):809-827, 2007.
25. S. Jin and S. Ryan. Capacity expansion in the integrated supply network for an electricity market. *IEEE Transactions on Power Systems*, 26(4):2275 - 2284, 2011.
26. S.J. Kazempour, A.J. Conejo, and C. Ruiz. Strategic generation investment using a complementarity approach. *IEEE Transactions on Power Systems*, 26(2):940-948, 2011.
27. D. M. Kreps and J. A. Scheinkman. Quantity precommitment and Bertrand competition yield Cournot outcomes. *The Bell Journal of Economics*, 14(2):326-337, 1983.
28. S. Leyffer and T. S. Munson. Solving multi-leader-common-follower games. *Optimization Methods and Software*, 25(4):601-623, 2010.
29. London Economics International, LLC, *A proposed methodology for evaluating the economic benefits in a restructured wholesale electricity market*. Prepared for the California Independent System Operator, Folsom, CA, February 28, 2003.
30. F.H. Murphy and Y. Smeers. Withholding investments in energy only markets: Can contracts make a difference? *Journal of Regulatory Economics*, to appear, 2012.
31. J. Nash. Equilibrium points in n-person games. *Proceedings of the National Academy of Sciences*, 36(1):48-49, 1950.
32. K. Neuhoff, J. Barquin, M.G. Boots, A. Ehrenmann, B.F. Hobbs, and F.A.M. Rijkers. Network-constrained Cournot models of liberalized electricity markets: The devil is in the details. *Energy Economics*, 27:495-525, 2005.
33. S.S. Oren. Economic inefficiency of passive transmission rights in congested electricity systems with competitive generation. *Energy Journal*, 18(1):63-83, 1997.
34. P.-O. Pineau and P. Murto. An oligopolistic investment model of the Finnish electricity market. *Annals of Operations Research*, 121:123-148, 2003.
35. J. Raffensperger, M. Milke, and E. Read. A Deterministic Smart Market Model for Ground Water. *Operations Research*, 57(6):1333-1346, 2009.
36. J-H. Ryu, D.J. Yoo, and I-B. Lee. Developing a cost minimizing power load prediction model for steelwork industries. *Ind. Eng. Chem. Res.*, 49(8):3952–3956, 2010.
37. S.W. Salant. Imperfect competition in the international energy market: a computerized Nash-Cournot model. *Operations Research*, 30(2):252-280, 1982.
38. R.E. Schuler and B.F. Hobbs. Spatial price duopoly under uniform delivered pricing. *Journal of Industrial Economics,* 31 (1,2):175-188, 1982.

39. H.D. Sherali. A multiple leader Stackelberg model and analysis. *Operations Research*, 32:390-405, 1984.
40. Y. Smeers. *Gas models and three difficult objectives.* CORE Discussion paper 2008/9, Université Catholique Louvain, Belgium, February 2008.
41. S. Stoft, Using game theory to study market power in simple networks. In: H. Singh, ed., *Game Theory Tutorial*, IEEE Winter Power Meeting, New York, February 1, 1999.
42. P.D. Straffin. *Game Theory and Strategy*, The Mathematical Association of America, Washington, 1993.
43. M. Ventosa, R. Denis, and C. Redondo. Expansion planning in electricity markets: two different approaches. In *Proceedings of the 14th PSCC Conference*, Seville, 2002.
44. J.Y. Wei & Y. Smeers. Spatial oligopolistic electricity models with Cournot generators and regulated transmission prices. *Operations Research*, 47(1):102-112, 1999.
45. S. Wogrin, B.F. Hobbs, D. Ralph, E. Centeno, and J. Barquin. Open versus closed loop capacity equilibria in electricity markets under perfect and oligopolistic competition. *Mathematical Programming B*, to appear, 2012.
46. S. Wogrin, E. Centeno, and J. Barquin. Generation capacity expansion in liberalized electricity markets: a stochastic MPEC approach. *IEEE Transactions on Power Systems*, 26(4):2526-2532, 2011.
47. J. Yao, I. Adler and S.S. Oren. Modeling and Computing Two-settlement Oligopolistic Equilibrium in Congested Electricity Networks. *Operations Research*, 56(1):34-47, 2008.

Chapter 8
Algorithms for LCPs, NCPs and VIs

8.1 Introduction

This chapter presents some of the key ideas in various algorithms that are used to solve some of the basic equilibrium models, namely, models in the form of the linear complementarity problem (LCP), the nonlinear complementarity problem (NCP)[1] and the variational inequality (VI) problem. We explain the main ideas of the algorithms, but avoid detailed discussions or proofs of important mathematical issues such as the conditions under which an algorithm is guaranteed to converge.

In order to simplify the discussions, we consider only pure LCPs and NCPs, and not mixed complementarity problems, i.e., the complementarity models do not have equality constraints with corresponding free variables. This is not unduly restrictive, as an MCP can be rewritten as an equivalent NCP, but it has the notational advantage of not having to make distinctions between inequalities and equalities.

We pay particular attention to the PATH algorithm for solving NCPs (or MCPs, more generally), because it can be used to solve LCPs and VIs, as well – the LCP is a special case of the NCP, and a VI has an equivalent MCP, as explained in Chapter 5.

Section 8.2 discusses pivoting and iterative methods for solving the LCP. Section 8.3 presents the key ideas of the PATH algorithm for solving NCPs. In Section 8.4, we outline some sequential optimization methods for solving VIs, in addition to a reminder that PATH can be used. Section 8.5 summarizes the chapter, and an appendix in Section 8.6 presents some of the theory for the PATH algorithm, as well as outlines of other methods to solve NCPs.

---

[1] The mixed complementarity problem (MCP) as described in Chapter 1 is equivalent to the NCP (see Section 8.3), but the Box-MCP as defined in Section 8.3 is more general than the NCP. We focus on the NCP because it is simpler to describe the algorithms for it than for either definition of the MCP.

Throughout this chapter, when discussing LCP, NCP or VI models, we use the symbol z for the variable vector, instead of x, to be consistent with literature on the PATH algorithm and solver. See the introductory remarks of the appendix to this chapter for details.

8.2 Algorithms for LCP Models

In a Linear Complementarity Problem (LCP), the n real variables to be solved for, z_i, $i = 1, ..., n$, are required to be non-negative, they must satisfy n constraints that are linear in the variables, and each constraint is complementary to a variable: for given parameters q_i, m_{ij}, $i, j = 1, ..., n$, the LCP is defined as

$$\text{LCP:} \qquad 0 \leq q_i + \sum_{j=1}^{n} m_{ij} z_j \perp z_i \geq 0\,,\; i = 1, ..., n \tag{8.1}$$

In vector-matrix form, with $z = (z_i)$, $q = (q_i)$ and $M = (m_{ij})$, the same LCP is written as

$$\text{LCP:} \qquad 0 \leq q + Mz \perp z \geq 0 \tag{8.2}$$

where complementarity is understood to be between each component of z and each row of $q + Mz$ individually.

Following the terminology of [6], we can distinguish two classes of solution methods for the LCP: pivoting methods, and iterative methods. Pivoting methods involve linear algebra computations on systems of linear equations that also occur in the simplex method for solving linear programs. Iterative methods involve solving a sequence of easy problems, each of which approximates the actual LCP, and each approximation depends on the solution of the last easy problem (or on a first guess, to get started). Pivoting methods involve a finite number of pivots to finish the algorithm (i.e., to find a solution, or to encounter a problem as described below), but iterative methods normally produce an infinite series of approximate solutions that approach a true solution in the limit of many iterations (the iterations are stopped finitely when the approximate solution is known to be close enough to a true solution).

Pivoting methods can encounter computational difficulties, especially for large problems. Roundoff errors at every pivot can accumulate rapidly, causing the solution produced by pivoting actually to be a solution to a somewhat different problem. Another computational difficulty with pivoting methods concerns sparsity of the data (q, M): large problems tend to have a very large percentage of zeroes among the data elements q_i and m_{ij}, and both storage and arithmetic can be done efficiently with sparse data. Unfortunately, just a few pivots can change a sparse system of linear equations into a system having many nonzero elements. Special methods must be used to control roundoff accumulation and destruction of sparsity.

In contrast, iterative methods do not have severe roundoff accumulation, and each approximate problem at an iteration uses the original problem data (q, M), so sparsity in the problem data can be exploited at every iteration. Iterative methods may therefore be helpful in solving large scale LCPs, if pivoting methods encounter severe difficulties.

8.2.1 Lemke's Pivoting Method for LCPs

Several pivoting-based algorithms for the LCP have been proposed. Here, we explain only the widely used method by Lemke, and refer readers to the book [6] for discussions of other pivoting methods.

We begin with a discussion of the algebra of pivoting, mainly through an example of pivoting applied to solve a linear programming problem, in the simplex method.

8.2.1.1 General Background on Pivoting

Consider a set of m linear equations in n variables, where $m < n$. In general, there can be an infinite number of solutions because the number of variables exceeds the number of equations. However, in algorithms to solve optimization models and complementarity models, some special solutions play an important role, namely solutions for which $n - m$ of the variables have the value 0, while the remaining m variables may be nonzero. The $n - m$ zero-valued variables are called non-basic variables, and the remaining m variables are called basic variables. As such algorithms proceed step by step, the classification of variables into m basic and $n - m$ non-basic variables changes. At each step, one non-basic variable becomes basic (and usually nonzero), and one basic variable becomes non-basic (with a value of zero), while the other non-basic variables remain at their zero values, and the other basic variables' values change in order to obey the m equations. The algebra that keeps track of the changing variable values at each step is called a pivot.

The following example solves a linear program by the simplex method, which involves pivoting.

8.2.1.2 Illustrative Example. Pivoting in Simplex Method for a Linear Program

Consider the following linear program in two variables, x_1 and x_2 (the symbol v is used simply as a shorthand reference for the value of the objective function).

$$\text{Maximize}_{x_1, x_2} \quad v = 2x_1 + 3x_2 \tag{8.3a}$$

s. t.

$$x_1 + x_2 \leq 3 \tag{8.3b}$$

$$x_1 + 2x_2 \leq 4 \tag{8.3c}$$

$$x_1 \geq 0, x_2 \geq 0 \tag{8.3d}$$

We first replace the inequality constraints (8.3b)-(8.3c) by equality constraints, by introducing two new non-negative variables, x_3 and x_4, called slack variables, which have zero coefficients in the objective function. The equivalent linear program is as follows.

$$\text{Maximize}_{x_1, x_2, x_3, x_4}$$

$$v = 2x_1 + 3x_2 + 0x_3 + 0x_4 \tag{8.4a}$$

s. t.

$$x_1 + x_2 + x_3 = 3 \tag{8.4b}$$

$$x_1 + 2x_2 \quad + x_4 = 4 \tag{8.4c}$$

$$x_1 \geq 0, x_2 \geq 0, x_3 \geq 0, x_4 \geq 0 \tag{8.4d}$$

Note that, for this example, the variable values $x_1 = 0$, $x_2 = 0$, $x_3 = 3$, $x_4 = 4$, form a feasible solution at which $v = 0$. This feasible solution is the starting point for two pivots, which will lead to an optimal solution of the linear program (8.3) or (8.4). For this feasible solution, the basic variables are x_3 and x_4, while the non-basic variables are x_1 and x_2. We rearrange the $m = 2$ equations (8.4b) and (8.4c), in $n = 4$ variables, to express the two basic variables in terms of the two non-basic variables, and we also keep track of the value of v, together with an expression that indicates how v would change if the non-basic variables were to increase:

$$x_3 = 3 - x_1 - x_2 \tag{8.5a}$$

$$x_4 = 4 - x_1 - 2x_2 \tag{8.5b}$$

$$v = 0 + 2x_1 + 3x_2 \tag{8.5c}$$

Now we increase one of the non-basic variables in order to increase v, and we keep on increasing this non-basic variable until one of the basic variables decreases to zero. At this point, the basic variable that hits zero becomes non-basic, and the non-basic variable that was increased becomes basic. Which of the two non-basic variables x_1 or x_2 should be increased? One answer is to choose x_2, based on observing that for each unit increase in x_2, the increase in v is 3, while for each unit increase in x_1, the increase in v is only 2. As we increase x_2, keeping $x_1 = 0$, both of x_3 and x_4 decrease towards zero, as can be seen in (8.5a) and (8.5b): x_3 would hit zero when $x_2 = 3$, and x_4 would hit zero when $x_2 = 2$ (but x_4 would be negative when $x_2 = 3$, which violates the non-negativity requirement). Therefore, we increase x_2 to the value 2, x_2 becomes basic, and x_4 becomes non-basic. From (8.5c), we can see that $v = 6$ now.

Next, we perform the first pivot: we rearrange (8.5a) and (8.5b) to have the new set of basic variables, x_2 and x_3, on the left, and the new set of non-basic variables, x_1 and x_4, on the right; we also replace (8.5c) with an equation to show how v would change from its current value of 6 due to changes in either of the non-basic variables, x_1 and x_4. The first of the new equations solves for x_2 in (8.5b), and the second and third equations substitute this expression for x_2 into (8.5a) and (8.5c), respectively:

$$x_2 = 2 - \frac{1}{2}x_1 - \frac{1}{2}x_4 \tag{8.6a}$$

$$x_3 = 3 - x_1 - (2 - \frac{1}{2}x_1 - \frac{1}{2}x_4) = 1 - \frac{1}{2}x_1 + \frac{1}{2}x_4 \tag{8.6b}$$

$$v = 0 + 2x_1 + 3(2 - \frac{1}{2}x_1 - \frac{1}{2}x_4) = 6 + \frac{1}{2}x_1 - \frac{3}{2}x_4 \tag{8.6c}$$

An increase in the non-basic variable x_1 would increase v, but an increase in x_4 would decrease v. Therefore, the next pivot involves an increase in x_1. As we increase x_1, the first basic variable to hit zero is x_3, which now becomes non-basic. Using (8.6b) to solve for x_1 in terms of the new non-basic variables x_3 and x_4, and substituting in (8.6a) and (8.6c), we have the result of the second pivot:

$$\begin{aligned} x_1 &= 2 - 2x_3 + x_4 \\ x_2 &= 2 - \frac{1}{2}(2 - 2x_3 + x_4) - \frac{1}{2}x_4 = 1 + x_3 - x_4 \\ v &= 6 + \frac{1}{2}(2 - 2x_3 + x_4) - \frac{3}{2}x_4 = 7 - x_3 - x_4 \end{aligned}$$

The new value of $v = 7$. Any increase in either of the non-basic variables x_3 or x_4 would decrease v; therefore, we are finished. An optimal solution is $x_1 = 2$ and $x_2 = 1$, where $v = 7$.

□

8.2.1.3 Lemke's Method

In solving an LCP, there is no objective function. Therefore, the decision about which non-basic variable to increase cannot be made by observing which non-basic variable causes the best rate of change of an objective function. A prominent feature of an LCP is the statement of complementarity conditions, and complementarity is at the heart of the decision about which non-basic variable to increase in a *complementary pivoting* method such as the widely used method due to Lemke [19], the first complementary pivoting method for solution of a general LCP. We introduce the method via the following example.

8.2.1.3.1 Illustrative Example. Simple LCP from Chapter 1

A small mixed complementarity problem from Chapter 1 (Illustrative Example 1.2.1) can be transformed to the following LCP by using the equation to eliminate the free variable y_1; we also change the variable symbol from x to z:

$$\begin{aligned} z_1 + z_2 \geq 0\ , \quad z_1 \geq 0\ , \quad (z_1 + z_2) \cdot z_1 = 0 \\ 2z_1 + z_2 - 2 \geq 0\ , \quad z_2 \geq 0\ , \quad (2z_1 + z_2 - 2) \cdot z_2 = 0 \end{aligned}$$

We introduce the variables w_1 and w_2 and express the LCP in an equivalent form, to set up Lemke's method:

$$\begin{aligned} w_1 &= \ \ 0 + z_1 + z_2\ , \quad z_1 \geq 0\ , \quad w_1 \geq 0\ , \quad w_1 z_1 = 0 \\ w_2 &= -2 + 2z_1 + z_2\ , \quad z_2 \geq 0\ , \quad w_2 \geq 0\ , \quad w_2 z_2 = 0 \end{aligned}$$

Note that there are now $m = 2$ equations and $n = 4$ variables, as well as the non-negativity and complementarity conditions. Lemke's method gets started by introducing an artificial variable z_0, for a total of $n = 5$ variables, so that the system becomes

$$w_1 = \ \ 0 + z_0 + z_1 + z_2\ , \quad z_1 \geq 0\ , \quad w_1 \geq 0\ , \quad w_1 z_1 = 0 \tag{8.7a}$$

$$w_2 = -2 + z_0 + 2z_1 + z_2\ , \quad z_2 \geq 0\ , \quad w_2 \geq 0\ , \quad w_2 z_2 = 0 \tag{8.7b}$$

Now choose a value of the artificial variable that makes both w variables non-negative while both of the original z variables are equal to zero: set $z_0 = \max(-(0), -(-2)) = 2$ to begin the algorithm. With this value of z_0, the values $z_1 = 0$, $z_2 = 0$, $w_1 = 2$, and $w_2 = 0$ satisfy the two equations, the non-negativity requirements, and the complementarity requirements. This is not yet a solution, because z_0 is not one of the variables of the problem that we want to solve; we must get rid of z_0 by pivoting to compute another set of values which have $z_0 = 0$, but still satisfy the equations, non-negativity and complementarity. At this first step, of the five variables, only two are nonzero, z_0 and w_1; these are the basic variables, while the other three (all equal to zero) are non-basic variables. We rearrange the two equations to place the basic variables on the left, expressed in terms of constants and the non-basic variables on the right, using the equation for w_2 (8.7b) to solve for z_0 in terms of the other variables (we rearrange the equation order of (8.7) by listing this new z_0 equation first, to make its substitution in the w_1 equation (8.7a) easier to follow):

$$\begin{aligned} z_0 &= 2 + w_2 - 2z_1 - z_2 \\ w_1 &= 0 + z_0 + z_1 + z_2 \\ &= 0 + (2 + w_2 - 2z_1 - z_2) + z_1 + z_2 \\ &= 2 + w_2 - z_1 \end{aligned}$$

The new system of equations (8.8) is equivalent to those of (8.7):

$$z_0 = 2 + w_2 - 2z_1 - z_2 \tag{8.8a}$$

$$w_1 = 2 + w_2 - z_1 \tag{8.8b}$$

Since w_2 has been forced to zero by our choice of z_0, we can increase its complement, z_2, to values greater than zero, but still maintain complementarity. When we increase z_2, we need to enforce the two equations, which will change the values of other variables, but which other variables will change? The procedure is to increase z_2, while keeping the other non-basic variables fixed at zero, until one of the basic variables is driven to zero, which "blocks" any further increase in z_2. In this example, at this step of the algorithm, the equation (8.8b) for w_1 shows no dependence on z_2, so the blocking occurs when z_0 is driven to zero: $z_0 = 0 = 2 + w_2 - 2z_1 - z_2 = 2 + 0 - 2(0) - z_2$; therefore, $z_2 = 2$ is the new value. The algorithm has finished, because the artificial variable $z_0 = 0$. The other variables' values are either zero (the non-basic variables z_1, w_2), or are given by an equation ($w_1 = 2 + w_2 - z_1 = 2$). To summarize, the solution to the original LCP is $z_1 = 0$, $z_2 = 2$ (as shown in Chapter 1), and the values of the linear expressions are $w_1 = 2$, $w_2 = 0$.
□

8.2.1.3.2 General Statement of Lemke's Method

Let us generalize the method of the first example, so that it can be applied to any LCP (8.2). The first step is to introduce n new variables, w_i, $i = 1, ..., n$, with $w=(w_i)$, and to write the LCP in the equivalent form with n equations and $2n$ variables

$$w = q + Mz\,,\, 0 \leq w \perp z \geq 0 \tag{8.9}$$

Before getting into the details, let us summarize the main ideas of Lemke's method, based on Illustrative Example 8.2.1.3.1:

- to introduce a single, scalar *artificial variable* $z_0 \geq 0$, [2] to define *a different but related problem*, for which the solution is known to be $z = 0$, and exactly one $w_i = 0$, so that $0 \leq w \perp z \geq 0$ at the start;
- to pivot repeatedly by choosing the non-basic variable that is to enter the basis (a z_i or a w_i) as *the complement* of the variable that was just driven to zero and is leaving the basis, thus maintaining complementarity of w and z; and
- to pivot repeatedly until z_0 is driven to zero, thus solving the original problem.

Next we present the details of Lemke's method. First, note that if, in the original problem (8.9), $q \geq 0$, then a solution is $z = 0$. Therefore, we confine our discussion to the interesting (and typical) problems, for which $q_i < 0$ for at least one i.

The "different-but-related" problem definition requires the user to supply a vector parameter d called the covering vector, with entries $d_i > 0$, $i = 1, ..., n$. Often, in applications, $d_i = 1$, $i = 1, ..., n$. The related problem to start the algorithm is

$$w = q + dz_0 + Mz\,, z_0 \geq 0\,, 0 \leq w \perp z \geq 0 \tag{8.10}$$

Note that problem (8.10) has $2n + 1$ variables (n variables in the w vector, n variables in the z vector, and the scalar variable z_0) and n equations to be used for pivoting. The initial value of z_0 is chosen to be large enough to make $q + dz_0 \geq 0$, which means that choosing $z = 0$ satisfies all the requirements of (8.10), including non-negativity and complementarity of w and z. However, if we choose z_0 to be too large, then we could have $w > 0$ at this first step of the algorithm, with no $w_i = 0$ to indicate that its complement z_i should be introduced into the basis at the first pivot. Therefore, the initial value of z_0 is chosen to be the *smallest value* such that $q + dz_0 \geq 0$, i.e.,

$$\text{initial value of } z_0 \equiv z_0^{init} = \text{Maximize}_i \quad \left(\frac{-q_i}{d_i}\right) \tag{8.11}$$

which means that $w_i = 0$ for the index i that achieves the maximum. Any value of z_0 greater than the value in (8.11) produces values $w > 0$ that, together with $z = 0$, satisfy (8.10); the set of all such solutions is called the *primary ray*, defined as $\{w \in \mathbb{R}^n, z \in \mathbb{R}^n | z = 0, w = q + dz_0 + Mz, z_0 \geq z_0^{init}\}$. If the maximization in (8.11) is achieved for index value $i = r$, then $w_r = 0$, making z_r the first non-basic variable to be pivoted into the basis. To summarize: before the first pivot, the n basic variables are z_0 and all the w_i except for w_r; the $n + 1$ non-basic variables are w_r and all of the z_i for $i = 1, ..., n$; and z_r is identified as the non-basic variable to be pivoted

[2] The scalar variable z_0 is not part of the original problem $z = (z_i)_{i=1}^n$.

into the basis at the first pivot. Before the first pivot, all of the equations $w = q + dz_0 + Mz$ should be rearranged such that the basic variables are on the left and equal to constants (their current values) plus linear expressions in the non-basic variables (which have been set equal to zero); that is, the equations are used to solve for the basic variables in terms of the non-basic variables.

Just before the first and later pivots, a variable has just become non-basic (w_r before the first pivot, or one of the w_i or z_i that was just driven to zero after the last pivot), and that variable's complement (currently non-basic) is to be increased and become a basic variable. There are two things that can happen when we attempt to increase a non-basic variable from zero to a positive value: (1) one or more of the currently basic variables decreases (because the coefficient of the increasing non-basic variable is negative), and the first one to hit zero blocks any further increase in the variable that is being made basic; or (2) none of the currently basic variables decreases (because the coefficients of the increasing non-basic variable are positive, for all basic variables), and the non-basic variable being introduced into the basis can be increased without bound. If the first possibility happens at a pivot, then the pivot is completed, and Lemke's algorithm proceeds to the next pivot (or stops, if z_0 just left the basis). If the second possibility happens, then the algorithm must stop because no variable has been identified to leave the basis; instead, a *secondary ray* has been found ("secondary" distinguishes it from the primary ray at the start of Lemke's algorithm).

If z_0 is the first variable to hit zero at a pivot, so that its value becomes zero, and it leaves the basis, then because the equality in (8.10) has been maintained throughout all pivots, it follows that the current values of w and z satisfy both (8.10) and the original LCP (8.9): we have found a solution to the LCP.

We continue this discussion of Lemke's method with another illustration, this time based on a simple economic equilibrium model.

8.2.1.3.3 Illustrative Example. Equilibrium of two commodities

Suppose that there are two commodities, with quantities $z_1 \geq 0$ and $z_2 \geq 0$, with inverse supply functions $\gamma_1 + \delta_1 z_1$ and $\gamma_2 + \delta_2 z_2$, respectively. The two commodities' demands are related: the inverse demand functions are $\alpha_1 + \beta_{11} z_1 + \beta_{12} z_2$ and $\alpha_2 + \beta_{21} z_1 + \beta_{22} z_2$. Equilibrium in the market for commodity 1 requires: (1) if price as given by the inverse supply function exceeds the price as given on the inverse demand function, then $z_1 = 0$; and (2) if $z_1 > 0$, then the inverse supply price must equal the inverse demand price. Similar requirements must hold for equilibrium in the market for commodity 2. These equilibrium conditions define the following LCP:

$$0 \leq (-\alpha_1 + \gamma_1) + (-\beta_{11} + \delta_1) z_1 - \beta_{12} z_2 \perp z_1 \geq 0$$
$$0 \leq (-\alpha_2 + \gamma_2) - \beta_{21} z_1 + (-\beta_{22} + \delta_2) z_2 \perp z_2 \geq 0$$

Here, we assign particular values to the parameters: $\alpha_1 = 40$, $\alpha_2 = 70$, $\beta_{11} = -4$, $\beta_{12} = -1$, $\beta_{21} = -2$, $\beta_{22} = -6$, $\gamma_1 = 8$, $\gamma_2 = 6$, $\delta_1 = 2$, $\delta_2 = 1$, which gives rise to the following LCP:

$$0 \le 6z_1 + z_2 - 32 \perp z_1 \ge 0 \tag{8.12a}$$
$$0 \le 2z_1 + 7z_2 - 64 \perp z_2 \ge 0 \tag{8.12b}$$

Rewriting in the form with equations and two w variables, and with the artificial variable z_0 (with covering vector d having all entries equal to 1) to start Lemke's algorithm, we have the system:

$$\begin{aligned} w_1 &= -32 + z_0 + 6z_1 + z_2 \;, \quad z_1 \ge 0 \;, \quad w_1 \ge 0 \;, \quad w_1 z_1 = 0 \\ w_2 &= -64 + z_0 + 2z_1 + 7z_2 \;, \quad z_2 \ge 0 \;, \quad w_2 \ge 0 \;, \quad w_2 z_2 = 0 \end{aligned}$$

Set $z_0 = \max(-(-32), -(-64)) = 64$, $z_1 = z_2 = w_2 = 0$, $w_1 = 32$, to begin the algorithm. The basic variables are z_0 and w_1, and the non-basic variables are z_1, z_2, and w_2. Using the equation for w_2 to solve for z_0 in terms of the other variables, we express the basic variables in terms of the non-basic variables:

$$\begin{aligned} z_0 &= 64 + w_2 - 2z_1 - 7z_2 \\ w_1 &= -32 + z_0 + 6z_1 + z_2 \\ &= -32 + (64 + w_2 - 2z_1 - 7z_2) + 6z_1 + z_2 \\ &= 32 + w_2 + 4z_1 - 6z_2 \end{aligned}$$

Because z_2 is complementary to w_2, we increase z_2 until further increase is blocked by one of the basic variables reaching zero: $z_2 = 64/7$ causes $z_0 = 0$, and $z_2 = 16/3$ causes $w_1 = 0$. Therefore, the greatest increase in z_1 that still leaves z_0 and w_1 non-negative is $z_2 = \min(64/7, 16/3) = 16/3$. The next system of two equations has basic variables z_0, z_2, and non-basic variables z_1, w_1, w_2. The system of two equations is rewritten by solving for z_2 in the equation in which z_2 was blocked, and substituting in the other equation:

$$\begin{aligned} z_2 &= 16/3 - 1/6w_1 + 1/6w_2 + 2/3z_1 \\ z_0 &= 64 + w_2 - 2z_1 - 7z_2 \\ &= 64 + w_2 - 2z_1 - 7(16/3 - 1/6w_1 + 1/6w_2 + 2/3z_1) \\ &= 80/3 + 7/6w_1 - 1/6w_2 - 20/3z_1 \end{aligned}$$

Since w_1 just became non-basic, we increase its complement, z_1, until it is blocked by a basic variable becoming zero. This happens when $z_1 = 4$, and $z_0 = 0$. The artificial variable has been driven to zero, so the algorithm is finished, and we have found a solution to the LCP:

$$z_1 = 4$$

$$z_2 = 16/3 - 1/6w_1 + 1/6w_2 + 2/3z_1 = 8$$

In fact, this is the unique solution to (8.12), because the matrix of coefficients of z, $\begin{pmatrix} 6 & 1 \\ 2 & 7 \end{pmatrix}$, is positive definite (see [6], Theorem 3.1.6 for this uniqueness result).

□

8.2.1.3.4 Convergence of Lemke's Method

There are important, basic questions to ask about any algorithm that is proposed for solving a problem. Is the algorithm guaranteed to find a solution when a solution exists, at least for some of the problems that it is intended to solve? Does the algorithm stop after a finite amount of calculation? The answers, for Lemke's method, depend in part on whether or not the problem is *nondegenerate.* Essentially, the problem is nondegenerate if there are never any ties with two or more basic variables hitting zero at the same time, when a non-basic variable is being increased during a pivot operation. One of the problems avoided when a problem is nondegenerate is cycling, i.e., revisiting the same basis after a few pivots, and going over and over the same sequence of bases, to pivot without end. Another matter to be concerned about is the possibility that Lemke's algorithm can end with a secondary ray (i.e., increasing the designated non-basic variable causes all basic variables to increase without bound). Thus, the best that can be said for Lemke's algorithm for a general LCP, in answer to the basic questions, is in the following theorem, paraphrased from Theorem 4.4.8 of [6].

Theorem 8.1. *When applied to a nondegenerate instance of the LCP (8.9), Lemke's algorithm will terminate in a finite number of steps with either a secondary ray or else a solution of the LCP.*

What does it mean if the algorithm stops with a secondary ray? Are there some LCPs for which it is impossible to get a secondary ray termination of the algorithm, and thus, the algorithm must stop with a solution? These questions, and related questions of the existence of solutions to LCPs, have been the subject of a great deal of mathematical research. We refer the reader to the book [6] for a comprehensive review, as of the time of its publication.

8.2.2 Iterative Methods for LCPs

An iterative method applied to the LCP (8.2) starts with a first guess z^0 of the solution, and generates a series of approximate solutions, $z^1, z^2, \ldots$ that converge in the limit to a solution z^* (assuming that a solution exists). Here, we describe the *matrix splitting* approach in some detail, and briefly mention some other approaches.

8.2.2.1 General Background on Matrix Splitting

Iterative methods for the solution of a system of linear equations have been used for a long time, and they have inspired many of the techniques for iterative solution of LCPs. We therefore review a few techniques for the iterative solution of a linear system

$$Ax = b \tag{8.13}$$

where $x, b \in \mathbb{R}^n$ and $A \in \mathbb{R}^{n \times n}$ is nonsingular. Many techniques involve the splitting of the matrix

$$A = B + C \tag{8.14}$$

where B is nonsingular. With the last iterate x^{k-1} already known, the next iterate x^k is calculated as the solution of

$$Bx + Cx^{k-1} = b \text{ , i.e.,} \tag{8.15a}$$

$$x^k = B^{-1}(b - Cx^{k-1}) \tag{8.15b}$$

If (8.15a) is easy to solve, then each iteration can be done efficiently.

One well-known example of the splitting approach (8.14), (8.15), is the Jacobi method, in which $B = D$, the diagonal of A, and C is the matrix of off-diagonal elements of A. The requirement that $B = D$ is nonsingular amounts to the requirement that the diagonal have nonzero elements. Therefore, $B^{-1} = D^{-1}$ is easy to calculate, and thus the step (8.15b) can be performed efficiently:

$$\begin{aligned} x^k &= D^{-1}(b - (A - D)x^{k-1}) \\ &\triangleq D^{-1}u^{k-1} \end{aligned}$$

where the vector u^{k-1} is the result of straightforward vector-matrix multiplication and addition, i.e., $u^{k-1} = b - (A - D)x^{k-1}$ involving the iterate x^{k-1}, a known quantity at iteration k. Since D has nonzero entries (a_{ii}) along the diagonal and zeroes elsewhere, the value of the ith entry in x^k is simple to calculate:

$$x_i^k = u_i^{k-1}/a_{ii}$$

Another well-known example is the Gauss-Seidel method, which is most easily described first in terms of the elements of the vectors and matrices. Given the results of the previous iteration $x^{k-1} = (x_1^{k-1}, ..., x_n^{k-1})^T$, the elements of the next iterate x^k are calculated one by one, starting with x_1^k, then x_2^k, x_3^k, etc. The value of x_i^k satisfies the following equation, in which the previous iterate x^{k-1} supplies the values x_j^{k-1} for $j > i$, and the current iterate's results, so far, supply x_j^k for $j < i$:

$$\sum_{j<i} a_{ij}x_j^k + a_{ii}x_i + \sum_{j>i} a_{ij}x_j^{k-1} = b_i \tag{8.16}$$

In matrix form, (8.16) may be written

$$Lx^k + Dx^k + Ux^{k-1} = b \tag{8.17a}$$

$$\text{i.e.,}\quad x^k = (L+D)^{-1}(b - Ux^{k-1}) \tag{8.17b}$$

where L is the strictly lower triangular part of A (containing the off-diagonal elements of A that are below the diagonal), and U is the strictly upper triangular part of A. Therefore, the Gauss-Seidel method corresponds to the splitting $B = D + L$, and $C = U$.

Several more examples of splitting may be found in the book [21].

8.2.2.2 Matrix Splitting for the LCP

In this approach, the square matrix M that is part of the definition of the LCP (8.2) is split into two matrices of the same dimension as M:

$$M = B + C \tag{8.18}$$

At each iteration k, a guess z^{k-1} is known (typically as the solution of the last iteration, or, at the beginning, as a guess provided by the user), and the following LCP is solved for the next iterate z^k:

$$\text{LCP}(z^{k-1}):\qquad 0 \le (q + Cz^{k-1}) + Bz \perp z \ge 0 \tag{8.19}$$

If the solution to LCP(z^{k-1}) repeats the last guess, $z^k = z^{k-1}$, then we have found a solution to the LCP (8.2) because

$$0 \le (q + Cz^k) + Bz^k \perp z^k \ge 0\text{ , by (8.19), and so}$$
$$0 \le q + Mz^k \perp z^k \ge 0\text{ , by (8.18)}$$

In practice, though, the algorithm is typically stopped when z^k is *very close to* z^{k-1}, as measured, for example, by a small value of the Euclidean distance norm

$$\|z^k - z^{k-1}\| = \left(\sum_{i=1}^{n}(z_i^k - z_i^{k-1})^2\right)^{1/2}$$

A good choice of B has the property that LCP(z^{k-1}) is easy to solve. For example, if B is a diagonal matrix, with diagonal entries b_i, then LCP(z^{k-1}) reduces to n independent, one-dimensional problems (where $\widetilde{q} = q + Cz^{k-1}$, for brevity):

$$\widetilde{q}_i + b_i z_i \ge 0\,,\ z_i \ge 0\,,\ (\widetilde{q}_i + b_i z_i)z_i = 0\,, i = 1, ..., n$$

If there is a solution to the ith problem, then either $z_i > 0$ and so $\widetilde{q}_i + b_i z_i = 0$, which implies that $z_i = -\widetilde{q}_i/b_i$, or $z_i = 0$ (which can happen only if $\widetilde{q}_i \ge 0$).

To summarize, if there is a solution, then $z_i = \max(0, -\widetilde{q}_i/b_i)$, which is a simple calculation to perform. If the matrix B is not just any diagonal matrix, but is the diagonal part of M, then we might expect that the algorithm will take only a few easy iterations, because the LCP at each iteration, (8.19), is related to the original problem; this might especially be true if the off-diagonal part of the splitting, C, is somehow less important than the diagonal part B. The notion of C being less important than B is made more precise in Section 8.2.2.2.3.

Another possible choice for a splitting of the matrix M is to choose a symmetric B. For this choice, the conditions of LCP(z^{k-1}), (8.19) are in fact the KKT necessary conditions for the following quadratic programming problem:

$$\begin{aligned} \text{QP}(z^{k-1})\text{: Minimize} \quad & (q + Cz^{k-1})^T z + \frac{1}{2} z^T B z \\ \text{s. t.} \quad & \\ & z \geq 0 \end{aligned}$$

If there is a reliable and efficient way to solve QP(z^{k-1}), and if B is somehow close to M, with C of less importance than B, then it seems intuitively plausible that the algorithm may converge quickly and efficiently to a small value of $\|z^k - z^{k-1}\|$. A precise statement of this notion is contained in the theorem of Section 8.2.2.2.3.

There are many other ways in which the matrix M can be split, and each way produces a different type of algorithm. See [6] for several other examples.

The first example below is a splitting with B as the diagonal of M, applied to the two-commodity equilibrium model (8.12). The second example below has a symmetric B, with a sequence of quadratic programs, for the same two-commodity problem (8.12).

8.2.2.2.1 Illustrative Example. Matrix splitting for two-commodity model: B = diagonal part of M

For this LCP model, the linear expression, in vector-matrix notation, is $q + Mz$, where $M = \begin{pmatrix} 6 & 1 \\ 2 & 7 \end{pmatrix}$ and $q = \begin{pmatrix} -32 \\ -64 \end{pmatrix}$. In this example, the matrix splitting $M = B + C$ has $B = \begin{pmatrix} 6 & 0 \\ 0 & 7 \end{pmatrix}$, and $C = \begin{pmatrix} 0 & 1 \\ 2 & 0 \end{pmatrix}$, i.e., B is the diagonal part of M. At any iteration k, the result from the previous iteration, z^{k-1} is known, and so the LCP to be solved at iteration k, with linear expression $(q + Cz^{k-1}) + Bz$, takes the form

$$\begin{aligned} w_1 &= (-32 + z_2^{k-1}) + 6z_1 \ , \ \ z_1 \geq 0 \ , \ \ w_1 \geq 0 \ , \ \ w_1 z_1 = 0 \\ w_2 &= (-64 + 2z_1^{k-1}) + 7z_2 \ , \ \ z_2 \geq 0 \ , \ \ w_2 \geq 0 \ , \ \ w_2 z_2 = 0 \end{aligned}$$

This LCP decomposes into two separate, single-variable LCPs, each of the form: $a + bx \geq 0$, $x \geq 0$, $(a + bx)x = 0$, where $b > 0$. It is easy to see that the solution to the single-variable LCP is: $x = -a/b$ if $a < 0$, or $x = 0$ if $a \geq 0$. Therefore, the LCP at each iteration can be solved very easily and quickly. The progress of the iterations, for two different starting guesses z^0, is shown in Table 8.1. Note that the iterations approach closer and closer to the unique solution $(4, 8)^T$ from both starting points. A more exact solution can be found with more than six iterations.

Table 8.1 Matrix splitting, diagonal B, from two starting points

k	z_1	z_2	$\|z^k - z^{k-1}\|$	z_1	z_2	$\|z^k - z^{k-1}\|$
0	1	1		2	50	
1	5.166667	8.857143	8.893582	0.000000	8.571429	41.47682
2	3.857143	7.666667	1.769770	3.904762	9.142857	3.946352
3	4.055556	8.040816	0.423504	3.809524	8.027211	1.119704
4	3.993197	7.984127	0.084275	3.995465	8.054422	0.187922
5	4.002646	8.001944	0.020167	3.990930	8.001296	0.053319
6	3.999676	7.999244	0.004013	3.999784	8.002592	0.008949

□

8.2.2.2.2 Illustrative Example. Matrix splitting for two-commodity model: symmetric B

For this splitting of $M = B + C$, $B = \begin{pmatrix} 6 & 1 \\ 1 & 7 \end{pmatrix}$, and $C = \begin{pmatrix} 0 & 0 \\ 1 & 0 \end{pmatrix}$. At iteration k, the LCP to be solved can be expressed as a quadratic minimization problem, because B is symmetric. The minimization problem has a unique solution because B is positive definite. The LCP at iteration k has parameter values that depend on the result $\mathbf{z}^{k-1}$ of iteration $k-1$. The LCP at iteration k is

$$w_1 = -32 + 6z_1 + z_2\ ,\quad z_1 \geq 0\ ,\quad w_1 \geq 0\ ,\quad w_1 z_1 = 0$$
$$w_2 = (-64 + z_1^{k-1}) + z_1 + 7z_2\ ,\quad z_2 \geq 0\ ,\quad w_2 \geq 0\ ,\quad w_2 z_2 = 0$$

and the equivalent minimization problem is

$$\text{Minimize} \quad -32z_1 + (-64 + z_1^{k-1})z_2 + 3z_1^2 + 2z_1z_2 + 3.5z_2^2$$

s. t.

$$z_1 \geq 0,\ z_2 \geq 0$$

Table 8.2 shows the iterations, which display more rapid convergence than for the splitting with a diagonal B.

Table 8.2 Matrix splitting, symmetric B, from two starting points

k	z_1	z_2	$\|z^k - z^{k-1}\|$	z_1	z_2	$\|z^k - z^{k-1}\|$
0	1	1		2	50	
1	3.926829	8.439024	7.994086	3.951220	8.292683	41.75293
2	3.998215	8.010708	0.434225	3.998810	8.007139	0.289483
3	3.999956	8.000261	0.010591	3.999971	8.000174	0.007061
4	3.999999	8.000006	0.000258	3.999999	8.000004	0.000172
5	4.000000	8.000000	0.000006	4.000000	8.000000	0.000004
6	4.000000	8.000000	0.000000	4.000000	8.000000	0.000000

□

8.2.2.2.3 Convergence of Matrix Splitting Algorithms for the LCP

Under what conditions does a matrix splitting algorithm converge? We cannot expect convergence to an exact solution in a finite number of iterations – even Table 8.2 would show nonzero values of $\|z^k - z^{k-1}\|$ at iteration 6 if an extra decimal of accuracy were displayed. The best that we can hope for is that $\|z^k - z^{k-1}\|$ approaches zero in the limit as the number of iterations k gets very large. The following theorem is Theorem 5.3.12 from [6], reworded for this book. The theorem requires that B should be positive definite – this is true of both examples of splitting, above. Readers are reminded of some basic facts about positive definite matrices: (1) if a positive definite matrix B is symmetrized by constructing $\frac{1}{2}(B + B^T)$, then the symmetrized matrix is also positive definite; and (2) any positive definite matrix has a nonsingular, square root matrix, i.e., there exists a nonsingular matrix $\widetilde{B}$ such that the original matrix equals $\widetilde{B}^T\widetilde{B}$.

We digress briefly, to explain some facts about the Euclidean norm of a square matrix, which appears in the theorem below. The Euclidean norm of a square matrix A is defined using the ordinary norm of a vector, as

$$\|A\| = \max_{\|x\|=1} \|Ax\| \tag{8.20}$$

the largest value that $\|Ax\|$ can take, when x is restricted to vectors of unit length. Two basic facts from matrix theory (see [21]) are that, for any two square matrices $A, B \in \mathbb{R}^{n \times n}$ and any vector $y \in \mathbb{R}^n$,

$$\|AB\| \le \|A\|\|B\| \quad \text{and}$$
$$\|Ay\| \le \|A\|\|y\|$$

Theorem 8.2. *([6], Theorem 5.3.12) Suppose that in the splitting $M = B + C$, the matrix B is positive definite. Let $\widetilde{B}$ be a nonsingular matrix such that $\widetilde{B}^T\widetilde{B} = \frac{1}{2}(B + B^T)$. Suppose also that*

$$\rho = \|(\widetilde{B}^{-1})^T C \widetilde{B}^{-1}\| < 1 \tag{8.21}$$

Then: (i) M is positive definite; (ii) for any starting guess z^0, the sequence of iterates $\{z^k\}$ defined by $LCP(z^{k-1})$ (8.19) satisfies

$$\|\widetilde{B}(z^{k+1} - z^k)\| \leq \rho \|\widetilde{B}(z^k - z^{k-1})\| \tag{8.22}$$

and converges to the unique solution of the LCP (8.2).

In (8.22), the theorem says that successive iterates z^{k+1} and z^k are closer together than z^k and z^{k-1} were at the previous iteration, because $\rho < 1$; this behavior of successive iterates is called *contraction*. The measure of distance between two successive iterates z^{k+1} and z^k is the Euclidean distance between $\widetilde{B}z^{k+1}$ and $\widetilde{B}z^k$, but keep in mind that if $\|\widetilde{B}(z^{k+1} - z^k)\|$ is getting smaller, then so is $\|(z^{k+1} - z^k)\| = \|(\widetilde{B})^{-1}\widetilde{B}(z^{k+1} - z^k)\| \leq \|(\widetilde{B})^{-1}\|\|\widetilde{B}(z^{k+1} - z^k)\|$. Note also that if we compare the closeness of successive iterates to how close successive iterates were two or more iterations before, the contraction parameter becomes a power of ρ, making it very small after sufficient iterations because $\rho < 1$. For example, applying (8.22) twice, we have

$$\begin{aligned} \|\widetilde{B}(z^{k+1} - z^k)\| &\leq \rho \|\widetilde{B}(z^k - z^{k-1})\| \\ &\leq \rho \left(\rho \|\widetilde{B}(z^{k-1} \quad z^{k-2})\|\right) \\ &= \rho^2 \|\widetilde{B}(z^{k-1} - z^{k-2})\| \end{aligned}$$

To help to understand the condition (8.21) in the theorem, let us apply it to the Illustrative Example of Section 8.2.2.2.1, with the splitting for which B is the diagonal of M. Note that B is positive definite, as required by the theorem, and in particular, all diagonal entries are positive. In this case, the matrix B is itself symmetric, because it is diagonal, and the square root matrix $\widetilde{B}$ is also a diagonal matrix with each diagonal entry equal to the square root of the corresponding entry in B. If b_{ij} (c_{ij}) is the entry in the ith row and jth column of B (C), then the entry in the ith row and jth column of the matrix $(\widetilde{B}^{-1})^T C \widetilde{B}^{-1}$ in (8.21) is

$$c_{ij}/(b_{ii}b_{jj})^{1/2}$$

With some effort, it can be shown that a sufficient condition for $\rho < 1$ in(8.21) is that

$$\sum_{j=1}^{n} \frac{|c_{ij}|}{(b_{ii}b_{jj})^{1/2}} < 1 \; \forall i = 1, ..., n \tag{8.23}$$

The sufficient condition (8.23 means that if, in the splitting, the off-diagonal entries of M are so much smaller than the diagonal entries that the sum of the ratios in (8.23) is less than 1, then the algorithm converges. Loosely stated, if C is less important than B, then the splitting algorithm converges.

We can easily see that the condition (8.23) is satisfied for the diagonal splitting of Section 8.2.2.2.1. Recalling that $n = 2$ and $c_{ii} = 0$,

$$\text{for } i = 1, \sum_{j=1}^{n} \frac{|c_{ij}|}{(b_{ii}b_{jj})^{1/2}} = \frac{1}{(6 \cdot 7)^{1/2}} < 1$$

$$\text{for } i = 2, \sum_{j=1}^{n} \frac{|c_{ij}|}{(b_{ii}b_{jj})^{1/2}} = \frac{2}{(6 \cdot 7)^{1/2}} < 1$$

8.2.2.3 Other Iterative Methods for LCPs

The damped Newton approach for solving an LCP is discussed in the next section, in the more general context of solving a nonlinear complementarity problem. This method relies on a reformulation of the LCP as a system of nonlinear equations $H(z) = 0$ whose defining functions are not differentiable everywhere. This reformulation can also be described in terms of choosing z to minimize the Euclidean norm of errors, $H(z)^T H(z)$, which leads to a method to choose a Newton direction at each iteration. See [6] and the next section for further discussion.

The very successful interior point methods for solving linear programs have stimulated much research into interior point methods for solving the LCP. For this approach to work, there must be a feasible, strictly interior solution to the inequalities in the definition of the LCP, i.e., there must exist a vector z such that $z > 0$ and $q + Mz > 0$. An interior point method starts from an interior solution, and proceeds, step by step, through a sequence of other interior solutions that get closer and closer to satisfaction of the complementarity requirement $(q + Mz)^T z = 0$. See [6] for an introduction to this approach.

8.3 Algorithms for NCP Models

Since about the 1980s, there has been a great deal of research into methods for solving the nonlinear complementarity problem (NCP). In this section, we introduce some of the main ideas of the algorithms that have been proposed for the NCP. The PATH algorithm [8], [9], [10], [25], is the focus of discussion because it is widely acknowledged to be the most successful of all of the algorithms for the NCP, i.e., it is robust (it usually solves most problems without breaking down) and efficient (it usually takes less time than other algorithms), and because a modeler using a modeling system such as GAMS can easily call the PATH solver [14] to calculate a solution to an NCP. One purpose of this section is to prepare the reader to be an informed user of the the PATH solver. Another purpose is to lay the groundwork for the reader to be able to read the literature about the PATH algorithm and about other algorithms to solve the NCP. An appendix to this chapter, Section 8.6, takes the reader through an overview of some theory about PATH and other agorithms for the NCP.

PATH is actually designed to solve a more general form of model than the NCP: it solves a mixed complementarity problem (MCP) defined on a *box*, i.e., each variable z_i has a lower bound l_i and an upper bound u_i, where $l_i \leq u_i$. We refer to this form of model as a *Box-MCP*. The following definition of the Box-MCP is more general than that presented in Chapter 1, i.e., any MCP of the Chapter 1 type can be represented as a Box-MCP of the following type, as discussed below.[3] As with the NCP, there is a mapping $f : \mathbb{R}^n \longrightarrow \mathbb{R}^n$ in the definition of the Box-MCP, as follows.

Box-MCP: find a vector $z \in \mathbb{R}^n$ that satisfies exactly one of the following conditions for each $i = 1, ..., n$

$$\begin{aligned} l_i \leq z_i \leq u_i &\text{ and } f_i(z) = 0 \\ z_i = l_i &\text{ and } f_i(z) > 0 \\ z_i = u_i &\text{ and } f_i(z) < 0 \end{aligned} \tag{8.24}$$

Note that $l_i = -\infty$ or $u_i = +\infty$ are allowed. If both of these bounds are infinite for the variable z_i, then z_i is a free variable as in the MCP definition of Chapter 1. If, for non-free variables, $l_i = 0$ and $u_i = +\infty$, then z_i is a non-negative variable as in the MCP definition of Chapter 1. These observations suggest the main ingredients to show that any MCP defined as in Chapter 1 can be represented in the Box-MCP form of (8.24).

If, in (8.24), $u_i = +\infty$ and $l_i = 0$ for *all* $i = 1, ..., n$, then the Box-MCP becomes the nonlinear complementarity problem (NCP):[4]

$$\begin{aligned} &\text{NCP: find a vector } z \in \mathbb{R}^n \text{ such that} \\ &0 \leq z_i,\ f_i(z) \geq 0 \text{ and } z_i f_i(z) = 0\ \forall\ i = 1, ..., n \end{aligned} \tag{8.25}$$

[3] In Chapter 1, the MCP is defined as follows: for the mapping $F : \mathbb{R}^n \longrightarrow \mathbb{R}^n$, find vectors $x \in \mathbb{R}^{n_1}, y \in \mathbb{R}^{n_2}, n_1 + n_2 = n$, such that

1. $F_i(x, y) \geq 0, x_i \geq 0, x_i \cdot F_i(x, y) = 0, i = 1, \ldots, n_1$
2. $F_{n_1+j}(x, y) = 0, y_j$ free, $j = 1, \ldots, n_2$.

This Chapter 1 form of the MCP can be represented as an NCP in the form (8.25), in either of two ways. If the equalities can be solved for the variables y_j, then these expressions can be substituted for y in $F_i(x, y)$. Alternatively, we can split each equality into two inequalities and we can split each variable y_j into its positive and negative parts, as $y_j = y_j^+ - y_j^-$, with $y_j^+ \geq 0, y_j^- \geq 0$; the details are left for the reader.

[4] More generally, if in (8.24) each variable has exactly one finite bound (upper or lower, but not both), then with a change of variables, the Box-MCP model (8.24) can be put into the NCP form (8.25). For example, if $l_i \leq z_i$, then define a new variable $w_i = z_i - l_i$, require that $w_i \geq 0$ and substitute $w_i + l_i$ wherever z_i appears in the Box-MCP model (8.24). The details are left for the reader.

In order to simplify the discussion, we describe the PATH algorithm for the NCP, and our remarks about other algorithms are also with respect to solving the NCP.

The PATH algorithm relies on the key ideas of Newton's method for solving a system of nonlinear equations, namely: generation of a sequence of approximate solutions by solving linear approximations to the nonlinear equations; constructing each linear approximation in the sequence as a first-order Taylor approximation about the previous approximate solution; and, when necessary to speed convergence, adjusting the generated sequence of approximate solutions by a procedure called *damping.* Newton's method must be modified for application to the solution of an NCP, but it results in: a sequence of LCPs that approximate the NCP; each LCP is formulated by a first-order Taylor approximation to the functions $f(z)$ about the approximate solution calculated by the previous LCP; and a type of damping is sometimes employed, to speed convergence.

In this section, we begin with an overview of the traditional Newton's method applied to a system of nonlinear equations. Then we give an easy, intuitive description of the modified Newton's method for NCPs – the PATH algorithm – with an introduction to the theory underlying the algorithm, in an appendix to this chapter in Section 8.6.

8.3.1 Newton's Method for Systems of Smooth Equations

One of the key ingredients of the PATH algorithm (and of many other algorithms) is Newton's method for the solution of smooth nonlinear equations. By "smooth" we mean that the functions that define the equations have continuous partial derivatives. Newton's method must be modified to be applicable to the NCP, because the NCP is not a system of smooth nonlinear equations. (However, it is possible to reformulate the NCP in various ways as a system of *nonsmooth* nonlinear equations, as discussed below in the theory underlying the PATH algorithm.)

We first provide a brief review of the traditional method, due to Newton, for solving a nonlinear equation (or a system of nonlinear equations) by solving a sequence of linear equations (or systems of linear equations) that approximate the nonlinear version of the problem. Each approximate linear problem is defined by using the derivative (gradient for systems of equations) evaluated at the most recent solution in the sequence, to determine the linear part of the approximation. Thus, the derivative (or gradient) of the nonlinear function must be well-defined, and preferably continuous – the nonlinear function is "smooth" – for this method to work. If all goes well, then the sequence of solutions to the linear approximations converges to the solution of the nonlinear problem. Convergence is not guaranteed in general, depending on whether the first guess of a solution – provided by the user, to determine the first linear approximation in the sequence – is far from the actual solution. Even with a bad starting guess, a modified version of the algorithm can converge nicely, namely if "damping" is employed.

We begin with a discussion of the pure Newton method, without damping, and then describe some damping procedures, for smooth equations.

8.3.1.1 Undamped Newton Method for Smooth Equations

Newton's method solves for a root of a system of nonlinear equation $f(z) = 0$, where $f : \mathbb{R}^n \longrightarrow \mathbb{R}^n$. It is an iterative method which starts with a guess of a solution, z^0, and then approximates the nonlinear function $f(z)$ by a linear function whose coefficients are in the gradient of f at z^0, $\nabla f(z^0)$. This gives an approximate equation $f(z^0) + \nabla f(z^0)(z - z^0) = 0$. The solution of this approximate, linear equation is the next iterate, z^1, where a new linear approximation is solved for z^2. The process is repeated until an iteration k, when $f(z^k)$ is sufficiently close to zero. Thus, the general iterative step solves for z^k in the equation

$$f(z^{k-1}) + \nabla f(z^{k-1})(z^k - z^{k-1}) = 0 \text{ , i.e.,} \tag{8.26a}$$

$$z^k = z^{k-1} - [\nabla f(z^{k-1})]^{-1} f(z^{k-1}) \tag{8.26b}$$

where $\nabla f(z^{k-1})$ is the matrix of partial derivatives of f, called the Jacobian, evaluated at z^{k-1}:

$$\nabla f(z) = \begin{pmatrix} \partial f_1/\partial z_1, & \cdots, \partial f_1/\partial z_n \\ \vdots & \vdots \\ \partial f_n/\partial z_1, & \cdots, \partial f_n/\partial z_n \end{pmatrix} \tag{8.27}$$

There are two important requirements on the mapping f: it must be continuously differentiable (i.e., ∇f must be well defined and continuous at all points z); and ∇f must be a nonsingular matrix (∇f^{-1} must exist) at every point z^{k-1} in the iterations of (8.26b).

8.3.1.1.1 Illustrative Example. Solving two equations in two unknowns by Newton's method

In this example, we wish to solve two equations in the two unknowns x and y. The definitions of the functions are:

$$f_1(x, y) = e^{-x} - 2x + y$$
$$f_2(x, y) = x^2 - xy + 1$$

For vector notation, define

$$z = \begin{pmatrix} x \\ y \end{pmatrix}, \quad f(z) = \begin{pmatrix} f_1(x,y) \\ f_2(x,y) \end{pmatrix}$$

The gradients of the two functions form the Jacobian matrix of f:

$$\nabla f(z) = \begin{pmatrix} \partial f_1/\partial x, \ \partial f_1/\partial y \\ \partial f_2/\partial x, \ \partial f_2/\partial y \end{pmatrix} = \begin{pmatrix} -e^{-x} - 2 & 1 \\ 2x - y & -x \end{pmatrix}$$

Let us start the iterations with the guess $(x^0, y^0) = (1, 1)$. Then with the updated estimate of z^k at the kth iteration given by

$$z^k = z^{k-1} - [\nabla f(z^{k-1})]^{-1} f(z^{k-1})$$

and the Euclidean norm of f by

$$\|f(z^k)\| = ((f_1(x^k, y^k))^2 + (f_2(x^k, y^k))^2)^{1/2}$$

the iterations proceed as follows:

$$z^0 = \begin{pmatrix} 1 \\ 1 \end{pmatrix}, \qquad f(z^0) = \begin{pmatrix} -0.63212 \\ 1 \end{pmatrix}, \ \|f(z^0)\| = 1.183037$$

$$z^1 = \begin{pmatrix} 1.268941 \\ 2.268941 \end{pmatrix}, \ f(z^1) = \begin{pmatrix} 0.012188 \\ -0.26894 \end{pmatrix}, \ \|f(z^1)\| = 0.269217$$

$$z^2 = \begin{pmatrix} 1.172404 \\ 2.036540 \end{pmatrix}, \ f(z^2) = \begin{pmatrix} 0.001353 \\ -0.01312 \end{pmatrix}, \ \|f(z^2)\| = 0.013186$$

$$z^3 = \begin{pmatrix} 1.167599 \\ 2.024089 \end{pmatrix}, \ f(z^3) = \begin{pmatrix} 3.58 \times 10^{-6} \\ -3.7 \times 10^{-5} \end{pmatrix}, \ \|f(z^3)\| = 3.69 \times 10^{-5}$$

□

8.3.1.1.2 Convergence of the Undamped Newton Method for Smooth Equations

Under not very restrictive conditions on f, if z^* stands for an exact solution to $f(z) = 0$, then if z^{k-1} is sufficiently close to z^*, it can be shown that for a positive constant c,

$$\|z^k - z^*\| \le c\|z^{k-1} - z^*\|^2 \tag{8.28}$$

This is called *quadratic convergence* because, when $\|z^{k-1} - z^*\|$ is very small (e.g., less than one), then the distance of the next iterate z^k from the solution z^* is proportional to the square of the very small number $\|z^{k-1} - z^*\|$. More precisely, (8.28) means that, when $\|z^{k-1} - z^*\| < 1/c$, the next iterate is closer to the solution than the previous iterate:

$$\|z^k - z^*\| \le c\|z^{k-1} - z^*\|^2$$

$$< c(1/c)\|z^{k-1} - z^*\|$$
$$= \|z^{k-1} - z^*\|$$

All of this means that, when the Newton method gets close to a solution, subsequent convergence is quite rapid. This is called local convergence. Unfortunately, getting close may be hard to achieve in some circumstances, but damping methods can help.

8.3.1.2 Damped Newton Methods for Smooth Equations

The quadratic convergence result (8.28) does not guarantee steady progress toward the solution, if the first guess z^0 is far from the solution. The iterates can wander wildly away from the solution, as illustrated in the next example. This wild behavior can often be tamed by a damping procedure, in which the next iterate z^k is not taken to be the result of the Newton calculation (8.26b), called the *Newton point*, but instead is a point between the current iterate z^{k-1} and the Newton point. The following example illustrates some damping procedures, applied to the previous illustration of two equations in two unknowns.

8.3.1.2.1 Illustrative Example. Damping Procedures to Accelerate Convergence

Consider again the problem of finding a solution of

$$f(z) = \begin{pmatrix} e^{-x} - 2x + y \\ x^2 - xy + 1 \end{pmatrix} = \begin{pmatrix} 0 \\ 0 \end{pmatrix}$$

Starting the undamped Newton method from a poor first guess can result in large oscillations for several iterations before getting close to the solution, e.g.:

$$z^0 = \begin{pmatrix} 0.2 \\ 19 \end{pmatrix}, \quad f(z^0) = \begin{pmatrix} 9.418731 \\ -0.76 \end{pmatrix}, \|f(z^0)\| = 9.449343$$

$$z^1 = \begin{pmatrix} 0.322630 \\ -0.073071 \end{pmatrix}, f(z^1) = \begin{pmatrix} 0.005912 \\ 1.127665 \end{pmatrix}, \|f(z^1)\| = 1.127680$$

$$z^2 = \begin{pmatrix} 7.356480 \\ 19.082928 \end{pmatrix}, f(z^2) = \begin{pmatrix} 4.370607 \\ -85.265377 \end{pmatrix}, \|f(z^2)\| = 85.37732$$

$$z^3 = \begin{pmatrix} 4.573886 \\ 9.145358 \end{pmatrix}, f(z^3) = \begin{pmatrix} 0.007903 \\ -19.909391 \end{pmatrix}, \|f(z^3)\| = 19.909392$$

$$z^4 = \begin{pmatrix} 2.412001 \\ 4.791377 \end{pmatrix}, f(z^4) = \begin{pmatrix} 0.057012 \\ -4.739058 \end{pmatrix}, \|f(z^4)\| = 4.739401$$

$$z^5 = \begin{pmatrix} 1.493084 \\ 2.814165 \end{pmatrix}, f(z^5) = \begin{pmatrix} 0.052675 \\ -0.972485 \end{pmatrix}, \|f(z^5)\| = 0.973911$$

Now let us modify the Newton procedure in a simple way that damps the oscillations: at iteration k, calculate the *Newton point* z_N^k as before from the previous iterate z^{k-1}, i.e., $z_N^k = z^{k-1} - [\nabla f(z^{k-1})]^{-1} f(z^{k-1})$, but instead of making the Newton point the next iterate, choose a point on the line joining z^{k-1} and z_N^k: $z^k = (1-\lambda)z^{k-1} + \lambda z_N^k$, where λ is a parameter between 0 and 1. A good damping procedure will choose λ differently at each iteration, such that there is some decrease in $\|f(z^k)\|$ compared with the previous iteration's value, $\|f(z^{k-1})\|$. To show that a dynamic choice of λ can improve over a constant choice, we first show a nondynamic approach and set $\lambda = 0.8$ for all iterations. Even this nondynamic choice produces a noticeable damping of the oscillations and gets close to a solution in few iterations:

$$z^0 = \begin{pmatrix} 0.2 \\ 19 \end{pmatrix}, \quad f(z^0) = \begin{pmatrix} 9.418731 \\ -0.76 \end{pmatrix}, \ \|f(z^0)\| = 9.449343$$

$$z^1 = \begin{pmatrix} 0.298104 \\ 1.741543 \end{pmatrix}, \ f(z^1) = \begin{pmatrix} 1.887560 \\ 0.569705 \end{pmatrix}, \ \|f(z^1)\| = 1.971661$$

$$z^2 = \begin{pmatrix} 0.759645 \\ 1.497146 \end{pmatrix}, \ f(z^2) = \begin{pmatrix} 0.445687 \\ 0.439761 \end{pmatrix}, \ \|f(z^2)\| = 0.626121$$

$$z^3 = \begin{pmatrix} 1.095758 \\ 1.970067 \end{pmatrix}, \ f(z^3) = \begin{pmatrix} 0.112836 \\ 0.041969 \end{pmatrix}, \ \|f(z^3)\| = 0.120388$$

$$z^4 = \begin{pmatrix} 1.152465 \\ 2.012168 \end{pmatrix}, \ f(z^4) = \begin{pmatrix} 0.023095 \\ 0.009222 \end{pmatrix}, \ \|f(z^4)\| = 0.024868$$

$$z^5 = \begin{pmatrix} 1.164531 \\ 2.021635 \end{pmatrix}, \ f(z^5) = \begin{pmatrix} 0.004642 \\ 0.001876 \end{pmatrix}, \ \|f(z^5)\| = 0.005006$$

However, $\lambda = 0.8$ was chosen after some experimentation with other values (which did not produce good convergence), so it is not at all clear that for different problems, this nondynamic choice will reduce the computational time to find an accurate solution.

Next we show the iterations when λ is chosen by a simple dynamic rule: try $\lambda^k = 1$ (making z^k correspond to the Newton point z_N^k), and if $\|f(z^k)\| < \|f(z^{k-1})\|$, then accept $\lambda^k = 1$; else set $\lambda^k = 0.9$, and if this makes $\|f(z^k)\| < \|f(z^{k-1})\|$, then accept $\lambda^k = 0.9$; etc., always decreasing λ by 0.1 if the last value of λ fails to reduce the norm. Thus, it is not necessary to specify a value of λ in advance. Here is what happens to the example:

$$z^0 = \begin{pmatrix} 0.2 \\ 19 \end{pmatrix}, \quad f(z^0) = \begin{pmatrix} 9.418731 \\ -0.76 \end{pmatrix}, \ \|f(z^0)\| = 9.449343$$

$$\lambda^1 = 1, \ z^1 = \begin{pmatrix} 0.32263 \\ -0.07307 \end{pmatrix}, \ f(z^1) = \begin{pmatrix} 0.00591 \\ 1.12767 \end{pmatrix}, \ \|f(z^1)\| = 1.12768$$

$$\lambda^2 = 0.1, \ z^2 = \begin{pmatrix} 1.02602 \\ 1.84253 \end{pmatrix}, \ f(z^2) = \begin{pmatrix} 0.14893 \\ 0.16224 \end{pmatrix}, \ \|f(z^2)\| = 0.22024$$

$$\lambda^3 = 1,\ z^3 = \begin{pmatrix} 1.16855 \\ 2.02976 \end{pmatrix},\ f(z^3) = \begin{pmatrix} 0.00347 \\ -0.0064 \end{pmatrix},\ \|f(z^3)\| = 0.00726$$

$$\lambda^4 = 1,\ z^4 = \begin{pmatrix} 1.16759 \\ 2.02406 \end{pmatrix},\ f(z^4) = \begin{pmatrix} 0.00000 \\ -0.00001 \end{pmatrix},\ \|f(z^4)\| = 0.00001$$

This simple dynamic rule accepts the Newton point at all but the second iteration, and it achieves an accuracy, in three iterations, almost as good as the nondynamic choice of λ achieves in five iterations. Furthermore, there is no need to experiment with different values for λ, as the rule does the experimentation, but only when necessary to reduce the norm. Note, however, that z^2 and $f(z^2)$ had to be evaluated 10 times to determine $\lambda^2 = 0.1$. As a general procedure, this way to determine λ is not a good one because, in other examples, the same-size steps in λ could fail to find a new iterate that reduces the norm.

There are other, more robust rules to do a *linesearch* for a norm-reducing point between z^{k-1} and z_N^k, requiring fewer evaluations of z^k and $f(z^k)$, such as an Armijo search – for a parameter $\rho \in (0,1)$, try $\lambda^k = \rho^m$ for $m = 0, 1, 2, ...$ (note that m is a power of ρ, which makes ρ^m smaller as m increases) and use the smallest value of m that produces a reduction in the norm.

Armijo Search for Next Iterate

Given $\rho \in (0,1)$ and Newton point z_N^k,

choose the next iterate as

$z^k = (1-\rho^m)z^{k-1} + \rho^m z_N^k$

for the smallest value of m such that

$\|f(z^k)\| < \|f(z^{k-1})\|$.

Applied to our example to determine λ^2, with $\rho = 0.5$, we would try $\lambda^2 = 1$; then $\lambda^2 = 0.5$ (if necessary); then $\lambda^2 = 0.25$ (if necessary); then $\lambda^2 = 0.125$ (if necessary); etc. For our example, this would have required only four evaluations of z^2 and $f(z^2)$ to choose $\lambda^2 = 0.125$.

□

8.3.1.2.2 Convergence of the Damped Newton Method for Smooth Equations

If the damping procedure (e.g., Armijo search) works as intended, then it guarantees that the norm of f is reduced at each iteration, which means that each iteration brings us closer to a solution, as measured by the norm of f. Eventually, the iterates get close enough to the solution that rapid quadratic convergence (8.28) begins. Therefore, if we can be sure that damping guarantees a reduction in the norm of f, then damping converts the pure Newton method into a method that has *global* convergence – from any starting guess

z^0, the algorithm eventually converges. Does damping guarantee a reduction in the norm of f? The answer is *yes*, under fairly general conditions on f – conditions that are not very restrictive, in practice. We next outline the main idea.

A reduction in $\|f(z)\|$ occurs if and only if a reduction in $\theta(z) = \frac{1}{2}\|f(z)\|^2 = \frac{1}{2}f(z)^T f(z) = \frac{1}{2}\sum_{i=1}^{n}(f_i(z))^2$ also occurs. We now argue that moving in the Newton direction $(z_N^k - z^{k-1}) = -[\nabla f(z^{k-1})]^{-1} f(z^{k-1})$ will produce a new value of θ that is smaller than $\theta(z^{k-1})$. Moving all the way to z_N^k from z^{k-1} may sometimes produce a reduction in θ, but we might have to keep close to the old iterate z^{k-1} along the line joining z^{k-1} and z_N^k, to find a new iterate that reduces θ. A line search such as the Armijo search finds a point in the Newton direction that reduces θ.

To show that there is a point in the Newton direction that reduces θ, we need to derive an expression for the gradient of θ:

$$\begin{aligned}\nabla\theta &= \big(\partial\theta/\partial z_1, \cdots, \partial\theta/\partial z_n\big)\\ &= \left(\sum_{i=1}^{n} f_i(z)\partial f_i/\partial z_1, \cdots, \sum_{i=1}^{n} f_i(z)\partial f_i/\partial z_n\right)\\ &= f(z)^T\nabla f(z)\ , \text{ using (8.27)}\end{aligned}$$

Points along the line joining z_N^k and z^{k-1} may be represented by $z = (1-\lambda)z^{k-1} + \lambda z_N^k$, for values of λ such that $0 \le \lambda \le 1$, which implies a step from z^{k-1} of $z - z^{k-1} = (z_N^k - z^{k-1})\lambda = -[\nabla f(z^{k-1})]^{-1} f(z^{k-1})\lambda$. The first-order Taylor approximation of θ is

$$\theta(z) = \theta(z^{k-1}) + \nabla\theta(z^{k-1})(z - z^{k-1}) + R$$

where the remainder term R is proportional to λ^2, as λ varies to produce different points on the line joining z_N^k and z^{k-1}. Therefore, for some constant number r, independent of z,

$$\begin{aligned}\theta(z) - \theta(z^{k-1}) &= \nabla\theta(z^{k-1})(z - z^{k-1}) + r\lambda^2\\ &= f(z^{k-1})^T\nabla f(z^{k-1})(z_N^k - z^{k-1})\lambda + r\lambda^2\\ &= -f(z^{k-1})^T\nabla f(z^{k-1})[\nabla f(z^{k-1})]^{-1} f(z^{k-1})\lambda + r\lambda^2\\ &= -f(z^{k-1})^T f(z^{k-1})\lambda + r\lambda^2\\ &= -\|f(z^{k-1})\|^2\lambda + r\lambda^2\end{aligned}$$

For small enough λ (i.e., for z close enough to z^{k-1}), the term linear in λ is larger in magnitude than the term $r\lambda^2$, which means that for small enough λ, $\theta(z) - \theta(z^{k-1}) < 0$, indicating a decrease in θ. Note that λ need not be reduced all the way to zero to bring about a decrease in θ – there is a positive value of λ that will reduce θ, i.e., we can find a new iterate that is different from and strictly better than the last iterate z^{k-1}.

The discussion above outlines the main idea to support the notion that damping converts the locally convergent pure Newton method into a globally convergent algorithm. However, there are some important technical requirements which are left out of the discussion above. A complete discussion may be found in [21].

8.3.2 Newton's Method for the NCP

The key idea of the undamped Newton method for smooth equations is to calculate a sequence of iterates z^k by solving a sequence of linear approximations to the nonlinear equations. The linear approximations are first-order Taylor approximations estimated at the previous value of the iterate. Early attempts to apply Newton's method to the NCP (see, e.g., [18], [20]) started with this sequential linear approximation idea, i.e., a sequence of approximate LCPs, and that is where we begin.

8.3.2.1 Constructing an Approximate LCP

The NCP (8.25), stated in vector form, is to find $z \in \mathbb{R}^n$ such that

$$0 \leq f(z) \perp z \geq 0 \tag{8.29}$$

Assuming that we have already calculated z^{k-1} at the previous iteration $k-1$, we construct a first-order Taylor approximation of the vector-valued function f about the point z^{k-1}:

$$f(z) \approx f(z^{k-1}) + \nabla f(z^{k-1})(z - z^{k-1}) \tag{8.30}$$

Collecting the constant terms together and introducing the compact notation

$$M^{k-1} = \nabla f(z^{k-1}) \quad \text{(an } n \times n \text{ matrix), and} \tag{8.31}$$

$$q^{k-1} = f(z^{k-1}) - M^{k-1}z^{k-1} \quad \text{(an } n- \text{ vector)} \tag{8.32}$$

we define the LCP approximation to (8.29) using (8.30), as

$$\text{LCP}(z^{k-1}): \qquad w = q^{k-1} + M^{k-1}z, \; 0 \leq w \perp z \geq 0 \tag{8.33}$$

In the undamped Newton method, (8.33) is solved to produce the next iterate, z^k. Damping procedures are discussed below, but first we must consider how to solve each approximate LCP (8.33).

8.3.2.2 Solving the Approximate LCP

Early attempts to apply Newton's method to (8.29) [20] turned to Lemke's pivoting method for solution of each approximate LCP in the sequence. Recall that Lemke's method to solve an LCP starts from an augmented system that includes an artificial variable at a value for which $z = 0$ and $w \geq 0$, so that complementarity is satisfied. Pivoting is done in a way that maintains complementarity of z and w with the goal of driving the artificial variable to zero, which corresponds to a solution to the LCP.

Because Lemke's method must start at the point $z = 0$, it is likely to be an inefficient method, compared to any method (if we could find one) that starts pivoting at the current iterate z^{k-1}. If the Newton method is gradually getting closer to the solution of the NCP (8.29), then, at least for later iterations k, z^{k-1} must also be close to the solution of the current approximation LCP(z^{k-1}). Therefore, it seems likely that a pivoting method that starts at z^{k-1} will have fewer pivots than Lemke's method, which starts pivoting at $z = 0$. The developers of the PATH algorithm [8], [9], [25], [26] invented a pivoting method that starts at z^{k-1}.

Like Lemke's method, the PATH method also constructs an augmented system with an artificial variable (called an *auxiliary* variable in the literature), but in a different way. The auxiliary variable $t \in \mathbb{R}^1$ starts at $t = 0$, and is driven to a special value, $t = 1$ during pivoting, to determine a solution to LCP(z^{k-1}). The key idea is to construct the augmented LCP such that, at $t = 0$, the previous iteration's result, (w^{k-1}, z^{k-1}) satisfies the augmented system, and after some pivoting, at $t = 1$, the (w, z) values satisfy the augmented system and also the requirements of LCP(z^{k-1}). Here is the augmented system which does the trick.

Augmented LCP(z^{k-1}) :

$$w = q^{k-1} + M^{k-1}z - (1-t)r^{k-1},\ 0 \leq w \perp z \geq 0 \tag{8.34a}$$

$$\text{where } r^{k-1} = f(z^{k-1}) - (q^{k-2} + M^{k-2}z^{k-1}) \tag{8.34b}$$

$$= f(z^{k-1}) - w^{k-1} \tag{8.34c}$$

In words, the vector r^{k-1} is the difference between the exact value of f at the previous iterate z^{k-1}, and the previous iteration's linear approximation (which used q^{k-2} and M^{k-2}) evaluated at z^{k-1}. The vector r^{k-1} can equally be defined as $f(z^{k-1}) - w^{k-1}$, because w^{k-1} is the value of the the last iteration's linear approximation, at z^{k-1}. Note that (w^{k-1}, z^{k-1}) satisfy the requirements $0 \leq w \perp z \geq 0$ because they are the solution to LCP(z^{k-2}) (except when $k = 0$, and z^0 must be a guess, to get the algorithm started – see below for a discussion of getting started).

For $t = 0$, we can evaluate w using (8.34a) at $z = z^{k-1}$:

$$w = q^{k-1} + M^{k-1}z^{k-1} - r^{k-1}$$

$$
\begin{aligned}
&= \{f(z^{k-1}) - M^{k-1}z^{k-1}\} + M^{k-1}z^{k-1} \\
&\qquad -\{f(z^{k-1}) - w^{k-1}\}, \text{ by (8.32), (8.34c)} \\
&= w^{k-1}
\end{aligned}
$$

and because (w^{k-1}, z^{k-1}) satisfy the requirements $0 \leq w \perp z \geq 0$, we can see that $(w^{k-1}, z^{k-1}, t = 0)$ is a valid complementary solution of (8.34a) where we can start the pivoting. The first pivot in the Augmented LCP(z^{k-1}) (8.34a) will be to increase t to a value larger than zero. After one or (typically) more pivots, when $t = 1$, then (8.34a) becomes

$$
\begin{aligned}
w &= q^{k-1} + M^{k-1}z - (1-1)r^{k-1} \\
&= q^{k-1} + M^{k-1}z, \; 0 \leq w \perp z \geq 0
\end{aligned}
$$

i.e., the values of w and z satisfy the requirements of LCP(z^{k-1}) (8.33), and so we have found the next iterate (w^k, z^k) (in the undamped Newton method), or the next Newton point (in the damped Newton method – see below).

In descriptions of the PATH algorithm and in output from the PATH solver, e.g., [14], the terms *major iteration* and *minor iteration* are used. A major iteration is the solution of an approximate LCP, and each pivot that is done to solve an LCP is a minor iteration.

8.3.2.3 Getting Started: Solving the First Approximate LCP

The user may have a reasonable first guess of the solution, $z^0 \geq 0$, or, in the absence of a user-provided guess, the solver software could provide a starting point such as $z^0 = 0$. At the first iteration, $k = 1$, the algorithm could construct a first approximation LCP(z^0) (8.33), but the question addressed here is: how do we solve LCP(z^0)? Should we construct an augmented system as in (8.34), and if so, what values should we assign to $q^{k-2} = q^{-1}$ and M^{-1}? This is difficult to answer, because there was no previous iteration at which there was an approximate LCP based on parameters q^{-2} and M^{-2}. Perhaps we could use the first guess z^0, and arbitrarily construct a complementary, positive vector w^0 – we do this in the two examples below.

Another idea is to avoid the augmented LCP (8.34) for the first iteration, and instead solve LCP(z^0) by Lemke's method. The linear approximation is made about the guess z^0, which may or may not equal 0, but Lemke's method sets up an augmented LCP with an artificial variable, and the starting value of z is $z = 0$.

In the implementation of the PATH algorithm, in the PATH solver [8], [9], [14], neither of these approaches is followed. Instead, a *crashing* heuristic technique starts from the user-provided first guess z^0, and seeks a good first approximation to the solution of the NCP. The first LCP approximation is constructed about the approximate solution found by the crashing technique. See [10] for details.

8.3.2.4 Two Examples Without Damping

The first example illustrates the PATH procedure as applied to a linear complementarity problem. No linear approximations are required, and consequently there is only one iteration $k = 1$. The point of the illustration is to show the set-up of the augmented problem, involving the vector $r^{k-1} = r^0$, and the pivots from $t = 0$ to $t = 1$, to find the Newton point, which in this case is the solution of the LCP.

8.3.2.4.1 Illustrative Example. PATH method for two-commodity LCP

Consider again the two-commodity equilibrium LCP (8.12):

$$w_1 = -32 + 6z_1 + z_2 \ , \quad z_1 \geq 0 \ , \quad w_1 \geq 0 \ , \quad w_1 z_1 = 0$$
$$w_2 = -64 + 2z_1 + 7z_2 \ , \quad z_2 \geq 0 \ , \quad w_2 \geq 0 \ , \quad w_2 z_2 = 0$$

Thus, $f(z) = Mz + q$, where $M = \begin{pmatrix} 6 & 1 \\ 2 & 7 \end{pmatrix}$ and $q = \begin{pmatrix} -32 \\ -64 \end{pmatrix}$. Because the functions are affine, the linear approximation never changes as the iterations proceed, i.e., the linear approximation is exactly the function itself: in (8.12), $M^{k-1} = M$ and $q^{k-1} = q$, for all k, but because we never update the linear approximation, there is only one iteration in this example, $k = 1$. Nevertheless, we can illustrate how PATH performs pivots from a starting guess that is different from zero. Suppose that the first guess of $z^0 \in \mathbb{R}^2$ is $z^0 = \begin{pmatrix} 1 \\ 0 \end{pmatrix}$. We must calculate the vector $r^0 = f(z^0) - w^0$, which requires that we have a vector w^0 which is non-negative and complementary to z^0. The PATH algorithm carries out a crashing procedure for this purpose, but here, we arbitrarily assign the value $w^0 = \begin{pmatrix} 0 \\ 1 \end{pmatrix}$, which satisfies the requirements of non-negativity and being complementary to z^0. Therefore, from (8.34c)

$$\begin{aligned} r^0 &= Mz + q - w^0 \\ &= \begin{pmatrix} 6 & 1 \\ 2 & 7 \end{pmatrix} \begin{pmatrix} 1 \\ 0 \end{pmatrix} + \begin{pmatrix} -32 \\ -64 \end{pmatrix} - \begin{pmatrix} 0 \\ 1 \end{pmatrix} = \begin{pmatrix} -26 \\ -63 \end{pmatrix} \end{aligned}$$

The augmented system of equations in (8.34a) is

$$w = Mz + q - (1 - t)r^0$$

$$= (q - r^0) + Mz + r^0 t$$

i.e.,

$$\begin{aligned} w_1 &= -6 + 6z_1 + z_2 - 26t \\ w_2 &= -1 + 2z_1 + 7z_2 - 63t \end{aligned}$$

When $t = 0$, at the beginning of the path, $z_1 = 1$, $z_2 = 0$, $w_1 = 0$, and $w_2 = 1$, which is a feasible and complementary solution to the augmented system. At this point, the basic variables are z_1 and w_2, while z_2, w_1, and t are the non-basic variables. Let us pivot, to express the basic variables (on the left) in terms of the non-basic variables:

$$\begin{aligned} z_1 &= 1 + (1/6)w_1 - (1/6)z_2 + (13/3)t \\ w_2 &= -1 + 2(1 + (1/6)w_1 - (1/6)z_2 + (13/3)t) + 7z_2 - 63t \\ &= 1 + (1/3)w_1 + (20/3)z_2 - (163/3)t \end{aligned}$$

We increase t to start the path, until increases are blocked, which in this case can happen either when $w_2 = 0$, or when $t = 1$ (z_1 remains basic because its t-coefficient is positive). Here, the blocking occurs when $w_2 = 0$, at which point $t = 3/163$, and $z_1 = 176/163$. Note that, as t increases to $3/163$, the variables z_1 and w_2 change linearly with respect to t on this piece of the path to the solution; we say that the whole path is *piecewise linear* because it is linear on each piece. Now the basic variables are z_1 and t, while z_2, w_1, and w_2 are the non-basic variables. After pivoting, the system of equations, with basic variables on the left and non-basic on the right becomes:

$$\begin{aligned} t &= (3/163) + (1/163)w_1 - (3/163)w_2 + (20/163)z_2 \\ z_1 &= (176/163) + (1/6 + 13/489)w_1 - (13/163)w_2 + (260/489 - 1/6)z_2 \end{aligned}$$

Since w_2 just became non-basic, we increase its complement, z_2, until increases are blocked, which in this case happens when $t = 1$, signalling the end of the iterations, and a solution to the LCP: $z_1 = 4$, $z_2 = 8$, $w_1 = 0$, $w_2 = 0$. In this last pivot, we observe that as z_2 increases, the basic variables t and z_1 change linearly with respect to z_2, on this second piece of the piecewise linear path to the solution.

□

8.3.2.4.2 Illustrative Example. PATH method for two-commodity NCP

In this example, f is nonlinear, and we show the linear approximation at the first iteration.

Let us add a nonlinear term to one of the functions in the LCP model of two-commodity equilibrium in the previous example (Section 8.3.2.4.1), which means we have an NCP:

$$f_1(z_1, z_2) = -32 + 6z_1 + z_2 + 0.1z_1^2 \ , \ 0 \le z_1 \perp f_1(z_1, z_2) \ge 0$$
$$f_2(z_1, z_2) = -64 + 2z_1 + 7z_2 \ , \quad 0 \le z_2 \perp f_2(z_1, z_2) \ge 0$$

We have a solution to the LCP which is very similar to the NCP, i.e., $z_1 = 4$, $z_2 = 8$, so it seems reasonable to start the PATH algorithm by linearizing about this point. That is, $z^0 = \begin{pmatrix} 4 \\ 8 \end{pmatrix}$, and $f(z^0) = \begin{pmatrix} 1.6 \\ 0 \end{pmatrix}$. To get started, we also need a non-negative vector $w^0 = (w_1^0, w_2^0)^T$ which is complementary to z^0, i.e., $4w_1^0 = 0$ and $8w_2^0 = 0$. The only possibility is $w^0 = \begin{pmatrix} 0 \\ 0 \end{pmatrix}$. Therefore,

$$\nabla f(z^0) = M^0 = \begin{pmatrix} (6 + 2 \cdot 0.1 \cdot 4) & 1 \\ 2 & 7 \end{pmatrix} = \begin{pmatrix} 6.8 & 1 \\ 2 & 7 \end{pmatrix}$$
$$q^0 = f(z^0) - M^0 z^0 = \begin{pmatrix} 1.6 \\ 0 \end{pmatrix} - \begin{pmatrix} 6.8 & 1 \\ 2 & 7 \end{pmatrix} \begin{pmatrix} 4 \\ 8 \end{pmatrix} = \begin{pmatrix} -33.6 \\ -64 \end{pmatrix}$$
$$r^0 = f(z^0) - w^0 = \begin{pmatrix} 1.6 \\ 0 \end{pmatrix}$$

Therefore, the augmented system (8.34a), Augmented LCP(z^0), for the first linearization is

$$\begin{aligned} w &= q^0 + M^0 z - (1 - t) r^0 \\ &= (q^0 - r^0) + M^0 z + r^0 t \ , \text{ i.e.,} \end{aligned}$$

$$\begin{aligned} w_1 &= -35.2 + 6.8z_1 + z_2 + 1.6t \\ w_2 &= -64 + 2z_1 + 7z_2 \end{aligned}$$

The initial basic variables are z_1,and z_2, while w_1, w_2, and t are non-basic. Expressing the system with the basic variables on the left, which involves inverting M^0, we have:

$$\begin{aligned} z_1 &= 4 + 0.154w_1 - 0.022w_2 - 0.245614t \\ z_2 &= 8 - 0.04386w_1 + 0.149123w_2 + 0.070175t \end{aligned}$$

Now we increase t until it is blocked, which happens for $t = 1$ in this case, for which $z_1 \simeq 3.754386$ and $z_2 \simeq 8.070175$. At this point, $f_1(z_1, z_2) \simeq 0.006032$ and $f_2(z_1, z_2) \simeq -0.000003$; this approximate solution satisfies $z \ge 0$, but not $f(z) \ge 0$ or complementarity. If we want a more accurate solution, then we can linearize $f(z)$ about the new iterate z^1, and compute the path from z^1 to the next iterate, z^2. We leave this second iteration as an exercise for the reader.

□

8.3.2.5 Damping in the Newton Method for NCPs

Damping takes a different form in the PATH algorithm, compared with the solution of smooth nonlinear equations, but it is an important element in the transformation of a locally convergent algorithm (the undamped Newton method described above) to a globally convergent algorithm. With smooth nonlinear equations $f(z) = 0$, damping involves a linesearch along the straight line joining the current iterate and the next Newton point (the solution of the current linear approximation), and a point on the line is chosen with a view to reduce $\|f(z)\|$. In the PATH algorithm, the current iterate and the next Newton point are connected by a piecewise linear path – a linear piece of the path corresponds to the changing values of the basic variables, as the next non-basic variable is increased. The end points of each line segment in the path are calculated when a pivot is made. This leads to the idea, first proposed by [25], that the path from the current iterate to the next Newton point can be searched for a good value of z as the next iterate.

However, $\|f(z)\|$ is not a good measure of how close we are to a solution of the NCP (8.29). To implement damping in the NCP context, we need another way to evaluate goodness of points along the piecewise linear path from the current iterate to the Newton point (the solution of the current LCP approximation). In order to make the notation clear, we use the symbol z_N^k for the Newton point, the solution of LCP(z^{k-1}), (8.33), and the symbol z^k will represent the next iterate, which may be the Newton point, or it may be another point along the piecewise linear path between z^{k-1} and z_N^k. Note that, even if $z^k \neq z_N^k$, i.e., z^k has been chosen as one of the points along the path from z^{k-1} to z_N^k, the pair (w^k, z^k) is non-negative and complementary, making it an acceptable starting point for pivoting to solve the augmented LCP(z^k) (8.34).

For the purpose of evaluating the goodness of points on the piecewise linear path, we need to define a *merit function*, which we denote by $\theta(z) \in \mathbb{R}^1$. By analogy with $\|f(z)\|$ as it is used in damping of Newton's method for smooth equations, we need to define $\theta(z)$ which has the following properties.

Merit Function Properties

1. $\theta(z) \geq 0$ for all z, and
2. $\theta(z) = 0$ if and only if z solves the NCP (8.29).

A merit function should also be easy to compute, in order to avoid long computation times. For example, one function that satisfies the requirements of a merit function is $\theta(z) = \frac{1}{2}H(z)^T H(z)$, where component i of the vector-valued function H is defined as $H_i(z) = \min(z_i, f_i(z))$. The function $\theta(z)$ is one-half of the (non-negative) squared length of the vector $H(z)$ (satisfying property 1). Furthermore, $\theta(z) = 0$ is equivalent to $H(z) = 0$, which means that for each i, either $z_i = 0$ and $f_i(z) \geq 0$, or $z_i \geq 0$ and $f_i(z) = 0$, i.e. z is a solution of the NCP (property 2). The min function is used for example in [15] and [23]. Many other merit functions have been defined, some of which are described in the appendix to this chapter, Section 8.6.

8.3.2.5.1 Illustrative Example. Min-based merit function for two-commodity NCP

Using the results of the previous example (Section 8.3.2.4.2) for the two-commodity NCP, we evaluate the min-function based merit function at the starting guess $z^0 = \begin{pmatrix} 4 \\ 8 \end{pmatrix}$ and at the first Newton point $z_N^0 = \begin{pmatrix} 3.754386 \\ 8.070175 \end{pmatrix}$. At these points, the functions evaluate to

$$f(z^0) = \begin{pmatrix} 1.6 \\ 0 \end{pmatrix} \text{ and } f(z_N^0) = \begin{pmatrix} 0.006032 \\ -0.000003 \end{pmatrix}$$

The vector-valued function values at each point are

$$H(z^0) = \begin{pmatrix} \min(4, 1.6) \\ \min(8, 0) \end{pmatrix} = \begin{pmatrix} 1.6 \\ 0 \end{pmatrix} \text{ and}$$
$$H(z_N^0) = \begin{pmatrix} \min(3.754386, 0.006032) \\ \min(8.070175, 0) \end{pmatrix} = \begin{pmatrix} 0.006032 \\ 0 \end{pmatrix}$$

The two values of the merit function are

$$\begin{aligned} \theta(z^0) &= \frac{1}{2}H(z^0)^T H(z^0) \\ &= \frac{1}{2}\{(1.6)^2 + (0)^2\} = 1.28, \text{ and} \\ \theta(z_N^0) &= \frac{1}{2}H(z_N^0)^T H(z_N^0) \\ &= \frac{1}{2}\{(0.006032)^2 + (-0.000003)^2\} = 1.81951 \times 10^{-5} \end{aligned}$$

There is a substantial decrease in the value of the merit function between the two points, indicating substantial progress towards a solution.

□

8.3.2.5.2 Path Search between Previous Iterate and Newton Point

Instead of searching along a line between z^{k-1} and z_N^k, as is done in the damped Newton method for smooth equations, the PATH algorithm carries out a *path search* along the piecewise linear path connecting the two points, with evaluations of the merit function at the end points of each piece of the path, i.e. at the bases when a pivot is completed. (Recall that in the LCP example of Section 8.3.2.4.1, there are two pivots to get from $t = 0$ to $t = 1$, and the two pieces of the piecewise linear path describe the linear changes in variables between the beginning and end of each pivot.) Two methods have been suggested for the path search: (1) evaluate the merit function in the forward direction, i.e., for each new basis as the pivoting proceeds; or (2) evaluate the merit function in the backward direction, after the Newton point has been found, i.e., evaluate the merit function at each basis, but in reverse order from the Newton point back to the previous iterate.

In the merit function evaluation, the first method (forward search) can use the z values which have just been computed in the latest pivot, so there is no need to store all z values from each pivot, a computational advantage. The second method (backward search) requires either the storing of all the z values from each pivot (so that the merit function can be evaluated for each pivot starting from the Newton point and working backwards), or the recomputation of z values from each pivot; either way, it appears that the second method is computationally expensive, in memory requirements or in computational time.

However, the usual procedure is not to find the very best merit function value for all pivots on the path, but just to find an *acceptable* decrease in the merit function. This means, in practice, that the calculation of the merit function is done only a few times. It is often observed that the Newton point itself produces an acceptable decrease in the merit function, and if not the Newton point, then only a few pivots before the Newton point has an acceptable decrease. Therefore, the second, backward search of merit function evaluation is the preferred method (and the one implemented in the PATH algorithm), because the forward method would typically require many merit function evaluations (one per pivot) until an acceptable decrease is found, and the evaluation of the merit function itself can be computationally expensive. In practice, there is often no path search at all, because the Newton point is accepted as the next iterate.

When a backward path search is done, the earlier bases are recalculated, starting from the Newton point for $t = 1$, and working backwards, but to aid this process, the index of the variable which leaves the basis at each pivot is stored during the pivoting procedure; the variables are put back into the basis, one by one, in reverse order, to recalculate the earlier bases, and evaluate the merit function.

8.3.2.6 Summary and Overview of Other Features of the PATH Algorithm

To this point in the discussion, the key steps in the PATH algorithm are as follows.

1. Starting from a user-supplied initial point, perform a crashing calculation for a good initial starting point z^0 as the first iterate.
2. From the current iterate (w^{k-1}, z^{k-1}), perform the path following pivoting procedure on the Augmented LCP(z^{k-1}), (8.34), ending at $t = 1$ with the next Newton point z_N^k.
3. If z_N^k reduces the merit function θ by an acceptable amount, then set the next iterate $z^k = z_N^k$, else perform a backward path search for a point with an acceptable decrease in θ and make this the new iterate z^k.
4. Repeat steps 2 and 3 until θ is acceptably close to zero.

However, this picture is too simple, and the actual PATH algorithm includes several other features that deal with things that can go wrong in this simple picture.

One thing that can go wrong in the simple picture is that t may fail to reach the value 1 during the pivoting to find a solution to LCP(z^{k-1}). This could happen because a secondary ray solution is encountered (the non-basic variable can be increased forever without forcing any basic variable to zero). In this case, a solution that has the non-basic variable slightly above zero is returned as the outcome of the LCP approximation, instead of an actual Newton point solution to the LCP. Another way in which t may fail to reach the value 1 is when cycling occurs during pivoting (the same sequence of bases is repeated over and over); in this situation, the solution z for the largest value of t that is found during pivoting is used as if it is the Newton point z_N^k.

Another thing that can go wrong is that a strict requirement that the merit function be reduced at each iteration (a *monotone* convergence requirement) can sometimes slow or prevent convergence of a damped Newton method, even for smooth nonlinear equations. Therefore, the PATH algorithm allows for *nonmonotone* reduction of θ. Several iterations in a row simply accept the Newton point (or z for the largest value of t) as the next iterate, as long as it is sufficiently close to the previous iterate in Euclidean distance; no check for reduction of θ is performed. A check for reduction in θ is performed periodically after a prescribed number of iterations, or if the next iterate would be too far away from the previous iterate; if the new point reduces θ sufficiently, then it becomes the next iterate, else a backward path search is performed for a point that does reduce θ sufficiently. Furthermore, early in the algorithm, θ is even allowed to increase somewhat when checks are performed. Another advantage of infrequent checks for reduction in θ (or checks for increases not being too large), apart from improved convergence, is that the computation of θ, which can be expensive in computer time, is performed infrequently, thus improving overall computational efficiency.

When various other convergence problems are detected, a gradient step is sometimes taken: from the iterate with the smallest value of θ found so far, a step is taken in a direction related to $\nabla\theta$, to try to find a new point that reduces θ. Of course, this requires that the gradient $\nabla\theta$ be well defined, and preferably continuous; in the chapter appendix, Section 8.6, we discuss other merit functions which have this property. Further details on many special features may be found in [14].

With all of the special features that allow the PATH algorithm to recover from difficulties, it has been found to be a robust and reliable algorithm for the solution of many NCPs (and bounded MCPs).

8.4 Algorithms for VI Models

In this section, we discuss several methods to solve variational inequality (VI) problems. The form of the VI that is considered is consistent with the form in Chapter 5. In particular, the form of the constraint functions ensures that the feasible set is convex, which is important for the basic definition of a VI; see Chapter 5.

As in Chapter 5, Section 5.3.2, let $G : \mathbb{R}^n \to \mathbb{R}^n$, let $g : \mathbb{R}^n \to \mathbb{R}^m$ be such that g_i is convex for all $i = 1, \ldots, m$, and let $h : \mathbb{R}^n \to \mathbb{R}^p$ be such that h_i is affine (i.e., a constant term plus a linear term) for all $i - 1, \ldots, p$. Assume further that g is continuously differentiable. Define the set $K = \{z \in \mathbb{R}^n | g(z) \leq 0, h(z) = 0\}$. We wish to solve the variational inequality problem

VI(K, G): find $z^* \in K$ such that $G^T(z^*)(z - z^*) \geq 0$ for all $z \in K$.

Here we discuss two general approaches to the solution of variational inequality models: solving an equivalent mixed complementarity problem; and sequential optimization methods. Other algorithms have been successfully applied to the solution of VI models, notably the simplicial decomposition methods (discussed in Chapter 9), for which an overview can be found in [24].

8.4.1 Solve Equivalent KKT System as MCP

The KKT conditions for VI(K, G) are in the form of the mixed complementarity problem denoted MCP-VI below, and MCP-VI is equivalent to VI(K, G) (see the discussion in Chapter 5, Section 5.3.2). The variable $v \in \mathbb{R}^m$ and $\beta \in \mathbb{R}^p$.

MCP-VI:

$$
\begin{aligned}
&G(z^*) + \nabla g^T(z^*)v^* + \nabla h^T(z^*)\beta^* = 0\,,\ z \text{ is free}\\
&0 \geq g(z^*) \perp v^* \geq 0\\
&h(z^*) = 0\,,\ \beta^* \text{ is free}
\end{aligned}
$$

For a modeler, the simplest procedure to follow for solution of VI(K, G) is to solve MCP-VI , using the robust PATH solver in GAMS. Furthermore, even the conversion of VI(K, G) into MCP-VI is automated by the Extended Mathematical Programming (EMP) interface in GAMS [13].

8.4.2 Iterative Methods: Sequential Optimization

When variational inequality models were first studied, successful algorithms existed for constrained optimization problems. Because of this, many of the first algorithms for VIs were designed to solve a sequence of constrained optimization problems, with each optimization problem being an approximation to the VI. Such an algorithm produces an iterate z^k at each iteration k as the solution to the optimization problem at that iteration; the algorithms are designed so that z^k converges towards the solution z^* of the VI.

One of the first such methods was developed at the United States Department of Energy in the mid 1970s (a little before finite-dimensional variational inequality models were studied at all) to solve their integrated energy market models known as the Project Independence Evaluation System (PIES); see [1], [2]. The sequential optimization algorithm became known as the PIES algorithm. The PIES model and algorithm was the impetus for a great deal of further research into algorithms for the solution of VIs and NCPs.

We start our overview of sequential optimization methods for VIs with the PIES model and algorithm. We then outline some other sequential optimization methods.

8.4.2.1 Project Independence Evaluation System (PIES)

In Chapter 1, the PIES model is presented, and in Chapter 5, a VI version of the PIES model is derived. Chapter 12 presents a multi-sector (oil and coal) PIES model.

Here, supply side decisions are in the vector x, and demand quantities are in $q \in \mathbb{R}^n$. At equilibrium, the total cost of meeting demand q is minimized according to a linear program in the variables x:

$$\text{Minimize}_x \quad c^T x \tag{8.35a}$$

s. t.

$$Ax \geq q \qquad (\pi) \tag{8.35b}$$

$$Bx \geq b \qquad (\gamma) \tag{8.35c}$$

$$x \geq 0 \tag{8.35d}$$

Furthermore, demand q is a function of a vector of demand prices p – with a slight abuse of notation, we write

$$q = q(p) \tag{8.36}$$

– and at equilibrium,

$$\pi = p \tag{8.37}$$

To summarize, equilibrium is defined by conditions (8.35), (8.36), (8.37) holding simultaneously.

In general, the multicommodity demand function (8.36) is not necessarily integrable, i.e., as discussed in Chapters 3 and 4, it is not possible to integrate the inverse demand function $p = p(q)$, except for the special case that the Jacobian of $p(q)$ is symmetric. Therefore, it is not possible, in general, to find an equilibrium by solving an optimization problem that maximizes consumers' plus producers' surplus.

The PIES algorithm instead solves a sequence of optimization problems whose solutions converge to an equilibrium of (8.35), (8.36), (8.37). Each optimization problem in the sequence is constructed by replacing the demand function (8.36) by an approximation of the demand function, and each approximation depends on the demand prices calculated in the immediately previous optimization problem (but for the first optimization problem in the sequence, the approximation depends on guesses of the equilibrium demand prices, provided by the modeler). For the kth optimization problem in the sequence, the approximation of the ith demand function is obtained by fixing the values of all of the *other* goods' prices at the values calculated in the $(k-1)^{th}$ optimization, p_j^{k-1} for $j \neq i$, to produce a function which depends only on the ith price:

$$\widehat{q}_i^k(p_i) = q_i(p_1^{k-1}, ..., p_{i-1}^{k-1}, p_i, p_{i+1}^{k-1}, ..., p_n^{k-1}) \tag{8.38}$$

where $q_i(\bullet)$ is the ith function in the vector-valued function (8.36). Note that, if the approximation (8.38) is inverted to give p as a function of quantities, then this approximate inverse demand function is integrable because the Jacobian is a diagonal matrix.

Thus, the kth optimization problem in the sequence can be formulated to maximize consumers' plus producers' surplus with the approximate demand functions (8.38): the objective function is the sum, over all demands i, of the integrals of the *inverse* of (8.38), minus the production costs $c^T x$, subject to the constraints (8.35b-8.35d).

The next example illustrates the PIES algorithm applied to a simple model involving two demand quantities.

8.4.2.1.1 Illustrative Example. Simple PIES model and algorithm

In this example, two commodities are demanded – electricity in the amount q_e and natural gas in the amount q_g. There are two firms that supply electricity in the amounts x_{e1}, x_{e2}, and two that supply natural gas in amounts x_{g1}, x_{g2}. There are upper capacity limits on each supply, and for each commodity, total supply must exceed or equal total demand. The cost minimizing supply side LP, for fixed q_e and q_g, is:

$$\begin{aligned}
&\text{Minimize}_x \quad 13x_{e1} + 15x_{e2} + 14x_{g1} + 19x_{g2} \\
&\text{s. t.} \\
&\quad x_{e1} \le 450,\ x_{e2} \le 100 \\
&\quad x_{g1} \le 350,\ x_{g2} \le 85 \\
&\quad x_{e1} + x_{e2} \ge q_e \\
&\quad x_{g1} + x_{g2} \ge q_g \\
&\quad x_{e1},\ x_{e2},\ x_{g1},\ x_{g2} \ge 0
\end{aligned}$$

Prices for the two commodities are represented by p_e and p_g. The demands are related to prices by the constant elasticity expression (the index i stands for e or g, and j is an alias for i):

$$q_i = \overline{q}_i \prod_j \left(\frac{p_j}{\overline{p}_j}\right)^{e_{ij}}$$

where the reference quantities and prices are $\overline{q}_e = 460$, $\overline{p}_e = 12$, $\overline{q}_g = 435$, $\overline{p}_g = 37$, and the elasticities are $e_{ee} = -.33$, $e_{eg} = .06$, $e_{ge} = .04$, $e_{gg} = -.52$. At each iteration k of the PIES algorithm, we need an estimate of the two prices, denoted by p_e^{k-1} and p_g^{k-1}; at the first iteration, $k = 1$, p_e^0 and p_g^0 are guesses supplied by the modeler, and at a later iteration k, p_e^{k-1} and $p_g^{k-1}$are the prices computed at the previous iteration. The approximations of the demand functions at an iteration are based on the idea that the main effect of prices on demands is the own-price effect. Therefore, the approximation of the electricity demand function at iteration k substitutes the number p_g^{k-1} in place of the variable p_g, and in the gas demand function, p_e^{k-1} replaces p_e. To summarize in a way that includes models with more than two commodities, the demand function approximations at iteration k are:

$$\widehat{q}_i^k(p_i) = \overline{q}_i \left(\frac{p_i}{\overline{p}_i}\right)^{e_{ii}} \prod_{j \ne i} \left(\frac{p_j^{k-1}}{\overline{p}_j}\right)^{e_{ij}}$$

We need the approximate inverse demand functions, which will be integrated for terms in the objective function at iteration k:

$$p_i = A_i^{k-1} \cdot (q_i)^{1/e_{ii}} \tag{8.39}$$

$$\text{where } A_i^{k-1} = \frac{\overline{p}_i}{(\overline{q}_i)^{1/e_{ii}}} \prod_{j \neq i} \left(\frac{p_j^{k-1}}{\overline{p}_j} \right)^{-e_{ij}/e_{ii}}$$

For this functional form, integration of (8.39) from zero to any positive quantity produces an infinite value if the own-price elasticity e_{ii} lies between -1 and 0. For example, suppose $e_{ii} = -1/2$; integration of (8.39) involves the factor

$$\int_0^{q_i} y^{1/(-1/2)} dy = \int_0^{q_i} y^{-2} dy$$
$$= -q_i^{-1} - \lim_{y \to 0} -y^{-1} = \infty$$

Therefore, we must integrate instead from a lower limit of an arbitrary positive value, up to q_i, which produces an expression in q_i, plus a constant term related to the lower limit of integration. In the objective function that we are constructing, the constant term can be ignored because it is unaffected by choice of variable values q_i; therefore we can drop the constant term. The resulting function is not the total value to consumers of consuming the amounts q_i. However, this function can be used to calculate the *difference* in total consumer value in comparing two consumption amounts. Ignoring the lower limit of integration, the integral that we use is:

$$\int^{q_i} A_i^{k-1} \cdot y^{1/e_{ii}} dy = \frac{A_i^{k-1}}{1 + 1/e_{ii}} \cdot (q_i)^{1+1/e_{ii}}$$

Therefore, the optimization problem to solve for supply *and* demand variables at iteration k is:

$$\text{Minimize}_{x,q} \quad 13x_{e1} + 15x_{e2} + 14x_{g1} + 19x_{g2}$$
$$- \frac{A_e^{k-1}}{1 + 1/e_{ee}} \cdot (q_e)^{1+1/e_{ee}} - \frac{A_g^{k-1}}{1 + 1/e_{gg}} \cdot (q_g)^{1+1/e_{gg}} \tag{8.40a}$$

s. t.

$$x_{e1} \leq 450, \; x_{e2} \leq 100 \tag{8.40b}$$
$$x_{g1} \leq 350, \; x_{g2} \leq 85 \tag{8.40c}$$
$$x_{e1} + x_{e2} \geq q_e \tag{8.40d}$$
$$x_{g1} + x_{g2} \geq q_g \tag{8.40e}$$
$$x_{e1}, x_{e2}, x_{g1}, x_{g2} \geq 0 \tag{8.40f}$$

The algorithm to follow can be stated precisely as follows.

Algorithm 8.1 (PIES).

Step 0. Set $k = 0$; choose starting guesses p_e^0, p_g^0 and convergence tolerance $\varepsilon > 0$.
Step 1. Update $k = k + 1$. Solve the NLP (8.40) to determine, in particular, q_e^k, q_g^k. Calculate new price estimates p_e^k, p_g^k using (8.39).
Step 2. If $|p_e^k - p_e^{k-1}| < \varepsilon$ and $|p_g^k - p_g^{k-1}| < \varepsilon$, then STOP; else go to Step 1.

With starting guesses of the prices, $p_e^0 = 22$ and $p_g^0 = 50$, the prices calculated at each iteration are given in Table 8.3. A solution is found at iteration $k = 3$, which is verified by noting that there is no change (to three decimals of accuracy) in the prices computed at iterations 3 and 4. (When there is no change in the prices, the approximate inverse demand functions (8.39) are consistent with the original, true price-quantity relationships, i.e., the equilibrium calculated for the most recent approximate problem is an equilibrium of the original problem).

Table 8.3 Prices by iteration for PIES example

k	p_e	p_g
0	22	50
1	13.548	38.766
2	13.000	37.347
3	13.000	37.229
4	13.000	37.229

□

8.4.2.1.2 PIES-q Algorithm

A variant of the PIES algorithm is also described in [1], [2], and is called the PIES-q algorithm. It is an instance of the more general *diagonalization* method to solve a VI model, described below in Section 8.4.2.2.

The PIES-q algorithm also solves a sequence of optimization problems, each of which maximizes consumers' plus producers' surplus, with an approximation of the demand-price relationship. In PIES-q, the demand function is inverted to give prices as functions of demands, $p(q)$, and in the kth optimization, the ith inverse demand function is approximated by fixing the *demands* of the other goods at the values obtained from the $(k-1)^{th}$ optimization:

$$\widehat{p}_i^k(q_i) = p_i(q_1^{k-1}, \ldots, q_{i-1}^{k-1}, q_i, q_{i+1}^{k-1}, \ldots, q_n^{k-1}) \tag{8.41}$$

8.4.2.1.3 Convergence of PIES and PIES-q Algorithms

In [1], [2], it is shown that the PIES algorithm converges to an equilibrium, under fairly general conditions, including a requirement that, loosely stated, means that the cross-price effects on demand are less important than the own-price effects on demand. To be more precise, if the demand vector q depends on the price vector p according to an affine relationship

$$q = b - Bp \tag{8.42}$$

then the key condition in [1], [2] for convergence of the PIES algorithm is that

$$\|I - D^{-\frac{1}{2}} B D^{-\frac{1}{2}}\| < 1 \tag{8.43}$$

where D is the diagonal of B, and $\| \bullet \|$ indicates the norm of the matrix (see (8.20) for the definition of a matrix norm). Note that the matrix $I - D^{-\frac{1}{2}} B D^{-\frac{1}{2}}$ in (8.43) has zeroes on the diagonal, and for the off-diagonal elements, the ith row and j^{th} column has the entry $-\frac{B_{ij}}{B_{ii}^{-1/2} B_{jj}^{-1/2}}$, where B_{ij} is the ij^{th} element of B. Thus, the off-diagonal elements of $I - D^{-\frac{1}{2}} B D^{-\frac{1}{2}}$ are small in magnitude if each off-diagonal element B_{ij} (corresponding to cross-price effects) is small compared to the diagonal elements B_{ii}, B_{jj} (corresponding to own-price effects). If the ratios $\frac{B_{ij}}{B_{ii}^{-1/2} B_{jj}^{-1/2}}$, are all small enough, then the norm of the matrix $I - D^{-\frac{1}{2}} B D^{-\frac{1}{2}}$ can be less than one, and the theorem in [2] says that the PIES algorithm converges.

For a general demand function, not necessarily affine in form, it is shown in [1], [2] that if an affine approximation (8.42) of the demand function is made at the equilibrium price vector, and if (8.43) is true for that affine approximation, then the PIES algorithm converges, provided the first guess of prices, p^0 (appearing in (8.38) for $k = 1$) is close enough to the equilibrium price vector.

The convergence results in [1], [2] for the PIES-q algorithm are similar to those for the PIES algorithm, but with respect to the inverse demand function $p(q)$: if the cross-goods effects in $p(q)$ are less important than the own-goods effects, then PIES-q converges.

8.4.2.2 A Nonlinear Approximation of G – Diagonalization Method

The diagonalization method for solving VIs is a generalization of the PIES-q algorithm: a sequence of optimization problems is solved, with their solutions converging to a solution of the original VI. At the kth iteration, the ith element of $G(z)$ is approximated by a function that depends only on z_i because the other z_j ($j \neq i$) variables are fixed at the values calculated at the $(k-1)^{th}$ iteration:

$$\widehat{G}_i^k(z_i) = G_i(z_1^{k-1}, ..., z_{i-1}^{k-1}, z_i, z_{i+1}^{k-1}, ..., z_n^{k-1}) \tag{8.44}$$

If G is nonlinear, then so is the approximating function $\widehat{G}^k$. Because it is a function of just one variable, $\widehat{G}_i^k(z_i)$ can normally be integrated, and the optimization problem solved at the kth iteration is

$$\text{Minimize} \quad \sum_i \int^{z_i} \widehat{G}_i^k(y)dy \tag{8.45a}$$

s. t.

$$z \in K \tag{8.45b}$$

where $K = \{z \in \mathbb{R}^n | g(z) \le 0, h(z) = 0\}$.

A diagonalization method is described in Chapter 7 to solve an EPEC, but there are differences from the method described above in (8.44), (8.45). In Chapter 7, the EPEC is viewed as the simultaneous solution of a collection of MPECs (one MPEC per leader, and each MPEC's constraints contain the same equilibrium problem of the follower). Each MPEC is influenced by other leaders' decisions in the MPEC's objective function and in the constraints for the follower's equilibrium problem. At iteration k, for each MPEC, the diagonalization method fixes the *other* leaders' variables at the values computed at iteration k, in the objective function *and* in the constraints for the follower's equilbrium problem. Thus, iteration k solves separate MPECs for each leader. In contrast, for diagonalization as described above in (8.44), (8.45), the fixing of *other* variables at iteration k occurs only in the mapping, and not in the constraints, which leads to only one optimization problem per iteration.

8.4.2.2.1 Illustrative Example. The PIES-q algorithm as diagonalization method on a VI

The PIES model, (8.35), (8.36), (8.37), can be rewritten slightly by using the inverse demand function $p(q)$ and combining (8.36), (8.37) as

$$\pi = p(q) \tag{8.46}$$

In Chapter 5, Section 5.2.3.3, it is shown that the PIES model, with form (8.46), may be written as a VI model:

$$\text{Let } K = \{x, q |\, Ax \ge q\,,\, Bx \ge b\,,\, x \ge 0\}. \tag{8.47a}$$

$$\text{Find } (x^*, q^*) \in K \text{ such that} \tag{8.47b}$$

$$c^T(x - x^*) - p(q^*)^T(q - q^*) \ge 0 \quad \forall (x, q) \in K \tag{8.47c}$$

In the diagonalization strategy (8.44) at the kth iteration, the ith component of $p(q^*)$ is replaced by an approximation that depends only on q_i as in (8.41), and the gradient-like vector-valued function $\begin{pmatrix} c \\ \widehat{p}(q) \end{pmatrix}$ is integrated to form the objective function. Therefore, the diagonalization method applied to the VI representation of the PIES model is the PIES-q algorithm.

□

8.4.2.3 Symmetric Linear Approximations of G

Newton methods for solving systems of nonlinear equations (Section 8.3.1) follow a strategy of linear approximation about a current iterate, using the gradients, and calculating the next iterate as the solution of the approximate linear equations (possibly with a damping line search step). Applying the general idea of linearization to the VI suggests an algorithm for which, at iteration k, the vector-valued function $G(z)$ is approximated by the affine function $G(z^{k-1})+\nabla G(z^{k-1})(z-z^{k-1})$. This still leaves us with a VI to solve at each step, and not an optimization problem, unless $\nabla G(z^{k-1})$ happens to be symmetric. It is perhaps surprising that other affine functions, with $\nabla G(z^{k-1})$ replaced by symmetric matrices in various ways, lead to useful algorithms that solve an optimization problem at each step.

One such method is the *linearized Jacobi* method, in which $G(z)$ is approximated at iteration k by the affine function $G(z^{k-1})+D(z^{k-1})(z-z^{k-1})$, where $D(z^{k-1})$ is a diagonal matrix equal to the diagonal part of $\nabla G(z^{k-1})$. Because a diagonal matrix has zeroes above and below the diagonal, it is a symmetric matrix, and the VI at iteration k can therefore be solved as an optimization problem.

The *symmetrized Newton* method replaces $\nabla G(z^{k-1})$ by a related matrix obtained by adding it to the transpose of itself, and dividing by 2: $G(z)$ is approximated at iteration k by the affine function $G(z^{k-1}) + \frac{1}{2}\{\nabla G(z^{k-1}) + \nabla G(z^{k-1})^T\}(z - z^{k-1})$. The symmetrized matrix has the same diagonal elements as $\nabla G(z^{k-1})$, and if the off-diagonal elements of the original Jacobian $\nabla G(z^{k-1})$ are close in value to the off-diagonal elements of the symmetrized matrix, then the optimization problem solved at each step is close to taking a Newton step.

The final example of a symmetric linear approximation is the *projection* method. In the affine approximation of G at iteration k, $\nabla G(z^{k-1})$ is replaced by a matrix M which need not be related to G at all, and M does not depend on the last iterate z^{k-1}. The only requirement on M is that it must be a symmetric, positive definite matrix. With no requirement for M to be related to G, it is surprising that any useful algorithm can be created, but in fact, the projection method has been quite successful. The method is called "projection" because each iteration can be interpreted as a mathematical calculation involving a projection onto the VI's feasible set K (but the normal minimization of Euclidean distance to K is replaced by minimization of an altered concept of distance, relying on the matrix M).

8.4.2.4 Convergence of Diagonalization and Symmetric Linear Approximation

All of the diagonalization and symmetric linear approximation schemes were grouped together in a general iterative framework in a paper by Dafermos [7].

A theorem provides a convergence condition that is a generalization of the PIES convergence condition (8.43), and that can be applied to the various iterative schemes described above.

8.5 Summary

In this chapter, we have surveyed some basic methods for solving LCPs, NCPs and VIs, with special attention paid to the PATH algorithm. PATH is for the NCP problem (MCP, more generally), but it can be used to solve LCPs and VIs, too, because an LCP is a special case of an NCP, and a VI has equivalent KKT conditions which form an MCP. Some further details on theory for the PATH algorithm, and other NCP algorithms, may be found in an appendix to this chapter, Section 8.6.

The chapter begins with Lemke's pivoting method for solving LCPs. Using an artificial variable technique, Lemke's method begins with the zero solution, which obeys the complementarity requirement of the LCP, and pivots through other bases, always maintaining complementarity, until the artificial variable is driven to zero. Some matrix splitting, iterative methods are also described, that approximate the matrix at each iteration by another matrix which is such that the LCP is easy to solve at each iteration, e.g., as a quadratic programming problem, for some types of matrix splitting.

The PATH algorithm for an NCP follows the general idea of Newton's iterative method for solving a system of nonlinear equations: approximate the nonlinear functions by affine functions, using a first-order Taylor approximation about the solution from the previous iteration; solve the resulting LCP for this iteration; and repeat in further iterations until convergence is achieved. Although Lemke's method could be used to solve the LCP at each iteration, it suffers from the fact that Lemke's method must start with the zero solution, which in general is far from the solution of the NCP and from the solution of the current iteration's LCP, causing many pivots and long computing times. The PATH algorithm employs a different method to start the complementary pivoting for each iteration's LCP, such that the solution of the LCP from the previous iteration is used as the first complementary solution, and pivoting proceeds until an auxiliary (artificial) variable reaches the special value of 1.

The idea of damping in Newton's method for a system of nonlinear equations is implemented in the PATH algorithm. After the LCP is solved at an iteration, a merit function is evaluated for each basis visited in the complementary pivoting calculations, starting with the solution, and working backwards through other bases, until an acceptable decrease in the merit function is found, compared with the merit function evaluated at the solution of the previous iteration.

In Section 8.4, we outline two main classes of methods for solving VIs. Often the most practical method is to use PATH to solve the MCP which is the KKT system equivalent to the VI. The modeler can write out the KKT system and code it, or the EMP software is available in GAMS to convert a VI model automatically to the equivalent KKT system, and call PATH for a solution. If PATH is not available, but an NLP solver is, or if PATH takes too long to find a solution, then sequential optimization algorithms can often be used to solve a VI. In a sequential optimization algorithm, the VI is approximated at an iteration by another VI which obeys the Principle of Symmetry (see Chapter 4), making the approximate model into an optimization problem. The approximation is updated at each iteration, using the solution of the previous iteration.

8.6 Appendix: Introduction to Theory for PATH and Other NCP Algorithms

In this appendix,we introduce some of the theory about the PATH algorithm, which also serves as an introduction to many other methods to solve NCPs. The purpose is to outline some of the key ideas, so that readers may pursue more reading of the literature about algorithms to solve NCPs. We show how an NCP can be understood as a problem to solve a system of nonlinear equations, which allows the successful machinery of Newton's method to be extended to the solution of NCPs. The most important complication is that the resulting formulation as a system of nonlinear equations involves nonsmooth equations: the derivatives are not continuous everywhere.

We begin by discussing the three key ingredients: projection mappings; the reformulation of the NCP as a system of nonsmooth equations, using projection mappings; and various merit functions that are useful for the damping steps of Newton's method. The two-volume survey [11] is a good place to start for readers interested in greater depth on these and other related topics. For a focus specifically on the PATH algorithm, [8] and [9] provide excellent overviews.

In this appendix, the reason for using the symbol z for the vector of variables, instead of x, becomes clear. In the literature that describes the PATH algorithm and solver, [8], [9], [25], it is customary to use the symbol x for the variable vector in the *normal equation* (NE, described below) which is equivalent to the NCP; the symbol z is used in the literature for the variable vector in the NCP. We have adopted the same symbols, to avoid confusing any readers who consult the literature for more detail.

8.6.1 Projection Mappings

Definition 8.1. Let B be a closed, convex set in $\mathbb{R}^n$. For $x \in \mathbb{R}^n$, the Euclidean projection of x on the set B, denoted $\pi_B(x)$ is the unique point $y \in B$ that minimizes the Euclidean distance $\|y - x\|$, i.e.

$$\pi_B(x) = \arg\min_{y \in B} \|y - x\| \tag{8.48}$$

The first example of a projection mapping is the one that is used in the transformation of an NCP into a system of nonlinear equations.

8.6.1.1 Illustrative Example. Projection Mapping for $B = \mathbb{R}^n_+$

In this example, we wish to define the mapping $\pi_B(x)$ that gives the point in $\mathbb{R}^n_+ = \{y \in \mathbb{R}^n \,|\, y_i \geq 0,\ i = 1, ..., n\}$ (the non-negative orthant of $\mathbb{R}^n$) that is closest to the point $x \in \mathbb{R}^n$. Therefore, we wish to find the point $y^* = \pi_B(x)$ which solves the following optimization problem:

$$\begin{aligned} &\text{Minimize}_{y_i} \quad (\sum_{i=1}^{n} (y_i - x_i)^2)^{1/2} \\ &\text{s. t.} \\ &\quad y_i \geq 0, \quad i = 1, ..., n \end{aligned}$$

It is clear that y^* also solves the optimization problem that minimizes the square of the distance from x to y

$$\begin{aligned} &\text{Minimize}_{y} \quad \sum_{i=1}^{n} (y_i - x_i)^2 \\ &\text{s. t.} \\ &\quad y_i \geq 0, \quad i = 1, ..., n \end{aligned}$$

and this problem splits into n separate problems, each of the form

$$\begin{aligned} &\text{Minimize}_{y_i} \quad (y_i - x_i)^2 \\ &\text{s. t.} \\ &\quad y_i \geq 0 \end{aligned}$$

The solution, therefore, is

$$y_i^* = \begin{Bmatrix} x_i, \text{ if } x_i \geq 0 \\ 0\ , \text{ if } x_i < 0 \end{Bmatrix}, \ i = 1, ..., n \tag{8.49}$$

□

The second example of a projection mapping is relevant to the transformation of the more general MCP (8.24) into a system of nonlinear equations. See [8], [9] for details about this transformation.

8.6.1.2 Illustrative Example. Projection Mapping for B as a Rectangular Box

In this example, the variables in the vector y are individually bounded above and below: the set $B = \{y \in \mathbb{R}^n \,|\, l_i \le y_i \le u_i\,,\ i = 1, ..., n\}$, where l_i and u_i are given, finite real numbers such that $l_i \le u_i$. Using the same approach as in the previous example, it is easy to see that the projection mapping $y^* = \pi_B(x)$ is given by

$$y_i^* = \left\{ \begin{array}{c} x_i, \text{ if } l_i \le x_i \le u_i \\ l_i\ , \text{ if } x_i < l_i \\ u_i\ , \text{ if } x_i > u_i \end{array} \right\}\ ,\ i = 1, ..., n$$

□

8.6.2 NCP Reformulated as Nonsmooth Equation Using Projection Mapping

We use the symbol x for a vector in $\mathbb{R}^n$ without any non-negativity condition, and the symbol z for a vector in $\mathbb{R}^n_+$, i.e., $z \ge 0$. We are interested to find a solution to the nonlinear complementarity problem

$$\text{NCP:} \qquad 0 \le f(z) \perp z \ge 0 \tag{8.50}$$

The next theorem shows that our search for a solution to NCP can be interpreted as a search for a solution of the set of n nonsmooth equations, a vector equation known in the literature as the *normal equation* (NE) – see, e.g., [26]. Furthermore, we show below that the application of Newton's method to NE leads naturally to an algorithm to find a solution to NCP.

$$\text{NE:} \qquad f_B(x) \equiv f(\pi_B(x)) + x - \pi_B(x) = 0, \text{ where } B = \mathbb{R}^n_+ \tag{8.51}$$

Two examples after the theorem illustrate the relations between the z and x variables, and between NCP and NE. The proof is left as an exercise for the reader, who may wish to consult the proof in [9] for the more general setting of the MCP (8.24).

Theorem 8.3. *(a) If x solves NE, then $z = \pi_B(x)$ solves NCP.*
(b) If z solves NCP, then $x = z - f(z)$ solves NE.

8.6.2.1 Illustrative Example. Illustration of Theorem 8.3 with $x \neq z$

This example uses an LCP (a special case of NCP) example from Chapter 1 that was used in the example of Section 8.2.1.3.1 (after converting it to a pure complementarity problem, i.e., without equations). The example was originally in the variables x_1, x_2, but we convert to z_1, z_2 to be consistent with the notation in this section, where z is used for the NCP and x for the NE. The NCP is:

$$0 \leq f(z) = \begin{pmatrix} 0 \\ -2 \end{pmatrix} + \begin{pmatrix} 1 & 1 \\ 2 & 1 \end{pmatrix} \begin{pmatrix} z_1 \\ z_2 \end{pmatrix} \perp z \geq 0$$

We first illustrate part (b) of the theorem. From Chapter 1 and the example of Section 8.2.1.3.1, a solution to the NCP is $z = (0, 2)^T$. The theorem states that the following value of x solves NE:

$$\begin{aligned} x &= z - f(z) \\ &= \begin{pmatrix} 0 \\ 2 \end{pmatrix} - \left\{ \begin{pmatrix} 0 \\ -2 \end{pmatrix} + \begin{pmatrix} 1 & 1 \\ 2 & 1 \end{pmatrix} \begin{pmatrix} 0 \\ 2 \end{pmatrix} \right\} \\ &= \begin{pmatrix} -2 \\ 2 \end{pmatrix} \end{aligned}$$

Let us now check that $x = (-2, 2)^T$ solves NE (8.51). For this, we need to evaluate the projection $\pi_B(x)$, which simply changes any negative component of x to zero: $\pi_B(x) = (0, 2)^T$. Therefore, evaluation of $f_B(x)$ proceeds as

$$\begin{aligned} f_B(x) &= f(\pi_B(x)) + x - \pi_B(x) \\ &= \left\{ \begin{pmatrix} 0 \\ -2 \end{pmatrix} + \begin{pmatrix} 1 & 1 \\ 2 & 1 \end{pmatrix} \begin{pmatrix} 0 \\ 2 \end{pmatrix} \right\} + \begin{pmatrix} -2 \\ 2 \end{pmatrix} - \begin{pmatrix} 0 \\ 2 \end{pmatrix} \\ &= \begin{pmatrix} 0 \\ 0 \end{pmatrix} \end{aligned}$$

as required, to satisfy NE (8.51).

Part (a) of the theorem starts from a known solution x to NE, and asserts that $z = \pi_B(x)$ solves NCP. Here, $x = (-2, 2)^T$ solves NE, and $\pi_B(x) = (0, 2)^T$, is a solution to the NCP, as noted above.

□

8.6.2.2 Illustrative Example. Illustration of Theorem 8.3 with $x = z$

In this example, we use the two-commodity LCP (8.12) that was first presented in Section 8.2.1.3.3:

$$0 \leq f(z) = \begin{pmatrix} -32 \\ -64 \end{pmatrix} + \begin{pmatrix} 6 & 1 \\ 2 & 7 \end{pmatrix} \begin{pmatrix} z_1 \\ z_2 \end{pmatrix} \perp z \geq 0$$

which has a solution $z = (4, 8)^T$. To illustrate part (b) of the theorem, we calculate $x = z - f(z)$ and check that x solves NE:

$$\begin{aligned} x &= z - f(z) \\ &= \begin{pmatrix} 4 \\ 8 \end{pmatrix} - \left\{ \begin{pmatrix} -32 \\ -64 \end{pmatrix} + \begin{pmatrix} 6 & 1 \\ 2 & 7 \end{pmatrix} \begin{pmatrix} 4 \\ 8 \end{pmatrix} \right\} \\ &= \begin{pmatrix} 4 \\ 8 \end{pmatrix} \end{aligned}$$

Inserting this value of x into the expression for $f_B(x)$ (noting that the projection $\pi_B(x) = x$ when $x \geq 0$) results in

$$\begin{aligned} f_B(x) &= f(\pi_B(x)) + x - \pi_B(x) \\ &= \left\{ \begin{pmatrix} -32 \\ -64 \end{pmatrix} + \begin{pmatrix} 6 & 1 \\ 2 & 7 \end{pmatrix} \begin{pmatrix} 4 \\ 8 \end{pmatrix} \right\} + \begin{pmatrix} 4 \\ 8 \end{pmatrix} - \begin{pmatrix} 4 \\ 8 \end{pmatrix} \\ &= \begin{pmatrix} 0 \\ 0 \end{pmatrix} \end{aligned}$$

as required by the theorem.

To illustrate part (a), we note again that, because $x \geq 0$, $z = \pi_B(x) = x = (4, 8)^T$, which is the known solution to NCP.

□

8.6.3 Some Useful Merit Functions and Corresponding Nonsmooth Equations

Recall from Section 8.3.2.5 that a merit function $\theta(z)$ has the following properties: $\theta(z) \geq 0$; and $\theta(z) = 0$ if and only if z solves the NCP (8.50). For the four examples of merit functions below, there is also a corresponding nonsmooth equation reformulation of the NCP, different from $f_B(x) = 0$. After these four merit functions are presented with their nonsmooth equations, we revert to using the nonsmooth equation $f_B(x) = 0$, because this is the basis of the PATH algorithm.

8.6.3.1 Merit Function Based on Min Function

In Section 8.3.2.5, we introduced the merit function that is based on the min function. We repeat the definition here, without discussion:

$$\theta(z) = \frac{1}{2} H(z)^T H(z) \tag{8.52}$$
$$\text{where } H_i(z) = \min(z_i,\, f_i(z))$$

As discussed in Section 8.3.2.5, the NCP can be reformulated as the nonsmooth equation $H(z) = 0$. Nonsmoothness is due to $H_i(z)$ taking two different functional forms for different ranges of z: where $z_i < f_i(z)$, $H_i(z) = z_i$ and $\partial H_i / \partial z_i = 1$; where $z_i > f_i(z)$, $H_i(z) = f_i(z)$ and $\partial H_i / \partial z_i = \partial f_i / \partial z_i$, which could, in general, take a very different value from 1. Therefore, if we move z from the region where $z_i < f_i(z)$ to the region where $z_i > f_i(z)$, then there will in general be a sudden change in the value of $\partial f_i / \partial z_i$ – nonsmoothness – as we cross the boundary where $z_i = f_i(z)$.

8.6.3.2 Merit Function Based on Norm of the Normal Map

This merit function is a function of x, the variable in the NE, instead of z, the variable in the NCP. At a solution of the NCP, or of the normal equation, $f_B(x) = 0$. Therefore, a merit function can be $\theta(x) = \|f_B(x)\|$. Alternatively, we could define $\theta(x) = \frac{1}{2}\|f_B(x)\|^2 = \frac{1}{2} f_B(x)^T f_B(x)$.

8.6.3.3 Merit Function Based on Fischer-Burmeister Function

The Fischer-Burmeister function of two real variables, $\phi : \mathbb{R}^2 \longrightarrow \mathbb{R}$ is defined as:

$$\phi(a,b) \equiv \sqrt{a^2 + b^2} - a - b \tag{8.53}$$

The Fischer-Burmeister function is an *NCP function* because requiring that $\phi(a,b) = 0$ is equivalent to requiring that $a \geq 0$, $b \geq 0$ and $ab = 0$. To prove that $\phi(a,b) = 0$ implies complementarity and non-negativity, observe that $\phi(a,b) = 0$ implies that $a^2 + b^2 = (a+b)^2$, which leads to $ab = 0$. Therefore, at least one of a or b is zero; if for, example, $a = 0$ then $\phi(a,b) = 0$ implies that $\sqrt{b^2} = b$, which shows that $b \geq 0$ because $\sqrt{\ }$ indicates the positive square root. Similarly, if $b = 0$, then $a \geq 0$. Readers can easily prove the implication in the opposite direction, i.e., that complementarity and non-negativity imply that $\phi(a,b) = 0$.

A merit function for the NCP, based on the Fischer-Burmeister function, relies on the mapping $\Phi : \mathbb{R}^n \longrightarrow \mathbb{R}^n$ whose i^{th} component is defined as

$$\Phi_i(z) = \phi(z_i, f_i(z)) \tag{8.54}$$

The NCP can be reformulated as a nonsmooth equation $\Phi(z) = 0$. Furthermore, for any z, $\|\Phi(z)\|^2 = \Phi(z)^T \Phi(z) = \sum_i \Phi_i^2(z) \geq 0$, and z^* is a solution of the NCP if and only if $\Phi(z^*)^T \Phi(z^*) = 0$. Therefore, the following function can be used as a merit function in the PATH algorithm (and in fact, it is used in PATH), to monitor progress towards a solution of the NCP:

$$\theta(z) = \frac{1}{2}\Phi(z)^T\Phi(z) \tag{8.55}$$

8.6.3.3.1 Illustrative Example. Fischer-Burmeister-based merit function for two-commodity NCP

Using the results of the path-following example for the two-commodity NCP of Section 8.3.2.4.2, we evaluate the Fischer-Burmeister-based merit function at the starting guess $z^0 = \begin{pmatrix} 4 \\ 8 \end{pmatrix}$ and at the first Newton point $z_N^0 = \begin{pmatrix} 3.754386 \\ 8.070175 \end{pmatrix}$. At these points, the functions evaluate to $f(z^0) = \begin{pmatrix} 1.6 \\ 0 \end{pmatrix}$ and $f(z_N^0) = \begin{pmatrix} 0.006032 \\ -0.000003 \end{pmatrix}$. The mapping

$$\Phi(z) = \begin{pmatrix} \sqrt{z_1^2 + f_1(z)^2} - z_1 - f_1(z) \\ \sqrt{z_2^2 + f_2(z)^2} - z_2 - f_2(z) \end{pmatrix}$$

and the merit function is $\theta(z) = \frac{1}{2}\Phi(z)^T\Phi(z) = \frac{1}{2}\{\Phi_1^2(z) + \Phi_2^2(z)\}$ yielding $\theta(z^0) - 0.834461664$, and $\theta(z_N^0) = 1.81658 \times 10^{-5}$.
□

8.6.3.4 Merit Function Based on Plus Function

The last example of a merit function is based on the *plus function*: for any $y \in \mathbb{R}^n$, and any component y_i,

$$(y_i)_+ = \text{Maximize} \quad \{0, y_i\} \tag{8.56}$$

and we write $(y)_+$ for the vector with the plus function applied to each component. In other words, the plus function is the same as the projection onto $B = \mathbb{R}_+^n$, as in (8.49). It is left as an exercise for the reader to show that $z \in \mathbb{R}^n$ solves NCP if and only if z solves the following nonsmooth equation:

$$z - (z - f(z))_+ = 0 \tag{8.57}$$

Note that in (8.57), the components of z can take positive or negative values in principle, but the equation itself limits z to non-negative values. The corresponding merit function is half the square of the residual (as in (8.55)):

$$\theta(z) = \frac{1}{2}\left(z - (z - f(z))_+\right)^T \left(z - (z - f(z))_+\right)$$

8.6.4 Damped Newton Method for NCP as Nonsmooth Equation

We now return to using the nonsmooth equation NE, $f_B(x) = 0$. Here, we restate the description of the PATH algorithm from Sections 8.3.2.1 and 8.3.2.2, but here in terms of the NE and its x vector.

Newton's method proceeds iteratively, with a key step being the calculation of the Newton point, i.e., the solution of the latest linear approximation of the system of nonlinear equations. In the linear approximation of NE corresponding to NCP, $\pi_B(x)$ (the projection onto the non-negative quadrant) is left as it is because (8.49) shows that it is already linear within the various regions of $\mathbb{R}^n$ defined by the signs of the components of x, i.e., $\pi_B(x)$ is piecewise linear. Given the most recent iterate x^{k-1}, the piecewise linear approximation is

$$f_B(x) \approx f(\pi_B(x^{k-1})) + \nabla f(\pi_B(x^{k-1}))(\pi_B(x) - \pi_B(x^{k-1})) + x - \pi_B(x)$$

Therefore, the next Newton point x_N^k is the solution of the following piecewise linearized nonsmooth equation

$$\begin{aligned} & q^{k-1} + M^{k-1}\pi_B(x) + x - \pi_B(x) = 0 \\ & \text{where } M^{k-1} = \nabla f(\pi_B(x^{k-1})) \\ & \text{and } q^{k-1} = f(\pi_B(x^{k-1})) - M^{k-1}\pi_B(x^{k-1}) \end{aligned} \tag{8.58}$$

The calculation of the next Newton point is actually done in terms of the vector variables z and w, by solving LCP(z^{k-1}) (8.33) as described in Section 8.3.2, with the following correspondence to x:

$$z = \pi_B(x), \; w = z - x$$

To show that a solution (z^k, w^k) of LCP(z^{k-1}) (8.33) provides a solution $x^k = z^k - w^k$ to (8.58), first note that the definitions of q^{k-1} and M^{k-1} in (8.58) above and in (8.31)-(8.32) in Section 8.3.2.1 are identical when $z^{k-1} = \pi_B(x^{k-1})$. Next observe that $z^k = \pi_B(x^k)$ because z^k and w^k are complementary: if a component $z_i^k = 0$, then $x_i^k = 0 - w^k \le 0$, which means that $(\pi_B(x^k))_i = 0 = z_i^k$; and if $z_i^k > 0$, then by complementarity, $w_i^k = 0$, which implies that $x_i^k = z_i^k - 0 > 0$, so $(\pi_B(x^k))_i = z_i^k$. It is now easy to see that a solution (z^k, w^k) of LCP(z^{k-1}) (8.33) provides a solution $x^k = z^k - w^k$ to (8.58): in the equality of (8.33), substitute $\pi_B(x^k)$ for z^k, and $z^k - x^k = \pi_B(x^k) - x^k$ for w^k, yielding

$$\pi_B(x^k) - x^k = q^{k-1} + M^{k-1}\pi_B(x^k)$$

and with a slight rearrangement, we have the equation (8.58). After the next Newton point is found, damping is done, if necessary, as described in Section 8.3.2.

The literature describes the PATH algorithm in terms of x and the normal equation, in spite of the fact that the actual calculations are done in terms of z and w, because the proof of convergence is more naturally shown in terms of x. We next provide some general comments about convergence.

8.6.5 Convergence of the PATH Algorithm

The most important theoretical result for any algorithm is to prove that it converges to a solution of the problem that it is intended to solve. Such convergence proofs typically depend on some characteristics of the problem (e.g., of the feasible region of the problem, and of functions that are used to define the problem), and also on values of parameters that control the algorithm (e.g., parameters that define whether a sufficient decrease in the merit function has been found during a backward path search). For the PATH algorithm, the proof of convegence is beyond the scope of this book, and even just the statement of the theorem on convergence is complex enough that it would take us too far afield. Readers who are interested may consult [8] or [25]. One important theme is that the sequence of iterates produced by the approximate LCPs without damping will converge locally, but damping (and other features of the PATH algorithm) bring about global convergence.

To summarize the convergence result, we can say that, for a large class of functions f, and for most "reasonable" first guesses x^0, the PATH algorithm converges to a solution of the normal equation $f_B(x) = 0$, i.e., to a solution of the NCP. The theoretical convergence result has been verified in practice – a great many NCP and MCP models have been successfully solved by PATH.

8.6.6 Other Methods to Solve NCPs

Most methods to solve the NCP involve a Newton approach, but the methods differ in the type of nonsmooth equation and merit function. They also differ in the way in which a direction is chosen for moving to the next iterate, and these differences are often due to differences in the type of nonsmooth equation and merit function. Most of the methods can be understood to be descent algorithms, i.e., they work to minimize a merit function. However, since the merit function often has local minima with a positive value (where the nonsmooth equation is *not* satisfied), these methods can get stuck at local minima.

We have seen in some detail how the PATH algorithm solves a linear complementarity problem to find the Newton point, and then does a backward search along the piecewise linear path for a point that sufficiently reduces the merit function. Another approach is taken in the NE/SQP (standing for nonsmooth equation, successive quadratic programming) algorithm [15], [23], based on the equation (8.52): a well-behaved quadratic program is solved at each step to find an approximate Newton *direction*, and then points along a line in that direction away from the current iterate are searched for another point that decreases the merit function sufficiently. The QPCOMP algorithm [3] built upon NE/SQP by devising a perturbation technique to move to a different point if NE/SQP gets stuck at a nonzero, local minimum of the merit function.

The BDIFF method [17] relies on a special kind of partial derivative of the function f_B (8.51), called the B-derivative, which is defined everywhere. It applies a damped Newton method directly with the B-derivative and a linesearch.

An example of an approach based in part on the merit function (8.55) derived from the Fischer-Burmeister function (8.53) is discussed in [12].

Some authors have adopted an altogether different strategy to deal with nonsmoothness in the system of equations: they approximate the nonsmooth equations by a family of smooth equations that depend on a parameter. As the parameter approaches zero, the approximate smooth equations approach the correct nonsmooth equations. The algorithms based on this idea solve the smooth equations by a standard Newton method, then adjust the parameter downwards and solve the more accurate smooth equations, starting from the solution obtained with the previous value of the parameter. For example, [4] and [5] approximate the plus function (8.56), which has a discontinuity in the derivative at $y_i = 0$, by various smooth functions.

Readers can find a thorough discussion of methods for solving NCPs in the two-volume book [11].

8.7 Exercises

1. Use Lemke's method to solve the following LCP, a modified form of the multicommodity equilibrium model of Section 8.2.1.3.3:

$$w_1 = -32 + 6z_1 - z_2 \ , \ \ z_1 \geq 0 \ , \ \ w_1 \geq 0 \ , \ \ w_1 z_1 = 0$$
$$w_2 = -64 - 2z_1 + 7z_2 \ , \ \ z_2 \geq 0 \ , \ \ w_2 \geq 0 \ , \ \ w_2 z_2 = 0$$

2. In Chapter 1, Section 1.2.4, a duopoly is described as an LCP, based on a linear inverse demand function $p = \alpha - \beta(z_1 + z_2)$, where z_1 and z_2 are the quantities produced by the firms 1 and 2, at unit costs γ_1 and γ_2,

respectively:

$$2\beta z_1 + \beta z_2 - \alpha + \gamma_1 \geq 0\ ,\quad z_1 \geq 0\ ,\quad (2\beta z_1 + \beta z_2 - \alpha + \gamma_1) \cdot z_1 = 0$$
$$\beta z_1 + 2\beta z_2 - \alpha + \gamma_2 \geq 0\ ,\quad z_2 \geq 0\ ,\quad (\beta z_1 + 2\beta z_2 - \alpha + \gamma_2) \cdot z_2 = 0$$

Here, we assign particular values to the parameters: $\alpha = 6, \beta = 1, \gamma_1 = 1$, $\gamma_2 = 2$, which gives rise to the following LCP:

$$2z_1 + z_2 - 5 \geq 0\ ,\quad z_1 \geq 0\ ,\quad (2z_1 + z_2 - 5) \cdot z_1 = 0$$
$$z_1 + 2z_2 - 4 \geq 0\ ,\quad z_2 \geq 0\ ,\quad (z_1 + 2z_2 - 4) \cdot z_2 = 0$$

a. Use Lemke's method to solve the LCP.
b. From the starting guess $z^0 = \begin{pmatrix} 1 \\ 0 \end{pmatrix}$, $w^0 = \begin{pmatrix} 0 \\ 1 \end{pmatrix}$, use the PATH method to solve the LCP (i.e., formulate the augmented LCP and perform the pivots to drive t to 1). Evaluate the merit function based on the Fischer-Burmeister function for the starting guess, and for each z value generated by pivoting.
c. Use the matrix splitting iterative method to solve the LCP, with B equal to the diagonal of the matrix. Use the first guess $z^0 = \begin{pmatrix} 1 \\ 1 \end{pmatrix}$, and stop when $\|z^k - z^{k-1}\| \leq 0.1$.
d. Use the matrix splitting iterative method to solve the LCP, with B equal to the best symmetric matrix that you can find, and the same first guess as in part (c). [Hint: you should only need one iteration.]

3. Consider the multicommodity equilibrium LCP that appears in Section 8.2.1.3.3 and in several other examples:

$$w_1 = -32 + 6z_1 + z_2\ ,\quad z_1 \geq 0\ ,\quad w_1 \geq 0\ ,\quad w_1 z_1 = 0$$
$$w_2 = -64 + 2z_1 + 7z_2\ ,\quad z_2 \geq 0\ ,\quad w_2 \geq 0\ ,\quad w_2 z_2 = 0$$

a. From the starting guess $z^0 = \begin{pmatrix} 0 \\ 2 \end{pmatrix}$, $w^0 = \begin{pmatrix} 1 \\ 0 \end{pmatrix}$, use the PATH method to solve the LCP. Evaluate the merit function based on the Fischer-Burmeister function for the starting guess, and for each z value generated by pivoting.
b. Use the matrix splitting iterative method to solve the LCP, with B equal to the diagonal part of the matrix. Use the first guess $z^0 = \begin{pmatrix} 2 \\ 5 \end{pmatrix}$, and stop when $\|z^k - z^{k-1}\| \leq 0.1$.
c. Use the matrix splitting iterative method to solve the LCP, with

$$B = \begin{pmatrix} 6 & 2 \\ 2 & 7 \end{pmatrix}, \text{ and } C = \begin{pmatrix} 0 & -1 \\ 0 & 0 \end{pmatrix}$$

Use the same first guess as in part (b), and stop when $\|z^k - z^{k-1}\| \leq 0.1$.

4. Consider the following modified form of the multicommodity equilibrium LCP of Section 8.2.1.3.3:

$$\begin{aligned} w_1 &= -32 + 6z_1 + z_2 \;, \quad z_1 \geq 0 \;, \quad w_1 \geq 0 \;, \quad w_1 z_1 = 0 \\ w_2 &= -2 + 2z_1 + 7z_2 \;, \quad z_2 \geq 0 \;, \quad w_2 \geq 0 \;, \quad w_2 z_2 = 0 \end{aligned}$$

a. Use Lemke's method to solve the LCP.
b. From the starting guess $z^0 = \begin{pmatrix} 0 \\ 1 \end{pmatrix}$, $w^0 = \begin{pmatrix} 1 \\ 0 \end{pmatrix}$, use the PATH method to solve the LCP. Evaluate the merit function based on the Fischer-Burmeister function for the starting guess, and for each z value generated by pivoting.
c. Use the matrix splitting iterative method to solve the LCP, with B equal to the diagonal of the matrix. Use the first guess $z^0 = \begin{pmatrix} 3 \\ 1 \end{pmatrix}$, and stop when $\|z^k - z^{k-1}\| \leq 0.1$.
d. Use the matrix splitting iterative method to solve the LCP, with

$$B = \begin{pmatrix} 6 & 1 \\ 1 & 7 \end{pmatrix}, \text{ and } C = \begin{pmatrix} 0 & 0 \\ 1 & 0 \end{pmatrix}.$$

Use the same first guess as in part (c), and stop when $\|z^k - z^{k-1}\| \leq 0.1$.
[Hint for all parts: the solution is $z_1 = 16/3$, $z_2 = 0$.]

5. Find the first linearization about the starting guess of $z^0 = \begin{pmatrix} 1 \\ 0 \end{pmatrix}$ for the two-commodity NCP from Section 8.3.2.4.2

$$\begin{aligned} f_1(z_1, z_2) &= -32 + 6z_1 + z_2 + 0.1z_1^2 \geq 0 \;, \quad z_1 \geq 0 \;, \quad f_1(z_1, z_2) z_1 = 0 \\ f_2(z_1, z_2) &= -64 + 2z_1 + 7z_2 \geq 0 \;, \quad z_2 \geq 0 \;, \quad f_2(z_1, z_2) z_2 = 0 \end{aligned}$$

a. Use Lemke's method to solve this first approximate LCP.
b. From the starting guess $z^0 = \begin{pmatrix} 1 \\ 0 \end{pmatrix}$, $w^0 = \begin{pmatrix} 0 \\ 1 \end{pmatrix}$, use the PATH method to solve the first approximate LCP. Evaluate the merit function based on the Fischer-Burmeister function for the starting guess, and for each z value generated by pivoting.
c. Choose the point from part (b) that provides the best merit function value, and then formulate the second approximate LCP about this point. Also formulate the augmented LCP for the PATH method to solve this second approximate LCP.

6. Prove the claim in Section 8.6.3.4 that $z \in \mathbb{R}^n$ solves NCP if and only if z solves the nonsmooth equation based on the plus function:

$$z - (z - f(z))_+ = 0$$

7. In this problem, you are to use the data and constant elasticity demand model in the PIES algorithm example of Section 8.4.2.1.1, and write three GAMS programs to solve it.

   a. Write out an NCP version of the model as in (8.47) by writing out the VI model's KKT conditions of Section 8.4.1 (use the full demand functions, not the approximate ones of the PIES algorithm), code it in GAMS, and solve it by calling the PATH solver. In the output from your program, you can see the prices p_e, p_g as the dual variables associated with the market clearing constraints for the electricity and gas; check that you get the same prices as in the text. [Hints: (i) inversion of the demand function, to get the inverse demand function, involves inverting the matrix of elasticities, which can be done in a spreadsheet such as MS-Excel; (ii) place small, positive lower bounds on the demand variables, in order ensure that PATH does not try to evaluate a demand raised to a negative power, which would cause an evaluation error.]
   b. Write out a VI version of the model as in (8.47), code it in GAMS, solve it by calling the EMP "solver" (which actually converts the VI to the equivalent NCP and calls PATH), and check that you get the same prices as in the text. [See hints for part (a).]
   c. Write a GAMS program that performs one optimization step of a diagonalization strategy applied to the PIES model as a VI, (8.47), i.e., one step of the PIES-q algorithm. Your program should take data from the previous solution, q^{k-1} (or from a first guess, q^0), and produce the next iterate (x^k, q^k). For first guesses q^0, use the reference quantities $\overline{q}_e = 460$, $\overline{q}_g = 435$. Run your program for 5 iterations, or until you observe convergence of the iterates, each time pasting the previous solution's value of q^{k-1} into the program as new data. (If you know how to code a loop in GAMS, then try to automate this procedure.) Compare the prices p_e and p_g at each iteration to the prices generated by the PIES algorithm, as shown in Table 8.3 (there should be a different sequence of prices, but converging to the same equilibrium prices).

8. For the Fischer-Burmeister function (8.53), prove that if $0 \leq a \perp b \geq 0$, then $\phi(a, b) = 0$.

References

1. B.H. Ahn. *Computation of Market Equilibria for Policy Analysis*, Garland Publishing Inc., New York, 1979.
2. B.H. Ahn and W.W. Hogan. On convergence of the PIES algorithm for computing equilibria. *Operations Research*, 30:281-300, 1982.
3. S.C. Billups and M.C. Ferris. QPCOMP: a quadratic programming based solver for mixed complementarity problems. *Mathematical Programming*, 76:533-562, 1997.
4. B. Chen and P.T. Harker. Smooth approximations to nonlinear complementarity problems. *SIAM Journal on Optimization*, 7:403-420, 1997.
5. C.H. Chen and O.L. Mangasarian. A class of smoothing functions for nonlinear and mixed complementarity problems. *Computational Optimization and Applications*, 5(2):97-138, 1996.
6. R.W. Cottle, J.-S. Pang, and R.E. Stone. *The Linear Complementarity Problem*, Academic Press, Inc., San Diego, 1992.
7. S. Dafermos. An iterative scheme for variational inequalities. *Mathematical Programming*, 26:40-47, 1983.
8. S.P. Dirkse and M.C. Ferris. The PATH solver: a nonmonotone stabilization scheme for mixed complementarity problems. *Optimization Methods and Software*, 5:123-156, 1995.
9. S.P. Dirkse and M.C. Ferris. A pathsearch damped Newton method for computing general equilibria. *Annals of Operations Research*, 68:211-232, 1996.
10. S.P. Dirkse and M.C. Ferris. Crash techniques for large-scale complementarity problems. In *Complementarity and Variational Problems: State of the Art*. M.C. Ferris and J.-S. Pang, eds., SIAM Publications, 40-61, 1997.
11. F. Facchinei and J.-S. Pang. *Finite-Dimensional Variational Inequalities and Complementarity Problems, Volumes I and II*, Springer, New York, 2003.
12. F. Facchinei and J. Soares. A new merit function for nonlinear complementarity problems and a related algorithm. *SIAM Journal on Optimization*, 7:225-247, 1997.
13. M.C. Ferris, S.P. Dirkse, J.-H. Jagla and A. Meeraus. An extended mathematical programming framework. *Computers and Chemical Engineering*, 33:1973-1982, 2009.
14. M.C. Ferris and T.S. Munson. PATH 4.6. In *GAMS: the Solver Manuals*. GAMS Development Corporation, Washington D.C., 2005. Available at: www.gams.com.
15. S.A. Gabriel and J.-S. Pang. An inexact NE/SQP method for solving the nonlinear complementarity problem. *Computational Optimization and Applications*, 1:67-91, 1992.
16. P.T. Harker. Accelerating the convergence of the diagonalization and projection algorithms for finite-dimensional variational inequalities. *Mathematical Programming*, 41:29-59, 1988.

17. P. T. Harker and B. Xiao. Newton's method for the nonlinear complementarity problem: a B-differentiable equation problem. *Mathematical Programming*, 48:339-357, 1990.
18. N.H. Josephy. Newton's method for generalized equations. Technical Summary Report 1965, Mathematics Research Center, University of Wisconsin-Madison, Madison, WI, 1979.
19. C.E. Lemke. Bimatrix equilibrium points and mathematical programming. *Management Science*, 11(7):681-689, 1965.
20. L. Mathiesen. Computational experience in solving equilibrium models by a sequence of linear complementarity problems. *Operations Research* 33(6):1225-1250, 1985.
21. J.M. Ortega and W.C. Rheinboldt. *Iterative Solution of Nonlinear Equations in Several Variables*, Academic Press, New York, 1970.
22. J.-S. Pang and D. Chan. Iterative methods for variational and complementarity problems. *Mathematical Programming*, 24:284-313, 1982.
23. J.-S. Pang and S.A. Gabriel. NE/SQP: a robust algorithm for the nonlinear complementarity problem. *Mathematical Programming*, 60:295-337, 1993.
24. M. Patriksson. Simplicial decomposition algorithms. *Encyclopedia of Optimization*, Springer, 2009.
25. D. Ralph. Global convergence of damped Newton's method for nonsmooth equations via the path search. *Mathematics of Operations Research*, 19(2):352-389, 1994.
26. S.M. Robinson. Newton's method for a class of nonsmooth functions. *Set-Valued Analysis*, 2:291-305, 1994.

Chapter 9
Some Advanced Algorithms for VI Decomposition, MPCCs and EPECs

9.1 Introduction

In this chapter, we present several advanced algorithms that can be useful for the solution of some of the models discussed in this book.

Section 9.2 discusses several decomposition algorithms for variational inequality (VI) models: an extension of the classical Dantzig-Wolfe decomposition algorithm [6] from the realm of optimization models to VIs; simplicial decomposition of VIs, which relies on the column generation idea of Dantzig-Wolfe decomposition; and an extension of Benders decomposition from optimization models to VIs, explained as an application of Dantzig-Wolfe decomposition to the dual VI of a given primal VI model. The cobweb decomposition algorithm is also briefly described.

Section 9.3 presents several methods to solve mathematical programs with complementarity constraints (MPCCs): regularization, which solves a sequence of NLPs with relaxed complementarity constraints that are tightened at each step of the sequence; penalization, which solves a sequence of NLPs that have no complementarity constraints but instead penalize violations of complementarity, in the objective function; and sequential quadratic programming, which solves a sequence of quadratic programs that approximate the given MPCC. Several other methods to solve MPCCs are briefly described.

Section 9.4 discusses two methods to solve equilibrium programs with equilibrium constraints (EPECs): a brief discussion of diagonalization, which is described more fully in Chapter 7; and the reformulation of an EPEC as an NLP.

9.2 Decomposition Algorithms for VIs

Sometimes, a VI model has a structure that is *close to* the structure of a VI that would be easier to solve. For example, it can happen that if a few complicating constraints were not present in the model, then the model would split apart into several smaller, independent subproblems that could be solved more easily than the original large model, and solution of the separate subproblems could perhaps be done simultaneously, in parallel on separate computers. Other VI models may have complicating variables, i.e., if these variables could be fixed at known values, then the model would be easier to solve, perhaps because it breaks apart into separate smaller subproblems. Decomposition algorithms have been developed for optimization problems – e.g., the famous method of Dantzig and Wolfe [6] for LPs – in order to take advantage of such structures, and some of these algorithms have been extended to the realm of VI models. See [5] for a thorough description of many decomposition algorithms as applied to optimization models.

There is sometimes another motivation to use decomposition techniques, even if decomposition does not speed up the solution process. Murphy and Mudrageda [19] explain that splitting the very large energy market model of the U.S. Department of Energy into separate submodels for different fuels (e.g., electricity, natural gas, etc.) allows different teams of experts to build and maintain the submodels; decomposition techniques are used to bring the separate pieces together into a consistent multi-fuel equilibrium.

In this section, we present three decomposition techniques for VIs – Dantzig-Wolfe, simplicial, and Benders decomposition – and we give an overview of cobweb decomposition. We show the details of Dantzig-Wolfe decomposition for VIs, and we explain simplicial and Benders techniques by deriving them based on Dantzig-Wolfe decomposition.

As an introduction and review of the basic ideas of Dantzig-Wolfe decomposition for optimization models, we begin with a simple illustrative example of Dantzig-Wolfe decomposition applied to an LP. Following the LP example, we present an illustrative example of a stochastic power VI model which is used to illustrate Dantzig-Wolfe, simplicial, and Benders decomposition in later subsections. The stochastic power model is a simplified version of a VI model from Chapters 4 and 5, and in [11].

9.2.1 Illustrative Example. Dantzig-Wolfe Decomposition of a Simple LP

Consider the following LP in four variables.

$$\text{Minimize} \quad 4x_1 + x_2 + 2x_3 + 6x_4 \tag{9.1a}$$

s. t.

$$x_1 + 2x_2 + 3x_3 + x_4 \geq 10 \quad (\beta) \tag{9.1b}$$

$$x_1 + 3x_2 \leq 5 \quad (\alpha_1) \tag{9.1c}$$

$$x_2 \leq 2 \quad (\alpha_2) \tag{9.1d}$$

$$2x_3 + x_4 \leq 6 \quad (\alpha_3) \tag{9.1e}$$

$$x_3 \leq 2 \quad (\alpha_4) \tag{9.1f}$$

$$x_1, x_2, x_3, x_4 \geq 0 \tag{9.1g}$$

If the *linking constraint* (9.1b) (also called a *complicating constraint*) were not present, then the LP would split into two smaller LPs – one in the variables x_1, x_2, and the other in x_3, x_4. Dantzig-Wolfe decomposition creates two subproblems by dropping the linking constraint (9.1b), but with a modification of the objective function coefficients, using an interpretation of the dual variable β. At an optimal solution of (9.1), β can be interpreted as the rate of change of the optimal objective value with respect to an increase in the right hand side of (9.1b). Let us move, say, the x_2 term to the right side of (9.1b), i.e., (9.1b) becomes

$$x_1 + 3x_3 + x_4 \geq 10 - 2x_2$$

and a one unit increase in x_2 causes a change of -2 on the right side. Thus the rate of change of the optimal objective value with respect to an increase in x_2 is -2β. If we have an estimate of the optimal value of β, then it is as if $-2\beta x_2$ is another component of cost associated with x_2 (but since $-2\beta x_2 \leq 0$ in this example, it might more properly be called a revenue). From this line of reasoning, the total cost associated with x_2 includes the original term of the objective function, and the new term involving β, i.e., $(1 - 2\beta)x_2$. The same reasoning can be followed for the other three variables, which gives the modified form of the objective functions of the two subproblems, described below.

At iteration k of Dantzig-Wolfe decomposition, an estimate β^{k-1} is available from the previous iteration's solution of the master problem (see below for a description of the master problem), or, if $k = 1$ and there was no previous master problem, then $\beta^{k-1} = \beta^0$ is a first guess provided by the modeler. The estimate β^{k-1} is the value of the dual variable of the master problem's linking constraint at the optimal solution of the master problem. Thus, the two subproblems, without the linking constraint, and with the modified objective functions, are

$$\text{Minimize} \quad (4 - \beta^{k-1})x_1 + (1 - 2\beta^{k-1})x_2 \tag{9.2a}$$

s. t.

$$x_1 + 3x_2 \leq 5 \quad (\alpha_1) \tag{9.2b}$$

$$x_2 \leq 2 \quad (\alpha_2) \tag{9.2c}$$

$$x_1, x_2 \geq 0 \tag{9.2d}$$

and

$$\text{Minimize} \quad (2 - 3\beta^{k-1})x_3 + (6 - \beta^{k-1})x_4 \tag{9.3a}$$

s. t.

$$2x_3 + x_4 \leq 6 \quad (\alpha_3) \tag{9.3b}$$

$$x_3 \leq 2 \quad (\alpha_4) \tag{9.3c}$$

$$x_3, x_4 \geq 0 \tag{9.3d}$$

The two subproblems (9.2), (9.3) are solved at the beginning of each iteration k. Here, we denote the optimal solutions calculated for (9.2), (9.3) at iteration k by $x_{1S}^k, x_{2S}^k, x_{3S}^k, x_{4S}^k$. Note that the "$S$" subscript denotes subproblem as will be explained below.

At the end of each iteration k, a master problem is solved. At iteration k, there have been k previous solutions of the subproblems, $(x_{1S}^j, x_{2S}^j, x_{3S}^j, x_{4S}^j)$, $j = 1, \ldots, k$, often called *proposals* in the decomposition literature. In the master problem, the variable vector $x = (x_1, x_2, x_3, x_4)^T$ is restricted to the set of *convex combinations* of $(x_{1S}^j, x_{2S}^j, x_{3S}^j, x_{4S}^j)^T$, $j = 1, \ldots, k$. For fractional weights $\lambda_j \geq 0$, $j = 1, \ldots, k$, with $\sum_{j=1}^{k} \lambda_j = 1$, the convex combination of previous subproblem proposals is

$$\sum_{j=1}^{k} \begin{pmatrix} x_{1S}^j \\ x_{2S}^j \\ x_{3S}^j \\ x_{4S}^j \end{pmatrix} \lambda_j = \begin{pmatrix} \sum_{j=1}^{k} x_{1S}^j \lambda_j \\ \sum_{j=1}^{k} x_{2S}^j \lambda_j \\ \sum_{j=1}^{k} x_{3S}^j \lambda_j \\ \sum_{j=1}^{k} x_{4S}^j \lambda_j \end{pmatrix} \tag{9.4a}$$

$$\text{where} \sum_{j=1}^{k} \lambda_j = 1 \tag{9.4b}$$

$$\text{and } \lambda_j \geq 0, j = 1, \ldots, k \tag{9.4c}$$

In the master problem, the variables are the weights λ_j. The reader can easily verify that any convex combination (9.4) satisfies constraints (9.1c)-(9.1g) of the original LP, because all of the previous subproblem solutions satisfied these constraints in the subproblems; therefore, the master problem does not need to list these constraints. However, the master problem enforces the linking constraint (9.1b).

For this example, (9.1), the zero vector $x = (0,0,0,0)^T$ is feasible with respect to the constraints (9.1c)-(9.1g) in the original LP and in the subproblems; therefore, we can always include the zero vector for the convex combination, as if it came from a solution of the subproblems. Denoting the zero vector by the index $j = 0$, the convex combination is thus $(\sum_{j=0}^{k} x_{1S}^j\lambda_j, \sum_{j=0}^{k} x_{2S}^j\lambda_j, \sum_{j=0}^{k} x_{3S}^j\lambda_j, \sum_{j=0}^{k} x_{4S}^j\lambda_j)^T$, where $\sum_{j=0}^{k} \lambda_j = 1$ and $\lambda_j \geq 0, j = 0,1,\ldots,k$; but the set of all vectors generated by such combinations is the same as the set generated by $(\sum_{j=1}^{k} x_{1S}^j\lambda_j, \sum_{j=1}^{k} x_{2S}^j\lambda_j, \sum_{j=1}^{k} x_{3S}^j\lambda_j, \sum_{j=1}^{k} x_{4S}^j\lambda_j)^T$, where $\sum_{j=1}^{k} \lambda_j \leq 1$ and $\lambda_j \geq 0, j = 1,\ldots,k$. The zero vector is not always feasible in the subproblems of every model, but when it is feasible, it can be used in this way to enlarge the feasible region represented by convex combinations, with no extra computational effort.

Therefore, the form of the master problem at iteration k, is

$$\text{Minimize} \quad 4\sum_{j=1}^{k}(x_{1S}^j\lambda_j)+\sum_{j=1}^{k}(x_{2S}^j\lambda_j)+2\sum_{j=1}^{k}(x_{3S}^j\lambda_j)+6\sum_{j=1}^{k}(x_{4S}^j\lambda_j) \tag{9.5a}$$

s. t.

$$\sum_{j=1}^{k}(x_{1S}^j\lambda_j)+2\sum_{j=1}^{k}(x_{2S}^j\lambda_j)+3\sum_{j=1}^{k}(x_{3S}^j\lambda_j)+\sum_{j=1}^{k}(x_{4S}^j\lambda_j)\geq 10 \quad (\beta) \tag{9.5b}$$

$$\sum_{j=1}^{k}\lambda_j \leq 1 \tag{9.5c}$$

$$\lambda_j \geq 0, \; j = 1,\ldots,k \tag{9.5d}$$

Because the master problem (9.5) is a *restriction* of the original LP[1] (9.1) – i.e., the feasible region of the master problem, in x space, is smaller than that of the original LP – the optimal objective value of the master problem is greater than or equal to the optimal value of the original LP. The master problem provides an *upper bound on the optimal value* of the original LP. Furthermore, at iteration k, the master problem's feasible region includes the feasible region of the master problem at iteration $k-1$, and so the optimal values of the master problems decrease (or possibly stay the same) from one iteration to the next, approaching the optimal value of the original LP.

Each time the subproblems are solved, at the beginning of an iteration k, a *lower bound on the optimal value* of the original problem can also be obtained. This can be seen from an examination of the dual of the original LP:

$$\text{Maximize} \quad 10\beta - 5\alpha_1 - 2\alpha_2 - 6\alpha_3 - 2\alpha_4 \tag{9.6a}$$

s. t.

$$\beta - \alpha_1 \leq 4 \tag{9.6b}$$

$$2\beta - 3\alpha_1 - \alpha_2 \leq 1 \tag{9.6c}$$

$$3\beta - 2\alpha_3 - \alpha_4 \leq 2 \tag{9.6d}$$

$$\beta - \alpha_3 \leq 6 \tag{9.6e}$$

$$\beta, \alpha_1, \alpha_2, \alpha_3, \alpha_4 \geq 0 \tag{9.6f}$$

It is well known that the optimal objective value (9.6a) of the dual LP is equal to the optimal objective value (9.1a) of the original, primal LP. As we explain below, after the subproblems are solved at the beginning of iteration k, a *feasible* solution to the dual problem (9.6) is available from the optimal dual

[1] Consider, for example, the values that the vector

$$x = \left(\sum_{j=1}^{k} (x_{1S}^j \lambda_j), \sum_{j=1}^{k} (x_{2S}^j \lambda_j), \sum_{j=1}^{k} (x_{3S}^j \lambda_j), \sum_{j=1}^{k} (x_{4S}^j \lambda_j) \right)^T$$

can take when $k = 1$: x would lie on a line between the origin and $x_S^1 = (x_{1S}^1, x_{2S}^1, x_{3S}^1, x_{4S}^1)^T$ because (9.5c)-(9.5d) restrict λ_1 to lie between 0 and 1; all of the points x on this line satisfy the subproblem constraints (9.1c)-(9.1g) because x_S^1 and the origin do, and because the feasible region of the subproblem is convex. Only the points on the line (if any) that satisfy (9.5b) are feasible in the master problem, and in the original problem (9.1). Since there are many points *off the line* that are feasible in the original problem, we can say that the feasible region of the master problem (9.5) is *restricted*, i.e. strictly included in the feasible region of the original problem (9.1).
Similarly, when $k = 2$, the points x that satisfy the subproblem constraints (9.1c)-(9.1g) lie on a plane bordered by a triangle whose vertices are x_S^1, x_S^2 and the origin. This triangle includes the line defined at iteration $k = 1$, which means that the feasible region of the master problem (9.5) has grown larger at the second iteration, but, depending on the particularities of a model, there could still be many points that are feasible in the original problem (9.1) but not in the master problem, i.e., the master problem is still a restriction of the original problem.

variables of the subproblems, together with the optimal dual variable β^{k-1} from the previous master problem. This feasible solution of the dual problem has an objective value that is less than or equal to the optimal value, and therefore provides a lower bound on the original problem's optimal objective value.

A feasible solution of the dual (9.6) can be constructed in the following way, after the subproblems are solved at the beginning of iteration k. First, the duals of the two subproblems (9.2), (9.3), are

$$\text{Maximize} \quad -5\alpha_1 - 2\alpha_2 \tag{9.7a}$$

s. t.

$$-\alpha_1 \leq 4 - \beta^{k-1} \tag{9.7b}$$

$$-3\alpha_1 - \alpha_2 \leq 1 - 2\beta^{k-1} \tag{9.7c}$$

$$\alpha_1, \alpha_2 \geq 0 \tag{9.7d}$$

and

$$\text{Maximize} \quad -6\alpha_3 - 2\alpha_4 \tag{9.8a}$$

s. t.

$$-2\alpha_3 - \alpha_4 \leq 2 - 3\beta^{k-1} \tag{9.8b}$$

$$-\alpha_3 \leq 6 - \beta^{k-1} \tag{9.8c}$$

$$\alpha_3, \alpha_4 \geq 0 \tag{9.8d}$$

When the subproblems (9.2), (9.3) are solved, dual values $(\alpha_1^*, \alpha_2^*, \alpha_3^*, \alpha_4^*)$ are available from the LP solver, and they satisfy the constraints of (9.7), (9.8). Moving the terms involving β^{k-1} from the right to the left sides of constraints (9.7b), (9.7c), (9.8b), (9.8c), we see that $(\beta^{k-1}, \alpha_1^*, \alpha_2^*, \alpha_3^*, \alpha_4^*)$ satisfies the constraints of the dual problem (9.6), as claimed.

To summarize the results about bounds on the optimal value of the original LP, at the end of iteration k, when the kth master problem has just been solved, the optimal value of the master problem is directly an upper bound on the optimal value of the original LP (9.1). When the subproblems are next solved at the beginning of iteration $k+1$, using the value β^k from the master problem of iteration k, the optimal dual variable values of the subproblems, $\alpha_1^*, \alpha_2^*, \alpha_3^*, \alpha_4^*$, together with β^k, can be used to evaluate the dual objective function (9.6a) to give a lower bound $10\beta^k - 5\alpha_1^* - 2\alpha_2^* - 6\alpha_3^* - 2\alpha_4^*$ on the optimal value of the original LP. The *gap* (upper bound minus lower bound) is often used as a stopping condition for the Dantzig-Wolfe algorithm: when the gap is small enough, the iterations cease, even if the gap is still nonzero.

For this problem, we started the iterations with a guess $\beta^0 = 8$. An optimal solution of the original problem is found by the second master problem, and this is recognized as optimal after the subproblems are solved for the third time. The results of the iterations are shown in Table 9.1.

Table 9.1 Iterations of Dantzig-Wolfe decomposition of small LP example

Iteration	Master Solution				Master	
k	x_1	x_2	x_3	x_4	Objective	Gap
1	0	1.471	1.765	1.765	15.588	8.882
2	0	1.667	2	0.667	9.667	0

□

9.2.2 Illustrative Example. Simplified Stochastic Power Model from Chapters 4 and 5

Here, we describe the model that is used to illustrate Dantzig-Wolfe, simplicial and Benders decomposition methods.

In this simplified version of a VI model from Chapters 4 and 5, Sections 4.6.3 and 5.2.3.4, and in [11], we assume that there are only two nodes $i \in \{1, 2\}$ and one line in the transmission system. We also assume that there is one generator at each node, and that the generators at the two nodes are owned by two different firms. Thus, we can eliminate the firm index f, because the node index i can be used to distinguish firms. Furthermore, because there is only one generator at a node, the index of different generating units, h, is also eliminated. There are two demand scenarios, m, for low demand and high demand. The variables in the simplified model are

$$s_{ijm} = \text{sales by firm at node } i \text{ to customers at node } j \text{ in scenario } m$$
$$x_i = \text{slow-ramping, baseload generation at node } i$$
$$r_{im} = \text{rapid-ramping generation at node } i \text{ in scenario } m$$
$$y_{im} = \text{net injection of power from the line into node } i \text{ in scenario } m$$

Because there is only one transmission line, the line limit constraints take a very simple form: the upper limit of T on power flow in scenario m from node 1 to node 2 is simply $y_{2m} \leq T$, and the upper limit on flow from node 2 to node 1 is $y_{1m} \leq T$.

With these simplifications, the feasible region is

$$K = \{s_{ijm}, x_i, r_{im}, y_{im} \mid$$
$$x_i - X_i \leq 0 \; \forall i \tag{9.9a}$$
$$r_{im} - R_i \leq 0 \; \forall i, m \tag{9.9b}$$
$$\sum_j s_{ijm} - x_i - r_{im} = 0 \; \forall i, m \tag{9.9c}$$
$$y_{im} \leq T \; \forall i, m \tag{9.9d}$$

$$-\sum_j s_{jim} + x_i + r_{im} + y_{im} = 0 \; \forall i, m \tag{9.9e}$$

$$x_i, r_{im}, s_{ijm} \geq 0, y_{im} \text{ unrestricted in sign}\} \tag{9.9f}$$

and for[2] variables arranged in a vector in this order

$$\begin{pmatrix} s_{ijm} \; \forall i, j, m \\ x_i \; \forall i \\ r_{im} \; \forall i, m \\ y_{im} \; \forall i, m \end{pmatrix}$$

the mapping, with elements in the same order, is

$$G(s_{ijm}) = \begin{pmatrix} prob_{jm} \left[-a_{jm} + b_j \left(s_{ijm} + \sum_{j'} s_{j'jm} \right) \right] \; \forall i, j, m \\ C_i \; \forall i \\ prob_{im} RC_i \; \forall i, m \\ 0 \; \forall i, m \end{pmatrix} \tag{9.10}$$

The values of all parameters are given in Table 9.2. Using GAMS with EMP[3] and PATH, a solution (calculated without decomposition techniques) is given in Table 9.3.

Table 9.2 Data for stochastic power model used in decomposition illustrations

Node i	1	2
baseload generation cost C_i	15	19
rapid ramping generation cost RC_i	20	25
baseload capacity X_i	400	180
rapid ramping capacity R_i	150	100
inverse demand price intercept $a_{im}, m = Lo$	25	25
inverse demand price intercept $a_{im}, m = Hi$	50	50
inverse demand slope	.08	.04
line limit T	175	
$prob_{im} = 0.5, \forall i, m$		

□

[2] Note that the summations in (9.9c) and (9.9e) are different, because the i and j subscripts are in different orders.

[3] The Extended Mathematical Programming (EMP) software converts a VI model to its equivalent KKT system, an MCP, and calls PATH for a solution.

Table 9.3 Solution to stochastic power model used in decomposition illustrations

Node i	1		2	
Scenario m	Lo	Hi	Lo	Hi
baseload generation x_i	243.47		141.48	
rapid ramping generation r_{im}	0	150.0	0	100.0
net injection at i, y_{im}	-115.15	-175.0	115.15	175.0
total sales at i, $\sum_j s_{jim}$	128.31	218.47	256.63	416.48
price at i, $a_{im} - b_i \sum_j s_{jim}$	14.74	32.52	14.74	33.34

9.2.3 Dantzig-Wolfe Decomposition of VIs

In this subsection, we summarize the key features and properties of Dantzig-Wolfe decomposition for VIs, as discussed in [4], [9] and [11]. We adopt the notation from these references, which introduces some slight differences compared with notation in Chapters 5 and 8.

We consider a VI defined over the set $K = \{z \in \mathbb{R}^n | g(z) \geq 0, h(z) \geq 0\}$[4], where $g : \mathbb{R}^n \to \mathbb{R}^{p_1}, h : \mathbb{R}^n \to \mathbb{R}^{p_2}$ with the constraints $h(z) \geq 0$ being the complicating ones in the sense that removing them would make the problem easier to solve, e.g., by breaking the overall problem into separate VIs. Given the function $G : \mathbb{R}^n \to \mathbb{R}^n$, we define the VI as:

$$\text{find } z^* \in K \text{ such that } G(z^*)^T (z - z^*) \geq 0 \, \forall z \in K \tag{9.11}$$

The LP model of the illustrative example in Section 9.2.1 can be represented as a VI, with $z = (x_1, x_2, x_3, x_4)^T$; $G(z) = (4, 1, 2, 6)^T$, the objective coefficients in (9.1a); $g(z) \geq 0$ corresponding to the constraints (9.1c)-(9.1g); and $h(z) \geq 0$ corresponding to the linking constraint (9.1b). We use this model, and the decomposition procedure described in Section 9.2.1, to illustrate the general development of Dantzig-Wolfe decomposition for VIs.

We distinguish two types of variables, $z_1 \in \mathbb{R}^{n_1}, z_2 \in \mathbb{R}^{n_2}$, with $n = n_1 + n_2$: z_1 is the vector of variables which appear only in the master problem, and z_2 is the vector of variables that appear in the subproblem.[5] The LP example of Section 9.2.1 does not have variables z_1 which appear only in the master problem, but the next illustrative example (using the stochastic power model) does have such variables. We assume that the vector-valued function $g(z)$ depends only on z_2 – we write $g(z_2)$, and $g : \mathbb{R}^{n_2} \to \mathbb{R}^{p_1}$ –

[4] Note that the inequalities are in the form ≥ 0, instead of the ≤ 0 form of Chapters 5 and 8, to be consistent with [4], [9] and [11].

[5] To speak of *the* subproblem may appear odd, in light of the *two* subproblems in the LP example of Section 9.2.1. However, there are some models for which decomposition produces a single subproblem that is easier to solve than the original problem, even without splitting the subproblem into several separate models – e.g., decomposition of an LP might produce a a network structure in the subproblem, for which very fast algorithms can be used. Furthermore, the two subproblems in Section 9.2.1 *could be* given to an LP solver as if they form just one problem. Therefore, to be general, we refer to *the* subproblem here.

while $h(z)$ depends on both z_1 and z_2, as a separable vector-valued function, $h(z) = h_1(z_1) + h_2(z_2)$. The function $G(z) = \begin{pmatrix} G_1(z_1) \\ G_2(z_2) \end{pmatrix}$ with $G_i(z_i)$ the same dimension as z_i for $i = 1, 2$. In the LP model of Section 9.2.1, there were no z_1 components hence, $z = z_2, h = h_2$ and $G = G_2$.

The Dantzig-Wolfe VI master problem keeps the complicating constraints $h(z) \geq 0$, and at each iteration computes (among other things – see below) an estimate of the dual variable vector β corresponding to the complicating constraints. The subproblem at iteration k drops the complicating constraints $h(z) \geq 0$ and adjusts the mapping G_2 based on the most recent master problem value of the dual variable [6] β^{k-1} as well as the previous master problem solution z_{2M}^{k-1}. Specifically, the subproblem VI is a problem in z_2 as follows:

$$\text{Sub-VI}^k : \text{ find } z_{2S}^k \in \bar{K} \text{ such that}$$

$$\left[G_2(z_{2S}^k) - \sum_{j=1}^{p_2} \nabla h_{2j}(z_{2M}^{k-1}) \beta_j^{k-1} \right]^T \left(z_2 - z_{2S}^k \right) \; \forall z_2 \in \bar{K} \tag{9.12}$$

where $\bar{K} = \{ z_2 \in \mathbb{R}^{n_2} | g(z_2) \geq 0 \}$.

The LP example of Section 9.2.1 could be expressed in VI terms: the feasible set is described by the constraints (9.1b)-(9.1g), and the mapping G is the vector of objective coefficients in (9.1a). In the LP example, there is only one complicating constraint (9.1b), i.e., the index $p_2 = 1$, and $\nabla h_{2j}(z_{2M}^{k-1}) = (1, 2, 3, 1)$, the vector of coefficients of the variables in the complicating constraint, which is independent of z_{2M}^{k-1} because h_2 is affine. The vector $G_2(z_{2S}^k) - \sum_{j=1}^{p_2} \nabla h_{2j}(z_{2M}^{k-1}) \beta_j^{k-1}$ in (9.12) has the subproblem objective function coefficients in (9.2a) and (9.3a). The feasible set $\bar{K}$ is specified by the subproblem constraints (9.2b)-(9.2d), (9.3b)-(9.3d).

The master problem at iteration k uses the previous subproblem solutions (also called proposals as in an optimization context) given by $\left[z_{2S}^1, z_{2S}^2, \ldots, z_{2S}^k \right]$ which are the columns of the matrix Z_2^k. The feasible set of weights and z_1, which are the variables in the master problem, are defined as follows, where e^k is the vector of all ones:

$$\Lambda^k = \left\{ z_1 \in \mathbb{R}^{n_1}, \lambda \in \mathbb{R}^k | h_1(z_1) + h_2\left(Z_2^k \lambda \right) \geq 0, e^{kT} \lambda = 1, \lambda \geq 0 \right\}$$

that is, z_1 and those convex combinations of subproblem proposals that are feasible for the complicating constraints. In the LP example of Section 9.2.1,

[6] The superscript that counts the iteration number of the previous master problem is $k-1$, not k, because from the definition of the algorithm below, the previous master problem was solved in the previous iteration.

the matrix-vector multiplication $Z_2^k\lambda$ is done in (9.4a), and the requirement $e^{kT}\lambda = 1$ is in (9.4b).

The master problem VI is given as follows:

$$\text{Master-VI}^k : \text{ find } (\, z_1^k, \lambda^k) \in \Lambda^k \text{ such that} \tag{9.13a}$$
$$G_1(z_1^k)^T(z_1 - z_1^k) + G_2(Z_2^k\lambda^k)^T Z_2^k(\lambda - \lambda^k) \geq 0 \, \forall (z_1, \lambda) \in \Lambda^k \tag{9.13b}$$

A solution $(\, z_1^k, \lambda^k)$ may be expressed in the original variables as $z_M^k = \begin{pmatrix} z_1^k \\ z_{2M}^k \end{pmatrix} \equiv \begin{pmatrix} z_1^k \\ Z_2^k\lambda^k \end{pmatrix}$; z_{2M}^k and the dual variables β^k are passed to the next subproblem, Sub-VI^{k+1}.

Convergence is monitored by calculating a scalar parameter at each iteration k, CG^k called the convergence gap, which is calculated from the kth master problem solution and the next subproblem solution (which is calculated at the beginning of the next iteration, $k+1$):

$$CG^k = \left[G_2(z_{2M}^k) - \sum_{j=1}^{p_2} \nabla h_{2j}(z_{2M}^k)\beta_j^k \right]^T \left(z_{2S}^{k+1} - z_{2M}^k \right) \tag{9.14}$$

Under certain assumptions (see [9], [11]), the values of CG^k are negative before convergence, they approach zero in the limit as $k \longrightarrow \infty$, and if $CG^k \geq 0$, then convergence has been achieved and the master problem solution is a solution to the original problem. In the LP example of Section 9.2.1, a gap was calculated as the difference between an upper bound on the optimal value (the optimal value of the master problem objective at iteration k) and a lower bound on the optimal value (the objective value of the dual LP, provided by the subproblem solved at iteration $k+1$, together with β^k). The gap in Section 9.2.1 is the negative of CG^k defined in (9.14); this is shown, for any LP, in [9].

For completeness, the following is a precise statement of the algorithm.

Algorithm 9.1 (Dantzig-Wolfe Algorithm for VIs).

Step 0: Set $k = 0$. Choose $\varepsilon > 0$. Set Z_2^0 to the null matrix (i.e., no columns), and set $\nabla h_2(z_{2M}^0)\beta^0$ = an estimate provided by the modeler. Go to Step 1.

Step 1: Solve Sub-VI^{k+1}. If $k = 0$ and the subproblem is infeasible, then STOP; else place the solution z_{2S}^{k+1} in the matrix $Z_2^{k+1} = [Z_2^k, z_{2S}^{k+1}]$. If $k = 0$, then go to Step 2; else {calculate CG^k from (9.14) and if $CG^k > -\varepsilon$, then STOP; else go to Step 2}.

Step 2: Increment $k \leftarrow k+1$. Solve Master-VI^k. Record $\nabla h_2(z_{2M}^k)\beta^k$. Go to Step 1.

□

9.2.3.1 Some Computational Enhancements to Dantzig-Wolfe Decomposition of VIs

Dantzig-Wolfe decomposition of VIs was generalized in [4], to allow for the subproblem mapping to be approximated in a way that relies on the previous master problem solution z_{2M}^{k-1}. In (9.12), if $G_2(z_{2S}^k)$ is replaced by any approximation $\widetilde{G}_2(z_{2S}^k; z_{2M}^{k-1})$ that evaluates to the original G_2 when $z_{2S}^k = z_{2M}^{k-1}$, then the proofs of convergence are still valid. For example, the subproblem could use the constant valued mapping $G_2(z_{2M}^{k-1})$, making the subproblem equivalent to an optimization problem; this approximation is used below to explain simplicial decomposition. Other examples of approximations $\widetilde{G}_2(z_{2S}^k; z_{2M}^{k-1})$ are discussed in [4] and [17]; both papers report improved decomposition computation times with some approximations, for some models. Furthermore, [17] reports that, for some large models, Dantzig-Wolfe decomposition (with some approximations $\widetilde{G}_2$) finds solutions faster than without decomposition, and in some cases, decomposition finds a solution when the standard, no-decomposition approach fails.

Another type of approximation is examined by Celebi in [2]. It is noted that, if many iterations are required for convergence, then the master problem VI (9.13) becomes very large, because the number of variables in λ^k equals the number of iterations; calling PATH to solve the master problem VI becomes very time consuming, especially when the iteration number k becomes large. Because of this, in several examples with many iterations, the time to solve all the master problems dominates the total computation time of the decomposition algorithm; the total computation time of all of the subproblems in all of the iterations is relatively small. To improve the master problem computation times, Celebi uses the diagonalization and symmetrization ideas of Chapter 8, Sections 8.4.2.2 and 8.4.2.3, to formulate *approximate* master problems which are optimization problems, and can be solved by NLP solvers which are much faster than PATH. The parameter z^{k-1} that is required for both the diagonalization and symmetrization strategies of Sections 8.4.2.2 and 8.4.2.3 comes from the previous master problem solution, z_{2M}^{k-1}. For almost all iterations, only one approximate master problem is solved, but occasionally for some iterations, it is necessary to take more than one step of the diagonalization or symmetrization algorithms to get a more accurate solution to the master problem. Celebi [2] reports improved computation times for Dantzig-Wolfe decomposition of two VIs, when the master problem is approximated.

9.2.4 Illustrative Example. Dantzig-Wolfe Decomposition of Simplified Stochastic Power Model

In this subsection, we apply Dantzig-Wolfe decomposition to the stochastic power model that is described in Section 9.2.2.

We must choose which constraints are complicating constraints, to be relaxed in the subproblem, and which variables (if any) are to appear only in the master problem but not in the subproblem. In principle, we could choose any constraints to be complicating constraints, as long as they refer to more than one variable. However, it is usually useful to identify complicating constraints such that, when they are removed, the problem's feasible region becomes simpler in some way, e.g., by breaking apart into separate pieces defined by constraints on separate subsets of the variables. In this example, we identify the nodal balance constraints (9.9e) as complicating, and we select the net injection variables y_{im} to appear only in the master problem, and not in the subproblem. This means that the subproblem feasible region can be split in two, i.e., one feasible set for the decision variables of the firm at node 1, and the other feasible set for the decision variables of the firm at node 2. As discussed below, instead of constraining the firms' decision variables by the nodal balance constraints, the subproblem mapping uses dual price information to influence the firms' decisions.

Below, we relate this problem's variables and functions to the general description of Dantzig-Wolfe decomposition in Section 9.2.3. However, there is one small difficulty: because the y_{im} variables do not appear in the subproblem, the lineflow limits (9.9d) must be included among the $h(z) = h_1(z_1) + h_2(z_2) \geq 0$ constraints of the general Dantzig-Wolfe description of Section 9.2.3; but the lineflow limits (9.9d) are not complicating constraints. Therefore, for the components of $h_1(z_1)$ that are for the lineflow limits (9.9d), we define the corresponding components of $h_2(z_2)$ to be zero.[7]

$$
\begin{aligned}
z_1 &= (y_{im}\ \forall i, m) \\
z_2 &= \begin{pmatrix} s_{ijm}\ \forall i, j, m \\ x_i\ \forall i \\ r_{im}\ \forall i, m \end{pmatrix} \\
G_1(z_1) &= (0\ \forall i, m)
\end{aligned}
$$

[7] In the general description of Section 9.2.3, all constraints are of the ≥ 0 form, but in the stochastic power model description, some constraints are in the form ≤ 0, some constraints are in the form $= 0$, and some have nonzero right sides. All of these forms can be converted to the ≥ 0 form – e.g., an equality can be written equivalently as two inequalities (≥ 0 and ≤ 0) and the ≤ 0 can be multiplied by -1 to convert to the ≥ 0 form. Alternatively, it is possible to describe the decomposition scheme when some of the constraints are equalities, without converting to the ≥ 0 form; we do this here. One consequence of the equality form of the complicating constraint is that its dual variable β_{im} is a free variable.

$$G_2(z_2) = \begin{pmatrix} prob_{jm}\left[-a_{jm} + b_j\left(s_{ijm} + \sum_{j'} s_{j'jm}\right)\right] \ \forall i,j,m \\ C_i \ \forall i \\ prob_{im} RC_i \ \forall i,m \end{pmatrix}$$

$$= G_2(s_{ijm} \ \forall i,j,m), \text{ i.e., no dependence on } x_i, r_{im}$$

$$h_1(z_1) = \begin{pmatrix} T - y_{im} \ \forall i,m \\ y_{im} \ \forall i,m \end{pmatrix}$$

$$h_2(z_2) = \begin{pmatrix} 0 \ \forall i,m \\ -\sum_j s_{jim} + x_i + r_{im} \ \forall i,m \end{pmatrix}$$

The feasible set and mapping for the subproblem are therefore given by

$$\begin{aligned} \overline{K} = \{ s_{ijm}, x_i, r_{im} \mid & \\ x_i - X_i &\le 0, \forall i \\ r_{im} - R_i &\le 0, \forall i,m \\ \sum_j s_{ijm} - x_i - r_{im} &= 0, \forall i,m \\ x_i, r_{im}, s_{ijm} &\ge 0 \} \end{aligned}$$

$$G_2(s_{ijm}) - \sum_{j=1}^{p_2} \nabla h_{2j}(z_{2M}^{k-1}) \left(\beta^{k-1}\right)_j$$

$$- \begin{pmatrix} prob_{jm}\left[-a_{jm} + b_j\left(s_{ijm} + \sum_{j'} s_{j'jm}\right)\right] + \beta_{jm}^{k-1} \ \forall i,j,m \\ C_i - \sum_m \beta_{im}^{k-1} \ \forall i \\ prob_{im} RC_i - \beta_{im}^{k-1} \ \forall i,m \end{pmatrix}$$

where β_{im}^{k-1} is the dual variable for the nodal balance constraint (9.9e) at node i in scenario m, as calculated by the previous master problem at iteration $k-1$ (but for the first subproblem, $\beta_{im}^{k-1} = \beta_{im}^0$ is an estimate provided by the modeler).

If the first k subproblems have solutions represented by the symbols $s_{ijm}^{k'}, x_i^{k'}, r_{im}^{k'}$ (i.e., column k' of the matrix Z_2^k) for $k' = 1, 2, \ldots, k$, then the kth master problem (9.13) here has the following feasible set and mapping

$$\begin{aligned} \Lambda^k = \{ y_{im} \ \forall i,m, \lambda \in \mathbb{R}^k | & \\ y_{im} &\le T \ \forall i,m \\ -\sum_j \sum_{k'=1}^{k} s_{jim}^{k'} \lambda_{k'} + \sum_{k'=1}^{k} x_i^{k'} \lambda_{k'} + \sum_{k'=1}^{k} r_{im}^{k'} \lambda_{k'} + y_{im} &= 0 \ \forall i,m \\ \sum_{k'=1}^{k} \lambda_{k'} &= 1 \end{aligned}$$

Table 9.4 Convergence gap values of Dantzig-Wolfe iterations, for stochastic power model

Iteration k	1	2	3
Gap CG^k	-1125.5	-17.4	0

$$y_{im} \text{ unrestricted in sign, } \lambda_{k'} \geq 0\}$$

$$\begin{pmatrix} G_1(y_{im}) \\ G_2(Z_2^k\lambda)^T Z_2^k \end{pmatrix} = \begin{pmatrix} 0, \forall i, m \\ G_2(Z_2^k\lambda)^T Z_2^k \end{pmatrix}, \text{ where} \tag{9.16a}$$

$$G_2(Z_2^k\lambda)^T Z_2^k =$$
$$\begin{pmatrix} \sum\limits_{i,j,m} prob_{jm} \left[-a_{jm} + b_j \left(\sum\limits_{\widehat{k}=1}^{k} s_{ijm}^{\widehat{k}} \lambda_{\widehat{k}} + \sum_{j'} \sum\limits_{\widehat{k}=1}^{k} s_{j'jm}^{\widehat{k}} \lambda_{\widehat{k}} \right) \right] s_{ijm}^{k'} \\ + \sum\limits_{i} C_i x_i^{k'} + \sum\limits_{i,m} prob_{im} RC_i r_{im}^{k'} \quad \forall k' \end{pmatrix} \tag{9.16b}$$

With the data as in Table 9.2, and starting guesses of the dual variables $\beta_{im}^0 = 0$, the algorithm takes $k = 3$ iterations to find the solution that is in Table 9.3, i.e., three subproblems, three master problems, and one more subproblem for the calculation of the convergence gap CG^3. The convergence gap values are listed in Table 9.4.

□

9.2.5 Simplicial Decomposition of VIs

Simplicial decomposition of a VI has a master problem and subproblem structure, as in Dantzig-Wolfe decomposition, but there are no complicating constraints that are relaxed to form the subproblem, and the subproblem is solved as an optimization problem. Although simplicial decomposition of VIs was developed before Dantzig-Wolfe decomposition of VIs (see [20] for an overview), it is nevertheless convenient to describe simplicial decomposition as a special case of Dantzig-Wolfe decomposition. To simplify the discussion, and to follow typical practice in simplicial decomposition, we assume that there are no variables z_1 that appear only in the master problem; thus, the variable vector is simply $z = z_2$. There are two specializations of the algorithm described in Section 9.2.3:

1. there are no complicating constraints that are relaxed in the subproblem; thus, the subproblem feasible region is the same as the original problem's feasible region; and

2. in the kth subproblem, the mapping has no term $-\sum_{j=1}^{p_2} \nabla h_{2j}(z_{2M}^{k-1})\beta_j^{k-1}$ (because there are no relaxed constraints), and the mapping is *approximated* by a constant vector (see the discussion about subproblem approximation in Section 9.2.3.1), equal in value to the previous master problem value, i.e., $\widetilde{G^{k-1}} = G(z_M^{k-1})$ (except for $k=1$, when an estimate $\widetilde{G^0}$ is provided by the modeler), making the subproblem an optimization problem.

Therefore, if the original VI is defined as

$$\text{find } z^* \in K = \{z \in \mathbb{R}^n | g(z) \geq 0\}$$
$$\text{such that } G(z^*)^T(z - z^*) \geq 0, \; \forall z \in K$$

then the kth subproblem is defined as

$$\begin{aligned} &\text{Minimize}_z \quad G(z_M^{k-1})^T z \\ &\text{s.t.} \\ &\quad g(z) \geq 0 \end{aligned}$$

If the constraint functions $g(z)$ are affine, then the subproblem is a linear program, which can be solved efficiently. As in Dantzig-Wolfe decomposition, the master problem forms convex combinations of proposals and searches over the weights $\lambda_{k'}$ on the proposals. With Z^k representing the $n \times k$ matrix whose columns are the first k subproblem proposals, $z^{k'}, k' = 1, 2, \ldots, k$, the kth master problem is defined as

$$\text{find } \lambda^* \in \Lambda^k = \{\lambda \in \mathbb{R}^k | \sum_{k'=1}^{k} \lambda_{k'} = 1, \lambda_{k'} \geq 0 \; \forall k'\}$$
$$\text{such that } G(Z^k\lambda^*)^T Z^k(\lambda - \lambda^*) \geq 0, \; \forall \lambda \in \Lambda^k$$

The solution to the kth master problem VI is expressed in terms of the original variables, $z_M^k = Z^k\lambda^*$, and passed to the subproblem to form the constant coefficients $G(z_M^k)$ in the objective function of the $(k+1)^{st}$ subproblem.

With these definitions of the master and subproblem, the iterations proceed as in Dantzig-Wolfe decomposition, and convergence is declared when the negative convergence gap, CG^k is sufficiently close to zero.

9.2.6 Illustrative Example. Simplicial Decomposition of Simplified Stochastic Power Model

The subproblem is a linear program, because all of the constraints are affine. The mapping G is evaluated from (9.10), using $s_{M,ijm}^{k-1}$ for the s_{ijm} variable values calculated by the $(k-1)^{st}$ master problem, to give the coefficients of the objective function. The next subproblem is thus defined as

$$\text{Minimize}_{s_{ijm},x_i,r_{im},y_{im}}$$

$$\sum_{i,j,m} prob_{jm}\left[-a_{jm}+b_j\left(s_{M,ijm}^{k-1}+\sum_{j'} s_{M,j'jm}^{k-1}\right)\right] s_{ijm} + \sum_{i} C_i x_i + \sum_{i,m} prob_{im} RC_i r_{im}$$

s. t.

all constraints in (9.9)

The mapping for the master problem is

$$G(Z^k\lambda^*)^T Z^k =$$

$$\left(\sum_{i,j,m} prob_{jm}\left[-a_{jm}+b_j\left(\sum_{\widehat{k}=1}^{k} s_{ijm}^{\widehat{k}}\lambda_{\widehat{k}}+\sum_{j'}\sum_{\widehat{k}=1}^{k} s_{j'jm}^{\widehat{k}}\lambda_{\widehat{k}}\right)\right] s_{ijm}^{k'} + \sum_{i} C_i x_i^{k'} + \sum_{i,m} prob_{im} RC_i r_{im}^{k'} \ \forall k'\right)$$

With the data as in Table 9.2, and starting guesses of the objective function coefficients of s_{ijm} for the first subproblem as

$$prob_{jm}\left[-a_{jm}+b_j\left(s_{M,ijm}^{0}+\sum_{j'} s_{M,j'jm}^{0}\right)\right] = \left\{\begin{array}{l} -10, \text{ for } m = Lo \ \forall i,j \\ -20, \text{ for } m = Hi \ \forall i,j \end{array}\right\}$$

the simplicial decomposition algorithm takes $k = 7$ iterations to find the solution that is in Table 9.3, i.e., seven subproblems, seven master problems, and one more subproblem for the calculation of the convergence gap CG^7. The convergence gap values are listed in Table 9.5.

□

Table 9.5 Convergence gap values of simplicial decomposition iterations, for stochastic power model

Iteration k	1	2	3	4	5	6	7
Gap CG^k	-4457.2	-2282.5	-1440.1	-699.8	-386.8	-77.5	0

9.2.7 Benders Decomposition of VIs

In Benders decomposition of an optimization problem, some complicating *variables* are identified, in the sense that, if the values of these variables were fixed, then the problem would be easy to solve. The subproblem in the Benders scheme has fixed estimates of the complicating variables, with these estimates coming from the most recent solution of the master problem. In the master problem of the Benders scheme, instead of a steadily growing number of variables as the iterations proceed (the weights $\lambda_{k'}$ on proposals), there is a steadily growing number of linear *constraints*, called cuts, whose coefficients are determined in part by *dual* variable solutions of the subproblems that preceded the current master problem.

For linear programming models, there is a close relationship between Dantzig-Wolfe and Benders decomposition methods, and this relationship can be understood through the relations between the LP primal problem and the corresponding LP dual problem. In a nutshell, Benders decomposition applied to an LP is equivalent to applying Dantzig-Wolfe decomposition to the dual of the LP. The complicating variables in the primal LP (that are fixed in the Benders subproblem) correspond to complicating constraints in the dual LP, i.e., a complicating primal variable is the dual variable of the corresponding complicating constraint in the dual LP. When, in Dantzig-Wolfe decomposition applied to the *dual* LP, the master problem passes a "dual" variable of the complicating constraints to the subproblem, it is actually passing an estimate of the corresponding complicating *primal variable.*

These ideas are clarified through the next illustrative example.

9.2.7.1 Illustrative Example. Benders Decomposition of a Simple LP

We use the same problem as in Section 9.2.1, but we reverse the roles of the primal and dual LPs. For the present example, suppose that the given LP that we wish to solve – the primal LP – is (9.6), repeated here for the reader's convenience:

$$\text{Maximize} \quad 10\beta - 5\alpha_1 - 2\alpha_2 - 6\alpha_3 - 2\alpha_4$$

s. t.

$$\beta - \alpha_1 \leq 4$$

$$2\beta - 3\alpha_1 - \alpha_2 \leq 1$$
$$3\beta - 2\alpha_3 - \alpha_4 \leq 2$$
$$\beta - \alpha_3 \leq 6$$
$$\beta, \alpha_1, \alpha_2, \alpha_3, \alpha_4 \geq 0$$

Note that if the *complicating variable* β were a fixed value, then the LP would split into two smaller LPs – one in the variables α_1, α_2, and the other in α_3, α_4. The dual of this LP is (9.1) of Section 9.2.1. Dantzig-Wolfe decomposition can be applied to the dual problem (9.1), as discussed in Section 9.2.1. In Section 9.2.1, the subproblems (9.2), (9.3) (in the variables x_1, x_2, x_3, x_4) can be re-expressed equivalently by their duals (9.7), (9.8), which are the two subproblems in Benders decomposition, in the variables $\alpha_1, \alpha_2, \alpha_3, \alpha_4$. Note that the Benders subproblems have modifications to the *right-hand sides of constraints*, involving the variable value β^{k-1} from the most recent solution of the Benders master problem.

To complete the description of Benders decomposition for this simple example, we next present the form of the Benders master problem. We do this by writing out the dual of the Dantzig-Wolfe master problem (9.5): for every variable λ_j in (9.5), there is a corresponding constraint (often called a *cut*) in the Benders master problem, and the number of cuts equals the number of iterations, k, that have occurred so far in the decomposition algorithm. In the following Benders master problem, θ is the dual variable corresponding to the constraint (9.5c), and $x_{1S}^j, x_{2S}^j, x_{3S}^j, x_{4S}^j$ are the dual variable values for the constraints (9.7b), (9.7c), (9.8b), (9.8c), respectively, calculated at the jth iteration in the Benders subproblems (9.7), (9.8). Here is the form of the Benders master problem:

$$\begin{aligned}
&\text{Maximize} \quad 10\beta - \theta \\
&\text{s. t.} \\
&\quad \{x_{1S}^j + 2x_{2S}^j + 3x_{3S}^j + x_{4S}^j\}\beta - \theta \\
&\qquad \leq \{4x_{1S}^j + x_{2S}^j + 2x_{3S}^j + 6x_{4S}^j\} \quad j = 1, \ldots, k \\
&\quad \beta, \theta \geq 0
\end{aligned}$$

□

9.2.7.2 General Development of Benders Decomposition for VIs

This primal-dual relationship is extended to a useful class of VIs in [10], and Benders decomposition is defined by applying Dantzig-Wolfe decomposition to the dual form of the VI. We first state the primal form of the VIs in this class, labeled VI_P, in which $q \in \mathbb{R}^{n_1}, F(q) \in \mathbb{R}^{n_1}, w \in \mathbb{R}^{n_2}, c \in \mathbb{R}^{n_2}, b \in \mathbb{R}^m$, A is an $m \times n_1$ matrix, and B is an $m \times n_2$ matrix. If a model requires nonnegativity of any variables, these are understood to be among the constraints listed in the definition of the feasible set.

$$\begin{aligned} \text{VI}_P: \quad & \text{Find } \begin{pmatrix} q^* \\ w^* \end{pmatrix} \in K_P = \left\{ \begin{pmatrix} q \\ w \end{pmatrix} \in \mathbb{R}^{n_1+n_2} | Aq + Bw \leq b \right\} \\ & \text{s.t. } F(q^*)^T(q - q^*) + c^T(w - w^*) \geq 0, \ \forall \begin{pmatrix} q \\ w \end{pmatrix} \in K_P \end{aligned}$$

Assuming that the mapping F is invertible, i.e., that we can write $f = F(q)$ and $q = F^{-1}(f)$ equivalently, and introducing $v \in \mathbb{R}^m$ as the vector of dual variables that correspond to the primal constraints $Aq + Bw \leq b$, we can define a dual VI in the variables f and v, denoted VI_D.

$$\begin{aligned} \text{VI}_D\text{: Find } & \begin{pmatrix} f^* \\ v^* \end{pmatrix} \in K_D = \left\{ \begin{pmatrix} f \\ v \end{pmatrix} \in \mathbb{R}^{n_1+m} | f + A^T v = 0, B^T v = -c, v \geq 0 \right\} \\ & \text{s.t. } F^{-1}(f^*)^T(f - f^*) + b^T(v - v^*) \geq 0, \ \forall \begin{pmatrix} f \\ v \end{pmatrix} \in K_D \end{aligned}$$

Note that the constraints in K_D have associated KKT multipliers, and as stated in the following theorem, the multipliers associated with $f + A^T v = 0$ can be represented by q, and the multipliers associated with $B^T v = -c$ can be represented by w.

Theorem 9.1. *The vectors q^*, w^* solve VI_P with KKT multipliers v^* if and only if the vectors $f^* = F(q^*), v^*$ solve VI_D with KKT multipliers q^*, w^*.*

Proof. See [10] for a proof relying on the equivalence of the KKT conditions for VI_P and VI_D.

To derive Benders decomposition of the primal problem VI_P, we follow the same general idea that leads to Benders decomposition for an LP problem: apply Dantzig-Wolfe decomposition to the dual problem, VI_D. If some of the constraints in $B^T v = -c$ are complicating constraints, then they can be relaxed to form a Dantzig-Wolfe subproblem of VI_D. The Dantzig-Wolfe subproblem of VI_D would be in the variables f and v, and the "dual" variable vector from the complicating constraints in the master problem would be some of the variables in the vector w, the complicating variables. However, we could write the subproblem of VI_D in an equivalent form with primal variables q and w, but the complicating variables in the vector w would be fixed at the values calculated in the previous master problem.

We illustrate the construction of the dual VI, and Benders decomposition, with the stochastic power VI.

9.2.7.3 Illustrative Example. Benders Decomposition of Simplified Stochastic Power Model

In this application of Benders decomposition, we identify the baseload generation variables x_i as the complicating variables since they do not depend on any scenario m. The Benders subproblem therefore has the x_i variables fixed at the values calculated by the previous master problem. This makes the subproblem easy to solve because it splits into two smaller problems, one for each scenario m.

But first, let us identify the variables and constraints that fit into the general form of VI$_P$, and then derive the dual problem, VI$_D$. In the general statement of VI$_P$ above, all constraints are in "$\leq$" form, and there are no explicit sign restrictions on variables; if a problem has such restrictions (as does the stochastic power problem), then they are understood to be among the constraints in $Aq + Bw \leq b$. Therefore, we must rewrite the definition of the feasible set (9.9) of the stochastic power model, making two types of changes: equality constraints are replaced by two inequality constraints, and sign restrictions on variables are represented as "$\leq$" constraints. The feasible set (9.9) is redefined below, and, for later developments, dual variables for each constraint are shown in parentheses.

$$K_P = \{s_{ijm}, x_i, r_{im}, y_{im} \mid$$

$$x_i \leq X_i, \quad (v_i^1)\forall i \tag{9.17a}$$

$$r_{im} \leq R_i, \quad (v_{im}^2)\forall i, m \tag{9.17b}$$

$$\sum_j s_{ijm} - x_i - r_{im} \leq 0, \quad (v_{im}^3)\forall i, m \tag{9.17c}$$

$$-\sum_j s_{ijm} + x_i + r_{im} \leq 0, \quad (v_{im}^4)\forall i, m \tag{9.17d}$$

$$y_{im} \leq T, \quad (v_{im}^5)\forall i, m \tag{9.17e}$$

$$-\sum_j s_{jim} + x_i + r_{im} + y_{im} \leq 0, \quad (v_{im}^6)\forall i, m \tag{9.17f}$$

$$\sum_j s_{jim} - x_i - r_{im} - y_{im} \leq 0, \quad (v_{im}^7)\forall i, m \tag{9.17g}$$

$$-x_i \leq 0, \quad (v_i^8)\forall i \tag{9.17h}$$

$$-r_{im} \leq 0, \quad (v_{im}^9)\forall i, m \tag{9.17i}$$

$$-s_{ijm} \leq 0, \quad (v_{ijm}^{10})\forall i, j, m\} \tag{9.17j}$$

To finish the identification of the structure of VI$_P$ in the stochastic power problem, we must classify the problem variables into the vectors q and w. The main distinction between q and w is that variables in q have corresponding mapping functions that are nonconstant, while variables in w have corresponding mapping functions that are constant; furthermore, the mapping functions corresponding to q depend only on q and not on w. From the form of the mapping (9.10), it is clear that

$$q = (s_{ijm}, \forall i, j, m)$$

$$w = \begin{pmatrix} x_i, \forall i \\ r_{im}, \forall i, m \\ y_{im}, \forall i, m \end{pmatrix}$$

$$F(q) = \left(prob_{jm} \left[-a_{jm} + b_j \left(s_{ijm} + \sum_{j'} s_{j'jm} \right) \right] \; \forall i, j, m \right)$$

$$c = \begin{pmatrix} C_i, \forall i \\ prob_{im} RC_i, \forall i, m \\ 0, \forall i, m \end{pmatrix}$$

Now we can write the feasible set of the dual problem VI_D for the stochastic power model, using the coefficients of the constraints (9.17) that describe the primal feasible set, K_P:

$$K_D = \{ f_{ijm}, v_i^1, v_{im}^2, \ldots, v_{ijm}^{10} \,|$$

$$f_{ijm} + v_{im}^3 - v_{im}^4 - v_{jm}^6 + v_{jm}^7 - v_{ijm}^{10} = 0 \;\forall i, j, m \qquad (9.18a)$$

$$v_i^1 + \sum_m (-v_{im}^3 + v_{im}^4 + v_{im}^6 - v_{im}^7) - v_i^8 = -C_i \;\forall i \qquad (9.18b)$$

$$v_{im}^2 - v_{im}^3 + v_{im}^4 + v_{im}^6 - v_{im}^7 - v_{im}^9 = -prob_{im} RC_i \;\forall i, m \qquad (9.18c)$$

$$v_{im}^5 + v_{im}^6 - v_{im}^7 = 0 \;\forall i, m \qquad (9.18d)$$

$$v_i^1, v_{im}^2, \ldots, v_{ijm}^{10} \geq 0 \; \} \qquad (9.18e)$$

For the mapping $F^{-1}(f)$, we need to invert the relation

$$f_{ijm} = prob_{jm} \left[-a_{jm} + b_j \left(s_{ijm} + \sum_{j'} s_{j'jm} \right) \right] \;\forall i, j, m$$

to solve for the s_{ijm} variables in terms of the f_{ijm} variables. For the present case of two nodes, $i \in \{1, 2\}$, readers can easily verify that

$$s_{1jm} = (2f_{1jm} - f_{2jm} + prob_{jm} a_{jm})/(3 prob_{jm} b_j) \;\forall j, m$$
$$s_{2jm} = (2f_{2jm} - f_{1jm} + prob_{jm} a_{jm})/(3 prob_{jm} b_j) \;\forall j, m.$$

The whole mapping for VI_D, $\binom{F^{-1}(f)}{b}$, is listed below, with the corresponding variable indicated after the symbol "$\sim$".

$$(2f_{1jm} - f_{2jm} + prob_{jm} a_{jm})/(3 prob_{jm} b_j) \sim f_{1jm} \;\forall j, m \qquad (9.19a)$$

$$(2f_{2jm} - f_{1jm} + prob_{jm} a_{jm})/(3 prob_{jm} b_j) \sim f_{2jm} \;\forall j, m \qquad (9.19b)$$

$$X_i \sim v_i^1 \; \forall i \tag{9.19c}$$
$$R_i \sim v_{im}^2 \; \forall i, m \tag{9.19d}$$
$$0 \sim v_{im}^3 \; \forall i, m \tag{9.19e}$$
$$0 \sim v_{im}^4 \; \forall i, m \tag{9.19f}$$
$$T \sim v_{im}^5 \; \forall i, m \tag{9.19g}$$
$$0 \sim v_{im}^6 \; \forall i, m \tag{9.19h}$$
$$0 \sim v_{im}^7 \; \forall i, m \tag{9.19i}$$
$$0 \sim v_i^8 \; \forall i \tag{9.19j}$$
$$0 \sim v_{im}^9 \; \forall i, m \tag{9.19k}$$
$$0 \sim v_{ijm}^{10} \; \forall i, j, m \tag{9.19l}$$

Now we apply Dantzig-Wolfe decomposition to the dual problem with feasible set (9.18) and mapping (9.19). The constraints (9.18b) can be treated as complicating constraints for the purpose of applying Dantzig-Wolfe decomposition to the dual VI. If constraints (9.18b) are relaxed, and the dual variables v_i^1, v_i^8 appear only in the master problem, then the subproblem can be split into two independent problems. However, for simplicity of presentation, we treat the subproblem as one VI problem, as in the general development of Dantzig-Wolfe decomposition in Section 9.2.3. Now we can relate the dual VI (9.18), (9.19), to the symbols for variables and functions in Section 9.2.3:

$$z_1 = \begin{pmatrix} v_i^1 \; \forall i \\ v_i^8 \; \forall i \end{pmatrix}$$

$$z_2 = \begin{pmatrix} f_{ijm} \; \forall i, j, m \\ v_{im}^2 \; \forall i, m \\ v_{im}^3 \; \forall i, m \\ v_{im}^4 \; \forall i, m \\ v_{im}^5 \; \forall i, m \\ v_{im}^6 \; \forall i, m \\ v_{im}^7 \; \forall i, m \\ v_{im}^9 \; \forall i, m \\ v_{ijm}^{10} \; \forall i, j, m \end{pmatrix}$$

$$G_1(z_1) = \begin{pmatrix} X_i \; \forall i \\ 0 \; \forall i \end{pmatrix} \tag{9.20a}$$

$$G_2(z_2)=\begin{pmatrix} (2f_{1jm}-f_{2jm}+prob_{jm}a_{jm})/(3prob_{jm}b_j)\ \forall j,m \\ (2f_{2jm}-f_{1jm}+prob_{jm}a_{jm})/(3prob_{jm}b_j)\ \forall j,m \\ R_i\ \forall i,m \\ 0\ \forall i,m \\ 0\ \forall i,m \\ T\ \forall i,m \\ 0\ \forall i,m \\ 0\ \forall i,m \\ 0\ \forall i,m \\ 0\ \forall i,m \end{pmatrix} \tag{9.20b}$$

and the constraint functions g, h_1, h_2 can be derived from the constraints in (9.18), keeping in mind that in the general development of Dantzig-Wolfe decomposition of Section 9.2.3, the constraints are all of the ≥ 0 form, but that ≤ 0 and $= 0$ forms can be converted to the ≥ 0 form, as noted in a footnote in Section 9.2.4.

$$\begin{aligned} h_1(z_1) &= v_i^1 - v_i^8 + C_i\ \forall i \\ h_2(z_2) &= \sum_m(-v_{im}^3 + v_{im}^4 + v_{im}^6 - v_{im}^7)\ \forall i \end{aligned}$$

$$g(z_2) = \begin{pmatrix} f_{ijm} + v_{im}^3 - v_{im}^4 - v_{jm}^6 + v_{jm}^7 - v_{ijm}^{10}\ \forall i,j,m \\ v_{im}^2 - v_{im}^3 + v_{im}^4 + v_{im}^6 - v_{im}^7 - v_{im}^9 + prob_{im}RC_i\ \forall i,m \\ v_{im}^5 + v_{im}^6 - v_{im}^7\ \forall i,m \\ v_{im}^2\ \forall i,m \\ \vdots \\ v_{im}^7\ \forall i,m \\ v_{im}^9\ \forall i,m \\ v_{ijm}^{10}\ \forall i,j,m \end{pmatrix}$$

The vector β^{k-1} that is passed from the $(k-1)^{th}$ master problem to the kth subproblem contains the duals of the complicating constraints (9.18b):

$$\beta^{k-1} = (x_i^{k-1}\ \forall i)$$

The feasible set of the dual subproblem is defined over the f and v variables, except v_i^1 and v_i^8, with constraints (9.18a), (9.18c), (9.18d), (9.18e). In the mapping for the dual subproblem, $G_2(z_{2S}^k) - \sum_{j=1}^{p_2} \nabla h_{2j}(z_{2M}^{k-1})\beta_j^{k-1}$, the second term is given by (with corresponding variables after the "$\sim$"symbol)

$$-\sum_{j=1}^{p_2} \nabla h_{2j}(z_{2M}^{k-1})\beta_j^{k-1} = \begin{pmatrix} 0 \sim f_{ijm}\ \forall i,j,m \\ 0 \sim v_{im}^2\ \forall i,m \\ x_i^{k-1} \sim v_{im}^3\ \forall i,m \\ -x_i^{k-1} \sim v_{im}^4\ \forall i,m \\ 0 \sim v_{im}^5\ \forall i,m \\ -x_i^{k-1} \sim v_{im}^6\ \forall i,m \\ x_i^{k-1} \sim v_{im}^7\ \forall i,m \\ 0 \sim v_{im}^9\ \forall i,m \\ 0 \sim v_{ijm}^{10}\ \forall i,j,m \end{pmatrix}$$

It is interesting and customary to present the dual of the dual subproblem, i.e., the subproblem VI expressed in primal variables. This primal form can be derived using Theorem 9.1, resulting in a primal subproblem that looks much like the original VI in primal variables, except that the complicating variables x_i are fixed at the previous master problem values. For the kth primal subproblem, the feasible region (with equalities recombined from pairs of inequalities) is

$$\overline{K}_P = \{s_{ijm}, r_{im}, y_{im} \mid$$

$$r_{im} \leq R_i, \forall i,m \tag{9.21a}$$

$$\sum_j s_{ijm} - r_{im} = x_i^{k-1}, \forall i,m \tag{9.21b}$$

$$y_{im} \leq T, \forall i,m \tag{9.21c}$$

$$-\sum_j s_{jim} + r_{im} + y_{im} = -x_i^{k-1}, \forall i,m \tag{9.21d}$$

$$-r_{im} \leq 0, \forall i,m \tag{9.21e}$$

$$-s_{ijm} \leq 0, \forall i,j,m\} \tag{9.21f}$$

and the mapping is the same as for the original primal problem, except for the absence of C_i elements, because x_i is not a variable.

The dual VI's master problem at iteration k has variables v_i^1, v_i^8, and weights $\lambda_{k'}$ for all previous subproblem solutions, $k' = 1,\ldots,k$, which delivered values of subproblem variables (proposals) to the master problem, $f_{ijm}^{k'}, v_{im}^{2k'}, \ldots, v_{im}^{7k'}, v_{im}^{9k'}, v_{ijm}^{10k'}$. The master problem constraints require that the weights sum to one, $\sum_{k'\leq k} \lambda_{k'} = 1$, the complicating constraint (9.18b) is satisfied, and non-negativity of v_i^1, v_i^8 and $\lambda_{k'}$. However, in (9.18b), subproblem variables z_2 are replaced by convex combinations of proposals, e.g., v_{im}^3 is replaced by $\sum_{k'\leq k} v_{im}^{3k'}\lambda_{k'}$. Using (9.13b), the mapping for the master problem is as in (9.19c), (9.19j), for the elements that correspond to v_i^1, v_i^8, i.e., $G_1(z_1^k)$ as defined in (9.20a). The mapping element $G_2(Z_2^k\lambda^k)^T Z_2^k$, corresponding to $\lambda_{k'}$, is more complex – a sum over all elements in (9.20b), with subproblem variable f_{ijm} replaced by the convex combination $\sum_{k''\leq k} f_{ijm}^{k''}\lambda_{k''}$, and with each term in the sum weighted by proposal number k' for the corresponding variable:

Table 9.6 Convergence gap values of Benders decomposition iterations, for stochastic power model

Iteration k	1	2	3	4	5	6	7
Gap CG^k	-10354.7	-717.2	-374.0	-83.2	-8.2	-0.1	0

$$\sum_{j,m}\{(2\sum_{k"\le k} f_{1jm}^{k"}\lambda_{k"} - \sum_{k"\le k} f_{2jm}^{k"}\lambda_{k"} + prob_{jm}a_{jm})/(3prob_{jm}b_j)\} f_{1jm}^{k'}$$
$$+\sum_{j,m}\{(2\sum_{k"\le k} f_{2jm}^{k"}\lambda_{k"} - \sum_{k"\le k} f_{1jm}^{k"}\lambda_{k"} + prob_{jm}a_{jm})/(3prob_{jm}b_j)\} f_{2jm}^{k'}$$
$$+\sum_{i,m}\{R_i v_{im}^{2k'} + T v_{im}^{5k'}\} \sim \lambda_{k'}$$

It is possible to convert this dual form of the master problem to a primal form. Whereas the dual master accumulates variables $\lambda_{k'}$, one for each subproblem solution, the primal master accumulates special constraints, often called cuts, one for each subproblem solution. The primal master problem has variables x_i and also sales variables s_{Mijm} (the M distinguishes them from the subproblem's sales variables). If we are trying to have all scenario-specific variables and constraints in the subproblem, for computational efficiency, then the variables s_{Mijm} present a difficulty, as the number of such variables is proportional to the number of scenarios. For this reason, it may be preferable to leave the master problem in dual form. See [11] for details.

With the data as in Table 9.2, and starting guesses $x_i = 0$ for the baseload generation for the first subproblem, Benders decomposition algorithm takes $k = 7$ iterations to find the solution that is in Table 9.3, i.e., seven subproblems, seven master problems, and one more subproblem for the calculation of the convergence gap CG^7. The convergence gap values are listed in Table 9.6.

□

9.2.8 Cobweb Decomposition Method – No Master Problem

Murphy and Mudrageda [19] discuss the properties of a decomposition algorithm that is used to solve the National Energy Modeling System (NEMS) at the United States' Energy Information Administration. The model can be viewed as an NCP or VI (see [12]), but it is maintained in separate submodels such as for demand, refining, electricity production, etc. The submodels are brought together for a coordinated equilibrium by solving one submodel at a time, with either prices or quantities fixed for variables that interact with other submodels. The values of variables that influence other submodels are passed to those submodels, which are then solved. The procedure is similar to the cobweb algorithm to find an equilibrium of supply and demand, as mentioned in many textbooks on microeconomics. In its simplest form, applied to a single commodity model where equilibrium is at the intersection of the supply and demand curves, one version of the cobweb algorithm starts with a guess of the quantity, calculates a price from the supply curve, which is passed to the demand curve for an updated estimate of the quantity, whereupon a new iteration begins. It is well known that the cobweb algorithm can fail to converge, but that convergence can be recovered through the use of relaxation techniques which, for example, pass not just the latest price from the supply curve, but instead a weighted average of the latest price and the price from the last iteration; relaxation is applied in the NEMS decomposition algorithm. Unlike other decomposition algorithms described above, there is no master problem in the NEMS calculations – subproblems pass information directly to each other.

9.3 Algorithms for Mathematical Programs with Complementarity Constraints (MPCCs)

In this section, we present some methods for solving a mathematical program with complementarity constraints (MPCC). This form arises, for example, in the Stackelberg leader-follower type of model. As explained in Chapter 6, if the lower-level optimization (e.g., the follower's problem) is represented by the KKT conditions of one or more optimization problems, then because the KKT conditions include complementarity conditions (assuming that the lower-level problem has some inequality constraints), the constraints of the upper-level problem (e.g., the leader's problem) must include complementarity conditions. A review of mathematical issues and algorithms for MPCCs may be found in [21]. Here, we focus on defining and illustrating three methods of solution that are accessible via GAMS, with NLPEC for reformulations of the MPCC as particular NLPs. Some other methods for solving MPCCs are briefly described at the end of this section.

The following problem is the general form of the MPCC, repeated from Chapter 6.

$$\text{Minimize}_{\{x\}\cup\{x^1,\ldots,x^n\}\cup\{\lambda^1,\ldots,\lambda^n,\mu^1,\ldots,\mu^n\}}$$

$$f(x, x^1, \ldots, x^n, \lambda^1, \ldots, \lambda^n, \mu^1, \ldots, \mu^n) \tag{9.22a}$$

s. t.

$$h(x, x^1, \ldots, x^n, \lambda^1, \ldots, \lambda^n, \mu^1, \ldots, \mu^n) = 0 \tag{9.22b}$$

$$g(x, x^1, \ldots, x^n, \lambda^1, \ldots, \lambda^n, \mu^1, \ldots, \mu^n) \leq 0, \tag{9.22c}$$

$$\nabla_{x^i} f^i(x, x^1, \ldots, x^n) + \lambda^{i^T} \nabla_{x^i} h^i(x, x^1, \ldots, x^n) + \mu^{i^T} \nabla_{x^i} g^i(x, x^1, \ldots, x^n) = 0, \quad i = 1, \ldots, n \tag{9.22d}$$

$$h^i(x, x^1, \ldots, x^n) = 0, \quad i = 1, \ldots, n \tag{9.22e}$$

$$0 \leq \mu^i \perp - g^i(x, x^1, \ldots, x^n) \geq 0, \quad i = 1, \ldots, n \tag{9.22f}$$

The complementarity conditions are indicated by the perp symbol $\perp$ in (9.22f), i.e., an extra set of equality constraints $-\mu^i g^i(x, x^1, \ldots, x^n) = 0$.

In this section, we use the illustrative example from Chapter 6, Section 6.2.9, of strategic offering into a pool type of electricity market to illustrate several methods to solve MPCCs. This example is repeated here for the reader's convenience:

$$\text{Minimize}_{\{\hat{c}_1^{\mathrm{g}}\} \cup \{g_1, g_2, g_3, d_1, d_2, d_3, \theta_1, \theta_2, \theta_3\} \cup \Xi_{\mathrm{SO}}}$$

$$c_1^{\mathrm{g}} g_1 - \alpha_1 g_1 \tag{9.23a}$$

s. t.

$$g_1 - d_1 = B_1(\theta_1 - \theta_2) + B_2(\theta_1 - \theta_3) \tag{9.23b}$$

$$g_2 - d_2 = B_1(\theta_2 - \theta_1) + B_3(\theta_2 - \theta_3) \tag{9.23c}$$

$$g_3 - d_3 = B_2(\theta_3 - \theta_1) + B_3(\theta_3 - \theta_2) \tag{9.23d}$$

$$\theta_3 = 0 \tag{9.23e}$$

$$\hat{c}_1^{\mathrm{g}} - \alpha_1 - \phi_1^{\mathrm{gmin}} + \phi_1^{\mathrm{gmax}} = 0 \tag{9.23f}$$

$$c_i^{\mathrm{g}} - \alpha_i - \phi_i^{\mathrm{gmin}} + \phi_i^{\mathrm{gmax}} = 0, \quad i = 2, 3 \tag{9.23g}$$

$$- c_i^{\mathrm{d}} + \alpha_i - \phi_i^{\mathrm{dmin}} + \phi_i^{\mathrm{dmax}} = 0, \quad i = 1, 2, 3 \tag{9.23h}$$

$$B_1(\alpha_1 - \alpha_2 - \rho_1^{\min} + \rho_1^{\max}) + B_2(\alpha_1 - \alpha_3 - \rho_2^{\min} + \rho_2^{\max}) = 0 \tag{9.23i}$$

$$B_1(-\alpha_1 + \alpha_2 + \rho_1^{\min} - \rho_1^{\max}) + B_3(\alpha_2 - \alpha_3 - \rho_3^{\min} + \rho_3^{\max}) = 0 \tag{9.23j}$$

$$B_2(-\alpha_1 + \alpha_3 + \rho_2^{\min} - \rho_2^{\max}) + B_3(-\alpha_2 + \alpha_3 + \rho_3^{\min} - \rho_3^{\max}) + \gamma = 0 \tag{9.23k}$$

$$0 \leq g_i \perp \phi_i^{\mathrm{gmin}} \geq 0, i = 1, 2, 3 \tag{9.23l}$$

$$0 \leq g_i^{\max} - g_i \perp \phi_i^{\mathrm{gmax}} \geq 0, \quad i = 1, 2, 3 \tag{9.23m}$$

$$0 \leq d_i \perp \phi_i^{\mathrm{dmin}} \geq 0, \quad i = 1, 2, 3 \tag{9.23n}$$

$$0 \leq d_i^{\max} - d_i \perp \phi_i^{\mathrm{dmax}} \geq 0, \quad i = 1, 2, 3 \tag{9.23o}$$

$$0 \leq B_1(\theta_1 - \theta_2) + f_1^{\max} \perp \rho_1^{\min} \geq 0 \tag{9.23p}$$

$$0 \leq f_1^{\max} - B_1(\theta_1 - \theta_2) \perp \rho_1^{\max} \geq 0 \tag{9.23q}$$

$$0 \leq B_2(\theta_1 - \theta_3) + f_2^{\max} \perp \rho_2^{\min} \geq 0 \tag{9.23r}$$

$$0 \leq f_2^{\max} - B_2(\theta_1 - \theta_3) \perp \rho_2^{\max} \geq 0 \tag{9.23s}$$

$$0 \le B_3(\theta_2 - \theta_3) + f_3^{\max} \perp \rho_3^{\min} \ge 0 \tag{9.23t}$$
$$0 \le f_3^{\max} - B_3(\theta_2 - \theta_3) \perp \rho_3^{\max} \ge 0 \tag{9.23u}$$

The generating company at bus 1 of a three-node electricity network chooses its offer price $\hat{c}_1^{\mathrm{g}}$ in order to influence the market operator to select the company's sales g_1 such that the company's profit is maximized. See Chapter 6, Section 6.2, for further details about the model.

9.3.1 Why Are MPCCs Difficult to Solve?

An MPCC is in fact a nonlinear program – in (9.22), replace (9.22f) by

$$0 \le \mu^i, -g^i(x, x^1, \ldots, x^n) \ge 0 \tag{9.24a}$$
$$-\mu_j^i g_j^i(x, x^1, \ldots, x^n) = 0 \; \forall j \in J^i, i = 1, \ldots, n \tag{9.24b}$$

where j indexes the components of the vector μ^i and the vector valued function g^i. Therefore, one might expect to be able to solve an MPCC with a standard NLP solver such as MINOS or CONOPT. However, there are two serious potential problems with this approach.

1. The feasible region is nonconvex, due to the equality constraints (9.24b). Therefore, since an NLP solver for a non-convex problem generally stops at a local optimum, there is no guarantee that a global optimum has been found.
2. NLP solvers generally stop when the KKT conditions are satisfied, but because of the equality constraints (9.24b), it is not even clear that KKT multipliers (dual variables) exist for the problem (9.22), due to violation of some standard constraint qualifications. Because of this, NLP solvers may not even be able to find a local optimum for some MPCCs. See [21] for a discussion.

The constraints (9.24b) require that either $\mu_j^i = 0$ or $g_j^i(x, x^1, \ldots, x^n) = 0$, for each $j \in J^i$ and $i = 1, \ldots, n$. The number of combinations of such possibilities is 2^N, where $N = \sum_i |J^i|$ is the number of variables μ_j^i. One might imagine trying to solve a restricted version of the MPCC 2^N times with an NLP solver, with the lower-level restricted each time to one of the possible combinations, and then to choose the best solution found. However, even for a moderate size of problem – e.g., $N = 100$ – the number of NLPs to solve, 2^N, becomes prohibitively time-consuming. In Chapter 6, we show essentially the same idea – that an MPCC could be converted into a mixed integer program, by the Fortuny-Amat and McCarl procedure [8] – with the warning that this is practical, computationally, only if there are not very many complementarity constraints.

9.3.2 Applying Standard NLP Algorithms to MPCCs

We discuss three methods to apply NLP algorithms to solve MPCCs: regularization, penalization and sequential quadratic programming.

9.3.2.1 Regularization of Complementarity Constraints

Although a particular MPCC may not satisfy a constraint qualification, as noted above, there is a very useful result in [22], which says that the linear independence constraint qualification (LICQ) is satisfied by most MPCCs which are almost the same as (9.22) – i.e., almost all MPCCs that are formed by slightly perturbing the data that define the functions f, h and g satisfy the LICQ.

The *regularization* method uses an NLP algorithm to solve a slightly perturbed version of the MPCC, and the size of this perturbation is gradually reduced in a sequence of NLPs.

The following non-negativity and complementarity requirements on two real variables, x and y,

$$x \geq 0, y \geq 0, xy = 0$$

are equivalent to

$$x \geq 0, y \geq 0, xy \leq 0 \tag{9.25}$$

because the non-negativity requirements on x and y ensure that $xy \geq 0$. For a given number $\varepsilon \geq 0$, we can relax the requirements (9.25) in a way that is controllable by choice of ε:

$$x \geq 0, y \geq 0, xy \leq \varepsilon \tag{9.26}$$

When $\varepsilon = 0$, (9.26) is the same as (9.25), and when $\varepsilon > 0$, (9.26) defines a larger feasible region than (9.25), i.e., (9.26) is a relaxation of (9.25). When ε is a very small positive number, the requirement $xy \leq \varepsilon$ can be described as enforcing near complementarity, and the smaller ε is, the closer any point (x, y) is to complementarity.

The same ideas can be applied if x and/or y are replaced by functions of one or more variables, e.g., in (9.22f) of the general MPCC form, the equivalent requirements (9.24) can be approximated by the relaxation

$$0 \leq \mu^i, -g^i(x, x^1, \ldots, x^n) \geq 0, -\mu^i_j g^i_j(x, x^1, \ldots, x^n) \leq \varepsilon_{ij} \; \forall j \in J^i, i = 1, \ldots, n \tag{9.27}$$

In a regularization approach to solving an MPCC, the relaxation (9.26) or (9.27) is applied not just once, but several times in succession, with progressively smaller values of ε at each iteration, until ε is close enough to zero that complementarity is deemed to hold closely enough for practical purposes. At each iteration k and corresponding value of $\varepsilon = \varepsilon^k$, a nonlinear program is solved. Since NLP solvers can be started from a user's guess of an approximate solution, the second and later NLPs in the sequence are started from the solution of the NLP with the previous value of ε. The user must specify the manner in which ε changes as the iteration number k increases.

9.3.2.2 Illustrative Example. Regularization Applied to the Strategic Offer MPCC

Applying the relaxations (9.26) and (9.27) to the complementarity conditions in the strategic offer model, we obtain the following modified form of (9.23l) - (9.23u). The objective function (9.23a) and other constraints (9.23b) - (9.23k) are unchanged. At each iteration k, the value of ε^k is assumed to be the same for all constraints, in this example.

$$0 \le g_i, \phi_i^{\text{gmin}} \ge 0, g_i\phi_i^{\text{gmin}} \le \varepsilon^k, i = 1,2,3 \tag{9.28a}$$
$$0 \le g_i^{\max} - g_i, \phi_i^{\text{gmax}} \ge 0, (g_i^{\max} - g_i)\phi_i^{\text{gmax}} \le \varepsilon^k, i = 1,2,3 \tag{9.28b}$$
$$0 \le d_i, \phi_i^{\text{dmin}} \ge 0, d_i\phi_i^{\text{dmin}} \le \varepsilon^k, i = 1,2,3 \tag{9.28c}$$
$$0 \le d_i^{\max} - d_i, \phi_i^{\text{dmax}} \ge 0, (d_i^{\max} - d_i)\phi_i^{\text{dmax}} \le \varepsilon^k, i = 1,2,3 \tag{9.28d}$$
$$0 \le B_1(\theta_1 - \theta_2) + f_1^{\max}, \rho_1^{\min} \ge 0, (B_1(\theta_1 - \theta_2) + f_1^{\max})\rho_1^{\min} \le \varepsilon^k \tag{9.28e}$$
$$0 \le f_1^{\max} - B_1(\theta_1 - \theta_2), \rho_1^{\max} \ge 0, (f_1^{\max} - B_1(\theta_1 - \theta_2))\rho_1^{\max} \le \varepsilon^k \tag{9.28f}$$
$$0 \le B_2(\theta_1 - \theta_3) + f_2^{\max}, \rho_2^{\min} \ge 0, (B_2(\theta_1 - \theta_3) + f_2^{\max})\rho_2^{\min} \le \varepsilon^k \tag{9.28g}$$
$$0 \le f_2^{\max} - B_2(\theta_1 - \theta_3), \rho_2^{\max} \ge 0, (f_2^{\max} - B_2(\theta_1 - \theta_3))\rho_2^{\max} \le \varepsilon^k \tag{9.28h}$$
$$0 \le B_3(\theta_2 - \theta_3) + f_3^{\max}, \rho_3^{\min} \ge 0, (B_3(\theta_2 - \theta_3) + f_3^{\max})\rho_3^{\min} \le \varepsilon^k \tag{9.28i}$$
$$0 \le f_3^{\max} - B_3(\theta_2 - \theta_3), \rho_3^{\max} \ge 0, (f_3^{\max} - B_3(\theta_2 - \theta_3))\rho_3^{\max} \le \varepsilon^k \tag{9.28j}$$

The initial value of ε is set at $\varepsilon^0 = 1$; for $k \ge 1$, $\varepsilon^k = 0.1\varepsilon^{k-1}$; and the iterations stop after $k = 5$ because complementarity is satisfied. Thus, there are six NLP problems solved altogether, for $k = 0, 1, \ldots, 5$, and for the last NLP solved, $\varepsilon^5 = 0.00001$. The model is coded in GAMS, using NLPEC to automate the specifications for ε^k, and to automate the production and solution of the sequence of NLPs, which are solved by the solver CONOPT.

An optimal solution reported in Chapter 6 is found (to three decimals of accuracy in the objective function) by the NLP for $k = 5$. The progression of objective function values is shown in the regularization row of Table 9.7, as positive values of profit for firm 1. Note that the optimal profits are larger than for the original model when ε^k is large (for early iterations), and the optimal profits decrease as ε^k decreases. This happens because each NLP

Table 9.7 Firm 1 profit, by iteration number for strategic offering MPCC

Iteration k	0	1	2	3	4	5
Profit, regularization	37.419	34.019	33.675	33.640	33.637	33.636
Profit, penalization	53.636	33.636				
Profit, SQP	33.636					

in the sequence has a larger feasible region than the original model, which allows for a larger profit than in the original model, and as ε^k decreases, the discrepancy between the approximate and true feasible regions is reduced.

□

9.3.2.3 Penalization of Complementarity Constraints

In this approach, the complementarity conditions are dropped from the list of constraints, but terms are added to the upper-level objective function, to penalize violations of the complementarity conditions. This removes the troublesome constraints that may be the cause of possible nonexistence of KKT multipliers.

In the general MPCC (9.22), the conditions $-\mu^i g^i(x, x^1, \ldots, x^n) = 0$ in (9.22f) can be replaced by $-\mu^i g^i(x, x^1, \ldots, x^n) \leq 0$ (because $0 \leq \mu^i$ and $-g^i(x, x^1, \ldots, x^n) \geq 0$), but the constraints $-\mu^i g^i(x, x^1, \ldots, x^n) \leq 0$ are dropped, and are moved into the objective function, with a large coefficient, as penalty terms to create extra costs whenever $-\mu^i g^i(x, x^1, \ldots, x^n) > 0$. The method solves a sequence of NLPs iteratively, as in the regularization method, with each NLP after the first started from the previous NLP's solution. The penalty terms at iteration k are proportional to $1/\varepsilon^k$, where ε^k becomes smaller as k increases, as in the regularization method – i.e., the penalty terms become more and more important in the objective as the iterations proceed, and eventually, the complementarity conditions $-\mu^i g^i(x, x^1, \ldots, x^n) = 0$ are satisfied. The form of the NLP for the general MPCC (9.22) at iteration k is as follows.

$$\begin{aligned}
&\text{Minimize}_{\{x\}\cup\{x^1,\ldots,x^n\}\cup\{\lambda^1,\ldots,\lambda^n,\mu^1,\ldots,\mu^n\}} \\
&\quad f(x, x^1, \ldots, x^n, \lambda^1, \ldots, \lambda^n, \mu^1, \ldots, \mu^n) \\
&+ \left(\frac{1}{\varepsilon^k}\right) \sum_i \{-\mu^i g^i(x, x^1, \ldots, x^n)\} \\
&\text{s. t.} \\
&(9.22\text{b})\text{-}(9.22\text{e}) \\
&\quad 0 \leq \mu^i, -g^i(x, x^1, \ldots, x^n) \geq 0, \quad i = 1, \ldots, n
\end{aligned}$$

9.3.2.4 Illustrative Example. Penalization Applied to the Strategic Offer MPCC

Application of the penalization approach to the strategic offer problem gives the following NLP at iteration k.

$$\text{Minimize}_{\{\hat{c}_1^{\text{g}}\}\cup\{g_1,g_2,g_3,d_1,d_2,d_3,\theta_1,\theta_2,\theta_3\}\cup\Xi_{\text{SO}}}$$

$$c_1^{\text{g}}g_1 - \alpha_1 g_1 \tag{9.29a}$$

$$+\left(\frac{1}{\varepsilon^k}\right)\{\sum_i (g_i\phi_i^{\text{gmin}} + (g_i^{\max} - g_i)\phi_i^{\text{gmax}}) \tag{9.29b}$$

$$+\sum_i (d_i\phi_i^{\text{dmin}} + (d_i^{\max} - d_i)\phi_i^{\text{dmax}}) \tag{9.29c}$$

$$+ (B_1(\theta_1 - \theta_2) + f_1^{\max})\rho_1^{\min} \tag{9.29d}$$

$$+ (f_1^{\max} - B_1(\theta_1 - \theta_2))\rho_1^{\max} \tag{9.29e}$$

$$+ (B_2(\theta_1 - \theta_3) + f_2^{\max})\rho_2^{\min} \tag{9.29f}$$

$$+ (f_2^{\max} - B_2(\theta_1 - \theta_3))\rho_2^{\max} \tag{9.29g}$$

$$+ (B_3(\theta_2 - \theta_3) + f_3^{\max})\rho_3^{\min} \tag{9.29h}$$

$$+ (f_3^{\max} - B_3(\theta_2 - \theta_3))\rho_3^{\max}\} \tag{9.29i}$$

s. t.

$$\text{(9.23b)-(9.23k)} \tag{9.29j}$$

$$0 \le g_i, \phi_i^{\text{gmin}} \ge 0, i = 1,2,3 \tag{9.29k}$$

$$0 \le g_i^{\max} - g_i, \phi_i^{\text{gmax}} \ge 0, \quad i = 1,2,3 \tag{9.29l}$$

$$0 \le d_i, \phi_i^{\text{dmin}} \ge 0, \quad i = 1,2,3 \tag{9.29m}$$

$$0 \le d_i^{\max} - d_i, \phi_i^{\text{dmax}} \ge 0, \quad i = 1,2,3 \tag{9.29n}$$

$$0 \le B_1(\theta_1 - \theta_2) + f_1^{\max}, \rho_1^{\min} \ge 0 \tag{9.29o}$$

$$0 \le f_1^{\max} - B_1(\theta_1 - \theta_2), \rho_1^{\max} \ge 0 \tag{9.29p}$$

$$0 \le B_2(\theta_1 - \theta_3) + f_2^{\max}, \rho_2^{\min} \ge 0 \tag{9.29q}$$

$$0 \le f_2^{\max} - B_2(\theta_1 - \theta_3), \rho_2^{\max} \ge 0 \tag{9.29r}$$

$$0 \le B_3(\theta_2 - \theta_3) + f_3^{\max}, \rho_3^{\min} \ge 0 \tag{9.29s}$$

$$0 \le f_3^{\max} - B_3(\theta_2 - \theta_3), \rho_3^{\max} \ge 0 \tag{9.29t}$$

The initial value of ε is set at $\varepsilon^0 = 1$; for $k \ge 1$, $\varepsilon^k = 0.1\varepsilon^{k-1}$; and the iterations stop after $k = 1$ when complementarity is satisfied, for two NLP problems altogether, and for the last NLP solved, $\varepsilon^1 = 0.1$. The model is coded in GAMS, using NLPEC to automate the specifications for ε^k, and to automate the production and solution of the sequence of NLPs, which are solved by the solver CONOPT. The two values of profit (i.e., (9.29a) but excluding (9.29b)-(9.29i)) are shown in Table 9.7. The optimal solution of Chapter 6 is found at iteration $k = 1$.

□

9.3.2.5 Sequential Quadratic Programming

One type of general purpose NLP solver employs a sequential quadratic programming (SQP) algorithm: at each iteration of SQP, a quadratic program is formed in which nonlinear constraints are approximated by linear expressions in the Newton fashion, and the nonlinear objective function is replaced by a quadratic approximation of the Lagrangian of the NLP. An optimal solution of the quadratic program provides the point around which nonlinear constraints are linearized and a new quadratic approximation of the Lagrangian is formed, for the next iteration.

When the complementarity conditions (9.22f) are formulated as (9.27) with $\varepsilon = 0$ and an SQP solver is applied to the resulting NLP, it is perhaps surprising that the solver is often able to find an optimal solution, without the user needing to reduce ε^k gradually to zero. However, the SQP solver sometimes does something similar, automatically – e.g., the SQP solver SNOPT enters an "elastic mode" which relaxes constraints and penalizes violations of constraints, when a quadratic program encounters computational difficulties.

The SQP method was used in [3] in a procedure to solve a very large MPCC (approximately 20,000 variables and 10,000 constraints) representing a Stackelberg leader-follower equilibrium in electric power and NO_x emissions allowances markets. First, a feasible solution was found by treating the leader as a follower, and solving an ordinary complementarity problem, using the PATH solver. Next, the SQP solver SNOPT solved a restricted version of the MPCC, with one of the leader's variables fixed, starting from the feasible solution found by PATH. Finally, the solution found by SNOPT was used as a starting point for another SQP solver, FILTER, to solve the MPCC without the restriction on the leader's variable.

9.3.2.6 Illustrative Example. SQP Applied to the Strategic Offer MPCC

In the strategic offer problem (9.23), we replace (9.23l)-(9.23u) by (9.28) with $\varepsilon^k = 0$, and call SNOPT (an SQP solver) to solve the NLP. As shown in Table 9.7, the solution of Chapter 6 is found in one application of the solver SNOPT.

□

9.3.2.7 Some Practical Advice

As mentioned above, the solution calculated by NLP methods may find a local optimal solution that is not a global optimum. To be sure that a solution is a global optimum, it is a good idea to try several different methods (e.g., regularization, penalization, SQP) with several different settings (e.g., different values of ε^0 and update factors, for regularization and penalization). In addition, since an NLP solver can be started from a user-defined starting point in its search for a local optimum, it is a good idea to calculate solutions from several different starting points (for the first NLP in the sequence, in the regularization or penalization methods, or for the single NLP in the SQP approach).

In the regularization and penalization methods, the choice of ε^0 can cause problems that can sometimes be fixed simply by making another choice. For example, in the penalization approach to solve the strategic offer problem, if $\varepsilon^0 = 4$, the CONOPT solver returns the message that the first NLP problem is unbounded, and no further NLPs are solved because there is no new starting point; but, as shown in the example, setting $\varepsilon^0 = 1$ leads to an optimal solution. In the regularization approach, setting $\varepsilon^0 = 0.00001$ produces an infeasible problem for the first NLP in the sequence (which stops the whole calculation), but as shown in the example, $\varepsilon^0 = 1$ leads to an optimal solution.

9.3.3 Some Other Methods for MPCCs

The study of solution methods for MPCCs continues to be an active area of investigation. Here, we briefly mention a few other methods that have been proposed.

The Fortuny-Amat and McCarl conversion to a mixed integer program (MIP) [8] is mentioned above and described in Chapter 6. It is useful if the number of complementarity constraints is small enough to lead to a MIP that can be solved in a reasonable amount of time, but because optimization software and computing hardware capabilities steadily improve, it is worthwhile to keep this technique in mind. There is an important advantage of the Fortuny-Amat and McCarl method: when any MIP is solved, typically to a near-optimal solution, the user obtains upper and lower bounds on the optimal objective function value; therefore this technique provides an error bound on its estimate of the optimal objective value, whereas the NLP methods can only provide an estimate, without an error bound, because the solution that is found may be a local but not a global optimum.

Interior point methods have been used successfully to solve linear and nonlinear programs. The main idea is to step iteratively through the interior of the feasible region towards an optimum. To stay in the interior, inequality constraints are removed from the list of constraints but violations of the inequality constraints are penalized in extra terms of the objective function. A parameter is adjusted at each iteration, to produce an approximate solution that is allowed to be closer to the boundary defined by inequality constraints, so that the approximate solutions can move closer to an optimum (which may lie on a boundary). One early algorithm for MPCCs used this approach, the Penalty Interior Point Algorithm; see [18]. More recent developments are discussed in [15], which is also summarized in [21].

Siddiqui and Gabriel [23] use Schur's decomposition and Special Ordered Sets of type 1 (SOS1) variables to express the complementarity conditions in an MPCC. To illustrate, for two real variables, x and y, the system

$$x \geq 0, y \geq 0, xy = 0$$

is equivlalent, by Schur's decomposition, to the system

$$\begin{aligned} & x \geq 0, y \geq 0 \\ & u = (x + y)/2 \\ & v = (x - y)/2 \\ & u^2 - v^2 = 0 \end{aligned}$$

Since x and y are non-negative, u is also non-negative, but the same cannot be said for v. Therefore, $u^2 - v^2 = 0$ can be replaced by $u - |v| = 0$. The absolute value of v is represented using the SOS1 variables (at most one can be nonzero) $v^+ \geq 0, v^- \geq 0$, where $v = v^+ - v^-$ and $|v| = v^+ + v^-$. The authors report substantially lower computing times for their method, compared with the MIP of [8], on small examples, and they successfully solve a large-scale MPCC model of the North American gas market which could not be solved by the MIP approach of [8].

Kazempour and Conejo [14] use Benders decomposition to split a large stochastic MPCC into independent, much smaller MPCC subproblems. The upper-level problem chooses a leader company's investment in new generating capacity that will be offered into a pool in a future year. The lower-level problem represents the social welfare maximization process of the pool in that year, with the leader and rival producers offering into the pool; the leader behaves strategically as in the strategic offer problem (9.23) above. However, conditions in the year in question are uncertain – rival producers' capacity, their offer prices and demand are uncertain – which leads to many scenarios, each with its own market outcome. The upper-level objective is to minimize the negative of expected profit (i.e., maximize expected profit) of the leader, i.e., the negative of expected operating profit from participation in the pool, plus the cost of new capacity. In the Benders decomposition framework, the complicating variable that ties together all lower-level scenarios is the leader's investment in new capacity; when this variable is fixed at an iteration, the large MPCC subproblem splits into many smaller subproblems, with one MPCC like (9.23) for each scenario. For Benders decomposition to work, the upper-level objective should be convex with respect to the complicating variable, but this is not true here because the lower-level problem is an MPCC. However, the authors demonstrate that, with strategic offering into the pool by the leader, and with many scenarios, the upper-level objective is *almost* convex. The use of Benders decomposition is shown, on two test problems, to lead to near optimal solutions in much less time than without decomposition.

9.4 Algorithms for Equilibrium Programs with Equilibrium Constraints (EPECs)

An EPEC is a model of simultaneous optimization of several MPECs, in which the lower-level problems are the same for all MPECs. Here, we assume that the individual MPECs take the form of MPCCs, and we extend the single MPCC (9.22) to several, to define an EPEC in a general form. As in the MPCC (9.22), the lower-level decision variables are the vectors x^i, and the dual variables from the lower-level optimization problems are the vectors λ^i, μ^i, $i = 1, \ldots, n$. For the EPEC, with several upper-level decision makers (or *leaders*), we need to replace the single upper-level decision variables vector x by several variable vectors, one for each leader; we use the symbol X^k for the vector of upper-level decision variables of the kth leader, and $k = 1, \ldots, m$, for m upper-level leaders altogether. The general EPEC is thus the simultaneous solution of m MPCCs:

EPEC: for each $k = 1, \ldots, m$,

Minimize$_{\{X^k\}\cup\{x^1,\ldots,x^n\}\cup\{\lambda^1,\ldots,\lambda^n,\mu^1,\ldots,\mu^n\}}$

$$f(X^1, \ldots, X^m, x^1, \ldots, x^n, \lambda^1, \ldots, \lambda^n, \mu^1, \ldots, \mu^n) \tag{9.30a}$$

s. t.

$$h^k(X^1, \ldots, X^m, x^1, \ldots, x^n, \lambda^1, \ldots, \lambda^n, \mu^1, \ldots, \mu^n) = 0 \tag{9.30b}$$

$$g^k(X^1, \ldots, X^m, x^1, \ldots, x^n, \lambda^1, \ldots, \lambda^n, \mu^1, \ldots, \mu^n) \leq 0, \tag{9.30c}$$

$$\begin{aligned} &\nabla_{x^i} f^i(X^1, \ldots, X^m, x^1, \ldots, x^n) + \\ &\quad {\lambda^i}^T \nabla_{x^i} h^i(X^1, \ldots, X^m, x^1, \ldots, x^n) + \\ &\quad {\mu^i}^T \nabla_{x^i} g^i(X^1, \ldots, X^m, x^1, \ldots, x^n) = 0, \quad i = 1, \ldots, n \end{aligned} \tag{9.30d}$$

$$h^i(X^1, \ldots, X^m, x^1, \ldots, x^n) = 0, \quad i = 1, \ldots, n \tag{9.30e}$$

$$0 \leq \mu^i \perp - g^i(X^1, \ldots, X^m, x^1, \ldots, x^n) \geq 0, \quad i = 1, \ldots, n \tag{9.30f}$$

Note that in (9.30), leader k assumes that the decision vectors of the other leaders, $X^1, \ldots, X^{k-1}$, $X^{k+1}, \ldots, X^m$, are beyond k's control, and they are treated as parameters in k's MPCC. Note also that the lower-level variables

x^i, λ^i, μ^i are assumed to take the same value in all MPCCs $k = 1, \ldots, m$ – e.g., we do *not* use symbols $x^{i,k}$, which would represent k's value assigned to the lower-level vector x^i.

A solution to the EPEC is values for the lower-level variables $x^i, \lambda^i, \mu^i, i = 1, \ldots, n$, and values for the upper-level variables $X^k, k = 1, \ldots, m$, that simultaneously solve the MPCCs (9.30). In particular, for each $k = 1, \ldots, m$, the kth leader uses as parameters the values $X^1, \ldots, X^{k-1}$, $X^{k+1}, \ldots, X^m$ that are optimal in the other leaders' MPCCs.

The EPEC is therefore a generalized Nash equilibrium – "Nash" because each leader takes the others' decisions as parameters, beyond its control, and "generalized" because there are constraints in each leader's problem that involve variables of the other leaders; see Chapters 1, 3 and 5 for discussions about generalized Nash models. This raises the issue that, for constraints that are identical across the MPCCs – e.g., the lower-level optimality conditions (9.30d), (9.30e), (9.30f) – the m leaders may or may not have the same dual variable values at an equilibrium. A modeler could formulate the EPEC in such a way as to force the dual variables of common constraints to be equal (see the example, below); the interpretation of this depends on the particular model.

As mentioned in Chapter 7, it is not always true that a solution exists for an EPEC problem. Therefore, any theoretical guarantees of convergence must necessarily be contingent on assuming the existence of an equilibrium, or conditions that amount to assuming existence.

Below, we mention the two main types of methods for solving EPECs – diagonalization and NLP reformulations. However, this is an active area of research, and other methods are under investigation.

9.4.1 Diagonalization Method for EPECs

Chapter 7 presents the diagonalization method for EPECs, which in essence solves the MPCC problem of one leader at a time, assuming fixed values for the strategic variables of the other leaders. The assumed values of the other leaders are updated at each iteration to the values most recently computed by the other leaders' MPCCs. Because the individual leaders' MPCCs are solved separately, one expects that the lower-level variables x^i, λ^i, μ^i take different values in the different leaders' problems. The diagonalization method is said to converge to an equilibrium if there is no change in the leaders' optimal variables $X^k, k = 1, \ldots, m$ from one iteration to the next, *and* if the lower-level variables x^i, λ^i, μ^i take the same values across all leaders' problems. Hu and Ralph [13] prove (in Proposition 7 of that paper) that if diagonalization converges to values of upper and lower variables that do not change from one iteration to the next, then an equilibrium has been found, but a key *assumption* is that the lower-level variables converge to the same values across all leaders' problems. In spite of this theoretical difficulty, diagonalization has successfully converged to an equilibrium for many models, such as the examples in Chapter 7.

The next subsection presents two related methods which avoid the potential difficulty of different values for lower-level variables, by enforcing equality of lower-level variables across all leaders.

9.4.2 NLP Reformulation of EPECs

Leyffer and Munson [16] present a method which is easy to describe in a brief overview. First, write out m sets of KKT conditions for the leaders' MPCCs. Second, solve the resulting complementarity problem by one of two methods: (a) use a complementarity solver such as the PATH solver [7]; or (b) remove the complementarity constraints from the KKT conditions, formulate an objective to minimize the sum of the removed complementarity expressions (which must be zero at an equilibrium), and solve the resulting optimization problem with an NLP solver. A model can be formulated and solved with or without equality of the dual variables of common constraints. Although the overview is easy to state, difficult complications (e.g., the potential nonexistence of KKT multipliers in the MPCCs) must be treated with great care.

In addition to the examples in [16], another energy example of the NLP approach may be found in [1]. In the next subsection, we present the NLP method through a very simple example, using the notation of [16] as far as possible. We do not assume equality of dual variables for common constraints.

9.4.3 Illustrative Example. A simple 2-Leader, 1-Follower EPEC

We suppose that there are two large firms (indexed by $k = 1, 2$) that generate electricity, and many small firms grouped together under the index $k = 3$. The small firms act as price-takers; we refer to the group of them as a competitive fringe. The generation amounts of the two large firms and the total of the fringe are represented by $g_k \geq 0, k = 1, 2, 3$, with cost functions of the following form:

$$k\text{'s cost} = c_k g_k + \frac{1}{2} d_k g_k^2$$

where c_k and d_k are positive parameters. For simplicity, we assume that there are no transmission constraints, and that the firms and fringe have no constraints (except for non-negativity).

The price-quantity relationship of electricity consumers is given by an inverse demand function (we assume that total demand equals total generation $\sum_{k=1}^{3} g_k$):

$$\text{price} = a - b\sum_{k=1}^{3} g_k$$

where a and b are positive parameters.

We assume further that firms 1 and 2 behave in a Cournot fashion with respect to each other (they use their knowledge of the inverse demand function and each assumes that the other's production is a parameter beyond its control). Additionally, we assume that they also behave as leaders with respect to the fringe – they anticipate the reactions of the follower to their profit maximizing decisions about g_1 and g_2.

As price-takers, the fringe produces a total amount g_3 such that price equals marginal cost $c_3 + d_3 g_3$, unless $g_3 = 0$, in which case marginal cost may exceed the price. This can be summarized in the complementarity model:

$$0 \leq g_3 \perp c_3 + d_3 g_3 - (a - b\sum_{k=1}^{3} g_k) \geq 0 \tag{9.31}$$

To stay close to the notation of [16], we introduce the surplus variable s in the right-most inequality constraint of (9.31) for an equivalent set of conditions, with the complementarity condition explicitly shown as a constraint:

$$c_3 + d_3 g_3 - (a - b\sum_{k=1}^{3} g_k) - s = 0$$
$$g_3 \geq 0, s \geq 0$$
$$g_3 s = 0$$

The complementarity condition $g_3 s = 0$ can be replaced by $g_3 s \leq 0$ because non-negativity of g_3 and s ensures that $g_3 s \geq 0$, as discussed above for MPCCs, in (9.25).

Therefore, the optimization problem for each leader $k = 1, 2$ is

$$\text{Minimize}_{g_k, g_3, s} \quad c_k g_k + \frac{1}{2} d_k g_k^2 - (a - b\sum_{j=1}^{3} g_j) g_k \tag{9.32a}$$

s. t.

$$c_3 + d_3 g_3 - (a - b\sum_{j=1}^{3} g_j) - s = 0 \quad (\mu_k) \tag{9.32b}$$

$$g_k \geq 0 \quad (\chi_k) \tag{9.32c}$$

$$g_3 \geq 0 \quad (\psi_k) \tag{9.32d}$$

$$s \geq 0 \quad (\sigma_k) \tag{9.32e}$$

$$-g_3 s \geq 0 \quad (\xi_k) \tag{9.32f}$$

where the dual variables for the constraints are in parentheses to the right of the constraints. The subscript k on the dual variables $\mu_k, \psi_k, \sigma_k, \xi_k$ allows them to have different values for each leader. That is, the dual variables may be different for the follower constraints that are the same in both leaders' problems, (9.32b), (9.32d), (9.32e) and (9.32f), which is an issue in a generalized Nash problem such as this.

Next, we write out the KKT conditions for each leader's problem, $k = 1, 2$:

$$c_k + d_k g_k - (a - b\sum_{j=1}^{3} g_j) + bg_k - b\mu_k - \chi_k = 0 \tag{9.33a}$$

$$bg_k - (b + d_3)\mu_k - \psi_k + s\xi_k = 0 \tag{9.33b}$$

$$\mu_k - \sigma_k + g_3\xi_k = 0 \tag{9.33c}$$

$$c_3 + d_3 g_3 - (a - b\sum_{j=1}^{3} g_j) - s = 0 \tag{9.33d}$$

$$0 \leq g_k \perp \chi_k \geq 0 \tag{9.33e}$$

$$0 \leq g_3 \perp \psi_k \geq 0 \tag{9.33f}$$

$$0 \leq s \perp \sigma_k \geq 0 \tag{9.33g}$$

$$0 \leq -g_3 s \perp \xi_k \geq 0 \tag{9.33h}$$

The conditions

$$g_3 \geq 0, s \geq 0, -g_3 s \geq 0, \xi_k \geq 0 \tag{9.34a}$$

$$-g_3 s\xi_k = 0 \tag{9.34b}$$

can be replaced by

$$g_3 \geq 0, s \geq 0, \xi_k \geq 0 \tag{9.35a}$$

$$g_3 s = 0 \tag{9.35b}$$

by the following argument. In (9.34), $g_3 \geq 0, s \geq 0, -g_3 s \geq 0$ imply that $g_3 s = 0$, and $-g_3 s\xi_k = 0$ may be dropped as it is a redundant condition, giving (9.35). Starting from (9.35), it can be concluded that $-g_3 s \geq 0$ and that $-g_3 s\xi_k = 0$, giving (9.34).

Therefore, the KKT conditions (9.33) for each leader's problem, $k = 1, 2$, may be written in the following form:

$$c_k + d_k g_k - (a - b\sum_{j=1}^{3} g_j) + bg_k - b\mu_k - \chi_k = 0$$

$$
\begin{aligned}
& bg_k - (b + d_3)\mu_k - \psi_k + s\xi_k = 0 \\
& \mu_k - \sigma_k + g_3\xi_k = 0 \\
& c_3 + d_3 g_3 - (a - b\sum_{j=1}^{3} g_j) - s = 0 \\
& 0 \le g_k \perp \chi_k \ge 0 \\
& 0 \le g_3 \perp \psi_k \ge 0 \\
& 0 \le s \perp \sigma_k \ge 0 \\
& 0 \le s \perp g_3 \ge 0 \\
& \xi_k \ge 0
\end{aligned}
$$

Concatenating the two firms' KKT conditions together forms a mixed complementarity problem, but it is not square – see [16].

Finally, we formulate an NLP by removing the complementarity conditions from the constraints, and minimizing the sum of the complementarity expressions, subject to the remaining constraints. Note that there is no need for a penalty factor in the objective function, because the only term in the objective is the sum of the complementarity expressions.

$$
\text{Minimize} \quad \sum_{k=1}^{2} (g_k \chi_k + g_3 \psi_k + s\sigma_k) + g_3 s
$$

s. t.

$$
\begin{aligned}
& c_k + d_k g_k - (a - b\sum_{j=1}^{3} g_j) + bg_k - b\mu_k - \chi_k = 0, \quad k = 1, 2 \\
& bg_k - (b + d_3)\mu_k - \psi_k + s\xi_k = 0 \quad k = 1, 2 \\
& \mu_k - \sigma_k + g_3\xi_k = 0, \quad k = 1, 2 \\
& c_3 + d_3 g_3 - (a - b\sum_{j=1}^{3} g_j) - s = 0 \\
& g_k, \chi_k, \psi_k, \sigma_k \ge 0, \quad k = 1, 2 \\
& g_3, s \ge 0
\end{aligned}
$$

The parameter values assumed for this example are given in Table 9.8.

Table 9.8 Parameter values for simple EPEC example

inverse demand parameter a		40
inverse demand parameter b		1
cost parameter c_k	$k = 1$	5
	$k = 2$	6
	$k = 3$	8
cost parameter d_k	$k = 1$	0.8
	$k = 2$	0.7
	$k = 3$	0.2

For comparison, we also solved a model in which firms 1 and 2 are Cournot competitors without anticipating the reaction of the follower fringe, while the fringe firms again play as price-takers. Because inverse demand is linear, the Cournot equilibrium can be represented by the following optimization model:[8]

$$\text{Minmize}_{g_k \geq 0}$$
$$\sum_{k=1}^{3}(c_k g_k + \frac{1}{2}d_k g_k^2) - (a\sum_{k=1}^{3} g_k - 0.5b(\sum_{k=1}^{3} g_k)^2) + 0.5b\sum_{k=1}^{2} g_k^2$$

(note that the last sum is over the two Cournot firms only).

The solutions calculated for the EPEC and Cournot models are given in Table 9.9.

Table 9.9 Solutions calculated for simple EPEC and Cournot models

		Cournot	EPEC
price		12.08	11.25
production	$k = 1$	3.934	6.463
	$k = 2$	3.577	6.054
	$k = 3$	20.407	16.236
profit	$k = 1$	21.668	23.667
	$k = 2$	17.276	18.939
	$k = 3$	41.645	26.360

In the solution found for the EPEC, the two leader firms increase their production compared with the Cournot solution, which causes the price to drop. A lower price causes the the production of the fringe to drop, but not by as much as the leaders' production increases. The net effect on the leaders – lower price but greater production – is an increase in profit for both leaders, while the fringe earns less profit.

□

[8] See Chapter 3, Section 3.4.2.7, for the optimization method for Cournot models with linear inverse demand.

9.5 Summary

In Section 9.2, we present the details of three decomposition methods for VIs – Dantzig-Wolfe, Benders and simplicial decomposition. Decomposition can sometimes improve computation times for large models, or make it possible to solve a large model when a direct method (without decomposition) fails. There is sometimes another advantage of decomposition of a large complex model: if the subproblem splits apart into several separate problems that can be developed and maintained by different teams of specialists, then model management can become more efficient than it would be without decomposition.

In Section 9.3, we describe in detail several NLP-based methods to solve mathematical programs with complementarity constraints (MPCCs). We also discuss several other methods for MPCCs.

Section 9.4 discusses two main methods to solve equilibrium programs with equilibrium constraints (EPECs) – the diagonalization method (which is described in detail in Chapter 7), and NLP reformulation of an EPEC.

9.6 Exercises

1. Modify the GAMS program for Dantzig-Wolfe decomposition of the stochastic power model of Section 9.2.4 (see Appendix B for the GAMS code), to implement the simplicial decomposition algorithm for the same model, as discussed in Section 9.2.6.
2. Modify the GAMS program for Dantzig-Wolfe decomposition of the stochastic power model of Section 9.2.4 (see Appendix B for the GAMS code), to implement Benders decomposition algorithm for the same model, as discussed in Section 9.2.7.3.
3. Modify the GAMS program for Dantzig-Wolfe decomposition of the stochastic power model of Section 9.2.4 (see Appendix B for the GAMS code), to implement subproblem approximation, as discussed in Section 9.2.3.1, with two different approximations of the mapping G_2 in the subproblem:
(a) $\widetilde{G}_2(z_{2S}^k; z_{2M}^{k-1}) = G_2(z_{2M}^{k-1})$, a constant, which allows the subproblem to be represented as an LP; and
(b) $\widetilde{G}_2(z_{2S}^k; z_{2M}^{k-1})$ is a *diagonalization* of G_2, which allows G_2 to be integrated, and the subproblem can be represented as an NLP.
4. In this problem, you are to revisit the coal market model of Chapter 3. In Section 3.5.1.2, an MPEC model is formulated, with firms 1 and 2 merged as one leader, and firm 3 as the follower. In this problem, you are asked to formulate an EPEC model with firms 1 and 2 as two *separate* leaders. Code the two MPCC leader problems in GAMS, and solve several iterations of the diagonalization method for EPECs.

5. Modify the GAMS code for the strategic offer problem (see Appendix B for the GAMS code) to implement the regularization, penalization and SQP methods as described in Section 9.3. (Hint for regularization and penalization: read the GAMS documentation about NLPEC, and study its use in the GAMS code for the Stackelberg MPEC, which can be found in Appendix B.) Experiment with different settings for parameters that control the algorithms, e.g., as discussed in Section 9.3.2.7.
6. Create GAMS code for the EPEC and Cournot models of Section 9.4.3. Check your calculated solutions with the results in Table 9.9. Modify some of the cost parameters in Table 9.8 to see the effect of different cost structures.
7. Modify the three firms' models of Section 9.4.3 to include capacity limits, i.e., positive upper bounds on the generation variables. Revise the formulation of the EPEC of this section to include these new constraints.

References

1. G. Bautista, M.F. Anjos and A. Vannelli. Formulation of oligopolistic competition in AC power networks: an NLP approach. *IEEE Transactions on Power Systems*, 22(1):105-115, 2007.
2. E. Celebi. *Decomposition of variational inequalities with applications to Nash-Cournot models in time of use electricity markets*. Ph.D. thesis, Department of Management Sciences, Faculty of Engineering, University of Waterloo, Waterloo ON, Canada, 2011.
3. Y. Chen, B.F. Hobbs, S. Leyffer and T.S. Munson. Leader-follower equilibria for electric power and NO_x allowances markets. *Computational Management Science*, 3:307-330, 2006.
4. W. Chung and J.D. Fuller. Subproblem approximation in Dantzig-Wolfe decomposition of variational inequality models with an application to a multicommodity economic equilibrium model. *Operations Research* 58(5):1318-1327, 2010.
5. A.J. Conejo, E. Castillo, R. Mínguez and R. García-Bertrand. *Decomposition Techniques in Mathematical Programming: Engineering and Science Applications*. Springer, New York, 2006.
6. G.B. Dantzig and P. Wolfe. The decomposition algorithm for linear programs. *Econometrica*, 29(4):767-778, 1961.
7. M.C. Ferris and T.S. Munson. PATH 4.6. In *GAMS: the Solver Manuals*. GAMS Development Corporation, Washington D.C., 2005. Available at www.gams.com.
8. J. Fortuny-Amat and B. McCarl. A representation and economic interpretation of a two-level programming problem. *The Journal of the Operational Research Society*, 32(9):783-792, 1981.
9. J.D. Fuller and W. Chung. Dantzig-Wolfe decomposition of variational inequalities. *Computational Economics,* 25:303-326, 2005.
10. J.D. Fuller and W. Chung. Benders decomposition for a class of variational inequalities. *European Journal of Operational Research*, 185(1):76-91, 2007.
11. S.A. Gabriel and J.D. Fuller. A Benders decomposition method for solving stochastic complementarity problems with an application in energy. *Computational Economics*, 35(4):301-329, 2010.
12. S.A. Gabriel, A.S. Kydes and P. Whitman. The National Energy Modeling System: a large-scale energy-economic equilibrium model. *Operations Research* 49(1):14-25, 2001.
13. X. Hu and D. Ralph. Using EPECs to model bilevel games in restructured electricity markets with locational prices. *Operations Research* 55(5):809-827, 2007.
14. S.J. Kazempour and A.J. Conejo. Strategic generation investment under uncertainty via Benders decomposition. *IEEE Transactions on Power Systems*, 17(1):424-432, 2011.

15. S. Leyffer, G. López-Calva and J. Nocedal. Interior methods for mathematical programs with complementarity constraints. *SIAM Journal on Optimization,* 17(1):52-77, 2006.
16. S. Leyffer and T. Munson. Solving multi-leader-common-follower games. *Optimization Methods and Software*, 25(4):601-623, 2010.
17. J.P. Luna, C. Sagastizabal and M. Solodov. A class of Dantzig-Wolfe type decomposition methods for variational inequality problems. Unpublished manuscript, 2011. Downloaded from pages.cs.wisc.edu/˜solodov/lss11DW.pdf.
18. Z.-Q. Luo, J.-S. Pang and D. Ralph. *Mathematical Programs with Equilibrium Constraints.* Cambridge University Press, Cambridge, 1996.
19. F.H. Murphy and M.V. Mudrageda. A decomposition approach for a class of economic equilibrium models. *Operations Research*, 46(3):368-377, 1998.
20. M. Patriksson. Simplicial Decomposition Algorithms. *Encyclopedia of Optimization*, Springer, 2009.
21. D. Ralph. Mathematical programs with complementarity constraints in traffic and telecommunications networks. *Philosophical Transactions of the Royal Society, A*, 366(1872):1973-1987, 13 June 2008
22. S. Scholtes and M. Stöhr. How stringent is the linear independence assumption for mathematical programs with complementarity constraints? *Mathematics of Operations Research*, 26(4):851-863, 2001.
23. S. Siddiqui and S.A. Gabriel. An SOS1-based approach for solving MPECs with a natural gas market application, accepted at *Networks and Spatial Economics*, May 2012.

Chapter 10
Natural Gas Market Modeling

10.1 Introduction

Natural gas is a key fuel in energy markets worldwide. It is produced from either onshore or offshore wells, processed to remove impurities, and then transported by either pipeline in gaseous form or cooled to about -260 degrees F (about -160 degrees C) and then transported as liquefied natural gas (LNG) to destinations around the world.[1] The main consuming sectors that use it are residential, commercial, industrial, electric power, and to some extent transportation. At present, the world has abundant gas supplies. According to [52], the global mean projected remaining recoverable resources is 16,200 trillion cubic feet (Tcf) or 150 times the current annual global consumption.[2] About 9,000 Tcf is gauged to be economically available at less than or equal to $4 per million British Thermal Units (Btu) [52].

In the last few years, natural gas markets have received considerable attention for several reasons. First, compared to other fossil fuels natural gas has the lowest carbon content. After combustion, natural gas produces 53.06 kilograms of carbon dioxide per million Btu as compared to 96.43 for lignite, 103.62 for anthracite coal, 73.15 for distillate fuel oil, and 78.80 for residential fuel oil as a way of comparison.[3] This means natural gas has the most environmentally advantageous position of the fossil fuels which is important given increasing societal and regulatory initiatives to reduce carbon dioxide (and other greenhouse gases).

Second, since the endowment of gas reserves is not equally distributed around the world and there is substantial infrastructure investments needed to transport and receive gas (e.g., LNG import terminals, pipelines), countries that have a surplus to sell can in some situations take advantage of this

[1] See [17] for more information about LNG.

[2] 1 trillion cubic feet (Tcf) of gas equals approximately 28.32 billion cubic meters (Bcm). Thus, 16,200 Tcf equates to about 458,784 Bcm.

[3] As described in [18] (Table B-3).

position to adversely influence prices and market conditions for downstream customers. This sort of phenomenon was observed in both 2006 and 2009 when Russian gas was curtailed through Ukraine due to a price dispute resulting in various European countries also feeling the negative effects.[4] Clearly with the rise of liquefied natural gas shipments around the world and various countries ramping up their LNG import infrastructure or diversifying their suppliers, such market power instances may be mitigated in the future. Nevertheless, countries dependent on natural gas from outside sources have begun to consider the security of supply aspects of this market especially given the recent formation of the Gas Exporting Countries Forum (GECF), a loose confederation of gas-supplying countries that collectively control over 70% of the world's natural gas reserves [11]. Several recent works have concentrated on counterfactual studies of a gas cartel (gas OPEC) forming and what might be the effects on the global gas market, e.g., [25]

Third, with the rise of renewable portfolio standards in the electric power sector in places like the U.S. and Europe, there is a current need for thermal back-up of intermittent renewable sources such as wind and solar. One very attractive option given the relatively favorable environmental footprint is to use combined cycle gas turbines that can ramp up quickly to produce power.and are also relatively inexpensive compared to other fossil-fuel power plants as well as other advantages [6]. However, gas and intermittent renewables has not always been a win-win situation with in some cases these fuels competing against each other [31].

Fourth, the rise of unconventional gas and in particular shale gas in the United States has sparked a number of substantial energy market changes. Due to the two techniques of horizontal drilling and hydraulic fracturing ("fracking") which have greatly increased the profit of low permeability geologic formations such as shale gas making this a sizeable resource in various parts of the world, especially the United States [15]. Indeed, the U.S. Energy Information Administration (EIA) projects in its *Annual Energy Outlook 2011* that the U.S. has some 862 Tcf (24,411 Bcm) of technically recoverable shale gas or about 34% of its total natural gas resource base of 2543 Tcf.[5] Moreover, as predicted by EIA, by 2035 shale gas production is estimated to account for 46% of the total U.S. natural gas production.

In its report, EIA (2011) estimates that technically recoverable resources for select basins in 32 countries including the U.S. amount to some 6,622 Tcf (187,535 Bcm). China is first with 1,275 Tcf followed by the U.S. at 862 Tcf and Argentina third at 774 Tcf. It should be noted that all these numbers are estimates at the time of their reporting and subject to in some cases dramatic changes. Indeed, as reported in [64], EIA revised down some 40% its 2011 estimate of 827 Tcf of U.S. shale gas to 482 total due to more drilling data. Despite the great increase of shale gas estimated for the U.S.,

---

[4] See for example [42] or [41] for related articles in the popular press or [60] for more information.

[5] From the AEO2011 Reference Case.

there are environmental questions related to contamination of the water table [62] and a possible link with fracking-induced earthquakes [20] and the U.S. Environmental Protection Agency (EPA) is considering how to best proceed with hydraulic fracturing given public health and environmental concerns [19]. These concerns are by no means limited to the U.S. as other countries need to balance energy vs. water (e.g., South Africa) taking into account in some cases dense populations but geopolitical energy considerations (e.g., Poland) [63].

Lastly, in a variety of countries natural gas is used for a significant percent of transportation vehicles and represents a cleaner-burning option to gasoline and diesel fuel. Some noteworthy examples are Bangladesh (61% of market share of total fleet represented by natural gas vehicles), Armenia (30%), Pakistan (26%), Bolivia (26%), Argentina (24%), Columbia (24%) [55]. In a number of studies both compressed natural gas (CNG) and LNG were shown to be substantially less polluting that petroleum-based fuels when considering a variety of pollutants.[6]

Thus, from the above discussion one can see that there are a number of distinct reasons why natural gas markets are important for the energy sector. In the rest of this chapter we discuss previous natural gas market and other gas models (Section 10.2), explore some of the engineering considerations (Section 10.3 and Appendix D), build a natural gas market model and discuss the functions of the players as well as provide illustrative numerical results (Section 10.4). Section 10.5 summarizes the chapter and exercises are provided in Section 10.6.

10.2 Natural Gas Market Models

The field of modeling natural gas operations and/or markets has been a particularly active one in recent years. In the literature one can find both descriptive models describing how prices in various regions are tending to converge [5], [54] analysis of pricing regulations [58] as well as prescriptive models. For the latter class of models, there are differing perspectives depending on where along the gas supply chain is the focus of study. An early example is [34] which considered supply, storage, and service reliability issues for a natural gas distribution utility. These sorts of models tend to be more specific to the company level and do not have the view of the whole market. Such models need not only consider operational aspects but may also investigate investment decisions for a company. For example, [8] considers the optimal dimensioning of pipelines from both an operations and investment perspective and [9] presents a model and Simplex-like algorithm for optimizing transmission for a gas company. Midthun [50] considers several other

[6] See [12] for a selection of studies.

models for optimal dispatch or optimal production as might be faced by a gas-producing company and includes a discussion of the engineering aspects as well (see also [51] for a discussion of some of these aspects as well as the next section and Appendix D of this book). Other gas engineering details can be found in [65] and [3].

Additionally, there has also been a plethora of market-level models in recent years. There are several ways in which these models can differ. First, some are regional concentrating on a country or number of countries within a region (e.g., Europe) with details on the gas infrastructure whereas others are more global in their purview. Another way these gas market models can differ is their assumptions about perfect or imperfect competition. Specifically, do certain market agents possess market power to directly influence prices or are all such agents price-takers? Lastly, these models can also differ in their incorporation or lack thereof of engineering laws (e.g., Weymouth equation for gas flows).

In what follows we briefly describe some of the more recent or well-known deterministic gas market models in the literature starting with the ones that don't depict market power and then discuss the ones that incorporate oligopolistic (imperfect) competition. After that we describe a few of the gas market models that considered stochastic elements (e.g., uncertain demand). This description of both deterministic and stochastic models is by no means exhaustive but rather meant to be illustrative. Besides the models described below, additional models explanations and studies can be found in [26] which gives an overview of the natural gas market from a theoretical perspective, [59] which analyzes a number of European gas models and three objectives of the European Commission, and "World Natural Gas Markets and Trade: A Multi-Modeling Perspective" from the Energy Modeling Forum (EMF) 23 with an overview by Huntington [39].

In [27], the Gas Systems Analysis Model (GSAM) is presented and applied to a study on Canadian carbon stabilization and its effect on North American gas markets. GSAM is a bottom-up type of market model with very fine detail on the gas reservoir side (over 17,000 reservoirs with explicit geologic and economic variables) that successively builds up supply curves based on market signals and technology levels based on the concept of a minimal acceptable supply price (break-even price) for exploration and production to occur. An integrating linear program (as an approximation of a nonlinear one based on maximizing total surplus less transportation costs) brings together both the supply, demand, and transportation sides of the market. In [24] a number of computational aspects of GSAM are investigated including for example, how to improve the convergence of the approximating linear program to a solution of the market equilibrium.

The U.S. Energy Information Administration (www.eia..gov) produces the *Annual Energy Outlook* (AEO) which is an important energy market report used by both public and private sector energy organizations. The National Energy Modeling System (NEMS) is one of the models used to produce the

AEO and is composed of many submodules and has been shown to be an example of an MCP/VI [22]. One of the components to NEMS is the Natural Gas Transmission and Distribution Module (NGTDM). This module, composed of several submodules uses a heuristic algorithm to establish a market equilibrium.

MAGELAN [48] [49] is a global gas supply model originally developed as a global extension of EUGAS [57]. MAGELAN optimizes long-term gas supply including investments in production and infrastructure capacity through 2035 via a linear programming approach. Its objective function includes both capital and operating costs for global gas production and transports and these costs are minimized subject to all relevant technical restrictions in production and transportation.

The TIGER (Transport Infrastructure for Gas with Enhanced Resolution) model [47], [46], [10] is another example with very fine granularity (daily) for Europe and surrounding production countries including a database on pipelines, LNG terminals, regasification and storage sites and has been used in a number of studies related to the European gas market. TIGER is a linear cost minimization model that also takes into account important technical constraints of the gas infrastructure such as capacity bounds on production and flow, material balance, and other ones as well. Since it is a linear program, it can handle many variables and constraints from a computational perspective. However, there is the overriding assumption that the market will be efficient so that cost-minimizing behavior is appropriate to characterize it. Given recent strategic behavior in European gas markets as mentioned above, other economic paradigms may also be appropriate depending on the questions being asked.

The Rice World Gas Trade Model (RWGTM) is another gas market model but with a global purview. RWGTM is a dynamic spatial equilibrium model of the global gas market based on economic theory and geologic data and constructed from the *MarketBuilder* software from Altos Management Partners. This model has extensive global coverage and subregional detail and for example has been recently used in a study of potential Russian gas exports [35].

Another modeling approach is to use a game theoretic representation of gas markets in the manner of Nash-Cournot or two-level models (e.g., MPEC, EPEC).

In [21], a Nash-Cournot model for a general gas market is considered where the traders are counterfactually depicted as having market power. Various existence and uniqueness results are proven and then the model is applied in [28] to a study of the North American natural gas market.

Another model is the World Gas Model (WGM), a large-scale, Nash-Cournot complementarity problem based on [21] and [28] but extended to many more countries and having much more functionality. WGM has been used in a number of studies for the U.S. Department of Energy as well as the Norwegian government on shale gas or cartels (e.g., [25]).

Another model with market power aspects is GASTALE (Gas Market System for Trade Analysis in a Liberalizing Europe) [4], [13], and more recently [45] with mostly a European market focus. In [4], a successive oligopoly is modeled in which oligopolistic producers compete against each other in the manner of Cournot, while anticipating the reactions of a lower level of oligopolistic within-country traders. The resulting model thus has an EPEC structure. The model described in [45] extends both [4] and [13] in several important ways including the dynamics of investment.

GASMOD [38] is another market power-oriented natural gas model. In this model, there is a two-stage game of successive natural gas exports to Europe (upstream model) and wholesale trade inside Europe (downstream model). In [38] three scenarios are compared, namely: Cournot competition in both markets, perfect competition in both markets, and perfect competition in the downstream market with Cournot competition in the upstream one.

Another example of a regional, gas model with representation of many market players is the recent work of Chyong and Hobbs [7] in which they endow market power to transit countries with a detailed representation of the Former Soviet Union (FSU) gas markets. This disaggregation includes transmission pipelines in Russia, Ukraine, Belarus and Central Asia as well as incorporating differentiation among gas production regions in Russia.

The last example of a deterministic, market-power gas model is GaMMES (Gas Market Modeling with Energy Substitution) [2], [1]. This model distinguishes itself from the others mentioned above in that it is a generalized Nash equilibrium (GNE) which depicts the major gas market participants with a European focus. Additionally, it has a sophisticated demand function based on a systems dynamic approach allowing for fossil fuel substitution between gas, coal, and oil. The generalized Nash aspects arise due to the modeling of independent gas traders who share market power with producers. Lastly, different from many other models, long-term contracts are endogenously determined.

While the above gas models have focused on either descriptive or prescriptive aspects of company or market-level aspects, they and most models like them assume that the inputs are deterministic. This is necessary in part due to the already big size of some of them and/or the computational difficulty due to non-convex terms. However, many engineering-economic problems have inherent uncertain elements such as costs, demand, or the like. Ignoring the uncertainty may be problematic in some cases and so another approach in which the stochasticity is endogenously treated may be preferred under certain circumstances. One early example of a stochastic operational model is [33]. Another more recent example of a stochastic operational or company-level gas models includes [43], [44] in which the authors considered the stochastic pooling problem for natural gas production.[7] Here both the

[7] The classic pooling problem as described in [36] and [37] models a process in which different inputs (e.g., in a refinery) are blended with a constraint on the pooled material.

design and operations of a gas production system are modeled taking into account the gas quality and the inherent non-convex pooling relationship.

In [67] and [68], a stochastic version of [21] and [28] is considered in which the demand curves are random. A scenario-tree approach was used to endogenize the uncertainty on the part of the market participants. It was noted that unlike in the case of stochastic optimization, the value of the stochastic solution (VSS) and the expected value of perfect information (EVPI) in a stochastic MCP may be negative for certain players.

In [29], a scenario-reduction approach for solving stochastic complementarity (or VI) problems is presented and tested on a stochastic duopoly, a stochastic gas market MCP for North America (but applicable to other regions), and a stochastic MCP for a subset of the countries in the European gas market as described in [14] with minor modifications. Lastly, in [1] a stochastic version of GaMMES (S-GaMMES) is presented in which a scenario-tree approach for different future oil price paths is considered.

In this chapter we do not assume any stochasticity but take all exogenous data to be deterministic. The rationale is that we shown an illustrative gas market model that includes the major features but leaves uncertainty for future considerations by the reader. In fact, one of the exercises asks the reader to reformulate the equilibrium problem considering stochastic demand elements. Besides ignoring stochasticity, we also omit non-convex engineering considerations in the equilibrium problem of Section 10.4. However, we do mention some important examples of these non-convexities in the next section as well as in the appendix of this chapter and ask the reader to include such consideration in one of the exercises.

10.3 Engineering Considerations

Natural gas operations inherently involve nonlinear and non-convex functions in equations and/or inequalities. These sorts of relationships can result in non-convex feasible regions for the associated problems. Consequently, solvers may have trouble finding global solutions. As a specific example, consider the Weymouth equation [56] that links the flow f_{ij} through an arc (i,j) and the pressures p_i and p_j, respectively, at nodes i and j:[8]

$$f_{ij} = sgn\left(p_i, p_j\right) C_{ij} \left|p_i^2 - p_j^2\right|^{1/2} \tag{10.1}$$

where

This leads to a non-convex constraint in an otherwise linear problem originally analyzed via recursion.

[8] Nodes here refer to locations in the gas network where gas is supplied, transferred, and/or removed. More generally, in this chapter a node could also refer to a particular market or country or other geographic entity depending on the model and application.

$$sgn\left(p_i, p_j\right) = \begin{cases} 1 & \text{if } p_i \geq p_j \\ -1 & \text{otherwise} \end{cases}$$

and C_{ij} is the conveyance of the pipe segment which is in general a function of the length, diameter, and the efficiency of the pipe and the properties of the gas that is flowing in the pipe segment. If for example (D.1) were just inserted as a constraint into an appropriate optimization problem, it is not clear, given its non-convex nature, that a solver could easily find a global solution. Combined with other challenging constraints (and possible non-convexities), and a large-scale problem setting, the end result would be a fairly challenging problem to solve. Thus, there is a need to approximate such a non-convex relationship or perhaps avoid it completely, depending on the questions that are to be answered by the model.

A natural question to ask is thus: When is it appropriate to use the Weymouth equation (D.1) or other nonlinear gas engineering ones? Essentially if the model needs to know the pressures and flows precisely given that these specific elements (e.g., pipe segments, compressors, valves) are explicitly being modeled then some form of the Weymouth equation is needed. In other instances, for example higher-level market models such as [2], [21], [28], [25] where only inter-regional flows are needed, these sorts of equations may be avoided. In addition, simplifying the above equations for large-scale problems or ones that already involve complex market aspects (e.g., non-convex terms, complementarity problems) may be sufficient motivation for simplication of these engineering laws. Of course, leaving these equations out may affect the results but that depends on the level of detail required. As shown in [50], including the Weymouth equation (or a linearization thereof) can make a difference versus using pressure-independent constraints on the flows when infrastructure expansion decisions are being considered. In Appendix D, we describe some other nonlinear gas engineering functions as well as some linearization approaches for using (D.1).

10.4 The Natural Gas Supply Chain and the Various Market Agents

In this section we expand on the simple network example originally presented in Chapter 1 (Illustrative Example 1.2.7). That model had four producers (two at each node), one network operator, and two demand curves (one at each node) that represent the supply and demand. To make it more realistic for natural gas markets worldwide, we need to consider other agents as well as more details on the models for these agents. In what follows we develop the extended version of the Chapter 1 model along the way pointing out how to make it more realistic as well as any simplifying assumptions imposed for presentation purposes. The purpose is not to present one model for all possible applications but to describe the key elements of gas models in an easy-to-undertand manner.

10.4.1 Sectoral and Seasonal Aspects and Gas Storage Operator

There are two regimes for consumers of natural gas: either "firm" or "interruptible" with the former paying more per unit of volume but having a more secure (and generally not interruptible) supply of gas. Another way to divide up the consumption is by sector. According to the Energy Information Administration, in 2011 in the U.S. the largest consumption sector for natural gas was electric power generation with some 34% of the total 22.3 Tcf. Next was the industrial sector with 30%, followed by the residential (21%), commercial (14%), and vehicle fuel (0.1%). First, it is important to note that the nature of the demand for natural gas differs in each of these sectors. For example, in the residential sector, natural gas for heating is an important factor in consumption levels [53]. However in the electric power sector, gas is at times a preferred fuel for power generation due to the relative capital cost advantage and lower emissions relative to other fossil fuels [53]. Based on serving the residential and commercial customers, the "high demand" season for gas is typically around October-April during which gas is extracted from reservoirs to meet this additional demand.[9] In the rest of the year, gas is often injected back into storage sites for use later on. Thus, to adequately answer questions about sectoral level or within-year demand, gas models should have at least two seasons corresponding to periods of high and low demand. Of course more than two seasons is also possible and may be desirable depending on the level of temporal detail required. Thus, we adjust the simple network problem from Chapter 1 to include two seasons: $s = 1$ (high demand) and $s = 2$ (low demand).

When a gas model includes seasons, it is also important to have a model of how the gas is stored to account for inter-temporal arbitrage and logistics. That is, typically the gas is bought when the price is lower (e.g., $s = 2$) and then stored for withdrawal when the demand (and price) is higher ($s = 1$). This allows the storage operator to arbitrage across time in order to take advantage of price differences. Consequently, in the extension to the Chapter 1 model, we add the functions of a storage operator. In the example shown below, for ease of presentation we assume that the producer also takes care of the storage functions as well as gas export.[10] The name we give to this

[9] Here and in the rest of this chapter, unless otherwise noted, we assume a geographical location whose weather pattern matches that of the U.S. East Coast.

[10] Who actually owns/controls the storage sites is an important question and for example it may be a power producer who wants a secure source of gas, an independent operator or other market participant.

market player is "supplier" to show that it's in general concerned with the supply of gas in several ways.[11]

In the extension to the Chapter 1 model, for simplicity we model only two sets of market agents, suppliers who handle production, storage, and gas export and network operators who manage the gas grid (pipeline or LNG). Alternatively, the storage and gas export functions could be handled with separate players who are modeled as optimizing profit subject to engineering/operational constraints as in [25]. The KKT conditions to their optimization problems would then be added to those of the producer and other players and combined with market-clearing conditions or explicit inverse demand equations to form the overall market equilibrium.

Such a stylized model while easier to understand necessarily misses some parts of the natural gas market which may or may not be important for an application in question. For example, pipeline nominations are not directly modeled. These forward notifications by a third party shipper to a pipeline company indicate intent to use part of the pipeline capacity for transportation of gas [61]. In the model to be presented, there is no separate shipper *per se* but rather this function is subsumed into the role of the supplier with the variable f_{ist}^{nm} (flow in volume/day) representing this nomination. Clearly, if the model needed to answer questions specific to the role of the shipper independent of production, a more complicated model than this one would be used. An example of such a study might be the analysis of capacity release in which firm transportation capacity is released by the shipper and sold to a third party some times through a closed bidding process [61]. Typically in these situations the third party would pay some or all of the reservation charge in addition to any commodity charge as well as other related charges.[12] Another set of aspects not considered in our illustrative model is the financial trading aspects of the natural gas market (e.g., futures contracts, swaps, take-or-pay contracts) that are described in [61].

Lastly, another example of an omitted feature in our model (which may not be important to include for market models), is that of other substances besides methane that can be produced/extracted during the drilling process. Hydrogen sulfide (H_2S) is one such substance which is very toxic in very small quantities. Besides avoiding exposure to it for human beings, it can also cause certain metal drilling equipment to fail. Consequently, there must be precautions taken in the cases when hydrogen sulfide may be encountered [40].

---

[11] Consequently, in our model there is no term for the purchasing of gas for storage.

[12] Under firm and interruptible transportation contracts, there are two types of main charges: commodity charge and reservation charge. The former only applies to gas that is actually transported. The latter, applied only to firm contracts is a fixed charge to reserve the pipeline capacity. There are also other charges as described in [61].

10.4.2 Capacity Expansion and Multi-Year Perspective

The simple model from Chapter 1 assumed that all production and transportation capacity was fixed. This is consistent with the assumption with a short-term model in which all these aspects would not be expanded on. However, for gas market models that cover approximately five years or more, it is important to consider capacity expansions in production, transportation and other elements (e.g., liquefaction/regasification). Generally it is not necessary to consider every year in a multi-year horizon but rather something on the order of five years corresponding roughly to how long it takes to add significant (e.g., pipeline) capacity in one of the gas market areas. Consequently, we extend the Illustrative Model 1.2.7 to include two time periods: the present ($t = 1$) and the future ($t = 2$) representing a time at least five years from the present where new capacity might be installed. It is important to note that capacity addition decisions made in time $t = 1$ don't get realized until time period $t = 2$. Of course, one can also consider capacity decisions for the second time period but then one needs to handle the "end-of-horizon" effect in some way. This phenomenon relates to the fact that the gas market doesn't cease to exist after time period $t = 2$, and so capacity decisions should be considered in that second period. However those decisions might never be selected by the model if the model stops at time $t = 2$. One way to handle this issue is to extend the model long enough so that decisions in the last period (e.g., $t = 2$) can be realized but then don't use the results from these extra time periods when analyzing the output. This is the approach in the World Gas Model [25] where results are generally reported in five-year increments through 2050 but the model is run additionally in 2055 and 2060 with the output from these last two years not used for reporting purposes. Lastly, it is important to consider the present value of capacity (and other) investments made in the future so a discount factor for later years should be applied. In the extension to the illustrative model from Chapter 1, for simplicity however, we ignore a discount factor and capacity expansion decisions but allow for two time periods with an analysis of binding resource availability.

10.4.3 Representation of Consumers and Strategic Versus Non-Strategic Players

In the simple example from Chapter 1, the producers were assumed to be price-takers, that is, they could not directly influence the price with their production decisions. As discussed above in the context of the Ukraine-Russia gas price disputes, there is some evidence of market power in gas markets. How best then to represent this market power over the consumers? One way is to use an inverse demand function of the form: $p^n_{st}(v^{n,prod}_{st})$ where $v^{n,prod}_{st}$ is the total production of gas in season s for node n at time t. This inverse demand function then replaces the fixed price that appeared in the producer's objective. Since producer i controls how much gas they produce and this amount is part of $v^{n,prod}_{st}$, this allows for producers to directly influence prices by adjusting their production quantity. Of course, the production from the other producers at that node are fixed and appear as a constant to producer i in this inverse demand function. Moreover, if imports and exports of gas into the region are also considered as part of the available supply which would be the case if the producer is modeled as managing both the production and import/exports, then $v^{n,prod}_{st}$ would be the total nodal production plus the net imports.

But what about if the gas market model covers regions of the world in which both perfect competition (price-taking) and imperfect competition (oligopolistic behavior) are present. There are at least two ways to handle this. First, one can just decide *a priori* whether each market in question is perfectly or imperfectly competitive and use either market-clearing conditions (as in the simple example) or an inverse demand function of the form just described, respectively. The complementarity format is sufficiently rich to include both these extremes. A second alternative which was adopted in the World Gas Model [25] is to use a convex combination of the price obtained from perfect and imperfect combination with the weights specific to the region in question. What this means is that if π^n_{st} represents the perfect competition price and p^n_{st} is the price obtained from an inverse demand function from Nash-Cournot players, then for $\delta^n_{st} \in [0, 1]$, the convex combination price is $\delta^n_{st}\ \pi^n_{st} + (1 - \delta^n_{st})p^n_{st}$. This latter approach has two advantages over the former one. First, it allows for a calibration "level" so as to better fit the markets being modeled. Second, it allows for calibration for the non-spot portion of the market which if significant could work towards mitigating market power. Lastly, in gas market models one must decide on which players (usually producers or their trading arms) should be modeled as strategic allowing them to influence the prices directly. For simplicity and consistency with the price-taking producers in Chapter 1, in the extension of Illustrative Example 1.2.7 discussed below we assume that each producer is non-strategic, i.e., price-taking and so the prices are computed via market-clearing conditions. If the market power modeling option were selected instead, then no market-clearing conditions would be needed. Instead, an inverse demand function would be used to compute the appropriate price. Note that one of the exercises asks the reader to convert the model so that at node 1, suppliers have market power and prices are determined via inverse demand curves instead of market-clearing conditions.

10.4.4 Additional Players and Engineering Aspects

Besides the producers and network operator depicted in the simple network example, a more extensive model could include the following other market agents: storage operators, traders, liquefiers, regasifiers, LNG tankers, and wholesale distributors. The first additional player has been briefly discussed above and the corresponding functions of this player will be spelled out in detail below. The traders are an entity that can either be dedicated to a producer and function as a trading arm as in [25] or also be an independent agent buying and selling gas by themselves as in [2]. Additionally, one could model the liquefaction process either with an optimization problem based for example on maximizing profit or simply with exogenous data related to costs, capacities, etc. The latter is easier to implement but more simplistic. The same sort of concept also applies to the regasification operations at LNG import terminals. The LNG tanker portion can be modeled separately as an optimization problem in which shipping profits are maximized or as in the case of the simple liquefaction/regasification approach, just as exogenous shipping costs and capacities. Lastly, depending on the extent of the supply chain being modeled, one could include wholesale distributors and their distribution network to the final users. Alternatively, one could just ignore this level and add exogenous adders to get the gas to the final consumers in the various end-use sectors. For ease of understanding, in the extension of the Chapter 1 simple network model we characterize liquefaction, regasification, and LNG shipping by exogenous factors and do not concentrate on retail-level consumption.

10.4.5 Suppliers

In what follows we build up the final optimization problem for supplier i located at node n starting with production.

10.4.5.1 Production

The main function of producers is to drill for natural gas and extract it from either onshore or offshore reservoirs. Producers have a number of constraints some of which involve engineering details (as previously outlined) and some logistical considerations. An example of the latter would be procuring a sufficient number of drilling rigs which as discussed in Illustrative Example 1.2.5 could lead to a generalized Nash equilibrium if two or more producers are competing for rigs. An example of the former would be the type of drilling machinery needed for the reservoir in question. Alternatively, a market for drilling rigs could be simulated with a market-clearing condition whose price would appear in each producer's objective function multiplied by the number of rigs that the producer engages. In many market models that are not concerned with specific reservoir or rig-level details, such elements can be left out of the formulation. Instead, production limits from both a daily or seasonal perspective could be included to better match a longer-term and higher-level perspective. This was the approach taken in [25]. Also, the cost of drilling for gas clearly may depend on the type of well (e.g., conventional or unconventional, offshore or onshore) that is being considered. We first describe the production cost curve in general terms and then give a specific example (the Golombek curve) for specificity whose parameters can be adjusted to suit the reservoir in question.

We let $q_{ist}^{n,prod}$ be the amount of gas produced by supplier i at node n in season s, time period t expressed in volume per day (either cubic feet or cubic meters usually). Since there are two seasons being modeled and the length of each season is not necessarily the same, we also include $Days_s$ to be the number of days in season s.[13] Consequently, the term $Days_s q_{ist}^{n,prod}$ represents the total volume of gas produced for that supplier in season s and time period t.[14] All the constraints to be shown will be expressed in the same units (volume) so as to have the associated multipliers in the same units ($/volume).

Producing a volume of $Days_s q_{ist}^{n,prod}$ generates a set of costs related to equipment, labor and materials. Of course, this depends on the reservoir in question but if we generalize a cost function $C_{ist}^{n,prod}(\cdot)$ to be the same for each reservoir in a node but varying perhaps by supplier and season and time period, we obtain the following expressed in dollars per volume extracted: $C_{ist}^{n,prod}\left(v_{ist}^{n,prod}\right)$ where the volume produced $v_{ist}^{n,prod} = Days_s q_{ist}^{n,prod}$.

Note that we have implicitly assumed that the lowest level of spatial aggregation is the (supplier, node) combination rather than the reservoir or well. This is a reasonable assumption if one can aggregate the production from the individual reservoirs and wells and/or if one is only interested in the total nodal production as is the case with many gas market models. Also, in some sense, if finer detail is needed additional suppliers can be included to represent different reservoirs.

There are a number of choices for the form of this cost function depending on where the gas is being extracted (onshore or offshore) as well as the type of gas (conventional or unconventional) to name just a few differentiating features. In general though, this function should be increasing in its argument (producing more gas costs more money) and for optimization purposes should be differentiable and (hopefully) convex so that the KKT conditions can be

13 Here and in the rest of the discussion, upper-case Roman letters correspond to exogenous parameters or functions and lower-case Roman ones to variables whose values are to be determined by the model. Greek variables are the Lagrange multipliers for their associated constraints.

14 Time periods are often years or five-year increments but depend on the model in question.

applied globally. One such choice is the Golombek cost function described in [32] and used in the World Gas Model [25] and GASTALE [45]. This Golombek marginal cost function has the form:

$$a + bv + c \ln\left(1 - \frac{v}{capacity}\right) \tag{10.2}$$

where a, b, c are parameters to be determined (with $c \leq 0$), v is the volume of gas produced, and *capacity* is the production capacity. Notice that if the volume produced exactly matches the capacity given that $c \leq 0$, the costs are infinite. Essentially what happens is that as this capacity is approached, the third term in (10.2) dominates and the costs rise exponentially. An example of this cost function when $a = b = 1$, $c = -1$, and $capacity = 10$ is shown in Figure 10.1. (In the exercises, the reader is asked to analyze the convexity of the Golombek cost function).

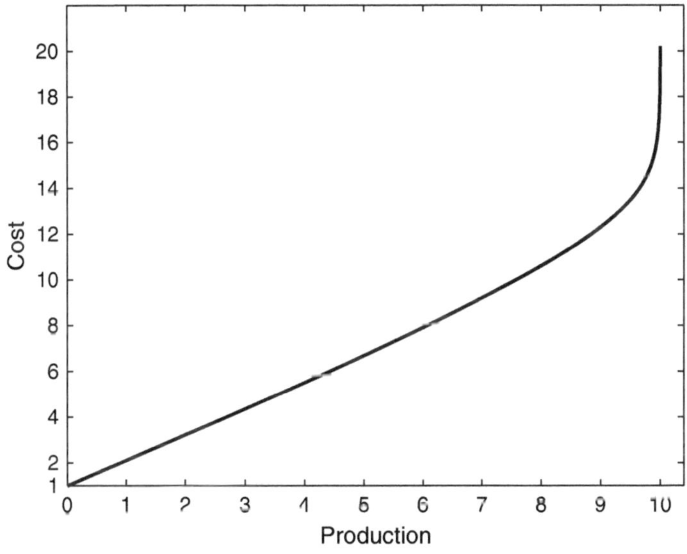

Fig. 10.1 Golombek marginal cost function: $a = b = 1, c = -1$, *capacity* $= 10$

Some reasonable constraints for a market-level model are to include the daily (or seasonal) production limits relating to rig and /or other equipment availability as well as ultimate recoverable volume limits for the entire life of the reservoirs. These two constraints are easy to implement and indirectly account for a number of more complicated engineering and logistical restrictions mentioned above. The production rate constraint is the following for supplier i at node n:[15]

$$Days_s q_{ist}^{n,prod} \leq Q_{ist}^{n,prod,\max}, \forall s, t \tag{10.3}$$

A simplification is to make $Q_{ist}^{n,prod,\max} = Q_i^{n,prod,\max}$ so that this maximum volume is independent of the season and time period.

The total volume constraint is the following for supplier i at node n:

$$\sum_t \sum_s Days_s q_{ist}^{n,prod} \leq V_i^{n,prod,\max} \tag{10.4}$$

where $V_i^{n,prod,\max}$ is the (estimated) total recoverable volume of natural gas by supplier i at node n over the time horizon. Taking into account the comments made above, (10.4) could also be construed as a total volume constraint for a reservoir if each supplier corresponds to exactly one reservoir. Of course, in this case some care must be exercised in interpreting the strategic aspects embodied in the inverse demand function which is usually at the supplier not reservoir level.

One could also consider capacity expansion decisions if the time period for the model is long enough. This might include for example a variable on how much capacity should be added to the right-hand side of (10.3). Given that the model to be presented in this section will only correspond to two time periods and has the goal of being simple enough to easily understand, these capacity decisions for the suppliers (and other players) are left out. However, they are important decisions that other models do consider especially in the context of congestion and infrastructure planning; see for example [25]) and for the effect of the Weymouth equations on capacity decisions [50].

10.4.5.2 Delivering Gas to the Market

The last point about modeling the suppliers concerns how to handle the activity of getting the produced gas to the market. This can be done as a separate activity but still under the control of the supplier or could involve another market agent such as a trader or arbitrageur. In [25] the traders are dedicated to the production companies but modeled as separate players for modeling advantages including representing gas cartels. In [2] and [1] both producers-traders as well as independent traders are modeled. In GASTALE [45], producers are their own bulk traders deciding on what to produce and then paying the pipeline system to ship it to consuming countries. The traders in that model are within country.

[15] The variable q is for quantities expressed in volume/day with the super and subscripts distinguishing type of quantity (production or storage). Similarly, v denotes the volume of gas.

In the gas market example in this chapter, to simplify the presentation as well as correspond more closely to Illustrative Example 1.2.7, we endow the producers directly with the trading capability. From that Chapter 1 example recall that only the producers at node 1 could send their production to the other node 2. Clearly, in a more general market model both directions should be possible. Thus, we need to consider two new index sets corresponding to the set of flows incoming to node n designated as $incoming(n)$ and those flows going out of node n given as $outgoing(n)$. We assume that $incoming(1) \supseteq \{n\}$ since suppliers can at least provide gas to their own node. From Illustrative Example 1.2.7 we see that

$$\begin{array}{ll} incoming(1) = \{1\} & outgoing(1) = \{2\} \\ incoming(2) = \{1,2\} & outgoing(2) = \emptyset \end{array}$$

We designate the flow from node n to node m by producer i as f_{ist}^{nm} expressed in volume/day (i.e., cubic feet/day or cubic meters/day) in season s and time t.

Also note that it is only the index of other nodes that matters in terms of flow in or out of the node n. This is because it is assumed that the flow from all the suppliers shipping gas from node m to node n is comingled on the same pipeline. This aggregation of pipelines and (supplier, customer) pairs is a simplification but one that is reasonable if a market-level perspective is considered and one in which individual shipments between a specific producer and customer are not needed to be individually tracked. If there are contracted volumes between a supplier and customer, one can always put them separately into the supplier's problem. We can take care of both the spot market volumes (described up to this point) as well as contracted volumes if we define some additional variables as follows:

$s_{ist}^{nm,contract}$ = total contracted sales going to market m from node n in volume/day

$s_{ist}^{nm,spot}$ = total spot market sales to market m from node n in volume/day

The following constraint links these variables to the ones discussed above:

$$\sum_{m \in outgoing(n)} Days_s s_{ist}^{nm,contract} + \sum_{m \in outgoing(n)} Days_s s_{ist}^{nm,spot} = Days_s q_{ist}^{n,prod}, \forall n, i, s, t \quad (10.5)$$

In a larger model for which computational aspects need to be carefully considered a set of arcs rather than all node pairs (n, m) could be considered. Additionally, one could include path flows as opposed to flows between nodes if the particular path that the gas took was important to track for cost or other (e.g., financial) reasons. However, since the number of paths grows quite quickly in even moderate size gas networks, this approach is usually not considered.[16] Thus, the flow f_{ist}^{nm} is constrained as follows:

$$Days_s f_{ist}^{nm} = Days_s s_{ist}^{nm,contract} + Days_s s_{ist}^{nm,spot} \tag{10.6}$$

10.4.5.3 Supplier's Problem (Version 1: Production and Export Functions)

Taking together all the parts of the supplier i's activities discussed above, the following represents an optimization problem to maximize the overall profits for that supplier located at node n where $P_{ist}^{nm,contract}$ is an exogenously determined price for shipping contracted gas from node n to node m, expressed in \$/volume.[17] The contract price $P_{ist}^{nm,contract}$ might vary by some indexing formulas based on other energy prices but is expressed here just as a constant for ease of presentation. Also, we assume for simplicity that contract sales can be determined separately for each season s and time period t. In practice, there is usually a longer-term arrangement but this can easily be handled by restricting the contract prices and sales accordingly with additional constraints.

Additionally, we use the exogenous factor $\tau_{st}^{nm,reg}$ as a regulated tariff for shipping gas between nodes n and m determined perhaps by the government, expressed in \$/volume. Lastly, $\tau_{st}^{nm,congest}$ is a congestion tariff for shipping gas between nodes n and m determined by the pipeline market-clearing conditions, expressed in \$/volume (see below for how it is determined).[18]

$$\text{Maximize}_{q_{ist}^{n,prod}, v_{ist}^{n,prod}, f_{ist}^{nm}, s_{ist}^{nm,contract}, s_{ist}^{nm,spot}}$$

$$\sum_t \sum_s \Big\{ \sum_{m \in outgoing(n)} \big[\big(P_{ist}^{nm,contract} Days_s s_{ist}^{nm,contract}\big) + \big(\pi_{st}^{m} Days_s s_{ist}^{nm,spot}\big) - \big(\tau_{st}^{nm,reg} + \tau_{st}^{nm,congest}\big) Days_s f_{ist}^{nm} \big] - C_{ist}^{n,prod}\left(v_{ist}^{n,prod}\right) \Big\} \tag{10.7a}$$

s. t.

$$Days_s q_{ist}^{n,prod} \le Q_{ist}^{n,prod,\max}, \forall s,t \qquad (\alpha 1_{ist}^{n}) \tag{10.7b}$$

---

[16] See [28] for a discussion of path flows in gas networks.

[17] There is an implicit assumption in this optimization problem that each producer is operating only at one node. This is easily overcome but summing the profits at all the nodes where the producer is operating but the former is preferred for clarity of presentation.

[18] In (10.7d) and other places, we have multiplied by $Days_s$ to convert the units for the equation to volume. This is done so that all the associated dual variables have the same units of \$/volume.

$$\sum_t \sum_s Days_s q_{ist}^{n,prod} \le V_i^{n,prod,\max} \quad (\alpha 2_i^n) \tag{10.7c}$$

$$Days_s f_{ist}^{nm} = Days_s s_{ist}^{nm,contract} + Days_s s_{ist}^{nm,spot}, \forall s,t \quad (\alpha 3_{ist}^{nm}) \tag{10.7d}$$

$$\sum_{m \in outgoing(n)} Days_s s_{ist}^{nm,contract} + \sum_{m \in outgoing(n)} Days_s s_{ist}^{nm,spot}$$

$$= Days_s q_{ist}^{n,prod}, \forall s,t \quad (\alpha 4_{ist}^n) \tag{10.7e}$$

$$v_{ist}^{n,prod} = Days_s q_{ist}^{n,prod} \quad (\alpha 5_{ist}^n) \tag{10.7f}$$

$$q_{ist}^{n,prod}, v_{ist}^{n,prod}, f_{ist}^{nm}, s_{ist}^{nm,contract}, s_{ist}^{nm,spot} \ge 0, \forall s,t \tag{10.7g}$$

Also, we have the following associated market-clearing conditions:

$$\sum_{m \in incoming(n)} \sum_{i \in supplier(m)} Days_s \left(s_{ist}^{mn,spot} + s_{ist}^{mn,contract} \right) = D_{st}^n \left(\pi_{st}^n \right) \quad (\pi_{st}^n) \tag{10.8}$$

where $supplier(n)$ is the set of suppliers at node n, $D_{st}^n(\pi_{st}^n)$ is the total demand for natural gas in market n and π_{st}^n is the nodal price.[19] If there are contract sales, i.e., $\sum_{m \in incoming(n)} \sum_{i \in supplier(m)} Days_s \left(s_{ist}^{mn,contract} \right) > 0$, then this amount can be subtracted from the right-hand side of (10.8) since the contract sales will have satisfied that portion of demand. The remaining demand can be construed as the spot market. For simplification, one can take $\tau_{st}^{nn,reg}, \tau_{st}^{nn,congest}$ both equal to zero consistent with the assumption that the costs of delivering gas within the same node is negligible compared to outside the node and thus these costs can be ignored. Other items that are not usually considered in models of this type but are part of the natural gas supply chain are for example: "lease and plant fuel" [17] "or storage at the production site." Clearly, these and other aspects of gas production may or may not be included at the discretion of the modeler. Note that we have explicitly included a constraint to convert from rates (volume/day) to volume and included separate volume variables $v_{ist}^{n,prod}$. The rationale for this is to be clear that the production cost curve $C_{ist}^{n,prod}$ is a function of volume. Clearly, instead we could just have used $C_{ist}^{n,prod}\left(Days_s q_{ist}^{n,prod}\right)$ but this is slightly less clear especially when taking the KKT conditions. In a large-scale model, to limit the number of variables and/or constraints, these extra volume variables and volume-rate constraints would be suppressed by either substituting $Days_s q_{ist}^{n,prod}$ for $v_{ist}^{n,prod}$ in the KKT conditions or beforehand in the optimization problem. A similar line of reasoning also applies to the storage and flow volumes and rates.

To summarize, we see that problem (10.7) models the production, sales, and gas tranportation parts of the gas suppliers. Also, (10.8) is the balancing

[19] The demand by individual sectors has been aggregated into a total demand for clarity of presentation. Also, any retail markups has been avoided to simplify the presentation.

of total sales at a node n with the total demand $D^n_{st}(\pi^n_{st})$ summed across all sectors. In the next section we expand on the supplier's problem to include storage as well as introduce the network operator's problem.

10.4.5.4 Storage Operations

In this section we describe the storage operations for the natural gas market and modify the supplier's optimization problem (10.7) appropriately. Typically gas is extracted from storage to supply consumers in the high demand season(s) when the supply is scarce after having injected it in the low demand season(s). Thus, storage operations represent a sort of temporal arbitrage where the gas is bought and injected when it is less expensive (low season) and withdrawn and sold in the higher demand season when it is more expensive.

A first question is "who owns the gas storage?" According to the U.S. Energy Information Administration [16], in the U.S., there are four types of main owners of gas storage:

1. Interstate pipeline companies
2. Intrastate pipeline companies
3. Local distribution companies
4. Independent storage services providers

Additionally, as noted in [16], it is important to note that the owners/operators of the storage facility may not correspond to those who own the natural gas. For simplicity though, in the model (10.13) to be presented below, we have made the owner and operator of the gas storage and the owner of the gas itself the same, i.e., the supplier.[20]

Another simplification that we have made is to take the maximum deliverability (volume extracted per day) as constant, e.g., the rate X^n_i differing only (potentially) by node and supplier. In practice this rate can depend on a host of factors such as: the amount of gas in the reservoir, the pressure in the reservoir, the compression capability in the reservoir, surface facilities, etc. However, it can be said that this deliverability is highest when the reservoir is almost full of gas and as a function of the amount of gas extracted. While deliverability varies directly with the total amount of gas in the reservoir, the injection rate varies inversely (it also varies by related factors to deliverability) [16]. For simplicity we have also assumed that this maximum injection rate is constant (i.e., I^n_i).

An example of this relationship is given by the following "backpressure" equation [30]:

$$q_g = C\left(p_R^2 - p_{wf}^2\right)^n \tag{10.9}$$

[20] The EIA categorizes three types of underground gas storage: 1. depleted reservoirs in oil and/or gas fields, 2. aquifers, 3. salt caverns as well as storing gas aboveground as LNG.

where q_g is the flow of gas in million cubic feet/day, p_R^2 is the pressure of the reservoir (taken away from the well), p_{wf}^2 is the bottom flowing pressure and C and n are constants to be determined for the reservoir in question; n is a value in between 0.5 and 1.0 [30]. When more gas is extracted from the reservoir, the value of p_R^2 will decrease and all else being equal (for $C > 0$) will make the flow rate decrease.

Besides the production and sales decisions shown above for the supplier problem (version 1), we now consider the decisions to inject or extract gas into/out of storage.[21] To keep it simple we assume that gas is only extracted in the high demand season ($s = 1$) and only injected in the low demand season ($s = 2$) but chronologically gas must of course be injected first. The resulting variables are thus defined as follows:

$q_{ist}^{n,extract}$ = the amount of gas extracted by supplier i in season s, time t in volume/day

$q_{ist}^{n,inject}$ = the amount of gas injected by supplier i in season s, time t in volume/day

with the understanding that $q_{i2t}^{n,extract} = q_{i1t}^{n,inject} = 0$.[22]

There are several sets of constraints that connect these variables where we have assumed that within one year, the storage injection/extraction must balance.[23] First, there is an upper bound on how much can be injected or extracted in a given season due to capacity of the compressors (injection) or other equipment. The associated constraints are the following:

$$Days_1 q_{i1t}^{n,extract} \leq Days_1 X_i^n, \forall t \tag{10.10a}$$

$$Days_2 q_{i2t}^{n,inject} \leq Days_2 I_i^n, \forall t \tag{10.10b}$$

where X_i^n, I_i^n represent, respectively, the extraction, injection maximum rates in volume/day (assumed to be constant for the entire season s). Additionally, the total injection (incuding losses) must balance with the total extraction for each year:

$$Days_1 q_{i1t}^{n,extract} = \left(1 - L_{it}^n\right) Days_2 q_{i2t}^{n,inject}, \forall i, t \tag{10.11}$$

[21] We assume somewhat reasonably that storage is only a local supply to the node. That is, the supplier only has access to the stored gas at its own node n.

[22] For ease of presentation, in the GAMS code, a seasonal index is used for the primal and dual variables and the appropriate extraction/injection activities set to zero. Using this seasonal index is done to reinforce the fact that these activities only occur in specific seasons and associating the correct $Days_s$ coefficients.

[23] This is reasonable in that gas not withdrawn corresponds to lost revenue hence there is a motivation to extract as much as possible by the supplier.

where L_{it}^{n} is the loss factor (%) to account for some of the injected gas fueling the compressors.[24] Lastly, we assume that there is a volumetric constraint as well related to how much can actually be taken out of the reservoir (working gas) to maintain a certain amount for technical reasons (cushion gas) [16]. This constraint is of the following form:

$$Days_1 q_{i1t}^{n,extract} \leq W_{it}^{n}, \forall t \tag{10.12}$$

where W_{it}^{n} is the total volume of working gas allowed to be extracted.[25] Lastly, we suppose that there are convex and differentiable cost functions $C_{it}^{n,extract}\left(v_{it}^{n,extract}\right)$ and $C_{it}^{n,inject}\left(v_{it}^{n,inject}\right)$ respectively, for storage extraction and injection in terms of the total extraction and injection volumes $v_{it}^{n,extract} = Days_1 q_{i1t}^{n,extract}$, $v_{it}^{n,inject} = Days_2 q_{ist}^{n,inject}$.[26]

10.4.5.5 Supplier's Problem (Version 2: Production, Export and Storage Functions)

Taking into account the storage constraints mentioned above, we see version 2 of the supplier's problem as follows:

$$\begin{aligned}
&\underset{q_{ist}^{n,prod}, v_{ist}^{n,prod}, f_{ist}^{nm}, s_{ist}^{nm,contract}, s_{ist}^{nm,spot}, q_{i1t}^{n,extract}, q_{i2t}^{n,inject}, v_{it}^{n,extract}, v_{it}^{n,inject}}{\text{Maximize}} \\
&\sum_t \left(\pi_{1t}^{n} Days_1 q_{i1t}^{n,extract}\right) \\
&+\sum_t \sum_s \Bigg\{ \sum_{m \in outgoing(n)} \Big[\left(P_{ist}^{nm,contract} Days_s s_{ist}^{nm,contract}\right) \\
&+ \left(\pi_{st}^{m} Days_s s_{ist}^{nm,spot}\right) \\
&- \left(\tau_{st}^{nm,reg} + \tau_{st}^{nm,congest}\right) Days_s f_{ist}^{nm} \Big] \\
&- C_{ist}^{n,prod}\left(v_{ist}^{n,prod}\right) - C_{it}^{n,extract}\left(v_{it}^{n,extract}\right) - C_{it}^{n,inject}\left(v_{it}^{n,inject}\right) \Bigg\}
\end{aligned} \tag{10.13a}$$

s. t.

$$Days_s q_{ist}^{n,prod} \leq Q_{ist}^{n,\max}, \forall s, t \qquad (\alpha 1_{ist}^{n}) \tag{10.13b}$$

[24] In the GAMS code relating to the gas market equilibrium presented later in this chapter, we assume for simplicity that the loss percentage is the same for each supplier i so that it becomes L_t^n.

[25] In the GAMS code for the gas market equilibrium we assume that the total working gas volume does not depend on time so it is represented by W_i^n.

[26] We do not use the seasonal index s for the volume since there are no $Days_s$ coefficients to match up. Alternatively we could use this index with restrictions on extraction volume being equal to 0 in $s = 2$ and injection volume equal to 0 in season $s = 1$.

$$\sum_t \sum_s Days_s q_{ist}^{n,prod} \leq V_i^{n,prod,\max} \quad (\alpha 2_i^n) \tag{10.13c}$$

$$Days_s f_{ist}^{nm} = Days_s s_{ist}^{nm,contract} + Days_s s_{ist}^{nm,spot}, \forall s,t \quad (\alpha 3_{ist}^{nm}) \tag{10.13d}$$

$$\sum_{m \in outgoing(n)} Days_s s_{ist}^{nm,contract} + \sum_{m \in outgoing(n)} Days_s s_{ist}^{nm,spot}$$

$$= Days_s q_{ist}^{n,prod}, \forall t \quad (\alpha 4_{i1t}^n) \quad s=1$$

$$\sum_{m \in outgoing(n)} Days_s s_{ist}^{nm,contract} + \sum_{m \in outgoing(n)} Days_s s_{ist}^{nm,spot}$$

$$+ Days_2 q_{i2t}^{n,inject} = Days_s q_{ist}^{n,prod}, \forall t \quad (\alpha 4_{i2t}^n) \quad s=2 \tag{10.13e}$$

$$v_{ist}^{n,prod} = Days_s q_{ist}^{n,prod}, \forall s,t \quad (\alpha 5_{ist}^n) \tag{10.13f}$$

$$Days_1 q_{i1t}^{n,extract} \leq Days_1 X_i^n, \forall t \quad (\alpha 6_{i1t}^n) \tag{10.13g}$$

$$Days_2 q_{i2t}^{n,inject} \leq Days_2 I_i^n, \forall t \quad (\alpha 7_{i2t}^n) \tag{10.13h}$$

$$Days_1 q_{i1t}^{n,extract} - (1 - L_{it}^n) Days_2 q_{i2t}^{n,inject} = 0, \forall t \quad (\alpha 8_{it}^n) \tag{10.13i}$$

$$Days_1 q_{i1t}^{n,extract} \leq W_{it}^n, \forall t \quad (\alpha 9_{it}^n) \tag{10.13j}$$

$$v_{it}^{n,extract} = Days_1 q_{i1t}^{n,extract}, \forall t \quad (\alpha 10_{it}^n) \tag{10.13k}$$

$$v_{it}^{n,inject} = Days_2 q_{i2t}^{n,inject}, \forall t \quad (\alpha 11_{it}^n) \tag{10.13l}$$

$$q_{ist}^{n,prod}, v_{ist}^{n,prod}, f_{ist}^{nm}, s_{ist}^{nm,contract}, s_{ist}^{nm,spot}, q_{i1t}^{n,extract}, q_{i2t}^{n,inject},$$

$$v_{it}^{n,extract}, v_{it}^{n,inject} \geq 0, \forall s,t \tag{10.13m}$$

In addition to adjusting the supplier problem to account for storage, we need to also modify the market-clearing conditions to obtain the following:

$$\begin{cases} \sum_{m \in incoming(n)} \sum_{i \in supplier(m)} Days_s \left(s_{ist}^{mn,contract} + s_{ist}^{mn,spot} \right) \\ \quad + \sum_{i \in supplier(n)} Days_s q_{i1t}^{n,extract} = D_{st}^n \left(\pi_{st}^n \right) \quad \left(\pi_{st}^n \right) \quad s=1 \\ \sum_{m \in incoming(n)} \sum_{i \in supplier(m)} Days_s \left(s_{ist}^{mn,contract} + s_{ist}^{mn,spot} \right) \\ \quad = D_{st}^n \left(\pi_{st}^n \right) \quad \left(\pi_{st}^n \right) \quad s=2 \end{cases} \tag{10.14}$$

To summarize, the problem (10.13) represents the production, export, and storage operations of each supplier i at node n with an objective of maximizing profit (10.13a). The equations constrain the production to be under a maximum rate (10.13b), within a total gas amount for the entire time horizon (10.13c), relate the spot and contract sales to the quantity to be exported (10.13d), divide up the production into sales or injection into storage (10.13e) or relate the production rate to the production volume (10.13f). Additionally, the constraints (10.13g)-(10.13l) restrict the storage injection and/or extraction activities. In the next section we develop an optimization problem for the network operator whose KKT conditions, combined with those of the suppliers and the marketing-clearing conditions (10.14) and similar ones for the pipeline capacity all taken together will constitute the gas market equilibrium.

10.4.6 Transportation

In this section we describe some of the operations of transporting natural gas with the two main choices being by pipeline or via LNG tankers. In the latter case both the liquefaction and then regasification of the gas needs to be considered. In some models (e.g., [25]), either the pipeline or LNG options are modeled the same as just inter-regional flows with the differences being the unit costs and the capacities. At the other extreme, one can directly model the LNG supply chain (called "train") with the liquefiers, regasifiers, LNG shipping all depicted with separate optimization problems for each player. For example, in GASTALE [45], liquefaction and regasification facilities are included. The choice is really based on the questions that need to be answered by the model. Clearly, if the focus is an LNG important terminal and how well it does as part of the overall gas supply chain, a separate optimization model for that player is probably best. On the other hand, if the level of detail is only regional then probably treating pipeline and LNG routes similarly with exogenous costs and capacities is more appropriate. For clarity of presentation we choose this last option.

We model the network operator whose purview is to maintain the smooth running of the gas grid by seeing that the flows along the links (pipeline or LNG) are within capacities and to do so at minimum cost. Other objectives such as maximizing social welfare are also possible. The network operator typically will only control a subset of the gas grid relating to specific links but for ease of presentation in the model below we assume that there is just one network operator with control of the entire grid. The main variables are g_{st}^{nm} which are the amount of volume/day flowed between nodes n and m in season s and time t. We could further distinguish the flow separately for pipelines and LNG as each would have a different supply chain for reaching the end-user. For simplicity and to avoid adding more variables related to how much of the sales by the supplier should be pipeline-based and how much should travel by LNG, we represent the flows in aggregate just by this one variable. There is a capacity constraint for the flows given as follows where G_s^{nm} is the link capacity (pipeline or LNG):

$$Days_s g_{st}^{nm} \leq G_s^{nm} \tag{10.15}$$

This inequality (for pipelines) is of course an approximation of the engineering aspects of transporting this form of gas. For example, as discussed above and in Appendix D, inclusion of the pressure-flow relationship given by the Weymouth equation would be more accurate but (10.15) may be sufficient given the modeling questions to answer. Associated with each unit of flow is a volume cost function designated as $C_{st}^{nm,flow}\left(v_{st}^{nm,flow}\right)$ which we assume to be convex, increasing, and differentiable for realism and computational considerations.[27] The network operator's problem is thus the following:

$$\text{Maximize}_{g_{st}^{nm},v_{st}^{nm,flow}}$$

$$\sum_t \sum_s \sum_n \sum_m \left(\tau_{st}^{nm,reg} + \tau_{st}^{nm,congest}\right) Days_s g_{st}^{nm} - C_{st}^{nm,flow}\left(v_{st}^{nm,flow}\right) \tag{10.16a}$$

s. t.

$$Days_s g_{st}^{nm} \leq G_s^{nm}, \quad \forall s,t \quad (\beta 1_{st}^{nm}) \tag{10.16b}$$

$$v_{st}^{nm,flow} = Days_s g_{st}^{nm}, \quad \forall s,t \quad (\beta 2_{st}^{nm}) \tag{10.16c}$$

$$g_{st}^{nm}, v_{st}^{nm,flow} \geq 0 \tag{10.16d}$$

The endogenous congestion price $\tau_{st}^{nm,congest}$ is calculated as the dual of the transportation market-clearing conditions given as follows:

$$\sum_{i \in supplier(n)} Days_s f_{ist}^{nm} = Days_s g_{st}^{nm} \quad \left(\tau_{st}^{nm,congest}\right) \tag{10.17}$$

where the left-hand side of (10.17) represents the demand for the particular link (n, m) in season s and time t and the right-hand side the supply. Note that the regulated tariff $\tau_{st}^{nm,reg}$ is an exogenous factor agreed upon by the regulatory agency and the network operator. By constrast, the congestion price $\tau_{st}^{nm,congest}$ is endogenously determined by the market-clearing conditions (10.17) and the rest of the market model. If the market in question has no actual congestion payments, this tariff can be considered as an economic mechanism for allocating scarce capacity on a link (pipeline or LNG).

10.4.7 A Model for the Whole Market

In this section we simultaneously take the KKT conditions to the supplier problem (version 2) (10.13) and the network operator problem (10.16), and the two sets of market-clearing conditions (10.14) and (10.17) to arrive at a market equilibrium.

The KKT conditions to the supplier problems (10.13) are the following where the first set are with respect to the primal variables and the second to the dual variables.

---

[27] For simplicity, we have characterized all links as (n, m) pairs where n, m are nodes in the network and $m = n$ means the same market.

First set:

$$0 \leq Days_s \left[\alpha 1_{ist}^n + \alpha 2_i^n - \alpha 4_{ist}^n - \alpha 5_{ist}^n\right] \perp q_{ist}^{n,prod} \geq 0, \forall n,i,s,t \quad (10.18a)$$

$$0 \leq \frac{dC_{ist}^{n,prod}\left(v_{ist}^{n,prod}\right)}{dv_{ist}^{n,prod}} + \alpha 5_{ist}^n \perp v_{ist}^{n,prod} \geq 0, \forall n,i,s,t \quad (10.18b)$$

$$0 \leq Days_s \left[\left(\tau_{st}^{nm,reg} + \tau_{st}^{nm,congest}\right) + \alpha 3_{ist}^{nm}\right] \perp f_{ist}^{nm} \geq 0, \forall n,m \in outgoing(n), i,s,t \quad (10.18c)$$

$$0 \leq Days_s \left[-P_{ist}^{nm,contract} - \alpha 3_{ist}^{nm} + \alpha 4_{ist}^n\right] \perp s_{ist}^{nm,contract} \geq 0, \forall n,m \in outgoing(n), i,s,t \quad (10.18d)$$

$$0 \leq Days_s \left[-\pi_{st}^m - \alpha 3_{ist}^{nm} + \alpha 4_{ist}^n\right] \perp s_{ist}^{nm,spot} \geq 0, \forall n,m \in outgoing(n), i,s,t \quad (10.18e)$$

$$0 \leq Days_1 \left[-\pi_{1t}^n + \alpha 6_{ist}^n + \alpha 8_{it}^n + \alpha 9_{it}^n - \alpha 10_{it}^n\right] \perp q_{ist}^{n,extract} \geq 0, \forall n,i,s,t \text{ for } s=1 \quad (10.18f)$$

$$0 \leq Days_2 \left[\alpha 4_{i2t}^n + \alpha 7_{ist}^n - (1 - L_{it}^n)\alpha 8_{it}^n - \alpha 11_{it}^n\right] \perp q_{ist}^{n,inject} \geq 0, \forall n,i,s,t \text{ for } s=2 \quad (10.18g)$$

$$0 \leq \frac{dC_{it}^{n,extract}\left(v_{it}^{n,extract}\right)}{dv_{it}^{n,extract}} + \alpha 10_{it}^n \perp v_{it}^{n,extract} \geq 0, \forall n,i,t \quad (10.18h)$$

$$0 \leq \frac{dC_{it}^{n,inject}\left(v_{it}^{n,inject}\right)}{dv_{it}^{n,inject}} + \alpha 11_{it}^n \perp v_{it}^{n,inject} \geq 0, \forall n,i,t \quad (10.18i)$$

Second set:

$$0 \leq Q_{ist}^{n,\max} - Days_s q_{ist}^{n,prod} \perp \alpha 1_{ist}^n \geq 0, \forall n,i,s,t \quad (10.19a)$$

$$0 \leq V_i^{n,prod,\max} - \sum_t \sum_s Days_s q_{ist}^{n,prod} \perp \alpha 2_i^n \geq 0, \forall n,i \quad (10.19b)$$

$$0 = Days_s f_{ist}^{nm} - Days_s s_{ist}^{nm,contract} - Days_s s_{ist}^{nm,spot}, \ \alpha 3_{ist}^{nm} \text{ free}, \forall n,m \in outgoing(n), i,s,t \quad (10.19c)$$

$$\begin{cases} 0 = \sum_{m \in outgoing(n)} Days_s s_{ist}^{nm,contract} + \sum_{m \in outgoing(n)} Days_s s_{ist}^{nm,spot} - Days_s q_{ist}^{n,prod}, \ \alpha 4_{ist}^n \text{ free}, \forall n,i,s,t \quad s=1 \\ 0 = \sum_{m \in outgoing(n)} Days_s s_{ist}^{nm,contract} + \sum_{m \in outgoing(n)} Days_s s_{ist}^{nm,spot} + Days_2 q_{ist}^{n,inject} - Days_s q_{ist}^{n,prod}, \ \alpha 4_{ist}^n \text{ free}, \forall n,i,s,t \quad s=2 \end{cases} \quad (10.19d)$$

$$0 = v_{ist}^{n,prod} - Days_s q_{ist}^{n,prod}, \ \alpha 5_{ist}^n \text{ free}, \forall n,i,s,t \quad (10.19e)$$

$$0 \le Days_1 X_i^n - Days_1 q_{i1t}^{n,extract} \perp \alpha 6_{ist}^n \ge 0, \forall n, i, t \text{ for } s = 1 \quad (10.19f)$$

$$0 \le Days_2 I_i^n - Days_2 q_{i2t}^{n,inject} \perp \alpha 7_{ist}^n \ge 0, \forall n, i, t \text{ for } s = 2 \quad (10.19g)$$

$$0 = Days_1 q_{i1t}^{n,extract} - (1 - L_{it}^n) Days_2 q_{i2t}^{n,inject}, \ \alpha 8_{it}^n \text{ free}, \forall n, i, t \quad (10.19h)$$

$$0 \le W_{it}^n - Days_1 q_{i1t}^{n,extract} \perp \alpha 9_{it}^n \ge 0, \ \forall n, i, t \quad (10.19i)$$

$$0 = v_{it}^{n,extract} - Days_1 q_{i1t}^{n,extract}, \ \alpha 10_{it}^n \text{ free}, \forall t \quad (10.19j)$$

$$0 = v_{it}^{n,inject} - Days_2 q_{i2t}^{n,inject}, \ \alpha 11_{it}^n \text{ free}, \forall t \quad (10.19k)$$

The KKT conditions to the network operator problem (10.16) are the following:

$$\begin{aligned} 0 \le & -Days_s \left(\tau_{st}^{nm,reg} + \tau_{st}^{nm,congest}\right) + Days_s \beta 1_{st}^{nm} \\ & - Days_s \beta 2_{st}^{nm} \perp g_{st}^{nm} \ge 0, \quad \forall n, m \in outgoing(n), s, t \end{aligned} \quad (10.20a)$$

$$0 \le \frac{dC_{st}^{nm,flow}\left(v_{st}^{nm,flow}\right)}{dv_{st}^{nm,flow}} + \beta 2_{st}^{nm} \perp v_{st}^{nm,flow} \ge 0, \quad \forall n, m, s, t \quad (10.20b)$$

$$0 \le G_s^{nm} - Days_s g_{st}^{nm} \perp \beta 1_{st}^{nm} \ge 0, \forall n, m, s, t \quad (10.20c)$$

$$0 = v_{st}^{nm,flow} - Days_s g_{st}^{nm}, \quad \beta 2_{st}^{nm} \text{ free }, \forall n, m, s, t \quad (10.20d)$$

So the full market equilibrium model is thus:

1. KKT conditions (10.18) and (10.19)
2. KKT conditions for the network operator, (10.20)
3. Spot market-clearing conditions (10.14) and
4. Network flow market-clearing conditions (10.17).

10.4.8 Illustrative Example. Small Natural Gas Network Equilibrium

10.4.8.1 Overview

In this section[28] we analyze the gas market equilibrium described in Sections 10.4.5 to 10.4.7 using the simple network from Chapter 1 (Illustrative Example 1.2.7) and consider several illustrative cases related to the effects of storage, total available gas, and contracted sales. The interested reader can consult the associated GAMS file found in Appendix B for the Base Case.

[28] We would like to thank Seksun Moryadee for his valuable analytical help on the numerical results as well as editorial suggestions in this chapter.

For all the results in this section, we have used a convergence tolerance of $1e{-}11$ and this setting as well as other ones are specified in the file "path.opt" included in the Base Case GAMS code in Appendix B.

In this chapter, the two producers (A,B) located at node 1 are now referred to, respectively, as 1 and 2 (for node 1). Likewise, the producers C and D at node 2 are now referred to, respectively, as 1 and 2 (at node 2) for overall consistency with the new, extended notation from this chapter. It is relevant to note that the original, simpler version of the gas market equilibrium (Illustrative Example 1.2.7) has a singular coefficient matrix. In particular, this coefficient matrix is as shown in Figure 10.2:

	s_1A	s_1B	s_2C	s_2D	q_1A	q_1B	q_2C	q_2D	f_12A	f_12B	lambda_1A	lambda_1B	lambda_2C	lambda_2D	g_12	eps_12	delta_1A	delta_1B	delta_2C	delta_2D	pi_1	pi_2	tau_12
1																	1				-1		
2											1						-1						
3																	1					-1	1
4					-1																		
5	1				-1				1														
6																		1			-1		
7												1						-1					
8																		1				-1	1
9						-1																	
10		1				-1				1													
11																			1			-1	
12													1						-1				
13							-1																
14			1				-1																
15																				1		-1	
16														1						-1			
17								-1															
18				1				-1															
19	1	1																			1		
20			1	1					1	1												2	
21																1							-1
22															-1								
23									-1	-1					1								

Fig. 10.2 Coefficient matrix from Illustrative Example 1.2.7

Notice that (column 1 for s_1A) + (column 10 for f_12B) = (column 2 for s_1B) + (column 9 for f_12A) so that these columns are linearly dependent and the whole matrix is therefore singular.[29] Methods like PATH (see Chapters 8 and 9) are able to handle such singularities but some care should be exercised when additionally fixing values in GAMS based on numerical experiments. In the exercises, the reader is asked to see if the coefficient matrix for the larger gas market equilibrium problem (from this chapter) is also singular and to investigate the solutions obtained with some numerical experiments.

To begin with, we first interpret these equilibrium conditions for the easy case of no storage, no contract sales, and original data from the Chapter 1 example. Just to reiterate, all cost functions are linear with the slope given by the symbol γ with appropriate superscripts and subscripts. Additionally, all demand functions are affine with the intercept given as a and the slope as

[29] Also rows 1+8 equal rows 3 + 6.

b. In particular, the following choices for exogenous data and functions were made to be consistent with Illustrative Example 1.2.7:

$$D^n_{st}\left(\pi^n_{st}\right) = a^n_{st} - b^n_{st}\pi^n_{st} \tag{10.21}$$

with $a^1_{st} = 20, a^2_{st} = 40, b^1_{st} = 1, b^2_{st} = 2$ for all s, t. Also, the following cost functions were chosen:

$$C^{n,prod}_{ist}\left(v^{n,prod}_{ist}\right) = \gamma^{n,prod}_{ist} \tag{10.22a}$$

$$C^{nm,flow}_{st}\left(v^{nm,flow}_{st}\right) = \gamma^{nm,flow}_{st} \tag{10.22b}$$

with $\gamma^{1,prod}_{1st} = 10$, $\gamma^{1,prod}_{2st} = 12$, $\gamma^{2,prod}_{1st} = 15$, $\gamma^{2,prod}_{2st} = 18$, $\gamma^{12,flow}_{st} = 1$ for all s, t. Also, consistent with Chapter 1's example, we take $\tau^{12,reg}_{st} = 0.5, \gamma^{12,flow}_{st} = 1$ for all s, t.

10.4.8.2 Base Case

Let's consider what happens at node $n = 1$ in terms of how the price at that node, π^n_{st}, is formed. For simplicity, let's assume that each of the two producers at that node (called A and B in Chapter 1, given respectively, as producers $i = 1, 2$ in this chapter) are producing positive amounts of gas ($q^{n,prod}_{ist} > 0, i = 1, 2$) so that by (10.19e), $v^{n,prod}_{ist} > 0$ as well. Since we assume no contracted sales and no storage activity, then by (10.19d) $\sum_{m\in outgoing(n)} Days_s s^{nm,spot}_{ist} > 0$ so that in particular for $n = 1$, $s^{11,spot}_{ist} > 0$ or $s^{12,spot}_{ist} > 0$ or both are positive. For simplicity let's assume that (at least) $s^{11,spot}_{ist} > 0$. Via (10.19c) the flow $f^{nn}_{ist} > 0$ for producer i at node $n = 1$. Then we have the following chain of logic:

$$q^{n,prod}_{ist} > 0 \Rightarrow \alpha 1^n_{ist} + \alpha 2^n_i = \alpha 4^n_{ist} + \alpha 5^n_{ist} \quad \text{by (10.18a)} \tag{10.23a}$$

$$v^{n,prod}_{ist} > 0 \Rightarrow \gamma^{n,prod}_{ist} = -\alpha 5^n_{ist} \quad \text{by (10.18b)} \tag{10.23b}$$

$$f^{nn}_{ist} > 0 \Rightarrow \left(\tau^{nn,reg}_{st} + \tau^{nn,congest}_{st}\right) = -\alpha 3^{nn}_{ist} \quad \text{by (10.18c)} \tag{10.23c}$$

$$s^{nn,spot}_{ist} > 0 \Rightarrow \pi^n_{st} = -\alpha 3^{nn}_{ist} + \alpha 4^n_{ist}, \quad \text{by (10.18e)} \tag{10.23d}$$

It is reasonable to assume that the flow within the region in the idealized model presented above has zero costs so that $\tau^{nn,reg}_{st} = 0$ for flow from node 1 to itself. Likewise, let's assume for simplicity that there are no congestion charges for transport within the node, i.e., $\tau^{nn,congest}_{st} = 0$ consistent with an infinite local capacity for transport. Then, the above shows that $\alpha 3^{nn}_{ist} = 0$ for $n = 1$ which in turn gives

$$\begin{aligned}\pi^1_{st} & \\ &= \alpha 4^1_{ist} \\ &= \alpha 1^1_{ist} + \alpha 2^1_i - \alpha 5^1_{ist} \\ &= \alpha 1^1_{ist} + \alpha 2^1_i + \gamma^{1,prod}_{ist}\end{aligned} \tag{10.24}$$

This relationship must hold for both producers $i = 1, 2$ or that

$$\pi^1_{st} = \alpha 1^1_{1st} + \alpha 2^1_1 + \gamma^{1,prod}_{1st} \tag{10.25a}$$

$$\pi^1_{st} = \alpha 1^1_{2st} + \alpha 2^1_2 + \gamma^{1,prod}_{2st} \tag{10.25b}$$

In words, these last two equations mean that the price at node $n = 1$ is determined by the marginal production costs for both producers $(\gamma^{1,prod}_{1st}, \gamma^{1,prod}_{2st})$ increased by the sum of the shadow price of the production capacity for the producer $(\alpha 1^1_{1st}, \alpha 1^1_{2st})$ and the shadow price of the producer's total available gas constraint $(\alpha 2^1_1, \alpha 2^1_2)$. If this latter constraint on the total available gas is not binding (i.e., not all the gas is produced out), then $\alpha 2^1_1 = \alpha 2^1_2 = 0$. In that instance, could both production amounts be under their limits? The answer is no from both a mathematical as well as economic perspective assuming different marginal costs of production.

Mathematically speaking if this situation happened, it would mean that $\pi^1_{st} = \gamma^{1,prod}_{1st} = \gamma^{1,prod}_{2st}$ but that $\gamma^{1,prod}_{1st} \neq \gamma^{1,prod}_{2st}$, a clear impossibility. From an economic perspective, assuming that $\gamma^{n,prod}_{1st} < \gamma^{n,prod}_{2st}$ (as is the case with this chapter's model), there would be no market incentive to start producing for the more expensive producer ($i = 2$) if the cheaper one ($i = 1$) hadn't reached its capacity.

Next, for ease of understanding, let's assume that the cheaper production (for producer $i = 1$) is at its capacity but the more expensive one is under its maximum level. Then, what happens in (10.25) is that the market price is the marginal production cost for the more expensive producer. Also, the shadow price $\alpha 1^1_{1st}$ is exactly the difference $(\gamma^{1,prod}_{2st} - \gamma^{1,prod}_{1st})$ reflecting the value of having more capacity with the cheaper producer. This relationship is confirmed in Table 10.1.

Now we consider how the price at node $n = 2$ is formed. Taking all the assumptions made so far but additionally assuming that $s^{12,spot}_{ist} > 0$ for $i = 1, 2$, let's see what results using similar logic to what was shown above.

$$f^{12}_{ist} > 0 \Rightarrow \left(\tau^{12,reg}_{st} + \tau^{12,congest}_{st}\right) = -\alpha 3^{12}_{ist} \tag{10.26a}$$

$$s^{12,spot}_{ist} > 0 \Rightarrow \pi^2_{st} = -\alpha 3^{12}_{ist} + \alpha 4^1_{ist} \tag{10.26b}$$

Taking into account (10.26) and the logic leading to (10.25) gives the following:

$$\pi^2_{st} \tag{10.27}$$

$$= \left(\tau_{st}^{12,reg} + \tau_{st}^{12,congest}\right) + \alpha 4_{ist}^{1}$$
$$= \left(\tau_{st}^{12,reg} + \tau_{st}^{12,congest}\right) + \alpha 1_{ist}^{1} + \alpha 2_{i}^{1} - \alpha 5_{ist}^{1}$$
$$= \left(\tau_{st}^{12,reg} + \tau_{st}^{12,congest}\right) + \alpha 1_{ist}^{1} + \alpha 2_{i}^{1} + \gamma_{ist}^{1,prod}$$
$$= \left(\tau_{st}^{12,reg} + \tau_{st}^{12,congest}\right) + \pi_{st}^{1}$$

showing that the price at node $n = 2$ is the price at node 1 plus the transportation charges $\left(\tau_{st}^{12,reg} + \tau_{st}^{12,congest}\right)$ to get the gas to node 2, consistent with economic theory. A relevant point about node 2 is that it can also get natural gas from producing it itself. Since $\gamma_{1st}^{1,prod} < \gamma_{2st}^{1,prod} < \gamma_{1st}^{2,prod} < \gamma_{2st}^{2,prod}$, how is the balance between imported and locally produced gas at node 2 determined? Assuming just producer $i = 1$ at node 2 (formerly producer C) is active, then reasoning similar to (10.24) shows that

$$\pi_{st}^{2} = \alpha 1_{1st}^{2} + \alpha 2_{1}^{2} + \gamma_{1st}^{2,prod} \tag{10.28}$$

or more simply

$$\pi_{st}^{2} = \gamma_{1st}^{2,prod} \tag{10.29}$$

if we assume that the production capacity and total available gas constraints are not binding. Combining (10.27) and (10.29) shows that

$$\gamma_{1st}^{2,prod} = \left(\tau_{st}^{12,reg} + \tau_{st}^{12,congest}\right) + \pi_{st}^{1} \tag{10.30}$$

or that the marginal production cost for local production (left-hand side) must equal the marginal delivered cost (right-hand side). This makes economic sense since if the delivered cost were greater than what could be produced locally, there would be no incentive for node 2 to import any gas from node 1. Note that both $\gamma_{1st}^{2,prod}$ and $\tau_{st}^{12,reg}$ are exogenous data so the model can endogenously adjust $\tau_{st}^{12,congest}$ and π_{st}^{1} to make the above equation hold. Since, $\gamma_{1st}^{2,prod} = 15$ and $\tau_{st}^{12,reg} = 0.5$, we see that $\tau_{st}^{12,congest} + \pi_{st}^{1}$ must equal $15 - 0.5 = 14.5$. This is the case since $\tau_{st}^{12,congest} = 2.5, \pi_{st}^{1} = 12$ as shown in Table 10.1 as well the fact that the difference in nodal prices is $3 = \tau_{st}^{12,reg} + \tau_{st}^{12,congest}$.

These values produce demand at nodes 1 and 2, respectively of 8 and 10 for all seasons and years. It is relevant to note that compared to the results from Chapter 1, there is still the same production level by producer (10,3,5, 0), the same total sales to node 1 (8), the same total exports from node 1 to 2 (5), and the same local sales at node 2 (5). What has changed is that the composition of these totals has shifted. In particular, from Chapter 1, of the total production of 10 by producer A, 7.439 went to local (spot) sales at node 1 and 2.561 was exported to node 2. In the Base Case for this chapter, the respective numbers changed to 5.000 and 0.000. A similar alternative

Table 10.1 Solution for base case-main variables (same values for all s, t)

Variable	Value (Chapter 10) (Base Case)
$v_{1st}^{1,prod}$	10.000
$v_{2st}^{1,prod}$	3.000
$v_{1st}^{2,prod}$	5.000
$v_{2st}^{2,prod}$	0.000
$Days_s f_{1st}^{12}$	5.000
$Days_s f_{2st}^{12}$	0.000
$Days_s s_{1st}^{11,spot}$	5.000
$Days_s s_{2st}^{11,spot}$	3.000
$Days_s s_{1st}^{12,spot}$	5.000
$Days_s s_{2st}^{12,spot}$	0.000
$Days_s s_{1st}^{22,spot}$	5.000
$Days_s s_{2st}^{22,spot}$	0.000
π_{st}^{1}	12.000
π_{st}^{2}	15.000
$\tau_{st}^{12,congest}$	2.500
$\alpha 1_{1st}^{1}$	2.000
$\alpha 1_{2st}^{1}$	0.000
$\alpha 1_{1st}^{2}$	0.000
$\alpha 1_{1st}^{2}$	0.000

division of production occurs for producer B, namely, previously 0.561 and 2.439 for local sales and exports, respectively, changed to 3.000 and 0.000, but still matching the total production of 3.

Why is this happening? First, recall that complementarity problems generalize nonlinear programs and nonlinear equations (and other problems) so that having more than one solution is not unexpected for certain applications. In this particular problem, there are an infinite number of solutions since the model does not distinguish between local and exported sales as long as the total net price received is the same. But $\pi_{st}^1 = 12$ if sold locally and $\pi_{st}^2 = 15$ less the transportation charges of $3 = \tau_{st}^{12,reg} + \tau_{st}^{12,congest}$ so the net prices are the same. In the exercises, the reader is asked to explain further the question of solution uniqueness for some of the variables.

Lastly, we turn to the calculations behind the congestion price. From (10.27) we know that the difference in the nodal prices is the sum of the regulated tariff and the congestion tariff, i.e., $\tau_{st}^{12,reg} + \tau_{st}^{12,congest}$. Since both these prices and the congestion price are calculated endogenously at the same time, one might ask how is the value of $\tau_{st}^{12,congest} = 2.5$ determined in this Base Case? Taking an economic perspective, the demand at node 2 can be served either locally at node 2 or exported from node 1. At node 1, the marginal supplier is producer B (since producer A is already at capacity).

The marginal cost of production for B is 12. To transport the gas to node 2, it has to be no more expensive than the marginal supply cost at node 2 which is 15 relating to the marginal production for producer C. Thus, $\tau_{st}^{12,congest} = 15 - 0.5 - 12 = 2.5$. This reasoning is confirmed by slightly adjusting the demand at node 2 by increasing the intercept from 40 to 40.1 in season 1, time 1. Since this is still less than the marginal production cost for D (18), what happens is that node 2's price in season 1, time 1 increases slightly to 15.050 and $\tau_{st}^{12,congest}$ rises to 2.55. This occurs since the inverse demand function at node 2 is given by

$$\begin{aligned}
&\pi_{11}^2 \qquad (10.31)\\
&= -\frac{1}{2}\,(\text{demand}) + \frac{1}{2}\,(40.1)\\
&= -\frac{1}{2}\,(10) + \frac{1}{2}\,(40.1)\\
&= 15.05
\end{aligned}$$

However, if the intercept changes to 47, the new price (still assuming a demand of 10 at node 2) would be $\pi_{11}^2 = 18.5$. However, at a price of 18 the previously inactive producer D (at node 2) becomes active since his marginal production cost is 18. Instead, what is picked is a demand of 11 resulting in $\pi_{11}^2 = 18$ and the production for A,B, and C unchanged but D producing one unit of gas. Under this scenario, $\tau_{st}^{12,congest}$ rises to 5.5 reflecting $18 - 12 - 0.5$ or backcalculating what production from node 1 to node 2 could charge at most to match the new node 2 price of 18.

10.4.8.3 Analysis of Storage

Unlike electric power, natural gas can be stored in large quantities for use at a later time. One of the main advantages of such an approach is to effectively increase the production level beyond its regular capacity. That is, some of the production in a low demand period (e.g., $s = 2$ in our example) is injected into storage rather than using it for meeting demand in that period. Then, in the high demand season ($s = 1$) this extra gas is extracted and available to the market. In effect, storage allows for seasonal arbitrage. To see the effect of storage, we ran the GAMS program for the Base Case but adjusted the following storage parameters:

- marginal storage extraction and injection costs changed to 0.5 (from 0 in the Base Case– since storage was made inactive)
- maximum extraction and injection rates set at 10 (from 0 in the Base Case)
- a 5% loss factor included for extracting the gas back from storage
- maximum volume of storage value still set at 1000 (to make it non-binding)
- demand intercept at node 1, season 1, time 1 changed from 20 to 35

This last point is especially important since the reason to store gas would be to handle higher demand. Also, it is important to note that chronologically season $s = 2$ when gas is injected would have to occur before season $s = 1$ when it is extracted. We have not focused on the chronological logistics of storage though. With these various adjustments made to the Base Case, the following key changes occur at node 1, season 1, time 1:

- demand increases from 8 to 21.342
- price rises from 12 to 13.658
- $\tau_{st}^{12,congest}$ decreases from 2.5 to 0.842

Also, producer B increases its production $v_{2s1}^{1,prod}$ from 3 in both seasons to 10 and 9.676, respectively for $s = 1, 2$. Lastly, producer B extracts and injects, respectively 6.342 and 6.676 in time 1. Let's examine each of these changes. The demand rises because the intercept has been shifted up by 15. It is interesting to note that the increase in the demand is only 13.342 though since the price has also increased slightly (by 1.658) depressing demand slightly. The congestion price backcalculated in the manner described above (also referred to as "netback") is $15 - 0.5 - 13.658 = 0.842$. Producer B then picks up the extra demand by increasing its production as indicated. Note that the amount of extraction is equal to the injection less losses or $6.676\,(1 - 0.05) = 6.342$. Also important to observe are what happens with the production from A and B at nodes 1 and 2. First, there is a demand of 21.342 in season 1, time 1. Given that both producers A and B can only each produce at most 10, extraction from storage is necessary to meet this elevated level of demand. The way that is chosen is to have A produce 5 and sell to the node 1 spot market (an additional 5 goes to node 2 which maxes out the capacity bound on the pipeline). Added to this 5 is B's production of 10 (B does not export to node 2 in season 1, time 1). This only gets the supply to 15 and so storage extraction is needed for the remaining $21.342 - 15 = 6.342$. Also, the composition of who sales what at node 1, season 2, time 1 changes. For example, in the Base Case producer A sold 5 locally and producer B sold 3. This changed to 8 and 0, respectively in this Storage Case. However, this is further manifestation of the multiple solutions for sales and less interesting since the demand is still 8 as in the Base Case. Lastly, the supply composition at node 2 doesn't change since there are still five units coming from node 1 (different amounts by producers A and B) and 5 produced locally (by C). Combined with no change in demand at node 2 means that the price at node 2 is unchanged from the Base Case.

10.4.8.4 Analysis of Total Gas Reserves Constraint

The only constraints in the gas market equilibrium model that span the different time periods $t1, t2$ are (10.4). Just to recall, these constraints limit the total amount of gas that is extracted over the time horizon. In this section

Table 10.2 New production for producer A under time horizon case 1

Season\Time	$t = 1$	$t = 2$
$s = 1$	3	7
$s = 2$	10	10

Table 10.3 New production for producer B under time horizon case 1

Season\Time	$t = 1$	$t = 2$
$s = 1$	10	6
$s = 2$	3	3

Table 10.4 Division of local use (spot sales to node 1) and exports (spot sales to node 2) for time horizon case 1

(Season,Time)\ Producer	A	B	A	B
	spot sales to node 1	spot sales to node 1	spot sales to node 2	spot sales to node 2
$(s = 1, t = 1)$	0	8	3	2
$(s = 2, t = 1)$	5	3	5	0
$(s = 1, t = 2)$	2	6	5	0
$(s = 2, t = 2)$	5	3	5	0

we analyze the effect of these constraints by comparison with the Base Case. Recall that in the Base Case, the production for producer A in each season, time combination was 10 or a total production of 40 for the time horizon. Thus, we make the first change to limit producer A's total volume to be 30 to see what changes. The demand, nodal prices, and congestion prices are all unchanged from the Base Case. What is new is that producer A cuts back on its production to satisfy the total of 30 being maintained as shown in Table 10.2.

Since the demand is still 8 at node 1 for both time periods, producer B has to make up the difference given that no storage extraction is allowed in this case. The resulting production for producer B is indicated in Table 10.3. We see that compared to the Base Case, producer B has increased its production from 3 to 10 and 6 in season 1, time periods 1 and 2, respectively so that total production (by producers A and B) at node 1 is 13 for all seasons and time periods. Since the demand is only 8 the extra 5 units are exported to node 2 (recall that the pipeline capacity is 5). The division by producer between local use (spot sales to node 1) and export spot sales to node 2) is given in Table 10.4.

The second case for analyzing the time horizon constraint is a further reduction in the total available gas– this time for producer B. From Table 10.3 we see that production for B was 22 over all seasons and time periods. Consequently, in addition to the restriction of 30 for producer A we make B's new total available amount of gas to be 20. The new results from this second case are shown in Tables 10.5, 10.6, and 10.7. Compared to Time Horizon Case 1, Producer A shifts its production in season 1 from 3 down to 2.5 in

Table 10.5 New production for producer A under time horizon case 2

Season\Time	$t = 1$	$t = 2$
$s = 1$	2.5	7.5
$s = 2$	10	10

Table 10.6 New production for producer B under time horizon case 2

Season\Time	$t = 1$	$t = 2$
$s = 1$	10	5
$s = 2$	2.5	2.5

Table 10.7 Division of local use and exports for time horizon case 2

(Season,Time)\Producer	A	B	A	B
	spot sales to node 1	spot sales to node 1	spot sales to node 2	spot sales to node 2
$(s = 1, t = 1)$	0	7.5	2.5	2.5
$(s = 2, t = 1)$	5	2.5	5	0
$(s = 1, t = 2)$	2.5	5	5	0
$(s = 2, t = 2)$	5	2.5	5	0

time 1 and up from 7 to 7.5 in time 2 but the total is still at the limit of 30. Producer B shifts down production in season 1, time 2 to 5 (from 6) and season 2 both years to 2.5 (from 3) to be at the limit of 20. Notice from Table 10.7 that there is some shifting of local use and export in this case but the end result is that at node 1, only 7.5 units of gas are available in each season and time period and 5 units of export are available at node 2.[30] Clearly this doesn't meet the previous demand of 8 at node 1 so demand shifts down to 7.5 causing the price at node 1 to go up to 12.5 (for all seasons and time periods) determined via the inverse demand function of 20−demand= $20 - 7.5 = 12.5$. Another interesting result is that now the price differential between the two nodes is $15 - 12.5 = 2.5$ so that $\tau_{st}^{12,reg} + \tau_{st}^{12,congest}$ can be 2.5 or that $\tau_{st}^{12,congest}$ drops to 2.0. Lastly, the shadow price on the constraint related to total gas, namely, $\alpha 2_i^1$ changes from 0 in the Base Case to 2.5 and 0.5, respectively, for producers A and B, reflecting the increased value of more gas if it were available.

10.4.8.5 Analysis of Contract Sales

In all the cases analyzed so far, only the spot sales were considered based on the market-clearing conditions (10.14) as well as the other parts of the equilibrium problem. In reality, a good deal of natural gas is also sold via pre-existing contracts so this case analyzes that part of the model. As com-

[30] In comparing cases when more than one solution is possible, care must be taken to attribute the differences to the case specifics alone.

Table 10.8 Exogenous contract prices for producer A at node 1

(Season,Time)\Producer	A
	Local Use
$(s=1,t=1)$	12.1
$(s=2,t=1)$	12.1
$(s=1,t=2)$	12.1
$(s=2,t=2)$	12.1

pared to the Base Case, the Contract Case allow for non-zero contract prices $P_{ist}^{nm,contract}$. In particular, as shown in Table 10.8, this exogenous contract price is now set just for producer A's local sales at node 1 (all other contract prices for all producers stay at 0):[31]

Under this setting, compared to the Base Case, the demand at node 1 increases from 8 to 10 (both seasons and time periods) with a nodal price decrease from 12 to 10.[32] One economic line of reasoning for this shift would be as follows. First, producer A, seeing the high contract prices, of 12.1 decides to produce at its maximum of 10 and sell all its production as contracted sales given its low marginal production cost of \$10.(In the Base Case producer A only sold to the spot market, 5 at node 1 and 5 at node 2). This results in a spot price of $20-(1)(10)=\$10$ at node 1 which is unfavorable to producer B who has a marginal cost of 12. As a result, producer B produces 5 and exports all of it to node 2 where a spot price of 15 is available. (In the Base Case, producer B sold all of its production of 3 to the node 1 spot market). There is also the possibility of multiple solutions so one must be careful about drawing conclusions and the reader is asked to consider uniqueness of solutions to this model in the exercises.

10.5 Summary

In this chapter we have presented a gas market equilibrium model. The players include two suppliers at each of the two nodes as well as a network operator who manages the pipeline from node 1 to node 2. In this model the suppliers take care of production as well as storage and gas exports (node 1 to node 2 only) and operate in perfect competition between each other at a given node. Besides the KKT conditions for these four suppliers and

[31] For a treatment of endogenously determined contract prices, see the generalized Nash gas market model described in [2].

[32] In the presence of positive contract and spot sales at node 1, nodal price π_{st}^{1} obtained from the market-clearing conditions ((10.8) or (10.14)) needs to be re-interpreted somewhat since it would not just be for spot sales. An alternative approach would be to treat the spot and contract markets separately with different demand curves.

the network operator, there are two sets of market-clearing conditions: one for the spot market at each node in each season and time period, one for the pipeline capacity. Several cases were run to highlight the importance of storage, contract sales, and the time horizon total gas constraints. Extensions to this model could include: oligopolistic competition between the suppliers, more general production cost functions (e.g., Golombek), separate players who handle storage, exports, liquefaction, regasification, separate residential sectors with potentially different demand functions, or two-level, Stackelberg games.

10.6 Exercises

1. Consider the Golombek cost function (10.2). Is it a convex function of the volume of gas v? If yes, prove this. If not, show a specific counter-example.
2. Use the GAMS code for the Base Case provided in Appendix B and modify the market equilibrium model as follows. Eliminate the market-clearing constraints (10.14) and replace them with an inverse demand function for the two suppliers at node 1 in effect making them Nash-Cournot players. You should come up with a demand function that matches the Base Case output. Then, compare your new model's output with the market equilibrium from this chapter using the data from the Storage Case. Report on the differences between your new model and what is presented for the Storage Case in this chapter.
3. Use the GAMS code for the Base Case provided in Appendix B and modify the market equilibrium model as follows. Use the Golombek cost function instead of the linear ones for production at node 1 (for both producers A and B). You should come up with parameters for the Golombek cost function that match the Base Case output. Then, compare your new model's output with the market equilibrium from this chapter using the data from the Storage Case. Report on the differences.
4. Consider the backpressure equation (10.9). Show if it is convex in p_R. You may assume that $C > 0$ and $n = 0.5$. If not convex, give a particular counter-example.
5. Consider the backpressure equation (10.9). Come up with a method to linearize it and add constraints (or a single constraint) representing this approximation into the GAMS code for the Base Case. Note: as part of this approximation you will have to come up with a relationship between p_R and volume extracted taking into account that the more that is extracted, the lower p_R will be. You may find [30] useful in this approximation exercise. Compare the results between the Base Case from this chapter and your new one.

6. Modify the Base Case GAMS code so that $\tau_{st}^{12,reg}$ are 10 different selected values (your choice) in [0,20]. How does the amount of gas exported from node 1 to node 2 change as a function of this tariff. Explain your results using economic theory as well as mathematically using the market equilibrium conditions.
7. Consider the Base Case market equilibrium model. Which endogenous variables will be unique if any? Prove your results or show a counter-example to uniqueness. Can you make minimal restrictions on the exogenous factors to ensure uniqueness of some of the endogenous variables? Explain your answer.
8. Modify the Base Case GAMS model and code so that there is a "supply security" constraint. That is, having no more than x % of total supply to node 2 coming from each node 1 producer. Let x vary in the range [0, 1] with $x = 1$ corresponding to no restriction. Analyze the export pattern by producers A and B to node 2 under differing values of x and also determine a reasonable value for the cost of supply security.
9. Consider the Base Case market equilibrium model. Reformulate this model assuming that the demand equations are of the form: $D_{st}^{n}(\pi_{st}^{n}) = a_{st}^{n} - b_{st}^{n}\pi_{st}^{n}$ where $a_{st}^{n}, b_{st}^{n} > 0$ are discrete random variables. Each of the players should optimize an appropriate expected objective function value derived from the original one stated above. Compare your answer with the Base Case.
10. Consider the coefficient matrix for the gas market equilibrium problem in this chapter. Determine if it is singular and perform numerical tests to see if: a. starting from different points if GAMS is able to produce a solution b. if using the ".fx" suffix for certain variables before solving will produce a valid solution. Report on your results.

References

1. I. Abada. *Modélisation des Marchés du Gaz Naturel en Europe en Concurrence Oligopolistique. Le Modèle GaMMES et Quelques Applications.* Ph.D. thesis, Université Paris Ouest, Nanterre–La Défense, 2012.
2. I. Abada, V. Briat, S.A. Gabriel, and O. Massol. A Generalized Nash-Cournot model for the European natural gas markets with a fuel substitution demand function: the GaMMES model. *Networks and Spatial Economics*, accepted, 2012.
3. J. André. *Optimization of Investments in Gas Networks*, Ph.D. Thesis, Université Lille Nord de France, 2010.
4. M.G. Boots, B.F. Hobbs, F.A.M. Rijkers. Trading in the downstream European gas market: a successive oligopoly approach. *The Energy Journal*, 25(3):73-102, 2004.
5. S.P.A. Brown and M.K. Yücel. Market arbitrage: European and North American natural gas prices, World natural gas markets and trade: A multi-modeling perspective. A Special Issue of *The Energy Journal*, 167-185, 2009.
6. U.C. Colpier and D. Cornland. The economics of the combined cycle gas turbine–an experience curve analysis. *Energy Policy*, 30(4):309-316, 2002.
7. C.-K. Chyong and B.H. Hobbs. Strategic Eurasian natural gas model for energy Security and Policy Analysis. EPRG Working Paper 1115, Cambridge Working Paper in Economics 1134, 2011.
8. D. de Wolf and Y. Smeers. Optimal dimensioning of pipe networks with application to gas transmission networks. *Operations Research,* 44(4):596-608, 1996.
9. D. de Wolf and Y. Smeers. The gas transmission problem solved by an extension of the simplex algorithm. *Management Science,* 46(11):1454-1465, 2000.
10. C. Dieckhöner. Simulating security of supply effects of the Nabucco and South Stream projects for the European natural gas market. *The Energy Journal*, 33(3):155-183, 2012.
11. M. Dietsch. The next global energy cartel, www.forbes.com, April 2012.
12. Department of Energy, *Natural Gas Vehicle Emissions*, www.afdc.energy.gov, April 2012.
13. R. Egging and S.A. Gabriel. Examining market power in the European natural gas market. *Energy Policy,* 34(17):2762-2778, 2006.
14. R. Egging, S.A. Gabriel, F. Holz, J. Zhuang J. A Complementarity model for the European natural gas market. *Energy Policy,* 36:2385-2414, 2008.
15. U.S. Energy Information Administration. World shale gas resources: an initial assessment of 14 regions Outside the United States, April 2011.
16. U.S. Energy Information Administration, The Basics of Underground Natural Gas Storage, www.eia.gov, April 2012.
17. Energy Information Administration, www.eia.gov, April 2012.
18. Direct Emissions from Stationary Combustion Sources, www.epa.gov, April 2012.

19. Natural Gas Extraction–Hydraulic Fracturing, www.epa.gov/hydraulicfracture, April, 2012.
20. H. Fountain. Ohio: sites of two earthquakes nearly identical. *The New York Times*, January 3, 2012.
21. S.A. Gabriel, S. Kiet, and J. Zhuang J. A mixed complementarity-based equilibrium model of natural gas markets. *Operations Research,* 53(5):799-818, 2005.
22. S.A. Gabriel, A.S. Kydes, and P. Whitman. The National Energy Modeling System: a large-scale energy-economic equilibrium model. *Operations Research*, 49(1):14-25, 2001.
23. S.A. Gabriel and F.U. Leuthold. Solving discretely-constrained MPEC problems with applications in electric power markets. *Energy Economics*, 32(1):3-14, 2010.
24. S.A. Gabriel, J. Manik, and S. Vikas. Computational experience with a large-scale, multi-period, spatial equilibrium model of the North American natural gas system. *Networks and Spatial Economics*, 3(2):97-122, 2003.
25. S.A. Gabriel, K.E. Rosendahl, R. Egging, H. Avetisyan, S. Siddiqui. Cartelization in gas markets: studying the potential for a "gas OPEC". *Energy Economics*, 34(1):137-152, 2012.
26. S.A. Gabriel and Y. Smeers. Complemenatarity problems in restructured natural gas markets. *Recent Advances in Optimization. Lecture Notes in Economics and Mathematical Systems*, Edited by A. Seeger, Vol. 563, Springer Berlin Heidelberg, 343-373, 2006.
27. S.A. Gabriel, S. Vikas, and D. Ribar. Measuring the influence of Canadian carbon stabilization programs on natural gas exports to the United States via a "bottom-up" intertemporal spatial price equilibrium model. *Energy Economics*, 22(5):497-525, 2000.
28. S.A. Gabriel, J. Zhuang, S. Kiet. A large-scale linear complementarity model of the North American natural gas market. *Energy Economics*, 27(4):639-665, 2005.
29. S.A. Gabriel, J. Zhuang, R. Egging. Solving stochastic complementarity problems in energy market modeling using scenario reduction. *European Journal of Operational Research,* 197(3):1028-1040, 2009.
30. M. Golan and C.H. Whitson. *Well Performance*, Prentice-Hall, New Jersey, 2nd edition, 1991.
31. R. Gold. Natural gas tilts at windmills in power feud. *The Wall Street Journal*, March 20, 2010.
32. R. Golombek , E. Gjelsvik, K.E. Rosendahl. Increased competition on the supply side of the western European natural gas market. *The Energy Journal*, 19(3):1-18, 1998.
33. J.-M. Guldmann. Supply, storage, and service reliability decisions by gas distribution utilities: a chance-constrained approach. *Management Science,* 29(8):884-906, 1983.
34. J.-M. Guldmann. A marginal-cost pricing model for gas distribution utilities. *Operations Research*, 34(6):851-863, 1986.
35. P.R. Hartley and K.A. Medlock III. Potential futures for Russian natural gas exports, World natural gas markets and trade: A multi-modeling perspective. A Special Issue of *The Energy Journal*, 73-95, 2009.
36. C.A. Haverly. Studies of the behavior of recursion for the pooling problem. *SIGMAP Bulletin*, 25:19-28, 1978.
37. C.A. Haverly. Behavior of recursion model– more studies. *SIGMAP Bulletin*, 26:22-28, 1979.
38. F. Holz, C. von Hirschhausen, C. Kemfert. A strategic model of European gas supply (GASMOD). *Energy Economics*, 30:766-788, 2008.
39. H.G. Huntington. Natural gas across country borders: an introduction and overview, World natural gas markets and trade: A multi-modeling perspective. A Special Issue of *The Energy Journal*, 1-8, 2009.
40. J.L. Kennedy. *Fundamentals of Drilling Technology and Economics*, PennWell Publishing Company, Tulsa, Oklahoma, 1983.
41. A. Kramer. Gazprom builds wealth for itself, but anxiety for others. *The New York Times*, January 13, 2006.

42. M. Landler. Gas halt may produce big ripples in European policy. *The New York Times*, January 4, 2006.
43. X. Li, E. Armagan, A. Tomasgard, and P.I. Barton. Stochastic pooling problem for natural gas production network design and operation under uncertainty. *American Institute of Chemical Engineers Journal*, 57(8):2120-2135, 2010.
44. X. Li, A. Tomasgard, P.I. Barton. Decomposition strategy for the stochastic pooling Problem, *Journal of Global Optimization*, Springer Netherlands, in press.
45. W. Lise and B.F. Hobbs. Future evolution of the liberalised European gas market: simulation results with a dynamic model. *Energy,* 33(7):989-1004, 2008.
46. S. Lochner. Identification of congestion and valuation of transport infrastructures in the European natural gas market. *Energy*, 36(5):2483-2492, 2011.
47. S. Lochner and D. Bothe D. From Russia with gas– an analysis of the Nord Stream pipeline's impact on the European gas transmission system with the TIGER model. EWI Working Paper 07/2, Institute of Energy Economics at the University of Cologne, Germany, 2007.
48. S. Lochner and D. Bothe D. The development of natural gas supply costs to Europe, the United States and Japan in a globalizing gas market–model-based analysis until 2030. *Energy Policy*, 37(4):1518-1528, 2009.
49. S. Lochner and J. Richter. The impact of recent gas market development on long-term projections for global gas supply. *Zeitschrift für Energiewirtschaft*, 34(1):61-69, 2010.
50. K. T. Midthun. *Optimization Models for Liberalized Natural Gas Markets*, Ph.D. Thesis, Norwegian University of Science and Technology, Trondheim, Norway, 2007.
51. K.T. Midthun, M. Bjørndal, and A. Tomasgard. Modeling optimal economic dispatch and system effects in natural gas networks. *The Energy Journal*, 30(4):155-180, 2009.
52. MIT Energy Initiatiave. *The Future of Natural Gas An Interdisciplinary MIT Study*, Interim Report, ISBN (978-0-9828008-0-5), 2010.
53. Natural Gas, www.naturalgas.org, April 2012.
54. A. Neumann. Linking natural gas markets – is LNG doing its job?, World natural gas markets and trade: A multi-modeling perspective. A Special Issue of *The Energy Journal*, 187-199, 2009.
55. M. Nijboer. The contribution of natural gas vehicles to sustainable transport. *International Energy Agency,* 2010.
56. R.P. O'Neill, M. Williard, B. Wilkins, and R. Pike. A mathematical programming model for allocation of natural gas. *Operations Research* 27(5):857-873, 1979.
57. J. Perner and A. Seeliger. Prospects of gas supplies to the European market until 2030– results from the simulation model EUGAS. *Utilities Policy*, 12(4):291-302, 2004.
58. J.C. Ramírez and J. Rosellón. Pricing natural gas distribution in Mexico. *Energy Economics*, 24(3):231-248, 2002.
59. Y. Smeers. Gas models and three difficult objectives. CORE, 2008.
60. J.P. Stern. *The Future of Russian Gas and Gazprom*, Oxford University Press, Oxford, 2005.
61. F. J. Sturm. *Trading Natural Gas Cash Futures Options and Swaps*, PennWell Publishing Company, Tulsa, Oklahoma, 1997.
62. I. Urbina. A tainted water well, and concern there may be more. *The New York Times*, August 3, 2012.
63. I. Urbina. Hunt for gas hits fragile soil, and South Africans fear risks. *The New York Times*, December 30, 2011.
64. I. Urbina. New report by agency lowers estimates of natural gas in U.S. *The New York Times*, January 28, 2012.
65. T. van der Hoeven. *Math in Gas and the Art of Linearization*, Energy Delta Institute, International Business School & Research Centre for Natural Gas, The Netherlands, 2004.
66. D.G. Victor, A.M. Jaffe, and M.H. Hayes. *Natural Gas and Geopolitics From 1970 to 2040*, Cambridge University Press, Cambridge, 2006.

67. J. Zhuang. *A Stochastic Equilibrium Model for the North American Natural Gas Market*, Ph.D. Thesis. Department of Civil & Environmental Engineering, University of Maryland, College Park, MD, USA, 2005.
68. J. Zhuang and S.A. Gabriel. A complementarity model for solving stochastic natural gas market equilibria. *Energy Economics,* 30(1):113-147, 2008.

Chapter 11
Electricity and Environmental Markets

11.1 Introduction

The purpose of this chapter is to provide a more in-depth exploration of applications of complementarity models to electricity markets. In doing so, we introduce two crucial features of energy markets. The first is transportation networks with capacity limits on links between different markets. The second is environmental restrictions, such as emissions markets. We address these in turn by building, analyzing, and solving models for electric power markets that incorporate these features.

Electric power differs from other commodities; its transport is governed by Kirchhoff's laws, which define an "unvalved" network in which power flow takes all paths between a source and sink. In contrast, water or gas networks have valves that allow flow to be directed. We consider how the effects of these laws on market outcomes can be modeled, and how they can affect the competitiveness of markets with oligopolistic producers. A power market with hundreds of power producers can still be vulnerable to the exercise of market power if some producers in some locations are shielded from competition by high transmission costs from other locations. The concept of a Nash-Cournot equilibrium is used to model oligopolistic competition on a network.

Meanwhile, the evolution of the electric power sector over the next few years and decades will be greatly shaped by increasingly stringent environmental demands. These take a variety of forms in the case of the electric utility industry. Examples include requirements for investment in renewable energy and cap-and-trade emissions markets in which power producers need to hold sufficient emissions permits (measured in tons) to cover their emissions of pollution. Those permits are tradable, and can be bought or sold by producers who are short or long or by independent speculators.

In presenting models of transmission and environmentally constrained power markets, we follow the following five general steps for building and analyzing complementarity models of markets:

1. *Agent Model Formulation.* Each agent (or player or participant) in the market is modeled as either an optimizer or someone who makes decisions following a rule. For the optimizers, this step consists of defining the optimization problem that an agent solves. For producers, this will be a profit-maximization problem, as it will be for the transmission system operator (TSO). For consumers, they could be modeled as maximizing their consumer surplus (integral of demand curves minus expenditures on energy); however, here we will assume that consumer behavior is defined by a demand function $q = f_d(p)$ or its inverse $p = f_d^{-1}(q)$. We note that electric power planning and market models have traditionally assumed that power demands are fixed constants. However, as we move towards "Smart Grids", consumers will assume their rightful place as equal partners in the market, modifying or shifting loads in order to manage costs and improve reliability [28]. Including price-responsive demand functions is one way to model such short-term demand response (see Section 12.3 for more on this topic).
2. *Agent First-Order Conditions.* In this step, the first-order conditions are defined for the optimization problem for each optimizing agent. For those agents whose primal decisions are instead modeled as following some (perhaps suboptimal) rules, functions embodying those rules (for instance, demand functions for consumers) are defined.
3. *Equilibrium Problem Formulation.* As explained in Section 3.4.1, above, a market equilibrium problem is created by concatenating the conditions and functions that define agent behavior from Step 2 together with supply-demand balances (i.e., market-clearing conditions, one per market), one per market. Each location in a spatial market is an individual market with its own clearing condition. If the model is also dynamic, then each location at each time considered is a separate market. Additional conditions that affect price formation or primal decisions by agents can also be defined; for example, a government agency might impose a condition that price for a particular market or service is based on some sort of average cost (perhaps including sunk costs) rather than marginal cost [5,31]. If correctly formulated, this step should result in a square system, in which the number of unknowns (primal and dual variables and market prices) equals the number of conditions. The resulting system, if it includes first-order conditions for inequality constrained problems, will be a complementarity problem.
4. *Analysis.* The model is analyzed mathematically for its properties, such as uniqueness and existence or its relationship to other models. General interpretations are made of the economic significance of the equilibrium conditions.
5. *Parameterization and Solution.* Parameter values are estimated for the model, which is then solved numerically with an algorithm appropriate to the structure of the problem.

We illustrate each of the steps of the modeling process with the models of this chapter. In Section 11.2, we apply these steps to the modeling of transmission constrained electricity markets, considering both the cases of perfect competition and Cournot oligopoly. Section 11.3 concerns environmental markets, where we contrast two different emissions trading systems using models built by the above step-by-step process.

11.2 Transmission-Constrained Electricity Markets

Electricity is an example of a commodity with spatially separated but connected markets. A producer at one location can sell to local customers, or can pay the transmission grid to have the power transmitted to customers at another location. Like other commodities with spatially distributed markets and transportation costs, the result is different prices in different locations.

A glance at the pattern of production of electricity in the US shows that the combined effect of heterogeneous distribution of fuel sources, the expense of fuel transportation, and transmission constraints results in a great deal of differentiation among the country's regions. In terms of types of fuels used, air pollution emission rates, and costs of electric power, the regions differ dramatically. The coal-rich southeast US has low power costs, while the northeastern states of New York and New England rely mainly on nuclear and natural gas fired plants, and pay significantly higher prices.

Furthermore, transmission capacity bottlenecks and costs can isolate local producers from competition, increasing their ability to exercise market power by raising prices above their marginal cost of production. Disregarding transmission can lead to erroneous conclusions about the competitiveness of markets, which California learned to its sorrow during its power crisis of 2000-2001.

Consequently, analyses of power markets, and energy markets in general, need to consider fuel and electricity transportation costs and the spatially distributed nature of power production and consumption in order to realistically characterize the impacts of policy changes upon the types of fuels used, the costs of power production, the ability to exercise market power, and environmental impacts. In this section, we focus on modeling transmission constraints and their effects on power production. First, we provide a simple example to illustrate the effect of transmission constraints under perfect competition, taking advantage of a linearized model of the transmission system, followed by a three-bus numerical example (Sections 11.2.1 and 11.2.2). Then we develop a model of an oligopolistic market in which Cournot producers compete to sell bilaterally to consumers and a grid operator provides transmission and arbitrage services (Section 11.2.3), followed by application to the same three bus system (Section 11.2.4).[1]

11.2.1 Short-Run, Perfectly Competitive Market

This model assumes that individual generating firms f can own and operate power plants in several network nodes i simultaneously, and sell power at those nodes to the market operator. However, in this short-run model, we do not consider investment. In fact, our simplified model considers just one hour, although a generalization to multiple hours is straightforward. A model node may be a single "bus" (an electrical engineer's term for an individual substation), or it can be a zone that consists of an aggregate of nearby buses. The TSO conveys power from the generators to consumers. In essence, the TSO acts as a market arbitrageur, purchasing power at nodes with excess supply and sell it where there is deficit. (In Section 11.2.3, we consider the case where the generator pays for transmission services in order to sell power itself in other markets, rather than allowing the TSO to resell the power.)

Another assumption of the model is that power flows on the TSO's network can be adequately approximated by a linearized direct current "dc" load flow, in which linear equations constrain those flows to conform to the "dc" versions of Kirchhoff's current and voltage laws [28]. The dc model is an increasingly common representation of transmission flows in power market models. Complementarity models can also be formulated using full nonlinear alternating current (ac) load flow models, but they are non-convex and more difficult to solve.

We will begin with Step 1, construction of optimization or rule models for each market player. However, first we introduce our notation:

Generating firm. We start by considering a firm which is a price-taker, rather than an exerciser of market power. As a result, it views price at each and every node as an unchangeable parameter in its objective function, even though from the market's point of view those prices are variable and are adjusted to balance supply and demand at each node. The generating company's optimization problem is given below.

$$\text{Maximize}_{x_{fi}} \quad \sum_i p_i x_{fi} - \sum_i C_{fi}(x_{fi}) \tag{11.1a}$$

s. t.

$$x_{fi} - K_{fi} \leq 0 \quad (\mu_{fi}) \quad \forall i \in I_f \tag{11.1b}$$

$$x_{fi} \geq 0 \tag{11.1c}$$

[1] For in-depth summaries of the literature on electric power market models, see [35, 39].

I_f is the set of nodes i at which firm f owns generation. The decision variable x_{fi} represents firm f's generation at node i, in MW, and $C_{fi}(x_{fi})$ and K_{fi} are the cost function and capacity, respectively, at i. The shadow price of the capacity constraint is μ_{fi}.

For simplicity, we disregard the possibility of more than one plant at each node, but such a generalization is easy to implement. We also disregard unit commitment constraints and costs, such as start-up and minimum load expenses and ramp rate limits. The latter are readily modeled by defining multiple periods in chronologic order, and adding constraints that limit the change in a generator's output from period to period. Start-up and minimum load costs in theory require the use of binary 0-1 variables to represent the decision whether or not to commit a unit and keep it on-line, which cannot be used in a complementarity equilibrium model. However, if the binary constraint is relaxed so that the commitment and on-line variables are continuous variables in the range [0,1], then they can be included in complementarity formulations, although this approximation will understate commitment costs if fractional values occur at a solution.

Consumer. Here, the response of the consumer is represented by its demand function $q_i = f_{d,i}(p_i)$.

Transmission provider. Turning to the TSO, we assume that it operates the network as a regulated but efficient entity. For example, it might be a non-profit "Independent System Operator" (ISO) that operates but does not own network or generation assets, or an incentive regulated "Transco" that owns and operates the network but does not own generation. In either case, its objective is to allocate transmission capacity to maximize the value that the market receives from those assets. We represent this behavior here by modeling the TSO as a market agent who believes it cannot affect the market price, even though in fact price is a variable that is endogenous to the market. Another way to view this model is that the TSO operates a market that auctions off the capacity of individual transmission components, similar to the so-called "flowgate" market proposed by Chao and Peck [8]. Viewed in this manner, the KKT conditions derived below for the TSO's problem ensure market clearing: a component's price is positive only if it is constraining, and in that case, the price is set at the level at which demand for the interface (the flow) just equals the supply (the capacity).

One possible version of the TSO's model has two types of variables and this is the model used here (and in Section 7.4.1.3). The first variable is the sales (or arbitrage) of power a_i (MW) to consumers at a node i that was generated elsewhere. This variable is unrestricted in sign, so that a negative value means that the TSO is instead buying power at the location.

The other variable is the flow t_k in a particular direction through a particular branch k of the network, where there are K branches. Branches may represent distinct components of the transmission network, such as power lines and transformers, or can model aggregations of lines or other constraints. This flow is also unrestricted in sign (with a negative value indicating reversal of direction), but subject to lower and upper bounds based on equipment and system characteristics. For instance, thermal or surge impedance loading considerations place limits on current, and thus power, flow through individual lines. So-called "n-1" constraints might limit the flow through a set of lines to a city to some total that can still be accommodated if one of the lines happens to fail. More complex "nomograms" can be linear or nonlinear inequality constraints that are functions not only of line flows across several lines but also the status of generators. Such constraints usually address system security concerns. For instance, the famous SCIT nomogram of Southern California is a nonlinear relationship between the maximum flows into that region and thermal generation within the region that can be used to keep the system stable if a line or other piece of equipment fails [7].

The flows t_k are related to the nodal power sales a_i as follows. The dc load flow model sets up a system of K independent linear equations that determines the K flows as a function of the TSO's power purchases at the I nodes. The model assumes that line resistance is much smaller than reactance (inductance and capacitance), and that actual voltage magnitudes are the same as the nominal voltage levels for all buses (e.g., the actual voltage hovers near 500 kV at buses on the 500 kV network) [28]. In the simplest case, in which there are only constraints on flows on individual lines, the K equations are derived from I-1 independent energy balances (analogous to Kirchhoff's current law) at the nodes, and K-I+1 independent loop equations that, in an analogy to Kirchhoff's voltage law, state that the sum, over all links in a loop in a network, of the MW link flow times the link impedance must equal zero. It can be proved that for a network such a set of K independent equations exists and uniquely determines the t_k as a function of the a_i. Because this system is linear in all variables, the principle of superposition applies, leading to the following relationship:

$$t_k = \sum_i PTDF_{ik} a_i, \quad \forall k \tag{11.2}$$

where $PTDF_{ik}$ is a linear response coefficient, translating net power withdrawals at the nodes into flow on a branch k in the network. These are called "swing factors" or "power transfer distribution factors" by power engineers. They can be interpreted as the MW flow through component k that results from a 1 MW withdrawal at node i (i.e., the TSO selling 1 MW to a consumer) and a matching 1 MW injection at the "hub" node, which is the node whose energy balance is not included in the K independent equations. Swing factors generally produce a good approximation of the actual flows, which are governed by nonlinear equations, except in extreme conditions of very high flows, high losses, and high reactive (or imaginary) power [28].[2] The resulting

[2] Reactive power is power passed back and forth in ac power systems among generators, transformers, motors, and other equipment that produce magnetic fields as a result of inductance or that store energy due to capacitance. Basically, with every switch of direction

pattern of flow is very different than what a transshipment or transportation network might produce [14], in which flow can be directed along different paths of a network, disregarding Kirchhoff's voltage law.

These K equalities can be used to eliminate the flow variables, so that the bounds on flows in the TSO's optimization model become bounds on $\sum_i PTDF_{ik}a_i$ instead. We can now write the TSO's model (repeating (7.5)):

$$\text{Maximize}_{a_i} \quad \sum_i p_i a_i \tag{11.3a}$$

s. t.

$$\sum_i a_i = 0 \quad (\gamma) \tag{11.3b}$$

$$\sum_i PTDF_{ik} a_i \le T_k \quad (\lambda_k), \quad \forall k \tag{11.3c}$$

A positive a_i increases the TSO's profit because it represents a net sale at that node to consumers (i.e., a withdrawal of power from the network), while a negative a_i is instead a net purchase by the TSO from generators there (i.e., power is injected injected into the network). The first constraint ensures that if the TSO sells power at one node, it has to buy a matching amount elsewhere; γ is its dual variable, which in equilibrium will equal the value of energy at the hub node. In the second constraint, T_k is the upper bound to the flow through component k, and λ_k is its nonnegative dual variable. For simplicity, we omit the lower bound (which we might call $-T_{k'}$) as it can be expressed as a constraint in the same form by noting that $-\sum_i PTDF_{ik}a_i \le T_{k'}$ can be converted into an inequality of the above form by multiplying the $PTDF_{ik}$by -1.

Market-clearing conditions and the equilibrium problem. In the second step, the first-order (KKT) conditions for firms' and TSO's optimization problems are derived. Then in Step 3, these conditions plus the demand functions that describe consumer behavior are combined with market-clearing conditions to create a square system of equilibrium conditions. The market-clearing conditions simply balance supply and demand at each node i, and in the complementarity problem are associated with the nodal price p_i:

$$q_i - a_i - \sum_{f \in F_i} x_{fi} = 0 \quad (p_i \text{ unrestricted}), \ \forall i \tag{11.4}$$

of current flow (120 times per second in North America and 100 times per second in Europe), the magnetic field from inductors or the state of charge in capacitors reverses, resulting in storage or release of energy, which then returns when the current direction reverses again. This sloshing to and fro of energy is called reactive power. Reactive power is closely related to voltage magnitude and it helps keep voltage magnitudes stable across the system, 110 V at the low voltage level in North America and 230 V in Europe. Voltage magnitude stability is needed for the proper functioning of most electrical appliances. See Schweppe et al. [28].

We define F_i as the set of firms that have a power plant at node i. Completing Step 3, we get the following complementarity problem for perfect competition among generators who sell their output to a TSO:

Find $\{x_{fi}, \mu_{fi}, q_i, a_i, \gamma, \lambda_k, p_i\}$ satisfying:

$$0 \leq x_{fi} \perp p_i - dC_{fi}/dx_{fi} - \mu_{fi} \leq 0, \quad \forall f, i \in G_f \tag{11.5a}$$

$$0 \leq \mu_{fi} \perp x_{fi} - K_{fi} \leq 0, \quad \forall f, i \in G_f \tag{11.5b}$$

$$q_i = f_{d,i}(p_i) \quad (q_i \text{ unrestricted}), \quad \forall i \tag{11.5c}$$

$$\sum_i a_i = 0 \quad (\gamma \text{ unrestricted}) \tag{11.5d}$$

$$p_i - \gamma - \sum_k PTDF_{ik} \lambda_k = 0 \quad (a_i \text{ unrestricted}), \quad \forall i \tag{11.5e}$$

$$0 \leq \lambda_k \perp \sum_i PTDF_{ik} a_i - T_k \leq 0, \quad \forall k \tag{11.5f}$$

$$q_i - a_i - \sum_f x_{fi} = 0 \quad (p_i \text{ unrestricted}), \quad \forall i \tag{11.5g}$$

We verify that the problem is square by noting that there are *3I* variables q_i, p_i, and a_i; $2G$ variables x_{fi} and μ_{fi}; K variables λ_k; and one variable γ, and that there is exactly one condition associated with each of those variables in the above mixed complementarity problem. The associated variable is either on the left side of the complementarity condition, or is shown as being unrestricted in parentheses to the right of the condition. G is the number of combinations of indices f and i that have generators.

Solution properties. In the fourth step, the modeler investigates the properties of this system. For instance, economic interpretations might be given for the equilibrium conditions. Conditions (11.5a)-(11.5b), for example, indicate that the price at a location i will equal the marginal cost of any generator whose output is positive but not at its capacity. The first condition is the classic "price equals marginal cost" result of microeconomic models of perfect competition; this model generalizes this result to spatial markets in which the commodity (here, electricity) is transported according to Kirchhoff's laws. The third from last condition shows that the price at each location i is equal to the "system energy price" (γ, equal to the hub node price) plus a weighted sum of shadow prices for binding transmission constraints. That sum represents the cost of bringing power from the hub node to i.

Further analysis of the equilibrium conditions can show that the TSO profit is equal to the difference between payments by consumers for power and payments to producers, which is in turn equal (by strong duality for the TSO problem) to the sum of the transmission constraints times their shadow prices [20]. This turns out to not be the case if quadratic power losses from electrical resistance are modeled. If commodity transport is instead modeled as a transportation, transshipment [32], or, say, pressurized natural gas network, then relationships of prices at different locations will differ from these.

Another set of questions that Step 4 addresses can be the uniqueness and existence of solutions of the equilibrium problem. In the case of (11.5), using the methods of Cottle et al. [11] (e.g., as in [24]), it can readily be shown that continuous and decreasing demand curves $f_{d,i}(p_i)$ and strictly convex cost functions $C_{fi}(x_{fi})$ are sufficient to ensure that a solution exists and that the quantities and prices q_i, x_{fi}, a_i, and p_i are unique. Multiple dual solutions μ_{fi} and λ_k are possible if their associated constraints are degenerate.[3] Uniqueness and existence are desirable properties for models that are used in policy analyses or regulatory proceedings, as multiple solutions make conclusions ambiguous. Multiple solutions for market operations models are also undesirable if different solutions imply different distributions of income among market players; political issues then arise [21].

Existence and uniqueness results can also be obtained by noting that this complementarity problem is the set of KKT conditions for a single convex optimization model in which values of the primal variables are chosen in order to maximize total social surplus (the sum of consumer, TSO, and producer surpluses) subject to generation and transmission flow bounds and market-clearing constraints [30]. The inverse of $f_{d,i}(p_i)$ is then interpreted as the derivative of the consumption benefit function used in the consumer surplus calculation. In general, however, market equilibrium problems do not necessarily correspond to the first-order conditions of a single optimization problem (as explained in Section 3.3, above). Examples of such problems are presented, for example, in Chapter 12.

11.2.2 Illustrative Example. Transmission-Constrained Perfect Competition Equilibrium

This simple example[4] (from [15]) illustrates Step 5 of the complementarity model construction procedure outlined at the start of this chapter. Competition in a single hour is considered, so MW and MWh are equivalent. There are three buses, i = 1,2,3, each of which has customers. However, generation

[3] Degeneracy occurs in a solution to a linear program if a basic variable has a value of zero [14]. A general property of, for instance, linear programs is that if the primal problem has multiple solutions, the dual is degenerate, and vice versa. As a simple example, imagine that a power system has two 100 MW power plants, one costing \$20/MWh and the other costing \$30/MWh, which serve a load of exactly 100 MW. The cheapest plant will be fully used, and the second plant will produce nothing in the optimal solution. Because the load and first plant's capacity are precisely equal, degeneracy occurs; either the slack variable for the first plant or the output variable of the second plant will be a basic variable but with a zero value. As a result, the dual variable for the first plant's capacity will be either zero or \$10/MWh, respectively.

[4] This numerical example and that in Section 11.2.4 are based upon [15].

occurs only at buses $i = 1{,}2$. Each pair of buses is interconnected by a single transmission line; all three lines have equal impedances (say, 2.7 units). Equal impedances on a triangular network imply that the linearized dc load flow equations (Appendix C) consist of one voltage law equation (corresponding to the only loop) and two out of the three energy balances (power sold by the TSO to a node equals physical flows to that node from its neighbors). These equations are as follows:

$$2.7t_{1\to 2} + 2.7t_{2\to 3} - 2.7t_{1\to 3} = 0$$

$$-t_{1\to 2} - t_{1\to 3} - a_1 = 0$$

$$+t_{1\to 2} - t_{2\to 3} - a_2 = 0$$

The first equation (the voltage law) was obtained by summing the product of each flow and its impedance while going clockwise around the one loop in the network, starting at node 1. Note the negative sign on $t_{1\to 3}$; this occurs because the flow variable is in the opposite direction of the clockwise tour around the loop. Solving these load flow equations for the flows yields:

$$t_{1\to 2} = -1/3a_1 + 1/3a_2$$

$$t_{1\to 3} = -2/3a_1 - 1/3a_2$$

$$t_{2\to 3} = -1/3a_1 - 2/3a_2$$

The coefficients on the power purchases are the swing factors. For instance, if the TSO sells 100 MW at node 1 and buys that power at the hub (node 3, whose energy balance was omitted from the flow equations), then two-thirds of that power will flow directly from 3 to 1, while the other one-third will flow through the path 3→2→1. The negative sign in front of many of the swing factors indicates that a sale of 1 MW at that node, which is balanced by a purchase of 1 MW at the hub, will decrease rather than increase the flow of interest.

The demand functions are $q_i = f_{d,i}(p_i) = 500–12.5p_i$, i=1,2 and $q_3 = f_{d,3}(p_3) = 620–19.38p_3$ \$/MWh. These functions imply that demand is more elastic at the demand-only node (bus 3). There are two producers f = 1,2, each with one generator. Firm 1's generator is sited at $i = 1$, while 2's is at i=2. Both generators have unlimited capacity, and a constant marginal cost: \$15/MWh for firm 1, and \$20/MWh for firm 2. The only transmission cost arises from congestion.

We set up this problem with two different transmission systems: one without congestion—infinite transmission capacity—and one with congestion on a single interface ($k = 1$) between buses 1 and 2 ($T_1 = 25$ MW in both directions). In both cases, we simplified the model, as explained above, by using the load flow equations just given to eliminate the flow variables t_k.

The resulting equilibrium model is shown below. In the first exercise at the end of the chapter, we ask the reader to implement and solve this model using a complementarity solver. Alternatively, the reader can check that the solutions in Tables 11.1 and 11.2 satisfy all conditions by inserting each condition as a separate cell in a spreadsheet, and then verifying that equality conditions are satisfied and that one side or another of each complementarity condition is zero.

Find $\{x_{1,1}, x_{2,2}, q_1, q_2, q_3, a_1, a_2, a_3, \gamma, \lambda_{1+}, \lambda_{1-}, p_1, p_2, p_3\}$ satisfying:

$$0 \le x_{1,1} \perp p_1 - 15 \le 0$$
$$0 \le x_{2,2} \perp p_2 - 20 \le 0$$
$$q_1 = 500 - 12.5 p_1 \quad (q_1 \text{ unrestricted})$$
$$q_2 = 500 - 12.5 p_2 \quad (q_2 \text{ unrestricted})$$
$$q_3 = 620 - 19.38 p_3 \quad (q_3 \text{ unrestricted})$$
$$a_1 + a_2 + a_3 = 0 \quad (\gamma \text{ unrestricted})$$
$$p_1 - \gamma + 1/3\lambda_{1+} - 1/3\lambda_{1-} = 0 \quad (a_1 \text{ unrestricted})$$
$$p_2 - \gamma - 1/3\lambda_{1+} + 1/3\lambda_{1-} = 0 \quad (a_2 \text{ unrestricted})$$
$$p_3 - \gamma = 0 \quad (a_3 \text{ unrestricted})$$
$$0 \le \lambda_{1+} \perp \quad -1/3 a_1 + 1/3 a_2 - T_1 \le 0$$
$$0 \le \lambda_{1-} \perp +1/3 a_1 - 1/3 a_2 - T_1 \le 0$$
$$q_1 - a_1 - x_{1,1} = 0 \quad (p_1 \text{ unrestricted})$$
$$q_2 - a_2 - x_{2,2} = 0 \quad (p_2 \text{ unrestricted})$$
$$q_3 - a_3 = 0 \quad (p_3 \text{ unrestricted})$$

Note that there are two transmission constraints, one for each direction on the transmission link between nodes 1 and 2. The transmission constraint index $k = 1+$ refers to the direction 1→2, while $k = 1-$ refers to the opposite direction. For the first model, T_1 is infinite, and in the second model, T_1 is set to 25 MW.

For each case, Table 11.1 shows the resulting primal solution, while Table 11.2 shows the prices, transmission dual, and surplus for each market party and p for perfect competition. As you see can see, the complementarity conditions are satisfied. For instance, the profit-maximization criterion (11.5a) for production is satisfied for firm 1 in both solutions in that a strictly positive output $x_{1,1}$ is accompanied by an equality between price and marginal cost (15 \$/MWh). For firm 2, however, price (\$15/MWh) is less than its marginal cost (\$20/MWh) in the infinite transmission capacity case. Due to the complementarity condition for $x_{2,2}$, that variable is zero in the solution. As another example, a positive transmission dual is possible only if the flow is binding, as it is for the flow between nodes 1 and 2 in the constrained transmission case. All the equality constraints (such as the sum of TSO purchases or market clearing at each bus) equal zero.

Table 11.1 Primal solutions for perfect competition on an electricity network

	Quantities demanded MWh			Generation MW	
Transmission Case	q_1	q_2	q_3	$x_{1,1}$	$x_{2,2}$
$T_1 = \infty$	312.5	312.5	329.5	954.5	0
$T_1 = 25$ MW	312.5	250	281.0	490.5	353.0

	TSO Sales MWh			Flow MW		
Transmission Case	a_1	a_2	a_3	$t_{1\to 2}$	$t_{1\to 3}$	$t_{2\to 3}$
$T_1 = \infty$	-642.0	312.5	329.5	318.2	323.8	5.7
$T_1 = 25$ MW	-178.0	-103.0	281.0	25	153.0	128.0

Table 11.2 Prices and surpluses for perfect competition on an electricity network

	Energy price $/MWh			Transmission dual λ_{1+} , λ_{1-} $/MWh
Transmission Case	p_1	p_2	$p_3(=\gamma)$	
$T_1 = \infty$	15	15	15	0,0
$T_1 = 25$ MW	15	20	17.5	7.5,0

	Generator profit $/hr		TSO profit $/hr	Consumer surplus $/hr	Social surplus $/hr
Transmission Case	f=1	f=2			
$T_1 = \infty$	0	0	0	10,613	10,613
$T_1 = 25$ MW	0	0	187	8444	8631

It is apparent that transmission constraints can drastically change the solution. Without the transmission limit between 1 and 2, prices are all the same at the buses and the cheaper producer (f=1) takes the entire market. With the constraint, prices separate, and firm 2 enters the market, even though its power is 33% more expensive. This is because the transmission constraint only allows 178.1 MW of firm 1's power to be shipped to other nodes (less than a third of what it exports in the no constraint case). Transmission constraints can increase profits in some situations, and decrease them in others; here, because there are no generation capacity constraints, price equals marginal cost for each generator, and they make no profit in either case. However, consumers are worse off, with their surplus (integral of demand function minus what they pay for power) shrinking by almost 20%. The transmission grid earns a small amount in the constrained case, selling power to consumers for $187/hour more than it paid producers for it. This difference is called the "congestion revenue" (often also called "congestion rent" or "congestion surplus"). It can be proven that the TSO can never lose money in this model, but if there are no binding transmission limits, it will make nothing either.

That the prices at the generation nodes should equal the marginal cost of active generation variables at those locations is obvious. But what about the 17.5 $/MWh at node 3 in the constrained solution? The physical reason for that can be understood if one considers how the dispatch has to change in order to accommodate a 1 MW increase in load at that node without violating the transmission constraint between nodes 1 and 2. If the entire load increase was met by incrementing upwards the cheap generator at node 1, then the load flow equations mean that 1/3 more MW will flow through the path 1→2→3, causing the flow $t_{1\rightarrow 2}$to grow from 25 MW to 25-1/3 MW, which is infeasible. Some generation must come instead from the costlier generator at node 2. It turns out that by incrementing generation at each of the generators by 1/2 MW (costing (0.5)(15)+(0.5)(20) =17.5 $/MWh), the increased load at node 3 can be met without violating the constraint. Thus, the price at a node in an electric network can represent a weighted average of the marginal costs of generators whose output variables are positive but not at their upper bounds. In other network configurations, prices at nodes can exceed the highest variable cost generator, or even be negative, depending on the location of transmission constraints relative to the basic generators. These and other counter-intuitive pricing results of Kirchhoff's laws are described by Wu et al. [36].

11.2.3 Oligopolistic Market: A Cournot Model

The previous section modeled two companies competing for demand at three locations. However, such a duopolistic situation in the real world would not result in prices equal to marginal cost. Market power—the ability of market players to cause prices to deviate systematically and persistently from marginal cost—is unfortunately a reality in power markets. In some places, most infamously California in 2000-2001, supply shortages and inability of consumers to adjust their consumption to prices that vary greatly from hour to hour have resulted in large price increases. During the California crisis, a large share of the increases experienced has been blamed on the ability of generators to manipulate output and bids. One way to exercise market power is to withhold output, for instance by declaring generators to be unavailable because of equipment failures. Prices can also be affected by strategic bidding, by manipulation of markets for needed inputs such as fuel or emissions allowances, and by deliberate congestion or decongestion of the network. However, despite the infamous "Get Shorty", "Death Star", and other sophisticated strategies detailed in the famous Enron memos from the 2000-2001 California power crisis, most of the market power problems experienced by California were due to simple economic and physical withholding of capacity. Economic withholding occurs when generation capacity is made available only at high prices. Meanwhile, physical withholding is when capacity is unavailable at any price, for example by declaring a maintenance outage when demand is high. It has been argued that economic withholding in California was facilitated by contrived shortages of natural gas and NO_x allowances at crucial times [22].

Exercise of market power affects not only prices, profits, and consumer economic welfare, but can also decrease productive efficiency (if prices increase, smaller firms with relatively high costs are encouraged to generate) while impact the environment (if outputs shift among generators with different emission rates). Market power is widely viewed as one of the most serious imperfections in new power markets. Therefore, models for projecting prices and other market outcomes should explicitly consider the potential for strategic behavior by power producers.

Many such models have been developed [35, 39]. Most are based on the calculation of Nash equilibria for a single time period for one commodity (electric energy). The most common Nash games represented are those in which the firm's strategic variable is either sales and/or production quantities (Nash-Cournot games, see Section 3.4.2.5, above) or bid functions (supply function games). Conjectural variation-based models have seen application [6, 12]. Such models can be viewed as a generalization of the Cournot model, in which a generator can conjecture that its rivals will change their decisions some way in reaction to quantity or price changes by that generator, rather than the Nash assumption of assuming that rivals will hold their strategies fixed.

Some models have gone beyond energy markets to consider market power in input markets (fuels, emissions markets) and other output markets (ancillary services such as operating reserves or green power markets) (e.g., [18]). Manipulation of transmission prices by changing outputs so as to deliberately force flow constraints to be binding or slack can enable strategic producers to isolate sub-markets from competition, or deprive TSOs of congestion revenues they would otherwise receive [26].

Here, we present an elementary model of oligopolistic competition among power generators. We generalize the perfect competition model of Section 11.2.1to represent a Cournot game with the following characteristics:

- Reflecting the predominance of bilateral trading of power between producers and consumers even in markets with a central buyer of power, the model has generators selling to consumers at locations throughout the network. Therefore, we define a new variable s_{fi}, the MW sales by generating company f to consumers at i. Thus both generators and the TSO (through its sales variable a_i) can sell directly to consumers, and quantity demanded $q_i = a_i + \sum_f s_{fi}$.
- Generators play a quantity game against other generators, in which each f assumes that the other firms $f\prime \neq f$ will not change the amount $s_{f\prime i}$ they sell. Firms recognize that if more or less power is sold at a node then price will

fall or rise, respectively, following the inverse demand curve. This provides an incentive to restrict sales and increase price.

- Generators pay the TSO for transmission services. The TSO's prices result from the same congestion pricing system as used in the previous section, in which the TSO rations scarce transmission capacity. A property of congestion pricing is the price of moving power from node j to node i is -1 times the price of moving it in the opposite direction. Furthermore, the price of moving power from j to i is equal to the price of moving it from j to an arbitrary node k and then from k to i, and is the same for all k. The price w_i \$/MWh represents the expense of transmission service, and is the price for delivering one MWh from the hub node to node i. All power is routed through the hub, so that a generator wanting to produce power at j and sell it at i pays $-w_j + w_i$.
- The generators adopt a Bertrand (price-taking) conjecture regarding the price of transmission services. That is, each generating company assumes that its sales and output actions will not affect the values of w_i. This is of course a strong assumption, in that a large generator might naturally assume that if it tried to send more power to another location, congestion would increase, causing the price of transmission services to that location to go up. In Sections 7.4 and 7.5, we consider another, more sophisticated assumption, in which the generator explicitly represents the response of the TSO using the latter's KKT conditions, yielding a MPEC or EPEC problem. Another approach to modeling transmission price responses to changes in generator decisions is to use a notion related to conjectural variations called conjectured price response [19].
- Complementing bilateral transactions, the TSO arbitrages among the markets in the same manner as the previous section.

As in the previous section, we undertake each of the steps of the model building process, starting with Step 1, the specification of the market party optimization models.

For a brief presentation of a stochastic version of this Cournot market model, see Section 4.6.3.

Consumers. As in the previous model, the response of the consumer is represented by its demand function $q_i = f_{d,i}(p_i)$. However, since generators consider how their sales affects price, it is more convenient to use the inverse demand curve:

$$p_i = f_{d,i}^{-1}(q_i) = f_{d,i}^{-1}(a_i + \sum_f s_{fi}). \tag{11.6}$$

The right-most equality arises from the identity that the quantity demanded at i equals the sum of sales by generators and the TSO at that location. This substitution allows us to simplify the model by eliminating the q_i variables.

Generating firm. The generator model is more complex than in the competitive model of the previous section in three ways. First, the generator recognizes that its sales affect the price, so $f_{d,i}^{-1}(a_i + \sum_f s_{fi})$ replaces p_i in the revenue expression in the objective. Second, because the generator can sell power at locations other than where generated, transmission charges w_i are applied to convey power from generators to the hub node, and to convey power from the hub to the consumers. Finally, a sales balance constraint limits the total sales to the amount of power generated.

$$\text{Maximize}_{x_{fi},s_{fi}} \quad \sum_i \left(f_{d,i}^{-1}(a_i + \sum_f s_{fi}) - w_i \right) s_{fi}$$

$$- \left(\sum_{i \in I_f} C_{fi}(x_{fi}) - w_i x_{fi} \right) \tag{11.7a}$$

s. t.

$$x_{fi} - K_{fi} \leq 0 \quad (\mu_{fi}) \quad \forall i \in I_f \tag{11.7b}$$

$$\sum_i s_{fi} - \sum_{i \in I_f} x_{fi} = 0 \quad (\phi_f) \tag{11.7c}$$

$$x_{fi} \geq 0 \quad \forall i \in I_f \tag{11.7d}$$

Transmission provider. The TSO is once again assumed to operate the network as a regulated but efficient entity. It provides y_i MW of transmission services for bilateral transactions to node i from the hub; if negative, that implies that it is, on net, delivering bilateral power sales from i. It also arbitrages the market using the variable a_i, buying power from generators in low value areas ($a_i < 0$) and reselling it to consumers in high value areas ($a_j > 0$), ensuring that in equilibrium the price difference between two locations is equal to the price of transmission between them:

$$p_j - p_i = w_j - w_i \tag{11.8}$$

The TSO's model is as follows:

$$\text{Maximize}_{a_i,y_i} \quad \sum_i p_i a_i + \sum_i w_i y_i \tag{11.9a}$$

s. t.

$$\sum_i a_i = 0 \quad (\gamma) \tag{11.9b}$$

$$\sum_i PTDF_{ik}(a_i + y_i) - T_k \leq 0 \quad (\lambda_k), \quad \forall k \tag{11.9c}$$

The difference compared to the competitive TSO is the addition of revenues from providing transmission services y_i, and the inclusion of those variables in the transmission flow constraint. Unlike arbitrage services, there is no need for a constraint that says that the sum of transmission services delivered is zero, as the generators' sale balance constraints will ensure that this is the case. Inclusion of a redundant constraint will cause two problems. The first problem is a linear dependency in the set of conditions, which can cause numerical headaches. The other problem is an ambiguity, in the form of multiple solutions, in the prices/dual variables associated with the extra constraint and the constraints it depends upon.

Market-clearing conditions and the equilibrium problem. The equilibrium problem consists of the first-order conditions for the generator and transmission problem, together with market-clearing conditions. It is not necessary to explicitly include a demand function for consumers, as that is already included in the generators objective function via the inverse demand function in the revenue term; however, we also append that function to the system so that we calculate price. Furthermore, the sales variables s_{fi} have been substituted in the generator's objective for quantity demanded q_i, so the latter variable does not need to be included in the equilibrium problem. For the same reason, the market-clearing condition 11.4 from the competitive model is unneeded. Instead, the only market-balancing constraint is for the provision of transmission services, in which the MW of transmission delivery y_i provided by the TSO to generators equals the net sales by those generators to at node i.

$$y_i - \sum_f (x_{fi} - s_{fi}) = 0 \quad (w_i), \quad \forall i \tag{11.10}$$

Note that the price associated with that market-clearing condition is the price of transmission services to that node.

Completing Step 3, we get the following complementarity problem for Cournot competition among generators who contract bilaterally in a market in which the TSO provides arbitrage and transmission services:

Find $\{x_{fi}, s_{fi}, \mu_{fi}, \phi_f, a_i, y_i, \gamma, \lambda_k, w_i, p_i\}$ satisfying:

$$0 \le x_{fi} \perp -(dC_{fi}/dx_{fi} - w_i + \mu_{fi}) + \phi_f \le 0, \quad \forall f, i \in I_f \tag{11.11a}$$

$$0 \le s_{fi} \perp (p_i + s_{fi}\, dp_i/ds_{fi}) - (w_i + \phi_f) \le 0, \quad \forall f, i \tag{11.11b}$$

$$0 \le \mu_{fi} \perp x_{fi} - K_{fi} \le 0, \quad \forall f, i \in I_f \tag{11.11c}$$

$$\sum_i s_{fi} - \sum_{i \in I_f} x_{fi} = 0 \quad (\phi_f \text{ unrestricted}) \quad \forall f \tag{11.11d}$$

$$p_i - \gamma - \sum_k PTDF_{ik}\lambda_k = 0 \quad (a_i \text{ unrestricted}), \quad \forall i \tag{11.11e}$$

$$w_i - \sum_k PTDF_{ik}\lambda_k = 0 \quad (y_i \text{ unrestricted}), \quad \forall i \tag{11.11f}$$

$$\sum_i a_i = 0 \quad (\gamma \text{ unrestricted}) \tag{11.11g}$$

$$0 \leq \lambda_k \perp \sum_i PTDF_{ik}(a_i + y_i) - T_k \leq 0, \quad \forall k \tag{11.11h}$$

$$y_i - \sum_f (x_{fi} - s_{fi}) = 0 \quad (w_i \text{ unrestricted}), \quad \forall i \tag{11.11i}$$

$$p_i - f_{d,i}^{-1}(a_i + \sum_f s_{fi}) = 0 \quad (p_i \text{ unrestricted}), \quad \forall i \tag{11.11j}$$

where:

$$p_i + dp_i/ds_{fi} = f_{d,i}^{-1}(a_i + \sum_f s_{fi}) + df_{d,i}^{-1}(a_i + \sum_f s_{fi})/ds_{fi} \tag{11.12}$$

which is the well-known marginal revenue term for Cournot oligopoly models. Exercise 4 at the end of the chapter asks the reader to verify that this system of conditions is indeed square, with one condition for each variable.

Solution properties. First, we indulge in a bit of Step 4 analysis of the economic implications of the model. We note that if both s_{fi} and x_{fj} are strictly positive for a given firm f and nodes i and j, then we can use the first two conditions to eliminate ϕ_f, resulting in (after some rearrangement):

$$p_i + s_{fi}dp_i/ds_{fi} = dC_{fj}/dx_{fj} + \mu_{fj} + (w_i - w_j) \tag{11.13}$$

This is the classic first-order condition of Cournot oligopoly models: that marginal revenue equals marginal cost, including the shadow price of capacity, if applicable, and the expensive of transporting the power. Note that because inverse demand curves are downward sloping, dp_i/ds_{fi} is negative. Therefore, in equilibrium:

$$p_i > dC_{fj}/dx_{fj} + \mu_{fj} + (w_i - w_j) \tag{11.14}$$

if s_{fi} is strictly positive. That is, price exceeds marginal cost. Furthermore, the greater a generator's sales, the greater the margin between price and marginal cost, which is a measure of the amount of market power that is exercised.

Turning to issues of existence and uniqueness, Metzler et al. [24] has established the uniqueness of the prices and profits resulting from the above model. Furthermore, it is shown that the results are equivalent to those of a market model in which every producer sells to the TSO at its point of production (as in the perfectly competitive model of the previous section), and anticipates that the TSO will adjust its arbitrage in order to maintain price differences $(p_i - p_j)$ between nodes.

11.2.4 Illustrative Example. Transmission-Constrained Cournot Equilibrium

We now execute Step 5 of the model construction process by inserting the parameters from the example in Section 11.2.2 into the Cournot model 11.11. The model is:

Find $\{x_{1,1}, x_{2,2}, s_{1,1}, s_{1,2}, s_{1,3}, s_{2,1}, s_{2,2}, s_{2,3}, \phi_1,\ \phi_2, a_1, a_2, a_3, y_1, y_2, y_3, \gamma, \lambda_{1+}, \lambda_{1-}, w_1, w_2, w_3, p_1, p_2, p_3\}$ satisfying:

$$
\begin{aligned}
&0 \le x_{1,1} \perp -15 + w_1 + \phi_1 \le 0\\
&0 \le x_{2,2} \perp -20 + w_2 + \phi_2 \le 0\\
&0 \le s_{1,1} \perp (p_1 + s_{1,1}(-0.08)) - (w_1 + \phi_1) \le 0\\
&0 \le s_{1,2} \perp (p_2 + s_{1,2}(-0.08)) - (w_2 + \phi_1) \le 0\\
&0 \le s_{1,3} \perp (p_3 + s_{1,3}(-0.0516)) - (w_3 + \phi_1) \le 0\\
&0 \le s_{2,1} \perp (p_1 + s_{2,1}(-0.08)) - (w_1 + \phi_2) \le 0\\
&0 \le s_{2,2} \perp (p_2 + s_{2,2}(-0.08)) - (w_2 + \phi_2) \le 0\\
&0 \le s_{2,3} \perp (p_3 + s_{2,3}(-0.0516)) - (w_3 + \phi_2) \le 0\\
&s_{1,1} + s_{1,2} + s_{1,3} - x_{1,1} = \ 0 \quad (\phi_1 \text{ unrestricted})\\
&s_{2,1} + s_{2,2} + s_{2,3} - x_{2,2} = \ 0 \quad (\phi_2 \text{ unrestricted})\\
&p_1 - \gamma + 1/3\lambda_{1+} - 1/3\lambda_{1-} = 0 \quad (a_1 \text{ unrestricted})\\
&p_2 - \gamma - 1/3\lambda_{1+} + 1/3\lambda_{1-} = 0 \quad (a_2 \text{ unrestricted})\\
&p_3 - \gamma = 0 \quad (a_3 \text{ unrestricted})\\
&w_1 + 1/3\lambda_{1+} - 1/3\lambda_{1-} = 0 \quad (y_1 \text{ unrestricted})\\
&w_2 - 1/3\lambda_{1+} + 1/3\lambda_{1-} = 0 \quad (y_2 \text{ unrestricted})\\
&w_3 = 0 \quad (y_3 \text{ unrestricted})\\
&a_1 + a_2 + a_3 = \ 0 \quad (\gamma \text{ unrestricted})\\
&0 \le \lambda_{1+} \perp -1/3(a_1 + y_1) + 1/3(a_2 + y_2) - T_1 \le 0\\
&0 \le \lambda_{1-} \perp +1/3(a_1 + y_1) - 1/3(a_2 + y_2) - T_1 \le 0\\
&y_1 - (s_{1,1} + s_{2,1}) + x_{1,1} = 0 \qquad (w_1 \text{ unrestricted})\\
&y_2 - (s_{1,2} + s_{2,2}) + x_{2,2} = 0 \qquad (w_2 \text{ unrestricted})\\
&y_3 - (s_{1,3} + s_{2,3}) = 0 \qquad (w_3 \text{ unrestricted})\\
&p_1 - (40 - 0.08(a_1 + s_{1,1} + s_{2,1})) = 0 \quad (p_1 \text{ unrestricted})\\
&p_2 - (40 - 0.08(a_2 + s_{1,2} + s_{2,2})) = 0 \quad (p_2 \text{ unrestricted})\\
&p_3 - (32 - 0.0516(a_3 + s_{1,3} + s_{2,3})) = 0 \quad (p_3 \text{ unrestricted})
\end{aligned}
$$

The solutions are shown in Tables 11.3 and 11.4. An obvious difference between these Cournot solutions and the competitive solutions of Tables 11.1 and 11.2 is the effect of imperfect competition. Perfect competition yields much lower prices and higher welfare (defined as the sum of consumer, TSO, and producer surpluses) than the Cournot solutions. Under perfect competition, without transmission constraints, the cheapest firm ($f = 1$) serves the entire demand at all nodes. But under imperfect competition, prices climb enough for the more costly firm ($f = 2$) to profitably enter.

Table 11.3 Primal solutions for Cournot competition on an electricity network

	Quantities demanded MWh			Generation MW	
Transmission Case	q_1	q_2	q_3	$x_{1,1}$	$x_{2,2}$
$T_1 = \infty$	202.1	202.1	158.2	392.1	170.2
$T_1 = 25$ MW	218.2	185.9	158.3	334.8	227.5

	Firm 1 sales MWh			Firm 2 sales MWh		
Transmission Case	$s_{1,1}$	$s_{1,2}$	$s_{1,3}$	$s_{2,1}$	$s_{2,2}$	$s_{2,3}$
$T_1 = \infty$	110.4	110.4	171.2	47.9	47.9	74.3
$T_1 = 25$ MW	94.3	94.3	146.2	64.1	64.1	99.4

	TSO Sales MWh			Flow MW		
Transmission Case	a_1	a_2	a_3	$t_{1\to 2}$	$t_{1\to 3}$	$t_{2\to 3}$
$T_1 = \infty$	43.7	43.7	-87.3	74	116.1	42.1
$T_1 = 25$ MW	59.8	27.5	-87.3	25	91.6	66.6

	Transmission service MW		
Transmission Case	y_1	y_2	y_3
$T_1 = \infty$	-233.7	-11.8	245.6
$T_1 = 25$ MW	-176.4	-69.1	245.6

Note: $q_i = a_i + s_{1,i} + s_{2,i}$

Table 11.4 Prices, dual variables, and surpluses for Cournot competition on an electricity network

Transmission Case	Energy price $/MWh			Transmission dual λ_{1+} , λ_{1-} $/MWh
	p_1	p_2	$p_{3,\gamma}$	
$T_1 = \infty$	23.84	23.84	23.84	0, 0
$T_1 = 25$ MW	22.54	25.13	23.84	3.87, 0

Transmission Case	Transmission price $/MWh			Marginal cost at hub $/MWh	
	w_1	w_2	w_3	ϕ_1	ϕ_2
$T_1 = \infty$	0	0	0	15	20
$T_1 = 25$ MW	-1.29	1.29	0	16.29	18.71

Transmission Case	Generator profit $/hr		TSO profit $/hr	Consumer surplus $/hr	Social surplus $/hr
	f=1	f=2			
$T_1 = \infty$	3465	653	0	3912	8030
$T_1 = 25$ MW	2526	1166	97	3933	7722

We now turn to the effect of transmission limits. Imposing the 25 MW transmission constraint between nodes 1 and 2 in the Cournot solution results in price dispersion among the three nodes, although not as much as under competition. The constraint has a differential effect on the profitability of the two firms. On one hand, firm 1 loses market share because its plant at node 1 cannot as readily access node 2's consumers, but on the other hand, firm 2's profits increase because it is insulated from competition. Illustrating the often counterintuitive effects of transmission on economic outcomes, consumer surplus actually increases slightly when the transmission constraint is imposed. This can happen even in competitive markets – for instance, a transmission constraint can bottle up cheap power in one market, and relieving it might cause the price there to go up without having much of an effect in the markets it is connected to. Whether increasing transmission capacity increases or decreases consumer surplus is parameter dependent; however, with rare exceptions [27], adding transmission will usually increase social surplus.

11.3 Environmental Markets: Emissions Trading

Environmental legislation impacts the electric power industry in many ways. Emissions can be capped or taxed, siting of new plants or lines may be restricted to areas considered less sensitive, pollution control equipment retrofits may be mandated, and investment in particular types of power plants may be ordered.

In this section, we consider how some of these restrictions can be modeled, starting with pollution cap-and-trade policies. These policies work by allocating or selling permits to pollute (or "allowances") to eligible pollution sources, who can then trade permits among themselves so that every polluter holds a number of allowances at least equal to their emissions. If the cap is below the emissions that would otherwise occur, the allowances have a positive market price. Under certain assumptions, such systems motivate least-cost control of the emissions [33]. The first large-scale system for emissions cap-and-trade was established by the US 1990 Clean Air Act Amendments for electric sector SO_2 emissions. Since then, the US has extended cap-and-trade to NO_x emissions. In Europe, cap-and-trade was implemented in 2005 for the greenhouse gas CO_2.

Cap-and-trade programs for CO_2 are anticipated to have much larger economic impacts than previous emissions trading programs. The cost of significant CO_2 reductions and the economic value of CO_2 allowances are likely to be about an order of magnitude greater than for NO_x and SO_2 [13]. Wholesale electricity price increases in Germany and the Netherlands of 40% or more in 2005 were blamed upon the introduction of CO_2 trading there [29]. However, these increases have moderated recently in the face of falling carbon prices. The Regional Greenhouse Gas Initiative, involving several northeastern states in the US, has not resulted in such price increases, however, because its cap is loose, while low gas prices and demand growth would have depressed coal plant output even without the cap. Models of the sort developed in this chapter can help understand how different policies for limiting emissions and allocating allowances to market participants might affect future prices, power plant operations and investment, imports and exports, and other market outcomes of interest to policy makers (e.g., [9,10,37,40]).

In this section, we introduce two simple cap-and-trade models. The first is the "source-based" type of system just mentioned, in which power producers are responsible for tracking emissions and holding enough allowances to cover their emissions. A model for source-based trading and an illustrative numerical application are presented in Sections 11.3.1 and 11.3.2.

However, in theory, other market participants at different levels of the electricity supply chain could instead be the responsible parties – for instance, fuel producers or even consumers (under personal carbon trading proposals, e.g., [25]). So we consider a second cap-and-trade proposal called "load-based trading" in which load-serving entities (LSEs), who buy bulk power supplies and resell them to consumers, instead keep track of emissions associated with their purchases and meet a target [4,16,37]. Examples of such LSEs are regulated distribution utilities in the US who are required by law to serve a particular region and provide power as cheaply as possible to its customers. Sections 11.3.3 and 11.3.4 present a load-based trading model and a numerical illustration. It has been claimed that load-based trading would cost consumers less than source-based trading. However we demonstrate in

Section 11.3.5 certain theoretical equivalences between the load- and source-based approaches.

11.3.1 A Simple Model of Emissions Trading among Producers

With a source-based pollution control system, the allowances that producers need can be initially distributed in several ways to the market—for instance, by free allocation by the government to the producers in proportion to their energy sales. The model here assumes that allowances are auctioned by the government, and the proceeds returned to consumers, perhaps as tax breaks. The emissions cap and, thus, the total number of allowances is RQ, where R is the target emissions rate (in tons/MWh) and Q is the total demand.

As in Section 11.2, we consider the five steps of model construction and analysis. First, we build optimization models for the producers and for the LSEs who buy power on behalf of consumers. We'll neglect transmission constraints in the model and for simplicity also ignore demand elasticity and market power. For simplicity, we consider energy and emissions trading only for a single hour We assume there is just one LSE that buys electricity and that each power company owns only one power plant; of course, more general models are readily developed. For simplicity, we just consider one representative hour both in this model and the load-based trading model of Section 11.3.3.

Generating firm. Starting with the producers, each solves the following profit-maximization problem for the single power plant we assume they own, which we assume has a constant ton/MWh emissions rate.

$$\text{Maximize}_{x_f} \quad p_f x_f - C_f(x_f) - p^{CO2} E_f x_f \tag{11.15a}$$

s. t.

$$x_f - K_f \leq 0 \quad (\mu_f) \tag{11.15b}$$

$$x_f \geq 0 \tag{11.15c}$$

This model shows that producer f sells its output x_f in the power market at price p_f (which might differ from what other generators get for their product) but must buy an input (in the form of CO_2 emissions allowances) from the allowance market at price p^{CO2}, in \$/ton. The tonnage of allowances that must be bought is the power plant's emissions rate E_f in tons/MWh times the output. Otherwise, the model is similar to 11.1, above.

Load-serving entity. The LSE also solves an optimization problem. We assume state regulation or ownership of the LSE, so that its problem is to choose which power plants to buy from in order to maximize an objective equal to revenues from the emissions allowance auction minus expenditures upon bulk power supplies. Instead of a single consumption variable q, one q_f is defined for each generation company f in order to be able to represent power as a differentiated commodity with different prices from different sources. Although this does not occur for this model, it does for the load-based equilibrium model, presented later in this section.

$$\text{Maximize}_{q_f} \quad p^{CO2}RQ - \sum_f p_f q_f \tag{11.16a}$$

s. t.

$$-\sum_f q_f + Q = 0 \quad (\beta) \tag{11.16b}$$

$$q_f \geq 0, \quad \forall f \tag{11.16c}$$

The first objective function term consists of the proceeds of the allowance auction. Since the decision variable q_f does not enter that term, the LSE views it as a constant. The second term is the expenditures on power purchases from generators. The objective is maximized subject to the constraint that power purchases equal the total electricity demand, which is assumed to be fixed. (Both sides of the constraint are multiplied by -1 so that β will usually be positive.)

Equilibrium model. In Step 2 of the modeling process, first-order conditions are defined for each of the models; in the third step, these first-order conditions are combined with market-clearing conditions to complete the equilibrium model. The model is shown below. In this market, there are two sets of clearing conditions. The first is for allowance trading, ensuring that total emissions do not exceed a cap, and allows the price to be positive only if the cap is binding. The second balances consumption and supply of energy for each producer. These conditions are shown as the last two conditions in the equilibrium model below:

Find $\{x_f, q_f, p_f, \mu_f, \beta, p^{CO2}\}$ satisfying:

$$0 \leq x_f \perp p_f - dC_f/dx_f - p^{CO2}E_f - \mu_f \leq 0, \quad \forall f \tag{11.17a}$$

$$0 \leq \mu_f \perp x_f - K_f \leq 0, \quad \forall f \tag{11.17b}$$

$$0 \leq q_f \perp - p_f + \beta \leq 0 \quad \forall f \tag{11.17c}$$

$$-\sum_f q_f + Q = 0 \quad (\beta \text{ unrestricted}) \tag{11.17d}$$

$$0 \leq p^{CO2} \perp \sum_f E_f x_f - RQ \leq 0 \tag{11.17e}$$

$$q_f - x_f = 0 \quad (p_f \text{ unrestricted}), \forall f \tag{11.17f}$$

This system is square, as it should be, with one variable corresponding to each condition.

Economic analysis. In Step 4, certain properties can be readily verified. For example, by the third and fifth conditions, the prices p_f received by different firms are all equal for firms with positive output x_f. However, this will not be the case with the load-based trading system of the next section. We undertake some further analysis of this model in Section 11.3.5.

11.3.2 Illustrative Example. Simple Source-Based Emissions Trading Equilibrium

The assumptions of this source-based example[5] are:

- The load-serving entity has a constant load $Q = 2000$ MW. The environmental regulator wants to reduce the average emissions per MWh to R = 0.55 tons/MWh, so that total emissions will be 1100 tons (= (0.55)(2000)). This number of allowances will be auctioned to generators, and the payments will be conveyed to consumers via the LSE.
- Generation type A (wind or hydro) has emissions E_A of 0 tons/MWh, constant marginal cost $dC_A/dx_A = 0$ \$/MWh, and capacity $K_A = 500$ MW.
- Generation type B (natural gas) emits $E_B = 0.6$ tons/MWh, incurs constant marginal cost $dC_B/dx_B = 80$\$/MWh, and has ample capacity K_B (no limit).
- Generation type C (coal) has emissions E_C of 1 ton/MWh, constant marginal cost $dC_C/dx_C = 40$ \$/MWh, and ample capacity K_C (no limit).

After setting up the optimization KKTs for each producer, combining them with the market-clearing conditions, and inserting the parameter values, the resulting complementarity problem is:

Find $\{x_A, x_B, x_C, q_A, q_B, q_C, p_A, p_B, p_C, \mu_A, \beta, p^{CO2}\}$ satisfying:

$$\begin{aligned}
&0 \le x_A \perp p_A - \mu_A \le 0 \\
&0 \le x_B \perp p_B - 80 - 0.6p^{CO2} \le 0 \\
&0 \le x_C \perp p_C - 40 - 1p^{CO2} \le 0 \\
&0 \le \mu_A \perp x_A - K_A \le 0 \\
&0 \le q_A \perp - p_A + \beta \le 0 \\
&0 \le q_B \perp - p_B + \beta \le 0 \\
&0 \le q_C \perp - p_C + \beta \le 0 \\
&q_A + q_B + q_C \ + \ Q = \ 0 \quad (\beta \text{ unrestricted}) \\
&0 \le p^{CO2} \perp 0.6x_B + 1x_C \ - (0.55)(2000) \le 0
\end{aligned}$$

[5] The numerical examples here and in Section 11.3.4 are based upon [16].

$$q_A - x_A = 0 \quad (p_A \text{ unrestricted})$$
$$q_B - x_B = 0 \quad (p_B \text{ unrestricted})$$
$$q_C - x_C = 0 \quad (p_C \text{ unrestricted})$$

The resulting solution is shown in Table 11.5. Basically, the market takes as much of the cheap and clean hydropower as it can ($x_A = K_A$); to meet the rest of the demand, as much generation as the 1100 ton CO_2 constraint allows is taken from the relatively inexpensive coal plant, with gas generation making up the rest. By the energy market-clearing condition, the outputs x_f also equal the MW purchases q_f by the LSE.

Table 11.5 Solution for source-based trading example

Primal Variable	MW	Energy Price/Dual	\$/MWh	Carbon Price	\$/ton
x_A	500	p_A	140	p^{CO2}	100
x_B	1000	p_B	140		
x_C	500	p_C	140		
q_A	500	β	140		
q_B	1000	μ_A	140		
q_C	500				

The price for power, \$140/MWh, is identical for all producers. The allowance price of \$100/ton times the limit of 1100 tons yields an allowances rent is \$110,000. This price means that the net marginal cost for B's output is \$140/MWh (= \$80/MWh for fuel + (0.6 ton/MWh)(\$100/ton) for allowances), which is the same as for C (= \$40/MWh for fuel + (1.0 ton/MWh)(\$100/ton)). Neither B nor C earns any operating profit; price equals marginal cost because each of their outputs are basic variables (being strictly positive and having no upper bound). On the other hand, A's marginal cost is \$0, as it has neither fuel costs nor emissions. As a consequence, it produces at its 500 MW capacity, and earn \$70,000 in profits (= (\$140/MWh)(500 MW)).

The LSE pays \$280,000 for its power (= (\$140/MWh)(2000 MW)). But since consumers get the proceeds of the allowances auction, they accrue the allowances rent ($p^{CO2}RQ$ = \$110,000), so the net cost to the consumers is \$170,000, or \$85/MWh.

11.3.3 A Simple Model of Emissions Trading among Load-Serving Entities

A load-based emissions control program frees generators of the responsibility of worrying about whether the allowances they hold are sufficient to cover their emissions. Instead, it is load-serving entities who must consider whether the MWh-weighted emissions of their portfolio of power purchases is less than a cap. This provides an incentive for them to buy power from cleaner producers who then get a higher price for their product, as is shown below. The model for a load-based system of emissions trading consists of submodels of LSE and producer decision making (built in Step 1). The first-order conditions of those submodels (derived in Step 2), when combined with a market-clearing condition, define the market equilibrium (Step 3). Unless otherwise indicated, the notation is the same as for the source-based trading model

Generating firm. We start with the producers, assuming one plant per producer, with a constant marginal cost and a fixed capacity as in the source-based system. Each producer is a price taker. The producer's problem is to maximize net revenues subject to the capacity constraint, without worrying about its emissions:

$$\text{Maximize}_{x_f} \quad p_f x_f - C_f\left(x_f\right) \tag{11.18a}$$

s. t.

$$x_f - K_f \leq 0 \quad (\mu_f) \tag{11.18b}$$

$$x_f \geq 0 \tag{11.18c}$$

Load-serving entity. Turning to the LSE, as in the source-based model, the LSE is regulated so that it pursues a cost minimization objective subject to a fixed demand for electricity Q. In particular, the LSE chooses which producers to buy power from in order to minimize the cost of meeting two constraints: the consumers' demand for power and a new constraint—the emissions constraint that says that the emissions associated with its power purchases cannot exceed the cap:

$$\text{Maximize}_{q_f} \quad -\sum_f p_f q_f \tag{11.19a}$$

s. t.

$$-\sum_f q_f + Q = 0 \quad (\beta) \tag{11.19b}$$

$$\sum_f E_f q_f \leq RQ \quad (\alpha) \tag{11.19c}$$

$$q_f \geq 0, \forall f \tag{11.19d}$$

Equilibrium model. As usual, the competitive equilibrium is achieved when each market party (LSE and producers) is maximizing their objective subject to the market prices, and markets clear. As elsewhere in this chapter, this is accomplished by forming an equilibrium model (Step 3) from the first-order (Kuhn-Tucker) conditions for each party's optimization problem derived in Step 2 together with the clearing condition that supply equals demand for each producer. No market clearing model is needed for allowances (although if there were several LSEs, such a market would be desirable and would likely arise for trade among the LSEs). The equilibrium problem is:

Find $\{x_f, q_f, p_f, \mu_f, \beta, \alpha\}$ satisfying:

$$0 \leq x_f \perp p_f - dC_f/dx_f - \mu_f \leq 0, \quad \forall f \tag{11.20a}$$

$$0 \leq \mu_f \perp x_f - K_f \leq 0, \quad \forall f \tag{11.20b}$$

$$0 \leq q_f \perp -p_f + \beta + \alpha E_f \leq 0 \quad \forall f \tag{11.20c}$$

$$-\sum_f q_f + Q = 0 \quad (\beta \text{ unrestricted}) \tag{11.20d}$$

$$0 \leq \alpha \perp \sum_f E_f q_f - RQ \leq 0 \tag{11.20e}$$

$$q_f - x_f = 0 \quad (p_f \text{ unrestricted}), \forall f \tag{11.20f}$$

The first two conditions are the KKT conditions for the producer, the second three are the KKTs for the LSE, and the last condition is market clearing for energy sales by each producer *f*.

Economic analysis. We now do some analysis of the economic meaning of these conditions, which is part of Step 4. If there is a plant with zero emissions and strictly positive output ($x_f = q_f > 0$), then the third equilibrium condition implies that its energy price, which we will call p^{ZeroE}, equals β. Further, that condition also implies that the price p_f for any generator f with positive output must equal $p^{ZeroE} - \alpha E_f$, where α is the LSE's shadow price of emissions. Thus, if the emissions constraint is binding with a positive dual multiplier α, then cleaner producers get a premium for their power, representing its value to the LSE for meeting its emissions constraint.

11.3.4 Illustrative Example. Simple Load-Based Market Equilibrium

This example uses the same parameter assumptions as the numerical example of source-based trading in Section 11.3.2 in which there are three producers A, B, and C with distinct emissions rates and marginal costs, and a single LSE. In the load-based model, the emissions constraint is in the LSE's optimization problem rather than a market-clearing condition for producer emissions trading. However, we assume the same emissions limit applies, so the LSE has to limit the emissions associated with its energy purchases to 1100 tons (the maximum emissions rate of 0.55 tons/MWh times the consumption of 2000 tons).

The resulting equilibrium problem is:

Find $\{x_A, x_B, x_C, q_A, q_B, q_C, p_A, p_B, p_C, \mu_A, \beta, \alpha\}$ satisfying:

$$\begin{aligned}
&0 \le x_A \perp p_A - \mu_A \le 0\\
&0 \le x_B \perp p_B - 80 \le 0\\
&0 \le x_C \perp p_C - 40 \le 0\\
&0 \le \mu_A \perp x_A - K_A \le 0\\
&0 \le q_A \perp -p_A + \beta \le 0\\
&0 \le q_B \perp -p_B + \beta - 0.6\alpha \le 0\\
&0 \le q_C \perp -p_C + \beta - 1\alpha \le 0\\
&q_A + q_B + q_C \;-\; Q = 0 \quad (\beta \text{ unrestricted})\\
&0 \le \alpha \perp 0.6x_B + 1x_C \;-\; (0.55)(2000) \le 0\\
&q_A - x_A = 0 \quad (p_A \text{ unrestricted})\\
&q_B - x_B = 0 \quad (p_B \text{ unrestricted})\\
&q_C - x_C = 0 \quad (p_C \text{ unrestricted})
\end{aligned}$$

Table 11.6 shows a solution to this model (Step 5 of the modeling process). In summary:

1. Load-based trading yields the same values of the primal variables as the source-based model.
2. However, prices p_f paid by the LSE for each type of generation differ, with the highest price being paid for the cleanest generation (p_A = \$140/MWh for hydropower) and the lowest price paid for the dirtiest (p_C = \$40/MWh for coal based generation).
3. As in the source-based system, only producer A makes a profit ((\$140 /MWh)(500 MW), or \$70,000).
4. The total paid for power by the LSE is \$170,000, or \$85/MWh. Thus, the two systems (load-based and source-based/consumer-owned allowances) result in the same cost to load.

Thus, cleaner generation earns a premium in this market which results from the value it provides by making it easier for the LSE to achieve its emissions target. Further, the shadow price α of the LSE's emissions constraint—\$100/ton—equals the price p^{CO2} of emissions allowances in the source-based example, above. That this is a general result is shown in the next subsection.

Table 11.6 Solution for load-based trading example

Primal Variable	MW	Energy Price/Dual	$/MWh	Carbon Dual	$/ton
x_A	500	p_A	140	α	100
x_B	1000	p_B	80		
x_C	500	p_C	40		
q_A	500	β	140		
q_B	1000	μ_A	140		
q_C	500				

11.3.5 Model Analysis: Equivalence of Source-Based and Load Based Trading

In debates concerning the design of state-level CO_2 trading programs in the U.S., load-based trading had been promoted as costing consumers less than source-based trading. In this section, we show an example of Step 4, the analysis of equilibrium conditions. In particular, we establish the equivalence of the equilibria defined by (11.17) and (11.20), as hinted at by the numerical examples above.

To establish equivalence, it is sufficient to note that a solution to one set of conditions also satisfies the other set, and vice versa. In particular, if $\{x_f, q_f, p_f - E_f p^{CO2}, \mu_f, \beta, p^{CO2}\}$is substituted for $\{x_f, q_f, p_f, \mu_f, \beta, \alpha\}$ in the load-based conditions (11.20), then the source-based equilibrium conditions (11.17) result. Going the other way, substituting $\{x_f, q_f, p_f, \mu_f, \beta, \alpha\}$ for $\{x_f, q_f, p_f - E_f p^{CO2}, \mu_f, \beta, p^{CO2}\}$ in the source-based conditions (11.17) yields the load-based equilibrium conditions (11.20). Thus, an equilibrium for one market is an equilibrium for the other if we note that the load-based energy price p_f is equivalent to the source-based price adjusted downwards by the cost of associated emissions allowances $p_f - E_f p^{CO2}$.

That the producer profits and consumer costs are the same in the load- and source-based systems can be established by substituting $p_f - E_f p^{CO2}$ for its equivalent $p_f$in the producer's profit objective under the load-based policy (11.18a). The result is the profit objective in the source-based case (11.15a):

$$\begin{aligned} \text{Load–based profit} &= p_f x_f - C_f(x_f) \\ &= p_f(x_f - p^{CO2} E_f x_f) - C_f(x_f) = \text{Source–based profit} \end{aligned} \tag{11.21}$$

A similar substitution for the consumer's objective in the load-based case in (11.19a) yields the consumer's objective for the source-based policy (11.16a):

$$\text{Load–based LSE objective} = -\sum_f p_f q_f$$
$$= -\sum_f (p_f - p^{CO2}E_f)q_f = \sum_f p^{CO2} q_f - \sum_f p_f q_f = p^{CO2}\sum_f q_f - \sum_f p_f q_f$$
$$= p^{CO2}RQ - \sum_f p_f q_f = \text{Source–based LSE ojective} \tag{11.22}$$

The first term on the last line results from assuming that the LSE's emissions constraint holds as an equality $\sum_f E_f q_f = RQ$. This implies that consumers pay the same amount for their power under a load-based system as under a source-based system, if the proceeds $p^{CO2}RQ$ from auctioning emissions allowances are returned to consumers.

11.4 Summary

In this chapter, we have introduced two features common to energy markets and, thus, to energy market models: geographic markets connected by capacity-limited transport, and policies that limit emissions. The particular examples were competitive and Cournot markets for electricity in which power transport is governed by a dc load flow model, and two different systems for capping carbon emissions: source-based caps and load-based caps.

The models presented embody some basic principles in modeling transport and emissions. Some of these are congestion pricing in transport; the role of system operators in moving energy and arbitraging price differences among markets; and trading of emissions allowances. The flexibility of complementarity approach to modeling markets makes possible many variants and extensions of these models.

For instance, in the case of transport, in Section 10.4.6, above, a natural gas network operator links gas producers and consumers, while in Section 12.4, international trade in coal and petroleum products is represented. Turning back to electricity, more elaborate load flow models can be embedded in competitive and oligopolistic electricity models, including quadratic resistance losses, phase shifters and other FACTS devices, a mix of dc and ac lines [17], and even full-blown ac load flow constraints ([1] and Ch. 6 of [39]). Oligopolistic competition in markets for transmission services [19], financial transmission rights [2], and operating reserves can also be modeled by complementarity methods [3].

The basic transmission model, such as that in Section 11.2, generates locational marginal prices, representing the marginal cost of providing power at each location at each time, and resulting in clearing of supply, demand, and net imports at each point. However, energy transport is often regulated, causing important deviations from that locational pricing model. This too can be modeled by complementarity approaches. For instance, regulators may impose average-cost based pricing for transmission, guaranteeing a return on capital for a regulated transport utility [31]. Alternatively, competition might take place in a system in which transmission is priced by auctioning off interfaces between neighboring countries in a path-based accounting scheme, while still respecting physical load flows [19]. Or a market design might define zones that impose the same price on all nodes within each zone [38].

The transport models just mentioned considered operations of a fixed transport infrastructure; but an important problem for the energy modeling community is the modeling of transport investment [23]. This problem poses several interesting challenges. Among them are model size, because investment models must encompass multiple years or decades, and model non-convexities, because investments are lumpy and, in some cases, introduce nonlinearities into the constraint sets. For instance, swing factors in the linearized dc load flow representation are a function of transmission capacity, and are multiplied by power injection variables in the constraint set of the system operator.

Likewise, with environmental policies, the complementarity framework makes it possible to capture many technical, economic, and institutional complications that are relevant to evaluating such policies. Section 12.4, below, illustrates the interactions of different energy sectors – coal and oil – who compete with each other for carbon allowances. As another example of a complexity modeled by complementarity approaches, emissions trading systems often involve free distribution of allowances by various rules. It has been shown that free distribution to new generation investment can distort the mix of those investments and increase the social cost of environmental cleanup, for instance if new dirtier plants are given more allowances [40]. Trade between markets, some of which have emissions caps and others which do not, can make it more difficult for the capped regions to achieve true reductions in emissions [9]. Unlike the models in Sections 7.4, 7.5 and 11.2, these models include generation investment as well as operations as decision variables, and could also represent investment in emissions controls such as carbon capture or SO_2 scrubbers. Renewable energy policies are another rich potential area of potential application of complementarity models [34], which can represent the goulash of tax subsidies, feed-in tariffs, and renewable energy credit policies that often simultaneously tug renewable investors in several ways at once.

11.5 Exercises

1. Set up and solve the perfect competition transmission equilibrium problem in Section 11.2.2 as a mixed complementarity problem. Confirm that the solution presented in that section is an equilibrium.
2. Set up and solve the oligopolistic transmission equilibrium problem in Section 11.2.4 as a mixed complementarity problem. Confirm that the solution presented in that section is an equilibrium.

3. Set up a version of the oligopolistic transmission equilibrium problem in Section 11.2.4, but omitting the arbitrage variables a_i from the TSO's problem and market clearing conditions. This represents a situation in which the TSO does not arbitrage the market, but only provides transmission services for generators. Implement and solve the resulting MCP. Verify that there is spatial price discrimination; that is, the differences in prices among the nodes of the network do not equal the differences in transmission charges w_i [15].
4. Verify that the MCP (11.11) has as many variables as conditions. Repeat for the MCP at the beginning of Section 11.2.4.
5. Set up a MCP model like that in Section 11.3.1 for an electric power market that is subject to a renewable portfolio standard rather than an emissions cap. In such a market, generating firms must either generate renewable energy or purchase renewable energy credits amounting to a given percentage (e.g., 20%) of their total MWh sales. Credits can be bought from other generating firms who generate more than 20% renewable energy credit. First, define the revised profit maximization problem for each generating company, including the net cost of credits bought (which will be negative if the company sells excess credits) and a constraint to account for the number of credits it must buy. Also include appropriate decision variables for the construction and operation of a renewable energy generator with an assumed fixed investment cost per MW and no operating costs. Second, define an appropriate market clearing constraint with a price. Third, define the complementarity problem for the market equilibrium. Fourth, assume values for model parameters and solve it numerically, noting the price for power and renewable energy credits as you increase the percent credits required.
6. Generalize the load-based model of Section 11.3.3 by allowing LSEs the opportunity to invest in energy efficiency (i.e., by investing money which will lower their energy requirements Q). Appropriately modify the LSE model with a decision variable for the amount to invest in energy efficiency. Write down the resulting market equilibrium problem. Assume values of the parameters (the cost of an energy efficiency program per MWh of load reduction, the maximum amount of load reduction that can take place), and solve the resulting model numerically.

References

1. G. Bautista, M.F. Anjos, and A. Vannelli. Formulation of oligopolistic competition in ac power networks: an NLP approach. *IEEE Transaction on Power Systems*, 22(1):105-115, 2006.
2. G. Bautista and V.H. Quintana. Complementarity-based models for financial transmission rights. *Proceedings, IEEE Power Engineering Society General Meeting*, San Francisco, June 12-16, 2005
3. G. Bautista, V.H. Quintana and J. Aguado. An oligopolistic model for an integrated market of energy and spinning reserve. *IEEE Transaction on Power Systems*, 21(1):132-142, 2006.
4. E.B. Bluemel. Regional regulatory initiatives addressing GHG leakage in the USA, In M. Faure and M. Peeters, Editors. *Climate Change and European Emissions Trading: Lessons for Theory and Practice*, Edward Elgar Publishing, 2009.
5. E. Celebi and D. Fuller. A model for efficient consumer pricing schemes in electricity markets. *IEEE Transactions on Power Systems*, 22(1):60-67, 2006.
6. E. Centeno, J. Reneses and J. Barquin. Strategic analysis of electricity markets under uncertainty: a conjectured-price-response approach. *IEEE Transactions on Power Systems*, 22(1):423-432, 2007.
7. R. Chakkapalli and P. Poonpun. System operating limit study report, California-Oregon Intertie, Pacific DC Intertie, Path 26 and SCIT Paths, Spring 2012. California Independent System Operator, Folsom. January 31, 2012.
8. H. P. Chao and S. Peck. A market mechanism for electric power transmission. *Journal of Regulatory Economics*, 10(1):25-59, 1996.
9. Y. Chen, A.L. Liu, and B.F. Hobbs. Economic and emissions implications of load-based, source-based and first-seller emissions trading programs under California AB32. *Operations Research*, 59:696-712, 2011.
10. Y. Chen, J. Sijm, B.F. Hobbs, and W. Lise. Implications of CO_2 emissions trading for short-run electricity market outcomes in northwest Europe. *Journal of Regulatory Economics*, 34(3):251-281, 2008.
11. R.W. Cottle, J.-S. Pang, and R.E. Stone. *The Linear Complementarity Problem*. Academic Press, 1992.
12. C.J. Day, B.F. Hobbs, and J.-S. Pang. Oligopolistic competition in power networks: a conjectured supply function approach. *IEEE Transactions on Power Systems*, 17(3):597-607, 2002.
13. M. Grubb and K. Neuhoff. Allocation and competitiveness in the EU emissions trading scheme: policy overview. *Climate Policy*, 6:7-30, 2006.
14. F.S. Hillier and G.J. Lieberman. *Introduction to Operations Research*. McGraw-Hill, New York, 8th edition, 2005.

15. B.F. Hobbs. Linear complementarity models of Nash-Cournot competition in bilateral and POOLCO power markets. *IEEE Transactions on Power Systems*, 16(2):194-202, May 2001.
16. B.F. Hobbs, J. Bushnell, and F.A. Wolak. Upstream vs. downstream CO_2 trading: a comparison in the electricity context. *Energy Policy*, 38(7):3632-3643, 2010.
17. B.F. Hobbs, G. Drayton, E.B. Fisher, and W. Lise. Improved transmission representations in oligopolistic market models: quadratic losses, phase shifters, and dc lines. *IEEE Transactions on Power Systems*, 23(3):1018-1029, 2008.
18. B.F. Hobbs and J.-S. Pang. Spatial oligopolistic equilibria with arbitrage, shared resources, and price function conjectures. *Mathematical Programming Series B*, 101(1):57-94, 2004,.
19. B.F. Hobbs and F.A.M. Rijkers. Modeling strategic generator behavior with conjectured transmission price responses in a mixed transmission pricing system I: formulation. *IEEE Transactions on Power Systems*, 19(2):707-717, 2004.
20. W.W. Hogan. Contract networks for electric power transmission. *Journal of Regulatory Economics*, 4(3):211-242, 1992.
21. R.B. Johnson, S.S. Oren, and A.J. Svoboda. Equity and efficiency of unit commitment in competitive electricity markets. *Utilities Policy*, 6(1):9-20, 1997.
22. J. Kolstad and F. Wolak. Using environmental emissions permit prices to raise electricity prices: evidence from the california electricity market. Technical report CSEM WP-113, University of California, Berkeley, May 2003.
23. W. Lise, B.F. Hobbs, and F. van Oostvoorn. Natural gas corridors between the EU and its main suppliers: simulation results with the dynamic GASTALE model. *Energy Policy*, 36(6):1890-1906, 2003.
24. C. Metzler, B.F. Hobbs, and J.-S. Pang. Nash-Cournot equilibria in power markets on a linearized dc network with arbitrage: formulations and properties. *Networks & Spatial Economics*, 3(2):123-150, 2003.
25. D. Niemeier, G. Gould, A. Karner, M. Hixson, B. Bachmann, C. Okma, Z. Lang, and D. Heres Del Valle. Rethinking downstream regulation: California's opportunity to engage households in reducing greenhouse gases. *Energy Policy*, 36(9):3436-3447, 2008.
26. S.S. Oren. Economic inefficiency of passive transmission rights in congested electricity systems with competitive generation. *Energy Journal*, 18(1):63-83, 1997.
27. E.E. Sauma and S.S. Oren. Proactive planning and valuation of transmission investments in restructured electricity markets, *Journal of Regulatory Economics*, 30:261-290, 2006.
28. F.C. Schweppe, M.C. Caramanis, R.D. Tabors, and R.E. Bohn. *Spot Pricing of Electricity*. Kluwer, Boston, 1988.
29. J. Sijm, K. Neuhoff, and Y. Chen. CO_2 cost pass-through and windfall profits in the power sector. *Climate Policy*, 6:49-72, 2006.
30. Y. Smeers and J. Boucher. Alternative models of restructured electricity systems, part 1: no market power. *Operations Research*, 49(6):821-838, 2001.
31. Y. Smeers and J.-Y. Wei. Spatial oligopolistic electricity models with Cournot generators and regulated transmission prices. *Operations Research*, 47(1):102-112, 1999.
32. T. Takayama and G. Judge. *Spatial and Temporal Price and Allocation Models*. North Holland Publishing Company, Amsterdam, 1971.
33. T. Tietenberg. *Emissions Trading—Principles and Practice*. Resources for the Future Press, Washington, DC, 2006.
34. T. Traber and C. Kemfert. Impacts of the German support for renewable energy on electricity prices, emissions, and firms. *Energy Journal*, 30(3):155-178, 2009.
35. M. Ventosa, A. Baillo, A. Ramos, and M. Rivier. Electricity market modeling trends. *Energy Policy*, 33(7):897-913, 2005.
36. F. Wu, P. Variya, P.T. Spiller and S.S. Oren. Folk theorems on transmission open access: proofs and counter examples, *Journal of Regulatory Economics*, 10:5-23, 1996.

37. H. Yang, C.Y. Chung, and K.P. Wong. Optimal fuel, power and load-based emissions trades for electric power supply chain equilibrium, *IEEE Transactions on Power Systems*, to appear, 2012.
38. J. Yao, S. Oren, and I. Adler. Two settlements electricity markets with price caps and Cournot generation firms. *European Journal of Operations Research*, 181(3):1279-1296, 2007.
39. X.-P. Zhang, Editor. *Restructured Electric Power Systems: Analysis of Electricity Markets with Equilibrium Models.* Wiley/IEEE Press, 2010.
40. J. Zhao, B.F. Hobbs, and J.-S. Pang. Long-run equilibrium modeling of alternative emissions allowance allocation systems in electric power markets. *Operations Research*, 58(3):529-548, 2010.

Chapter 12
Multicommodity Equilibrium Models: Accounting for Demand-Side Linkages

12.1 Introduction

Several of the models previously introduced in this book have focused on the market for a single commodity with a single price, such as power at a particular location in a particular hour. However, many of this book's models instead considered several markets simultaneously, recognizing that linkages among them imply that equilibrium prices in one market cannot be calculated without considering how they affect, and are affected by, prices in other markets. In the earlier chapters, linkages among markets were mainly through the supply-side, in which the cost of providing commodity in one market depends in part on prices in other markets. For instance, a power generator with only a small amount of capacity with low running costs might experience a rise in its marginal cost of serving one part of the network if it also sells a lot of power elsewhere, thereby exhausting its cheap capacity. The purpose of this chapter is to introduce the modeling of multiple energy markets in which it is instead the behavior of consumers that links the markets. In particular, the amount that final consumers buy of one commodity affects how much they are willing to pay for other commodities.

In this chapter, we model the demand-side linkages among multiple energy markets. We use two extensive case studies to illustrate the ideas, and to show how disregarding cross-price elasticities can distort the results of energy models. Before introducing the case studies, however, we provide in Section 12.2 an overview of some important concepts and applications in modeling demand-linkages in energy markets. Then in Section 12.3, we consider our first case study, which illustrates the modeling of markets in which the cross-price elasticities occur across time. In particular, we present an example that considers the effect of real-time electricity pricing on the time profile of demand and the costs of meeting it. Cross-price elasticities are used in that case to represent the ability of consumers to shift their use of electricity from one hour to another. Consideration of this capability significantly affects the estimated amount of power that consumers consume in the market equilibrium.

In Section 12.4, we turn to multicommodity energy models that are used to model entire energy economies, accounting for interactions of different types of fuels and, ultimately, how consumers choose among different ways to meet their demands for energy services. The numerical example is inspired by the famous PIES (Project Independence Evaluation System), one of the first national energy policy models, which we have already discussed in Chapters 1, 5, 9 and 10.

In presenting the modeling of demand linkages through cross-price elasticities, we also introduce the reader to some other important concepts in modeling and modeling practice. One is the use of linear (Leontief) production functions in the modeling of energy conversion (Section 12.4.2.3). In Section 12.4.4, we then present some useful approaches to interpreting model results. We show how price formation can be understood by identifying basic primal variables and studying the associated KKT conditions. For instance, final product prices might be influenced most by marginal production costs, or instead bottlenecks in supply or transport might mean that consumer willingness-to-pay determines prices. We also illustrate how multiple runs of a model under alternative policy assumptions can be used to understand the cost of a policy, and who bears it. It is important that the modeler formulate some hypotheses ahead of time about the possible impacts of the policy being studied for three reasons: to specify the experimental design (which model runs to undertake), to aid in the checking of results, and to structure the discussion of the model results. Our PIES application shows that assumptions about cross-price elasticities can make a large difference in conclusions about the expense of achieving environmental goals, and about who bears the brunt of that expense.

12.2 Linkages among Multiple Energy Markets

In this section, we discuss some general considerations in modeling linked energy markets. Markets may be distinguished by type of energy, timing, or location. They can be linked through various types of supply-side linkages, but in this chapter we emphasize demand-side linkages. We introduce the crucial modeling tool of cross-price elasticities in demand functions, but also note that there are alternative ways to represent complementary and substitution relationships among the energy commodities that a consumer buys. We close the section with some observations on the importance of considering demand linkages in energy models, which indeed were a major motivation for development of the original PIES model [13].

The various types of markets that might be considered simultaneously in a multimarket model, might be for fundamentally distinct commodities, such as coal, natural gas, emissions allowances, and electricity. Technical relationships on the supply side may mean that two commodities can substitute for each other (e.g., gas for coal in plants). Or, two commodities can be complements (e.g., if coal is burned in a power station, the generator must also buy emissions allowances). In either case, what a producer would be willing to pay for one input depends on the price of the other. Or one commodity might be transformed into another through the production process, so that the expense of inputs (say, gas) will affect the cost of producing outputs (in that case, electricity). Changes in input prices then affect the supply of the output.

Alternatively, the multiple markets may be for essentially the same commodity (e.g., natural gas), but because of transport costs or storage limitations commodities might be differentiated over space (e.g., different prices in different countries) or time (e.g., higher prices for peak electricity than for off-peak). In a spatial market equilibrium, transport costs or constrained transmission capacity limit how much energy can be exported by one market or imported by another. These spatial linkages can be quite complicated, as in power load flow models or gas transportation pipeline systems. In these circumstances, prices at different locations diverge. In competitive markets, these price differences will equal the marginal cost of energy transport, if there are no binding capacity limits in the transport system, as competitive producers or traders will arbitrage away non-cost based price differences (see Section 12.4.4.1). When transport constraints bind, differences in prices will include a shadow price component accounting for the marginal value of transmission capacity, and competitive markets will ensure that the price paid to those providing transport services equals price differences among markets (see Section 11.2 as an example). In oligopolistic markets, however, this arbitrage may not occur, and price differences may reflect the ability of large producers to price discriminate among consumers at different locations.

Such trade and arbitrage can be viewed as taking place through time as well, if an energy form can be put into storage in one time period and then withdrawn and consumed later when price is higher (as in the natural gas model featured in Chapter 10). In this case, there are multiple markets for the same commodity, but at different times. Just as in spatial arbitrage, competitive conditions will determine whether or not the price paid for storage and the price differences at different times are based on underlying costs.

In nearly all of the previous models of this book, the linkages among multiple markets have been on the supply-side, through transport or storage of commodities, or through substitution or complementary relationships among different inputs when transforming one energy form to another. Final consumers, in contrast, have been represented in previous chapters simply as agents who buy just one commodity, following a demand curve in which the amount they buy q_i of commodity i is a function only of its price p_i, $q_i = f_{d,i}(p_i)$. However, there are in fact very important linkages among markets on the demand side. A final consumer's willingness-to-pay for one commodity can depend on the price of the other, as we explained with a simple coal-wood example in Section 3.2.5. Commodities can be complements (natural gas and gas-fired furnaces, electricity and light-bulbs) or substitutes (different types of light bulbs, alternative fuels for heating, and gasoline versus diesel and, increasingly, natural gas or electricity for vehicle transport).

Commodities at different times can also have these relationships. A key feature of the "Smart Grid" is the use of real-time electricity prices (or other signals) to encourage consumers to be more flexible in the timing and amount of their energy use. As a result, the price of peak energy will affect off-peak demands because a residential consumer may decide to delay the timing of doing the wash or dishes, or may have an energy storage device in the home. Use at different times can be substitutes (when energy-using activities can be rescheduled to take advantage of low prices at certain times). They can also be complements (in the short run, if a dishwasher takes two adjacent hours to complete its operation, while in the long run, consistently low prices at some times during the day triggers purchase of energy-using equipment that increases energy use in all hours).

These relationships can be captured in a more general demand function $q_i = f_{d,i}(p)$, with p being a vector of prices $\{p_1,\ldots, p_I\}$. The partial derivatives $\partial q_i/\partial p_j$, $i{\neq}j$ signal whether different commodities are complements or substitutes (see Section 3.2.6). If increasing the price for commodity j (such as diesel or peak power) increases the quantity demanded of i (say, gasoline or off-peak power), then there is a substitution relationship between the commodities. A negative partial derivative, however, indicates that the commodities are complements.

When describing these relationships, economists more commonly refer to the dimensionless cross-price elasticity, which Section 3.2.6.1 defined as:

$$\varepsilon_{ij} = (\partial q_i/\partial p_j)(p_j/q_i) \tag{12.1}$$

Generally, own-price elasticities ε_{ii} (almost always negative) are larger in magnitude than cross-price elasticities $\varepsilon_{ij}, i{\neq}j$. For instance, short-run own elasticities for electricity demand in a particular hour are on the order of -0.1 to -0.2, while elasticities with respect to demand in other hours are positive but significantly smaller in magnitude (see, for instance, [6]).

An alternative view of the final consumer is the household production framework, which represents the consumer as a kind of "producer." That view recognizes that the final consumer is really interested in the ultimate amount of energy services that are consumed (light, heat, transport, . . .), and uses capital, electricity, labor, and other inputs to produce those services. In Amory Lovins' famous phrasing, what consumers want is cold beer and warm showers, not electricity or gas. In this framework, a household "production function" links household inputs (including energy) with the desired energy services. An economist's characterization of the consumer's decision process would then be of an agent who chooses the least-cost mix of inputs to meet her demand for energy services, and that the amount of energy services demanded depends on the implicit price of those services, based on the price of inputs and their marginal productivity in producing the services. The demand function is then derived from the optimizing behavior of consumers, rather than assumed to have particular elasticities and cross-price elasticities. If one builds an energy model in which energy suppliers optimize some objective subject to the optimizing behavior of consumers with household production functions, the resulting model has an EPEC structure.

As an example of the use of a household production framework in a complementarity-based energy model, Hobbs and Nelson [12] present an EPEC in which an electric utility (the upper level, or Stackelberg leader) anticipates how consumers (the lower level, or Stackelberg followers) will react to two different types of incentives: electricity prices and subsidies for investments in efficient energy-using equipment. The consumers behave according to the household production framework, choosing how much energy to buy and how much to invest in energy using equipment in order to maximize the net benefits of consuming energy services.

Although the household production function representation of consumer decision making may seem both over simplified (assuming optimizing behavior) and difficult to model (because of the lack of data on household production functions and demands for energy services), it can still provide useful insights. For instance, from this perspective, the phenomenon of energy "rebound" or "takeback", in which energy savings from home insulation or efficient lighting are less than anticipated due to behavioral changes can be understood as a rational response to a decrease in the implicit price of energy services. In particular, a consumer will see that the marginal cost of cooling her home or keeping her lights on has fallen because less energy is now required to do so, and she may decide that the worth of a more comfortable temperature setting or leaving the lights on longer exceeds the now-lower incremental energy cost. This behavior is exhibited in the model in [12]. The household production function approach even allows an economist to quantify the economic value (in terms of consumer surplus for the energy service) of this "rebound" [10]. Thus, although the existence of rebound may disappoint the engineer who calculated the theoretical energy savings of a more efficient home or energy-using equipment, that rebound actually increases the net benefit to the consumer of that investment.

Models that capture the interactions of final demand for different energy commodities are increasingly useful for many policy and planning activities. Recognizing these interactions allows policies to exploit substitution possibilities in order to lower the cost and environmental impact of energy use. Smart Grids, for instance, can lower the amount of power plant and transmission capacity needed by encouraging consumers to shift their uses off-peak. Policies to promote electric or natural-gas vehicles can reduce oil imports. Pollution can be lessened by electric vehicles as well, if the fuels used to generate electricity are cleaner than gasoline and diesel fuels. On the other hand, disregarding such interactions can cause policies to fail to meet their objective. As an example, taxing residential natural gas use to encourage conservation may simply motivate consumers to use more electric portable or baseboard heaters. Then if that electricity is produced by gas fired power plants, the resulting total gas use may be higher, not lower because most of the energy value of the fuel is lost when generating and transmitting electricity. As another example, promoting biofuels as substitutes for oil in transportation may or may not lower carbon emissions or oil imports, depending on how the biofuels are produced.

For these reasons, representing demand linkages was considered to be a critical feature in the original PIES model [13], which interfaced a linear programming supply-side model with an econometric demand model with nonzero cross-price elasticities (see Section 1.2.3 for a brief overview of PIES). Today, its successor, the National Energy Modeling System, is used by the US Energy Information Agency for energy projection and policy analysis purposes, and a distinguishing feature of that model is its capturing of cross-price elasticities in the demand for energy by final consumers [8, 16, 18]. However, not all economy-wide multicommodity energy models include cross-price elasticities in final demand [4]. For instance, the popular MARKAL model [7] (and its related models EFOM [5] and TIMES [17]) assumed a fixed demand for energy services in a linear programming formulation. In those models, substitution relationships are partially but not completely captured when consumers choose among different energy forms to meet those fixed demands at least cost.

12.3 Demand Relations over Time

In this section,[1] we illustrate the modeling of demand interdependencies over time by considering a simple case in which there are two electricity demand periods – off-peak (or base) and peak. Consumption in one period can substitute for another, a relationship that is represented by positive cross-price elasticities. In the model of Section 12.3.1, we also illustrate the use of regulatory rules to calculate prices. That example represents a regulated vertically integrated utility; in the case of Section 12.3.2, we instead consider how interdependencies affect the outcome of a competitive generation market in which retail prices also include a regulated transmission and distribution (T&D) component.[2] Comparing the two models also allows us to explore how time-

[1] The contributions of Dr. Cedric De Jonghe of KU Leuven, who performed the calculations of this section, are gratefully acknowledged.

[2] Celebri and Fuller [1, 2] present more realistically sized electricity market equilibrium models that include such intertemporal elasticities. An interesting feature of their models is their resolution of the different time scales for generator price variations (e.g., hourly

differentiated prices can result in beneficial shifts in the timing of electricity demand.

In both models there are two demand periods per year of equal length (4380 hrs each), with period $t = 1$ being the off-peak period and period $t = 2$ being the peak period. Prices p_t \$/MWh are differentiated by period t, while quantity demanded q_t MW in each period is a function of both of the prices as follows:

$$q_1 = 3500 - 14.86p_1 + 2.972p_2 \tag{12.2a}$$

$$q_2 = 5600 + 4.756p_1 - 23.78p_2 \tag{12.2b}$$

(The particular parameter values assumed in these functions, and in the rest of the model of this section, are hypothetical ones for the purposes of illustrating the methods.) In this case, the response of each quantity (e.g., q_1) to its own price (p_1) is negative and five times larger in magnitude than its response to the other period's price (in that case, p_2). For instance, an increase of 1 \$/MWh in p_1 decreases load in that period q_1 by 14.86 MW (or 14.86 MW $\times$ 12 hrs = 178.32 MWh), but an increase in the other period's price p_2 by the same amount increases q_1 by 2.972 MW (35.66 MWh). The load-shifting effect of price changes is apparent if we consider the two effects of increasing the peak price p_2 by \$1/MWh: peak load q_2 goes down by 23.78 MW (12.2b), but is partially offset by a 2.972 MW increase in off-peak load (12.2a).

Note that (12.2) allows quantities demanded to be negative; if they must be nonnegative, then a more general version of the demand relationship involving a complementarity rather than equality condition would be used. Section 12.4.2.1 provides details on how those complementarity conditions would be formulated.

Next, we solve a complementarity model that represents a regulated monopoly's decisions in the face of these demand functions, where regulators require that the two periods' prices are equal. Then in Section 12.3.2, we consider an unbundled electricity market in which prices in different periods can differ.

12.3.1 Regulated Vertically Integrated Utility Model

First consider a combined generation-transmission-distribution utility that is a regulated monopoly. In the U.S., such vertically integrated utilities are still the rule, for example, in the southeast. The regulator ensures that all generation and distribution costs (including a regulated return on capital) are recovered from consumers by charging them a uniform regulated price p

---

in the North American day-ahead markets) and retail consumer price variations (e.g., time-of-day, with two periods per day and the same prices in each day of the month).

that is the same in all periods ($p_1 = p_2 = p$). We assume that T&D costs are independent of quantity demanded and arbitrarily assume they amount to \$854.1×10^6/year.

The model has three components or submodels: the demand functions (12.2), generation investment and operations, and a set of regulatory rules.

In the generation submodel, power x_{it} is produced from either of the monopolist's two power plants $i \in \{A,B\}$; we will consider a long-run problem with variable capacity, so the plants' capacity k_i is also a decision variable. We consider a single year, rather than a dynamic (multi-year) investment problem in which the objective of the regulated utility is to minimize its total generation cost,[3] including both variable generation costs and annualized investment costs, subject to meeting the demand at the regulated prices. Its optimization model, which includes in the objective function both the annualized cost for capacity k_i and fuel costs for generation x_{it}, is as follows:

$$\begin{aligned} \text{Minimize}_{x_{it},k_i} \quad & 180,000k_A + 4380(30x_{A1}) + 4380(30x_{A2}) \\ & + 70,000k_B + 4380(50x_{B1}) + 4380(50x_{B2}) \end{aligned} \qquad (12.3a)$$

s. t.

$$\sum_{i \in \{A,B\}} x_{it} = q_t \qquad (\lambda_t), \quad t \in \{1,2\} \qquad (12.3b)$$

$$-x_{it} + k_i \geq 0 \qquad (\mu_{it}), \quad i \in \{A,B\}, t \in \{1,2\} \qquad (12.3c)$$

$$x_{it}, k_i \geq 0, \quad i \in \{A,B\}, \quad t \in \{1,2\} \qquad (12.3d)$$

The capacity constraint is written as a greater than or equal to constraint so that its dual multiplier will be nonnegative in this minimization problem. A more sophisticated alternative to the annualized cost formulation is to minimize the present worth of costs over all years in a planning time horizon, considering how demand, fuel prices, etc. evolve from year to year.[4] This more complex model would certainly be more realistic, but is not necessary for the purposes of this chapter.

The regulator's submodel has two parts: first, a cost-recovery rule that can be written by setting the utility's revenue equal to the fixed T&D cost plus generation cost (12.4a) and, second, a rule that prices in the two periods are equal (12.4b):

$$4380p_1q_1 + 4380p_2q_2 = 854,100,000$$

---

[3] Why such a utility would be motivated to do so when the regulator guarantees cost recovery is a mystery as well as a heroic assumption, but we will assume that it does! However, in the absence of a widely accepted alternative behavioral model for regulated utilities, the cost-minimization assumption is frequently made, at least as a base case against which other solutions are compared.

[4] The classic Turvey and Anderson [21] text shows how such a multiyear model can be formulated; or see Hobbs [11].

$$+ [180,000k_A + 4380(30x_{A1}) + 4380(30x_{A2})$$
$$+ 70,000k_B + 4380(50x_{B1}) + 4380(50x_{B2})] \quad (12.4a)$$
$$p_1 = p_2 \quad (12.4b)$$

Now the three components can be put together to form the regulated vertically integrated utility model, which we use to calculate the equilibrium between the consumers and the utility. The model is created by concatenating the demand functions (12.2), regulatory conditions (12.4), and the KKTs for the generation problem (12.3), yielding the following square LCP:

Find $\{q_1, q_2, p_1, p_2, k_A, k_B, x_{A1}, x_{A2}, x_{B1}, x_{B2}, \lambda_1, \lambda_2, \mu_{A1}, \mu_{A2}, \mu_{B1}, \mu_{B2}\}$ satisfying:

Demand functions 12.2

Regulatory cost-recovery and price rules 12.4

KKTs for the utility's generation cost minimization 12.3:

$$0 \leq k_A \perp 180,000 - \mu_{A1} - \mu_{A2} \geq 0 \quad (12.5a)$$
$$0 \leq k_B \perp 70,000 - \mu_{B1} - \mu_{B2} \geq 0 \quad (12.5b)$$
$$0 \leq x_{A1} \perp (4380)(30) + \mu_{A1} - \lambda_1 \geq 0 \quad (12.5c)$$
$$0 \leq x_{A2} \perp (4380)(30) + \mu_{A2} - \lambda_2 \geq 0 \quad (12.5d)$$
$$0 \leq x_{B1} \perp (4380)(50) + \mu_{B1} - \lambda_1 \geq 0 \quad (12.5e)$$
$$0 \leq x_{B2} \perp (4380)(50) + \mu_{B2} - \lambda_2 \geq 0 \quad (12.5f)$$
$$\sum_{i \in \{A,B\}} x_{i1} = q_1 \quad (p_1) \quad (12.5g)$$
$$\sum_{i \in \{A,B\}} x_{i2} = q_2 \quad (p_2) \quad (12.5h)$$
$$0 \leq \mu_{A1} \perp - x_{A1} + k_A \geq 0 \quad (12.5i)$$
$$0 \leq \mu_{A2} \perp - x_{A2} + k_A \geq 0 \quad (12.5j)$$
$$0 \leq \mu_{B1} \perp - x_{B1} + k_B \geq 0 \quad (12.5k)$$
$$0 \leq \mu_{B2} \perp - x_{B2} + k_B \geq 0 \quad (12.5l)$$

An important characteristic of this model is that the above linear MCP (consisting of (12.2), (12.4), (12.5) does *not* correspond to the KKT conditions of any known single optimization model. That is, it is not possible to obtain the solution by solving a single equivalent optimization model. This is in contrast to some of the other market models we've presented elsewhere in this book. For instance, in the case of a competitive market with no regulatory constraints and zero cross-price elasticities, the Samuelson [20] principle can be used to formulate a single optimization problem whose KKTs are the equilibrium conditions of the competitive market. As another example, for Cournot markets with linear demand and neither price regulations nor cross-price elasticities, Salant's [19] and Hashimoto's [9] equivalent single-optimization problem yields the same solution as the mixed complementarity problem that defines the Cournot equilibrium (see for example Section 3.4.2.5). However, in the case here, the cross-price elasticities prevent use of a single equivalent optimization model to find the equilibrium. (An exception, pointed out later in this chapter, occurs in the special case in which the cross-price coefficients in the linear demand functions are symmetric.) Further, even if there were no cross-price elasticities, the regulatory conditions (12.4) would rule out the existence of such an equivalent optimization problem.

Solving the above linear MCP with, say, the PATH solver yields a regulated utility equilibrium:

$$\begin{aligned}&\{q_1, q_2, p_1, p_2, k_A, k_B, x_{A1}, x_{A2}, x_{B1}, x_{B2}, \lambda_1, \lambda_2, \mu_{A1}, \mu_{A2}, \mu_{B1}, \mu_{B2}\}\\&= \{2500 \text{ MW}, 4000 \text{ MW}, 84.11 \text{ \$/MWh}, 84.11 \text{ \$/MWh},\\&2500 \text{ MW}, 1500 \text{ MW},\\&2500 \text{ MW}, 2500 \text{ MW}, 0 \text{ MW}, 1500 \text{ MW},\\&(4380)(35.11) \text{ \$/MW-yr}, (4380)(65.98) \text{ \$/MW-yr},\\&(4380)(5.11) \text{ \$/MW-yr}, (4380)(35.98) \text{ \$/MW-yr},\\&0 \text{ \$/yr}, (4380)(15.98) \text{ \$/MW-yr}\}.^5\end{aligned}$$

The price of 84.11 \$/MWh is charged for each MWh sold in both periods, even though the marginal cost of generation is different in each period (35.11 \$/MWh in t=1 and 65.98 \$/MWh in t=2).[6] The gap between the generation

[5] Note that the units of the dual variables are \$/MW-yr, not \$/MWh, because the objective function is annual cost, in \$/yr, and the constraint right-hand side is in MW. Thus the units of the dual variables are the ratio of the two sets of units (\$/yr)/(MW), or \$/MW-yr. If we divide the dual variables by the number of hours in the demand block, in this case 4380hr/yr, the duals can be expressed in terms of \$/MWh. For instance, the dual λ_1 is 153,800 \$/yr, or 35.11 \$/MWh.

[6] The on-peak marginal cost \$65.98/MW equals the fuel cost of the peaker (plant type B, at \$50/MWh) plus its capital cost divided by the 4380 hrs/yr duration of the peak period (\$70,000/MW/yr ÷ 4380 hr/yr, or \$15.98/MWh). This is because 1 more MW demand during that period would result in 1 more MW of peaker capacity being built and then operated for that length of time.

However, the off-peak marginal cost is more complicated, involving an interaction between the baseload capacity and peaker decisions. If load increases by 1 MW off-peak, 1 more MW of baseload capacity (plant type A) would be built and operated for all 8760 hours, while 1 MW less of peaker would then be needed, which saves both its capital cost and 4380 hours of operation. This is a total cost increase of ((\$180,000/MW/yr) + (30 \$/MWh)(4380 hours/yr)) minus ((\$70,000/MW/yr) + (50 \$/MWh)(4380 hrs/yr)) = \$153,800/MW/yr. Dividing that by the 4380 hr/yr duration of the off-peak period gives the off-peak marginal cost of \$35.11/MWh.

Notice that these calculations do not account for the feedback between the stated 1 MW change in one period's load, equilibrium prices (because of the need to recover T&D costs and the elasticity demand) and thus load in both periods. For instance, an increase in peak load would allow the fixed T&D costs to be spread among more MWh, lowering the T&D surcharge in the unbundled generator model, which would then lead to a further peak load increase as well as an off-peak load increase. The dual variables do not reflect this feedback marginal costs and price is in part necessary to offset the fixed T&D cost of $854.1×10^6/year—and in fact, by (12.4a), produces exactly enough extra revenue to just cover those costs.

At these prices and quantities, the price elasticities $\epsilon_{t,t'} = (\partial q_t/\partial p_{t'})(p_{t'}/q_t)$ for the demand functions (12.2) can be calculated. The own elasticities $\epsilon_{t,t} = \epsilon_{1,1} = \epsilon_{2,2} =$ -0.5; thus, a 1% increase in price in period t decreases load in that period by 0.5%. The cross-price elasticities are positive: $\epsilon_{1,2} = \epsilon_{2,1} =$ 0.1. This means that power consumption in different periods are economic substitutes. This arises if in response to a price increase in one period, consumers react in part by shifting some energy using activities from that time to the other period. This is somewhat analogous to the supply shifting done in natural gas markets via storage injection and extraction, as discussed in Chapter 10.

12.3.2 Unbundled Power Market with and without Cross-Price Elasticities

We now consider an unbundled electricity market consisting of a regulated T&D company that provides "wire" services for a competitive generation sector whose bulk power prices vary over time. The total price paid by the consumer is the competitive energy price plus a constant per MWh surcharge that the regulator sets at a level sufficient to recover the T&D costs of $854.1×10^6/year. This example allows us to examine how the presence or absence of cross-elasticities affect the results, including the loads, generation mix, and prices.

The model has four components: demand, T&D pricing, generation, and market clearing. Each is presented in turn.

In this unbundled market, we let p_t, $t \in \{1,2\}$ be the retail prices paid by consumers, which include the surcharge, while p_t^b, $t \in \{1,2\}$ are the bulk power prices charged by generators, which exclude the surcharge. To represent consumer behavior in the unbundled power market, we again consider the demand functions (12.2) in which, as we just noted, the cross-price elasticities are positive, but considerably smaller in magnitude than the own-price elasticities. These functions show the relationship between retail power prices and quantity demanded.

The connection between these retail prices and the bulk power market is provided by the T&D submodel. The regulated T&D company (which can also be viewed as the load-serving entity, confusedly called the supplier in the UK) buys power at the bulk power price p_t^b and then adds the T&D

because it is internal neither to the regulated utility's submodel (which minimizes cost subject to what the utility thinks is an exogenous load) nor the competitive generators' submodels (which maximize profit subject to what the generators think are exogenous prices).

charge $p^{T\&D}$ to yield the retail price p_t. The charge $p^{T\&D}$ is endogenous to the market model. The regulated T&D company model consists of these price relationships, plus the requirement that $p^{T\&D}$ be set at a level to recover all the T&D costs:

$$p_t^b + p^{T\&D} = p_t, t \in \{1,2\} \tag{12.6a}$$

$$4380p^{T\&D}q_1 + 4380p^{T\&D}q_2 = \$854.1 \times 10^6\$/\text{yr} \tag{12.6b}$$

The third component of the model is the profit-maximizing generator sub-model. As long as we represent the firms as price-takers (i.e., the bulk power prices are treated as exogenous) and assume no scale economies (constant per unit cost coefficients), then it doesn't matter how many firms we represent. Whether we have one monopolist firm who controls both technologies A and B, two firms (one for A, and one for B), or multiple firms each being able to choose from both technologies or just one, the energy produced and prices will be the same because the applicable KKT conditions in each case are the same. We arbitrarily consider here two firms, one per technology, whose profit-maximization problems are:

$$\begin{aligned} \text{Maximize}_{x_{At},k_A} \quad & 4380p_1^b x_{A1} + 4380p_2^b x_{A2} \\ & - [180,000k_A + 4380(30x_{A1}) + 4380(30x_{A2})] \end{aligned} \tag{12.7a}$$

s. t.

$$-x_{At} + k_A \geq 0 \qquad (\mu_{At}), \quad t \in \{1,2\} \tag{12.7b}$$

$$x_{At}, k_A \geq 0, \quad t \in \{1,2\} \tag{12.7c}$$

$$\begin{aligned} \text{Maximize}_{x_{Bt},k_B} \quad & 380p_1^b x_{B1} + 4380p_2^b x_{B2} \\ & - [70,000k_B + 4380(50x_{B1}) + 4380(50x_{B2})] \end{aligned} \tag{12.8a}$$

s. t.

$$-x_{Bt} + k_B \geq 0 \qquad (\mu_{Bt}), \quad t \in \{1,2\} \tag{12.8b}$$

$$x_{Bt}, k_B \geq 0, \quad t \in \{1,2\} \tag{12.8c}$$

The final component of the unbundled market model is market clearing: what's sold by the generators must equal the quantities bought by the consumers in each period:

$$\sum_{i \in \{A,B\}} x_{it} = q_t \quad (p_t), \quad t \in \{1,2\} \tag{12.9}$$

We can now assemble the equilibrium model for the unbundled market as follows:

Find

$\{q_1, q_2, p_1, p_2, p_1^b, p_2^b, p^{T\&D}, k_A, k_B, x_{A1}, x_{A2}, x_{B1}, x_{B2}, \mu_{A1}, \mu_{A2}, \mu_{B1}, \mu_{B2}\}$

satisfying:

Demand functions (12.2)

Regulated T&D company pricing (12.6)

KKTs for generator A's problem (12.7):

$0 \le k_A \perp -180,000 + \mu_{A1} + \mu_{A2} \le 0$

$0 \le x_{A1} \perp 4380(p_1^b - 30) - \mu_{A1} \le 0$

$0 \le x_{A2} \perp 4380(p_2^b - 30) - \mu_{A2} \le 0$

$0 \le \mu_{A1} \perp -x_{A1} + k_A \ge 0$

$0 \le \mu_{A2} \perp -x_{A2} + k_A \ge 0$

KKTs for generator B's problem (12.8):

$0 \le k_B \perp -70,000 + \mu_{B1} + \mu_{B2} \le 0$

$0 \le x_{B1} \perp 4380(p_1^b - 50) - \mu_{B1} \le 0$

$0 \le x_{B2} \perp 4380(p_2^b - 50) - \mu_{B2} \le 0$

$0 \le \mu_{B1} \perp -x_{B1} + k_B \ge 0$

$0 \le \mu_{B2} \perp -x_{B2} + k_B \ge 0$

Market clearing 12.9

The reader can readily verify that this system is square.[7] Similar to the regulated utility model, the presence of cross-price elasticities (with asymmetric cross-price terms) in (12.2) and the T&D surcharge rule (12.6) means that we can't define a single optimization problem whose KKT conditions are the same as the above mixed MCP (and thus could be used to calculate the equilibrium solution for the market).

A solution to this system is:

$\{q_1, q_2, p_1, p_2, p_1^b, p_2^b, p^{T\&D}, k_A, k_B, x_{A1}, x_{A2}, x_{B1}, x_{B2}, \mu_{A1}, \mu_{A2}, \mu_{B1}, \mu_{B2}\}$
= {2814.0 MW, 3621.7 MW, 65.41 \$/MWh, 96.28 \$/MWh,
35.11 \$/MWh, 65.98 \$/MWh, 30.3 \$/MWh,
2814.0 MW, 807.6 MW,
2814.0 MW, 2814.0 MW, 0 MW, 807.6 MW,
(4380)(5.11) \$/MW-yr, (4380)(35.98) \$/MW-yr,

[7] The reader can also verify that if instead of two separate generating companies there was just one that built both types A and B, then the above linear MCP would not be changed. Note that when numerically solving such a complementarity problem, the software will generally require that a variable be associated with each and every condition, even equality ones. For equality constraints within a market agent's problem, these would be their dual variables. In the case of equality conditions that do not occur within an agent's optimization problem, such as market clearing conditions (12.9) or demand functions, the logical variable to associate with the condition would often be the price of the commodity in question.

0 \$/yr, (4380)(15.98) \$/MW-yr}

Before discussing this solution, let's also solve a version of the model using demand functions that have zero cross-price elasticities.

$$q_1 = 3750 - 14.86p_1 \tag{12.10a}$$
$$q_2 = 6000 - 23.78p_2 \tag{12.10b}$$

Like (12.2), the above demand functions also pass through the following price-quantity pairs for the regulated utility solution (Section 12.3.1): $\{p_1,q_1\} = \{84.11$ \$/MWh, 2500 MW$\}$ and $\{p_2,q_2\} = \{84.11$ \$/MWh, 4000 MW$\}$, in which the two period's prices are equal and recover all costs incurred to meet the assumed loads. Consequently, those demand functions would yield the same solution as (12.2) if used in the regulated utility model. Also like (12.2), these demand functions have own-price elasticities of -0.5, but comparing them with (12.2) shows that the cross-price elasticity terms are now absent.

Substituting these demand functions into the unbundled market LCP just described ((12.2), (12.6), (12.9), and the KKTs for (12.7) and (12.8)), we can then solve for an equilibrium solution for the zero cross-price elasticity case:

$\{q_1, q_2, p_1, p_2, p_1^b, p_2^b, p^{T\&D}, k_A, k_B, x_{A1}, x_{A2}, x_{B1}, x_{B2}, \mu_{A1}, \mu_{A2}, \mu_{B1}, \mu_{B2}\}$
= {2782.3 MW, 3717.7 MW, 65.11 \$/MWh, 95.98 \$/MWh,
35.11 \$/MWh, 65.98 \$/MWh, 30 \$/MWh,
2782.3 MW, 935.4 MW,
2782.3 MW, 2782.3 MW, 0 MW, 935.4 MW,
(4380)(5.11) \$/MW-yr, (4380)(35.98) \$/MW-yr,
0 \$/yr, (4380)(15.98) \$/MW-yr}

We now contrast the results of the two unbundled models (with and without cross-price elasticities) with each other and with the regulated vertically integrated utility which is forced by the regulator to charge a single price in both periods that is just high enough to recover all costs. Figure 12.1 shows the effects on each period's electricity prices, while Figure 12.2 shows the quantity effects. Three values are shown in each figure for each period: one for the regulated utility model (which, by design, is the same under either positive cross-price elasticities (12.2) or zero cross-elasticities (12.10)), and one each for the unbundled competitive generator model with and without positive cross-price elasticities.

We see from Figure 12.1 that the effect of relaxing the regulatory rule that prices in all periods must be the same is to allow prices to fall in the off-peak period and to increase in the peak period. These changes reflect the fact that marginal costs of generating power off-peak are about half the cost of generating on-peak (\$35.11/MWh vs. \$65.98/MWh in both models).

However, the retail prices paid by the final consumer in the unbundled models are about \$30/MWh higher because of the T&D surcharge. Meanwhile, whether or not cross-price elasticities are included does not affect prices in the unbundled model by more than 0.5%, as the figure shows.

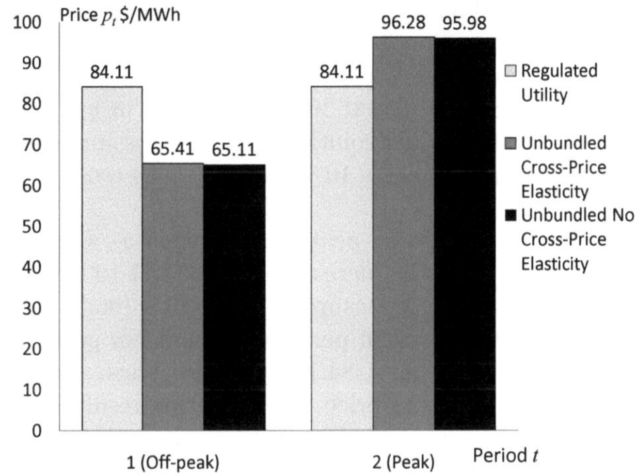

Fig. 12.1 Off-peak and peak power retail prices for solutions to the regulated utility model and unbundled model

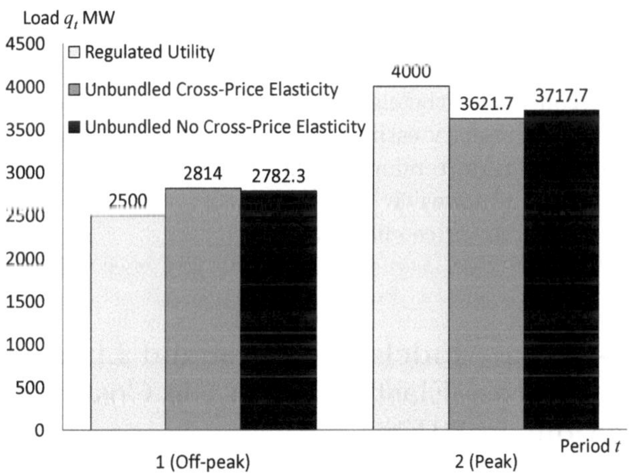

Fig. 12.2 Off-peak and peak quantities demanded for solutions to the regulated utility model and unbundled model

The effect of the change from uniform (across time) regulated prices to time-differentiated prices is to dampen peak demand and boost off-peak demand, as Figure 12.2 shows. In the cross-elasticity case, dropping the off-peak price from \$84.11 to \$65.41 per MWh as well as increasing the on-peak price from \$84.11 to \$96.28 both encourage 314 MW more off-peak consumption. Most of that off-peak consumption increase is due to the own price effect (by (12.2), the \$18.70/MWh decrease in p_1 yields an increase in q_1 of 278 MW), but a significant portion is due to the load-shifting effect of the positive cross-elasticity with peak prices ((12.2) implies that the \$12.17/MWh increase in p_2 induced an additional 36 MW increase in q_1). This load-shift effect is absent in the unbundled solution without cross-price elasticities: as a result, the off-peak load increases 10% less than with cross-elasticities (282 MW vs 314 MW).

The reverse effect is seen in on-peak consumption q_2. Using the demand functions (12.2), we see that the increase in p_2 (\$84.11 to \$96.28 per MWh) causes consumers to adjust peak consumption by -23.78(96.28-84.11) = -289.4 MW (by (12.2)), while the lower off-peak price p_1 pushes peak demand lower by a further 89.9 MW (4.756(65.41-84.11) = -89.9). Thus, about a quarter of the peak demand change due to price differentiation results from the cross-price effect. This can also be seen in Figure 12.2, where the unbundled model without cross-price elasticities shows an appreciably smaller response of peak demand to time-differentiated prices (q_2 decreasing by only 282 MW rather than 378 MW from the regulated value of 4000 MW).

Thus, disregarding cross-price elasticities can result in appreciable distortions in estimates of the response of time-differentiated markets to price changes. In this case, their omission resulted in an underestimate of quantity changes in each period. Price effects are less, although whether prices or quantities are affected more will, in general, depend on the particular assumptions made concerning the elasticities of demand and supply. In general, although cross-price elasticity estimates are not as available as own-price estimates and their inclusion renders market models more complex, modelers should, at a minimum, do sensitivity analyses to gauge the possible bias that results from leaving cross-price effects out.

12.4 Multi-Sector Models with Demand Linkages: The Effect of Cross-Elasticities on the Cost of CO_2 Reductions in PIES

12.4.1 The Project Independence Evaluation System

The Project Independence Evaluation System (PIES) was created by the US Federal Energy Administration in the mid-1970s to provide projections of the cost and effectiveness of possible government strategies to reduce imports of oil, accounting for interactions with other fuel markets [13]. History might judge the Project Independence policy itself to be a failure because oil imports to the US have increased greatly since that time both in absolute quantity as well as a percentage of end use. However, PIES and its successor energy-economy models have been a success in terms of providing relatively objective information that has beneficially influenced policy in the oil, natural gas, and electricity sectors [18].

In this section we present a highly simplified pedagogical version of PIES based on the version in Hogan [13]. The purpose of Hogan's article was to illustrate the structure of the model as well as the PIES iterative algorithm for its solution (see Chapter 8, Section 8.4.2.1 above for our summary of that approach). We modify Hogan's demand functions in our example, and then extend it to consider CO_2 trading. Our goal here is to explore the effect of considering demand linkages, in the form of cross-price elasticities, upon the conclusions of a multisector energy policy model, in particular conclusions concerning the cost of CO_2 reductions. It turns out that having positive cross-elasticities (i.e., making fuels substitutes) means that CO_2 price increases become less efficient in lowering emissions through the route of demand reductions. This is because the CO_2 price inflates the price of coal and reduces its consumption, as anticipated, but consumers then switch to oil in the model, and oil's CO_2 emissions would then rise compared to the situation of zero cross-elasticities.

Figure 12.3 shows a schematic of the markets in our version of the PIES model and their connections. Seventeen distinct markets and their interconnections are modeled. Oil quantities are in barrels of oil equivalent per day (bbl/d), while coal and CO_2 quantities are in English tons/day (tons/d). Coal and petroleum are each extracted in two different regions. Coal is directly shipped to the consuming regions (which are also two in number). Oil, however, is first transported to refineries, where it can be used to produce either light or heavy oil, which are then sent to the two consumption markets. The quality of coal or oil is not differentiated (in real markets, they would differ in sulfur content and other characteristics that would affect the prices they would fetch in the market). There is an international carbon dioxide emissions market as well, in which producers of coal and oil are required to buy allowances in proportion to the carbon content of the fuel they produce.[8] Consequently, the 17 markets differ in either the type of commodity (of which there are five, including CO_2) or location (of which there are eight).

Next, we will introduce each of the components (submodels) of our PIES model. The parameter values assumed are based loosely on [13], and are arbitrary. Along the way, we introduce the idea of a fixed proportions (or Leontief) production function. The components are assembled into the overall PIES model in Section 12.4.3, and solution approaches discussed. Our

[8] Carbon could instead be regulated at the final point of use, or even at an intermediate point, such as the refineries in this example. For example, the US greenhouse gas legislation passed by the House of Representatives in 2009 (H.B. 2454) regulated carbon emissions at various points in the supply chain. For the electricity industry, this was done at an intermediate point where the fuel was burned for generating electricity. On the other hand, producers or importers of coal for direct industrial use would have had to buy allowances, while distributors of natural gas would have been responsible for emissions for their customers.

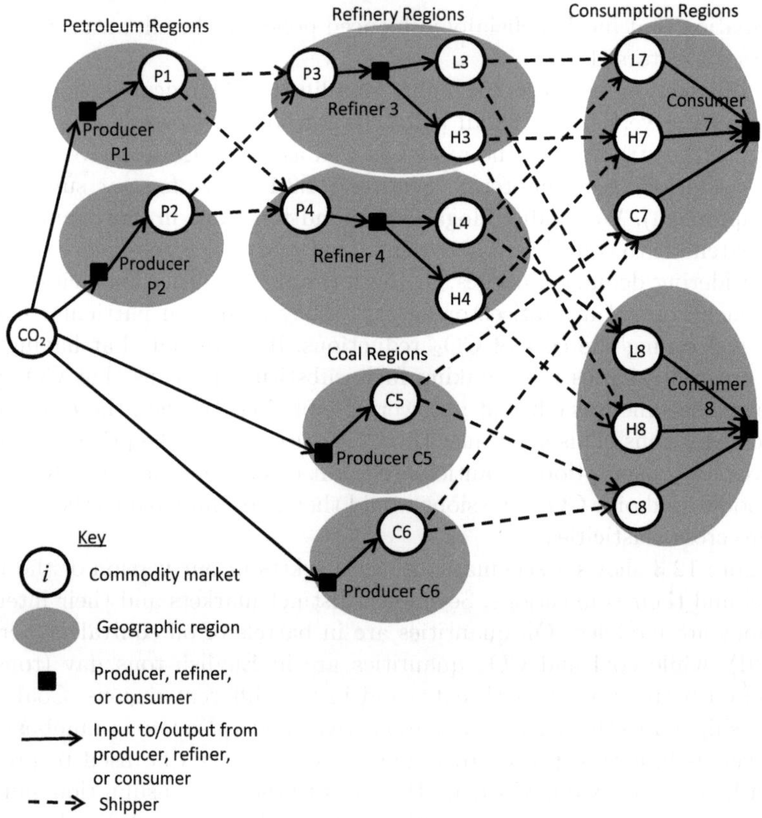

Fig. 12.3 Commodity markets, market participants, and energy flows in PIES example

application and conclusions concerning the effects of demand linkages then follow in Section 12.4.4.

12.4.2 PIES Model Components

The notation of our version of the PIES model is as follows:

x_{is} : Quantity of supply into market i from supply source s. Suppliers can either be raw fuel producers or refineries, depending on the product. The market indices include $i \in \{P1, P2, P3, P4$(petroleum), $C5, C6, C7, C8$(coal), $L3, L4, L7, L8$(light oil), $H3, H4, H7, H8$(heavy oil), $CO_2\}$. At the coal and oil production regions, there are up to three sources (or tranches) of supply s with various costs. These tranches have different marginal costs and are used to represent the steps of a stair-step (or step-wise) supply curve for market i.

q_i : Quantity purchased from market i (either by refineries or final consumers).

t_{ij} : Quantity transported from market i to market j.

p_i : Price of commodity in market i.

Market participants include producers, refiners, shippers, and consumers. Below, we describe demand functions and the producers, refiner, and shipper profit-maximization problems, followed by a list of the market clearing conditions. Combining the demand functions, KKTs for the profit-maximization problems, and the clearing conditions yields a mixed linear complementarity problem that we then solve under different demand and carbon assumptions.

12.4.2.1 Consumers

The consumers' demand functions for final fuel products (light oil, heavy oil, and coal) are identical in the two consumption regions (7 and 8). Prices for each final commodity (coal, light oil, and heavy oil) in each region affect the quantity demanded at that location only, but there are cross-price elasticities between the different commodity types within the region.

In region 7, the inverse demand functions $p_i = f_{d,i}^{-1}(q)$ (which express marginal willingness-to-pay for a commodity as a function of all the quantities produced) are assumed to be:

$$p_{L7} = 68 - 0.03q_{L7} - 0.01q_{H7} - 0.006q_{C7} \tag{12.11a}$$

$$p_{H7} = 63 - 0.01q_{L7} - 0.03q_{H7} - 0.009q_{C7} \tag{12.11b}$$

$$p_{C7} = 47.2 - 0.006q_{L7} - 0.009q_{H7} - 0.019q_{C7} \tag{12.11c}$$

The inverse demand functions for region 8 are the same, except they involve that location's prices and quantities instead.

An alternative representation recognizes that quantities cannot be negative, and that quantity demanded is zero if the price is above a "choke" level (the price intercept of the demand curve, which is a function of the levels of consumption of other goods). This is a complementarity condition in which quantity demanded is strictly positive only if price and marginal willingness-to-pay are equal; if price exceeds that amount, then the consumer buys nothing. Here, the demand curves in (12.11) would then be stated as the following complementarity conditions:

$$0 \leq q_{L7} \perp p_{L7} - (68 - 0.03q_{L7} - 0.01q_{H7} - 0.006q_{C7}) \geq 0 \tag{12.12a}$$

$$0 \leq q_{H7} \perp p_{H7} - (63 - 0.01q_{L7} - 0.03q_{H7} - 0.009q_{C7}) \geq 0 \tag{12.12b}$$

$$0 \le q_{C7} \perp p_{C7} - (47.2 - 0.006 q_{L7} - 0.009 q_{H7} - 0.019 q_{C7}) \ge 0 \qquad (12.12c)$$

If we instead solve (12.11) for quantities as a function of prices $q_i = f_{d,i}(p)$, we obtain region 7's demand functions:

$$q_{L7} = 1600.31 - 38.3 p_{L7} + 10.65 p_{H7} + 7.048 p_{C7} \qquad (12.13a)$$

$$q_{H7} = 1134.06 + 10.65 p_{L7} - 38.3 p_{H7} + 16.44 p_{C7} \qquad (12.13b)$$

$$q_{C7} = 1441.66 + 7.048 p_{L7} + 16.44 p_{H7} - 62.647 p_{C7} \qquad (12.13c)$$

The same functions also apply to region 8, substituting in that region's prices and quantities. If we instead consider the complementarity version of inverse demand, which limits quantities to be nonnegative, then the demand functions would instead define each q_i as the maximum of the relevant right hand side of (12.13) and zero.

At prices of $\{p_{L7}, p_{H7}, p_{C7}\} = \{\$12/\text{bbl}, \$12/\text{bbl}, \$12/\text{ton}\}$, demand functions (12.13) yield $\{q_{L7}, q_{H7}, q_{C7}\} = \{1200 \text{ bbl/day}, 1000 \text{ bbl/day}, 1000 \text{ tons/day}\}$. At those price-quantity pairs, we use (12.1) to obtain the price elasticities ε_{ij}:

$$\begin{bmatrix} \varepsilon_{L,L} & \varepsilon_{L,H} & \varepsilon_{L,C} \\ \varepsilon_{H,L} & \varepsilon_{H,H} & \varepsilon_{H,C} \\ \varepsilon_{C,L} & \varepsilon_{C,H} & \varepsilon_{C,C} \end{bmatrix} = \begin{bmatrix} -0.51 & 0.106 & 0.07 \\ 0.17 & -0.50 & 0.197 \\ 0.113 & 0.197 & -0.753 \end{bmatrix} \qquad (12.14)$$

12.4.2.2 Fuel Producers

Oil or coal producers in market i have several tranches of supply, indexed by s, which allows creation of a stepped supply curve in that market. Associated with each tranche is the quantity supplied x_{is} from that step (bbl/d or tons/d, respectively), which has constant marginal cost CX_{is} (\$/bbl or \$/ton) and a maximum production rate X_{is}. Fuel producers also have to buy emissions allowances to cover the carbon dioxide emissions that will ultimately result from combustion of their fuel; E_{is} is the CO_2 emissions rate per unit of fuel (tons/bbl or tons/ton, respectively, for oil and coal). This model does not account for effects of resource depletion, in that production in one period does not affect the amount available in subsequent periods; more complex models can be devised, in which case a dynamic, multiperiod formulation is needed (as, for example, in [22]).

In each case, assuming one producer per market and assuming competitive (price-taking) behavior, producer i's problem becomes:

$$\text{Maximize}_{x_{is}} \quad \sum_s (p_i - CX_{is} - p_{CO2} E_{is}) x_{is} \qquad (12.15a)$$

s. t.

$$x_{is} \leq X_{is} \qquad (\mu_{is}), \quad \forall s \tag{12.15b}$$

Profit equals the revenue from selling x_{is} minus the two cost components: the cost of production (CX_{is} being the per unit cost) and the expense of CO_2 allowances. Price-taking behavior means that the producer believes that the p_i are exogenous and won't change no matter what the producer does. More general models could be formulated as oligopoly models introduced in other chapters, in which each producer recognizes that the price p_i it receives is a downward sloping function of the amount that it supplies to the market. Obviously, oligopoly models are relevant to the cartelized international oil market.

The values of CX_{is} and X_{is} for each of the producers are shown in Table 12.1. E_{is} is assumed to be 0.49 tons CO_2/bbl for oil and 2.2 tons CO_2/ton for coal. We assume that only CO_2 associated with fuel produced must acquire permits, and not downstream products (light and heavy oil in this model, electricity in other models). Other CO_2 market designs are possible, as noted in Section 11.3.

Table 12.1 Producer cost and capacity data (after Hogan [13])

Commodity	Producer i	Tranche s	Cost CX_{is} [a]	Capacity X_{is} [b]
Petroleum	P1	1	1	1100
		2	1.5	1200
	P2	1	1.25	1300
		2	1.5	1100
Coal	C5	1	5	300
		2	6	300
		3	8	400
	C6	1	4	200
		2	5	300
		3	7	600

a. Units are \$/bbl for petroleum, \$/ton for coal
b. Units are bbl/d for petroleum, tons/d for coal

As an example of a producer model, there is one crude oil producer in petroleum market i = P1. That producer has two supply tranches s = 1,2. Its profit-maximization model is:

$$\begin{aligned} \text{Maximize}_{x_{P1,1},x_{P1,2}} \quad & (p_{P1} - 1.00 - 0.49p_{CO2})x_{P1,1} \\ & + (p_{P1} - 1.50 - 0.49p_{CO2})x_{P1,2} \end{aligned} \tag{12.16a}$$

s. t.

$$x_{P1,1} \leq 1100 \qquad (\mu_{P1,1}) \tag{12.16b}$$

$$x_{P1,2} \leq 1200 \qquad (\mu_{P1,2}) \tag{12.16c}$$

$$x_{P1,1}, x_{P1,2} \geq 0. \tag{12.16d}$$

12.4.2.3 Oil Refiners

Refiners in a region convert crude oil, which we model as having been purchased in that region's crude oil market j, to refined products, either light or heavy refined oil, which are modeled as being sold in the region's products markets i. In general, models of oil refineries can be formulated as highly complicated LPs or NLPs in which several different feedstocks (different qualities of oil) can be converted into a large number of product streams. The constraint sets of those math programs can capture substitution relationships in which more of one product can be produced at the expense of less of another, for a given set of inputs.

Such models are complicated examples of what economists call "production functions" (Chapter 3). They represent one of the first and most famous areas of application of linear programming (e.g., [14]).

Here, however, a very simple type of linear relationship between inputs and outputs is used, called a Leontief production function. It is named after the famous economist Wassily Leontief who received a Nobel prize for his invention of the input-output model which is extensively used for modeling flows of goods and services in the economy [15]. The Leontief, or "fixed proportions", production function has the convenient advantage of yielding linear constraints in a profit-maximization problem. The Leontief function is like a cake recipe: to make a single cake, you might need 1.25 cups of sugar, 1 cup of flour, 1/2 teaspoon of salt, 10 egg whites, 2 tablespoons of water, and 1/2 teaspoon of vanilla. If you want to make two cakes, you would need to double the amounts of ingredients. The "fixed proportions" in this case means that the ratios of the inputs required are constant, and that multiplying all inputs by X will result in X more output.

The cake example is a case of many inputs to one output. This "recipe" type of production function is the basis of the Leontief input-output model. Another fixed proportions production function relates one input to many outputs: consider wool and lamb chop production. The input is sheep, and the outputs are in fixed proportion (fiber and meat) to the number of sheep. The most general case could be multiple inputs to multiple outputs, all of which are in fixed proportions. For example, feed, water, and labor may be the inputs that a sheep farmer uses to produce wool and lamb chops; a fixed proportions production function results if the inputs have to be provided in fixed proportion yielding a fixed ratio of wool/lamb, and doubling the inputs results in doubling the outputs.

For the refinery, the production function is of the single input/several output type of Leontief function. Crude oil is the input, and the each refinery splits it, in fixed proportion, into two products: light and heavy oil, at a cost. The refiner's optimization problem is to maximize profit subject to the production function, which ensures that nothing is produced unless the requisite input is purchased. Again, we assume price-taking (competitive) behavior:

$$\text{Maximize}_{x_{j1},q_i} \quad \sum_{j\in R(i)} p_j x_{j1} - (CR_i + p_i)q_i \tag{12.17a}$$

s. t.

$$x_{j1} - F_{ij}q_i = 0 \qquad (\phi_{ij}), \quad \forall j \in R(i) \tag{12.17b}$$

$$x_{j1} \geq 0, \quad \forall j \in R(i); \quad q_i \geq 0 \tag{12.17c}$$

where CR_i is the cost per bbl of refining crude oil, F_{ij} is the fixed ratio of output per unit input, and $R(i)$ is the set of product markets j in which the refinery located at i sells its products. Note that there is only one tranche of supply (the subscript s in x_{is} just 1). In the below, we omit that subscript s in the refiner's model for simplicity. We could also have imposed an upper bound X_{j1} upon x_{j1}, just as in the producer's model, but again for the sake of brevity we omit it.

As Figure 12.3 shows, there are two refiners in two different regions. The parameterized linear programs for each (based on Hogan, 1975) are as follows for the refineries located in regions 3 and 4:

$$\text{Maximize}_{x_{L3},x_{H3},q_{P3}} \quad p_{L3}x_{L3} + p_{H3}x_{H3} - (6.5 + p_{P3})q_{P3} \tag{12.18a}$$

s. t.

$$x_{L3} - 0.6q_{P3} = 0 \qquad (\phi_{L3,P3}) \tag{12.18b}$$

$$x_{H3} - 0.4q_{P3} = 0 \qquad (\phi_{H3,P3}) \tag{12.18c}$$

$$x_{L3}, x_{H3}, q_{P3} \geq 0 \tag{12.18d}$$

and:

$$\text{Maximize}_{x_{L4},x_{H4},q_{P4}} \quad p_{L4}x_{L4} + p_{H4}x_{H4} - (5 + p_{P4})q_{P4} \tag{12.19a}$$

s. t.

$$x_{L4} - 0.5q_{P4} = 0 \qquad (\phi_{L4,P4}) \tag{12.19b}$$

$$x_{H4} - 0.5q_{P4} = 0 \qquad (\phi_{H4,P4}) \tag{12.19c}$$

$$x_{L4}, x_{H4}, q_{P4} \geq 0 \tag{12.19d}$$

Note that the refinery in region 3 splits each barrel of crude into 0.6 bbl of light and 0.4 bbl of heavy oil product. Region 4's refinery is less effective at producing the more valuable light oil, and splits each barrel of crude evenly into light and heavy products. However, region 4's refinery has lower costs, incurring 5 \$/bbl in refining expenses as opposed to the 6.5 \$/bbl cost in the region 3 refinery.

12.4.2.4 Shippers

Here we model the energy shipper as a price-taking arbitrageur between the markets, buying in one market i and shipping to another market j. The cost of shipping here is represented as a simple model of the "transportation" type, in which a fixed per unit of cost is incurred to ship a commodity from i to j and there are no limits to the amount shipped. Of course, shipment costs can be much more complicated. For instance, there might be alternative routes that can be chosen through various intermediate nodes; this situation might be represented as a transshipment problem. The gas and electricity networks introduced in other chapters involve more complex constraints and even nonlinear and nonconvex relationships.

We could define a separate shipping company for each commodity or for different shipping routes. This is indeed the case in the real world, where oil shipping is done by different companies than liquefied natural gas shipping, etc. However, if we assume that all shippers are price-takers and there are no economies of scale or scope in shipping, the model results will be the same whether we have separate companies for each route or commodity or one single shipper. In contrast, if there are economies of scale (doubling the amount shipped results in less than a doubling of shipping cost) or scope (coordinated shipping of more than one product results in cost savings relative to separate, uncoordinated shipping), then it would matter whether we have separate or a single combined shipping company. Likewise, if shippers had market power, then the number and sizes of shippers would also matter.

The resulting shipper transportation-based model can be phrased generally as:

$$\text{Maximize}_{t_{ij}} \quad \sum_{ij \in IJ} (p_j - p_i - CT_{ij}) t_{ij} \tag{12.20a}$$

s. t.

$$t_{ij} \leq T_{ij} \qquad (\gamma_{ij}), \quad \forall ij \in IJ \tag{12.20b}$$

$$t_{ij} \geq 0, \quad \forall ij \in IJ \tag{12.20c}$$

CT_{ij} is the per unit cost of shipping from i to j. Set IJ is the set of possible shipment routes (origin-destination pairs). This formulation includes upper bounds T_{ij} on flows t_{ij}; however, in our case study, we assume that these can be ignored.

Referring to Figure 12.3, we see that there are 16 such origin-destination pairs (assuming that CO_2 allowances do not require shipping). We use the costs assumed by Hogan [13], and insert them into the above model, resulting in:

$$\text{Maximize}_{t_{ij}} \quad (p_{P3} - p_{P1} - 2)t_{P1,P3} + (p_{P4} - p_{P1} - 3)t_{P1,P4}$$

$$+ (p_{P3} - p_{P2} - 4)t_{P2,P3} + (p_{P4} - p_{P2} - 2)t_{P2,P4}$$
$$+ (p_{L7} - p_{L3} - 1)t_{L3,L7} + (p_{L8} - p_{L3} - 1.2)t_{L3,L8}$$
$$+ (p_{L7} - p_{L4} - 1)t_{L4,L7} + (p_{L8} - p_{L4} - 1.5)t_{L4,L8}$$
$$+ (p_{H7} - p_{H3} - 1)t_{H3,H7} + (p_{H8} - p_{H3} - 1.2)t_{H3,H8}$$
$$+ (p_{H7} - p_{H4} - 1)t_{H4,H7} + (p_{H8} - p_{H4} - 1.5)t_{H4,H8}$$
$$+ (p_{C7} - p_{C5} - 1)t_{C5,C7} + (p_{C8} - p_{C5} - 2.5)t_{C5,C8}$$
$$+ (p_{C7} - p_{C6} - 0.75)t_{C6,C7} + (p_{C8} - p_{C6} - 2.75)t_{C6,C8} \qquad (12.21a)$$

s. t.

$$t_{ij} \geq 0, \quad \forall ij \in IJ \qquad (12.21b)$$

12.4.2.5 Market Clearing

There are 17 markets (Figure 12.3), yielding an equal number of market clearing conditions and prices. For commodities other than CO_2, the market clearing condition states that consumption minus production plus net exports equals zero. In general terms:

$$q_i - \sum_{s} x_{is} + \Big(\sum_{j \in JE(i)} t_{ij} - \sum_{j \in JI(i)} t_{ji}\Big) = 0 \qquad (p_i) \qquad (12.22)$$

where $JE(i)$ is the set of markets j that market i exports to, while $JI(i)$ is the set of markets j that market i imports from. Meanwhile for CO_2, the market instead satisfies the following complementarity condition:

$$0 \leq p_{CO2} \perp \sum_{is \in BF} E_{is} x_{is} - \overline{CO_2} \leq 0 \qquad (12.23)$$

where BF is the set $\{is\}$ of oil and coal supply tranches that must acquire CO_2 permits. This is not an equality condition because the demand for CO_2 allowances can be strictly less than the (regulator-specified) supply, in which case the price would be zero. In particular, if the total emissions associated with petroleum and coal production fall short of the carbon dioxide cap of $\overline{CO_2}$, then the price of CO_2 allowances must be zero.

For our 17 markets, we get the following market clearing conditions:

$$-(x_{P1,1} + x_{P1,2}) + t_{P1,P3} + t_{P1,P4} = 0 \qquad (p_{P1}) \qquad (12.24a)$$
$$-(x_{P2,1} + x_{P2,2}) + t_{P2,P3} + t_{P2,P4} = 0 \qquad (p_{P2}) \qquad (12.24b)$$
$$q_{P3} - t_{P1,P3} - t_{P2,P3} = 0 \qquad (p_{P3}) \qquad (12.24c)$$
$$q_{P4} - t_{P1,P4} - t_{P2,P4} = 0 \qquad (p_{P4}) \qquad (12.24d)$$
$$-x_{L3} + t_{L3,L7} + t_{L3,L8} = 0 \qquad (p_{L3}) \qquad (12.24e)$$
$$-x_{L4} + t_{L4,L7} + t_{L4,L8} = 0 \qquad (p_{L4}) \qquad (12.24f)$$

$$-x_{H3}+t_{H3,H7}+t_{H3,H8}=0 \qquad (p_{H3}) \tag{12.24g}$$
$$-x_{H4}+t_{H4,H7}+t_{H4,H8}=0 \qquad (p_{H4}) \tag{12.24h}$$
$$q_{L7}-t_{L3,L7}-t_{L4,L7}=0 \qquad (p_{L7}) \tag{12.24i}$$
$$q_{L8}-t_{L3,L8}-t_{L4,L8}=0 \qquad (p_{L8}) \tag{12.24j}$$
$$q_{H7}-t_{H3,H7}-t_{H4,H7}=0 \qquad (p_{H7}) \tag{12.24k}$$
$$q_{H8}-t_{H3,H8}-t_{H4,H8}=0 \qquad (p_{H8}) \tag{12.24l}$$
$$-(x_{C5,1}+x_{C5,2}+x_{C5,3})+t_{C5,C7}+t_{C5,C8}=0 \qquad (p_{C5}) \tag{12.24m}$$
$$-(x_{C6,1}+x_{C6,2}+x_{C6,3})+t_{C6,C7}+t_{C6,C8}=0 \qquad (p_{C6}) \tag{12.24n}$$
$$q_{C7}-t_{C5,C7}-t_{C6,C7}=0 \qquad (p_{C7}) \tag{12.24o}$$
$$q_{C8}-t_{C5,C8}-t_{C6,C8}=0 \qquad (p_{C8}) \tag{12.24p}$$
$$\begin{aligned}0 \le p_{CO2} \perp\, & 0.49(x_{P1,1}+x_{P1,2}+x_{P2,1}+x_{P2,2}) \\ & +2.2(x_{C5,1}+x_{C5,2}+x_{C5,3}+x_{C6,1}+x_{C6,2}+x_{C6,3})-\overline{CO_2}\le 0\end{aligned} \tag{12.24q}$$

With the exception of the CO_2 market, all of the market clearing conditions are equalities (supply = demand).

12.4.3 Assembling and Solving the PIES Model

12.4.3.1 Market Equilibrium LCP

The market equilibrium conditions consist of a concatenation of the demand functions, first-order (KKT) conditions of the profit-maximization functions for producers, refiners, and shippers, and the market clearing conditions. In general, this gives the following mixed linear complementarity problem:

Find $\{x_{is},\ x_i,\ q_i,\ t_{ij},\ \mu_{is},\phi_{ij},\gamma_{ij}\}$ that satisfy:

- Demand functions (12.13) or inverse demand functions (12.11) or (12.12) for all consumers of final products
- KKT conditions for producer problem (12.15) for all producers of raw fuels
- KKT conditions for refinery problem (12.17) for all refineries
- KKT conditions for shipper problem (12.20)
- Market clearing conditions (12.22) and (12.23)

Properly formulated, the resulting system should be square, with as many variables as conditions. In this particular case:
Problem PIES: Find $\{x_{is},\ x_i,\ q_i,\ t_{ij},\ \mu_{is},\phi_{ij}\}$ that satisfy:

- Demand functions (12.13) or inverse demand functions (12.11) or (12.12) for locations 7 and 8

- KKT conditions for problems for producers P1, P2, C5, C6 (e.g., (12.16)):
  $0 \leq x_{P1,1} \perp (p_{P1} - 1.00 - 0.49p_{CO2}) - \mu_{P1,1} \leq 0$
  $0 \leq x_{P1,2} \perp (p_{P1} - 1.50 - 0.49p_{CO2}) - \mu_{P1,2} \leq 0$
  $0 \leq \mu_{P1,1} \perp x_{P1,1} - 1100 \leq 0$
  $0 \leq \mu_{P1,2} \perp x_{P1,2} - 1200 \leq 0$
  $0 \leq x_{P2,1} \perp (p_{P2} - 1.25 - 0.49p_{CO2}) - \mu_{P2,1} \leq 0$
  $0 \leq x_{P2,2} \perp (p_{P2} - 1.50 - 0.49p_{CO2}) - \mu_{P2,2} \leq 0$
  $0 \leq \mu_{P2,1} \perp x_{P2,1} - 1300 \leq 0$
  $0 \leq \mu_{P2,2} \perp x_{P2,2} - 1100 \leq 0$
  $0 \leq x_{C5,1} \perp (p_{C5} - 5 - 2.2p_{CO2}) - \mu_{C5,1} \leq 0$
  $0 \leq x_{C5,2} \perp (p_{C5} - 6 - 2.2p_{CO2}) - \mu_{C5,2} \leq 0$
  $0 \leq x_{C5,3} \perp (p_{C5} - 8 - 2.2p_{CO2}) - \mu_{C5,3} \leq 0$
  $0 \leq \mu_{C5,1} \perp x_{C5,1} - 300 \leq 0$
  $0 \leq \mu_{C5,2} \perp x_{C5,2} - 300 \leq 0$
  $0 \leq \mu_{C5,3} \perp x_{C5,3} - 400 \leq 0$
  $0 \leq x_{C6,1} \perp (p_{C6} - 4 - 2.2p_{CO2}) - \mu_{C6,1} \leq 0$
  $0 \leq x_{C6,2} \perp (p_{C6} - 5 - 2.2p_{CO2}) - \mu_{C6,2} \leq 0$
  $0 \leq x_{C6,3} \perp (p_{C6} - 7 - 2.2p_{CO2}) - \mu_{C6,3} \leq 0$
  $0 \leq \mu_{C6,1} \perp x_{C6,1} - 200 \leq 0$
  $0 \leq \mu_{C6,2} \perp x_{C6,2} - 300 \leq 0$
  $0 \leq \mu_{C6,3} \perp x_{C6,3} - 600 \leq 0$
- KKT conditions for the two refineries' problems (12.18) and (12.19):
  $0 \leq x_{L3} \perp p_{L3} - \phi_{L3,P3} \leq 0$
  $0 \leq x_{H3} \perp p_{H3} - \phi_{H3,P3} \leq 0$
  $0 \leq q_{P3} \perp -(p_{P3} + 6.5) + 0.6\phi_{L3,P3} + 0.4\phi_{H3,P3} \leq 0$
  $x_{L3} - 0.6q_{P3} = 0 \qquad (\phi_{L3,P3})$
  $x_{H3} - 0.4q_{P3} = 0 \qquad (\phi_{H3,P3})$
  $0 \leq x_{L4} \perp p_{L4} - \phi_{L4,P4} \leq 0$
  $0 \leq x_{H4} \perp p_{H4} - \phi_{H4,P4} \leq 0$
  $0 \leq q_{P4} \perp -(p_{P4} + 5) + 0.5\phi_{L4,P4} + 0.5\phi_{H4,P4} \leq 0$
  $x_{L4} - 0.5q_{P4} = 0 \qquad (\phi_{L4,P4})$
  $x_{H4} - 0.5q_{P4} = 0 \qquad (\phi_{H4,P4})$

- KKT conditions for the shipper's problem (12.21) :
  $0 \leq t_{P1,P3} \perp (p_{P3} - p_{P1} - 2) \leq 0$
  $0 \leq t_{P1,P4} \perp (p_{P4} - p_{P1} - 3) \leq 0$
  $0 \leq t_{P2,P3} \perp (p_{P3} - p_{P2} - 4) \leq 0$
  $0 \leq t_{P2,P4} \perp (p_{P4} - p_{P2} - 2) \leq 0$
  $0 \leq t_{L3,L7} \perp (p_{L7} - p_{L3} - 1) \leq 0$
  $0 \leq t_{L3,L8} \perp (p_{L8} - p_{L3} - 1.2) \leq 0$
  $0 \leq t_{L4,L7} \perp (p_{L7} - p_{L4} - 1) \leq 0$
  $0 \leq t_{L4,L8} \perp (p_{L8} - p_{L4} - 1.5) \leq 0$
  $0 \leq t_{H3,H7} \perp (p_{H7} - p_{H3} - 1) \leq 0$
  $0 \leq t_{H3,H8} \perp (p_{H8} - p_{H3} - 1.2) \leq 0$
  $0 \leq t_{H4,H7} \perp (p_{H7} - p_{H4} - 1) \leq 0$
  $0 \leq t_{H4,H8} \perp (p_{H8} - p_{H4} - 1.5) \leq 0$
  $0 \leq t_{C5,C7} \perp (p_{C7} - p_{C5} - 1) \leq 0$
  $0 \leq t_{C5,C8} \perp (p_{C8} - p_{C5} - 2.5) \leq 0$
  $0 \leq t_{C6,C7} \perp (p_{C7} - p_{C6} - 0.75) \leq 0$
  $0 \leq t_{C6,C8} \perp (p_{C8} - p_{C6} - 2.75) \leq 0$
- Market clearing conditions (12.24) .

There are 69 variables and 69 conditions, so the system is indeed square.

12.4.3.2 Solution Approaches

The above linear MCP can be solved with a complementarity solver such as PATH. However, in a very narrow set of circumstances, the complementarity problem is also the KKT conditions for a *single* quadratic program, which would allow use of a NLP or quadratic programming (QP) solver to obtain the equilibrium. (See Chapter 4 for an in-depth discussion on the Principle of Symmetry, also called Integrability.) This narrow condition is symmetry of the matrix B in the set of demand functions (12.13) written in vector form:

$$Q = A + BP$$

Or, equivalently, symmetry in the matrix D in the set of inverse demand functions (12.11) when written as:

$$P = C + DQ$$

with P and Q being the column vector of prices and quantities consumed of the commodities for which demand functions are defined.

If B or D are symmetric, the PIES model can be formulated as the below QP. (We ask the reader to be forgiving of our mix of vector and scalar notation.)

$$\text{Maximize}_{q_i, x_{is}, t_{ij}} \quad CQ + Q^T DQ/2$$

$$-\Big(\sum_{is\in IS} CX_{is}x_{is} + \sum_{i\in R} CR_i q_i + \sum_{ij\in IJ} CT_{ij}t_{ij}\Big) \quad (12.25a)$$

s. t.

$$x_{is} \le X_{is} \qquad (\mu_{is}), \quad \forall i, s \quad (12.25b)$$

$$x_{j1} - F_{ij}q_i = 0 \qquad (\phi_{ij}), \quad \forall i, j \in R(i) \quad (12.25c)$$

$$t_{ij} \le T_{ij} \qquad (\gamma_{ij}), \quad \forall ij \in IJ \quad (12.25d)$$

$$q_i - \sum_s x_{is} + \Big(\sum_{j\in JE(i)} t_{ij} - \sum_{j\in JI(i)} t_{ji}\Big) = 0 \qquad (p_i), \forall i \quad (12.25e)$$

$$\sum_{is\in BF} E_{is}x_{is} \le \overline{CO_2} \quad (p_{CO2}) \quad (12.25f)$$

$$x_{is} \ge 0, \forall i, s; \quad q_i \ge 0, \forall i; \quad t_{ij} \ge 0, \forall ij \in IJ \quad (12.25g)$$

The KKT conditions for optimization problem (12.25) yield the linear MCP that describes the market equilibrium for the PIES example, if the D matrix of coefficients of the demand functions is symmetric. The objective function might be loosely interpreted as a measure of market efficiency (benefits to consumers minus costs of production, refining, and shipping). However, the "benefit" term $CQ+Q^TDQ/2$ should not, strictly speaking, be interpreted literally as benefit component of consumer surplus, in the way that the integral of a single commodity demand curve can be (see Chapter 3). However, its derivative does yield the original inverse demand curve (12.11), which can be interpreted as the marginal willingness-to-pay. The resulting KKT condition for q_i can then be interpreted as marginal benefit = marginal cost of supplying that demand, if q_i is strictly positive; but if q_i is zero, then marginal cost can exceed marginal benefit. This is the complementarity form of the inverse demand function we previously introduced (12.12).

If we examine the inverse demand relationships (12.11), (12.12) or the demand equations (12.13), we see that they indeed satisfy the symmetry requirement. Therefore, we can solve our particular PIES equilibrium problem either with a complementarity solver or as a QP. However, in the more general case of asymmetry in the matrices B or D, the QP approach cannot be used. Either the complementarity method or an iterative approach (as described by Hogan [13] or Section 8.4.2.1 of this book) would need to be applied.

12.4.4 PIES Equilibrium: Interpreting the Solutions of a Multicommodity Model with Demand Linkages

In this section, we present two equilibrium solutions. One solution has no CO_2 cap. In the other solution, such a cap is imposed, which requires fossil fuel producers to buy allowances to cover the CO_2 emissions that would

Table 12.2 PIES Solution under no CO_2 limit

Production

i	s	x_{is}
P1	1	1100
	2	1195.07
P2	1	1300
	2	1100
C5	1	300
	2	300
	3	400
C6	1	200
	2	300
	3	600

Refining Input

i	q_i
P3	2245.61
P4	2449.46

Refining Output

i	x_i
L3	1347.37
L4	1224.73
H3	898.25
H4	1224.73

Consumption

i	q_i
L7	1281.04
H7	1056.49
C7	1100
L8	1291.06
H8	1066.49
C8	1000

Prices

i	p_i
P1	1.5
P2	2.5
P3	3.5
P4	4.5
L3	11.403
L4	11.404
L7	12.404
L8	12.603
H3	7.895
H4	7.595
H7	8.595
H8	9.095
C5	8.355
C6	8.355
C7	9.105
C8	10.855
CO_2	0

Transport

i	j	t_{ij}
P1	P3	2245.61
P1	P4	49.46
P2	P3	0
P2	P4	2400
L3	L7	56.31
L3	L8	1291.05
L4	L7	1224.73
L4	L8	0

Transport

i	j	t_{ij}
H3	H7	0
H3	H8	898.24
H4	H7	1056.49
H4	H8	168.24
C5	C7	0
C5	C8	1000
C6	C7	1100
C6	C8	0

Emissions

CO_2	6921

Of which 4620 is from coal

ultimately result from combustion of the coal they mine and the oil they pump. We dig into the solution in some depth in order to help the reader learn how to interpret solutions of multicommodity models, in particular the factors that determine the formation of prices. We then consider how the inclusion or omission of cross-price elasticities alters the equilibrium and policy conclusions.

12.4.4.1 Interpreting Solutions: Where Do Prices Come From?

We first consider just the market equilibrium without a CO_2 cap (yielding 6921 tons of CO_2) in Table 12.2, and consider the impact of a CO_2 policy in a later subsection. All primal variables and all commodity prices are shown in that table. For brevity, however, we omit the shadow prices (dual variables) of capacity and refining constraints, which the interested reader can readily recover from the commodity prices using the equilibrium conditions for those duals in the PIES linear MCP (Section 12.4.3.1).

However, because of the arbitrary nature of our assumptions, we realize the specific numbers we obtain won't interest the reader so much as some general rules about what characteristics of the solutions to such models are likely to be of interest when examining them. One set of characteristics concerns the setting of prices: is it supply costs that determine prices, or the value to demand as expressed by the demand function? Our solutions exhibit both phenomena, as we explain in this subsection. The second set concerns the differences caused by assumptions or, as here, a change in policy (imposition of a CO_2 cap), which we'll turn to in subsequent subsections.

In studying solutions of these or other market equilibrium models, the most economic insight is often obtained by examining the primal variables (physical quantities) and dual variables (prices) together to understand how the prices are formed. Often, but not always, prices in competitive markets will be determined by either supply's marginal cost or, if supply is producing at capacity, the marginal willingness of demand to pay. This is illustrated in Figures 12.4(a) and 12.4(b), which show a stepped supply curve and a downward sloping demand curve. Dashed lines indicate the equilibrium price p and quantity q where the two curves intercept. Each step of the supply curve is a different tranche of supply with a marginal cost and limited production capability. The inverse demand curve shows price as a function of only the associated quantity. More generally, though, the functional form of (12.11)-(12.13) means that the intercept of this curve (but not the slope) depends on how much is consumed of other (substitute or complementary) goods.

(However, as pointed out in Chapter 3, prices in oligopolistic markets result from a complicated interplay of additional factors, including number of firms and demand elasticity. We limit our discussion here to competitive markets in which suppliers and consumers both assume that prices are exogenous to their maximization problems, i.e., are assumed by each player to be uninfluenced by their individual decisions.[9])

In Figure 12.4(a), the demand curve cuts a horizontal portion of the supply curve. In such a situation, the primal variable representing the output of that step will be basic, as its value lies strictly between its lower and upper bounds. Further, the price will equal that supply tranche's marginal cost.

Figure 12.4(b) shows a different situation: there, the demand curve intercepts a vertical portion of the supply curve, representing a situation where the last producing tranche is at capacity. Price is then set not by marginal production cost, but by the willingness of consumers to pay for the good. This situation is sometimes called "scarcity pricing" or "ration pricing", as price is raised above the marginal cost of the last unit of production in order to ration consumption to what can be produced. The difference between the ration price and the last production tranche's marginal cost equals the shadow

---

[9] For an analysis of how prices for energy and CO_2 are formed in more complicated oligopoly models, together with solutions from a large market equilibrium model, see Chen et al. [3].

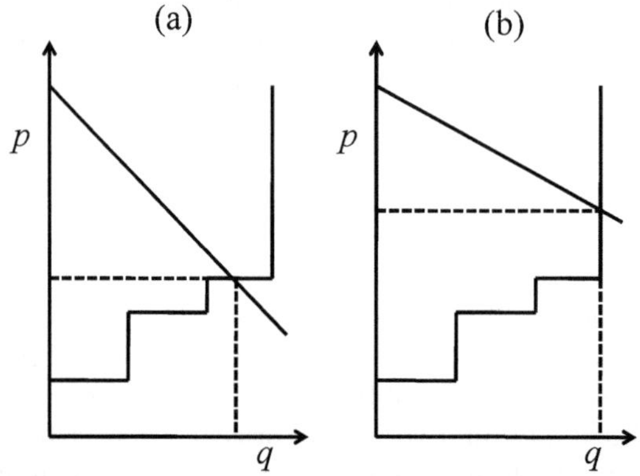

Fig. 12.4 Schematic of equilibria between stepped supply curves and linear demand: (a) supply cost sets price; (b) demand willingness-to-pay alone sets price

price (dual variable) of the associated capacity constraint in the producer's model. It contributes to the short-run producers' surplus (difference between revenues and variable costs, excluding the fixed cost associated with capital investments). That surplus is also called "economic rent" by economists and "gross margin" by accountants. In real markets, such rents are necessary in order to provide a return to capital—otherwise investment would not be profitable to producers. If rents are too low to maintain or replace existing plant, then eventually capacity is retired, shrinking supply and raising prices further until returns to capital rise sufficiently to justify investment in maintenance or new production facilities.

In our PIES solutions, we find both types of price setting situations. As an example of the situation in Figure 12.4(a), consider petroleum and coal production. What's perhaps most notable about the no-CO_2 limit solution is that nearly all the possible petroleum and coal is produced, with the exception of a small amount of the most expensive tranche of oil in market P1. The latter tranche is basic (greater than its lower bound but short of its upper bound). The existence of this basic production variable indicates that at least some market prices will be determined by that variable's cost coefficient. This results from complementary slackness: the marginal cost of a production variable strictly between its lower and upper bounds must equal the price in its market.

For this case of the basic tranche in market P1, that condition is the KKT conditions for $x_{P1,2}$ and $\mu_{P1,2}$ in (12.16) (which are therefore conditions in the PIES MCP). There, the \$1.5/bbl marginal cost of producing P1's last tranche of oil sets the price for petroleum in that region (p_{P1} = \$1.5/bbl). Furthermore, when marked up by the transport cost, that marginal cost also sets the price of petroleum in downstream refining regions, as a result of the

KKT conditions for positive oil transport variables from region 1 to elsewhere. For instance, in the case of transport from region 1 to region 3:

$$0 \leq t_{P1,P3} \perp (p_{P3} - p_{P1} - 2) \leq 0$$

Since a positive amount of oil is shipped from region 1 to both refining regions 3 and 4, the latter prices p_{P3} and p_{P4} are 3.5 and 4.5 \$/bbl, respectively, as a result of adding the \$1.5/bbl production cost to the respective costs of transport from region 1 (2 and 3 \$/bbl respectively). In the case of p_{P3}, a strictly positive $t_{P1,P3}$ means that the right side of the above complementarity condition forces $p_{P3} - p_{P1} - 2$ to equal zero, driving p_{P3} to \$3.5/bbl.

The basic production variable for oil in region 1 not only affects downstream prices, but also prices for competing producers elsewhere. In particular, the marginal cost in P1 propagates back to production market P2, where the price is \$2.5/bbl. The reason is that the producer there competes with P1's producer in the refinery market in P4, where the prices equal P1's marginal cost plus its transport cost. Because the transport variable $t_{P2,P4}$ is positive, its KKT condition

$$0 \leq t_{P2,P4} \perp (p_{P4} - p_{P2} - 2) \leq 0$$

(derived from (12.21)) will force the price at P2 to equal the downstream price P4 minus the \$2/bbl cost of transport from P2 to P4. Note that P2's transport costs to P4 are \$1/bbl less than P1's \$3/bbl transport cost to P4, so the producer in region 2 receives a \$1/bbl higher price than region 1's producer.

However, the prices of oil products aren't as easily interpreted as Figure 12.4(a). The very low petroleum prices don't translate into equally low product prices (light and heavy oil) because of the cost of refining and transporting refined fuels to markets. Those costs are pancaked onto the crude oil price through the KKT conditions to problems (12.18), (12.19), yielding the final product prices in markets L7, L8, H7, H8 (which are still amazingly cheap by 21st century standards!). The relationship between those prices and P1's marginal cost is made a bit more complex by the combination of (a) the higher prices that consumers are generally willing to pay for light oil than heavy oil and (b) each refinery's splitting of petroleum into products. This higher valuation of light oil is reflected in the inverse demand functions (price as a function of quantity) (12.11). They also show higher intercepts for the light oil price, and furthermore, also exhibit larger negative cross coefficient for coal consumption for the heavy oil price, meaning that coal's substitution possibilities depress the consumers' willingness-to-pay for heavy oil more than light oil, as one might expect. Meanwhile, the first-order conditions for the refineries allow the prices for light and heavy oil to separate; we see from the KKT conditions from the refiner's problem(12.17) that the only restriction that refining operations place on light and heavy oil prices is that their weighted average equals the price of input petroleum plus the refining cost.

For instance, consider the refinery in region 3. If it buys a positive amount of crude oil q_{P3} and produces positive amounts of light and heavy oil products (x_{L3}, x_{H3}), then we can derive $(p_{P3} + 6.5) = 0.6p_{L3,P3} + 0.4p_{H3,P3}$ from the following KKT conditions from refinery 3's profit-maximization problem (12.18):

$$0 \leq x_{L3} \perp p_{L3} - \phi_{L3,P3} \leq 0$$

$$0 \leq x_{H3} \perp p_{H3} - \phi_{H3,P3} \leq 0$$

$$0 \leq q_{P3} \perp - (p_{P3} + 6.5) + 0.6\phi_{L3,P3} + 0.4\phi_{H3,P3} \leq 0$$

In words, P3's price plus the 6.5\$/bbl refining cost must equal a weighted sum of the product prices, with the price of light oil getting 60% of the weight (since 0.6 bbl of light oil is produced for each barrel of crude input) and the price of heavy oil getting 40%. That is, 3.5+6.5 = (0.6)(11.403)+(0.4)(7.895). Because the downstream consumers push light oil prices up, due to their higher valuations of that commodity, that weighted sum equation then ensures that heavy oil prices must go down in compensation. (Note the complication that downstream prices to the consumer also include the price of transport from the refinery.)

Thus, the final prices of oil products reflect a more complex picture than Figure 12.4(a), involving an interplay of production costs (here, at P1), transport costs, refining technology (which determines the weights of light and heavy oil), and consumers' differing relative valuations of the different products. However, in addition to providing economic insights, tracing the prices in the above manner provides a useful check on the solution: is the solution reported by the software really an equilibrium, or was there a computational or data input error?

The price relationships would be even more complicated if the transport system had capacity limitations or other constraints. In that case, price differences between markets sending and receiving a commodity could exceed the cost coefficient in the transporter's problem (12.20). Indeed, in the history of energy markets, such capacity limits have been experienced at various times, for instance for oil tankers during times of vigorous international economic growth, or for rail shipments of low sulfur coal from western US markets when the US acid rain law was first implemented in the mid-1990s. However, we suspect that the reader will be grateful that we've omitted that complication here!

Let's now turn to the case of demand-determined prices, as in Figure 12.4(b). In the No CO_2 limit PIES solution (Table 12.2), the price of coal is determined, in large measure, by consumer valuations and not at all by production costs. This is because there are no basic production variables – the mines are producing at full capacity, a total of 2100 tons/d. In such circumstances, prices in an equilibrium model such as ours must rise to incent consumers to reduce their consumption to the point where the production capacity can meet the demand, so that markets clear. At that point, generally, the price will be higher than the marginal cost of production for all produced coal.

Here, the prices in markets C7 and C8 (9.105 and 10.855 $/ton, respectively) are high enough such that total consumption is precisely 2100 tons. The level of these ration prices is not unique, however, for two reasons. One is consumer substitution: the prices of light and heavy oil, which are partial substitutes for coal, also determine how much coal is demanded at different prices (via the demand functions (12.13)). However, even given the prices of other commodities, the C7 and C8 prices still won't be unique—there are an infinite number of combinations of the two prices that yield the same 2100 tons/d total demand, but with different splits of that consumption between the two regional markets.

The exact coal prices result from the interaction of demand functions and transport, accounting for transport costs from the locations of production. The equilibrium conditions imply, for example, that if two coal producers are serving the same consumption market, then the price each receives at their point of production equals the price the consumer pays minus transport costs. However, in Table 12.2's solution, two coal producers never serve the same consumer market – the coal producer at 5 ships its entire production of 1000 tons to consumers at 8, while the producer at 6 sends its 1100 tons of coal only to consumers at 7. By the KKT condition for coal transport variable $t_{C5,C8}$ from the shippers optimization problem (12.21)

$$0 \leq t_{C5,C8} \perp (p_{C8} - p_{C5} - 2.5) \leq 0,$$

this implies that the price in market C5 is 2.5 $/t less than that paid by consumers in C8. Meanwhile, for the same general reason, $t_{C6,C7}$ being positive implies that 6's producer (in market C6) receives 0.75 $/t less payment than what 7's consumer pays (in market C7). The prices shown in Table 12.2 indeed satisfy these conditions.

That the situation in Figure 12.4(b) applies to our PIES coal market in Table 12.2's solution can be determined by considering, for instance, demand at C7. If we insert the purchases of node 7's consumers of light and heavy oil (1281.04 and 1056.49 bbl/d, respectively) into the inverse demand function for coal in market C7 (12.11), we get the equation:

$$p_{C7} = 47.2 - (0.006)(1281.04) - (0.009)(1056.49) - 0.019\ q_{C7}$$

In order for q_{C7} to be rationed to the 1100 tons/d available to that market (as in Figure 12.4(b)), the resulting coal price has to be 9.105 $/t. This is indeed what occurs in Table 12.2's solution. The seller of these 1100 tons/d (in market C6) earns an economic rent on its last tranche of production equal to the price it receives at C6 (equal to C7's prices 9.105 $/t minus the 0.75 $/t transport cost from C6) minus that tranche's marginal cost of $7/ton, or 1.35 $/ton. This is, as it should be, the shadow price $\mu_{C6,3}$ on that tranche's capacity constraint (not shown in the table).

Table 12.3 PIES Solution under CO_2 limit with cross-price elasticities

Production

I	s	x_{is}
P1	1	1100
	2	1200
P2	1	1300
	2	1100
C5	1	300
	2	300
	3	0
C6	1	200
	2	300
	3	125.91

Refining Input

i	q_i
P3	2136.99
P4	2563.02

Refining Output

i	x_i
L3	1282.19
L4	1281.51
H3	854.80
H4	1281.5

Consumption

i	q_i
L7	1281.51
H7	1064.16
C7	654.43
L8	1282.19
H8	1072.15
C8	571.48

Prices

i	p_i
P1	4.678
P2	5.678
P3	6.678
P4	7.678
L3	14.184
L4	13.987
L7	14.987
L8	15.384
H3	11.67
H4	11.37
H7	12.37
H8	12.87
C5	16.5
C6	16.75
C7	17.45
C8	19
CO_2	4.432

Transport

I	J	t_{ij}
P1	P3	2136.99
P1	P4	163.02
P2	P3	0
P2	P4	2400
L3	L7	0
L3	L8	1282.19
L4	L7	1282.51
L4	L8	0

Transport

i	j	t_{ij}
H3	H7	0
H3	H8	854.79
H4	H7	1064.16
H4	H8	217.35
C5	C7	28.52
C5	C8	571.48
C6	C7	625.91
C6	C8	0

Emissions

CO_2	5000

Of which 2697 is from coal

12.4.4.2 Interpreting Solutions: Effects of Policy

We have just explored how insight into price formation can be obtained by studying a particular solution. We now compare different solutions in order to understand how changes in policies and other assumptions can affect results.[10] In Table 12.3, we show the market equilibrium resulting from imposing a cap of 5000 tons/d upon CO_2 emissions. This represents a decrease of 1921 tons/d from the no limit solution of Table 12.2.

[10] Another potential issue that readers should be aware of is that a solution to a particular complementarity problem might not be unique. Therefore, differences between alternative policy runs of a model might arise not because of fundamental differences in the models, but because the different solutions might simply be alternative solutions to what are actually equivalent models. Or the differences between policies might not be unique because there may be more than one distinct solution for a given model. Uniqueness results can be established by the analytical approaches in Section 4.5.

In this subsection, we illustrate what can be learned by a comparison of runs based on different policies, in this case an environmental policy. We consider both changes in primal variables and the associated changes in prices.[11]

In interpreting solutions, it is helpful to form hypotheses *before* calculating solutions. Developing hypotheses about how a policy or other change in assumptions will change solutions will aid you to think about aspects of the solution that are likely to be most interesting to you and your audience. This gives you a structure for your presentation of results to others. Finally, it is particularly useful when the solutions contradict your hypotheses. Either some mistake in formulation, data input, or computation has been made (which you can then debug), or the model's surprising result is teaching you something you didn't expect. If the latter is true, the resulting insights will be particularly interesting to you and your audience. (Economists in particular love conclusions that contradict conventional wisdom, and systems theorists such as Jay Forrester have long warned us to expect counterintuitive behavior from complex systems.)

Our first hypothesis is that *the imposition of the cap will result in a positive price for CO_2, a reduction in CO_2 emissions, and reductions in energy use.* This is confirmed. (Perhaps this disappoints the surprise-loving economists among our readers!) In a sense, the price of CO_2 is a "scarcity" or "ration price" situation (Figure 12.4(b)) in which a capacity limit causes the price to be set not by the marginal cost of supply, but instead by the value to consumers. In particular, because the supply of CO_2 allowances is perfectly inelastic, the price is determined by where the effective demand curve for allowances intersects the vertical demand curve. However, explaining why a 4.43 \$/ton price for CO_2 results and not some other level is not straightforward, as the effective demand by fossil fuel producers for allowances results from two intertwined phenomena: consumers restricting their consumption and switching from high- to low-carbon fuels. (In a more realistic situation, other actions can also reduce the effective demand for allowances, such as production, transport, and conversion efficiency improvements, carbon sequestration, and generation of carbon credits by planting trees. The resulting complexity is precisely why comprehensive energy market models are needed to study carbon policy.)

A second hypothesis is that *CO_2 costs would affect the prices of coal more than they do oil because of the greater carbon content of coal.* And the tables show this to be the case. Coal prices nearly double when the CO_2 cap is imposed (compare tables 12.2 and 12.3 prices for consumers in markets C7 and C8). This is because the prices at the mine-mouth (markets C5 and C6) go up by 2.2 tons CO_2/ton coal times the shadow price of CO_2 allowances, which is 4.43 \$/ton of CO_2. The reason for this can be seen from the KKT condition for the basic production variable $x_{C6,3}$:

[11] Subject to the caveat that the individual solutions might not be unique, so therefore the differences between them might not be unique either.

$$0 \le x_{C6,3} \perp (p_{C6} - 7 - 2.2p_{CO2}) - \mu_{C6,3} \le 0$$

Since $x_{C6,3}$ is strictly between its upper and lower bound, the price p_{C6} for coal in region 6 must equal the production cost for that tranche (7 \$/ton) plus 220% of the price of CO2 ((2.2)(4.43) = 9.75), or \$16.75/ton. (This is because the shadow price $\mu_{C6,3}$ will be zero if the upper bound is not binding, and the inequality on the right side of $x_{C6,3}$'s complementarity condition will be forced to be an equality by its complementarity with $x_{C6,3}$.) This amount is indeed the price of coal in region 6 in Table 12.3. Meanwhile, oil prices also go up, but not as much percentage-wise because of oil's lower CO_2 content.

A third hypothesis is that *significant decreases in fossil fuel consumption would accompany the carbon and fuel price increases, with more decreases experienced for carbon-intensive coal.* This is mostly, but not entirely true. Tables 12.2 and 12.3 show that the greater price increase for coal had the effect of decreasing coal consumption, as well as its associated CO_2 emissions, just as we would hypothesize. Without a CO_2 cap, coal is the source of two-thirds of the emissions, but with the cap, coal's share drops to just over half, as the tables indicate. Total coal consumption falls from 2100 tons/d to just over 1200 tons/d.

We also expected that higher oil prices would also cause oil product consumption to go down. Yet, in fact, this did *not* happen. Petroleum production is actually (slightly) higher in the CO_2 cap case, despite the higher oil prices. (So if you were hoping for a surprise, we were able to muster one for you!) Table 12.3 shows now that the oil producers are pumping as much as their capacity allows (2700 tons/d), about 5 bbl/day more than when there is no CO_2 cap.

The higher oil demand occurs despite higher oil prices largely because oil is a substitute for coal, and the high coal prices have driven consumers to switch (ever so slightly) to oil. The demand functions (12.13) show that every \$1/ton increase in coal price causes light and heavy oil demand to go up by 7 and 16 bbl/d, respectively, in each of the two consumer markets (nodes 7 and 8). Since coal prices rose roughly \$9/ton (almost doubling), coal price increases by themselves would induce an increase of about 400 bbl/d in oil demand (2 regions multiplied by (7+16 (bbl/d)/(\$/t))(9 \$/t)). Nearly, but not all of this demand increase is offset by the dampening effect of oil price increases, with the net effect being a small increase in oil consumption.

12.4.4.3 Comparison with Own-Elasticity Only Results

We finally turn to the question that is the heart of this chapter: does consideration of cross-price elasticities materially affect the solution? After all, those elasticities are much smaller in magnitude than the own-price elasticity, and most energy models disregard them.

We answer the question by, first, defining inverse demand curves that pass through the no CO_2 limit solution of Table 12.2, have the same own-quantity slope as the original demand curves (12.13), but zero coefficients for cross effects. Then, second, we simulate the same CO_2 cap policy as considered in the last section, and examine the differences in prices and quantities. The detailed procedures used in each step are described below.

In the first step, we define own-elasticity only demand curves for regions 7 and 8 in the market so that they satisfy two conditions. The first condition is the demand curves must yield the same equilibrium prices and quantities as the original demand curves (12.11), (12.12), (12.13) in the absence of a CO_2 limit. This allows us to examine how simplifying the demand functions affects the PIES model's assessment of the impact of imposing a CO_2 cap. The second condition is that the new demand functions have same own-price elasticities at that equilibrium as the original demand curves, but zero cross-price elasticities. (We did the same thing for the electricity model of Section 12.3, so that the two sets of demand relationships considered there yielded the same regulated utility base case and would have the same own-price elasticities at that solution.) The resulting functions are:

$$p_{L7} = 45.846 - 0.02611 q_{L7} \tag{12.26a}$$

$$p_{H7} = 36.186 - 0.02611 q_{H7} \tag{12.26b}$$

$$p_{C7} = 26.662 - 0.01596 q_{C7} \tag{12.26c}$$

$$p_{L8} = 46.306 - 0.02611 q_{L8} \tag{12.26d}$$

$$p_{H8} = 36.945 - 0.02611 q_{H8} \tag{12.26e}$$

$$p_{C8} = 26.813 - 0.01596 q_{C8} \tag{12.26f}$$

The calibrated own-elasticity demand curves now differ for the two consumption markets (7 and 8) because the equilibrium prices and quantities in the two markets diverge in the no CO_2 case in Table 12.2.[12]

Table 12.4 shows the equilibrium solution that results from substituting these inverse demand functions (12.26) for the original ones (12.11) in the PIES mixed complementarity problem of Section 12.4.3.1, and simultaneously imposing a CO_2 limit of 5000 tons. (By the way we defined the own-elasticity

[12] For instance, consider demand for light oil at node 7. Table 12.2 shows that the No-CO_2 equilibrium base case was $\{p_{L7}, q_{L7}\} = \{12.40\ \$/\text{bbl}, 1281.04\ \text{bbl/d}\}$. Let the own-elasticity only linear demand function be $q_{L7} = A + Bp_{L7}$. To preserve the same price elasticity as in (12.13) for this commodity, the slope B has to be the same in the own-elasticity only function as in (12.13). Since elasticity for this linear function equals $(d\, q_{L7} / d\, p_{L7})\ (p_{L7} / q_{L7}) = B(\, p_{L7} / q_{L7}\,)$, so $B = -38.3$, Then we can solve for $A = q_{L7} - Bp_{L7} = 1281.04 - (-38.3)12.5 = 1760.79$. Inverting $q_{L7} = 1760.79 - 38.3\, p_{L7}$ yields $p_{L7} = 45.9 - 0.002611\, q_{L7}$, which is the inverse demand function for L7 shown in (12.26a). Note that the condition that the own elasticities of the original demand function and the own elasticity-only function are equal means that the own-variable slopes of the demand functions (12.13) are preserved, not the slopes of the inverse demand functions (12.11).

only demand functions, the no CO_2 limit solution is the same as Table 12.2 , and so it's unnecessary to show it.)

This solution differs in some important ways from Table 12.3 's solution, which imposes the same CO_2 limit, but considers cross-price elasticities. The most important implication is that omitting cross-price elasticities means that reducing quantity consumed of one commodity no longer increases the consumer's willingness-to-pay for (and use) substitutes. Consider the example of coal; raising coal's price in an attempt to meet the carbon limit by dampening demand for that high-carbon fuel will be partially frustrated if oil is a substitute (positive cross-price elasticity). In that case, some consumers will simply switch to the alternative fuel, whose CO_2 emissions will then offset some of the reductions achieved by burning less coal. As we see in the demand function (12.13), a \$1/ton increase in the price of coal decreases coal demand by 63 tons/d, but under our cross-price elasticity assumptions, consumers will also expand their use of light oil by 7 bbl/d and heavy oil by 16 bbl/d. The solution in Table 12.4 doesn't consider those cross-price effects, unlike that in Table 12.3 . Consequently, the no cross-effects model finds it easier to achieve the CO_2 target, as evidenced by that model's CO_2 price (which, at 3.6 \$/ton, is around 19% lower than Table 12.3's price).

Table 12.4 PIES Solution under CO_2 limit, no cross elasticities

Production

i	s	x_{is}
P1	1	1100
	2	931.73
P2	1	1300
	2	1100
C5	1	300
	2	300
	3	0
C6	1	200
	2	300
	3	185.66

Refining Input

i	q_i
P3	2031.73
P4	2400

Refining Output

i	x_i
L3	1219.04
L4	1200
H3	812.69
H4	1200

Consumption

i	q_i
L7	1204.52
H7	1001.34
C7	685.66
L8	1214.51
H8	1011.34
C8	600

Transport

i	j	t_{ij}
P1	P3	2031.73
P1	P4	0
P2	P3	0
P2	P4	2400
L3	L7	4.53
L3	L8	1214.51
L4	L7	1200
L4	L8	0.00

Transport

i	j	t_{ij}
H3	H7	0
H3	H8	812.69
H4	H7	1001.35
H4	H8	198.65
C5	C7	0
C5	C8	600.00
C6	C7	685.66
C6	C8	0

Emissions

CO_2	5000

Prices

i	p_i
P1	3.274
P2	4.218
P3	5.274
P4	6.218
L3	13.401
L4	13.401
L7	14.401
L8	14.601
H3	9.334
H4	9.034
H7	10.034
H8	10.534
C5	14.74
C6	14.968
C7	15.718
C8	17.24
CO_2	3.622

In our PIES model, the only way to reduce CO_2 is use less fuel (inducing conservation by jacking up prices) or to substitute cleaner oil for dirtier coal. Considering the first mechanism, achieving the CO_2 standard by raising fuel prices to consumers is less effective when one fuel substitutes for the other (positive cross-price elasticities); consequently, Table 12.3's fuel prices are significantly higher than Table 12.4's. With positive cross-price elasticities (Table 12.3), consumers at nodes 7 and 8 pay about 10% more for coal, 20% more for heavy oil, and 5% more for light oil. For instance, coal costs 14.7 $/ton at node 5 without cross price terms (Table 12.4), but 16.5 $/ton with them (Table 12.3).

The fuel mix impacts of the CO_2 limit also change if cross-price elasticities are dropped. With cross-price elasticities (Table 12.3), all emission reductions are achieved by reducing coal use, with oil consumption essentially unchanged despite the increases in all commodity prices. Ignoring the cross-price terms (Table 12.4) causes the model to instead project that the oil sector will contribute to the CO_2 reductions, with total oil consumption dropping 6% relative to demand without CO_2 limits (Table 12.2). With the oil sector requiring fewer CO_2 allowances in the simplified model, the coal sector gets a bit more room to breathe, expanding 5% compared to the CO_2-constrained cross-price case (Table 12.3).

Thus, disregarding substitution possibilities (represented by positive cross-price elasticities) in multifuel models can change conclusions about policies. In this particular case, the cost of a policy of reducing greenhouse gasses is underestimated by 28%, and the impacts upon different fuels are distorted.

12.5 Summary

This chapter is not only about modeling demand linkages, but also about how to interpret model outputs, especially prices and differences among policy runs of a model.

We show how complementarity models can readily be formulated to account for how consumption in one energy market can affect demand in other markets through demand functions with cross-price elasticities. We provide two fairly detailed examples of models with cross-price elasticities whose results can be replicated by the reader. The first example, which looks at the effect of time varying pricing in electricity markets, considers linkages among consumption at different times. In that case, positive cross-price elasticities between peak and off-peak consumption yield larger estimates for the impact of time-differentiated pricing than would result if only own-price elasticities are considered. This is because peak demand is dampened by both an increase in on-peak price and (because of the cross effect) a decrease in off-peak price. Ignoring cross-price elasticities in electricity modeling means that the second effect will be missed.

The second example instead considers cross-price elasticities among consumer demand for distinct energy commodities, such as natural gas and electricity. This is in the context of a simple PIES-type model of oil and coal trade that we use to estimate the impact of imposing a CO_2 cap upon the energy economy. We show how prices from a complementarity model can be analyzed to understand whether supply costs, capacity limits, or consumer willingness-to-pay for energy determine prices. We also show that cross-price elasticities result in appreciably higher estimates for the cost of lowering CO_2 production, mainly because raising prices on coal to decrease its emissions would partially frustrated by the boost in oil demand that would occur from the higher coal prices. Disregarding cross-price elasticities distorts the resulting estimates of the cost of environmental policy.

12.6 Exercises

1. Village X's residents heat their homes using either fuel oil or natural gas. In a given year, the village's retail market is observed to have the following price-quantity pairs for oil and gas, respectively: $\{p_{fo} = \$4/\text{gallon}, q_{fo} = 20{,}000 \text{ gal/yr}\}$ and $\{p_{ng} = \$8/\text{MBTU}, q_{ng} = 16{,}000 \text{ MBTU/yr}\}$. From other studies of retail energy markets, you estimate that the own-elasticities for each fuel type should be -0.5, while the cross-price elasticities should be +0.2. Calibrate linear demand functions for these two commodities that are consistent with those observed price-quantity pairs, assuming that those elasticities apply at those points. Does this demand system satisfy the Integrability condition?
2. Now consider the supply situation in the retail energy market for Village X. Say that the supply of fuel oil is perfectly elastic (horizontal) at a price of \$3.5/gal, but that the supply of natural gas is upward sloping as follows: $p_{ng} = \$6/\text{MBTU} + 0.0001 q_{ng}$. However, because of pipeline capacity limits into the town, the supplier of natural gas can provide no more than q_{ng}= 14,000 MBTU/yr. Define the retail energy market equilibrium problem for that village as a complementarity problem, and then solve the problem for the equilibrium.
3. Repeat Problem 2 under the assumption that the natural gas capacity limit has been relieved by construction of another pipeline into the town, so now there is no limit at all to q_{ng}. Hence, the supply function for gas is now $p_{ng} = \$6/\text{MBTU} + 0.0001 q_{ng}$ for all quantities. How does that change in the market affect the equilibrium quantity demanded of fuel oil? (To answer this question, you will need to calculate the new market equilibrium for both energy commodities.) Why (or why doesn't) this change in oil demand make sense to you?
4. The government that regulates the distribution utility in Section 12.3 decides that it needs revenue, and so it imposes an annual property tax of $\$100\times10^6$/yr upon the utility. Consequently, the utility's annual fixed cost is increased to $\$954.1\times10^6$/yr. Recalculate the regulated and unbundled market equilibria (Sections 12.3.1 and 12.3.2, respectively). Describe the incidence of the tax for each off the two models. In particular, calculate how much more consumers pay for power in each equilibrium, and how much the profits change for the unregulated power generators A and B in the unbundled solution (Section 12.3.2). (Assume non-zero cross-price elasticities.)
5. Consider the PIES model with a CO_2 trading system described in Section 12.4.2. Solve the version with non-zero cross-price elasticities for CO_2 limits of 1000, 2000, 3000, 4000, and 6000 tons/day. Plot the resulting prices of CO_2 permits against these quantities, and also plot the the price-quantity pairs from Tables 12.2 and 12.3. These points can be viewed as tracing out a demand curve for CO_2 emissions permits. From this curve, estimate how many CO_2 emissions permits would be demanded if instead

of a tradable emissions permit system, a tax of \$7/ton was imposed on carbon emissions.

References

1. E. Celebi and D. Fuller. A model for efficient consumer pricing schemes in electricity markets. *IEEE Transactions on Power Systems*, 22(1):60-67, 2007.
2. E. Celebi and D. Fuller. Time of use pricing in electricity markets under different market structures. *IEEE Transactions on Power Systems*, to appear, 2012.
3. Y. Chen, J. Sijm, B.F. Hobbs, and W. Lise. Implications of CO_2 emissions trading for short-run electricity market outcomes in Northwest Europe. *Journal of Regulatory Economics*, 34(3):251-281, Dec. 2008.
4. D. Connolly, H. Lund, B.V. Mathiesen, and M. Leahy. A review of computer tools for analysing the integration of renewable energy into various energy systems. *Applied Energy*, 87(4):1059–1082, 2010
5. C. Cormio, M. Dicorato, A. Minoia, and M. Trovato. A regional energy planning methodology including renewable energy sources and environmental constraints. *Renewable and Sustainable Energy Reviews*, 7(2):99–130, 2003.
6. A. Faruqui and S. Sergici. Household response to dynamic pricing of electricity - a survey of the experimental evidence. Manuscript, Brattle Group, San Francisco, 2009.
7. L.G. Fishbone and H. Abilock. MARKAL, A linear-programming model for energy-systems analysis – technical description of the BNL Version. *International Journal of Energy Research*, 5(4): 353-375, 1981.
8. S. Gabriel, A. Kydes, and P. Whitman. The National Energy Modeling System: a large-scale energy-economic equilibrium model. *Operations Research*, 49(1):14-25, 2001.
9. H. Hashimoto. A spatial Nash equilibrium model. In P. Harker, Editor. *Spatial Price Equilibrium: Advances in Theory, Computation, and Application.* Springer, Berlin, 1982.
10. B.F. Hobbs. The "Most Value" criterion: economic evaluation of utility demand-side management programs considering customer value. *The Energy Journal*, 12(2):67-91, 1991.
11. B.F. Hobbs. Optimization methods for electric utility resource planning. *European J. Operational Research*, 83(1):1-20, 1995
12. B.F. Hobbs and S.K. Nelson. A nonlinear bilevel model for analysis of electric utility demand-side planning issues. *Annals of Operations Research*, 34:255-274, 1992.
13. W.W. Hogan. Energy policy models for Project Independence, *Computers and Operations Research*, 2:251-271, 1975.
14. D. Klingman, N. Phillips, D. Steiger, R. Wirth, and W. Young. The challenges and success factors in implementing an integrated products planning system for Citgo. *Interfaces*, 16(3):1-19, 1986.
15. W.W. Leontief. *Input-Output Economics.* Oxford University Press, New York, 2nd edition, 1986.

16. S.C. Morris, G.A. Goldstein, and V.M. Fthenakis. NEMS and MARKAL-MACRO models for energy-environmental-economic analysis: a comparison of the electricity and carbon reduction projections. *Environmental Modeling & Assessment*, 7(3):207-216, 2002.
17. R. Loulou and M. Labriet. ETSAP-TIAM: the TIMES integrated assessment model, part I: model structure. *Computational Management Science*, 5(1-2):7-40, 2008.
18. F.H. Murphy and S.H. Shaw. The evolution of energy modeling at the Federal Energy Administration and the Energy Information Administration. *Interfaces*, 25(5):173-193, 1995.
19. S.W. Salant. Imperfect competition in the international energy market: a computerized Nash-Cournot model. *Operations Research*, 30(2):252-280, 1982.
20. P.A. Samuelson. Spatial price equilibrium and linear programming. *The American Economic Review*, 42:283-303, 1952.
21. R. Turvey and D. Anderson. *Electricity Economics*.The Johns Hopkins University Press, Baltimore, 1977.
22. G. Zwart and M. Mulder. NATGAS: a model of the European natural gas market. CPB Memorandum 144, CPB Netherlands Bureau for Economic Policy Analysis, 2006.

Appendix A
Convex Sets and Functions

This appendix provides a brief summary of the main concepts associated with convex sets and convex functions. Clarity and conciseness are preferred to mathematical formality. The interested reader can find extended technical material in references [7], [8] and [5] regarding convex analysis, and in references [1], [2], [3], [6] and [4] regarding the role that convexity plays in linear and nonlinear programming.

Definition A.1. A set $\Omega \in \mathbb{R}^n$ is convex if for any $\alpha \in [0,1]$ and $x^1, x^2 \in \Omega$, $\alpha x^1 + (1-\alpha)x^2 \in \Omega$.

Note that $\alpha x^1 + (1-\alpha)x^2$ with $\alpha \in [0,1]$ corresponds to the line segment connecting x^1 and x^2 in two-dimensional space.

Figure A.1 illustrates Definition A.1 in $\mathbb{R}^2$. Note that for convex sets (Figure A.1(a)) all points that lie on the line segment connecting any two points within the set, are also in the set. This is not true for nonconvex sets (Figure A.1(b)) since the line segment may contain points that do not belong to the set.

Two relevant properties pertaining to convex sets can be directly derived from Definition A.1:

1. If sets $\{\Omega_j\}_{j\in J}$ are convex, then $\cap_{j\in J}\Omega_j \equiv \{x \mid x \in \Omega_j \ \forall j \in J\}$ is also convex, i.e., the set resulting from the intersection of several convex sets, is also convex.
2. If Ω_1 and Ω_2 are two convex sets then the set $\Omega_1 + \Omega_2 \equiv \{x^1 + x^2 \mid x^1 \in \Omega_1 \text{ and } x^2 \in \Omega_2\}$ is also convex.

The three illustrative examples below present some representative convex sets: hyperplane and half-space, polyhedral set and polyhedral cone.

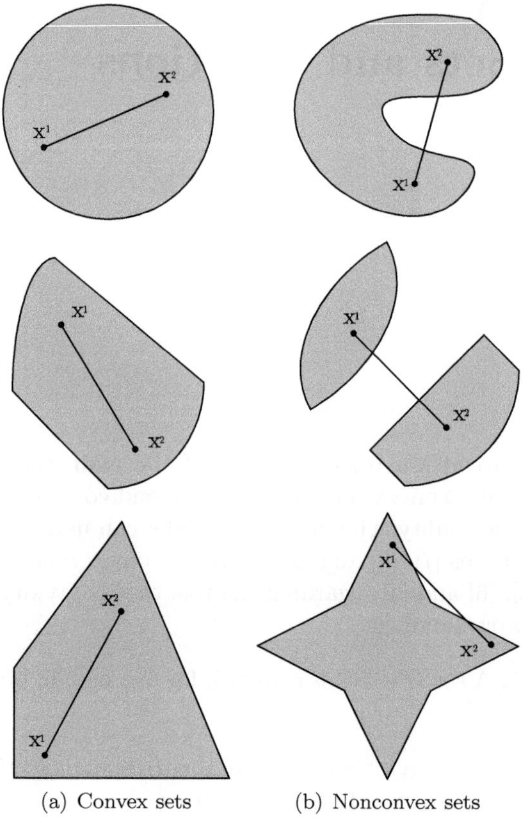

Fig. A.1 Instances of convex and nonconvex sets in two dimensions

Illustrative Example. Hyperplane and half-space

A hyperplane is a set defined as $\Omega = \{x \in \mathbb{R}^n \mid p^T x = \beta\}$ where β is a scalar and p is a nonzero vector in $\mathbb{R}^n$. Similarly, a half-space is defined as $\Omega = \{x \in \mathbb{R}^n \mid p^T x \leq \beta\}$. From Definition A.1 it is easy to check the convexity of both hyperplanes and half-spaces.

For instance, consider that x^1 and x^2 belong to the hyperplane $p^T x = \beta$ so that $p^T x^1 = \beta$ and $p^T x^2 = \beta$. Then $\alpha x^1 + (1-\alpha)x^2$ belongs also to the hyperplane since $p^T(\alpha x^1 + (1-\alpha)x^2) = \alpha p^T x^1 + (1-\alpha)p^T x^2 = \alpha\beta + (1-\alpha)\beta = \beta$. This derivation can be similarly applied to show the convexity of the half-space set.

Figure A.2 depicts two half-spaces in two dimensions and their corresponding supporting hyperplane, where $p = (2 \ -1)^T$ and $\beta = 1$. Note that in $\mathbb{R}^2$ every straight line is a hyperplane that divides $\mathbb{R}^2$ in two half-spaces.

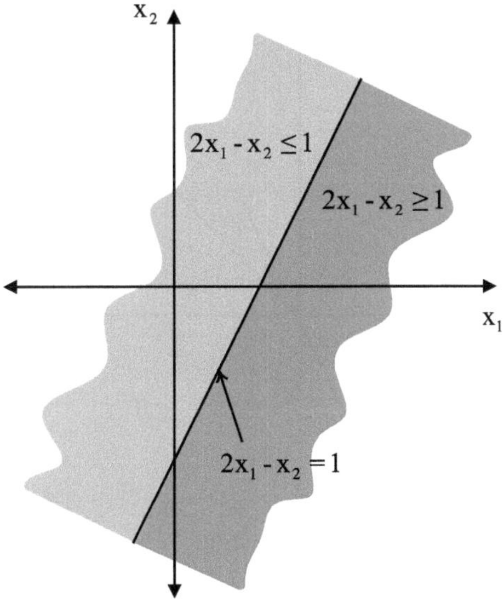

Fig. A.2 Half-spaces separated by a hyperplane

Illustrative Example. Polyhedral cone

A polyhedral cone is a set defined as $\Omega = \{x \in \mathbb{R}^n \mid Ax \leq 0\}$ where A is an $m \times n$ matrix. Note that $Ax \leq 0$ represents the intersection of several half-spaces whose supporting hyperplanes contain the origin of $\mathbb{R}^n$.

Figure A.3 provides an instance of a polyhedral cone in $\mathbb{R}^2$ where $A = \begin{pmatrix} 1 & -1 \\ -2 & 1 \end{pmatrix}$. The polyhedral cone is the darker area of the graph that results from the intersection of two half-spaces, defined as $x_1 - x_2 \leq 0$ and $-2x_1 + x_2 \leq 0$.

Illustrative Example. Polyhedral set

A polyhedral set is defined as $\Omega = \{x \in \mathbb{R}^n \mid Ax \leq b\}$ where A is a $m \times n$ matrix. Observe that the polyhedral cone can be interpreted as a particular case of a polyhedral set i.e., the case for $b = 0$.

An example of a polyhedral set in $\mathbb{R}^2$ is depicted in Figure A.4 (darker area). This set is defined as

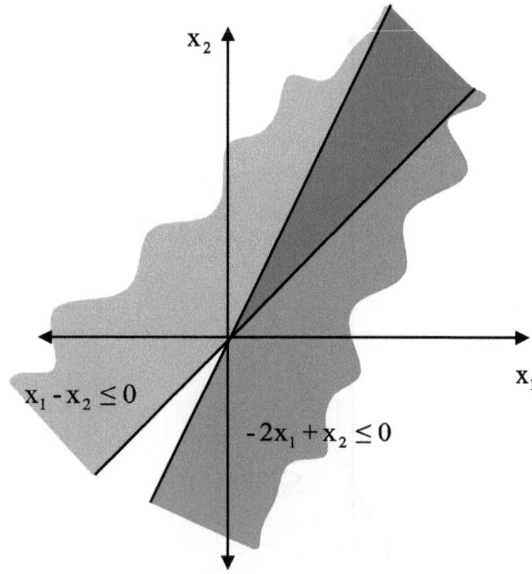

Fig. A.3 Polyhedral cone

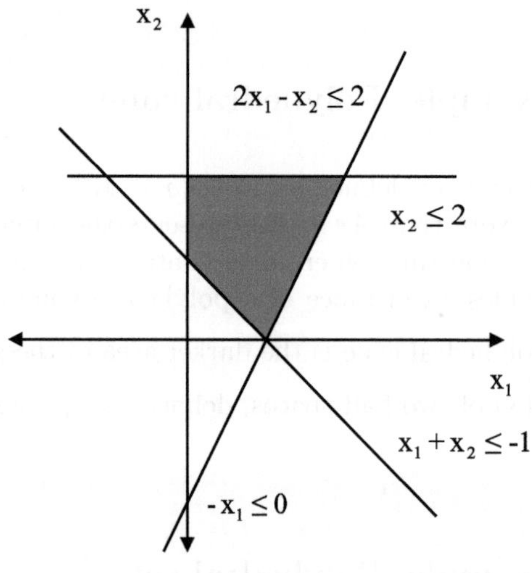

Fig. A.4 Polyhedral set

$$\begin{pmatrix} 2 & -1 \\ 1 & 1 \\ -1 & 0 \\ 0 & 1 \end{pmatrix} \begin{pmatrix} x_1 \\ x_2 \end{pmatrix} \leq \begin{pmatrix} 2 \\ 1 \\ 0 \\ 2 \end{pmatrix} \tag{A.1}$$

and is the intersection of four half-spaces.

Definition A.2. Suppose that Ω is a convex set. Then a function $f(x) : \Omega \subseteq \mathbb{R}^n \to \mathbb{R}$ is convex if $\forall \alpha \in [0, 1]$ and every $x^1, x^2 \in \Omega$, $f(\alpha x^1 + (1 - \alpha)x^2) \leq \alpha f(x^1) + (1 - \alpha)f(x^2)$. It is strictly convex if for all x^1 not equal to x^2, $f(\alpha x^1 + (1 - \alpha)x^2) < \alpha f(x^1) + (1 - \alpha)f(x^2)$. Similarly, a function $f(x)$ is said to be concave (strictly concave) if $-f(x)$ is convex (strictly convex).

Figures A.5-A.7 provide a geometric interpretation of Definition A.2 for $n = 1$. Note that for a convex function as shown in Figure A.5, the value of $f(\alpha x^1 + (1 - \alpha)x^2)$ is always below the value of $\alpha f(x^1) + (1 - \alpha)f(x^2)$ for any pair of points x^1 and x^2 and all $\alpha \in [0, 1]$. On the contrary, a concave function as shown in Figure A.6, is characterized by $f(\alpha x^1 + (1 - \alpha)x^2) \geq \alpha f(x^1) + (1 - \alpha)f(x^2)$ for all x^1, x^2 and all $\alpha \in [0, 1]$. Finally, Figure A.7 presents a function that is neither convex nor concave since the values of $f(\alpha x^1 + (1 - \alpha)x^2)$ can be above or below $\alpha f(x^1) + (1 - \alpha)f(x^2)$ depending on the selected α (as well as x^1 and x^2 values).

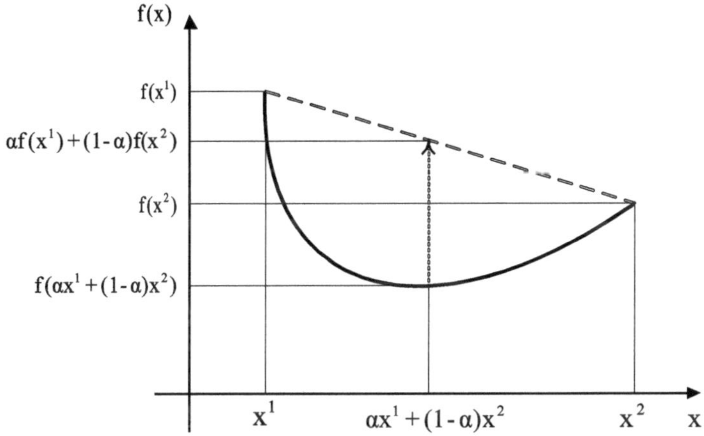

Fig. A.5 Convex function

We present below some relevant properties derived from considering that $\Omega \subset \mathbb{R}^n$ is a convex set and $f(x) : \Omega \to \mathbb{R}$ is a convex function. For clarity, proofs are omitted although they can be easily derived from definitions A.1 and A.2. Let $f(x)$ be continuous on Ω.

1. The sets $\{x \in \Omega \mid f(x) \leq \beta\}$ and $\{x \in \Omega \mid f(x) < \beta\}$ are convex for all scalars β.
2. All local minima of $f(x)$ on Ω are also global minima. Moreover, if $f(x)$ is strictly convex then there exists only one global minimum.
3. If $f(x)$ is differentiable over Ω then $f(x)$ is convex if and only if $f(x^1) \geq f(x^2) + \nabla f(x^1)^T(x^2 - x^1) \quad \forall x^1, x^2 \in \Omega$. This property is illustrated in Figure A.8.

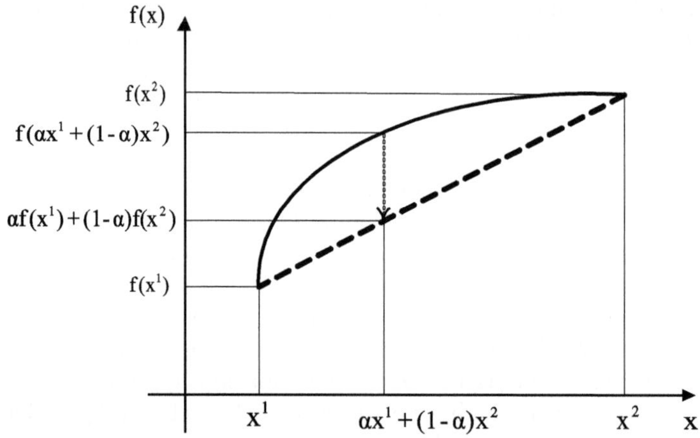

Fig. A.6 Concave function

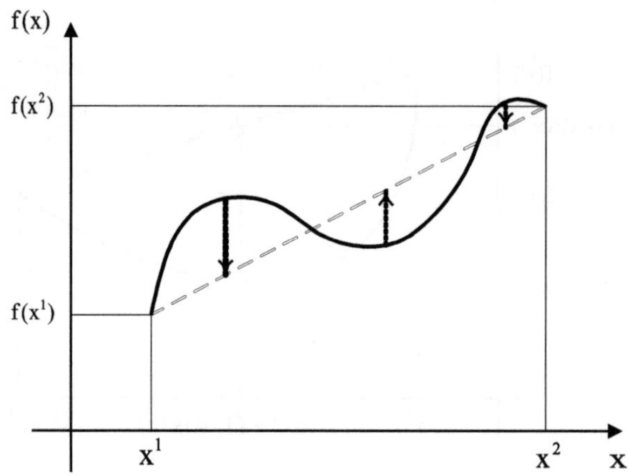

Fig. A.7 Neither convex nor concave function

4. If $f(x)$ is twice continuously differentiable on Ω then $f(x)$ is convex if and only if $y^T\nabla^2 f(\tilde{x})y \geq 0 \quad \forall \tilde{x} \in \Omega,\ \forall y \in \mathbb{R}^n$.
5. The above property is equivalent to the following one: If $f(x)$ is twice continuously differentiable on Ω then $f(x)$ is convex if and only if the Hessian matrix $(\nabla^2 f(x))$ is positive semidefinite $\forall x \in \Omega$.
6. Let $f_j(x),\ j = 1, \ldots, J$ be a family of convex functions and $\lambda_j,\ j = 1, \ldots, J$ be positive scalars. Then $h(x) = \sum_{j=1}^{J} \lambda_j f_j(x)$ is a convex function as well.
7. If $k(x)$ is a linear function then it is convex.

8. If $g(x)$ is a convex function and $k(x)$ is a linear function then $h = g(k(x))$ is a convex function.

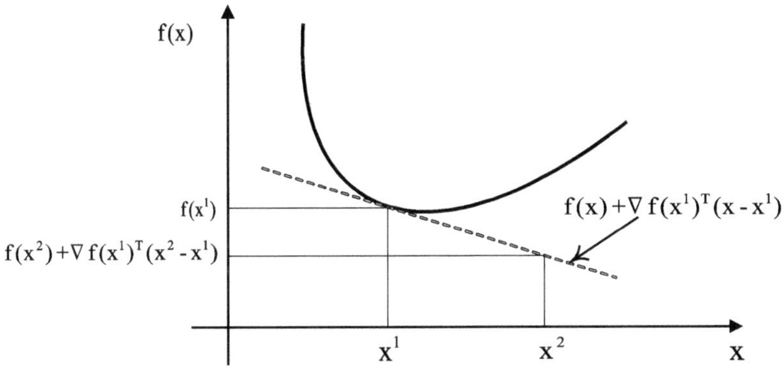

Fig. A.8 Convexity characterization of a differentiable function

Two relevant instances of convex functions are presented below.

Illustrative Example. Affine function

An affine function is defined as $f(x) = b^T x + a$ where $a \in \mathbb{R}$ and $b, x \in \mathbb{R}^n$.

The convexity of this function can be analyzed according to Definition A.2 by comparing $f(\alpha x^1 + (1-\alpha)x^2)$ and $\alpha f(x^1) + (1-\alpha)f(x^2)$.

In particular, observe that $f(\alpha x^1 + (1-\alpha)x^2) = b^T(\alpha x^1 + (1-\alpha)x^2) + a = \alpha(b^T x^1 + a) + (1-\alpha)(b^T x^2 + a) = \alpha f(x^1) + (1-\alpha)f(x^2)$, which satisfies the definition of a convex function. Moreover, since $-f(x)$ is also convex then function $f(x)$ is both convex and concave.

This result can also be determined by analyzing the Hessian matrix. Observe that $\nabla^2 f(x) = \nabla^2(-f(x)) = 0$ which shows that $f(x)$ is both convex and concave.

Illustrative Example. Quadratic function

Consider the quadratic function $f(x) = cx^2 + bx + c$ where for simplicity it is assumed that $c, b, a, x \in \mathbb{R}$.

The sign of c determines the convexity/concavity of $f(x)$. Note that:

- If $c \geq 0$ then $f(x)$ is convex since $f(\alpha x^1 + (1-\alpha)x^2) \leq \alpha f(x^1) + (1-\alpha)f(x^2)$.

- If $c \leq 0$ then $f(x)$ is concave since $f(\alpha x^1 + (1-\alpha)x^2) \geq \alpha f(x^1) + (1-\alpha)f(x^2)$.

Again, this result can be directly obtained from the analysis of the Hessian matrix. Observe that $\nabla^2 f(x) = 2c$ so that if $c \geq 0$ ($c \leq 0$) then $\nabla^2 f(x) \geq 0$ ($\nabla^2 f(x) \leq 0$) and $f(x)$ is convex (concave).

Similarly, in the general case for which $f(x) = x^T Cx + b^T x + a$ with $a, b, x \in \mathbb{R}^n$ and $C \in \mathbb{R}^{n\times n}$ is a symmetric matrix, $f(x)$ is convex if and only if the matrix $\nabla^2 f(x) = C$ is positive semidefinite and concave if negative semidefinite.

Exercises

1. Consider the following sets and determine, by applying Definition A.1, whether or not they are convex.

   a. $\Omega_1 = \{(x_1, x_2) \in \mathbb{R}^2 \mid x_1 \geq 0, x_1 - x_2 \geq 0\}$.
   b. $\Omega_2 = \{(x_1, x_2) \in \mathbb{R}^2 \mid x_1^2 + x_2^2 \leq 1\}$.
   c. $\Omega_3 = \{(x_1, x_2) \in \mathbb{R}^2 \mid x_1^2 + x_2^2 = 1\}$.
   d. $\Omega_4 = \{(x_1, x_2) \in \mathbb{R}^2 \mid x_1^2 + x_2^2 \geq 1\}$.
   e. $\Omega_5 = \{(x_1, x_2) \in \mathbb{R}^2 \mid x_1 \geq -1, x_2 \geq -1, x_1 \leq 1, x_2 \leq 1\}$.

2. Determine if the folowing functions are convex, concave or neither convex nor concave.

   a. $f_1(x_1, x_2) = x_1 + 2x_2$.
   b. $f_2(x) = |x|$.
   c. $f_3(x) = x^3 + 3x + 1$.
   d. $f_4(x_1, x_2) = {x_1}^2 + 2x_1x_2{x_2}^2$.
   e. $f_5(x) = x^{1/2}$.
   f. $f_6(x) = \sin(x)$ where $x \in (0, \pi)$.
   g. $f_7(x) = \sin(x)$ where $x \in (\pi, 2\pi)$.

References

1. M. Avriel. *Nonlinear programming: analysis and methods.* Dover Publishing, New York, 2003.
2. M.S. Bazaraa, H.D. Sherali, and C.M. Shetty. *Nonlinear programming: theory and algorithms.* Wiley, New Jersey, 2006.
3. D.P. Bertsekas. *Nonlinear programming.* Athena Scientific, Belmont, Massachusetts, 1999.
4. E. Castillo, A.J. Conejo, P. Pedregal, R. Garcia, and N. Alguacil. *Building and Solving Mathematical Programming Models in Engineering and Science.* Wiley, New York, 2001.
5. I. Ekeland and R. Temam. *Convex analysis and variational problems.* SIAM, New York, 1987.
6. D.G. Luenberger. *Introduction to linear and nonlinear programming.* Addison-Wesley publishing company, New York, 1973.
7. R.T. Rockafellar. *Convex analysis.* Princeton Univ Press, New Jersey, 1997.
8. J. Stoer and C. Witzgall. *Convexity and optimization in finite dimensions.* Springer, New Jersey, 1970.

Appendix B
GAMS codes

This appendix provides GAMS codes for some of the illustrative examples presented throughout this book. These codes can be implemented and modified by the interested readers to further understand the functioning of the proposed models. In particular, the GAMS codes included correspond to the following illustrative examples:

1. Energy production duopoly example (Sections 1.2.5 and 1.4.3).
2. Energy network with multiple players example (Sections 1.2.7 and 1.4.4).
3. Three price-taking firms: social welfare maximization model (Section 3.4.2.1).
4. Three price-taking firms: complementarity model (Section 3.4.2.2).
5. Nash-Cournot model of three firms: complementarity model (Section 3.4.2.6).
6. Generalized Nash model for coal market: limit on coal yard, with auctioning of shares and unequal marginal utilities (Section 3.4.3.3).
7. Stackelberg MPEC with firm 2 as leader problem (Section 3.5.1.1).
8. Energy production bimatrix game, version 1 example (Section 4.2.4.1).
9. Energy production bimatrix game, version 2 example (Section 4.2.4.2).
10. SPE for energy products example (Section 4.4.1.1).
11. Spatial price equilibrium problem, version 2 (Section 4.8.2).
12. Electric capacity market model with green price a multiple of conventional price (Section 5.4.1.6).
13. Strategic offering example (Section 6.5.4).
14. Vulnerability assessment example (Section 6.3.3).
15. Transmission investment example (Section 6.3.4).
16. Dantzig-Wolfe decomposition of the stochastic power model (Section 9.2.4).
17. Small natural gas network equilibrium problem (Section 10.4.8).

Additional information related to optimization environment GAMS can be found in [3]. Tools similar to GAMS are AIMMS [1] and AMPL [2].

GAMS Code for the Energy Production Duopoly (Sections 1.2.5 and 1.4.3)

```
**********************************************************************
*        DATA
**********************************************************************
SET p /p1, p2/;

PARAMETERS gamma(p)
/p1 1
p2  2/,
beta /5/,
alpha /10/
q_TOT /1.06 /;

**********************************************************************
*        DECLARATION OF VARIABLES
**********************************************************************
POSITIVE VARIABLE q1, q2,lambda_TOT;

**********************************************************************
*        EQUATIONS
**********************************************************************
EQUATIONS
equ_1,equ_2,equ_3;

equ_1..      2*beta*q1+beta*q2-alpha+gamma('p1')+lambda_TOT=g=0;
equ_2..      2*beta*q2+beta*q1-alpha+gamma('p2')+lambda_TOT=g=0;
equ_3..      q_TOT-q1-q2=g=0;

**********************************************************************
*        MODEL
**********************************************************************
MODEL problem2 /equ_1.q1,equ_2.q2, equ_3.lambda_TOT/;
solve problem2 using mcp;

PARAMETER price;
price = alpha-beta*(q1.l+q2.l);

option decimals=8;
display q1.l, q2.l, lambda_TOT.l,price;
```

GAMS Code for the Energy Network with Multiple Players Example (Sections 1.2.7 and 1.4.4)

```
**********************************************************************
*       DATA
**********************************************************************
option mcp=path;
PARAMETERS
tau_12reg          /0.5/
gamma_1A           /10/
gamma_1B           /12/
gamma_2C           /15/
gamma_2D           /18/
a_1                /20/
b_1                /1/
```

```
a_2             /40/
b_2             /2/
q_1A_max        /10/
q_1B_max        /10/
q_2C_max        /5/
q_2D_max        /5/
g_12_max        /5/
gamma_TSO       /1/;

****************************************************************************
*       DECLARATION OF VARIABLES
****************************************************************************
POSITIVE VARIABLES
s_1A
s_1B
s_2C
s_2D
q_1A
q_1B
q_2C
q_2D
f_12A
f_12B
lambda_1A
lambda_1B
lambda_2C
lambda_2D
g_12
eps_12;

VARIABLES
delta_1A
delta_1B
delta_2C
delta_2D
pi_1
pi_2
tau_12;


****************************************************************************
*       EQUATIONS
****************************************************************************
EQUATIONS
Eq_s_1A,Eq_q_1A,Eq_f_12A,Eq_lambda_1A,Eq_eq1A,Eq_s_1B,Eq_q_1B,Eq_f_12B,
Eq_lambda_1B,Eq_eq1B,Eq_s_2C,Eq_q_2C,Eq_lambda_2C,Eq_eq2C,Eq_s_2D,Eq_q_2D,
Eq_lambda_2D,Eq_eq2D,Eq_MC1,Eq_MC2,Eq_g_12,Eq_eps_12,Eq_tau_12;

Eq_s_1A..        -pi_1+delta_1A =g= 0;
Eq_q_1A..        gamma_1A+lambda_1A-delta_1A =g= 0;
Eq_f_12A..       -pi_2+(tau_12reg+tau_12)+delta_1A =g= 0;
Eq_lambda_1A..   q_1A_max-q_1A =g= 0;
Eq_eq1A..        s_1A-q_1A+f_12A =e=0;

Eq_s_1B..        -pi_1+delta_1B =g= 0;
Eq_q_1B..        gamma_1B+lambda_1B-delta_1B =g= 0;
Eq_f_12B..       -pi_2+(tau_12reg+tau_12)+delta_1B =g= 0;
Eq_lambda_1B..   q_1B_max-q_1B =g= 0;
Eq_eq1B..        s_1B-q_1B+f_12B =e=0;

Eq_s_2C..        -pi_2+delta_2C =g= 0;
Eq_q_2C..        gamma_2C+lambda_2C-delta_2C =g= 0;
Eq_lambda_2C..   q_2C_max-q_2C =g= 0;
Eq_eq2C..        s_2C-q_2C=e=0;

Eq_s_2D..        -pi_2+delta_2D =g= 0;
Eq_q_2D..        gamma_2D+lambda_2D-delta_2D =g= 0;
```

```
Eq_lambda_2D..   q_2D_max-q_2D =g= 0;
Eq_eq2D..        s_2D-q_2D=e=0;

Eq_MC1..         (s_1A+s_1B)-(a_1-b_1*pi_1) =e= 0;
Eq_MC2..         (s_2C+s_2D+f_12A+f_12B)-(a_2-b_2*pi_2) =e= 0;

Eq_g_12..        -tau_12reg-tau_12+gamma_TSO+eps_12 =g= 0;
Eq_eps_12..      g_12_max-g_12 =g= 0;

Eq_tau_12..      g_12-(f_12A+f_12B) =e= 0;

***************************************************************************
*       MODEL
***************************************************************************
MODEL compl1 /Eq_s_1A.s_1A,Eq_q_1A.q_1A,Eq_f_12A.f_12A,Eq_lambda_1A.lambda_1A,
Eq_eq1A,Eq_s_1B.s_1B,Eq_q_1B.q_1B,Eq_f_12B.f_12B,
Eq_lambda_1B.lambda_1B,Eq_eq1B,Eq_s_2C.s_2C,Eq_q_2C.q_2C,Eq_lambda_2C.lambda_2C,
Eq_eq2C,Eq_s_2D.s_2D,Eq_q_2D.q_2D,Eq_lambda_2D.lambda_2D,Eq_eq2D,Eq_MC1,Eq_MC2,
Eq_g_12.g_12,Eq_eps_12.eps_12,Eq_tau_12/

solve compl1 using mcp
```

GAMS Code for the Three Price-taking Firms: Social Welfare Maximization Model (Section 3.4.2.1)

```
***************************************************************************
*       DATA
***************************************************************************
* Social Welfare Maximization Model of Chapter 3
* Linear inverse demand function with 3 firms each with 2 mines
SETS
i firms / i1, i2, i3 /
b production process / b1, b2 /
TABLE C(i,b) variable costs
   b1   b2
i1 0.55 0.81
i2 0.62 1.25
i3 0.78 1.35
TABLE K(i,b) process capacities
    b1    b2
i1 21000 16000
i2 17000 22000
i3 18000 14000
Parameter alpha inverse demand intercept / 2.5 /
beta inverse demand slope / 0.0000166666667 / ;

***************************************************************************
*       DECLARATION OF VARIABLES
***************************************************************************
POSITIVE VARIABLES
x(i,b) production by i from process b
q demand quantity ;
VARIABLES
NegSocWel negative of social welfare ;

***************************************************************************
*       EQUATIONS
***************************************************************************
EQUATIONS
```

```
SocWelDef social welfare definition
Capacity(i,b) upper limit on generation output
MarkClear market clearing ;
SocWelDef.. sum((i,b), C(i,b)* x(i,b))
- (alpha*q - 0.5*beta*q**2) - NegSocWel =e= 0 ;
Capacity(i,b).. x(i,b) - K(i,b) =l= 0 ;
Markclear.. q - sum((i,b), x(i,b)) =e= 0 ;


****************************************************************************
*       MODEL
****************************************************************************
MODEL SocWelMax /all/ ;
Solve SocWelMax minimizing NegSocWel using nlp ;
Parameters Rev(i), VarCost(i), ProdSurp(i) ;
Parameters Price, ConPmt, ConSurp, TotProdSurp, SocWel ;
Price = alpha - beta*q.l ;
ConPmt = Price*q.l ;
ConSurp = alpha*q.l - 0.5*beta*q.l**2 - Price*q.l ;
Rev(i) = Price*sum(b,x.l(i,b));
VarCost(i) = sum(b, C(i,b)* x.l(i,b)) ;
ProdSurp(i) = Rev(i) - VarCost(i) ;
TotProdSurp = sum(i,ProdSurp(i)) ;
SocWel = ConSurp + TotProdSurp ;
Display x.l, Rev, VarCost, ProdSurp ;
Display q.l, Price, ConPmt, ConSurp, TotProdSurp, SocWel ;
```

GAMS Code for the Three Price-taking Firms: Complementarity Model (Section 3.4.2.2)

```
****************************************************************************
*       DATA
****************************************************************************
* Complementarity Model that is equivalent to
* Social Welfare Maximization Model of Chapter 3
* Linear inverse demand function with 3 firms each with 2 mines
SETS
i firms / i1, i2, i3 /
b production process / b1, b2 /
TABLE C(i,b) variable costs
    b1   b2
i1 0.55 0.81
i2 0.62 1.25
i3 0.78 1.35
TABLE K(i,b) process capacities
    b1      b2
i1 21000 16000
i2 17000 22000
i3 18000 14000
Parameter alpha inverse demand intercept / 2.5 /
beta inverse demand slope / 0.0000166666667 / ;

****************************************************************************
*       DECLARATION OF VARIABLES
****************************************************************************
POSITIVE VARIABLES
x(i,b) production by i from process b
q demand quantity
gamma(i,b) dual of capacity constraint ;
```

```
VARIABLES lamda price ;

***************************************************************************
*       EQUATIONS
***************************************************************************
EQUATIONS
PriceDef price definition as inverse demand
Mx(i,b) KKT condition for derivative w.r.t. x
Capacity(i,b) upper limit on generation output
MarkClear market clearing ;
PriceDef.. lamda - (alpha - beta*q) =e= 0 ;
Mx(i,b).. c(i,b) - lamda + gamma(i,b) =g= 0 ;
Capacity(i,b).. - x(i,b) + K(i,b) =g= 0 ;
Markclear.. q - sum((i,b), x(i,b)) =e= 0 ;

***************************************************************************
*       MODEL
***************************************************************************
MODEL CompetComplem /PriceDef.q, Mx.x, Capacity.gamma, MarkClear.lamda/ ;
Solve CompetComplem using mcp ;
Parameters Rev(i), VarCost(i), ProdSurp(i) ;
Parameters Price, ConPmt, ConSurp, TotProdSurp, SocWel ;
Price = alpha - beta*q.l ;
ConPmt = Price*q.l ;
ConSurp = alpha*q.l - 0.5*beta*q.l**2 - Price*q.l ;
Rev(i) = Price*sum(b,x.l(i,b));
VarCost(i) = sum(b, C(i,b)* x.l(i,b)) ;
ProdSurp(i) = Rev(i) - VarCost(i) ;
TotProdSurp = sum(i,ProdSurp(i)) ;
SocWel = ConSurp + TotProdSurp ;
Display x.l, Rev, VarCost, ProdSurp ;
Display q.l, Price, ConPmt, ConSurp, TotProdSurp, SocWel ;
```

GAMS Code for the Nash-Cournot Model of Three Firms: Complementarity Model (Section 3.4.2.6)

```
***************************************************************************
*       DATA
***************************************************************************
* Complementarity Model version of Nash-Cournot Model of Chapter 3
* with all 3 firms acting in Cournot fashion
* Linear inverse demand function with 3 firms each with 2 mines
SETS
i firms / i1, i2, i3 /
b production process / b1, b2 /
alias(b,bb)
TABLE C(i,b) variable costs
    b1   b2
i1 0.55 0.81
i2 0.62 1.25
i3 0.78 1.35
TABLE K(i,b) process capacities
     b1     b2
i1 21000 16000
i2 17000 22000
i3 18000 14000
Parameter alpha inverse demand intercept / 2.5 /
beta inverse demand slope / 0.0000166666667 / ;
```

```
***************************************************************************
*       DECLARATION OF VARIABLES
***************************************************************************
POSITIVE VARIABLES
x(i,b) production by i from process b
q demand quantity
gamma(i,b) dual of capacity constraint;
VARIABLES
lamda price ;

***************************************************************************
*       EQUATIONS
***************************************************************************
EQUATIONS
PriceDef price definition as inverse demand
Mx(i,b) KKT condition for derivative w.r.t. x
Capacity(i,b) upper limit on generation output
MarkClear market clearing ;
PriceDef.. lamda - (alpha - beta*q) =e= 0 ;
Mx(i,b).. c(i,b)-lamda+beta*sum(bb,x(i,bb))+gamma(i,b) =g= 0;
Capacity(i,b).. - x(i,b) + K(i,b) =g= 0 ;
Markclear.. q - sum((i,b), x(i,b)) =e= 0 ;

***************************************************************************
*       MODEL
***************************************************************************
MODEL CournotComplem
/PriceDef.q, Mx.x, Capacity.gamma, MarkClear.lamda/ ;
Solve CournotComplem using mcp ;
Parameters Rev(i), VarCost(i), ProdSurp(i) ;
Parameters Price, ConPmt, ConSurp, TotProdSurp, SocWel ;
Price = alpha - beta*q.l ;
ConPmt = Price*q.l ;
ConSurp = alpha*q.l - 0.5*beta*q.l**2 - Price*q.l ;
Rev(i) = Price*sum(b,x.l(i,b));
VarCost(i) = sum(b, C(i,b)* x.l(i,b)) ;
ProdSurp(i) = Rev(i) - VarCost(i) ;
TotProdSurp = sum(i,ProdSurp(i)) ;
SocWel = ConSurp + TotProdSurp ;
Display x.l, Rev, VarCost, ProdSurp ;
Display q.l, Price, ConPmt, ConSurp, TotProdSurp, SocWel ;
```

GAMS Code for the Generalized Nash Model for Coal Market: Limit on Coal Yard, with Auctioning of Shares and Unequal Marginal Utilities (Section 3.4.3.3)

```
***************************************************************************
*       DATA
***************************************************************************
* Generalized Nash version of Nash-Cournot Model of Chapter 3
* with all 3 firms acting in Cournot fashion, limit on total
* coal, and dual weights method.
* Linear inverse demand function with 3 firms each with 2 mines
SETS
i firms / i1, i2, i3 /
b production process / b1, b2 /
alias(b,bb)
TABLE C(i,b) variable costs
    b1   b2
```

```
i1 0.55 0.81
i2 0.62 1.25
i3 0.78 1.35
TABLE K(i,b) process capacities
    b1     b2
i1 21000 16000
i2 17000 22000
i3 18000 14000
Parameter w(i) weight on yard rental rate / i1 1, i2 2, i3 2 /
Parameter Lim upper limit on yard capacity / 60000 /
Parameter alpha inverse demand intercept / 2.5 /
beta inverse demand slope / 0.0000166666667 / ;

***************************************************************************
*        DECLARATION OF VARIABLES
***************************************************************************
POSITIVE VARIABLES
x(i,b) production by i from process b
q demand quantity
gamma(i,b) dual of capacity constraint
phi dual of YardLimit ;
VARIABLES
lamda price ;

***************************************************************************
*        EQUATIONS
***************************************************************************
EQUATIONS
PriceDef price definition as inverse demand
Mx(i,b) KKT condition for derivative w.r.t. x
Capacity(i,b) upper limit on generation output
MarkClear market clearing
YardLimit limit on yard capacity ;
PriceDef.. lamda - (alpha - beta*q) =e= 0 ;
Mx(i,b)..
c(i,b)-lamda+beta*sum(bb,x(i,bb))+gamma(i,b)+w(i)*phi =g= 0;
Capacity(i,b).. - x(i,b) + K(i,b) =g= 0 ;
Markclear.. q - sum((i,b), x(i,b)) =e= 0 ;
YardLimit.. - sum((i,b), x(i,b)) =g= -Lim ;

***************************************************************************
*        MODEL
***************************************************************************
MODEL CournotGenNash
/PriceDef.q,Mx.x,Capacity.gamma,MarkClear.lamda,YardLimit.phi/;
Solve CournotGenNash using mcp ;
Parameters Rev(i), VarCost(i), ProdSurp(i) ;
Parameters Price, ConPmt, ConSurp, TotProdSurp, SocWel ;
Price = alpha - beta*q.l ;
ConPmt = Price*q.l ;
ConSurp = alpha*q.l - 0.5*beta*q.l**2 - Price*q.l ;
Rev(i) = Price*sum(b,x.l(i,b));
VarCost(i) = sum(b, C(i,b)* x.l(i,b)) ;
ProdSurp(i) = Rev(i) - VarCost(i) ;
TotProdSurp = sum(i,ProdSurp(i)) ;
SocWel = ConSurp + TotProdSurp ;
Display x.l, Rev, VarCost, ProdSurp ;
Display q.l, Price, ConPmt, ConSurp, TotProdSurp, SocWel ;
Display phi.l, w ;
```

GAMS Code for the Stackelberg MPEC with Firm 2 as Leader (Section 3.5.1.1)

```
***************************************************************************
*        DECLARATION OF VARIABLES
***************************************************************************
POSITIVE VARIABLES
x2(b) production by leader from process b
x(ifol,b) production by i from process b
q demand quantity
gamma(ifol,b) dual of capacity constraints for followers ;
VARIABLES
NegStackObj negative of Stackelberg objective ;

***************************************************************************
*        EQUATIONS
***************************************************************************
EQUATIONS
StackObjDef Stackelberg objective definition
Capacity2(b) upper limit on generation output of Leader
Capacity(ifol,b) upper limit on generation output of Followers
MarkClear market clearing
Mx(ifol,b) KKT condition for derivative w.r.t. x of follower;
StackObjDef.. sum(b,C('i2',b)*x2(b) - (alpha - beta*q)*x2(b))
- NegStackObj =e= 0 ;
Capacity2(b).. x2(b) - K('i2',b) =l= 0 ;
Capacity(ifol,b).. -x(ifol,b) + K(ifol,b) =g= 0 ;
Markclear.. q - sum((ifol,b), x(ifol,b)) - sum(b,x2(b)) =e= 0 ;
Mx(ifol,b).. c(ifol,b)-(alpha-beta*q)+gamma(ifol,b) =g= 0 ;

***************************************************************************
*        MODEL
***************************************************************************
MODEL Stackelberg
/StackObjDef,Capacity2,Markclear,Capacity.gamma,Mx.x / ;
Stackelberg.optfile = 1 ;
Solve Stackelberg minimizing NegStackObj using mpec ;
Parameter Complem ;
Parameters Rev(i), VarCost(i), ProdSurp(i) ;
Parameters Price, ConPmt, ConSurp, TotProdSurp, SocWel ;
Complem = sum((ifol,b),
(c(ifol,b) - (alpha - beta*q.l) + gamma.l(ifol,b))*x.l(ifol,b)
+ (-x.l(ifol,b) + K(ifol,b))*gamma.l(ifol,b)) ;
Price = alpha - beta*q.l ;
ConPmt = Price*q.l ;
ConSurp = alpha*q.l - 0.5*beta*q.l**2 - Price*q.l ;
Rev('i2') = Price*sum(b,x2.l(b));
Rev(ifol) = Price*sum(b,x.l(ifol,b));
VarCost('i2') = sum(b, C('i2',b)* x2.l(b)) ;
VarCost(ifol) = sum(b, C(ifol,b)* x.l(ifol,b)) ;
ProdSurp(i) = Rev(i) - VarCost(i) ;
TotProdSurp = sum(i,ProdSurp(i)) ;
SocWel = ConSurp + TotProdSurp ;
Display Complem ;
Display x2.l, x.l, Rev, VarCost, ProdSurp ;
Display q.l, Price, ConPmt, ConSurp, TotProdSurp, SocWel ;
```

Note: the line "Stackelberg.optfile = 1" directs the solver to a file in the project directory called "nlpec.opt" which contains the following user-chosen settings to control the algorithm (please see the GAMS NLPEC solver manual for details):

```
reftype mult
aggregate none
```

```
constraint inequality
initmu 1.0
numsolves 4
```

GAMS Code for the Energy Production Bimatrix Game, Version 1 (Section 4.2.4.1)

```
***************************************************************************
*       DATA
***************************************************************************
SETS
p /p1, p2/
m /1,2,3/
n /1,2/;

TABLE A(m,n)
    1 2
1  50 2
2  10 3
3  20 4;

TABLE B(m,n)
    1  2
1  10  5
2  7   11
3  130 17;

***************************************************************************
*       DECLARATION OF VARIABLES
***************************************************************************
POSITIVE VARIABLE x(m), y(n);

***************************************************************************
*       EQUATIONS
***************************************************************************
EQUATIONS
player1(m)
player2(n);

player1(m)..       sum(n,A(m,n)*y(n))-1=g=0;
player2(n)..       sum(m,B(m,n)*x(m))-1=g=0;

***************************************************************************
*       MODEL
***************************************************************************
MODEL bimatrix /player1.x, player2.y/;
solve bimatrix using mcp;
option decimals=8;
display x.l,y.l;
```

GAMS Code for the Energy Production Bimatrix Game, Version 2 (Section 4.2.4.2)

```
***************************************************************************
*        DATA
***************************************************************************
SETS
p /p1, p2/
m /1,2,3/
n /1,2,3/;

TABLE A(m,n)
    1     2    3
1  1600  1600 1600
2  800   900  1000
3  200   400  1100 ;


TABLE B(m,n)
   1     2   3
1  1600  750 100
2  1600  850 1100
3  1600  950 600 ;

***************************************************************************
*        DECLARATION OF VARIABLES
***************************************************************************
POSITIVE VARIABLE x(m), y(n);

***************************************************************************
*        EQUATIONS
***************************************************************************
EQUATIONS
player1(m)
player2(n);

player1(m)..        sum(n,A(m,n)*y(n))-1=g=0;
player2(n)..        sum(m,B(m,n)*x(m))-1=g=0;

***************************************************************************
*        MODEL
***************************************************************************
MODEL bimatrix /player1.x, player2.y/;
solve bimatrix using mcp;
option decimals=8;
display x.l,y.l;
```

GAMS Code for the SPE for Energy Products (Section 4.4.1.1)

```
***************************************************************************
*        DATA
***************************************************************************
SETS
i /1,2,3,4/
j /1,2,3,4,5/;

TABLE costs(i,j)
    1         2         3         4         5
1   10        20        30        5         6
2   3         15        12        5         6
3   17        40        5         3         7
4   50        23        28        12        4;
```

```
PARAMETER Supply(i)
 /1 20
  2 30
  3 35
  4 15/;

PARAMETER Demand(j)
 /1 20
  2 25
  3 10
  4 10
  5 35/;

****************************************************************************
*        DECLARATION OF VARIABLES
****************************************************************************
POSITIVE VARIABLES F(i,j);
VARIABLES z;

****************************************************************************
*        EQUATIONS
****************************************************************************
EQUATIONS
obj
supplies(i)
demands(j);

obj.. z=e=sum((i,j), costs(i,j)*F(i,j));
supplies(i).. supply(i)-sum(j,F(i,j))=g=0;
demands(j)..  sum(i,F(i,j))-demand(j)=g=0

****************************************************************************
*        MODEL
****************************************************************************
MODEL transportation /all/;
solve transportation minimizing z using lp;

option decimals=8;
```

GAMS Code for the Price Equilibrium Problem, Version 2 (Section 4.8.2)

```
****************************************************************************
*        DATA
****************************************************************************
File spe_out;
Put spe_out;

SETS
i /1,2,3,4/
j /1,2,3,4,5/
;
alias (i,ii);
alias (j,jj);

TABLE c(i,j)
    1        2         3         4        5
1   10       20        30        5        6
```

```
2   3       15        12        5       6
3   17      40        5         3       7
4   50      23        28        12      4;

PARAMETER rho_supply(i)
 /1 0.20
  2 0.20
  3 0.20
  4 0.20/;
PARAMETER kappa_supply(i)
 /1 0.00
  2 0.00
  3 -7.0
  4 0.0/;

PARAMETER rho_demand(j)
 /1 1.00
  2 1.00
  3 1.00
  4 1.00
  5 1.00/;
PARAMETER kappa_demand(j)
 /1 29.00
  2 46.00
  3 15.00
  4 13.00
  5 42.00/;

*************************************************************************
*        DECLARATION OF VARIABLES
*************************************************************************
POSITIVE VARIABLES x(i,j);

x.l('1','1')=100.0;
x.l('1','2')=100.0;
x.l('1','3')=100.0;
x.l('1','4')=100.0;
x.l('1','5')=100.0;
x.l('2','1')=100.0;
x.l('2','2')=100.0;
x.l('2','3')=100.0;
x.l('2','4')=100.0;
x.l('2','5')=100.0;
x.l('3','1')=100.0;
x.l('3','2')=100.0;
x.l('3','3')=100.0;
x.l('3','4')=100.0;
x.l('3','5')=100.0;
x.l('4','1')=100.0;
x.l('4','2')=100.0;
x.l('4','3')=100.0;
x.l('4','4')=100.0;
x.l('4','5')=100.0;


*************************************************************************
*        EQUATIONS
*************************************************************************
EQUATIONS
pairs(i,j);
pairs(i,j).. rho_supply(i)*sum(jj,x(i,jj))+ kappa_supply(i)+ c(i,j)
+rho_demand(j)*sum(ii,x(ii,j))-kappa_demand(j)=g=0;


*************************************************************************
*        MODEL
*************************************************************************
```

```
MODEL spe /pairs.x/;
spe.optfile=1;
solve spe using mcp;
loop((i,j),
put x.l(i,j):10:10 /;
);
putclose;
option decimals=8;
display x.l;
```

GAMS Code for the Electric Capacity Market Model with Green Price a Multiple of Conventional Price (Section 5.4.1.6)

```
*************************************************************************
*       DATA
*************************************************************************
**VI model for Illustrative Example 5.4.1.6 in Chapter 5
** Green price is a multiple of conventional price
SETS
f firms /f1,f2/
h generation types /exist, convent, green/;
Parameter Mult(h) /exist 1, convent 1, green 1.25 /;
Parameter CapReq capacity requirement /400/;
TABLE MaxCap(f,h) maximum capacities
   exist convent green
f1 140   90      50
f2 100   75      80 ;
TABLE  MC_Const(f,h) marginal cost constant term
    exist convent green
f1  180   1000    1800
f2  110   900     2000 ;
TABLE  MCostSlope(f,h) marginal cost slope
    exist convent green
f1  0       15    90
f2  0       18    70;

*************************************************************************
*       DECLARATION OF VARIABLES
*************************************************************************
POSITIVE VARIABLES
x(f,h) capacity of firm i type h
mu dual for capacity requirement in multiplier VI model ;

*************************************************************************
*       EQUATIONS
*************************************************************************
EQUATIONS
MCostMult(f,h) marginal cost expression for multiplier VI model
Capacity total capacity requirement
CapLim(f,h) limit on capacity of firm i type h ;
MCostMult(f,h)..MC_Const(f,h)+MCostSlope(f,h)*x(f,h)
-Mult(h)*mu =N= 0 ;
Capacity.. -CapReq+sum((f,h),x(f,h)) =N= 0 ;
CapLim(f,h).. MaxCap(f,h)-x(f,h) =g= 0;

*************************************************************************
*       MODEL
*************************************************************************
```

```
**VI model for multiplier case
MODEL MultMod / MCostMult,Capacity,CapLim/ ;
* define the pairing of mapping elements and variables for VI
file fx /"%emp.info%"/;
put fx 'vifunc MCostMult x'/;
putclose fx 'vifunc Capacity mu';
solve MultMod using emp ;
Parameter TotPmt;
Parameter TotCostOut;
Parameter CapByType(h);
Parameter CostByType(h);
Parameter ProfitByType(h);
TotPmt=sum((f,h),Mult(h)*mu.l*x.l(f,h));
TotCostOut=sum((f,h),MC_Const(f,h)*x.l(f,h)
+0.5*MCostSlope(f,h)*x.l(f,h)**2);
CapByType(h)=sum(f,x.l(f,h));
CostByType(h)= sum(f,MC_Const(f,h)*x.l(f,h)
+0.5*MCostSlope(f,h)*x.l(f,h)**2);
ProfitByType(h)= sum(f,Mult(h)*mu.l*x.l(f,h))-CostByType(h);
display TotPmt, CapByType ;
```

GAMS Code for the Strategic Offering Example (Section 6.5.4)

```
***************************************************************************
*        DATA
***************************************************************************
PARAMETERS
fmax_1   /5/
fmax_2   /10/
fmax_3   /10/
c_g_1    /16/
c_g_2    /19/
c_g_3    /15/
c_d_1    /18/
c_d_2    /20/
c_d_3    /21/
gmax_1   /20/
gmax_2   /10/
gmax_3   /15/
dmax_1   /5/
dmax_2   /20/
dmax_3   /15/
B_1      /100/
B_2      /125/
B_3      /150/
M        /10000/;

***************************************************************************
*        DECLARATION OF VARIABLES
***************************************************************************
POSITIVE VARIABLES
g_1                Power produced by unit 1
g_2                Power produced by unit 2
g_3                Power produced by unit 3
d_1                Power consumed by demand 1
d_2                Power consumed by demand 2
d_3                Power consumed by demand 3
phi_gmin_1     Dual variable for the minimum generation capacity unit 1
```

```
phi_gmin_2      Dual variable for the minimum generation capacity unit 2
phi_gmin_3      Dual variable for the minimum generation capacity unit 3
phi_dmin_1      Dual variable for the minimum level of demand 1
phi_dmin_2      Dual variable for the minimum level of demand 2
phi_dmin_3      Dual variable for the minimum level of demand 3
phi_gmax_1     Dual variable for the maximum generation capacity of unit 1
phi_gmax_2     Dual variable for the maximum generation capacity of unit 2
phi_gmax_3     Dual variable for the maximum generation capacity of unit 3
phi_dmax_1     Dual variable for the maximum level of demand 1
phi_dmax_2     Dual variable for the maximum level of demand 2
phi_dmax_3     Dual variable for the maximum level of demand 3
rho_min_1       Dual variable for the minimum lower bound capacity of line 1
rho_min_2       Dual variable for the minimum lower bound capacity of line 2
rho_min_3       Dual variable for the minimum lower bound capacity of line 3
rho_max_1      Dual variable for the maximum upper bound capacity of line 1
rho_max_2      Dual variable for the maximum upper bound capacity of line 2
rho_max_3      Dual variable for the maximum upper bound capacity of line 3;

VARIABLES
alpha_1            Dual variable for power balance at node 1
alpha_2            Dual variable for power balance at node 2
alpha_3            Dual variable for power balance at node 3
cc_g_1             Strategic price offer for unit 1
theta_1            Angle bus for node 1
theta_2            Angle bus for node 2
theta_3            Angle bus for node 3
gamma              Dual variable for the slack bus equality
L_1                Power flow through line 1
L_2                Power flow through line 2
L_3                Power flow through line 3;

BINARY VARIABLES
*Used for the linearization of the complementarity constraints
v_gmin_1,v_gmin_2,v_gmin_3,v_gmax_1,v_gmax_2,v_gmax_3,v_dmin_1,v_dmin_2,v_dmin_3
v_dmax_1,v_dmax_2,v_dmax_3,v_min_1,v_min_2,v_min_3,v_max_1,v_max_2,v_max_3;

FREE VARIABLES
profit_1             Profit for the strategic producer 1
X                  Auxiliary variable
SW                 Social welfare
SW_real            True social welfare;

***************************************************************************
*       EQUATIONS
***************************************************************************
EQUATIONS
Eq_profit,Eq_X,Eq_bal1,Eq_bal2,Eq_bal3,Eq_slack,Eq_dual1,Eq_dual2,Eq_dual3,
Eq_dual4,Eq_dual5,Eq_dual6,Eq_dual7,Eq_dual8,Eq_dual9,Eq_compl1a,Eq_compl1b,
Eq_compl2a,Eq_compl2b,Eq_compl3a,Eq_compl3b,Eq_compl4a,Eq_compl4b,Eq_compl4c,
Eq_compl5a,Eq_compl5b,Eq_compl5c,Eq_compl6a,Eq_compl6b,Eq_compl6c,Eq_compl7a,
Eq_compl7b,Eq_compl8a,Eq_compl8b,Eq_compl9a,Eq_compl9b,Eq_compl10a,Eq_compl10b,
Eq_compl10c,Eq_compl11a,Eq_compl11b,Eq_compl11c,Eq_compl12a,Eq_compl12b,
Eq_compl12c,Eq_compl13a,Eq_compl13b,Eq_compl13c,Eq_compl14a,Eq_compl14b,
Eq_compl14c,Eq_compl15a,Eq_compl15b,Eq_compl15c,Eq_compl16a,Eq_compl16b,
Eq_compl16c,Eq_compl17a,Eq_compl17b,Eq_compl17c,Eq_compl18a,Eq_compl18b,
Eq_compl18c,Eq_L1,Eq_L2,Eq_L3,Eq_SW_real;

** PROFIT FOR THE STRATEGIC PRODUCER 1
Eq_profit..      profit_1 =e= X-c_g_1*g_1;

** LINEAR FORM OF alpha_1*g_1
Eq_X..           X =e= -c_g_2*g_2-c_g_3*g_3+c_d_1*d_1+c_d_2*d_2+c_d_3*d_3-
                 gmax_2*phi_gmax_2-gmax_3*phi_gmax_3-
                 dmax_1*phi_dmax_1-dmax_2*phi_dmax_2-dmax_3*phi_dmax_3-
                 fmax_1*rho_max_1-fmax_2*rho_max_2-fmax_3*rho_max_3-
                 fmax_1*rho_min_1-fmax_2*rho_min_2-fmax_3*rho_min_3;
```

```
** BALANCE EQUATION FOR EACH NODE
Eq_bal1..        g_1-d_1 =e= B_1*(theta_1-theta_2)+B_2*(theta_1-theta_3);
Eq_bal2..        g_2-d_2 =e= B_1*(theta_2-theta_1)+B_3*(theta_2-theta_3);
Eq_bal3..        g_3-d_3 =e= B_2*(theta_3-theta_1)+B_3*(theta_3-theta_2);

** NODE 3 SLACK BUS DEFINITION
Eq_slack..       theta_3 =e= 0;

** DUAL EQUATIONS OF THE LOWER-LEVEL PROBLEM
Eq_dual1..       cc_g_1-alpha_1-phi_gmin_1+phi_gmax_1 =e= 0;
Eq_dual2..       c_g_2-alpha_2-phi_gmin_2+phi_gmax_2 =e= 0;
Eq_dual3..       c_g_3-alpha_3-phi_gmin_3+phi_gmax_3 =e= 0;
Eq_dual4..       -c_d_1+alpha_1-phi_dmin_1+phi_dmax_1 =e= 0;
Eq_dual5..       -c_d_2+alpha_2-phi_dmin_2+phi_dmax_2 =e= 0;
Eq_dual6..       -c_d_3+alpha_3-phi_dmin_3+phi_dmax_3 =e= 0;
Eq_dual7..       B_1*(alpha_1-alpha_2-rho_min_1+rho_max_1)+
                 B_2*(alpha_1-alpha_3-rho_min_2+rho_max_2) =e= 0;
Eq_dual8..       B_1*(-alpha_1+alpha_2+rho_min_1-rho_max_1)+
                 B_3*(alpha_2-alpha_3-rho_min_3+rho_max_3) =e= 0;
Eq_dual9..       B_2*(-alpha_1+alpha_3+rho_min_2-rho_max_2)+
                 B_3*(-alpha_2+alpha_3+rho_min_3-rho_max_3) + gamma =e= 0;

** LINEARIZATION OF THE COMPLEMENTARITY CONSTRAINTS
Eq_compl1a..     g_1 =l= v_gmin_1*M;
Eq_compl1b..     phi_gmin_1 =l= (1-v_gmin_1)*M;

Eq_compl2a..     g_2 =l= v_gmin_2*M;
Eq_compl2b..     phi_gmin_2 =l= (1-v_gmin_2)*M;

Eq_compl3a..     g_3 =l= v_gmin_3*M;
Eq_compl3b..     phi_gmin_3 =l= (1-v_gmin_3)*M;

Eq_compl4a..     gmax_1-g_1 =l= v_gmax_1*M;
Eq_compl4b..     phi_gmax_1 =l= (1-v_gmax_1)*M;
Eq_compl4c..     gmax_1-g_1 =g= 0;

Eq_compl5a..     gmax_2-g_2 =l= v_gmax_2*M;
Eq_compl5b..     phi_gmax_2 =l= (1-v_gmax_2)*M;
Eq_compl5c..     gmax_2-g_2 =g= 0;

Eq_compl6a..     gmax_3-g_3 =l= v_gmax_3*M;
Eq_compl6b..     phi_gmax_3 =l= (1-v_gmax_3)*M;
Eq_compl6c..     gmax_3-g_3 =g= 0;

Eq_compl7a..     d_1 =l= v_dmin_1*M;
Eq_compl7b..     phi_dmin_1 =l= (1-v_dmin_1)*M;

Eq_compl8a..     d_2 =l= v_dmin_2*M;
Eq_compl8b..     phi_dmin_2 =l= (1-v_dmin_2)*M;

Eq_compl9a..     d_3 =l= v_dmin_3*M;
Eq_compl9b..     phi_dmin_3 =l= (1-v_dmin_3)*M;

Eq_compl10a..     dmax_1-d_1 =l= v_dmax_1*M;
Eq_compl10b..     phi_dmax_1 =l= (1-v_dmax_1)*M;
Eq_compl10c..     dmax_1-d_1 =g= 0;

Eq_compl11a..     gmax_2-d_2 =l= v_dmax_2*M;
Eq_compl11b..     phi_dmax_2 =l= (1-v_dmax_2)*M;
Eq_compl11c..     dmax_2-d_2 =g= 0;

Eq_compl12a..     dmax_3-d_3 =l= v_dmax_3*M;
Eq_compl12b..     phi_dmax_3 =l= (1-v_dmax_3)*M;
Eq_compl12c..     dmax_3-d_3 =g= 0;

Eq_compl13a..    B_1*(theta_1-theta_2)+fmax_1 =l= v_min_1*M;
Eq_compl13b..    rho_min_1 =l= (1-v_min_1)*M;
```

```
Eq_compl13c..     B_1*(theta_1-theta_2)+fmax_1 =g= 0;

Eq_compl14a..     B_2*(theta_1-theta_3)+fmax_2 =l= v_min_2*M;
Eq_compl14b..     rho_min_2 =l= (1-v_min_2)*M;
Eq_compl14c..     B_2*(theta_1-theta_3)+fmax_2 =g= 0;

Eq_compl15a..     B_3*(theta_2-theta_3)+fmax_3 =l= v_min_3*M;
Eq_compl15b..     rho_min_3 =l= (1-v_min_3)*M;
Eq_compl15c..     B_3*(theta_2-theta_3)+fmax_3 =g= 0;

Eq_compl16a..     fmax_1-B_1*(theta_1-theta_2) =l= v_max_1*M;
Eq_compl16b..     rho_max_1 =l= (1-v_max_1)*M;
Eq_compl16c..     fmax_1-B_1*(theta_1-theta_2) =g= 0;

Eq_compl17a..     fmax_2-B_2*(theta_1-theta_3) =l= v_max_2*M;
Eq_compl17b..     rho_max_2 =l= (1-v_max_2)*M;
Eq_compl17c..     fmax_2-B_2*(theta_1-theta_3) =g= 0;

Eq_compl18a..     fmax_3-B_3*(theta_2-theta_3) =l= v_max_3*M;
Eq_compl18b..     rho_max_3 =l= (1-v_max_3)*M;
Eq_compl18c..     fmax_3-B_3*(theta_2-theta_3) =g= 0;

** POWER FLOW THROUGH EACH LINE
Eq_L1..           L_1 =e= B_1*(theta_1-theta_2);
Eq_L2..           L_2 =e= B_2*(theta_1-theta_3);
Eq_L3..           L_3 =e= B_3*(theta_2-theta_3);

** TRUE SOCIAL WELFARE
Eq_SW_real..      SW_real =e= -c_g_1*g_1-c_g_2*g_2-c_g_3*g_3+
                              c_d_1*d_1+c_d_2*d_2+c_d_3*d_3;

****************************************************************************
*       MODEL
****************************************************************************
MODEL strategic /all/;
solve strategic maximizing profit_1 using mip
```

GAMS Code for the Vulnerability Assessment Example (Section 6.3.3)

```
****************************************************************************
*       DATA
****************************************************************************
PARAMETERS
x_1       /0/
x_2       /0/
x_3       /1/
y_1       /0/
y_2       /0/
y_3       /1/
d_1       /5/
d_2       /20/
d_3       /15/
B_1       /100/
B_2       /125/
B_3       /150/
g_min_1  /10/
g_min_2  /5/
g_min_3  /12.5/
g_max_1  /20/
```

```
g_max_2  /10/
g_max_3  /25/
f_max_1  /5/
f_max_2  /10/
f_max_3  /10/;

**************************************************************************
*       DECLARATION OF VARIABLES
**************************************************************************
POSITIVE VARIABLES
r_1                     Unserved energy at bus 1
r_2                     Unserved energy at bus 2
r_3                     Unserved energy at bus 3
phi_min_1               Dual variable for the minimum power output of unit 1
phi_min_2               Dual variable for the minimum power output of unit 2
phi_min_3               Dual variable for the minimum power output of unit 3
phi_max_1               Dual variable for the maximum power output of unit 1
phi_max_2               Dual variable for the maximum power output of unit 2
phi_max_3               Dual variable for the maximum power output of unit 3
nu_min_1                Dual variable for the minimum unserved energy at bus 1
nu_min_2                Dual variable for the minimum unserved energy at bus 2
nu_min_3                Dual variable for the minimum unserved energy at bus 3
nu_max_1                Dual variable for the maximum unserved energy at bus 1
nu_max_2                Dual variable for the maximum unserved energy at bus 2
nu_max_3                Dual variable for the maximum unserved energy at bus 3
rho_min_1               Dual variable lower bound capacity of line 1
rho_min_2               Dual variable lower bound capacity of line 2
rho_min_3               Dual variable lower bound capacity of line 3
rho_max_1               Dual variable upper bound capacity of line 1
rho_max_2               Dual variable upper bound capacity of line 2
rho_max_3               Dual variable upper bound capacity of line 3;

VARIABLES
alpha_1                 Dual variable for the power balance at node 1
alpha_2                 Dual variable for the power balance at node 2
alpha_3                 Dual variable for the power balance at node 3
theta_1                 Angle bus at node 1
theta_2                 Angle bus at node 2
theta_3                 Angle bus at node 3
gamma                   Dual variable for the slack bus equation
g_1                     Power produced by unit 1
g_2                     Power produced by unit 2
g_3                     Power produced by unit 3
f_1                     Power flow through line 1
f_2                     Power flow through line 2
f_3                     Power flow through line 3
beta_1                  Dual variable for the power flow through line 1
beta_2                  Dual variable for the power flow through line 2
beta_3                  Dual variable for the power flow through line 3;

FREE VARIABLES
tot_r                   Total unserved energy;

**************************************************************************
*       EQUATIONS
**************************************************************************
EQUATIONS
Eq_unserved,Eq_bal_1,Eq_bal_2,Eq_bal_3,Eq_f_1,Eq_f_2,Eq_f_3,Eq_theta_3,
Eq_g_min_1,Eq_g_min_2,Eq_g_min_3,Eq_g_max_1,Eq_g_max_2,Eq_g_max_3,Eq_r_max_1,
Eq_r_max_2,Eq_r_max_3,Eq_f_max_1,Eq_f_max_2,Eq_f_max_3,Eq_f_min_1,Eq_f_min_2,
Eq_f_min_3,Eq_dual_1,Eq_dual_2,Eq_dual_3,Eq_dual_f_1,Eq_dual_f_2,Eq_dual_f_3,
Eq_dual_r_1,Eq_dual_r_2,Eq_dual_r_3,Eq_dual_theta_1,Eq_dual_theta_2,
Eq_dual_theta_3,Eq_SDT;

** TOTAL UNSERVED ENERGY
Eq_unserved..     tot_r =e= r_1+r_2+r_3;
```

```
** BALANCE EQUATION FOR EACH NODE
Eq_bal_1..        g_1+r_1-d_1-f_1-f_2 =e= 0;
Eq_bal_2..        g_2+r_2-d_2+f_1-f_3 =e= 0;
Eq_bal_3..        g_3+r_3-d_3+f_2+f_3 =e= 0;

** POWER FLOW FOR EACH LINE
Eq_f_1..          f_1-(1-y_1)*B_1*(theta_1-theta_2) =e= 0;
Eq_f_2..          f_2-(1-y_2)*B_2*(theta_1-theta_3) =e= 0;
Eq_f_3..          f_3-(1-y_3)*B_3*(theta_2-theta_3) =e= 0;

** NODE 3 IS THE REFERENCE BUS
Eq_theta_3..      theta_3 =e= 0;

** MINIMUM POWER OUTPUT FOR GENERATING UNITS
Eq_g_min_1..      (1-x_1)*g_min_1 =l= g_1;
Eq_g_min_2..      (1-x_2)*g_min_2 =l= g_2;
Eq_g_min_3..      (1-x_3)*g_min_3 =l= g_3;

** MAXIMUM POWER OUTPUT FOR GENERATING UNITS
Eq_g_max_1..      (1-x_1)*g_max_1 =g= g_1;
Eq_g_max_2..      (1-x_2)*g_max_2 =g= g_2;
Eq_g_max_3..      (1-x_3)*g_max_3 =g= g_3;

** UNSERVED ENERGY LIMITS
Eq_r_max_1..      r_1 =l= d_1;
Eq_r_max_2..      r_2 =l= d_2;
Eq_r_max_3..      r_3 =l= d_3;

** POWER FLOW BOUNDS FOR EACH LINE
Eq_f_max_1..      f_1 =l= f_max_1;
Eq_f_max_2..      f_2 =l= f_max_2;
Eq_f_max_3..      f_3 =l= f_max_3;
Eq_f_min_1..      f_1 =g= -f_max_1;
Eq_f_min_2..      f_2 =g= -f_max_2;
Eq_f_min_3..      f_3 =g= -f_max_3;

** DUAL EQUATIONS OF THE LOWER-LEVEL PROBLEM
Eq_dual_1..       alpha_1-phi_min_1+phi_max_1 =e= 0;
Eq_dual_2..       alpha_2-phi_min_2+phi_max_2 =e= 0;
Eq_dual_3..       alpha_3-phi_min_3+phi_max_3 =e= 0;
Eq_dual_f_1..     -alpha_1+alpha_2+beta_1-rho_min_1+rho_max_1 =e= 0;
Eq_dual_f_2..     -alpha_1+alpha_3+beta_2-rho_min_2+rho_max_2 =e= 0;
Eq_dual_f_3..     -alpha_2+alpha_3+beta_3-rho_min_3+rho_max_3 =e= 0;
Eq_dual_r_1..     1+alpha_1-nu_min_1+nu_max_1 =e= 0;
Eq_dual_r_2..     1+alpha_2-nu_min_2+nu_max_2 =e= 0;
Eq_dual_r_3..     1+alpha_3-nu_min_3+nu_max_3 =e= 0;
Eq_dual_theta_1..         -beta_1*(1-y_1)*B_1-beta_2*(1-y_2)*B_2 =e= 0;
Eq_dual_theta_2..         beta_1*(1-y_1)*B_1-beta_3*(1-y_3)*B_3 =e= 0;
Eq_dual_theta_3..         beta_2*(1-y_2)*B_2-beta_3*(1-y_3)*B_3 + gamma =e= 0;

** STRONG DUALITY EQUALITY
Eq_SDT..          tot_r =e= -alpha_1*d_1-alpha_2*d_2-alpha_3*d_3+
                  phi_min_1*(1-x_1)*g_min_1+phi_min_2*(1-x_2)*g_min_2
                  +phi_min_3*(1-x_3)*g_min_3-
                  phi_max_1*(1-x_1)*g_max_1+phi_max_2*(1-x_2)*g_max_2
                  +phi_max_3*(1-x_3)*g_max_3-
                  nu_max_1*d_1-nu_max_2*d_2-nu_max_3*d_3-
                  f_max_1*rho_min_1-f_max_2*rho_min_2-f_max_3*rho_min_3-
                  f_max_1*rho_max_1-f_max_2*rho_max_2-f_max_3*rho_max_3;

***************************************************************************
*       MODEL
***************************************************************************
model Vulnerability /All/;
solve Vulnerability maximizing tot_r using mip;
```

GAMS Code for the Transmission Investment Example (Section 6.3.4)

```
***************************************************************************
*        DATA
***************************************************************************
PARAMETERS
e        /10/
c_1      /500/
c_2      /1000/
c_3      /1000/
c_U_1    /20/
c_U_2    /18/
c_U_3    /25/
c_g_1    /16/
c_g_2    /19/
c_g_3    /15/
d_1      /8/
d_2      /10/
d_3      /45/
B_N_1    /100/
B_N_2    /125/
B_N_3    /150/
B_E_1    /100/
B_E_2    /125/
B_E_3    /150/
g_min_1  /10/
g_min_2  /15/
g_min_3  /5/
g_max_1  /25/
g_max_2  /30/
g_max_3  /10/
t_max_1  /5/
t_max_2  /8/
t_max_3  /5/
f_max_1  /10/
f_max_2  /15/
f_max_3  /15/
x_1      /0/
x_2      /1/
x_3      /1/;

***************************************************************************
*        DECLARATION OF VARIABLES
***************************************************************************
POSITIVE VARIABLES
r_1                     Unserved energy at bus 1
r_2                     Unserved energy at bus 2
r_3                     Unserved energy at bus 3
phi_min_1               Dual variable for the minimum power output of unit 1
phi_min_2               Dual variable for the minimum power output of unit 2
phi_min_3               Dual variable for the minimum power output of unit 3
phi_max_1               Dual variable for the maximum power output of unit 1
phi_max_2               Dual variable for the maximum power output of unit 2
phi_max_3               Dual variable for the maximum power output of unit 3
nu_min_1                Dual variable for the minimum unserved energy at bus 1
nu_min_2                Dual variable for the minimum unserved energy at bus 2
nu_min_3                Dual variable for the minimum unserved energy at bus 3
nu_max_1                Dual variable for the maximum unserved energy at bus 1
nu_max_2                Dual variable for the maximum unserved energy at bus 2
nu_max_3                Dual variable for the maximum unserved energy at bus 3
psi_min_1             Dual variable for the lower bound capacity of existing line 1
psi_min_2             Dual variable for the lower bound capacity of existing line 2
psi_min_3             Dual variable for the lower bound capacity of existing line 3
```

```
psi_max_1           Dual variable for the upper bound capacity of existing line 1
psi_max_2           Dual variable for the upper bound capacity of existing line 2
psi_max_3           Dual variable for the upper bound capacity of existing line 3
rho_min_1              Dual variable for the lower bound capacity of new line 1
rho_min_2              Dual variable for the lower bound capacity of new line 2
rho_min_3              Dual variable for the lower bound capacity of new line 3
rho_max_1              Dual variable for the upper bound capacity of new line 1
rho_max_2              Dual variable for the upper bound capacity of new line 2
rho_max_3              Dual variable for the upper bound capacity of new line 3;

VARIABLES
g_1                    Power produced by unit 1
g_2                    Power produced by unit 2
g_3                    Power produced by unit 3
f_1                    Power flow through new line 1
f_2                    Power flow through new line 2
f_3                    Power flow through new line 3
t_1                    Power flow through existing line 1
t_2                    Power flow through existing line 2
t_3                    Power flow through existing line 3
theta_1                Angle bus at node 1
theta_2                Angle bus at node 2
theta_3                Angle bus at node 3
alpha_1                Dual variable for the power balance at node 1
alpha_2                Dual variable for the power balance at node 2
alpha_3                Dual variable for the power balance at node 3
beta_1                 Dual variable for the power flow through new line 1
beta_2                 Dual variable for the power flow through new line 2
beta_3                 Dual variable for the power flow through new line 3
delta_1                Dual variable for the power flow through existing line 1
delta_2                Dual variable for the power flow through existing line 2
delta_3                Dual variable for the power flow through existing line 3
gamma                  Dual variable for the reference bus equation;

FREE VARIABLES
tot_cost               investment cost plus unserved energy cost
invest_cost            investment cost
unserv_ener            unserved energy cost
gen_cost               generation cost;

***************************************************************************
*       EQUATIONS
***************************************************************************
EQUATIONS
Eq_cost,Eq_bal_1,Eq_bal_2,Eq_bal_3,Eq_f_1,Eq_f_2,Eq_f_3,Eq_t_1,Eq_t_2,Eq_t_3,
Eq_theta_slack_3,Eq_gmin_1,Eq_gmin_2,Eq_gmin_3,Eq_gmax_1,Eq_gmax_2,Eq_gmax_3,
Eq_max_r_1,Eq_max_r_2,Eq_max_r_3,Eq_tmin_1,Eq_tmin_2,Eq_tmin_3,Eq_tmax_1,
Eq_tmax_2,Eq_tmax_3,Eq_fmin_1,Eq_fmin_2,Eq_fmin_3,Eq_fmax_1,Eq_fmax_2,Eq_fmax_3,
Eq_dual_g_1,Eq_dual_g_2,Eq_dual_g_3,Eq_dual_r_1,Eq_dual_r_2,Eq_dual_r_3,
Eq_dual_f_1,Eq_dual_f_2,Eq_dual_f_3,Eq_dual_t_1,Eq_dual_t_2,Eq_dual_t_3,
Eq_theta_1,Eq_theta_2,Eq_theta_3,Eq_SDT;

** INVESTMENT COST + UNSERVED ENERGY COST
Eq_cost..        tot_cost =e= c_1*x_1+c_2*x_2+c_3*x_3
                              +e*(c_U_1*r_1+c_U_2*r_2+c_U_3*r_3);

** BALANCE EQUATION FOR EACH NODE
Eq_bal_1..       g_1+r_1-d_1-f_1-f_2-t_1-t_2 =e= 0;
Eq_bal_2..       g_2+r_2-d_2+f_1-f_3+t_1-t_3 =e= 0;
Eq_bal_3..       g_3+r_3-d_3+f_2+f_3+t_2+t_3 =e= 0;

** POWER FLOW THROUGH EACH NEW LINE
Eq_f_1..         f_1-x_1*B_N_1*(theta_1-theta_2) =e= 0;
Eq_f_2..         f_2-x_2*B_N_2*(theta_1-theta_3) =e= 0;
Eq_f_3..         f_3-x_3*B_N_3*(theta_2-theta_3) =e= 0;
```

```
** POWER FLOW THROUGH EACH EXISTING LINE
Eq_t_1..        t_1-B_E_1*(theta_1-theta_2) =e= 0;
Eq_t_2..        t_2-B_E_2*(theta_1-theta_3) =e= 0;
Eq_t_3..        t_3-B_E_3*(theta_2-theta_3) =e= 0;

** NODE 3 IS THE REFERENCE BUS
Eq_theta_slack_3..     theta_3 =e= 0;

** MINIMUM POWER OUTPUT FOR GENERATING UNITS
Eq_gmin_1..     g_min_1 =l= g_1;
Eq_gmin_2..     g_min_2 =l= g_2;
Eq_gmin_3..     g_min_3 =l= g_3;

** MAXIMUM POWER OUTPUT FOR GENERATING UNITS
Eq_gmax_1..     g_max_1 =g= g_1;
Eq_gmax_2..     g_max_2 =g= g_2;
Eq_gmax_3..     g_max_3 =g= g_3;

** UNSERVED ENERGY LIMITS
Eq_max_r_1..    r_1 =l= d_1;
Eq_max_r_2..    r_2 =l= d_2;
Eq_max_r_3..    r_3 =l= d_3;

** POWER FLOW BOUNDS FOR EXISTING LINES
Eq_tmin_1..     -t_max_1 =l= t_1;
Eq_tmin_2..     -t_max_2 =l= t_2;
Eq_tmin_3..     -t_max_3 =l= t_3;
Eq_tmax_1..     t_max_1 =g= t_1;
Eq_tmax_2..     t_max_2 =g= t_2;
Eq_tmax_3..     t_max_3 =g= t_3;

** POWER FLOW BOUNDS FOR NEW LINES
Eq_fmin_1..     -f_max_1 =l= f_1;
Eq_fmin_2..     -f_max_2 =l= f_2;
Eq_fmin_3..     -f_max_3 =l= f_3;
Eq_fmax_1..     f_max_1 =g= f_1;
Eq_fmax_2..     f_max_2 =g= f_2;
Eq_fmax_3..     f_max_3 =g= f_3;

** DUAL EQUATIONS OF THE LOWER-LEVEL PROBLEM
Eq_dual_g_1..   c_g_1+alpha_1-phi_min_1+phi_max_1 =e= 0;
Eq_dual_g_2..   c_g_2+alpha_2-phi_min_2+phi_max_2 =e= 0;
Eq_dual_g_3..   c_g_3+alpha_3-phi_min_3+phi_max_3 =e= 0;
Eq_dual_r_1..   c_U_1+alpha_1-nu_min_1+nu_max_1 =e= 0;
Eq_dual_r_2..   c_U_2+alpha_2-nu_min_2+nu_max_2 =e= 0;
Eq_dual_r_3..   c_U_3+alpha_3-nu_min_3+nu_max_3 =e= 0;
Eq_dual_f_1..   -alpha_1+alpha_2+beta_1-rho_min_1+rho_max_1 =e= 0;
Eq_dual_f_2..   -alpha_1+alpha_3+beta_2-rho_min_2+rho_max_2 =e= 0;
Eq_dual_f_3..   -alpha_2+alpha_3+beta_3-rho_min_3+rho_max_3 =e= 0;
Eq_dual_t_1..   -alpha_1+alpha_2+delta_1-psi_min_1+psi_max_1 =e= 0;
Eq_dual_t_2..   -alpha_1+alpha_3+delta_2-psi_min_2+psi_max_2 =e= 0;
Eq_dual_t_3..   -alpha_2+alpha_3+delta_3-psi_min_3+psi_max_3 =e= 0;
Eq_theta_1..    -beta_1*x_1*B_N_1-beta_2*x_2*B_N_2
                      -delta_1*B_E_1-delta_2*B_N_2 =e= 0;
Eq_theta_2..    beta_1*x_1*B_N_1-beta_3*x_3*B_N_3
                      +delta_1*B_E_1-delta_3*B_N_3 =e= 0;
Eq_theta_3..    beta_2*x_2*B_N_2+beta_3*x_3*B_N_3
                      +delta_2*B_E_2+delta_3*B_N_3 + gamma =e= 0;

** STRONG DUALITY EQUALITY
Eq_SDT..        c_g_1*g_1+c_g_2*g_2+c_g_3*g_3+c_U_1*r_1+c_U_2*r_2+c_U_3*r_3 =e=
                -alpha_1*d_1-alpha_2*d_2-alpha_3*d_3+
                phi_min_1*g_min_1+phi_min_2*g_min_2+phi_min_3*g_min_3-
                phi_max_1*g_max_1-phi_max_2*g_max_2-phi_max_3*g_max_3-
                nu_max_1*d_1-nu_max_2*d_2-nu_max_3*d_3-
                t_max_1*psi_min_1-t_max_2*psi_min_2-t_max_3*psi_min_3-
                t_max_1*psi_max_1-t_max_2*psi_max_2-t_max_3*psi_max_3-
```

```
              f_max_1*rho_min_1-f_max_2*rho_min_2-f_max_3*rho_min_3-
              f_max_1*rho_max_1-f_max_2*rho_max_2-f_max_3*rho_max_3;

*************************************************************************
*       MODEL
*************************************************************************
model transm /All/;
solve transm minimizing tot_cost using mip;
invest_cost.l =  c_1*x_1+c_2*x_2+c_3*x_3;
unserv_ener.l =  c_U_1*r_1.l+c_U_2*r_2.l+c_U_3*r_3.l;
gen_cost.l = g_1.l*c_g_1+g__2.l*c_g_2+g_3.l*c_g_3;
Display invest_cost.l,unserv_ener.l,gen_cost.l,tot_cost.l
```

GAMS Code for the Dantzig-Wolfe Decomposition of the Stochastic Power Model (Section 9.2.4)

```
*************************************************************************
*       DATA
*************************************************************************
* This is the stochastic power VI model of Chapter 9,
* solved by Dantzig-Wolfe decomposition.

Sets
i       nodes (and firms)   / 1, 2 /
m       scenarios / Lo, Hi /  ;
Alias( i, j, jj );
**
* For decomposition, we also need an iteration label, kk, which is a set, and a
* subset k(kk) to label iterations from the start, up to and including kk, in
* the process. The maximum number of iterations is given by the second number,
* i.e., the number after the * .
**
sets  kk  master problem labels for proposals  / 1*10 /
      k(kk)  proposals received since beginning of iterations ;
alias(k,ka);

**
* We need a criterion to decide whether to stop the iterations before the
* maximum number of iterations has been reached. When the value of
* the relative convergence gap RGap (look in the loop below for definition) is
* less than the parameter RGapMax, the iterations stop.  If RGap never falls
* below RGapMax, then the iterations continue until the maximum number of
* iterations is reached.
Parameter RGapMax convergence criterion / 0.000001 / ;

* For decomposition, we also need a first guess of the dual variables of
* the linking constraints (which are the nodal market clearing constraints),
* in order to solve the first subproblem.  This is set equal to the wG
* parameter later in this program; wG is the linking dual variable, and is
* equal to the actual value of the dual in the master problem, i.e.,
* it includes Pr(m), so it is the expected value of the wheeling fee in
* scenario m.
Table FirstGuess(i,m) first guess of wG for first subproblem
        Lo      Hi
1       0       0
2       0       0 ;

* The model parameters are listed next.
Parameter Cour Cour=0 for price-taker behavior or =1 for Cournot  / 1 / ;
```

```
Parameter C(i)       baseload generation costs /1 15, 2 19 / ;
Parameter RC(i) rapid ramping generation costs /1 20, 2 25 / ;
Parameter XK(i)        baseload capacities /1 400, 2 180 / ;
Parameter RK(i)  rapid ramping capacities /1 150, 2 100 / ;
Table A(m,i)  inverse demand price intercept
           1     2
 Lo       25    25
 Hi       50    50
Table B(m,i)  inverse demand slope (absolute value)
           1     2
 Lo       .08   .04
 Hi       .08   .04 ;
Parameter  T   line limit / 175 / ;
Parameter  Pr(m)  scenario probabilities    / Lo 0.5,  Hi 0.5 / ;

**
* Set up the subproblem. In particular, we need a parameter for the latest dual
* variables of linking constraints, from the last master problem. This
* parameter is given an initial value, a guess, for the first subproblem.
**
Parameter wG(i,m) guess (or from master) of dual for market clearing constraint;
* Note: wG(i,m) is the actual value of the dual in the master problem, i.e.,
* it includes Pr(m), so it is the expected value of the wheeling fee in
* scenario m.
wG(i,m)=FirstGuess(i,m);

************************************************************************
*        DECLARATION OF VARIABLES
************************************************************************

Positive Variables
s(i,j,m)  sales of firm at node i to load at node j in scenario m
x(i)      baseload generation of firm at node i
r(i,m)    rapid ramping generation of firm at node i in scenario m ;

* Note: no y variables in subproblem; they are in the master problem.
* Also: no constraints for bounds on y, or for market clearing.


************************************************************************
*        EQUATIONS
************************************************************************
Equations
BaseCap(i)  baseload capacity of firm at node i
RapidCap(i,m) rapid ramp capacity of firm at node i
Sales(i,m)  sales equals generation for firm at i in scenario m
Gs(i,j,m)   VI mapping element for sales from i to j in scenario m
Gx(i)       VI mapping element for baseload generation of firm at i
Gr(i,m)     VI mapping for rapid ramp generation of firm at i in scenario m ;

BaseCap(i).. x(i) =l= XK(i);
RapidCap(i,m)..r(i,m) =l= RK(i);
Sales(i,m).. sum(j,s(i,j,m))-x(i)-r(i,m) =e= 0;
Gs(i,j,m)..Pr(m)*(-A(m,j)+B(m,j)*(Cour*s(i,j,m)+sum(jj,s(jj,j,m))))+wG(j,m)=n=0;
Gx(i)..      C(i)-sum(m,wG(i,m)) =n= 0;
Gr(i,m)..    Pr(m)*RC(i)-wG(i,m) =n= 0;

************************************************************************
*        MODEL
************************************************************************

model SubStochpower /BaseCap,RapidCap,Sales,Gs,Gx,Gr / ;
file myinfo / '%emp.info%' /;
put myinfo '* complementarity pairs for stochastic power sub model' /;
put 'viFunc Gs s';
put 'viFunc Gx x';
putclose 'viFunc Gr r';
```

```
* solve subproblem with first guesses of wheeling fees wG.
solve SubStochpower using emp;

* Declare parameters that hold proposals from subproblem and assign values
* from first subproblem that was just solved. These are used in the first
* master problem of the loop defined below.
Parameter sP(i,j,m,kk) proposals for s variable
          xP(i,kk) proposals for x variable
          rP(i,m,kk) proposals for r variable    ;
sP(i,j,m,'1')=s.l(i,j,m);
xP(i,'1')=x.l(i);
rP(i,m,'1')=r.l(i,m);

**
* Define master problem.
**
Positive Variables
lam(kk) lamda weights on proposals ;

Variables
y(i,m)    power injected into node i in scenario m ;

Equations
LineLim(i,m) line limit in both directions in scenario m
Market(i,m) market clearing at node i in scenario m
convex   sum of lamda weights <= one
Glam(kk) VI mapping for lamda variable ;

LineLim(i,m).. y(i,m) =l= T;
Market(i,m)..-sum(j,sum(k,sP(j,i,m,k)*lam(k)))+sum(k,xP(i,k)*lam(k))
          +sum(k,rP(i,m,k)*lam(k))+y(i,m) =e= 0;
convex..     sum(k,lam(k)) =l= 1 ;
* Note the =l= form of the convex constraint.  This is valid because
* the zero solution is feasible in the subproblem.
Glam(k)..   sum((i,j,m),Pr(m)*(-A(m,j)+B(m,j)*(Cour*sum(ka,sP(i,j,m,ka)*lam(ka))
          +sum(jj,sum(ka,sP(jj,j,m,ka)*lam(ka)))))*sP(i,j,m,k))
          +sum(i,C(i)*xP(i,k)) + sum((i,m),Pr(m)*RC(i)*rP(i,m,k)) =n= 0;
;

model MastStochpower /LineLim,Market,convex,Glam /;

* Define parameters to be assigned within the loop.
Parameter sM(i,j,m,kk) sales calculated by master problem
          xM(i,kk)  baseload generation calculated by master problem
          rM(i,m,kk) rapid ramping generation calculated by master problem
          TotalSales(i,m) total sales at node i as calculated by master
          TotalGen(i,m) total generation at node i as calculated by master
          Price(i,m) price at node i as calculated by master
          Gap(kk)  convergence gap at iteration kk
          RGap(kk) relative convergence gap at iteration kk-1 ;

* Define subset k for first loop in master problem, and set RGap to a large
* value for kk=1, to start.
k('1')=yes;
RGap(kk)=0;
RGap('1')=100*RGapMax;

* now enter the loop to  iterate between master and subproblem.

loop(kk$(RGap(kk)>=RGapMax),
        put myinfo '* complementarity pairs for stochastic power master model' /;
        putclose 'viFunc Glam lam';
        solve MastStochpower using emp;
        wG(i,m)=Market.m(i,m) ;
        sM(i,j,m,kk)=sum(ka,sP(i,j,m,ka)*lam.l(ka)) ;
        xM(i,kk)=sum(ka,xP(i,ka)*lam.l(ka)) ;
```

```
        rM(i,m,kk)=sum(ka,rP(i,m,ka)*lam.l(ka));
        TotalSales(i,m)=sum(j,sM(j,i,m,kk));
        TotalGen(i,m)=xM(i,kk)+rM(i,m,kk);
        Price(i,m)=A(m,i)-B(m,i)*TotalSales(i,m);
        display TotalSales,TotalGen,Price ;

        put myinfo '* complementarity pairs for stochastic power sub model' /;
        put 'viFunc Gs s';
        put 'viFunc Gx x';
        putclose 'viFunc Gr r';
        solve SubStochpower using emp;
        sP(i,j,m,kk+1)=s.l(i,j,m) ;
        xP(i,kk+1)=x.l(i);
        rP(i,m,kk+1)=r.l(i,m);
        Gap(kk)= sum((i,j,m),(Pr(m)*(-A(m,j)+B(m,j)*(Cour*sM(i,j,m,kk)
                +sum(jj,sM(jj,j,m,kk))))+wG(j,m))*(s.l(i,j,m)-sM(i,j,m,kk)))
                +sum(i,(C(i)-sum(m,wG(i,m)))*(x.l(i)-xM(i,kk)))
                +sum((i,m),(Pr(m)*RC(i)-wG(i,m))*(r.l(i,m)-rM(i,m,kk))) ;
        RGap(kk+1)=abs(Gap(kk))/(1+abs(
                sum((i,j,m),(Pr(m)*(-A(m,j)+B(m,j)*(Cour*sM(i,j,m,kk)
                +sum(jj,sM(jj,j,m,kk))))+wG(j,m))*sM(i,j,m,kk))
                +sum(i,(C(i)-sum(m,wG(i,m)))*xM(i,kk))
                +sum((i,m),(Pr(m)*RC(i)-wG(i,m))*rM(i,m,kk)))) ;
        k(kk+1) = yes ;
);
display Gap,RGap,sM,xM,rM,TotalSales,TotalGen,Price ;
```

GAMS Code for the Small Natural Gas Network Equilibrium Problem (Section 10.4.8)

```
*****************************************************************************
*        DATA
*****************************************************************************
OPTION ITERLIM = 100000;
option limrow = 1000;
option limcol = 1000;


SETS
s seasons       /s1*s2/
t time periods  /t1*t2/
n nodes         /n1*n2/

* suppliers are relative to the node.
* @ node 1, suppliers 1, 2 are A, B, respectively
* @ node 2, suppliers 1, 2 are C, D, respectively
i suppliers     /i1,i2/
;


alias(n,m);
alias(i,ii);
alias(s,ss);


$ontext
Assume for now that every (n,m) is possible
Will restrict with .FX later

volume typically will be in cubic feet or cubic meters
```

```
rate typically will be in cubic feet/day or cubic meters/day
$offtext

PARAMETERS

Days(s) # of days in season s (s=1:Oct-Feb S=2:Mar-Sept) /s1 151, s2 214/
*** from a node to itself, no regulated tariff
*** from node 2 to node 1 flow will be restricted to zero
*** (could alternatively have very high tariff)

TABLE Tau_reg(n,m,s,t)  regulated tariff
      s1.t1      s2.t1       s1.t2      s2.t2
n1.n1 0.0        0.0         0.0        0.0
n1.n2 0.5        0.5         0.5        0.5
n2.n1 0.0        0.0         0.0        0.0
n2.n2 0.0        0.0         0.0        0.0
;

TABLE P_contract(n,m,i,s,t) contract price
         s1.t1  s1.t2  s2.t1 s2.t2
n1.n1.i1 0      0      0     0
n1.n1.i2 0      0      0     0
n1.n2.i1 0      0      0     0
n1.n2.i2 0      0      0     0
n2.n1.i1 0      0      0     0
n2.n1.i2 0      0      0     0
n2.n2.i1 0      0      0     0
n2.n2.i2 0      0      0     0
;

TABLE gamma_prod(n,i,s,t) slope of linear production function
      s1.t1 s1.t2 s2.t1 s2.t2
n1.i1  10    10    10    10
n1.i2  12    12    12    12
n2.i1  15    15    15    15
n2.i2  18    18    18    18
;


TABLE gamma_extract(n,i) slope of linear storage extraction cost function
     i1  i2
n1   0   0
n2   0   0
;

TABLE gamma_inject(n,i) slope of linear storage injection cost (same for  t)
    i1  i2
n1  0   0
n2  0   0
;


*** from a node to itself, no transportation cost
*** from node 2 to node 1 flow will be restricted to zero
***(could alternatively have very high tariff)

TABLE gamma_flow(n,m,s) slope of linear flow cost function (same for  s t)
      s1  s2
n1.n1 0   0
n1.n2 1   1
n2.n1 0   0
n2.n2 0   0
;

TABLE L(n,i) storage loss %'s assume the same over time
    i1   i2
n1  0.0  0.0
```

```
n2  0.0  0.0
;

***  Days(s)*q LE Max Daily Volume in season s
TABLE Q_prod_max(n,i,s) maximum production quantity (volume)
      s1 s2
n1.i1 10 10
n1.i2 10 10
n2.i1 5  5
n2.i2 5  5
;

*** choose large value if don't want to be binding
TABLE V_prod_max(n,i) maximum production volume over the time horizon
    i1   i2
n1  1000 1000
n2  1000 1000
;

TABLE X_extract_max(n,i) maximum extraction rate
    i1   i2
n1  0    0
n2  0    0
;

TABLE I_inject_max(n,i) maximum injection rate
    i1   i2
n1  0    0
n2  0    0
;

TABLE W_volume_max(n,i) maximum volume
    i1   i2
n1  1000 1000
n2  1000 1000
;

***  unrestricted from node 1 to itself (set bound high)
***  unrestricted from node 2 to itself (set bound high)
***  no flow from node 2 to node 1
TABLE G_flow_max(n,m,s) maximum volume
       s1         s2
n1.n1  10000      10000
n1.n2  5          5
n2.n1  0          0
n2.n2  10000      10000
;

*** demand equation of the form intercept_n -slope_n* pi_n

TABLE intercept(n,s,t) intercept for linear demand equation
   s1.t1  s2.t1  s1.t2  s2.t2
n1 20.0   20.0   20.0   20.0
n2 40.0   40.0   40.0   40.0
;

TABLE slope(n,s,t) slope for linear demand equation
    s1.t1  s2.t1   s1.t2  s2.t2
n1  1.0    1.0     1.0    1.0
n2  2.0    2.0     2.0    2.0;

******************************************************************************
*        DECLARATION OF VARIABLES
******************************************************************************
*************** SUPPLIER PROBLEM VARIABLES ***************
POSITIVE VARIABLES
*Nonnegative primal variables
```

```
q_prod(n,i,s,t)          production rate (volume per day)
v_prod(n,i,s,t)          production volume (volume)
f(n,m,i,s,t)             flow rate (volume per day)
s_contract(n,m,i,s,t)    contract sales rate (volume per day)
s_spot(n,m,i,s,t)        spot sales rate (volume per day)
q_extract(n,i,s,t)       storage extraction rate (volume per day)
q_inject(n,i,s,t)        storage injection rate (volume per day)
v_extract(n,i,t)         storage extraction volume (volume)
v_inject(n,i,t)          storage injection volume (volume)

*Nonnegative Lagrange multipliers
alpha1(n,i,s,t)
alpha2(n,i)
alpha6(n,i,s,t)
alpha7(n,i,s,t)
alpha9(n,i,t)
;

VARIABLES
* Free Lagrange multipliers
alpha3(n,m,i,s,t)
alpha4(n,i,s,t)
alpha5(n,i,s,t)
alpha8(n,i,t)
alpha10(n,i,t)
alpha11(n,i,t)
;

*************** END SUPPLIER PROBLEM VARIABLES ***************

********** NETWORK OPERATOR PROBLEM VARIABLES **********
POSITIVE VARIABLES
*Nonnegative primal variables
g(n,m,s,t)               flow rate (volume per day)
v_flow(n,m,s,t)          flow volume (volume)

* Nonnegative Lagrange multipliers
beta1(n,m,s,t)
;
VARIABLES
*Free Lagrange multipliers
beta2(n,m,s,t)
;
********** END NETWORK OPERATOR PROBLEM VARIABLES **********


********** MARKET-CLEARING CONDITIONS VARIABLES **********
VARIABLES
pi(n,s,t)                prices ($ per volume)
tau_congest (n,m,s,t)    congestion tariff ($ per volume)
;

***************************************************************************
*       EQUATIONS
***************************************************************************
EQUATIONS

* intermediate equation

* First Set, Supplier Problems
KKT_q_prod(n,i,s,t)
KKT_v_prod(n,i,s,t)
KKT_f(n,m,i,s,t)
KKT_s_contract(n,m,i,s,t)
KKT_s_spot(n,m,i,s,t)
KKT_q_extract(n,i,s,t)
KKT_q_inject(n,i,s,t)
```

```
KKT_v_extract(n,i,t)
KKT_v_inject(n,i,t)

* Second Set , Supplier Problems
Production_rate(n,i,s,t)
Production_horizon(n,i)
Flow_balance(n,m,i,s,t)
Production_balance(n,i,s,t)
Production_v_q(n,i,s,t)
Extraction_rate(n,i,s,t)
Injection_rate(n,i,s,t)
Storage_balance(n,i,t)
Storage_volume(n,i,t)
Extraction_v_q(n,i,t)
Injection_v_q(n,i,t)


* Network Operator
KKT_g(n,m,s,t)
KKT_v(n,m,s,t)

Flow_rate(n,m,s,t)
Flow_v_q(n,m,s,t)

* Market-Clearing Conditions
MCC_spot(n,s,t)
MCC_transportation(n,m,s,t)

;

* Definitions
*Supplier Problem: First Set
KKT_q_prod(n,i,s,t)..Days(s)*(alpha1(n,i,s,t)+alpha2(n,i)
                             -alpha4(n,i,s,t)-alpha5(n,i,s,t))=g=0;
KKT_v_prod(n,i,s,t)..gamma_prod(n,i,s,t)+alpha5(n,i,s,t)=g=0;
KKT_f(n,m,i,s,t).. Days(s)*(tau_reg(n,m,s,t)+tau_congest(n,m,s,t)
                         +alpha3(n,m,i,s,t))=g=0;
KKT_s_contract(n,m,i,s,t)..Days(s)*(-P_contract(n,m,i,s,t)
                           -alpha3(n,m,i,s,t)+alpha4(n,i,s,t))=g=0;
KKT_s_spot(n,m,i,s,t)..Days(s)*(-pi(m,s,t)-alpha3(n,m,i,s,t)
                           +alpha4(n,i,s,t))=g=0;
KKT_q_extract(n,i,s,t)..Days('s1')*(-pi(n,'s1', t)+alpha6(n,i,s,t)
                            +alpha8(n,i,t)+alpha9(n,i,t)-alpha10(n,i,t))=g=0;
KKT_q_inject(n,i,s,t)..Days('s2')*(alpha4(n,i,'s2',t)+alpha7(n,i,s,t)
                            -(1-L(n,i))*alpha8(n,i,t)-alpha11(n,i,t))=g=0;
KKT_v_extract(n,i,t)..gamma_extract(n,i)+alpha10(n,i,t)=g=0;
KKT_v_inject(n,i,t)..gamma_inject(n,i)+alpha11(n,i,t)=g=0;

*Supplier Problem: Second Set
Production_rate(n,i,s,t)..(Q_prod_max(n,i,s)-Days(s)*q_prod(n,i,s,t))=g=0;
* note: needed to divide Q_prod_max by Days(s)
* to get GAMS to compute the factors exactly.

Production_horizon(n,i)..V_prod_max(n,i)
                                  -sum((s,t), Days(s)*q_prod(n,i,s,t))=g=0;
Flow_balance(n,m,i,s,t)..Days(s)*(f(n,m,i,s,t)-s_contract(n,m,i,s,t)
                                  -s_spot(n,m,i,s,t))=e=0;
* was =g=0 should have been =e=0, fixed 10 April 2012;

Production_balance(n,i,s,t).. Days(s)*(sum(m,(s_contract(n,m,i,s,t)
                                          +s_spot(n,m,i,s,t))))
                                          +Days(s)*q_inject(n,i,s,t)
                                          -Days(s)*q_prod(n,i,s,t)=e=0;
Production_v_q(n,i,s,t)..v_prod(n,i,s,t)-Days(s)*q_prod(n,i,s,t)=e=0;
Extraction_rate(n,i,s,t).. Days('s1')*X_extract_max(n,i)
                                   -Days('s1')*q_extract(n,i,s,t)=g=0;
Injection_rate(n,i,s,t).. Days('s2')*I_inject_max(n,i)
```

```
                                    -Days('s2')*q_inject(n,i,s,t)=g=0;
Storage_balance(n,i,t)..Days('s1')*q_extract(n,i,'s1',t)
                                    -(1-L(n,i))*Days('s2')*q_inject(n,i,'s2',t)=e=0;
Storage_volume(n,i,t).. W_volume_max(n,i)-Days('s1')*q_extract(n,i,'s1',t)=g=0;

* storage volumes (extraction or inject no "s" index
*but yes for q_extract and q_inject-easier that way)
Extraction_v_q(n,i,t).. v_extract(n,i,t)-Days('s1')*q_extract(n,i,'s1',t)=e=0;
Injection_v_q(n,i,t)..  v_inject(n,i,t)-Days('s2')*q_inject(n,i,'s2',t)=e=0;


* Network Operator
KKT_g(n,m,s,t)..-Days(s)*(tau_reg(n,m,s,t)+tau_congest(n,m,s,t))
                             +Days(s)*beta1(n,m,s,t)-Days(s)*beta2(n,m,s,t)=g=0;
KKT_v(n,m,s,t)..gamma_flow(n,m,s)+beta2(n,m,s,t)=g=0;
Flow_rate(n,m,s,t)..G_flow_max(n,m,s)-Days(s)*g(n,m,s,t)=g=0;
Flow_v_q(n,m,s,t)..v_flow(n,m,s,t)-Days(s)*g(n,m,s,t)=e=0;


* Market-Clearing Conditions
* Spot market, node n
MCC_spot(n,s,t)..sum((m,i),Days(s)*((s_spot(m,n,i,s,t)
                         +s_contract(m,n,i,s,t))))
                         -(intercept(n,s,t)-slope(n,s,t)*pi(n,s,t))
                          +sum(i,Days(s)*q_extract(n,i,s,t))=e=0;
MCC_transportation(n,m,s,t)..sum(i,Days(s)*f(n,m,i,s,t))-Days(s)*g(n,m,s,t)=e=0;
;

*************************************************************************
*       MODEL
*************************************************************************
MODEL Gas_MCP
/
KKT_q_prod.q_prod,
KKT_v_prod.v_prod,
KKT_f.f,
KKT_s_contract.s_contract,
KKT_s_spot.s_spot,
KKT_q_extract.q_extract,
KKT_q_inject.q_inject,
KKT_v_extract.v_extract,
KKT_v_inject.v_inject,

Production_rate.alpha1,
Production_horizon.alpha2,
Flow_balance.alpha3,
Production_balance.alpha4,
Production_v_q.alpha5,
Extraction_rate.alpha6,
Injection_rate.alpha7,
Storage_balance.alpha8,
Storage_volume.alpha9,
Extraction_v_q.alpha10,
Injection_v_q.alpha11,

KKT_g.g,
KKT_v.v_flow,
Flow_rate.beta1,
Flow_v_q.beta2


MCC_spot.pi
MCC_transportation.tau_congest


/ ;
```

```
*** Fix certain variables to zero to match problem description

*** Don't allow flow from node 2 to node 1
f.fx('n2','n1',i,s,t)=0;

*** Don't allow spot or contract sales from node 2 to node 1
s_contract.fx('n2','n1',i,s,t)=0;
s_spot.fx('n2','n1',i,s,t)    =0;

*** No extraction in season 2
q_extract.fx(n,i,'s2',t)=0;


*** No injection in season 1
q_inject.fx(n,i,'s1',t) =0;




*** check with solution from Chapter 1
*s_spot.fx('n1','n1','i1',s,t)=7.43945984/Days(s);
*s_spot.fx('n1','n1','i2',s,t)=0.56054016/Days(s);
*s_spot.fx('n1','n2','i1',s,t)=2.56054016/Days(s);
*s_spot.fx('n1','n2','i2',s,t)=2.43945984/Days(s);


*alpha1.fx('n1','i1',s,t)=2;
*tau_congest.fx('n1','n2','s1',t)=2.5;
*tau_congest.fx('n1','n2','s2',t)=2.5;

*** Initial starting points
*s_spot.l('n1','n1','i1',s,t)=7.439/Days(s);
*s_spot.l('n1','n1','i2',s,t)=0.561/Days(s);
*s_spot.l('n1','n2','i1',s,t)=2.561/Days(s);
*s_spot.l('n1','n2','i2',s,t)=2.439/Days(s);

* set up options file
Gas_MCP.optfile =1;
option mcp=path;

Solve Gas_MCP using mcp;


*** Checking Output
parameter
demand(n,s,t), Production_A(s,t),Production_B(s,t),Production_C(s,t),
Production_D(s,t),sales_spot1_1(i,s,t),sales_spot1_2(i,s,t),
sales_spot2_2,sales_contract1_1(i,s,t),sales_contract1_2(i,s,t),
flow1_2(i,s,t);

demand(n,s,t)=intercept(n,s,t)-slope(n,s,t)*pi.l(n,s,t);
production_A(s,t)=Days(s)*q_prod.l('n1','i1',s,t);
production_B(s,t)=Days(s)*q_prod.l('n1','i2',s,t);
production_C(s,t)=Days(s)*q_prod.l('n2','i1',s,t);
production_D(s,t)=Days(s)*q_prod.l('n2','i2',s,t);

sales_spot1_1(i,s,t)=Days(s)*s_spot.l('n1','n1',i,s,t);
sales_spot1_2(i,s,t)=Days(s)*s_spot.l('n1','n2',i,s,t);
sales_spot2_2(i,s,t)=Days(s)*s_spot.l('n2','n2',i,s,t);
sales_contract1_1(i,s,t)=Days(s)*s_contract.l('n1','n1',i,s,t);
sales_contract1_2(i,s,t)=Days(s)*s_contract.l('n1','n2',i,s,t);

flow1_2(i,s,t)=Days(s)*f.l('n1','n2',i,s,t);

display demand, production_A, production_B, production_C, production_D,
sales_spot1_1,sales_spot1_2,sales_spot2_2,
sales_contract1_1,sales_contract1_2,flow1_2
```

```
v_extract.l,
v_inject.l
;
```

Note: the line "Gas_MCP.optfile =1;" directs the solver to a file in the project directory called "path.opt" which contains the following user-chosen settings to control the algorithm (please see the GAMS PATH solver manual for details):

```
convergence_tolerance = 1e-11
preprocess =0
output_factorization_singularities =yes
```

References

1. Advanced Interactive Multidimensional Modeling System. AIMMS, June 2011. www.aimms.com.
2. A Modeling Language for Mathematical Programming. AMPL, June 2011. www.ampl.com.
3. General Algebraic Modeling System. GAMS, June 2011. www.gams.com.

Appendix C
DC Power Flow

The power flow P_{ij} through a transmission line ij, connecting the sending node i of the transmission line to its receiving node j, can be expressed by the general equation below, [2], [1],

$$P_{ij} = V_i V_j \left[G_{ij} \cos(\theta_i - \theta_j) + B_{ij} \sin(\theta_i - \theta_j) \right] - V_i^2 G_{ij}. \tag{C.1}$$

where V_i and θ_i are the state variables characterizing the state of nodes i and are respectively, the voltage magnitude and voltage angle. Additionally, G_{ij} and B_{ij} are two parameters that characterize the physical structure of the transmission line called conductance and susceptance, respectively. Expression (C.1) is know as alternative current (AC) power flow.

Considering a transmission line, not a distribution one, the following simplifications generally hold:

1. $V_i = V_j = 1$ if voltage magnitudes are expressed in per unit terms.
2. $(\theta_i - \theta_j) < 2\pi/9$ radians under normal operating conditions and thus, [2], $\cos(\theta_i - \theta_j) \approx 1$ and $\sin(\theta_i - \theta_j) \approx (\theta_i - \theta_j)$.

Therefore, (C.1) reduces to:

$$P_{ij} = B_{ij}(\theta_i - \theta_j). \tag{C.2}$$

The expression (C.2) is know as the direct current (DC) power flow.

Denoting $\frac{1}{X_{ij}} = B_{ij}$, we obtain the two equivalent equations below:

$$P_{ij} = \frac{1}{X_{ij}}(\theta_i - \theta_j) \tag{C.3a}$$

$$(\theta_i - \theta_j) = X_{ij} P_{ij}. \tag{C.3b}$$

We consider a network with n nodes and m transmission lines and define the following terms:

A the node-branch incidence matrix $(n-1 \times m)$, not including the row corresponding to the reference node. The reference node is the node in which the phase angle is fixed to 0. Note that the selection of a reference node is convenient to handle equations (C.3) because only angle differences appear in these equations.

X a diagonal matrix $(m \times m)$ containing the reactances of the lines.

P_F the vector $(m \times 1)$ containing the power flow through the lines.

θ the vector $(n-1 \times 1)$ of state variables (phase angles), not including the reference node.

All equations (C.3b), i.e., one for each transmission line, can be expressed in matrix form by either of the two equivalent matrix equations below:

$$A^T\theta = XP_F \tag{C.4a}$$

$$P_F = \left(X^{-1}A^T\right)\theta \tag{C.4b}$$

Denoting by P the $(n-1 \times 1)$ vector of the power injections at all nodes other than the reference one, we can express the relationship between the power injection vector and the power flow vector as

$$P = AP_F \tag{C.5}$$

and

$$P = \left(AX^{-1}A^T\right)\theta = B\theta \tag{C.6}$$

where $B = AX^{-1}A^T$ is called the susceptance matrix.

Considering (C.4b) and (C.6), the power flow vector can be expressed as a function of the power injection vector:

$$P_F = \left(X^{-1}A^TB^{-1}\right)P = SP \tag{C.7}$$

where $S = X^{-1}A^TB^{-1}$ is the sensitivity matrix (it relates changes in power injections with changes in power flows) or "PTDF" (power transmission distribution factor) matrix.

To represent the state of a power transmission network and the flows through lines, two options are generally used:

1. Equation (C.3b) for all lines, which allows for explicitly considering the state variables (phase angles), power injections and power flows.
2. Matrix equation (C.7), which allows for considering just power injections and power flows.

Both of the above representations are used throughout this book. Further details can be found in [2].

Illustrative Example. DC Power Flow

In this example we analyze the 3-node network depicted in Figure C.1. The power injections are P_1, P_2 and P_3, located at nodes 1, 2 and 3, respectively. Additionally, each transmission line is characterized by its associated reactance, i.e., X_{12}, X_{13} and X_{23}.

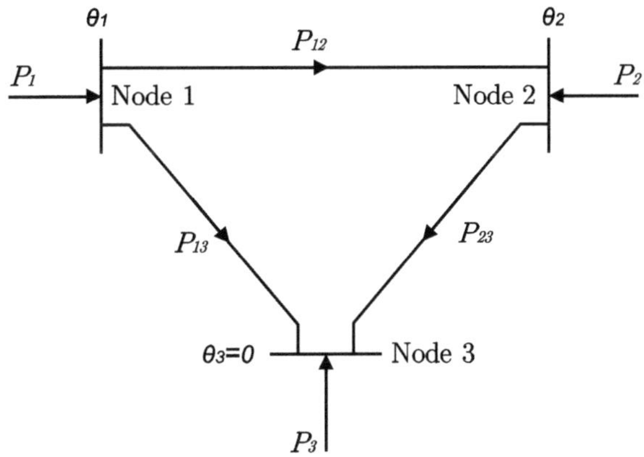

Fig. C.1 Illustrative Example: Three-node network

The power flow equations (C.3a) corresponding to this 3-node network are:

$$P_{12} = \frac{1}{X_{12}}(\theta_1 - \theta_2) \tag{C.8a}$$

$$P_{13} = \frac{1}{X_{13}}(\theta_1 - \theta_3) \tag{C.8b}$$

$$P_{23} = \frac{1}{X_{23}}(\theta_2 - \theta_3) \tag{C.8c}$$

If we consider node 3 to be the reference one ($\theta_3 = 0$), equations (C.8) can be rearranged in the form of the matrix equation (C.4b), i.e.,

$$\begin{aligned}\begin{pmatrix} P_{12} \\ P_{13} \\ P_{23} \end{pmatrix} &= \begin{pmatrix} \frac{1}{X_{12}} & 0 & 0 \\ 0 & \frac{1}{X_{13}} & 0 \\ 0 & 0 & \frac{1}{X_{23}} \end{pmatrix} \begin{pmatrix} 1 & -1 \\ 1 & 0 \\ 0 & 1 \end{pmatrix} \begin{pmatrix} \theta_1 \\ \theta_2 \end{pmatrix} \\ &= \begin{pmatrix} \frac{1}{X_{12}} & \frac{-1}{X_{12}} \\ \frac{1}{X_{13}} & 0 \\ 0 & \frac{1}{X_{23}} \end{pmatrix} \begin{pmatrix} \theta_1 \\ \theta_2 \end{pmatrix},\end{aligned} \tag{C.9}$$

where

$$A = \begin{pmatrix} 1 & 1 & 0 \\ -1 & 0 & 1 \end{pmatrix}, \; X^{-1} = \begin{pmatrix} \frac{1}{X_{12}} & 0 & 0 \\ 0 & \frac{1}{X_{13}} & 0 \\ 0 & 0 & \frac{1}{X_{23}} \end{pmatrix}, \; P_F = \begin{pmatrix} P_{12} \\ P_{13} \\ P_{23} \end{pmatrix}, \; \theta = \begin{pmatrix} \theta_1 \\ \theta_2 \end{pmatrix}.$$

Similarly to (C.5), the relationship between the power flow vector and the power injections vector can be expressed through the matrix A as:

$$\begin{pmatrix} P_1 \\ P_2 \end{pmatrix} = \begin{pmatrix} 1 & 1 & 0 \\ -1 & 0 & 1 \end{pmatrix} \begin{pmatrix} P_{12} \\ P_{13} \\ P_{23} \end{pmatrix}, \tag{C.10}$$

which taking into account (C.9) allows deriving the matrix equation (C.6) as follows:

$$\begin{aligned} \begin{pmatrix} P_1 \\ P_2 \end{pmatrix} &= \begin{pmatrix} 1 & 1 & 0 \\ -1 & 0 & 1 \end{pmatrix} \begin{pmatrix} \frac{1}{X_{12}} & 0 & 0 \\ 0 & \frac{1}{X_{13}} & 0 \\ 0 & 0 & \frac{1}{X_{23}} \end{pmatrix} \begin{pmatrix} 1 & -1 \\ 1 & 0 \\ 0 & 1 \end{pmatrix} \begin{pmatrix} \theta_1 \\ \theta_2 \end{pmatrix} \\ &= \begin{pmatrix} \frac{1}{X_{12}} + \frac{1}{X_{13}} & \frac{-1}{X_{12}} \\ \frac{-1}{X_{12}} & \frac{1}{X_{12}} + \frac{1}{X_{23}} \end{pmatrix} \begin{pmatrix} \theta_1 \\ \theta_2 \end{pmatrix} \end{aligned} \tag{C.11}$$

where $B = \begin{pmatrix} \frac{1}{X_{12}} + \frac{1}{X_{13}} & \frac{-1}{X_{12}} \\ \frac{-1}{X_{12}} & \frac{1}{X_{12}} + \frac{1}{X_{23}} \end{pmatrix}$.

Finally, the matrix equation (C.7) can be derived from (C.9) and (C.11):

$$\begin{aligned} \begin{pmatrix} P_{12} \\ P_{13} \\ P_{23} \end{pmatrix} &= \begin{pmatrix} \frac{1}{X_{12}} & 0 & 0 \\ 0 & \frac{1}{X_{13}} & 0 \\ 0 & 0 & \frac{1}{X_{23}} \end{pmatrix} \begin{pmatrix} 1 & -1 \\ 1 & 0 \\ 0 & 1 \end{pmatrix} \begin{pmatrix} \frac{1}{X_{12}} + \frac{1}{X_{13}} & \frac{-1}{X_{12}} \\ \frac{-1}{X_{12}} & \frac{1}{X_{12}} + \frac{1}{X_{23}} \end{pmatrix}^{-1} \begin{pmatrix} P_1 \\ P_2 \end{pmatrix} \\ &= \frac{X_{12}X_{13}X_{23}}{X_{12}+X_{13}+X_{23}} \begin{pmatrix} \frac{1}{X_{12}X_{23}} & \frac{-1}{X_{12}X_{13}} \\ \frac{1}{X_{13}}\left(\frac{1}{X_{12}} + \frac{1}{X_{23}}\right) & \frac{1}{X_{12}X_{13}} \\ \frac{1}{X_{12}X_{23}} & \frac{1}{X_{23}}\left(\frac{1}{X_{12}} + \frac{1}{X_{13}}\right) \end{pmatrix} \begin{pmatrix} P_1 \\ P_2 \end{pmatrix}, \end{aligned} \tag{C.12}$$

where $S = \frac{X_{12}X_{13}X_{23}}{X_{12}+X_{13}+X_{23}} \begin{pmatrix} \frac{1}{X_{12}X_{23}} & \frac{-1}{X_{12}X_{13}} \\ \frac{1}{X_{13}}\left(\frac{1}{X_{12}} + \frac{1}{X_{23}}\right) & \frac{1}{X_{12}X_{13}} \\ \frac{1}{X_{12}X_{23}} & \frac{1}{X_{23}}\left(\frac{1}{X_{12}} + \frac{1}{X_{13}}\right) \end{pmatrix}$ is the sensitivity matrix or "PTDF" matrix associated with the 3-node network considered in this illustrative example. Observe that (C.12) shows how to obtain the power flow vector directly from the power injections vector.

References

1. A.R. Bergen and V. Vittal. *Power Systems Analysis*. Prentice Hall, Upper Saddle River, New Jersey, 2nd edition, 1999.
2. A. Gómez-Expósito, A.J. Conejo, and C. Cañizares. *Electric Energy System. Analysis and Operation*. CRC Press, Boca Raton, 2009.

Appendix D
Natural Gas Engineering Considerations

The flow of gas between two points as well as the extraction/injection of it from wells can be described by nonlinear and non-convex relationships. In order to simplify things, many models of natural gas assume a linearized form or ignore certain components all together if appropriate. In this appendix, we delve more into the nonlinear engineering aspects and illustrate some linearization techniques and other approximations that have been used in the literature.

We focus on the following three nonlinear equations as described in [4] that all involve the flow f_{ij} through an arc (i,j) and the pressures p_i and p_j, respectively, at nodes i and j.[1] First is the flow along a pipe segment given by the following relationship known as the Weymouth equation:

$$f_{ij} = sgn\,(p_i, p_j)\, C_{ij} \left|p_i^2 - p_j^2\right|^{1/2} \tag{D.1}$$

where

$$sgn\,(p_i, p_j) = \begin{cases} 1 & \text{if } p_i \geq p_j \\ -1 & \text{otherwise} \end{cases}$$

and C_{ij} is the conveyance of the pipe segment which is in general a function of the length, diameter, and the efficiency of the pipe and the properties of the gas that is flowing in the pipe segment. Figure D.1 depicts the nonlinear, non-convex nature of (D.1).

The next engineering equation considered is the compressor activity given by the following equation:[2]

$$f_{ij} = HP_{ij} / \left(k_1\, (p_j/p_i)^{k_3} - k_2\right) \tag{D.2}$$

where HP_{ij} is the horsepower of the compressor and the constants k_1, k_2, k_3 are empirical values for the compressor in question. And the last equation

[1] Nodes here refer to locations in the gas network where gas is supplied, transferred, and/or removed.

[2] Compressors are used to increase the pressure of the gas.

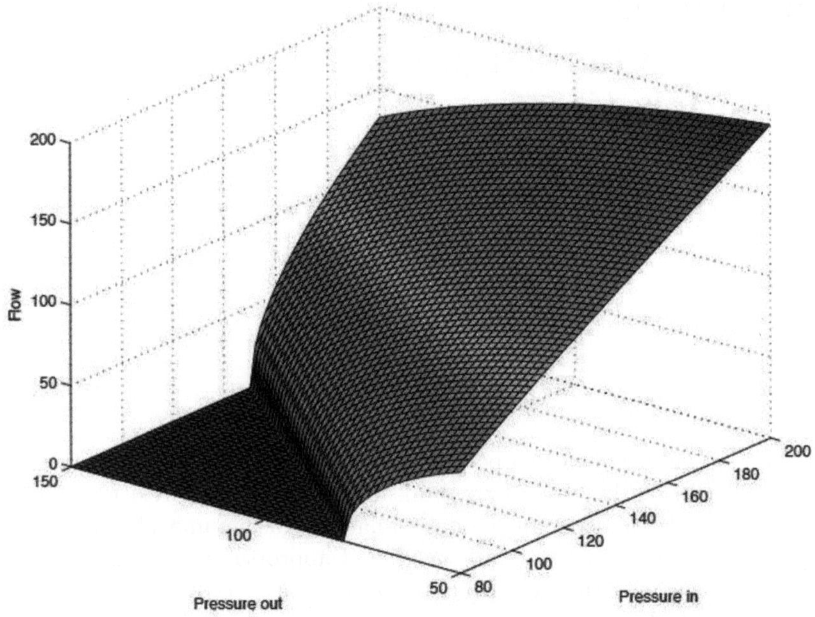

Fig. D.1 Example of Weymouth equation from [3]; flow [SM^3], pressure [bar]

refers to the valve activity[3] given by the following equation:

$$f_{ij} = sgn\,(p_i, p_j)\, K_{ij}\, [|p_i - p_j| \min(p_i, p_j)]^{1/2} \tag{D.3}$$

where K_{ij} is a constant that depends on the valve flow area and the loss coefficient (and other things).

In principle, one could approximate any of the nonlinear engineering equations described above by a first-order, hence linear, Taylor Series approximation. This was the approach taken in [4] and [3] for the nonlinear Weymouth equation (D.1). Of course, care must be exercised to insure that this approximation is close to the original equation for the values of interest and this can be done in a number of ways. In the rest of this section we focus mostly on the Weymouth equation and consider several ways to approximate it and we assume that the pressure at node i is greater than or equal to that at node j so that $sgn\,(p_i, p_j) = 1$. In addition, it is assumed that the inlet pressure is given as p_i and the outlet pressure as p_j.

As described in [4], after linearizing (D.1), the resulting equation becomes:

$$f_{ij} = a_i p_i - a_j p_j \tag{D.4}$$

[3] Used to open/close the pipeline for various reasons.

where $a_i = a_0 p_{i0}, a_j = a_0 p_{j0}$, with

$$a_0 = \frac{f_0}{p_{i0}^2 - p_{j0}^2}$$

with f_0, p_{i0}, p_{j0} being the test values around which the linearization is done for f_{ij}, p_i, p_j, respectively.[4] The linearlized Weymouth equation (D.4) can then be inserted as a constraint into the optimization problem under consideration representing the gas network. O'Neill et al. [4] use two inequalities instead of (D.4) resulting in:

$$-\varepsilon_{ij} \leq -f_{ij} + a_i p_i - a_j p_j \leq \varepsilon_{ij} \tag{D.5}$$

where ε_{ij} is a parameter that is determined at each iteration of an iterative process described in [4]. Upper and lower bounds on the pressure differences $p_i - p_j$ are also handled by O'Neill et al. [4] in a similar way to (D.5) but with different ε_{ij} values that also are determined at each iteration of an iterative process. The purpose is to simplify the constraints for compressors (D.2) and valves (D.3).

Midthun [3] uses a related but different approach in his linearization. Specifically, the Weymouth equation (D.1) is approximated by a first-order, linear Taylor series approximation of the form

$$f_{ij} \leq b_{il} p_i - b_{jl} p_j \tag{D.6}$$

where $l = 1, \ldots, L$ are the points where the pressures p_i and p_j are sampled and

$$b_{il} = C_{ij} \frac{p_{il}}{\sqrt{p_{il}^2 - p_{jl}^2}} \tag{D.7a}$$

$$b_{jl} = C_{ij} \frac{p_{jl}}{\sqrt{p_{il}^2 - p_{jl}^2}} \tag{D.7b}$$

As noted by Midthun [3], for any given pipeline flow, at most one of the L constraints given in (D.6) will be binding, corresponding to which one best approximates the flow.

An alternative approach is to take the Weymouth equation along with other constraints for gas networks after possibly making some transformations of variables, and show why at optimality that the Weymouth equation (and other gas engineering equations) hold. This approach for example is used in [1] and [2]. The advantage is that no linearization is needed, the disadvantage is that the results are specific (perhaps) to the gas network optimization problem form that is studied.

---

[4] In [4], the equation $0 = -f_{ij} + a_i p_i - a_j p_j$ is solved via a modified Newton method, hence the need to test at a flow and two pressure values and not just the pressures.

References

1. D. de Wolf and Y. Smeers. Optimal dimensioning of pipe networks with application to gas transmission networks. *Operations Research,* 44(4):596-608, 1996.
2. D. de Wolf and Y. Smeers. The gas transmission problem solved by an extension of the simplex algorithm. *Management Science,* 46(11):1454-1465, 2000.
3. K. T. Midthun. *Optimization Models for Liberalized Natural Gas Markets* Ph.D. Thesis, Norwegian University of Science and Technology, Trondheim, Norway, 2007.
4. R.P. O'Neill, M. Williard, B. Wilkins, and R. Pike. A mathematical programming model for allocation of natural gas. *Operations Research*, 27(5):857-873, 1979.

List of Tables

List of Figures

Index

MIX
Papier aus verantwortungsvollen Quellen
Paper from responsible sources
FSC® C105338

If you have any concerns about our products,
you can contact us on
ProductSafety@springernature.com

In case Publisher is established outside the EU,
the EU authorized representative is:
Springer Nature Customer Service Center GmbH
Europaplatz 3, 69115 Heidelberg, Germany

Printed by Libri Plureos GmbH
in Hamburg, Germany